General Relativity and its Applications

General Relativity and its Applications

Black Holes, Compact Stars and Gravitational Waves

Valeria Ferrari
Leonardo Gualtieri
Paolo Pani

CRC Press
Taylor & Francis Group
Boca Raton London New York

CRC Press is an imprint of the
Taylor & Francis Group, an **informa** business

CRC PressFirst edition published 2021
by CRC Press
6000 Broken Sound Parkway NW, Suite 300, Boca Raton, FL 33487-2742

and by CRC Press
2 Park Square, Milton Park, Abingdon, Oxon, OX14 4RN

Library of Congress Cataloging-in-Publication Data

Names: Ferrari, Valeria, 1952- author. | Gualtieri, Leonardo, 1971- author. | Pani, Paolo, 1984- author.
Title: General relativity and its Applications : Black Holes, Compact Stars and Gravitational Waves
Valeria Ferrari, Leonardo Gualtieri, Paolo Pani.
Description: Boca Raton : CRC Press, 2020. | Includes bibliographical references and index.
Identifiers: LCCN 2020028364 | ISBN 9780367625320 (paperback) | ISBN 9781138589773 (hardback) | ISBN 9780429491405 (ebook)
Subjects: LCSH: General relativity (Physics) | Gravitation. | Gravitational waves.
Classification: LCC QC173.6 .F45 2020 | DDC 530.11--dc23
LC record available at https://lccn.loc.gov/2020028364

ISBN: 978-1-138-58977-3 (hbk)
ISBN: 978-0-367-62532-0 (pbk)
ISBN: 978-0-429-49140-5 (ebk)

Typeset in Computer Modern font
by KnowledgeWorks Global Ltd.

Contents

Preface

The General Theory of Relativity (General Relativity, for short) was formulated by Albert Einstein more than a hundred years ago. At the time of writing (2020), all experiments and astrophysical observations have firmly confirmed the validity of its foundations. Einstein's seminal idea, based on the Equivalence Principle, that the gravitational interaction can be interpreted in terms of spacetime curvature has dramatically changed our view of the Universe.

This book originates from a series of lectures held at the University of Chicago in 1993-1994 and in 1997-1998, and from the classes taught by one of the authors (Valeria Ferrari) over the past twenty years at Sapienza University of Rome. During these years, few scientific disciplines have experienced a revolution as dramatic and far-reaching as the one that gravitational physics is still undergoing. The very concept of a black hole – which is now mainstream within both the scientific community and the society – was not so popular and well-accepted just a few years ago. The quest for gravitational waves, more than half a century long, culminated on September 14th, 2015 by the historical LIGO detection, marking the dawn of gravitational wave astronomy. This new discipline is only five years old, but is steadily growing in diverse directions that were unforeseeable just a decade ago. The original lecture notes have been re-elaborated and largely expanded by the authors, including new chapters and new sections on recent discoveries: the detection of gravitational waves from binary black holes and neutron stars, and the first observation of the shadow of a supermassive black hole.

In parallel to the prominent role acquired within the scientific community in the last few years, gravitational physics has also attracted an ever-growing attention from students at all levels. While historically limited to a niche audience, nowadays the teaching of General Relativity and of its phenomenological applications is (with full rights) an essential part of any course in Physics or Astronomy.

In this context, this book is addressed to third-year undergraduate and especially to graduate students in Physics or Astrophysics, who want to learn the basics of General Relativity and its diverse phenomenological consequences. There exist excellent textbooks on the foundations of General Relativity and on its mathematical aspects, as well as outstanding monographs on advanced topics such as black-hole perturbation theory, post-Newtonian theory, gravitational wave modelling, etc. Our main goal was to write a self-contained book that – starting from the basic pillars of the theory – would present in details its applications and, at the same time, could serve as a bridge towards more advanced topics. While we always opted for mathematical rigor, we tried to limit the most technical parts related to differential geometry (which might conceal the physical consequences of the theory) to a few introductory chapters. We adopted the view that General Relativity is the *physical* theory of gravity, and a great deal of attention was spent on its phenomenology and on its observational (both classical and modern) tests.

As a prerequisite, we have assumed that the reader is familiar with the theory of Special Relativity, and with classical Newtonian physics. We also decided to focus on the *astrophysical* consequences of General Relativity, and we do not treat cosmological applications. We believe that cosmology has equally developed into a broad independent field, which deserves a dedicated course and textbooks.

In each chapter we have included several boxes; their purpose is manifold: to solve exercises, to provide examples, to expand on some calculations, to stress some issue of particular relevance, to summarize the main results of a section or of a chapter, or to shortly introduce advanced topics which go beyond the scope of the book.

We start with an introductory chapter, where we discuss why the Newtonian theory of gravity, which had successfully survived for three centuries after its formulation, was challenged in the middle of the 19th century by some unexplained astrophysical observation (Mercury's precession) and, on a theoretical side, by its lack of invariance under Lorentz's transformations. Furthermore, we explain why the Principle of Equivalence between inertial and gravitational mass is central to the theory of General Relativity.

The basics of differential geometry, needed to formulate Einstein's equations, are introduced in a rigorous way in Chapter 2 to Chapter 4. All statements are demonstrated and supported with examples, unless trivial.

After introducing the stress-energy tensor and discussing its properties in Chapter 5, Einstein's equations are derived using a heuristic approach in Chapter 6 and, in a more formal way, using a variational approach in Chapter 7. The role of symmetries and the conservation laws associated to them are discussed in Chapter 8, which concludes the first part of the book on the theory of General Relativity.

The next twelve chapters are devoted to the study of the physical consequences of the theory. We derive the Schwarzschild solution and describe the structure of the black hole horizon and of the singularities, also discussing their role in General Relativity. We study the equations of motion of test particles in the Schwarzschild geometry, and present the most important kinematical tests of General Relativity, which include the "classical tests" (the gravitational redshift of spectral lines, the deflection of light, the periastron precession, the Shapiro time delay) and the most recent one, namely the "image" of a black hole through its shadow.

Four chapters are devoted to gravitational waves. In addition to the standard material common to most textbooks (e.g., the derivation of the wave equation from Einstein's equations linearized about Minkowski's metric and the quadrupole approximation), we describe the basic principles of functioning of a laser interferometric detector of gravitational waves, and derive the formula which allows to evaluate the luminosity of a gravitational wave source by introducing the stress-energy pseudo-tensor of the gravitational field. Particular effort is devoted to derive and explain the features of the signals emitted by coalescing binary systems composed of black holes or neutron stars, with an overview on the signals detected by the LIGO and Virgo experiments since 2015 and including the first two observational runs. Our goal in this part of the book was to give the reader all the tools needed to understand these milestones in a self-contained way. We conclude the part of the book on gravitational waves with a chapter on black-hole perturbation theory, in which we show how to derive the Regge-Wheeler and the Zerilli equations governing the non-radial oscillations of a Schwarzschild black hole, and introduce the notion of quasi-normal modes.

Chapter 16 is devoted to the study of white dwarfs and neutron stars. The structure of white dwarfs is discussed in the framework of Newtonian gravity; we show how the notion of critical mass (the Chandrasekhar limit) emerges when the matter in the interior of these stars is modeled as a fully degenerate gas of fermions. We then discuss the structure of neutron stars, generalizing the laws of thermodynamics and deriving the equations of stellar structure in General Relativity. We discuss how these equations can be solved numerically in the general case, and solve them analytically in the simple case of a star with uniform density. Finally, we discuss some general properties of compact stars: the theoretical bounds on their compactness, their equilibrium, and stability.

The next three chapters are devoted to the study of rotating bodies in General Relativity. We start by deriving the metric of an isolated, stationary object in the far-field limit; using the stress-energy pseudo-tensor, we define the mass and the angular momentum of a rotating object, we derive the Lense-Thirring and the geodesic precession of gyroscopes, and we briefly discuss the results of experiments which have measured these effects. We then introduce the Kerr solution and discuss in detail the structure of the horizons, the peculiar properties of the ergoregion, and the structure of the singularity. The geodesic equations are derived in the general case introducing the Carter constant, and are studied in detail on the equatorial plane. Penrose's process and superradiant scattering are also discussed.

The last chapter is a brief overview on black-hole thermodynamics. We show that it is possible to associate an entropy to a black hole, and discuss that, when quantum mechanics is taken into account, black holes are not totally black, but rather emit radiation with a black-body spectrum. We discuss the black hole area theorem and introduce the laws of black hole thermodynamics. Chapter 20 ends with a qualitative description of Hawking's radiation and black hole evaporation.

We believe that this book presents some novelties with respect to standard textbooks on General Relativity: it provides – in a self-contained way – the tools to understand the most important discoveries of recent years, in particular the detection of gravitational waves; it extends the standard perturbative approach of Minkowski's spacetime to the perturbations of a non-flat solution of Einstein's equations, and applies the theory to the Schwarzschild solution; finally, it provides a detailed description of rotating black holes and of some of the most interesting phenomena which occur in their neighbourhood.

Thus, from its very premise, the book can be approached at different levels, and different paths can be followed across the chapters, depending on the focus and the level of the course. Traditionally, the material presented in this book is taught at Sapienza University of Rome, in two master's classes and an advanced PhD course. Most of the material contained in Chapters 1 to 12 is presented in an introductory course (60 hours, 6 credits) on General Relativity. A second, advanced course on "Gravitational waves, Black holes and Compact Stars" (60 hours, 6 credits) includes Chapters 13, 14, and 16-19, together with some topics from the first twelve chapters not included in the introductory course. Finally, Chapters 15 and 20 and the most advanced topics of Chapter 18 (e.g., Sec. 18.8) are covered in a PhD course (30 hours, 3 credits). Different paths are possible; for instance the introductory chapters can be taught in an undergraduate course, leaving the more advanced chapters to master's or PhD courses. Furthermore, since the three main advanced topics – gravitational waves, compact stars, and rotating black holes – are treated independently, it is possible to skip any of them without affecting the presentation of the others.

General Relativity is often considered the most beautiful of the fundamental theories of nature, but it is actually much more than that: it is the key to explore new territories of knowledge. In this book we tried to convey our enthusiasm and to provide a useful tool to teach and learn this fascinating subject.

Rome, May 2020 Valeria Ferrari, Leonardo Gualtieri, Paolo Pani

ACKNOWLEDGMENTS

We are indebted to Omar Benhar, Emanuele Berti, Marco Bruni, and Luciano Rezzolla, who took the time to go through the various chapters of the book, suggesting improvements and correcting misprints. We also thank our students Vania Vellucci, Margherita Fasano, and Francesco Iacovelli for a careful reading of parts of the book.

Notation and conventions

Up to Chapter 9, we always use physical units – adopting either the cgs or the International System of Units (SI). In particular, we denote Newton's constant by G and the speed of light by c. From Chapter 9, we shall also adopt geometrized units in which $G = c = 1$ (see Box 9-B), although we will occasionally restore physical units when needed.

All vectors (both four-vectors and n-dimensional vectors defined in generic spaces, as those introduced in Chapter 2) will be indicated by an arrow, as in $\vec{V}$, except Euclidean vectors which will be denoted by boldface characters, as in $\mathbf{V}$.

Unless otherwise specified, we shall use Greek indices for the components of four-vectors, V^μ with $\mu = 0, \ldots, 3$. Latin indices will be used for the components of Euclidean three-vectors, V^i with $i = 1, \ldots, 3$, and for the components of vectors in generic n-dimensional spaces, V^i with $i = 1, \ldots n$, as those introduced in Chapter 2.

As customary we shall adopt Einstein's convention for the sum, i.e., the product of two quantities having the same index, appearing once in the lower and once in the upper case ("dummy index"), implies summation over the repeated index. For example,

$$\begin{aligned} v_\alpha V^\alpha &\equiv \sum_{\alpha=0}^{3} v_\alpha V^\alpha = v_0 V^0 + v_1 V^1 + v_2 V^2 + v_3 V^3 \,, \\ v_i V^i &\equiv \sum_{i=1}^{3} v_i V^i = v_1 V^1 + v_2 V^2 + v_3 V^3 \,. \end{aligned}$$

Manifolds, spaces, and points on manifolds will be denoted by the standard boldface symbol (e.g., $\mathbf{M}$, $\mathbf{T}$, $\mathbf{p} \in \mathbf{M}$), whereas tensors will be denoted by math-boldface symbols.

The components of the partial derivative of (say) a scalar quantity f will be denoted by $\frac{\partial f}{\partial x^\mu} \equiv \partial_\mu f \equiv f_{,\mu}$. The components of the covariant derivative will be denoted as $\nabla_\mu f \equiv f_{;\mu}$. Sometimes, for brevity, we shall indicate a function of the coordinates $f(x^0, \ldots, x^3)$ as $f(x)$.

We use a mostly positive metric signature, so that Minkowski's metric in 3+1 dimensions reads $\eta_{\mu\nu} = \mathrm{diag}(-1, 1, 1, 1)$.

Our conventions for the Christoffel symbols and for the components of the Riemann tensor are

$$\begin{aligned} \Gamma^\sigma_{\lambda\mu} &= \frac{1}{2} g^{\nu\sigma} \left(\frac{\partial g_{\mu\nu}}{\partial x^\lambda} + \frac{\partial g_{\lambda\nu}}{\partial x^\mu} - \frac{\partial g_{\lambda\mu}}{\partial x^\nu} \right) , \\ R^\alpha{}_{\beta\mu\nu} &= \Gamma^\alpha{}_{\beta\nu,\mu} - \Gamma^\alpha{}_{\beta\mu,\nu} - \Gamma^\alpha{}_{\kappa\nu}\Gamma^\kappa{}_{\beta\mu} + \Gamma^\alpha{}_{\kappa\mu}\Gamma^\kappa{}_{\beta\nu} \,. \end{aligned}$$

Table A: Fundamental constants and other useful quantities used in this book, in cgs units.

Newton's constant	G	6.674×10^{-8}	$\frac{\text{cm}^3}{\text{g s}^2}$
speed of light	c	2.998×10^{10}	$\frac{\text{cm}}{\text{s}}$
Planck's constant	h	6.626×10^{-27}	$\text{cm}^2\ \text{g s} = \text{erg s}$
reduced Planck's constant	$\hbar$	1.055×10^{-27}	$\text{cm}^2\ \text{g s} = \text{erg s}$
Planck's length	l_{P}	1.616×10^{-33}	cm
Boltzmann's constant	k_{B}	1.381×10^{-16}	$\frac{\text{cm}^2\ \text{g}}{\text{s}^2\ \text{K}} = \frac{\text{erg}}{\text{K}}$
electric charge of the electron	e	1.602×10^{-19}	C
electron mass	m_e	9.109×10^{-28}	g
proton mass	m_p	1.673×10^{-24}	g
neutron mass	m_n	1.675×10^{-24}	g
mass of the Sun	$M_{\odot}$	1.989×10^{33}	g
radius of the Sun	$R_{\odot}$	6.960×10^{10}	cm
mass of the Earth	$M_{\oplus}$	5.972×10^{27}	g
equatorial radius of the Earth	$R_{\oplus}$	6.378×10^{8}	cm
light year	ly	9.461×10^{17}	cm
parsec	pc	3.086×10^{18}	cm

CHAPTER 1

Introduction

The General Theory of Relativity (or "General Relativity" for short) is the theory of gravity formulated by Albert Einstein in 1915. It is based on two fundamental principles: the *Equivalence Principle of gravitation and inertia*, which establishes a relation, central to the theory, between the gravitational field and the geometry of the spacetime, and the *Principle of General Covariance*, according to which the laws of physics have the same form in any reference frame. General Relativity has changed dramatically our understanding of space, time, and gravity. In this book we shall investigate its fascinating and far-reaching predictions, such as the existence of black holes, and the emission of gravitational waves in the most violent and energetic phenomena occurring in our universe, as the collision of black holes and of neutron stars.

The language of General Relativity is that of differential geometry. There is no way to understand Einstein's theory of gravity without mastering the concept of manifold, or of tensor field. Therefore, we shall dedicate a few chapters to the mathematical tools that are essential to describe the theory and its physical consequences. This first chapter, however, will be a pedagogical introduction aimed at answering the following questions:

1. Why did Newtonian theory become inappropriate to describe the gravitational field?
2. Why do we need to introduce geometrical objects, like the metric tensor or the Christoffel symbols, to describe the gravitational field; and what is the role of the Equivalence Principle in this new geometrical framework?

In the next chapters we shall rigorously define manifolds, vectors, tensors, and then, after introducing the principle of general covariance, we will formulate Einstein's equations. But first of all, since as anticipated there exists a relation between the gravitational field and the spacetime geometry, let us introduce non-Euclidean geometries, which are in some sense the precursors of General Relativity.

1.1 NON-EUCLIDEAN GEOMETRIES

In the pre-relativistic years, the arena of all physical theories was the flat space of Euclidean geometry, which is based on the five Euclid's postulates. Among them, the fifth postulate had been the object of a millenary dispute: for over two thousand years mathematicians tried to show, without succeeding, that the fifth postulate is a consequence of the other four. The fifth postulate states the following: *Consider two straight lines and a third straight line crossing the two. If the sum of the two internal angles (see Fig. 1.1) is smaller than 180°, the two lines will meet at some point on the side of the internal angles.*

The problem was solved independently by three eminent scientists, Gauss (1824, Germany), Bolyai (1832, Hungary), and Lobachevski (1826, Russia), who discovered a geometry

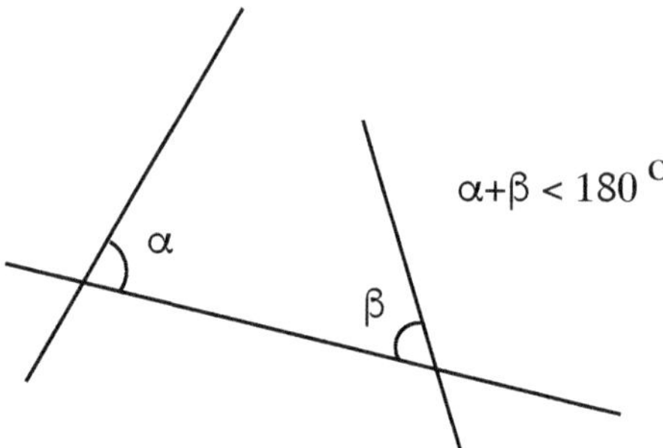

Figure 1.1: Geometrical illustration of Euclid's fifth postulate.

that satisfies all Euclid's postulates except the fifth (for a more detailed historic account, see Weinberg's book [117]). The analytic representation of this geometry was discovered by Felix Klein in 1870. He found that a point in this geometry can be represented by a pair of real numbers, $\boldsymbol{x} = (x^1, x^2)$, with

$$\boldsymbol{x} \cdot \boldsymbol{x} \equiv (x^1)^2 + (x^2)^2 < 1\,, \tag{1.1}$$

and the distance $d(\boldsymbol{x}, \boldsymbol{y})$ between two points $\boldsymbol{x}$ and $\boldsymbol{y}$ is defined as

$$d(\boldsymbol{x}, \boldsymbol{y}) = L \cosh^{-1}\left[\frac{1 - \boldsymbol{x} \cdot \boldsymbol{y}}{\sqrt{1 - \boldsymbol{x} \cdot \boldsymbol{x}}\sqrt{1 - \boldsymbol{y} \cdot \boldsymbol{y}}}\right], \tag{1.2}$$

where L has the dimensions of a length. This space is infinite, because $d(\boldsymbol{x}, \boldsymbol{y}) \to \infty$ as $|\boldsymbol{x}| \to 1$ or $|\boldsymbol{y}| \to 1$. The logical independence of Euclid's fifth postulate was thus established. As we shall see shortly, this geometry describes a *two-dimensional space of constant negative curvature*.

In 1827 Gauss, in his *Disquisitiones generales circa superficies curvas*, distinguished for the first time the inner, or *intrinsic*, properties of a surface from the outer, or *extrinsic*, properties: the former can be measured by an observer belonging to the surface; the latter arise from embedding the surface in a higher-dimensional space. Gauss realized that *the fundamental inner property is the distance between two points, defined as the length of the shortest path between them on the surface.*

For example, a cylinder has the same inner properties as a plane. The reason is that it can be obtained by a flat piece of paper suitably rolled, without distorting its *metric* relations, i.e., without stretching or tearing. This means that the distance between any two points on the surface is the same as it was in the original piece of paper, and parallel lines remain parallel. Thus the *intrinsic geometry* of a cylinder is flat. This is not true in the case of a sphere, since a sphere cannot be mapped onto a plane without distortions: the inner properties of a sphere are different from those of a plane. It is noteworthy that the intrinsic geometry of a surface only involves relations between points on the surface itself.

On the other hand, since a cylinder is "round" in one direction, we think of it as a curved surface. This is due to the fact that we consider the cylinder as a two-dimensional surface in a three-dimensional space, and we intuitively compare lines on the surface with straight lines in the flat three-dimensional space. Thus, the *extrinsic curvature* relies on the notion of a higher-dimensional space. In the following, we shall mostly be concerned with the intrinsic properties of curved spaces. Indeed, we shall see that the spacetime of General Relativity has the intrinsic geometry of a curved space, but without an extrinsic geometry, because *there is no higher-dimensional space* in which it is embedded.

The distance between two points can be defined in a variety of ways, and consequently we can construct different metric spaces (as we shall discuss in the next chapter, a *metric space* is a space endowed with a notion of length). Following Gauss' original idea, we shall select those metric spaces for which, given any sufficiently small region of space, it is possible to choose – in the case of a two-dimensional space – a system of coordinates (ξ^1, ξ^2) such that the *distance*[1] between two neighbouring points, $\mathbf{p}= (\xi^1, \xi^2)$ and $\mathbf{p}' = (\xi^1 + d\xi^1, \xi^2 + d\xi^2)$, satisfies Pythagoras' law

$$ds^2 = (d\xi^1)^2 + (d\xi^2)^2 . \tag{1.3}$$

If the space has n dimensions, the coordinates are $(\xi^1, \dots, \xi^n)$, and the distance between $\mathbf{p}$ and $\mathbf{p}'$ is $ds^2 = (d\xi^1)^2 + \cdots + (d\xi^n)^2$. These spaces are called *locally Euclidean spaces* and the coordinates $\{\xi^i\}_{i=1,n}$ are called *locally Euclidean coordinates.*

The possibility of setting up a locally Euclidean coordinate system is a local property, which deals only with the inner metric relations for infinitesimal neighbourhoods. Thus, unless the space is globally Euclidean, the coordinates (ξ^1, ξ^2) have only a local meaning. Let us now consider some other coordinate system (x^1, x^2); how do we express the distance between two neighbouring points in this new system? If we explicitly evaluate $d\xi^1$ and $d\xi^2$ in terms of the new coordinates, we obtain

$$\begin{aligned}
\xi^1 = \xi^1(x^1, x^2) \quad &\rightarrow \quad d\xi^1 = \frac{\partial \xi^1}{\partial x^1} dx^1 + \frac{\partial \xi^1}{\partial x^2} dx^2 , \\
\xi^2 = \xi^2(x^1, x^2) \quad &\rightarrow \quad d\xi^2 = \frac{\partial \xi^2}{\partial x^1} dx^1 + \frac{\partial \xi^2}{\partial x^2} dx^2 ,
\end{aligned} \tag{1.4}$$

and therefore

$$\begin{aligned}
ds^2 &= \left[\left(\frac{\partial \xi^1}{\partial x^1} \right)^2 + \left(\frac{\partial \xi^2}{\partial x^1} \right)^2 \right] (dx^1)^2 + \left[\left(\frac{\partial \xi^1}{\partial x^2} \right)^2 + \left(\frac{\partial \xi^2}{\partial x^2} \right)^2 \right] (dx^2)^2 \\
&+ 2 \left[\left(\frac{\partial \xi^1}{\partial x^1} \right) \left(\frac{\partial \xi^1}{\partial x^2} \right) + \left(\frac{\partial \xi^2}{\partial x^1} \right) \left(\frac{\partial \xi^2}{\partial x^2} \right) \right] dx^1 dx^2 \\
&= g_{11}(dx^1)^2 + g_{22}(dx^2)^2 + 2g_{12} dx^1 dx^2 = \sum_{i=1}^{2} \sum_{j=1}^{2} g_{ij} dx^i dx^j \equiv g_{ij} dx^i dx^j .
\end{aligned} \tag{1.5}$$

In the last term of the above equation, we have adopted Einstein's convention for the sum (i.e., a sum over repeated indices is left implicit), and we have defined the following quantities:

$$g_{ij} \equiv \left(\frac{\partial \xi^1}{\partial x^i} \right) \left(\frac{\partial \xi^1}{\partial x^j} \right) + \left(\frac{\partial \xi^2}{\partial x^i} \right) \left(\frac{\partial \xi^2}{\partial x^j} \right) \qquad i, j = 1, 2 . \tag{1.6}$$

In other words, we have introduced a metric tensor[2] g_{ij}, which is an object that *allows us to compute the distance in any coordinate system.* Thus, the notion of metric associated to a space emerges in a natural way from the change of the coordinate system. From Eq. 1.6 it follows that g_{ij} is a symmetric tensor, i.e., $g_{ij} = g_{ji}$. The discussion above can be easily generalized to a space of arbitrary dimensions.

---

[1] Throughout this book, with "distance between two points" we will always refer to points that are infinitely close, unless otherwise specified.

[2] We shall define tensors in a rigorous way in Chapter 2.

Box 1A

How to compute g_{ij}: an example

Given a two-dimensional, locally Euclidean coordinate system (ξ^1, ξ^2), let us introduce polar coordinates $(x^1 = r, x^2 = \theta)$ such that $\xi^1 = r\cos\theta$ and $\xi^2 = r\sin\theta$. Then

$$\begin{aligned} d\xi^1 &= \cos\theta dr - r\sin\theta d\theta\,, \\ d\xi^2 &= \sin\theta dr + r\cos\theta d\theta\,, \end{aligned} \tag{1.7}$$

and

$$ds^2 = (d\xi^1)^2 + (d\xi^2)^2 = dr^2 + r^2 d\theta^2\,. \tag{1.8}$$

In this case, the *components* of metric tensor read

$$g_{11} = 1, \qquad g_{22} = r^2, \qquad g_{12} = g_{21} = 0\,. \tag{1.9}$$

1.1.1 The metric tensor in different coordinate frames

We shall now see how the metric tensor transforms under an arbitrary coordinate transformation. Let us assume that g_{ij} is given in terms of the coordinates (x^1, x^2). In the previous section we have shown that, for example, the component g_{11} is defined as (see Eq. 1.6)

$$g_{11} = \left(\frac{\partial\xi^1}{\partial x^1}\right)^2 + \left(\frac{\partial\xi^2}{\partial x^1}\right)^2\,, \tag{1.10}$$

where (ξ^1, ξ^2) are the coordinates of the locally Euclidean reference frame. If we now change from (x^1, x^2) to a new reference frame $(x^{1'}, x^{2'})$, where $x^1 = x^1(x^{1'}, x^{2'})$ and $x^2 = x^2(x^{1'}, x^{2'})$, the metric tensor becomes

$$\begin{aligned} g_{1'1'} &= \left(\frac{\partial\xi^1}{\partial x^{1'}}\right)^2 + \left(\frac{\partial\xi^2}{\partial x^{1'}}\right)^2 \\ &= \left(\frac{\partial\xi^1}{\partial x^1}\frac{\partial x^1}{\partial x^{1'}} + \frac{\partial\xi^1}{\partial x^2}\frac{\partial x^2}{\partial x^{1'}}\right)^2 + \left(\frac{\partial\xi^2}{\partial x^1}\frac{\partial x^1}{\partial x^{1'}} + \frac{\partial\xi^2}{\partial x^2}\frac{\partial x^2}{\partial x^{1'}}\right)^2 \\ &= \left[\left(\frac{\partial\xi^1}{\partial x^1}\right)^2 + \left(\frac{\partial\xi^2}{\partial x^1}\right)^2\right]\left(\frac{\partial x^1}{\partial x^{1'}}\right)^2 + \left[\left(\frac{\partial\xi^1}{\partial x^2}\right)^2 + \left(\frac{\partial\xi^2}{\partial x^2}\right)^2\right]\left(\frac{\partial x^2}{\partial x^{1'}}\right)^2 \\ &+ 2\left(\frac{\partial\xi^1}{\partial x^1}\frac{\partial\xi^1}{\partial x^2} + \frac{\partial\xi^2}{\partial x^1}\frac{\partial\xi^2}{\partial x^2}\right)\left(\frac{\partial x^1}{\partial x^{1'}}\frac{\partial x^2}{\partial x^{1'}}\right) \\ &= g_{11}\left(\frac{\partial x^1}{\partial x^{1'}}\right)^2 + g_{22}\left(\frac{\partial x^2}{\partial x^{1'}}\right)^2 + 2g_{12}\left(\frac{\partial x^1}{\partial x^{1'}}\frac{\partial x^2}{\partial x^{1'}}\right). \end{aligned}$$

In general, using a compact notation we can write

$$g_{i'j'} = g_{kl}\frac{\partial x^k}{\partial x^{i'}}\frac{\partial x^l}{\partial x^{j'}}\,. \tag{1.11}$$

As discussed more rigorously in the next chapter, this is *how a tensor transforms under an arbitrary coordinate transformation.* Eq. 1.11 can be generalized to a space of arbitrary dimensions.

In conclusion, given a locally Euclidean space and an arbitrary reference frame, the knowledge of g_{ij} allows us to compute the distance between any two points, i.e., the fundamental intrinsic property of the space.

1.1.2 The Gaussian curvature

In the case of two-dimensional spaces, Gauss showed that it is possible to define a scalar function which depends on the inner properties of the space. This function is called, after him, the *Gaussian curvature*

$$
\begin{aligned}
k(x^1,x^2) &= \frac{1}{2g}\left[2\frac{\partial^2 g_{12}}{\partial x^1 \partial x^2} - \frac{\partial^2 g_{11}}{\partial x^{2^2}} - \frac{\partial^2 g_{22}}{\partial x^{1^2}}\right] \\
&- \frac{g_{22}}{4g^2}\left[\left(\frac{\partial g_{11}}{\partial x^1}\right)\left(2\frac{\partial g_{12}}{\partial x^2} - \frac{\partial g_{22}}{\partial x^1}\right) - \left(\frac{\partial g_{11}}{\partial x^2}\right)^2\right] \\
&+ \frac{g_{12}}{4g^2}\left[\left(\frac{\partial g_{11}}{\partial x^1}\right)\left(\frac{\partial g_{22}}{\partial x^2}\right) - 2\left(\frac{\partial g_{11}}{\partial x^2}\right)\left(\frac{\partial g_{22}}{\partial x^1}\right) + \left(2\frac{\partial g_{12}}{\partial x^1} - \frac{\partial g_{11}}{\partial x^2}\right)\left(2\frac{\partial g_{12}}{\partial x^2} - \frac{\partial g_{22}}{\partial x^1}\right)\right] \\
&- \frac{g_{11}}{4g^2}\left[\left(\frac{\partial g_{22}}{\partial x^2}\right)\left(2\frac{\partial g_{12}}{\partial x^1} - \frac{\partial g_{11}}{\partial x^2}\right) - \left(\frac{\partial g_{22}}{\partial x^1}\right)^2\right],
\end{aligned}
\tag{1.12}
$$

where g is the determinant of the two-dimensional metric g_{ij}

$$g = g_{11}g_{22} - g_{12}^2 . \tag{1.13}$$

The Gaussian curvature is an *invariant*, i.e., its value at a given point is independent of the particular choice of the coordinate system: if we choose a different coordinate system, the components of the metric tensor g_{ij} and the functional dependence on the coordinates of k will change, but the value of k at that point will remain the same.

For example, given a spherical surface of radius L, in polar coordinates (θ, φ) the distance between two close points is $ds^2 = L^2 d\theta^2 + L^2 \sin^2\theta d\varphi^2$, and the components of the metric tensor are $g_{11} = L^2$, $g_{22} = L^2 \sin^2\theta$, $g_{12} = 0$. Its Gaussian curvature is

$$k = \frac{1}{L^2}, \tag{1.14}$$

i.e., a spherical surface has a constant, positive curvature.

For the Gauss-Bolyai-Lobachewski geometry introduced above, whose metric tensor components are

$$g_{11} = \frac{L^2\left[1-(x^2)^2\right]}{\left[1-(x^1)^2-(x^2)^2\right]^2}, \qquad g_{22} = \frac{L^2\left[1-(x^1)^2\right]}{\left[1-(x^1)^2-(x^2)^2\right]^2}, \qquad g_{12} = \frac{L^2 x^1 x^2}{\left[1-(x^1)^2-(x^2)^2\right]^2}, \tag{1.15}$$

the Gaussian curvature is

$$k = -\frac{1}{L^2}, \tag{1.16}$$

i.e., this geometry has a constant, negative curvature. It is also important to stress that $k = 0$ if and only if the space is flat.

1.1.3 Pseudo-Euclidean geometries and spacetime

In the previous sections, following Gauss we have selected a class of two-dimensional spaces where, in the neighbourhood of any point, it is possible to set up a coordinate system (ξ^1, ξ^2) such that the distance between two close points is given by Pythagoras' law. Then, we have defined the metric tensor g_{ij}, which allows us to compute the distance between points in an arbitrary coordinate system, and we have derived the law according to which g_{ij} transforms when changing coordinates. Finally, we have seen that there exists a scalar quantity, the Gaussian curvature, which describes the inner properties of the space: it is a function of g_{ij} and of its first and second derivatives, and is invariant under coordinate transformations.

These results can be extended to an arbitrary D-dimensional space. In particular the space of General Relativity is a four-dimensional space, called the *spacetime*, such that in the neighbourhoods of any point it is possible to set up a coordinate system $\{\xi^\mu\}_{\mu=0,\dots,3}$ [3] such that the distance between two close points is locally prescribed by Special Relativity (see the detailed discussion in Sec. 1.3)

$$ds^2 = -(d\xi^0)^2 + (d\xi^1)^2 + (d\xi^2)^2 + (d\xi^3)^2 . \tag{1.17}$$

The coordinates $\{\xi^\mu\}$ are called *locally Minkowskian coordinates.* Such a spacetime is called locally *pseudo-Euclidean*, since the only (albeit crucial) difference from a $D = 4$ Euclidean space is the sign of the first term in the above equation. As known from Special Relativity, this accounts for giving a *causal structure* to the spacetime. While the metric tensor of a four-dimensional Euclidean space in locally Euclidean coordinates is simply $g_{\mu\nu} \equiv \delta_{\mu\nu} = \text{diag}(1,1,1,1)$, where $\delta_{\mu\nu}$ is Kronecker's delta, from Eq. 1.17 we see that *Minkowski's metric tensor* in the locally Minkowskian coordinates $\{\xi^\mu\}$ reads

$$g_{\mu\nu} \equiv \eta_{\mu\nu} = \text{diag}(-1,1,1,1) . \tag{1.18}$$

Henceforth, $\eta_{\mu\nu}$ will denote the metric tensor of the flat, Minkowski spacetime in the Minkowskian coordinates $\{\xi^\mu\}$.

To describe the inner properties of a four-dimensional space we need more than one scalar function. Indeed, since $g_{\mu\nu}$ is symmetric, it has 10 independent components. In addition, we can make a coordinate transformation, and impose 4 functional relations among them. Therefore, the number of independent functions that describe the inner properties of the spacetime is $N = 6$. If $D = 2$, as we have seen, $N = 1$ and the Gaussian curvature 1.12 is sufficient to describe the geometry.

1.2 NEWTONIAN THEORY AND ITS SHORTCOMINGS

In this section we shall discuss why the Newtonian theory of gravity became inappropriate to correctly describe the gravitational field. Newton's theory was published in 1685 in the *Philosophiae Naturalis Principia Mathematica*, which contains an incredible variety of fundamental results and, among them, the cornerstones of classical physics:

- *Newton's law*

$$\boldsymbol{F} = m_{\text{I}}\boldsymbol{a} , \tag{1.19}$$

  which gives the acceleration $\boldsymbol{a}$ of a point particle with *inertial* mass m_{I}, given a force $\boldsymbol{F}$ acting on it;

- *Newton's law of gravitation*

$$\boldsymbol{F}_{\text{G}} = m_{\text{G}}\boldsymbol{g} , \tag{1.20}$$

  describing the gravitational force acting on a point particle with *gravitational* mass m_{G}; the gravitational acceleration,

$$\boldsymbol{g} = -\frac{G\sum_i M_{\text{G}i}(\boldsymbol{r} - \boldsymbol{r}'_i)}{|\boldsymbol{r} - \boldsymbol{r}_i|^3} , \tag{1.21}$$

[3] Recall that Greek indices run from 0 to 3.

depends on the position of the particle with respect to the other masses, M_{Gi}, that generate the field. The gravitational acceleration (and hence the force) decreases as the inverse square of the distance, $g \sim 1/r^2$.

The two laws combined together clearly show that a body falls in a gravitational field with acceleration given by

$$\boldsymbol{a} = \left(\frac{m_{\rm G}}{m_{\rm I}}\right)\boldsymbol{g}\,. \tag{1.22}$$

If $m_{\rm G}/m_{\rm I}$ is a constant independent of the body, then the acceleration is the same for every infalling body, and independent of their mass. Galileo had already experimentally discovered this law, and Newton himself tested the *Equivalence Principle* between the gravitational and the inertial mass, by studying the motion of several pendulums of different composition and equal length, finding no difference in their oscillation periods. The validity of the Equivalence Principle was the core of Newton's arguments for the universality of his law of gravitation; indeed, after describing his experiments with different pendulums in the *Principia* he writes:

But, without all doubt, the nature of gravity towards the planets is the same as towards the Earth.

Since then, several experiments confirmed this crucial result. Among them, in 1889 Eötvös' experiment established the equality of gravitational and inertial mass with the accuracy of 1 part in 10^9; subsequently Dicke's experiment (1964, 1 part in 10^{11}), Braginsky's (1972, 1 part in 10^{12}), and more recently the Lunar-Laser Ranging experiment (1 part in 10^{13}) have confirmed the Equivalence Principle with increasing accuracy (for a detailed account, see Clifford Will's review [119]). As of 2020, the most stringent constraint comes from the MICROSCOPE satellite mission, which reached the outstanding accuracy of 1 part in 10^{15} [20]. Thus, all experiments up to date confirm the Principle of Equivalence of gravitational and inertial mass.

Before describing why at a certain point the Newtonian theory fails to be a satisfactory description of gravity, let us briefly recall the reasons of its great success, that remained untouched for more than 200 years. In the *Principia*, Newton formulated the universal law of gravitation, developed the theory of lunar motion and tides and that of planetary motion around the Sun which, at those times, were the most elegant descriptions of these phenomena.

After Newton's outstanding accomplishment, the law of gravitation was used to investigate in more detail the solar system; its application to the study of the perturbations of Uranus' orbit around the Sun led, in 1846, J.C. Adams (England) and U. Le Verrier (France) to predict the existence of a new planet which was named Neptune. A few years later, the discovery of Neptune was a triumph of Newton's theory of gravitation.

However, already in 1845 Le Verrier had observed anomalies in the motion of Mercury. He found that the perihelion precession of 35 arcsec per century exceeded the value due to the perturbation introduced by the other planets as predicted by Newton's theory. In 1882, Newcomb confirmed this discrepancy, updating Mercury's perihelion precession to its actual value, roughly 43 arcsec per century larger than the Newtonian value. In order to explain this effect, scientists developed models that predicted the existence of some interplanetary matter, and in 1896 Seelinger showed that an ellipsoidal distribution of matter surrounding the Sun could explain the observed precession. In 1859, Le Verrier attempted to use the same logic as for the prediction of Neptune and postulated the existence of a new hypothetical planet, dubbed Vulcan, in an orbit between Mercury and the Sun. We know today that these models were wrong, and that the reason for the exceedingly high precession of Mercury's perihelion has a relativistic origin [4]. In any event, we can say that the Newtonian theory of

[4]It is amusing to note that Vulcan is one of the first examples of *dark matter*, namely some hypothetical

gravity worked remarkably well to explain planetary motion, but already in 1845 the fact that something did not work perfectly had some experimental evidence.

Let us turn now to a more philosophical problem of the theory. The equations of Newtonian mechanics are invariant under Galileo's transformations:

$$\boldsymbol{x}' = \boldsymbol{M}\boldsymbol{x} + \boldsymbol{v}t + \boldsymbol{x}_0\,, \tag{1.23}$$

$$t' = t + T\,, \tag{1.24}$$

where $\boldsymbol{v}$ is the relative velocity of the two frames, $\boldsymbol{x}_0$ is the initial distance between the two origins, T is a constant, and $\boldsymbol{M}$ is the orthogonal, constant matrix expressing how the second frame is rotated with respect to the first. For a generic rotation around the x, y, and z axes (in this order),

$$\boldsymbol{M} = \begin{pmatrix} \cos\theta & -\sin\theta & 0 \\ \sin\theta & \cos\theta & 0 \\ 0 & 0 & 1 \end{pmatrix} \begin{pmatrix} \cos\varphi & 0 & \sin\varphi \\ 0 & 1 & 0 \\ -\sin\varphi & 0 & \cos\varphi \end{pmatrix} \begin{pmatrix} 1 & 0 & 0 \\ 0 & \cos\zeta & -\sin\zeta \\ 0 & \sin\zeta & \cos\zeta \end{pmatrix}, \tag{1.25}$$

where (ζ, φ, θ) are the three Euler angles. The ten parameters (3 Euler angles, plus 3 components of $\boldsymbol{v}$ and of $\boldsymbol{x}_0$, plus the time shift T) identify the ten-dimensional Galileo group.

The invariance of Newton's equations of motion with respect to Galileo's transformations implies the existence of *inertial frames*, where the laws of Mechanics hold in their standard form (i.e., Eqs. 1.19, 1.20). What then determines which frames are inertial? According to Newton there exists an absolute space, as the result of the famous experiment of the rotating vessel would show [5]: inertial frames are those in uniform relative motion with respect to the absolute space. However, this idea was rejected by Leibniz who claimed that there is no philosophical need for such a notion, and the debate on this issue continued during the next centuries. One of the major opponents to the concept of absolute space was Mach, who argued that if the masses in the entire universe rigidly rotate with respect to the vessel, the water surface will bend in exactly the same way as when the vessel rotates with respect to them. This is because the inertia is a measure of the gravitational interaction between a body and the matter content of the rest of the Universe.

The problems just described (the discrepancy in the advance of Mercury's perihelion and the postulated absolute space) were however only small clouds: the Newtonian theory remained *the* theory of gravity until the end of the 19th century. The big storm approached when Maxwell developed the theory of electrodynamics in 1864. Maxwell's equations establish that the velocity of light is a universal constant. It was soon understood that these equations are not invariant under Galileo's transformations; indeed, according to Eqs 1.23 and 1.24, if the velocity of light is c in a given coordinate frame, it would have a different value in another frame moving with respect to the first with assigned velocity $\boldsymbol{v}$. To justify this discrepancy, Maxwell formulated the hypothesis that light does not really propagate

matter distribution that is postulated in order to reconcile observations with the known laws of gravity. While this approach had worked for Neptune, the existence of Vulcan turned out to be wrong. In modern terms, we might say that Mercury's anomalous precession was a hint of new physics. An analogous debate still exists nowadays in cosmology. In the standard paradigm, in order to reconcile cosmological observations at various scales with Einstein's theory, one has to invoke the existence of a dominant component of *dark matter* and *dark energy* in the universe, but there exist also attempts to explain cosmological observations in terms of corrections to General Relativity.

[5] In this experiment, a vessel is filled with water and rotates with a given angular velocity about the symmetry axis. After some time the surface of the water assumes the typical shape of a paraboloid, being in equilibrium under the action of the gravity force, the centrifugal force, and the fluid forces. At the beginning, when the vessel moves with respect to the water, the surface of the water is still flat; it is only when the water moves together with the vessel that it assumes the paraboloid shape. This shows that the surface of the water is not affected by its relative motion with respect to the vessel: the shape changes only when the water moves with respect to the *absolute space*.

in vacuum: electromagnetic waves are carried by a medium, the *luminiferous ether*, and the equations are invariant only with respect to a set of Galilean inertial frames that are at rest with respect to the ether. However in 1887 Michelson and Morley showed that the velocity of light is the same, within 5 km/s (today the accuracy is of about 10^{-12} km/s [61]), along the direction of the Earth orbital motion, and transverse to it. How can this result be justified? One possibility was that the Earth is at rest with respect to the ether; but this hypothesis was totally unsatisfactory, since it would have been a coming back to an antropocentric picture of the world (moreover, it would contradict the observations of the aberration of light from stars). Another possibility was that the ether simply does not exist, and one has to accept the fact that the speed of light is the same in all directions, and whatever the velocity of the source is. If this was the case, one had to find the coordinate transformation with respect to which Maxwell's equations are invariant. The problem was solved by Einstein in 1905 with the formulation of the Special Theory of Relativity. He showed that Galileo's transformations have to be replaced by the *Lorentz transformations*

$$x^{\alpha\prime} = L^{\alpha}{}_{\gamma} x^{\gamma}\,, \tag{1.26}$$

where

$$L^0{}_0 = \gamma, \quad L^0{}_j = L^j{}_0 = \frac{\gamma}{c} v^j, \qquad L^i{}_j = \delta^i{}_j + \frac{\gamma - 1}{v^2} v^i v_j, \quad i, j = 1, \ldots, 3\,, \tag{1.27}$$

$\gamma = 1/\sqrt{1 - \frac{v^2}{c^2}}$, and v^i are the components of the relative velocity of the two frames (boost). A precise definition of covariant (lower) and contravariant (upper) indices will be provided in the next chapter.

As it was immediately realized, however, while Maxwell's equations are invariant with respect to Lorentz's transformations, Newton's equations are not. Consequently, one had to face a new problem: how to modify the equations of mechanics and gravity in such a way that they become invariant with respect to Lorentz's transformations? It is at this point that Einstein made his fundamental observation.

1.3 THE ROLE OF THE EQUIVALENCE PRINCIPLE

Let us consider a non-relativistic particle moving in a *constant* and *uniform* gravitational field $\boldsymbol{g}$. Note that constant means that $\boldsymbol{g}$ does not change with time, and uniform means that it is the same at each space point. Let $\boldsymbol{F}$ be the vector sum of other forces acting on the particle. According to Newtonian mechanics, the equation of motion reads

$$m_{\mathrm{I}} \frac{d^2 \boldsymbol{x}}{dt^2} = m_{\mathrm{G}} \boldsymbol{g} + \boldsymbol{F}\,. \tag{1.28}$$

Let us now jump on an elevator which is freely falling in the same gravitational field, i.e., let us make the following coordinate transformation,

$$\boldsymbol{x}' = \boldsymbol{x} - \frac{1}{2} \boldsymbol{g} t^2, \qquad t' = t. \tag{1.29}$$

In this new frame Eq. 1.28 becomes

$$m_{\mathrm{I}} \left[\frac{d^2 \boldsymbol{x}'}{dt^2} + \boldsymbol{g} \right] = m_{\mathrm{G}} \boldsymbol{g} + \boldsymbol{F}\,. \tag{1.30}$$

Since by the Equivalence Principle $m_{\mathrm{I}} = m_{\mathrm{G}}$, and since this is true for *any* particle, this equation reduces to

$$m_{\mathrm{I}} \frac{d^2 \boldsymbol{x}'}{dt^2} = \boldsymbol{F}\,. \tag{1.31}$$

Let us compare Eq. 1.28 and Eq. 1.31. It is clear that an observer O' who is in the elevator, i.e., in free fall in the gravitational field, sees the same laws of physics as the initial observer O, but (s)he does not feel the gravitational field. This follows from the (experimentally tested) equivalence between inertial and gravitational mass. It is important to note that since the ratio between m_G and m_I is the same for all bodies, the coordinate transformation 1.29, which allows to eliminate the effects of gravity in the freely falling frame, is valid for all bodies independently of their mass and composition. Consider now a particle at rest in this frame and no force $\boldsymbol{F}$ acting on it. Under this assumption, Eq. 1.31 shows that the particle remains at rest forever. Therefore the frame which is in free fall in a constant and uniform gravitational field is an inertial frame. This would not be true if $\boldsymbol{g}$ were not constant, since additional terms depending on the derivatives of $\boldsymbol{g}$ would appear in Eq. 1.30. In addition, would $\boldsymbol{g}$ not be uniform, the coordinate transformation 1.29 would be different in different space points. However, we can always take an interval of time so short, and a region of space so small, that $\boldsymbol{g}$ can be considered constant and uniform; in this case the freely falling reference frame defined by the coordinate transformation 1.29 will be called **locally inertial frame** (henceforth LIF), where "locally" means that, although it can be set up at any given spacetime point, it holds only in a small neighbourhood of that point. As we shall later discuss, locally inertial frames play an important role in the development of the theory of gravity.

Gravity is distinguished from all other forces because in a gravitational field, all bodies, given the same initial velocity, follow the same trajectory, regardless of their internal composition. This is not the case, for example, for electromagnetic forces which act on charged particles but not on neutral bodies; moreover, the trajectories of charged particles depend on their charge-to-mass ratio, which is not the same for all of them. Similarly, other forces, like the strong and weak interactions, affect different particles in a different way. It is this distinctive feature that makes it possible to describe the effects of gravity in terms of curved geometry, as we shall see in the next chapters.

We are now in a position to state the Principle of Equivalence, of which there are (at least) two formulations:

- **The strong Principle of Equivalence:** *In an arbitrary gravitational field, at any given spacetime point, it is possible to choose a locally inertial reference frame such that, in a sufficiently small region surrounding that point, all laws of physics take the same form as they would take in absence of gravity, namely the form prescribed by Special Relativity.*

- **The weak Principle of Equivalence:** *A weaker version of the principle states the same as above, but it refers only to the laws of motion of freely falling bodies, rather than to all laws of physics.*

Thus, the strong (weak) version of the Equivalence Principle states that in a locally inertial frame all laws of physics (the laws of motion) must coincide, locally, with those of Special Relativity. In particular, Special Relativity is constructed axiomatically upon the experimentally verified assumption that the speed of light is the same in every reference frame. This assumption has remarkable consequences, such as the relativity of simultaneity and the emergence of spacetime as a single entity. An *event* in spacetime is identified by the coordinates $\{\xi^\mu\} = (ct, x, y, z)$, where c is the speed of light, and in going from one inertial frame to another these coordinates change according to the Lorentz transformations. As a consequence the following quantity, known as *Minkowski's line element*,

$$ds^2 = -c^2dt^2 + dx^2 + dy^2 + dz^2 = -(d\xi^0)^2 + (d\xi^1)^2 + (d\xi^2)^2 + (d\xi^3)^2 \equiv \eta_{\mu\nu} d\xi^\mu d\xi^\nu \,, \quad (1.32)$$

is invariant under Lorentz's transformations and, as such, is interpreted as the distance between two (infinitely) close events. Thus, according to the Equivalence Principle, this must be true also in a LIF and, in this frame, the distance between two close points (events) must coincide with the Minkowski expression above.

This statement, which - we stress it again - follows from the Equivalence Principle, resembles very much the axiom that Gauss chose as a basis for non-Euclidean geometries, namely that at any given point in space, there exists a locally Euclidean reference frame such that, in a sufficiently small region surrounding that point, the distance between two points is given by Pythagoras' law. In our case this can be reformulated as follows: *at any given spacetime point, there exists a locally Minkowskian reference frame such that, in a sufficiently small spacetime region surrounding that point, the distance between two points is given by the Minkowski line element 1.32.*

1.4 GEODESIC EQUATIONS AS A CONSEQUENCE OF THE EQUIVALENCE PRINCIPLE

Let us start exploring what are the consequences of the Equivalence Principle. We wish to find the equations of motion of a particle moving under the sole action of a gravitational field, when this motion is observed in an arbitrary reference frame.

We start by analysing the motion in a LIF, the one in free fall with the particle, and let $\{\xi^\mu\}$, $\mu = 0, \ldots, 3$ be the coordinates of the LIF. According to the Equivalence Principle, in this frame the distance between two neighbouring points is

$$ds^2 = -(d\xi^0)^2 + (d\xi^1)^2 + (d\xi^2)^2 + (d\xi^3)^2 = \eta_{\mu\nu} d\xi^\mu d\xi^\nu \,, \tag{1.33}$$

and the equations of motion are those prescribed by Special Relativity, i.e.,

$$\frac{d^2\xi^\alpha}{d\tau^2} = 0 \,, \tag{1.34}$$

where τ/c is the particle proper time. We remark that, as discussed in Sec. 1.3, a LIF is defined in a sufficiently small region around a given spacetime point $\mathbf{p}$. This means that Eq. 1.34 holds exactly only in $\mathbf{p}$, whereas in a point $\mathbf{p}' \neq \mathbf{p}$ it holds with corrections of the order of the distance between $\mathbf{p}'$ and $\mathbf{p}$.

We now change to a frame where the coordinates are labeled $x^\alpha = x^\alpha(\xi^\gamma)$, i.e., we assign a transformation law which allows to express the new coordinates as functions of the old ones[6]. In the new frame the distance between two neighbouring points is

$$ds^2 = \eta_{\alpha\beta} \frac{\partial \xi^\alpha}{\partial x^\mu} dx^\mu \frac{\partial \xi^\beta}{\partial x^\nu} dx^\nu = g_{\mu\nu} dx^\mu dx^\nu \,, \tag{1.35}$$

where we have defined the metric tensor $g_{\mu\nu}$ as

$$g_{\mu\nu} = \frac{\partial \xi^\alpha}{\partial x^\mu} \frac{\partial \xi^\beta}{\partial x^\nu} \eta_{\alpha\beta} \,. \tag{1.36}$$

This formula is the four-dimensional generalization of Eq. 1.6, derived in Sec. 1.1 for two-dimensional spaces. In the new frame the particle equation of motion 1.34 becomes (see the detailed calculations in Box 1-B)

$$\frac{d^2x^\alpha}{d\tau^2} + \left(\frac{\partial x^\alpha}{\partial \xi^\lambda} \frac{\partial^2 \xi^\lambda}{\partial x^\mu \partial x^\nu} \right) \left(\frac{dx^\mu}{d\tau} \frac{dx^\nu}{d\tau} \right) = 0 \,. \tag{1.37}$$

[6]In the following chapter we shall clarify and introduce rigorously all concepts that we are now using, such us metric tensor, coordinate transformations, etc.

If we now define the following quantities

$$\Gamma^{\alpha}_{\mu\nu} = \frac{\partial x^{\alpha}}{\partial \xi^{\lambda}} \frac{\partial^2 \xi^{\lambda}}{\partial x^{\mu} \partial x^{\nu}} , \tag{1.38}$$

Eq. 1.37 can be written as

$$\frac{d^2 x^{\alpha}}{d\tau^2} + \Gamma^{\alpha}_{\mu\nu} \frac{dx^{\mu}}{d\tau} \frac{dx^{\nu}}{d\tau} = 0 . \tag{1.39}$$

The $\Gamma^{\alpha}_{\mu\nu}$ defined in Eq. 1.38 are called *Christoffel's symbols*, the properties of which will be discussed in Chapter 3. Eq. 1.39 is the **geodesic equation**, i.e., the equation of motion of a freely falling particle when observed in an arbitrary coordinate frame.

Let us analyse this equation in more detail. We have seen that according to the Equivalence Principle if we choose a locally inertial frame the particle equation of motion reduces to that of a free particle, Eq. 1.34. If we change the frame the particle will feel the gravitational field, as well as all apparent forces, like centrifugal, Coriolis', and dragging forces. In the new frame the geodesic equation becomes Eq. 1.39 and therefore the term

$$\Gamma^{\alpha}_{\mu\nu} \frac{dx^{\mu}}{d\tau} \frac{dx^{\nu}}{d\tau} \tag{1.40}$$

includes the gravitational force per unit mass acting on the particle, plus the additional apparent accelerations. Thus, in Newtonian mechanics the term 1.40 is $\boldsymbol{g}$, which is the gradient of the gravitational potential, plus the apparent accelerations. Note that the Christoffel symbols $\Gamma^{\alpha}_{\mu\nu}$ defined in Eq. 1.38 are constructed from the second derivatives of ξ^{α} with respect to the new coordinates x^{μ}. This suggests that, since the metric tensor 1.36 contains the first derivatives of ξ^{α}, ${\Gamma^{\alpha}}_{\mu\nu}$ should involve *first* derivatives of $g_{\mu\nu}$. In Chapter 3 we will show explicitly that this is the case, and we will derive the important relation between Christoffel's symbols and the metric tensor, which we here anticipate

$$\Gamma^{\sigma}_{\lambda\mu} = \frac{1}{2} g^{\nu\sigma} \left(\frac{\partial g_{\mu\nu}}{\partial x^{\lambda}} + \frac{\partial g_{\lambda\nu}}{\partial x^{\mu}} - \frac{\partial g_{\lambda\mu}}{\partial x^{\nu}} \right) . \tag{1.41}$$

Thus, in analogy with Newtonian theory, we might consider Christoffel's symbols as the generalization of the Newtonian gravitational field, and the metric tensor as the generalization of the Newtonian gravitational potential. Their role emerges in a natural way from the Equivalence Principle.

This interpretation is based on a *physical* analogy, namely on the comparison of Eq. 1.39, which describes the motion of freely falling bodies in an arbitrary frame, with the Newtonian equation of motion.

In addition, the metric tensor $g_{\mu\nu}$ is also a *geometrical* entity since, through the notion of distance, it characterizes the spacetime geometry. This double role, physical and geometrical, of the metric tensor is a direct consequence of the Principle of Equivalence, a point that one could hardly overstress.

Box 1-B

Given the equation of motion of a free particle in a LIF with coordinates $\{\xi^\mu\}$, $\mu = 0,\ldots,3$

$$\frac{d^2\xi^\alpha}{d\tau^2} = 0\,,$$

let us make a coordinate transformation to an arbitrary frame $\{x^\alpha\}$, $\alpha = 0,\ldots,3$

$$\xi^\alpha = \xi^\alpha(x^\gamma), \quad \rightarrow \quad \frac{d\xi^\alpha}{d\tau} = \frac{\partial\xi^\alpha}{\partial x^\gamma}\frac{dx^\gamma}{d\tau}\,.$$

Thus, $d^2\xi^\alpha/d\tau^2 = 0$ can be written as

$$\frac{d}{d\tau}\left(\frac{\partial\xi^\alpha}{\partial x^\gamma}\frac{dx^\gamma}{d\tau}\right) = \frac{d^2x^\gamma}{d\tau^2}\frac{\partial\xi^\alpha}{\partial x^\gamma} + \frac{\partial^2\xi^\alpha}{\partial x^\beta\partial x^\gamma}\frac{dx^\beta}{d\tau}\frac{dx^\gamma}{d\tau} = 0\,.$$

Multiplying the above equation by $\frac{\partial x^\sigma}{\partial\xi^\alpha}$ and reminding that

$$\frac{\partial\xi^\alpha}{\partial x^\gamma}\frac{\partial x^\sigma}{\partial\xi^\alpha} = \frac{\partial x^\sigma}{\partial x^\gamma} = \delta^\sigma_\gamma\,,$$

where $\delta^\sigma_\gamma = \mathrm{diag}(1,1,1,1)$ is the Kronecker symbol ($\delta^\sigma_\gamma = 1$ if $\sigma = \gamma$ and $\delta^\sigma_\gamma = 0$ otherwise), we find

$$\frac{d^2x^\gamma}{d\tau^2}\delta^\sigma_\gamma + \frac{\partial x^\sigma}{\partial\xi^\alpha}\frac{\partial^2\xi^\alpha}{\partial x^\beta\partial x^\gamma}\frac{dx^\beta}{d\tau}\frac{dx^\gamma}{d\tau} = 0\,.$$

Finally, the above equation can be written as

$$\frac{d^2x^\sigma}{d\tau^2} + \left[\frac{\partial x^\sigma}{\partial\xi^\alpha}\frac{\partial^2\xi^\alpha}{\partial x^\beta\partial x^\gamma}\right]\frac{dx^\beta}{d\tau}\frac{dx^\gamma}{d\tau} = 0\,,$$

which is Eq. 1.37.

1.5 LOCALLY INERTIAL FRAMES

Let us consider a reference frame with coordinates $\{x^\mu\}$ ($\mu = 0,\ldots,3$); we shall now show that if we know the metric tensor $g_{\mu\nu}$ and the Christoffel symbols $\Gamma^\alpha_{\mu\nu}$ (i.e., $g_{\mu\nu}$ and its first derivatives) at a point $\mathbf{p}$ with coordinates $x^\mu = X^\mu$, we can always determine the coordinates of a locally inertial frame $\{\xi^\alpha(x^\mu)\}$ in the neighbourhood of $\mathbf{p}$ in the following way.

Multiply $\Gamma^\beta_{\mu\nu}$ given in Eq. 1.38 by $\frac{\partial\xi^\beta}{\partial x^\lambda}$

$$\frac{\partial\xi^\beta}{\partial x^\lambda}\Gamma^\lambda_{\mu\nu} = \frac{\partial\xi^\beta}{\partial x^\lambda}\frac{\partial x^\lambda}{\partial\xi^\alpha}\frac{\partial^2\xi^\alpha}{\partial x^\mu\partial x^\nu} = \delta^\beta_\alpha\frac{\partial^2\xi^\alpha}{\partial x^\mu\partial x^\nu} = \frac{\partial^2\xi^\beta}{\partial x^\mu\partial x^\nu}\,, \tag{1.42}$$

i.e.,

$$\frac{\partial^2\xi^\beta}{\partial x^\mu\partial x^\nu} = \frac{\partial\xi^\beta}{\partial x^\lambda}\Gamma^\lambda_{\mu\nu}\,. \tag{1.43}$$

This is a differential equation for ξ^β, and we shall solve it by a series expansion about $\mathbf{p}$; for brevity, we shall indicate any function of the coordinates $f(x^0,\ldots,x^3)$ as $f(x)$, and, in

p, any $f(X^0, \ldots, X^3)$ as $f(X)$:

$$\begin{aligned} \xi^\beta(x) &= \xi^\beta(X) + [\frac{\partial \xi^\beta(x)}{\partial x^\lambda}]_{x=X}(x^\lambda - X^\lambda) \\ &+ \frac{1}{2}[\frac{\partial \xi^\beta(x)}{\partial x^\lambda}\Gamma^\lambda_{\mu\nu}]_{x=X}(x^\mu - X^\mu)(x^\nu - X^\nu) + \ldots, \end{aligned} \tag{1.44}$$

i.e.,

$$\xi^\beta(x) = a^\beta + b^\beta_\lambda(x^\lambda - X^\lambda) + \frac{1}{2}b^\beta_\lambda[\Gamma^\lambda_{\mu\nu}]_{x=X}(x^\mu - X^\mu)(x^\nu - X^\nu) + \ldots, \tag{1.45}$$

where, to simplify the notation, we have introduced

$$a^\beta \equiv \xi^\beta(X) \qquad \text{and} \qquad b^\beta_\nu \equiv \left[\partial\xi^\beta(x)/\partial x^\nu\right]_{x=X}.$$

In addition, by Eq. 1.36 we know that $g_{\mu\nu} = \frac{\partial \xi^\alpha}{\partial x^\mu}\frac{\partial \xi^\beta}{\partial x^\nu}\eta_{\alpha\beta}$, which, evaluated in X, gives

$$g_{\mu\nu}(X) = \eta_{\alpha\beta}\frac{\partial \xi^\alpha(x)}{\partial x^\mu}|_{x=X}\frac{\partial \xi^\beta(x)}{\partial x^\nu}|_{x=X} = \eta_{\alpha\beta}b^\alpha_\mu b^\beta_\nu\,. \tag{1.46}$$

This equation allows us to compute the coefficients b^β_μ.

Thus, given $g_{\mu\nu}$ and $\Gamma^\alpha_{\mu\nu}$ at a given point X, we can determine the coordinates of the locally inertial frame in the neighbourhood of X, to order $|x - X|^2$ by using Eq. 1.45. This equation defines the $\{\xi^\alpha\}$, but its solution is not unique. Indeed, Eq. 1.46 has ten independent components (those of the symmetric 4×4 matrix $g_{\mu\nu}$) and sixteen unknowns b^β_μ, which are then determined modulo six free independent parameters. With the four constants a^μ, the general solution for the coordinates of the locally inertial frame depends on ten free parameters. This ambiguity corresponds to the freedom to make an inhomogeneous Lorentz transformation, under which the new frame is still locally inertial: a four-dimensional translation (four degrees of freedom) and a Lorentz transformation (six degrees of freedom, see Box 1-C).

Box 1-C

The degrees of freedom of Lorentz's transformations

Given a locally inertial frame $\{\xi^\alpha\}$, where

$$ds^2 = \eta_{\mu\nu} d\xi^\mu d\xi^\nu \,, \tag{1.47}$$

let us consider the Lorentz transformation

$$\xi^\alpha = L^\alpha{}_{\mu'} \xi^{\mu'} \,, \tag{1.48}$$

where (assuming as usual that Greek indices run from 0 to 3 and Latin indices run from 1 to 3)

$$L^i_j = \delta^i_j + v^i v_j \frac{\gamma - 1}{v^2}, \qquad L^0_j = \frac{\gamma v_j}{c}, \qquad L^0_0 = \gamma, \qquad \gamma = (1 - \frac{v^2}{c^2})^{-\frac{1}{2}} \,. \tag{1.49}$$

The distance can now be written as

$$ds^2 = \eta_{\alpha\beta} d\xi^\alpha d\xi^\beta = \eta_{\alpha\beta} \frac{\partial \xi^\alpha}{\partial \xi^{\mu'}} \frac{\partial \xi^\beta}{\partial \xi^{\nu'}} d\xi^{\mu'} d\xi^{\nu'} \,. \tag{1.50}$$

Since

$$\frac{\partial \xi^\alpha}{\partial \xi^{\mu'}} = L^\alpha{}_{\nu'} \delta^{\nu'}{}_{\mu'} = L^\alpha{}_{\mu'} \,, \tag{1.51}$$

it follows that

$$ds^2 = \eta_{\alpha\beta} L^\alpha{}_{\mu'} L^\beta{}_{\nu'} d\xi^{\mu'} d\xi^{\nu'} \,. \tag{1.52}$$

Since $L^\alpha_{\mu'}$ is a Lorentz transformation,

$$\eta_{\alpha\beta} L^\alpha{}_{\mu'} L^\beta{}_{\nu'} = \eta_{\mu'\nu'} \,, \tag{1.53}$$

consequently the new frame is still a locally inertial frame, and

$$ds^2 = \eta_{\mu'\nu'} d\xi^{\mu'} d\xi^{\nu'} \,. \tag{1.54}$$

We note that Eq. 1.53 has ten independent components (those of a symmetric 4×4 matrix), therefore it determines the sixteen unknown $L^\alpha{}_{\mu'}$ modulo six free independent parameters: these are the six degrees of freedom of Lorentz's transformations, corresponding to the three components of the velocity of the boost and to the three angles of an arbitrary rotation.

CHAPTER 2

Elements of differential geometry

In Chapter 1 we have shown that the Principle of Equivalence allows us to establish a relation between the metric tensor and the gravitational field. We have used the notions of vectors and tensors, but we did not rigorously define these geometrical objects, nor did we discuss whether we are entitled to use these notions. In this chapter we shall now define the basic elements of differential geometry (vector, one-form, and tensor fields, coordinate transformations, metric spaces, etc.) in a more rigorous way.

2.1 TOPOLOGICAL SPACES, MAPPING, MANIFOLDS

2.1.1 Topological spaces

In General Relativity we deal with *topological spaces.* The word *topology* has two distinct meanings: local (in which we are mainly interested), and global, which involves the study of the large-scale features of a space.

Before introducing the general definition of topological space, let us recall some properties of $\mathbb{R}^n$, which is a particular case of topological space; this will help us in the understanding of the general notion.

Given a point $y = (y^1, y^2, \dots, y^n) \in \mathbb{R}^n$, a neighbourhood of y is the collection of points x such that

$$|x - y| \equiv \sqrt{\sum_{i=1}^{n}(x^i - y^i)^2} < r\,, \tag{2.1}$$

where r is a real number.

A set of points $\mathbf{S} \subset \mathbb{R}^n$ is *open* if every point $x \in \mathbf{S}$ has a neighbourhood entirely contained in $\mathbf{S}$. This implies that an open set does not include the points on its boundary. For instance, an open ball is an open set; a closed ball, defined by $|x - y| \leq r$, is not an open set, because the points of the boundary, i.e. $|x - y| = r$, do not admit a neighbourhood contained in the set.

Intuitively, this is a **continuum space**, i.e. there are points of $\mathbb{R}^n$ arbitrarily close to any given point, and the line joining two points can be subdivided into arbitrarily many pieces which also join points of $\mathbb{R}^n$. A non-continuous space is, for example, a lattice. A formal characterization of a continuum space is the *Hausdorff criterion*: for any two points of a continuum space which do not coincide, there exist two neighbourhoods which do not intersect (see Fig. 2.1).

Figure 2.1: Representation of Hausdorff's criterion.

The open sets of $\mathbb{R}^n$ satisfy the following properties:

(1) if $\mathbf{O}_1$ and $\mathbf{O}_2$ are open sets, their intersection is also an open set;

(2) the union of any collection (possibly infinite in number) of open sets is open.

We have introduced these concepts (open sets, neighbourhood, continuum space, etc.) in the space $\mathbb{R}^n$, but, as we are going to show, they can be extended to a *general* set, without the need of using the properties of $\mathbb{R}^n$.

Let us consider a *general set* $\mathbf{T}$, and a collection of subsets of $\mathbf{T}$, say $\mathbf{O}=\{\mathbf{O}_i\}$, and call them *open sets*, such that $\mathbf{T}$ itself and the empty set $\emptyset$ belong to the collection. We say that the pair $(\mathbf{T},\mathbf{O})$, consisting of the set and the collection of subsets, is a **topological space**, if it satisfies the properties (1) and (2) above. We remark that the set $\mathbf{T}$ can be any kind of set; the only specification we give is the collection of subsets $\mathbf{O}$, which are by definition the open sets, and that satisfy the properties (1), (2). In particular, in a topological space the notion of distance is a structure which has not been introduced: all definitions only require the notion of open sets.

In a topological space $\mathbf{T}$, a neighbourhood $\mathbf{B}(\mathbf{x})$ of a point $\mathbf{x} \in \mathbf{T}$ is an open set which contains the point $\mathbf{x}$, i.e. $\mathbf{x} \in \mathbf{B}(\mathbf{x})$ and $\mathbf{B}(\mathbf{x}) \in \mathbf{O}$. With these definitions, the Hausdorff criterion (and thus the notion of continuum space) can be applied to a topological space: the criterion is satisfied if $\forall\, \mathbf{x}, \mathbf{y} \in \mathbf{T}$ such that $\mathbf{x} \neq \mathbf{y}$, there exist two neighbourhoods $\mathbf{B}(\mathbf{x})$, $\mathbf{B}(\mathbf{y})$ which do not intersect.

2.1.2 Mapping

A map f from a space $\mathbf{M}$ to a space $\mathbf{N}$ is a rule which associates an element $\mathbf{x}$ of $\mathbf{M}$ to a unique element $\mathbf{y} = f(\mathbf{x})$ of $\mathbf{N}$ (see Fig. 2.2). The spaces $\mathbf{M}$ and $\mathbf{N}$ need not to be different. For example, the simplest maps are ordinary real-valued functions on $\mathbb{R}$, such as:

$$y = x^3, \qquad x \in \mathbb{R}, \qquad y \in \mathbb{R}\,. \tag{2.2}$$

In this case $\mathbf{M}$ and $\mathbf{N}$ coincide. It is also said that f maps a points $\mathbf{x} \in \mathbf{M}$ *into* a point $f(\mathbf{x}) \in \mathbf{N}$.

Surjective, injective, and bijective mappings

A map gives a unique $f(\mathbf{x})$ for every $\mathbf{x}$, but not necessarily a unique $\mathbf{x}$ for every $f(\mathbf{x})$ (see, e.g., Fig. 2.3). If f maps $\mathbf{M}$ into $\mathbf{N}$, then for any set $\mathbf{S}$ in $\mathbf{M}$ we have an *image* in $\mathbf{N}$, i.e. the set $\mathbf{T}= f(\mathbf{S})$ of all points mapped by f from $\mathbf{S}$ into $\mathbf{N}$. Conversely the set $\mathbf{S}$ is the *inverse image* of $\mathbf{T}$ (Fig. 2.4)

$$\mathbf{S} = f^{-1}(\mathbf{T})\,.$$

The statement "f maps $\mathbf{M}$ into $\mathbf{N}$" is indicated as

$$f : \mathbf{M} \to \mathbf{N}\,.$$

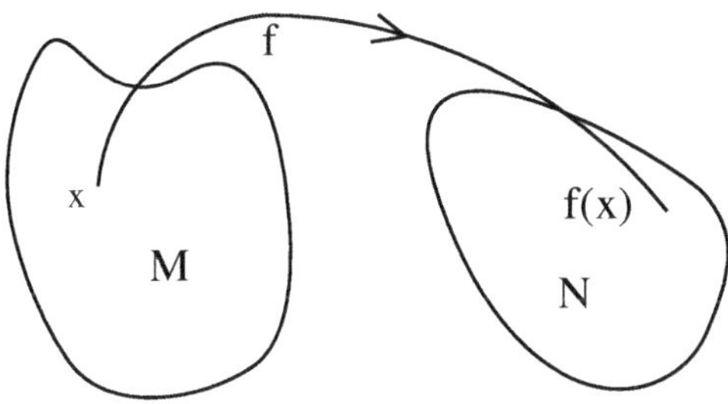

Figure 2.2: Mapping from **M** to **N**.

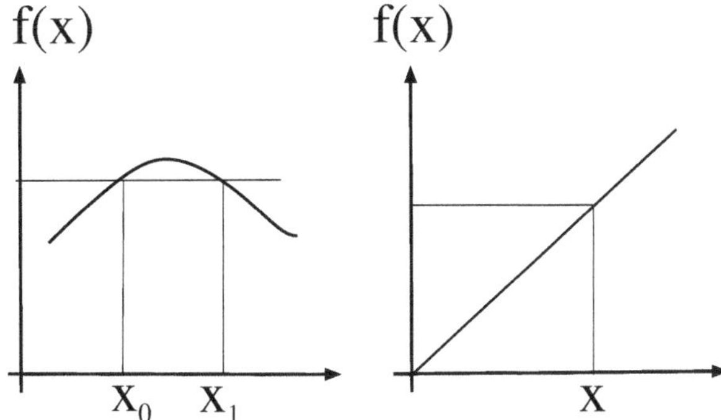

Figure 2.3: A given value of $f(\mathbf{x})$ may correspond either to different values (left panel) or to a single value (right panel) of $\mathbf{x}$.

In addition, f maps a particular element $\mathbf{x} \in \mathbf{M}$ into $\mathbf{y} \in \mathbf{N}$; the mapping is indicated as

$$f : \mathbf{x} \mapsto \mathbf{y} \,. \tag{2.3}$$

The image of a point $\mathbf{x}$ is $f(\mathbf{x})$. We can introduce the following definitions:

- If every point of **N** has an inverse image (but not necessarily a unique one), it is a map from **M onto N**, and it is called a **surjective** map.
- If, for any point of **N** which has an inverse image in **M**, the inverse image is *unique*, the map is said to be **injective**.
- A map of **M** into **N**, which is both surjective and injective, is called **bijective**, or a **one-to-one** map.
- If the map is one-to-one, every point of **M** corresponds to one and only one point of **N**, and vice versa; therefore, bijective maps are *invertible*.
- Inverse mapping is possible only in the case of one-to-one mapping. The *inverse map* to f is indicated as f^{-1}.

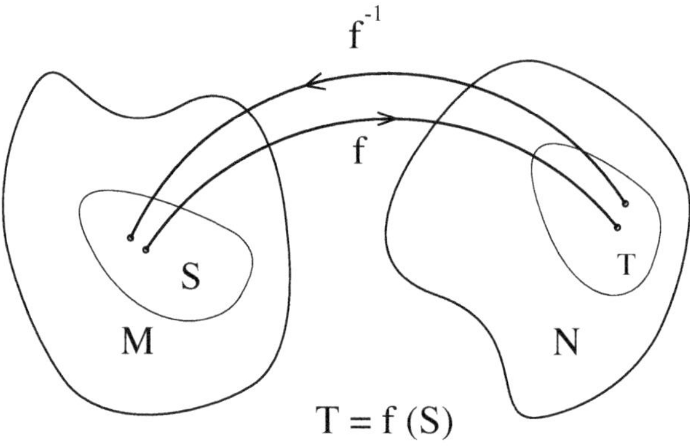

Figure 2.4: Image and inverse image of a mapping.

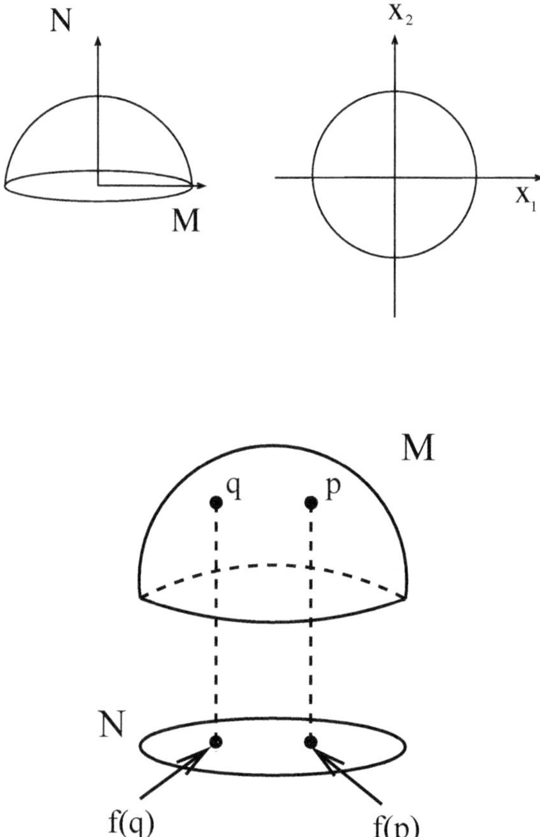

Figure 2.5: The sphere and its mapping to a disc.

Box 2-A

Example of a one-to-one mapping

Let $\mathbf{N}$ be the unit open disc in $\mathbb{R}^2$, i.e. the set of all points in $\mathbb{R}^2$ such that the distance from the center is less than unity, $d(0, x) < 1$. Let $\mathbf{M}$ be the surface of a hemisphere $\theta < \frac{\pi}{2}$ belonging to the unit sphere. There exists a one-to-one mapping f from $\mathbf{M}$ to $\mathbf{N}$ (see Fig. 2.5).

Composition of maps

Given two maps $f : \mathbf{M} \to \mathbf{N}$ and $g : \mathbf{N} \to \mathbf{P}$, there exists a map $g \circ f$ that maps $\mathbf{M}$ to $\mathbf{P}$

$$g \circ f : \mathbf{M} \to \mathbf{P} \,. \tag{2.4}$$

This means the following: take a point $\mathbf{x} \in \mathbf{M}$ and find the image $f(\mathbf{x}) \in \mathbf{N}$, then use g to map this point to a point $g\left(f(\mathbf{x})\right) \in \mathbf{P}$ (Fig. 2.6).

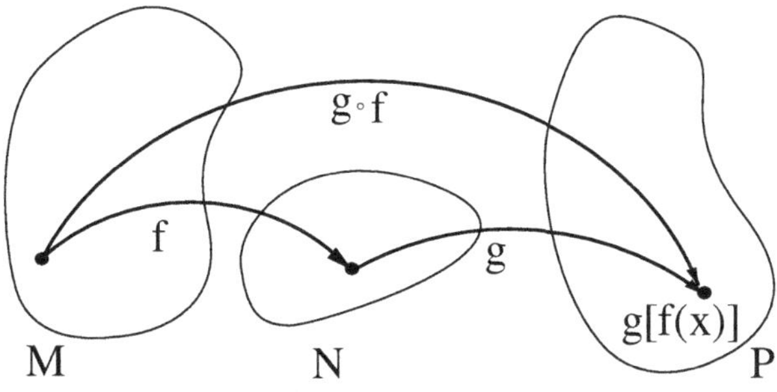

Figure 2.6: Composition of maps.

Box 2-B

Example: composition of maps

$$\begin{aligned} f : x &\mapsto y & \qquad y &= x^3 \\ g : y &\mapsto z & \qquad z &= y^2 \\ g \circ f : x &\mapsto z & \qquad z &= x^6 \end{aligned}$$

Continuous mapping

Given two topological spaces $\mathbf{M}$ and $\mathbf{N}$, a map $f : \mathbf{M} \to \mathbf{N}$ is continuous at $\mathbf{x} \in \mathbf{M}$ if any open set of $\mathbf{N}$ containing $f(\mathbf{x})$ contains the image of an open set of $\mathbf{M}$.

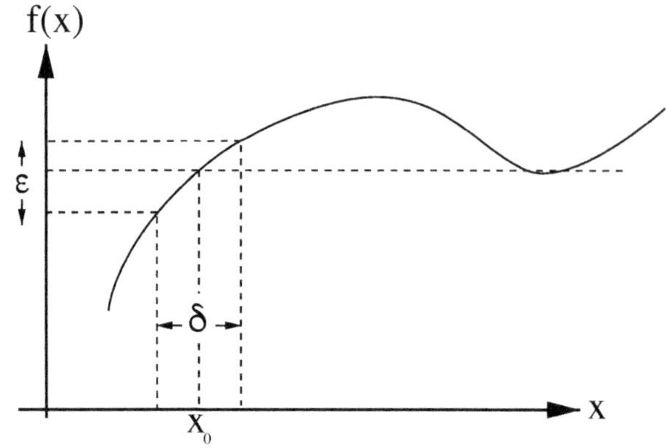

Figure 2.7: Continuous function.

This definition is related to the familiar notion of continuous functions. Suppose that f is a real-valued function of one real variable, i.e. f is a map of $\mathbb{R}$ to $\mathbb{R}$

$$f : \mathbb{R} \to \mathbb{R}\,. \tag{2.5}$$

In the elementary calculus f is said to be continuous at a point x_0 if, for every $\epsilon > 0$, there exists a $\delta > 0$ for which

$$\forall x \quad \text{such that} \quad |x - x_0| < \delta, \qquad\qquad |f(x) - f(x_0)| < \epsilon \tag{2.6}$$

(see Fig. 2.7). This definition can be translated in terms of open sets. We see from Fig. 2.7 that any open set containing $f(\mathbf{x_0})$, i.e. $|f(\mathbf{x}) - f(\mathbf{x_0})| < \epsilon$ (for any ϵ), contains an image of an open set of **M**. This is true at least in the domain of definition of f. This definition is more general than that of continuous functions, because it is based on the notion of open sets, and not on the notion of distance.

2.1.3 Manifolds and differentiable manifolds

A **manifold M** *is a topological space, which satisfies the Hausdorff criterion, and such that each point of* M *has an open neighbourhood which has a continuous one-to-one map to an open set of* $\mathbb{R}^n$*, where n is the dimension of the manifold.*

In this definition we have used the concepts previously defined: the space must be topological, continuous, and we want to associate an n-tuple of real numbers, i.e. *a set of coordinates*, to each point. For example, when we consider the diagram in Fig. 2.8, we are just using the notion of manifold: we take a point **p**, and map it to the point $(x_1, y_1) \in \mathbb{R}^2$, and this operation can be done for any open neighbourhood of **p**. It should be stressed that the definition of manifold involves open sets and not the whole **M** and $\mathbb{R}^n$, because we do not want to restrict the global topology of **M**. Moreover, at this stage we only require the map to be one-to-one. We have not yet introduced any geometrical notion such as length, angles, etc.

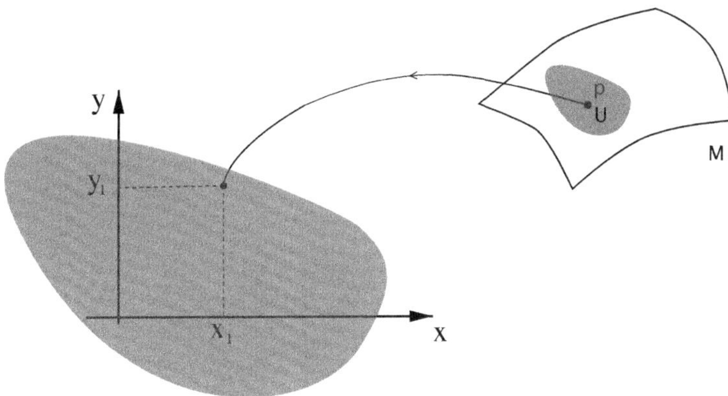

Figure 2.8: A coordinate frame in $\mathbb{R}^2$.

Coordinate systems and coordinate transformations

A coordinate system, or coordinate frame or **chart**, is a pair consisting of an open set of **M** and its map to an open set of $\mathbb{R}^n$. The open set does not necessarily include all **M**; thus there will be other open sets with the associated maps, and each point of **M** must lie in at least one of such open sets.

Let **U** and **V** be two overlapping open sets of **M** with two distinct maps into $\mathbb{R}^n$. The overlap $\mathbf{U} \cup \mathbf{V}$ is open (since it is the intersection of two open sets), and is given two different coordinate systems by the two maps. Thus, there must exist some relation between these maps, which we want to find.

As shown in the lower panel of Fig. 2.9, pick a point $(x^1, \ldots, x^n)$ in the image of the overlapping region belonging to $f(\mathbf{U})$. The map f has an inverse f^{-1} which brings to the point $\mathbf{p}$ of $\mathbf{U} \cup \mathbf{V}$. Then, use the map g to go from $\mathbf{p}$ to its image belonging to $g(\mathbf{V})$, i.e. to the point $(y^1, \ldots, y^n)$ in $\mathbb{R}^n$,

$$g \circ f^{-1} : \mathbb{R}^n \to \mathbb{R}^n. \tag{2.7}$$

The result of this operation is a functional relation between the two sets of coordinates, called **coordinate transformation** between the two charts:

$$\begin{cases} y^1 = y^1(x^1, \ldots, x^n) \\ . \\ . \\ . \\ y^n = y^n(x^1, \ldots, x^n). \end{cases} \tag{2.8}$$

If the partial derivatives of order $\leq k$ of all the functions $\{y^i\}$ with respect to all $\{x^i\}$ exist and are continuous, then the charts $(\mathbf{U}, f)$ and $(\mathbf{V}, g)$ are said to be C^k-related. If it is possible to construct a system of charts such that each point of **M** belongs at least to one of the open sets, and every chart is C^k related to every other one it overlaps with, then the manifold is said to be a C^k manifold. If $k \geq 1$, the manifold is called **differentiable**.

The notion of differentiable manifold is crucial, because it allows to add "structure" to the manifold, i.e. to define vectors, tensors, differential forms, Lie derivatives, etc. In order to complete the definition of coordinate transformation we still need to clarify the following.

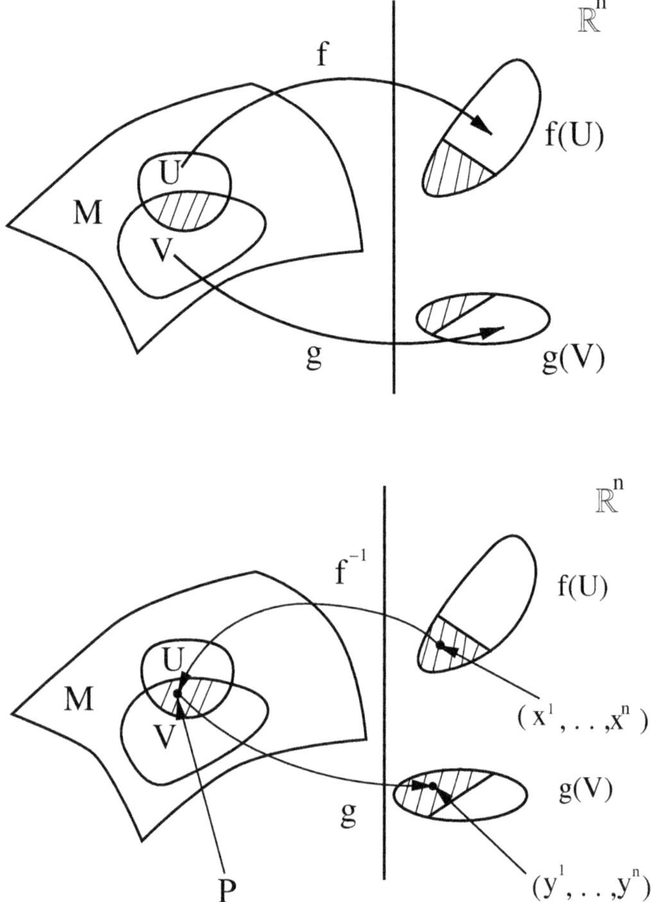

Figure 2.9: Two overlapping open sets of a manifold, with different maps f, g into $\mathbb{R}^n$ (upper panel). Transformation between two coordinate systems (lower panel).

Let us write Eqs. 2.8 in the form

$$y^i = f^i(x^1, \ldots, x^n), \qquad i = 1, \ldots, n, \tag{2.9}$$

where f^i are C^k-differentiable. Let J be the Jacobian of the transformation

$$J = \frac{\partial(f^1, \ldots, f^n)}{\partial(x^1, \ldots, x^n)} = \det \begin{pmatrix} \frac{\partial f^1}{\partial x^1} & \frac{\partial f^1}{\partial x^2} & \cdot & \cdot & \cdot & \frac{\partial f^1}{\partial x^n} \\ \frac{\partial f^2}{\partial x^1} & \frac{\partial f^2}{\partial x^2} & \cdot & \cdot & \cdot & \frac{\partial f^2}{\partial x^n} \\ \cdot & \cdot & \cdot & \cdot & \cdot & \cdot \\ \cdot & \cdot & \cdot & \cdot & \cdot & \cdot \\ \frac{\partial f^n}{\partial x^1} & \frac{\partial f^n}{\partial x^2} & \cdot & \cdot & \cdot & \frac{\partial f^n}{\partial x^n} \end{pmatrix}. \tag{2.10}$$

If J is non-zero at some point $\mathbf{p}$, then the inverse function theorem ensures that the map f is one-to-one in some neighbourhood of $\mathbf{p}$. If J is zero in $\mathbf{p}$ the transformation is singular.

Since a coordinate transformation must be one-to-one in its domain, J must not vanish in this domain.

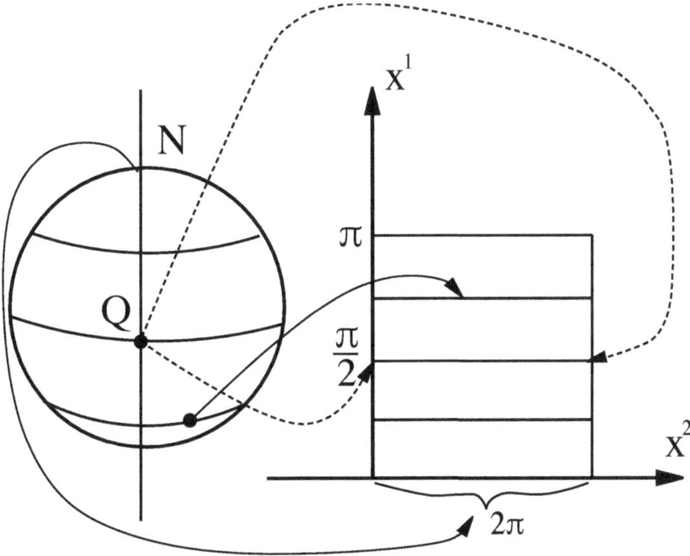

Figure 2.10: Two-sphere as a manifold.

Box 2-C

Example: the two-sphere as a manifold

Consider the two-sphere $\mathbf{S^2}$. It is defined as the set of all points in $\mathbb{R}^3$ such that $x^2+y^2+z^2 = const$ (see Fig. 2.10). Suppose that we want to map the *whole* sphere to $\mathbb{R}^2$ by using a single chart, for example using spherical coordinates $\theta \equiv x^1$, and $\varphi \equiv x^2$. The sphere will be mapped onto the rectangle $0 \le x^1 \le \pi, \quad 0 \le x^2 \le 2\pi$.
However, this is not a one-to-one mapping, and, strictly speaking, it is not even a mapping! Indeed, every point of the sphere should correspond to one *and only one* point of $\mathbb{R}^n$, while the north pole $\theta = 0$ is mapped to the entire line

$$x^1 = 0, \quad 0 \le x^2 \le 2\pi\,. \tag{2.11}$$

In addition all points of the semicircle $\varphi = 0$ are mapped in two places

$$x^2 = 0, \qquad \text{and} \qquad x^2 = 2\pi\,. \tag{2.12}$$

The only way to avoid these problems and have a well-defined (and one-to-one) mapping is to restrict the map to open regions (indeed, the definition of a manifold requires the maps to be defined between open sets):

$$0 < x^1 < \pi, \quad 0 < x^2 < 2\pi\,. \tag{2.13}$$

The two poles and the semicircle $\varphi = 0$ are left out. We may consider a second map, again in spherical coordinates but "rotated" in such a way that the line $\varphi = 0$ coincides with the equator of the old system. Then, every point of the sphere would be covered by one of the two charts, and the coordinate transformation for the overlapping region (which is simply a rotation) can be found. It is interesting to note that these mappings do not preserve angles and lengths. The two-sphere is an example of manifold that cannot be covered by a single chart, i.e. by a single coordinate frame, which is the reason why a globe cannot be mapped entirely into a two-dimensional map.

2.1.4 Diffeomorphisms

Given a differentiable manifold $\mathbf{M}$, let us consider an invertible (one-to-one) mapping of $\mathbf{M}$, or more generally of an open set $\mathbf{U} \subset \mathbf{M}$, into itself:

$$\begin{aligned} \Psi :\ & \mathbf{U} \to \mathbf{U} \\ & \mathbf{p} \mapsto \mathbf{p}' = \Psi(\mathbf{p})\,. \end{aligned} \tag{2.14}$$

In a coordinate frame $\mathbf{U} \to \mathbb{R}^n$, it defines n real functions of (an open set of) $\mathbb{R}^n$:

$$x^i \mapsto x'^i = \psi^i(x^j) \in \mathbb{R}^n\,. \tag{2.15}$$

If these are differentiable functions, the mapping Ψ is called a **diffeomorphism**.

Note that Eq. 2.15 is formally similar to a coordinate transformation, $x^i \mapsto x^{i'}$, but its meaning is very different: in a coordinate transformation the chart changes, but the point $\mathbf{p}$ on the manifold remains the same, while in a diffeomorphism the chart is the same, but the point $\mathbf{p}$ on the manifold is mapped to a different point $\mathbf{p}'$ (see Fig. 2.11). As we shall discuss in Chapter 8, diffeomorphisms are very important in General Relativity, since they are associated to spacetime symmetries.

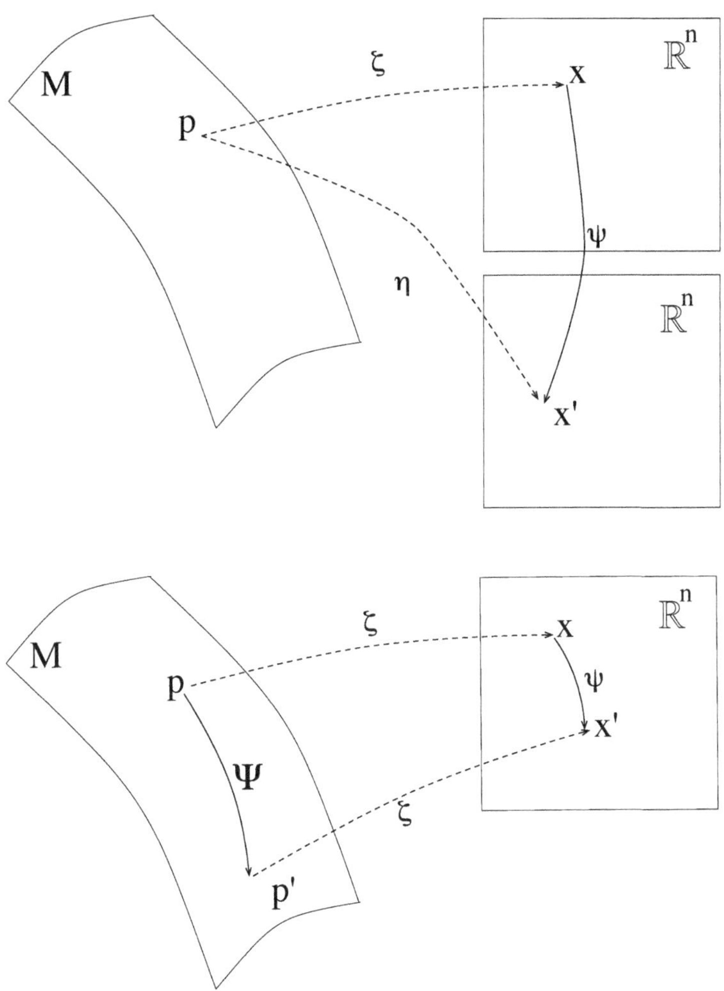

Figure 2.11: In the top panel we show a coordinate transformation: x and x' describe the same point $\mathbf{p}$ of the manifold in two different charts. In the bottom panel we show a diffeomorphism: x and x' describe different points $\mathbf{p}$, $\mathbf{p}'$ of the manifold, in the same chart.

2.2 VECTORS

2.2.1 The traditional definition of a vector

Let us consider an n-dimensional manifold, and a generic coordinate transformation

$$x^{i'} = x^{i'}(x^j), \qquad i', j = 1, \ldots, n\,. \tag{2.16}$$

A *contravariant* vector

$$\vec{V} \to_O \{V^i\}_{i=1,\ldots n}\,, \tag{2.17}$$

where the symbol $\to_O$ indicates that $\vec{V}$ has components $\{V^i\}$ with respect to a given frame O, can be defined as a collection of n numbers which transform under the coordinate transformation (2.16) as follows:

$$V^{i'} = \sum_{j=1,\dots,n} \frac{\partial x^{i'}}{\partial x^j} V^j \equiv \frac{\partial x^{i'}}{\partial x^j} V^j \,. \tag{2.18}$$

Notice that, in writing the last term, we have used Einstein's convention. $V^{i'}$ are the components of the vector in the new frame. If we now define the $n \times n$ matrix

$$(\Lambda^{i'}{}_j) = \begin{pmatrix} \frac{\partial x^{1'}}{\partial x^1} & \frac{\partial x^{1'}}{\partial x^2} & \cdots \\ \cdot & \cdot & \cdots \\ \cdot & \cdot & \cdots \\ \cdot & \cdot & \cdots \\ \frac{\partial x^{n'}}{\partial x^1} & \frac{\partial x^{n'}}{\partial x^2} & \cdots \end{pmatrix}, \tag{2.19}$$

the transformation law can be written in the general form

$$V^{i'} = \Lambda^{i'}{}_j V^j. \tag{2.20}$$

In addition, *covariant vectors* are defined as objects that transform according to the following rule

$$A_{i'} = \frac{\partial x^j}{\partial x^{i'}} A_j = \Lambda^j{}_{i'} A_j, \tag{2.21}$$

where $\Lambda^j{}_{i'}$ is the inverse matrix of $\Lambda^{i'}{}_j$ (see Box 2-D).

This definition of a vector relies on the choice of a coordinate system. However, we know that in $\mathbb{R}^n$ a vector can be defined as an oriented segment joining two points, without introducing a coordinate frame. We shall now show that in a *general manifold* it is possible to define a vector as a **geometrical object**, i.e. one which exists regardless of the coordinate system. Of course, once a coordinate system is given we can associate to a vector its components with respect to that system and, if the frame changes, the vector components transform as in Eq. 2.18. However, the vector itself – as a geometrical object – does not change.

Box 2-D

The matrices $\Lambda^i{}_{j'}$ and $\Lambda^{i'}{}_j$

Given a coordinate transformation $x^{i'} = x^{i'}(x^j)$, or the inverse $x^j = x^j(x^{i'})$, $i', j = 1, \dots, n$, the matrices

$$\Lambda^{i'}{}_j = \frac{\partial x^{i'}}{\partial x^j} \tag{2.22}$$

and

$$\Lambda^i{}_{j'} = \frac{\partial x^i}{\partial x^{j'}}, \tag{2.23}$$

are the inverse of each other. Indeed

$$\Lambda^{i'}{}_k \Lambda^k{}_{j'} = \frac{\partial x^{i'}}{\partial x^k} \frac{\partial x^k}{\partial x^{j'}} = \frac{\partial x^{i'}}{\partial x^{j'}} = \delta^{i'}_{j'} \,. \tag{2.24}$$

Note that, when we write $\Lambda^{i'}{}_j$ or $\Lambda^i{}_{j'}$, the first index (i.e. the one on the left) always refers to the row of the matrix, the second to the column.

Box 2-E

A comment on notation

Here and in the following, we shall use indices with and without primes to refer to different coordinate frames. Strictly speaking, Eq. 2.16 should be written as

$$x'^{i'} = x'^{i'}(x^j), \qquad i', j = 1, \ldots, n, \tag{2.25}$$

because the coordinate with (say) $i' = 1$ belongs to the new frame, and is then different from the coordinate with $j = 1$, belonging to the old frame. However, for brevity of notation, we will omit the primes in the coordinates, keeping only the primes in the indices, as in Eq. 2.16.

2.2.2 A geometrical definition

In order to define vectors as geometrical objects, we need to go by steps: firstly we shall introduce the notions of *paths* and *curves* to define the tangent vector to a curve at a given point $\mathbf{p}$. Then we shall introduce the directional derivative along a curve in $\mathbf{p}$, which will be shown to be in a one-to-one correspondence with the vector tangent to the same curve at the same point. This will allow us to give a definition of vectors that is independent of the coordinate system.

Paths and curves

A **path** $\mathcal{C}$ is a connected sequence of points in a manifold. An example of a path is shown in Fig. 2.12.

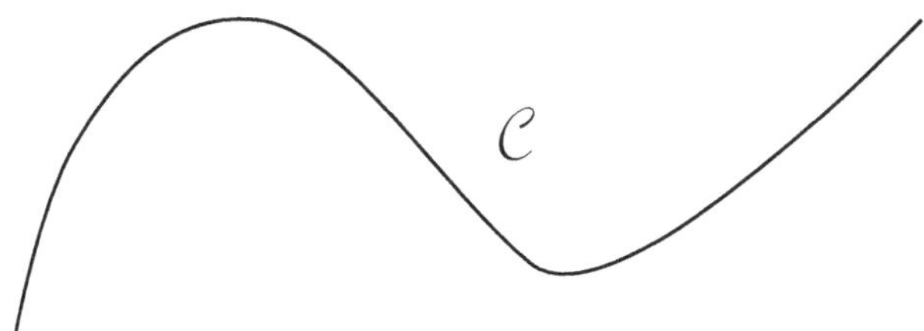

Figure 2.12: Path on a manifold.

A **curve** is a mapping from an interval $I = [a, b] \subset \mathbb{R}$ to a path in a manifold,

$$\gamma\ :\ s \in [a, b] \mapsto \gamma(s) \in \mathcal{C} \subset\ \mathbf{M}\,. \tag{2.26}$$

Thus, a curve γ associates a real number to each point of the path. We say that the curve is a *parametrization* of the path $\mathcal{C}$, and the variable $s \in [a, b]$ is called **parameter** of the curve. The path is then the image of the real interval I in the manifold.

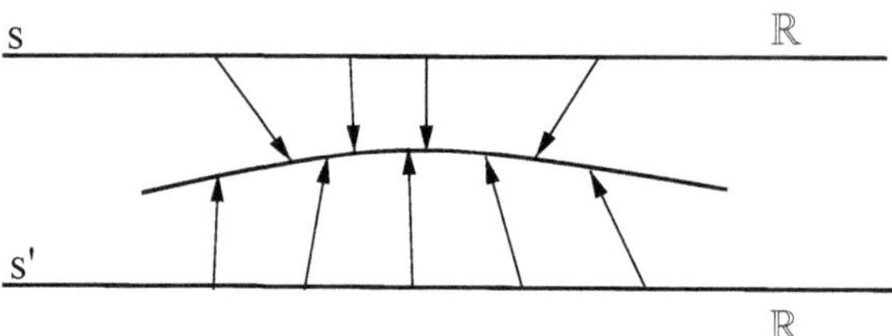

Figure 2.13: Different parametrizations of the same path.

Given a coordinate system $(x^1, \ldots, x^n)$ defined in the open set of the manifold containing the path, we can express the curve γ as a set of n real functions $(x^1(s), \ldots, x^n(s))$

$$\gamma \ : \ s \in [a, b] \mapsto (x^1(s), x^2(s), \ldots, x^n(s)) \,. \tag{2.27}$$

We say that the curve is C^k if the n functions are C^k.

If we do a reparametrization $s' = s'(s)$, the number associated to a given point of the path changes, i.e. the curve changes (see Fig. 2.13); therefore we get

$$\gamma' \ : \ s' \in [a', b'] \mapsto (x^1(s(s')), x^2(s(s')), \ldots, x^n(s(s'))) = (x'^1(s'), x'^2(s'), \ldots, x'^n(s')) \,, \tag{2.28}$$

where x'^1, x'^2 are new functions of s'. This is a *new curve*, although the path is the same.

Box 2-F

Example: paths and curves

The trajectory of a bullet shot by a gun in the three-dimensional space is a path; when we associate the parameter t (time) at each point of the trajectory, we define a curve; if we change the parameter, for instance we choose the curvilinear abscissa, we define a new curve.

Tangent vector to a curve

Let us consider a regular (i.e., C^1) curve γ on a differentiable manifold $\mathbf{M}$, with parameter λ, and a point $\mathbf{p}$ belonging to the curve. Given a coordinate system $(x^1, \ldots, x^n)$, as shown in Eq. 2.27 we can express the curve as a set of n real, C^1 functions $(x^1(\lambda), \ldots, x^n(\lambda))$ (see Fig. 2.14).

The set of numbers $\left\{\frac{dx^1}{d\lambda}, \ldots, \frac{dx^n}{d\lambda}\right\}$ are the components of the **tangent vector** to γ in $\mathbf{p}$:

$$\vec{V} \to_O \left\{\frac{dx^i}{d\lambda}\right\}_{i=1,\ldots,n} \,. \tag{2.29}$$

One must be careful not to confuse the curve with the path. In fact a path has, at any given point, an infinite number of tangent vectors, all parallel, but with different lengths, corresponding to the different possible parametrizations of the path. A curve, instead, has a *unique tangent vector* in any given point. Note that there are paths that are tangent to

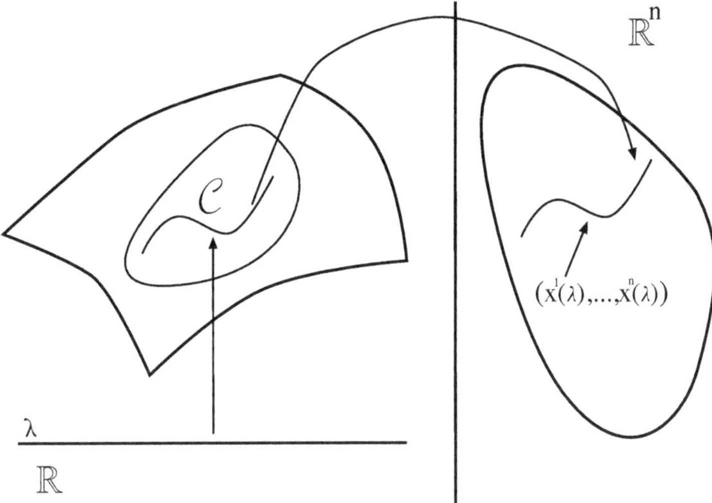

Figure 2.14: Curve on a manifold expressed as n real functions.

Figure 2.15: Different curves having the same tangent vector.

one another in $\mathbf{p}$ (see Fig. 2.15); we can choose suitable parametrizations of the paths such that the resulting curves have the same tangent vector in $\mathbf{p}$.

Let us now consider a different coordinate system $(x^{1'}, \ldots, x^{n'})$. Since $\mathbf{M}$ is a differentiable manifold, the functions $(x^{1'}(x^1, \ldots, x^n), \ldots, x^{n'}(x^1, \ldots, x^n))$ are regular and invertible in their domain. The components of the tangent vector in the new coordinate frame become

$$V^{i'} = \frac{dx^{i'}}{d\lambda} = \frac{\partial x^{i'}}{\partial x^j}\frac{dx^j}{d\lambda} = \Lambda^{i'}{}_j \frac{dx^j}{d\lambda}, \tag{2.30}$$

where $\Lambda^{i'}{}_j$ are the components of the matrix associated to the coordinate transformation defined in Eq. 2.19. As expected, Eq. 2.30 is the same transformation given in Eq. 2.20, which was used to define contravariant vectors in Sec. 2.2.1:

$$V^{i'} = \Lambda^{i'}{}_j V^j. \tag{2.31}$$

The definition 2.29 of vector tangent to a curve in a given point still depends on the choice of the coordinate system. In order to show that vectors are geometrical objects, i.e. objects that do not depend on the coordinate frame, we need to define the directional derivatives along a curve.

Directional derivatives along a curve

Let us consider a regular curve γ on a differentiable manifold $\mathbf{M}$, with parameter λ, and a point $\mathbf{p}$ belonging to the curve. Let $\mathbf{U}$ be a neighbourhood of $\mathbf{p}$. Let us also consider a real, differentiable function Φ defined in $\mathbf{U}$

$$\Phi : \mathbf{U} \to \mathbb{R}. \tag{2.32}$$

Given a coordinate system $(x^1, \ldots, x^n)$, we can express Φ as a function on $\mathbb{R}^n$, $\Phi = \Phi(x^1, \ldots, x^n)$, and the curve γ as a set of n real, C^1 functions $(x^1(\lambda), \ldots, x^n(\lambda))$.

We define the **directional derivative of Φ** in $\mathbf{p}$ along the curve γ as the real number

$$\frac{d\Phi}{d\lambda} = \frac{\partial \Phi}{\partial x^i}\frac{dx^i}{d\lambda}. \tag{2.33}$$

Since the function Φ is totally arbitrary, we can rewrite this expression as

$$\frac{d}{d\lambda} = \frac{dx^i}{d\lambda}\frac{\partial}{\partial x^i}, \tag{2.34}$$

where $\frac{d}{d\lambda}$ is the *directional derivative operator* in $\mathbf{p}$, acting on the space of the C^1 functions in $\mathbf{U}$:

$$\frac{d}{d\lambda} : C^1(\mathbf{U}) \to \mathbb{R}. \tag{2.35}$$

We have just found an important result: *Eq. 2.34 establishes a one-to-one relation between the directional derivative $\frac{d}{d\lambda}$ along a curve in $\mathbf{p}$, and the components of the tangent vector to the same curve in $\mathbf{p}$, $\frac{dx^i}{d\lambda}$.*

Let us consider a different coordinate frame defined in $\mathbf{U}$, $(x^{1'}, \ldots, x^{n'})$. As discussed in the previous section, the functions $(x^{1'}(x^1, \ldots, x^n), \ldots, x^{n'}(x^1, \ldots, x^n))$ are regular and invertible in their domain. Then we can write

$$\frac{\partial \Phi}{\partial x^{i'}}\frac{dx^{i'}}{d\lambda} = \left(\frac{\partial \Phi}{\partial x^j}\frac{\partial x^j}{\partial x^{i'}}\right)\left(\frac{\partial x^{i'}}{\partial x^k}\frac{dx^k}{d\lambda}\right) = \frac{\partial \Phi}{\partial x^k}\frac{dx^k}{d\lambda} = \frac{d\Phi}{d\lambda}. \tag{2.36}$$

Therefore *the directional derivative of a function does not depend on the choice of the coordinate system, i.e. the directional derivative operator is a geometrical object.*

We shall now show that the space of directional derivatives along a curve on a differentiable manifold satisfies the axiomatic definition of vector spaces, which is the following[1].

A vector space is a set V on which two operations are defined:

1. *Vector sum*
$$(\vec{v},\vec{w}) \rightarrow \vec{v}+\vec{w} \tag{2.37}$$

2. *Multiplication by a real number:*
$$(a,\vec{v}) \rightarrow a\vec{v} \tag{2.38}$$

where $\vec{v},\vec{w} \in V$, $a \in \mathbb{R}$, which satisfy the following properties:

- *Associativity and commutativity of vector sum*
$$\vec{v}+(\vec{w}+\vec{u}) = (\vec{v}+\vec{w})+\vec{u} \tag{2.39}$$
$$\vec{v}+\vec{w} = \vec{w}+\vec{v}\,, \qquad \forall\,\vec{v},\vec{w},\vec{u} \in V\,. \tag{2.40}$$

- *Existence of a zero vector, i.e. of an element $\vec{0} \in V$ such that*
$$\vec{v}+\vec{0}=\vec{v} \qquad \forall\,\vec{v} \in V\,. \tag{2.41}$$

- *Existence of the opposite element: for any $\vec{w} \in V$ there exists an element $\vec{v} \in V$ such that*
$$\vec{v}+\vec{w}=\vec{0}\,. \tag{2.42}$$

- *Associativity and distributivity of multiplication by real numbers:*
$$\begin{aligned} a(b\vec{v}) &= (ab)\vec{v} \\ a(\vec{v}+\vec{w}) &= a\vec{v}+a\vec{w} \\ (a+b)\vec{v} &= a\vec{v}+b\vec{v} \qquad \forall\vec{v} \in V\,, \forall\, a,b \in \mathbb{R}. \end{aligned} \tag{2.43}$$

- *Finally, the real number 1 must act as an identity on vectors:*
$$1\,\vec{v}=\vec{v} \qquad \forall\,\vec{v}\,. \tag{2.44}$$

Let us now go back to directional derivatives, and consider two curves on a differentiable manifold $\mathbf{M}$ passing through the same point $\mathbf{p}$, with parameters λ and μ, respectively. Given the coordinate system $(x^1,\ldots,x^n)$, the curves are described by the functions $x^i = x^i(\lambda)$ and $x^i = x^i(\mu)$. The directional derivatives in $\mathbf{p}$ along these curves are
$$\frac{d}{d\lambda} = \frac{dx^i}{d\lambda}\frac{\partial}{\partial x^i}\,, \qquad \frac{d}{d\mu} = \frac{dx^i}{d\mu}\frac{\partial}{\partial x^i}\,. \tag{2.45}$$

Let a be a real number. We define the following two operations on the space of directional derivatives along the curves passing through $\mathbf{p}$.

[1]To be precise, what we are defining here is a *real* vector space, but we will omit this specification, because in this book we will only deal with real vector spaces.

- *Sum* of two directional derivatives

$$\frac{d}{d\lambda}+\frac{d}{d\mu}\equiv\left(\frac{dx^i}{d\lambda}+\frac{dx^i}{d\mu}\right)\frac{\partial}{\partial x^i}. \tag{2.46}$$

  The numbers $\left(\frac{dx^i}{d\lambda}+\frac{dx^i}{d\mu}\right)$ are the components of a new vector, which is tangent to some curve through $\mathbf{p}$. Therefore, there must exist a curve with a parameter, say, s, such that in $\mathbf{p}$

$$\frac{dx^i}{ds}=\left(\frac{dx^i}{d\lambda}+\frac{dx^i}{d\mu}\right), \qquad \text{and} \qquad \frac{d}{ds}=\frac{dx^i}{ds}\frac{\partial}{\partial x^i}=\frac{d}{d\lambda}+\frac{d}{d\mu}. \tag{2.47}$$

- *Multiplication* of the directional derivative $\frac{d}{d\lambda}$ by a real number a

$$a\frac{d}{d\lambda}\equiv\left(a\frac{dx^i}{d\lambda}\right)\frac{\partial}{\partial x^i}. \tag{2.48}$$

  The numbers $\left(a\frac{dx^i}{d\lambda}\right)$ are the components of a new vector, which is tangent to some curve in $\mathbf{p}$. Therefore, there must exist a curve with parameter, say, s', such that in $\mathbf{p}$

$$\frac{dx^i}{ds'}=\left(a\frac{dx^i}{d\lambda}\right), \qquad \text{and} \qquad \frac{d}{ds'}=\frac{dx^i}{ds'}\frac{\partial}{\partial x^i}=a\frac{d}{d\lambda}. \tag{2.49}$$

It is easy to verify that the operations of sum and multiplication by a real number defined in Eqs. 2.46 and 2.48, respectively, satisfy the properties required for a vector space. For instance:

- Commutativity of the sum:

$$\frac{d}{d\lambda}+\frac{d}{d\mu}=\left(\frac{dx^i}{d\lambda}+\frac{dx^i}{d\mu}\right)\frac{\partial}{\partial x^i}=\left(\frac{dx^i}{d\mu}+\frac{dx^i}{d\lambda}\right)\frac{\partial}{\partial x^i}=\frac{d}{d\mu}+\frac{d}{d\lambda}. \tag{2.50}$$

- Associativity of multiplication by real numbers:

$$\begin{aligned} a\left(b\frac{d}{d\lambda}\right) &= a\left(\left(b\frac{dx^i}{d\lambda}\right)\frac{\partial}{\partial x^i}\right) \\ &= \left(a\left(b\frac{dx^i}{d\lambda}\right)\right)\frac{\partial}{\partial x^i}=\left(ab\frac{dx^i}{d\lambda}\right)\frac{\partial}{\partial x^i}=(ab)\frac{d}{d\lambda}. \end{aligned} \tag{2.51}$$

- Distributivity of multiplication by real numbers:

$$\begin{aligned} a\left(\frac{d}{d\lambda}+\frac{d}{d\mu}\right) &= a\left(\frac{dx^i}{d\lambda}\frac{\partial}{\partial x^i}+a\frac{dx^i}{d\mu}\frac{\partial}{\partial x^i}\right)=\left(a\frac{dx^i}{d\lambda}+a\frac{dx^i}{d\mu}\right)\frac{\partial}{\partial x^i} \\ &= \left(a\frac{dx^i}{d\lambda}\right)\frac{\partial}{\partial x^i}+\left(a\frac{dx^i}{d\mu}\right)\frac{\partial}{\partial x^i}=a\frac{d}{d\lambda}+a\frac{d}{d\mu}. \end{aligned} \tag{2.52}$$

- The zero element is the vector tangent to the curve $x^\mu\equiv const$, i.e. the point $\mathbf{p}$.
- The opposite of the vector $\vec{v}$ tangent to a given curve is obtained by changing sign to the parameter

$$\lambda\to-\lambda. \tag{2.53}$$

The proof of the remaining properties is analogous. *Therefore, the directional derivatives along all the curves through a point* $\mathbf{p}$ *of a manifold form a vector space.* We call this space the **tangent space in p to the manifold M**, and we shall denote it by $\mathbf{T_p}$. The directional derivative operator $\frac{d}{d\lambda}\in\mathbf{T_p}$ is then a **vector**.

A basis for the space of directional derivatives

In any coordinate system $(x^1, \dots, x^n)$ there are special curves, the *coordinate lines.* Along these lines one of the coordinates is taken as parameter, while the others are constant (think for example to the grid of Cartesian coordinates). The directional derivatives along these lines are

$$\frac{d}{dx^i} = \frac{\partial x^k}{\partial x^i}\frac{\partial}{\partial x^k} = \delta^k_i \frac{\partial}{\partial x^k} = \frac{\partial}{\partial x^i} \,. \tag{2.54}$$

Thus, the operator of directional derivative along the coordinate lines coincides with the operator of partial derivative. Since, as shown in Eq. 2.34,

$$\frac{d}{d\lambda} = \frac{dx^i}{d\lambda}\frac{\partial}{\partial x^i} \,, \tag{2.55}$$

the generic directional derivative $\frac{d}{d\lambda}$ is a linear combination of the directional derivatives along the coordinate lines, $\frac{\partial}{\partial x^i}$; therefore, these form a basis of the tangent space $\mathbf{T_p}$, which is the **coordinate basis** associated with the coordinate system $(x^1, \dots, x^n)$. The quantities $\{\frac{dx^i}{d\lambda}\}$ are the *components* of the vector $\frac{d}{d\lambda}$ in this basis.

Vectors as geometrical objects

As previously remarked, Eq. 2.34 establishes a one-to-one correspondence between the directional derivatives along the curves through $\mathbf{p}$ and the tangent vectors to the same curves in $\mathbf{p}$. Therefore, the tangent space $\mathbf{T_p}$ is also the space of the tangent vectors to the curves in $\mathbf{p}$. Since the directional derivative is independent of the choice of the coordinate system, this correspondence shows that vectors are *geometrical objects*, i.e.

$$\vec{V} = \frac{d}{d\lambda} \,. \tag{2.56}$$

In a coordinate system $(x^1, \dots, x^n)$ we can express this vector in the corresponding coordinate basis using Eq. 2.55:

$$\vec{V} = \frac{dx^i}{d\lambda}\frac{\partial}{\partial x^i} = V^i \frac{\partial}{\partial x^i} \tag{2.57}$$

where $V^i = \frac{dx^i}{d\lambda}$ are the components of $\vec{V}$ in the coordinate basis $\{\frac{\partial}{\partial x^i}\}$.

If we now apply $\frac{d}{d\lambda}$, i.e. $\vec{V}$, to a generic function Φ we find

$$\frac{d\Phi}{d\lambda} = \vec{V}(\Phi) = V^i \frac{\partial \Phi}{\partial x^i} \,, \tag{2.58}$$

and this is the directional derivative of Φ along $\vec{V}$.

Thus, *vectors map real, C^1 functions to real numbers.*

Note that this mapping is **linear**; indeed, given a function $\Phi = a\Phi_1 + b\Phi_2$, with Φ_1, Φ_2 functions on an open set $\mathbf{U}$ of the manifold $\mathbf{M}$, and a, b real numbers, from the linearity of the partial differentiation operator and from Eq. 2.58 it follows

$$V^i \frac{\partial \Phi}{\partial x^i} = aV^i \frac{\partial \Phi_1}{\partial x^i} + bV^i \frac{\partial \Phi_2}{\partial x^i} = a\vec{V}(\Phi_1) + b\vec{V}(\Phi_2) \,, \tag{2.59}$$

i.e.

$$\vec{V}(a\Phi_1 + b\Phi_2) = a\vec{V}(\Phi_1) + b\vec{V}(\Phi_2) \quad \forall \Phi_1 \,, \Phi_2 \text{ functions on } \mathbf{U} \,, \;\; \forall\, a, b \in \mathbb{R} \,. \tag{2.60}$$

In conclusion, we have shown that *a vector is a linear map which associates to any function*

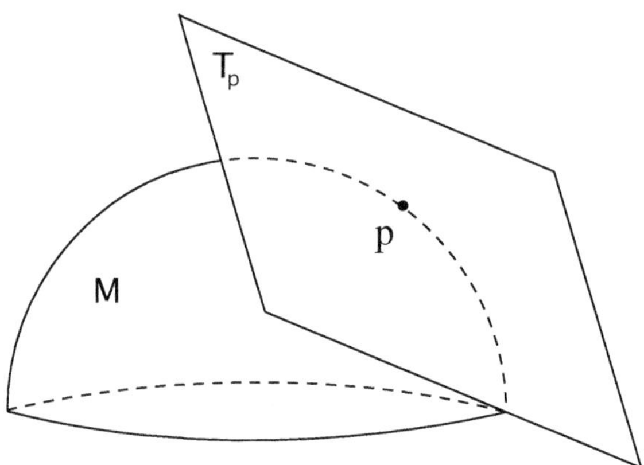

Figure 2.16: Tangent space $\mathbf{T_p}$ to a manifold $\mathbf{M}$.

Φ *the real number* $V^i \frac{\partial \Phi}{\partial x^i}$. In the rest of the book, we shall consider this as the definition of a vector on a differentiable manifold.

It should be stressed that vectors *do not belong to the manifold* $\mathbf{M}$: they belong to the tangent space to $\mathbf{M}$ in $\mathbf{p}$, namely $\mathbf{T_p}$. If the manifold is $\mathbb{R}^n$ this distinction may be overlooked, because the tangent space (at any point) coincides with $\mathbb{R}^n$, but for a general manifold $\mathbf{M}$ the two spaces are different. Indeed, a generic manifold is *not* a vector space. For example, if the manifold is a sphere, we can not define the vectors as "arrows" on the sphere: they lie in the tangent space, which is the plane tangent to the sphere at a given point. For more general manifolds it is not easy to visualize $\mathbf{T_p}$. In any case $\mathbf{T_p}$ *has the same dimensions as the manifold* $\mathbf{M}$.

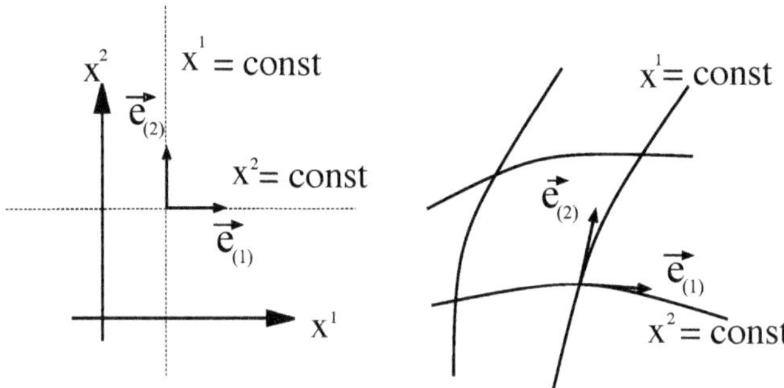

Figure 2.17: Coordinate basis of the tangent space.

We shall denote the vectors of the coordinate basis, associated with the coordinate system $\{x^i\}$, as

$$\vec{e}_{(i)} \equiv \frac{\partial}{\partial x^i}, \tag{2.61}$$

see, e.g., Fig 2.17. Hereafter, we shall enclose within ( ) the index that indicates the vector of a given basis, not to be confused with the index that indicates the components of the vector. The only exception is the operator of partial derivative, $\frac{\partial}{\partial x^i}$, for which we shall omit the ( ) around the index and the superscripted arrow, to follow the standard notation of textbooks. For instance $e^1_{(2)}$ indicates the component 1 of the basis vector $\vec{e}_{(2)}$.

Any vector $\vec{A}$ at a point $\mathbf{p}$ can be expressed as a linear combination of the basis vectors

$$\vec{A} = A^i \vec{e}_{(i)} \,, \quad \text{i.e.} \;\; \vec{A} \to_O \{A^i\} \qquad i = 1, \ldots, n \tag{2.62}$$

where the numbers A^i are the components of $\vec{A}$ with respect to the chosen basis O. If we make a coordinate transformation, the new set of coordinates $(x^{1'}, x^{2'}, \ldots, x^{n'})$ defines a new coordinate basis O', $\{\vec{e}_{(i')}\} \equiv \{\frac{\partial}{\partial x^{i'}}\}$. The vector $\vec{A}$ in the new basis is

$$\vec{A} = A^{j'} \vec{e}_{(j')} \,, \quad \text{i.e.} \;\; \vec{A} \to_{O'} \{A^{j'}\} \tag{2.63}$$

where $A^{j'}$ are the components of $\vec{A}$ with respect to the new basis vectors $\vec{e}_{(j')}$. Since the vector $\vec{A}$ is a geometrical object, i.e. it is independent on the coordinate frame, the following equality must hold

$$A^i \vec{e}_{(i)} = A^{i'} \vec{e}_{(i')}. \tag{2.64}$$

From Eq. 2.31 we know how to express $A^{i'}$ as functions of the components of $\vec{A}$ in the old basis, i.e. $A^{i'} = \Lambda^{i'}{}_j A^j$, and replacing these expressions into Eq. 2.64 we find

$$A^i \vec{e}_{(i)} = \Lambda^{i'}{}_j A^j \vec{e}_{(i')} \tag{2.65}$$

where $\Lambda^{i'}{}_j = \frac{\partial x^{i'}}{\partial x^j}$. By relabeling the dummy indices this equation can be written as

$$\left[\vec{e}_{(j)} - \Lambda^{i'}{}_j \vec{e}_{(i')}\right] A^j = 0\,. \tag{2.66}$$

Since Eq. 2.66 must be satisfied for any non-vanishing vector $\vec{A}$, the term in square brakets must vanish, i.e.

$$\vec{e}_{(j)} = \Lambda^{i'}{}_j \vec{e}_{(i')}\,. \tag{2.67}$$

Multiplying both members by $\Lambda^j{}_{k'}$ and remembering that $\Lambda^j{}_{k'}\Lambda^{i'}{}_j = \delta^{i'}_{k'}$ (see Eq. 2.24), we find

$$\Lambda^j{}_{k'} \vec{e}_{(j)} = \Lambda^j{}_{k'}\Lambda^{i'}{}_j \vec{e}_{(i')} = \delta^{i'}_{k'} \vec{e}_{(i')}\,, \tag{2.68}$$

i.e.

$$\vec{e}_{(k')} = \Lambda^j{}_{k'} \vec{e}_{(j)}. \tag{2.69}$$

It should be noted that we do not need to choose necessarily a coordinate basis. We may choose a set of independent basis vectors that are not tangent to the coordinate lines. In this case the matrix which transforms one basis to another has to be assigned, since it can not be written in terms of partial derivatives of a coordinate transformation. This will be further discussed in Sec. 3.10.

Box 2-G

Exercise

Let us consider the two-dimensional Euclidean space, where we choose a Cartesian coordinate frame (x^1, x^2), and let $\vec{V}$ be a vector at a point $(1, 1)$, whose components in this frame are $V^i = (1, 2)$, $i = 1, 2$. Given the coordinate transformation $(x^1, x^2) \to (r, \theta)$

$$\begin{cases} x^1 = r\cos\theta \\ x^2 = r\sin\theta\,, \end{cases} \tag{2.70}$$

with $x^{1'} = r$, $x^{2'} = \theta$, compute

- the new coordinate basis for vectors;
- the components of $\vec{V}$ in the same point, in the new frame.

The coordinate basis for vectors, $\{\vec{e}_{(i)}\} = \{\frac{\partial}{\partial x^i}\}$, and the components of a vector, transform according to Eq. 2.69 and Eq. 2.31, respectively, i.e.

$$\vec{e}_{(k')} = \Lambda^i{}_{k'}\vec{e}_{(i)}\,, \qquad \text{and} \qquad V^{k'} = \Lambda^{k'}{}_i V^i\,. \tag{2.71}$$

Thus, we need to compute the matrices

$$\Lambda^i{}_{k'} = \frac{\partial x^i}{\partial x^{k'}}\,, \qquad \text{and} \qquad \Lambda^{k'}{}_i = \frac{\partial x^{k'}}{\partial x^i}\,,$$

associated to the coordinate transformation 2.70. The matrix $\Lambda^i{}_{k'}$ can easily be computed

$$\Lambda^i{}_{k'} = \begin{pmatrix} \cos\theta & -r\sin\theta \\ \sin\theta & r\cos\theta \end{pmatrix}. \tag{2.72}$$

Remember that the first index (i in this case) indicates the row of the matrix, and the second (k') indicates the column. Thus, using Eqs. 2.71 and 2.72 the new basis vectors are

$$\begin{cases} \vec{e}_{(1')} \equiv \vec{e}_{(r)} = \Lambda^i{}_{1'}\ \vec{e}_{(i)} = \cos\theta\ \vec{e}_{(1)} + \sin\theta\ \vec{e}_{(2)} \\ \vec{e}_{(2')} \equiv \vec{e}_{(\theta)} = \Lambda^i{}_{2'}\ \vec{e}_{(i)} = -r\sin\theta\ \vec{e}_{(1)} + r\cos\theta\ \vec{e}_{(2)}\,. \end{cases} \tag{2.73}$$

To compute $\Lambda^{k'}{}_i$, needed to compute the components of $\vec{V}$ in the new frame, we can either invert the coordinate transformation, i.e.

$$\begin{cases} x^{1'} \equiv r = \sqrt{(x^1)^2 + (x^2)^2} \\ x^{2'} \equiv \theta = \arctan(x^2/x^1)\,, \end{cases} \tag{2.74}$$

and compute the derivatives $\frac{\partial x^{k'}}{\partial x^i}$ or, we can directly compute the matrix inverse to 2.72. The result is

$$\Lambda^{k'}{}_i = \begin{pmatrix} \cos\theta & \sin\theta \\ -\sin\theta/r & \cos\theta/r \end{pmatrix}. \tag{2.75}$$

Thus, the components of $\vec{V}$ in the new frame are

$$\begin{cases} V^{1'} \equiv V^r = \Lambda^{1'}{}_j V^j = \Lambda^{1'}{}_1 V^1 + \Lambda^{1'}{}_2 V^2 = \cos\theta + 2\sin\theta \\ V^{2'} \equiv V^\theta = \Lambda^{2'}{}_j V^j = \Lambda^{2'}{}_1 V^1 + \Lambda^{2'}{}_2 V^2 = \frac{1}{r}(-\sin\theta + 2\cos\theta)\,. \end{cases} \tag{2.76}$$

The point $(1, 1)$ in the new coordinates is $(\sqrt{2}, \pi/4)$; therefore in that point

$$V^{i'} = (V^r, V^\theta) = \left(\frac{3}{2}\sqrt{2}, \frac{1}{2}\right).$$

2.3 ONE-FORMS

2.3.1 One-forms as geometrical objects

A **one-form** is a linear, real-valued function of vectors:

$$\begin{aligned} \tilde{q} \,:\, \mathbf{T_p} &\rightarrow \mathbb{R} \\ \vec{V} &\mapsto \tilde{q}(\vec{V}) \,. \end{aligned} \tag{2.77}$$

This means that the one-form $\tilde{q}$ at the point $\mathbf{p}$ *associates to any vector $\vec{V}$ at* $\mathbf{p}$ *a real number, which we call $\tilde{q}(\vec{V})$.*

Hereafter a tilde on the top shall indicate one-forms (or 1-forms), whereas an arrow indicates vectors. Since one-forms are linear functions, then

$$\tilde{q}(a\vec{V} + b\vec{W}) = a\tilde{q}(\vec{V}) + b\tilde{q}(\vec{W}) \,, \tag{2.78}$$

for any pair of vectors $\vec{V}, \vec{W}$, any pair of real numbers a, b, and any one-form $\tilde{q}$. We define two operations on the space of one-forms:

- *Sum*: given two one-forms $\tilde{q}, \tilde{\sigma}$, we define the new one-form $\tilde{q} + \tilde{\sigma}$ such that, for any vector $\vec{V}$,
$$(\tilde{q} + \tilde{\sigma})(\vec{V}) = \tilde{q}(\vec{V}) + \tilde{\sigma}(\vec{V}) \,. \tag{2.79}$$
- *Multiplication* by real numbers: given a one-form $\tilde{q}$ and a real number a, we define the new one-form $a\tilde{q}$ such that, for any vector $\vec{V}$,
$$(a\tilde{q})(\vec{V}) = a\tilde{q}(\vec{V}) \,. \tag{2.80}$$

These operations satisfy the properties 2.40-2.44, which define a vector space. For example:

- Commutativity of the sum. Given two one-forms $\tilde{q}, \tilde{\sigma}$, Eq. 2.79 shows that
$$(\tilde{q} + \tilde{\sigma})(\vec{V}) = \tilde{q}(\vec{V}) + \tilde{\sigma}(\vec{V}) = \tilde{\sigma}(\vec{V}) + \tilde{q}(\vec{V}) = (\tilde{\sigma} + \tilde{q})(\vec{V}) \,; \tag{2.81}$$
since this is true for any $\vec{V}$, it follows that $\tilde{\sigma} + \tilde{q} = \tilde{q} + \tilde{\sigma}$.
- Distributivity of the multiplication with real numbers. Given two one-forms $\tilde{q}, \tilde{\sigma}$ and a real number a, Eq. 2.80 gives
$$\begin{aligned} \left[a\,(\tilde{q} + \tilde{\sigma})\right](\vec{V}) &= a\left[(\tilde{q} + \tilde{\sigma})\,(\vec{V})\right] = a[\tilde{q}(\vec{V}) + \tilde{\sigma}(\vec{V})] \\ &= a[\tilde{q}(\vec{V})] + a[\tilde{\sigma}(\vec{V})] = (a\tilde{q})(\vec{V}) + (a\tilde{\sigma})(\vec{V}) \\ &= [(a\tilde{q}) + (a\tilde{\sigma})](\vec{V}) \,; \end{aligned} \tag{2.82}$$
since this is true for any $\vec{V}$, it follows that $a(\tilde{q} + \tilde{\sigma}) = (a\tilde{q}) + (a\tilde{\sigma})$.
- Existence of the zero element. We define the one-form $\tilde{0}$ such that, for any $\vec{V}$,
$$\tilde{0}(\vec{V}) = 0 \,. \tag{2.83}$$

The other properties can be proved in a similar way. Therefore, *one-forms form a vector space*, $\mathbf{T^*_p}$, which is called the **dual** vector space to $\mathbf{T_p}$, or the **cotangent space in p**. $\mathbf{T^*_p}$ is the space of the maps (the one-forms) that associate to any given vector a number, i.e. that map $\mathbf{T_p}$ on $\mathbb{R}$.

Basis one-forms

We define a basis for the one-forms, $\{\tilde{\omega}^{(i)}\}_{i=1,\dots,n}$, as follows: *the basis one-form* $\tilde{\omega}^{(i)}$ *applied to* $\vec{V}$ *gives as a result the i-th component of the vector,* V^i

$$\tilde{\omega}^{(i)}(\vec{V}) = V^i\,. \tag{2.84}$$

As for vector bases, the index (i) in parenthesis selects the i-th one-form of the basis. In particular, if $\tilde{\omega}^{(i)}$ is applied to the basis vector $\vec{e}_{(j)}$, whose only non-vanishing component is $e^j_{(j)} = 1$, from Eq. 2.84 we find

$$\tilde{\omega}^{(i)}(\vec{e}_{(j)}) = \delta^i_j\,. \tag{2.85}$$

In order to show that $\{\tilde{\omega}^{(i)}\}_{i=1,\dots,n}$ is a basis of the space of one-forms, let us consider an arbitrary one-form $\tilde{q}$ acting on an arbitrary vector $\vec{V}$. By expressing $\vec{V}$ as a linear combination of the basis vectors $\vec{e}_{(j)}$, and using the linearity of one-forms, we find

$$\begin{aligned}\tilde{q}(\vec{V}) &= \tilde{q}(V^j\vec{e}_{(j)}) = V^j\tilde{q}(\vec{e}_{(j)}) = \\ &= \tilde{\omega}^{(j)}(\vec{V})\tilde{q}(\vec{e}_{(j)})\,,\end{aligned} \tag{2.86}$$

where the last equality follows from Eq. 2.84. This equation holds for any vector $\vec{V}$; therefore we can write

$$\tilde{q} = \tilde{q}(\vec{e}_{(j)})\tilde{\omega}^{(j)}\,. \tag{2.87}$$

Since $\tilde{q}(\vec{e}_{(j)})$ are real numbers, this equation shows that any one-form $\tilde{q}$ can be written as a linear combination of the $\{\tilde{\omega}^{(j)}\}$, i.e. $\{\tilde{\omega}^{(j)}\}_{j=1,\dots,n}$ is a basis for the one-forms. The quantities

$$q_j \equiv \tilde{q}(\vec{e}_{(j)}) \tag{2.88}$$

are the *components* of $\tilde{q}$ in the basis $\{\tilde{\omega}^{(i)}\}_{i=1,\dots,n}$

$$\tilde{q} = q_j\,\tilde{\omega}^{(j)}\,, \quad \text{i.e.} \quad \tilde{q} \to_O \{q_j\}\,. \tag{2.89}$$

We are now in a position to make clear which is the number resulting from the application of the one-form $\tilde{q}$ to a vector $\vec{V}$. Since $\vec{V} = V^i\vec{e}_{(i)}$, using Eqs. 2.89 and 2.85 and the linearity of one-forms, we find

$$\tilde{q}(\vec{V}) = q_j\,\tilde{\omega}^{(j)}(V^i\vec{e}_{(i)}) = q_jV^i\,\tilde{\omega}^{(j)}(\vec{e}_{(i)}) = q_jV^i\delta^j_i = q_jV^j\,. \tag{2.90}$$

Due to the symmetry of the expression $\tilde{q}(\vec{V}) = q_jV^j$ in the components of the one-form and of the vector, it is also possible, given a vector $\vec{V}$, to define a map which associates to any one-form $\tilde{q}$ a real number, which we call $\vec{V}(\tilde{q})$, as follows

$$\vec{V}(\tilde{q}) = \tilde{q}(\vec{V}) = q_jV^j\,, \tag{2.91}$$

an equation which will be very useful in the next chapters. This is the reason why we call the space $\mathbf{T}^*_\mathbf{P}$ the dual vector space to $\mathbf{T}_\mathbf{P}$. Similarly, the basis for the one-forms $\{\tilde{\omega}^{(i)}\}_{i=1,\dots,n}$ is called *dual basis.*

Transformation rules of one-forms

Let us consider an open set $\mathbf{U}$ of the manifold $\mathbf{M}$, and choose a coordinate system $\{x^i\}$. We have seen that this defines a *coordinate basis for vectors,* $\{\vec{e}_{(i)}\} \equiv \left\{\frac{\partial}{\partial x^i}\right\}$ and the *dual coordinate basis for one-forms* $\{\tilde{\omega}^{(i)}\}_{i=1,\dots,n}$. If we make a coordinate transformation

$x^{k'} = x^{k'}(x^i)$, the new coordinate basis vectors $\vec{e}_{(j')}$ are related to the old ones by (see Eq. 2.69)

$$\vec{e}_{(i')} = \Lambda^k{}_{i'}\vec{e}_{(k)}, \tag{2.92}$$

where $\Lambda^k{}_{i'} = \frac{\partial x^k}{\partial x^{i'}}$. Consequently

$$q_{j'} = \tilde{q}(\vec{e}_{(j')}) = \tilde{q}(\Lambda^k{}_{j'}\vec{e}_{(k)}) = \Lambda^k{}_{j'}\tilde{q}(\vec{e}_{(k)}) \tag{2.93}$$

and, using Eq. 2.88, the components of $\tilde{q}$ with respect to the new coordinate basis are

$$q_{j'} = \Lambda^k{}_{j'}q_k \,. \tag{2.94}$$

By comparing this result with Eq. 2.21, we immediately recognize that this is the way covariant vectors transform; thus *covariant vectors are one-forms.*

Let us now derive how the basis one-forms transform. Since a one-form is a geometric object, it is independent of the coordinate system. Consequently

$$\tilde{q} = q_j \,\tilde{\omega}^{(j)} = q_{k'} \,\tilde{\omega}^{(k')} \tag{2.95}$$

and, using Eq. 2.94, we find

$$q_j\tilde{\omega}^{(j)} = \Lambda^i{}_{k'}q_i \,\tilde{\omega}^{(k')}\,, \qquad \rightarrow \qquad \left[\Lambda^i{}_{k'}\tilde{\omega}^{(k')} - \tilde{\omega}^{(i)}\right]q_i = 0\,. \tag{2.96}$$

Since this equation must be satisfied for any non-vanishing one-form $\tilde{q}$, the term in square brackets must vanish, i.e.

$$\tilde{\omega}^{(i)} = \Lambda^i{}_{k'}\tilde{\omega}^{(k')}\,. \tag{2.97}$$

The matrix $\Lambda^i{}_{k'}$ is the inverse of $\Lambda^{k'}{}_i$, thus $\Lambda^{k'}{}_j\Lambda^j{}_{i'} = \delta^{k'}_{i'}$, and $\Lambda^{k'}{}_j\Lambda^i{}_{k'} = \delta^i_j$. Multiplying both sides of Eq. 2.97 by $\Lambda^{j'}{}_i$ we find

$$\Lambda^{j'}{}_i \,\tilde{\omega}^{(i)} = \Lambda^{j'}{}_i\Lambda^i{}_{k'} \,\tilde{\omega}^{(k')} = \delta^{j'}_{k'} \,\tilde{\omega}^{(k')}, \tag{2.98}$$

hence

$$\tilde{\omega}^{(j')} = \Lambda^{j'}{}_i \,\tilde{\omega}^{(i)}\,. \tag{2.99}$$

Note that the transformation matrix is the inverse of that used to transform the basis vectors in Eq. 2.69.

Differentials as one-forms

The differential $d\Phi$ of a real function Φ is the variation of the function in an *unspecified direction*, at first order in the displacement. If we specify the direction as given by an arbitrary vector $\vec{V}$, we can explicitly compute the specific variation of the function, which is a real number, i.e.[2]

$$d\Phi(\vec{V}) = V^j \frac{\partial \Phi}{\partial x^j}\,. \tag{2.100}$$

This expression is linear in $\vec{V}$; thus, according to the definition 2.77, *the differential of a function Φ is a one-form.* The components of $d\Phi$ are

$$d\Phi_i = d\Phi(\vec{e}_{(i)}) = e^j_{(i)}\frac{\partial \Phi}{\partial x^j} = \delta^j_i \frac{\partial \Phi}{\partial x^j} = \frac{\partial \Phi}{\partial x^i}\,. \tag{2.101}$$

[2]Note that the right-hand side of Eq. 2.100 coincides with that of Eq. 2.58, i.e. with the directional derivative of the function Φ along $\vec{V}$. However, while the directional derivative $d/d\lambda$ maps C^1 functions to $\mathbb{R}$, the differential of a function $d\Phi$ maps vectors to $\mathbb{R}$.

Thus, *the components of the one-form $d\Phi$ are the components of the gradient of the function.* Hereafter, we shall omit the superscripted tilde over the differential $d\Phi$ to follow the standard notation of textbooks.

According to the definition, the differential of any coordinate x^i is the one-form dx^i such that

$$dx^i(\vec{V}) = V^j \frac{\partial x^i}{\partial x^j} = V^j \delta^i_j = V^i \,, \tag{2.102}$$

i.e. the one-form dx^i associates to any vector $\vec{V}$ the component V^i. This is precisely the definition of the one-forms of the coordinate basis given in Eq. 2.84, i.e. $\tilde{\omega}^{(i)}(\vec{V}) = V^i$; therefore, *the differentials dx^i are the coordinate basis one-forms*:

$$\tilde{\omega}^{(i)} = dx^i \,, \tag{2.103}$$

whose components are

$$\omega^{(i)}_j = dx^i(\vec{e}_{(j)}) = \delta^i_j \,. \tag{2.104}$$

First-order displacement along a curve

Let us consider a manifold $\mathbf{M}$ with a coordinate system $(x^1, \ldots, x^n)$ and a path joining two nearby points $\mathbf{p}$ and $\mathbf{p}'$. Let λ be a parameter chosen on the path and $\{x^i(\lambda)\}$ and $\{x^i(\lambda + \Delta\lambda)\}$ the coordinates of the two points. The coordinate separation between $\mathbf{p}$ and $\mathbf{p}'$ can be expanded in $\Delta\lambda$ as follows

$$x^i(\lambda + \Delta\lambda) - x^i(\lambda) = \frac{dx^i}{d\lambda}\Delta\lambda + \mathcal{O}(\Delta\lambda^2) = \delta x^i + \mathcal{O}(\Delta\lambda^2) \quad (i = 1, \ldots, n) \,, \tag{2.105}$$

where

$$\delta x^i = \frac{dx^i}{d\lambda}\Delta\lambda \tag{2.106}$$

are the components of a vector $\vec{\delta x}$, which is the first-order displacement between $\mathbf{p}$ and $\mathbf{p}'$. Note that the vector $\vec{\delta x}$ does not depend on the parametrization of the path joining the two points. If the terms $\mathcal{O}(\Delta\lambda^2)$ can be neglected, $\vec{\delta x}$ is the *infinitesimal displacement* from $\mathbf{p}$ to $\mathbf{p}'$. If we apply the basis one-forms dx^i to $\vec{\delta x}$, we get

$$dx^i(\vec{\delta x}) = \delta x^i \,; \tag{2.107}$$

thus, the basis one-forms dx^i can be considered as the components of the infinitesimal displacement along a generic direction.

Box 2-H

Exercise: Lorentz's transformations and one-forms

Let us consider the four-dimensional flat spacetime of Special Relativity, restricted to the $(x - y)$ plane, where we choose the coordinates $(ct, x, y) \equiv (x^0, x^1, x^2)$. Let us consider a Lorentz boost in the x direction

$$\begin{cases} x^{0'} = \gamma\left(x^0 - \frac{vx^1}{c}\right) \\ x^{1'} = \gamma\left(x^1 - \frac{vx^0}{c}\right) \\ x^{2'} = x^2 \end{cases} \tag{2.108}$$

where $\gamma = 1/\sqrt{1 - v^2/c^2}$ is the Lorentz factor and v the velocity of the boost. Compute the coordinate basis for one-forms and the components of a generic one-form $\tilde{q}$ in the new frame.

The coordinate basis for one-forms, $\{\tilde{\omega}^{(\alpha)}\} \equiv \{dx^\alpha\}$, transforms according to Eq. 2.99

$$\tilde{\omega}^{(\alpha')} = \Lambda^{\alpha'}{}_{\mu}\tilde{\omega}^{(\mu)}\,, \tag{2.109}$$

whereas the components of $\tilde{q}$ transform according to Eq. 2.94

$$q_{\mu'} = \Lambda^{\alpha}{}_{\mu'}q_\alpha\,. \tag{2.110}$$

The matrices $\Lambda^{\alpha'}{}_{\mu}$ and $\Lambda^{\alpha}{}_{\mu'}$ associated to the coordinate transformation 2.108 are

$$\Lambda^{\alpha'}{}_{\mu} = \frac{\partial x^{\alpha'}}{\partial x^\mu} = \begin{pmatrix} \gamma & -\gamma\, v/c & 0 \\ -\gamma\, v/c & \gamma & 0 \\ 0 & 0 & 1 \end{pmatrix}, \tag{2.111}$$

and

$$\Lambda^{\alpha}{}_{\mu'} = \frac{\partial x^{\alpha}}{\partial x^{\mu'}} = \begin{pmatrix} \gamma & \gamma v/c & 0 \\ \gamma v/c & \gamma & 0 \\ 0 & 0 & 1 \end{pmatrix}. \tag{2.112}$$

Thus, the new basis one-forms are

$$\begin{cases} \tilde{\omega}^{(0')} = \Lambda^{0'}{}_{\mu}\,\tilde{\omega}^{(\mu)} = \gamma\tilde{\omega}^{(0)} - \frac{\gamma v}{c}\tilde{\omega}^{(1)} \\ \tilde{\omega}^{(1')} = \Lambda^{1'}{}_{\mu}\,\tilde{\omega}^{(\mu)} = -\frac{\gamma v}{c}\tilde{\omega}^{(0)} + \gamma\tilde{\omega}^{(1)} \\ \tilde{\omega}^{(2')} = \Lambda^{2'}{}_{\mu}\,\tilde{\omega}^{(\mu)} = \tilde{\omega}^{(2)} \end{cases} \tag{2.113}$$

and the new components of $\tilde{q}$ are

$$\begin{cases} q_{0'} = \Lambda^{\alpha}{}_{0'}q_\alpha = \gamma q_0 + \frac{\gamma v}{c}q_1 \\ q_{1'} = \Lambda^{\alpha}{}_{1'}q_\alpha = \frac{\gamma v}{c}q_0 + \gamma q_1 \\ q_{2'} = \Lambda^{\alpha}{}_{2'}q_\alpha = q_2\,. \end{cases} \tag{2.114}$$

2.3.2 Vector fields and one-form fields

Vectors and one-forms are defined *at a point* $\mathbf{p}$ of the manifold, and belong to the vector spaces $\mathbf{T_p}$ and $\mathbf{T^*_p}$, respectively, which also are defined in $\mathbf{p}$; to make this explicit, we could denote a vector in $\mathbf{p}$ as $\vec{V}_{\mathbf{p}}$, a one-form in $\mathbf{p}$ as $\tilde{q}_{\mathbf{p}}$, but we shall often leave the point implicit in order to simplify the notation. We shall now define *vector fields* and *one-form fields.*

Given an open set $\mathbf{S}$ of a differentiable manifold $\mathbf{M}$, we define the vector spaces

$$\mathbf{T_S} \equiv \bigcup_{\mathbf{p}\in\mathbf{S}} \mathbf{T_p}$$
$$\mathbf{T^*_S} \equiv \bigcup_{\mathbf{p}\in\mathbf{S}} \mathbf{T^*_p},$$

i.e., the union of the tangent spaces in the points $\mathbf{p} \in \mathbf{S}$, and the union of the cotangent spaces in the points $\mathbf{p} \in \mathbf{S}$. A vector field $\vec{V}$ is a mapping

$$\begin{aligned} \vec{V} : \mathbf{S} &\rightarrow \mathbf{T_S} \\ \mathbf{p} &\mapsto \vec{V}_{\mathbf{p}} \end{aligned}$$

which associates to every point $\mathbf{p} \in \mathbf{S}$, a vector $\vec{V}_{\mathbf{p}}$ defined on the tangent space in $\mathbf{p}$, $\mathbf{T_p}$. A one-form field $\tilde{q}$ is a mapping

$$\begin{aligned} \tilde{q} : \mathbf{S} &\rightarrow \mathbf{T^*_S} \\ \mathbf{p} &\mapsto \tilde{q}_{\mathbf{p}} \end{aligned}$$

which associates to every point $\mathbf{p} \in \mathbf{S}$, a one-form $\tilde{q}_{\mathbf{p}}$ defined on the cotangent space in $\mathbf{p}$, $\mathbf{T^*_p}$. If a coordinate system (a chart) $\{x^i\}$ is defined on $\mathbf{S}$, we can indicate the vector field and the one-form field as $\vec{V}(x)$, $\tilde{q}(x)$, respectively.

In the following, we will mainly consider vector fields and one-form fields; however, for brevity of notation, we will usually refer to them as vectors and one-forms.

Box 2-I

Summary on vectors and one-forms

- Vectors are linear, real-valued functions of one-forms: a vector $\vec{V}$ maps a generic one-form $\tilde{q}$ to the real number $\vec{V}(\tilde{q})$

$$\begin{aligned} \vec{V} : \mathbf{T}^*_{\mathbf{p}} &\to \mathbb{R} \\ \tilde{q} &\mapsto \vec{V}(\tilde{q}) . \end{aligned} \tag{2.115}$$

- One-forms are linear, real valued functions of vectors: a one-form $\tilde{q}$ maps a generic vector $\vec{V}$ to the real number $\tilde{q}(\vec{V}) = \vec{V}(\tilde{q})$

$$\begin{aligned} \tilde{q} : \mathbf{T}_{\mathbf{p}} &\to \mathbb{R} \\ \vec{V} &\mapsto \tilde{q}(\vec{V}) . \end{aligned} \tag{2.116}$$

- The number $\tilde{q}(\vec{V}) = \vec{V}(\tilde{q})$ is given by

$$\tilde{q}(\vec{V}) = \vec{V}(\tilde{q}) = V^i q_i , \tag{2.117}$$

  where V^i and q_i are the components of $\vec{V}$ and $\tilde{q}$ with respect to a chosen basis.

Vectors and one-forms are "attached" to the point $\mathbf{p}$, since they belong to the tangent space $\mathbf{T}_{\mathbf{p}}$ and to the cotangent space $\mathbf{T}^*_{\mathbf{p}}$ to the manifold $\mathbf{M}$ in $\mathbf{p}$.

Coordinate basis for vectors and one-forms

Given a coordinate system $(x^1, \ldots, x^n)$ the coordinate basis for vectors is the set of n vectors tangent to the coordinate lines

$$\{\vec{e}_{(i)}\} \equiv \left\{\frac{\partial}{\partial x^i}\right\} , \qquad \text{where} \qquad e^j_{(i)} = \left(\frac{\partial}{\partial x^i}\right)^j = \delta^j_i . \tag{2.118}$$

The dual coordinate basis for the one-forms is the set of n one-forms

$$\{\tilde{\omega}^{(i)}\} = \{dx^i\} , \qquad \text{where} \qquad \omega^{(i)}_j = (dx^i)_j = \delta^i_j . \tag{2.119}$$

The two coordinate bases are related by the following equation

$$\tilde{\omega}^{(i)}(\vec{e}_{(j)}) = \vec{e}_{(j)}(\tilde{\omega}^{(i)}) = \delta^i_j . \tag{2.120}$$

Components of vectors and one-forms

Given the coordinate bases $\{\vec{e}_{(i)}\}$ and $\{\tilde{\omega}^{(i)}\}$ for vectors and one-forms, respectively, any vector $\vec{V}$ and any one-form $\tilde{q}$ can be written as

$$\vec{V} = V^i \vec{e}_{(i)} \qquad \text{and} \qquad \tilde{q} = q_i \, \tilde{\omega}^{(i)} , \qquad i = 1, \ldots, n , \tag{2.121}$$

where

$$\begin{aligned} V^i &= \tilde{\omega}^{(i)}(\vec{V}) \equiv \vec{V}(\tilde{\omega}^{(i)}) \\ q_i &= \tilde{q}(\vec{e}_{(i)}) \equiv \vec{e}_{(i)}(\tilde{q}) . \end{aligned} \tag{2.122}$$

Box 2-J

Transformation rules for vectors and one-forms

Given the coordinate transformation $x^i = x^i(x^{j'})$, or its inverse $x^{j'} = x^{j'}(x^i)$, $i, j' = 1, \dots, n$, the components of vectors and one-forms transform as follows

$$\begin{cases} V^i = \Lambda^i{}_{j'}\, V^{j'} \\ V^{j'} = \Lambda^{j'}{}_i\, V^i \,, \end{cases} \quad \text{and} \quad \begin{cases} q_i = \Lambda^{j'}{}_i\, q_{j'} \\ q_{j'} = \Lambda^i{}_{j'}\, q_i \,, \end{cases} \tag{2.123}$$

where $\Lambda^i{}_{j'} = \frac{\partial x^i}{\partial x^{j'}}$, and $\Lambda^{j'}{}_i = \frac{\partial x^{j'}}{\partial x^i}$ is the inverse matrix of $\Lambda^i{}_{j'}$. The basis vectors and the basis one-forms transform as

$$\begin{cases} \vec{e}_{(i)} = \Lambda^{j'}{}_i\, \vec{e}_{(j')} \\ \vec{e}_{(j')} = \Lambda^i{}_{j'}\, \vec{e}_{(i)} \,, \end{cases} \quad \text{and} \quad \begin{cases} \tilde{\omega}^{(i)} = \Lambda^i{}_{j'}\, \tilde{\omega}^{(j')} \\ \tilde{\omega}^{(j')} = \Lambda^{j'}{}_i\, \tilde{\omega}^{(i)} \,. \end{cases} \tag{2.124}$$

2.4 TENSORS

2.4.1 Geometrical definition of a tensor

The following definition of a tensor is a generalization of the definition of vectors and one-forms. Let us consider a point $\mathbf{p}$ of an n-dimensional manifold $\mathbf{M}$.

A **tensor** *of type* $\binom{N}{N'}$ *in* $\mathbf{p}$ *is a linear, real valued function, which associates to N one-forms and N′ vectors a real number.* The sum of N and N' is called *rank* of the tensor.

For example if $\boldsymbol{F}$ is a $\binom{2}{2}$ tensor this means that

$$\boldsymbol{F}(\tilde{q}, \tilde{g}, \vec{V}, \vec{W}) \tag{2.125}$$

is a real number, which we will specify in the following, and the linearity implies that $\forall\, a, b \in \mathbb{R}$,

$$\boldsymbol{F}(a\tilde{q} + b\tilde{g}, \tilde{\sigma}, \vec{V}, \vec{W}) = a\boldsymbol{F}(\tilde{q}, \tilde{\sigma}, \vec{V}, \vec{W}) + b\boldsymbol{F}(\tilde{g}, \tilde{\sigma}, \vec{V}, \vec{W}) \tag{2.126}$$

and

$$\boldsymbol{F}(\tilde{q}, \tilde{g}, a\vec{V}_1 + b\vec{V}_2, \vec{W}) = a\boldsymbol{F}(\tilde{q}, \tilde{g}, \vec{V}_1, \vec{W}) + b\boldsymbol{F}(\tilde{q}, \tilde{g}, \vec{V}_2, \vec{W}) \tag{2.127}$$

and similarly for the other arguments. This definition of tensors is rather abstract, but we shall see how to make it concrete with specific examples. We remark that the order in which the arguments appear is important; actually, this is true for any function of real variables: for example, if $f(x, y) = 4x^3 + 5y$, then $f(1, 5) \neq f(5, 1)$. In the same way

$$\boldsymbol{F}(\tilde{q}, \tilde{g}, \vec{V}, \vec{W}) \neq \boldsymbol{F}(\tilde{g}, \tilde{q}, \vec{V}, \vec{W}).$$

Given a coordinate system $(x^1, \dots, x^n)$, we can define the *components* of a $\binom{N}{N'}$-tensor, generalizing the definitions of components of vectors and one-forms given in Eqs. 2.122: they are the real numbers obtained by applying the tensor to the one-forms and to the vectors of the coordinate basis:

$$T^{i_1 \cdots i_N}{}_{j_1 \cdots j_{N'}} = T(\tilde{\omega}^{(i_1)}, \dots, \tilde{\omega}^{(i_N)}, \vec{e}_{(j_1)}, \dots, \vec{e}_{(j_{N'})}) \,. \tag{2.128}$$

Box 2-K

Examples of tensors

- A $\binom{0}{1}$ tensor is a function that takes a vector as argument, and returns a number. This is precisely what one-forms do: *a* $\binom{0}{1}$ *tensor is a one-form,*

$$\tilde{q}(\vec{V}) = q_i V^i \,. \tag{2.129}$$

- A $\binom{1}{0}$ tensor is a function that takes a one-form as argument, and returns a number. This is what vectors do: *a* $\binom{1}{0}$ *tensor is a vector,*

$$\vec{V}(\tilde{q}) = q_i V^i \,. \tag{2.130}$$

- A $\binom{0}{2}$ tensor is a function that takes 2 vectors as arguments and returns a number,

$$\vec{V}, \vec{W} \rightarrow \boldsymbol{F}(\vec{V}, \vec{W}) \in \mathbb{R} \,. \tag{2.131}$$

As shown by Eq. 2.128, the components of the tensor $\boldsymbol{F}$ are

$$F_{ij} = \boldsymbol{F}(\vec{e}_{(i)}, \vec{e}_{(j)}) \,. \tag{2.132}$$

Since there are n basis vectors, the quantities F_{ij} are the components of an $n \times n$ real matrix. Using the linearity properties of tensors, we find that given two arbitrary vectors $\vec{A}$ and $\vec{B}$,

$$\boldsymbol{F}(\vec{A}, \vec{B}) \;=\; F(A^i \vec{e}_{(i)}, B^j \vec{e}_{(j)}) = A^i B^j \boldsymbol{F}(\vec{e}_{(i)}, \vec{e}_{(j)}) = F_{ij} A^i B^j \,. \tag{2.133}$$

Thus, $F_{ij} A^i B^j$ is the real number which results from the application of the $\binom{0}{2}$ tensor $\mathbf{F}$ to any pair of vectors $\vec{A}$ and $\vec{B}$.

Basis of the tensor space

The coordinate basis $\{\boldsymbol{\omega}^{(i)(j)}\}$ for $\binom{0}{2}$ tensors can be constructed as follows. We want to write the tensor as a linear combination of the basis elements, i.e.

$$\boldsymbol{F} = F_{ij} \boldsymbol{\omega}^{(i)(j)} \,, \tag{2.134}$$

where F_{ij} are the components of the tensor defined in Eq. 2.132. If we apply $\boldsymbol{F}$ to any two vectors $\vec{A}$ and $\vec{B}$, Eq. 2.134 gives

$$\boldsymbol{F}(\vec{A}, \vec{B}) = F_{ij} \boldsymbol{\omega}^{(i)(j)}(\vec{A}, \vec{B}). \tag{2.135}$$

On the other hand, from Eq. 2.133 we know that $\boldsymbol{F}(\vec{A}, \vec{B}) = F_{ij} A^i B^j$, and since $A^i = \tilde{\omega}^{(i)}(\vec{A})$ and $B^i = \tilde{\omega}^{(i)}(\vec{B})$ (see Eq. 2.122), we get

$$\boldsymbol{F}(\vec{A}, \vec{B}) = F_{ij} \tilde{\omega}^{(i)}(\vec{A}) \tilde{\omega}^{(j)}(\vec{B}) \,. \tag{2.136}$$

By equating Eq. 2.135 and Eq. 2.136 we find

$$\boldsymbol{\omega}^{(i)(j)}(\vec{A}, \vec{B}) = \tilde{\omega}^{(i)}(\vec{A}) \tilde{\omega}^{(j)}(\vec{B}) \,, \tag{2.137}$$

and this equation holds for *any* pair of vectors. We now define $\boldsymbol{\omega}^{(i)(j)}$ as the "outer product", indicated with the symbol $\otimes$, of the two basis one-forms, i.e.

$$\boldsymbol{\omega}^{(i)(j)} = \tilde{\omega}^{(i)} \otimes \tilde{\omega}^{(j)} . \tag{2.138}$$

This equation means that when we apply $\boldsymbol{\omega}^{(i)(j)}$ to the vectors $\vec{A}$ and $\vec{B}$, we get as a result a number which is the product of two numbers: the first is given by the first one-form, $\tilde{\omega}^{(i)}$, applied to the first vector, $\vec{A}$; the second is given by the second one-form $\tilde{\omega}^{(j)}$, applied to the second vector, $\vec{B}$, as in Eq. 2.137. It should be stressed that the order of $\vec{A}$ and $\vec{B}$ cannot be changed, i.e.

$$\boldsymbol{\omega}^{(i)(j)}(\vec{A}, \vec{B}) \neq \tilde{\omega}^{(i)}(\vec{B})\tilde{\omega}^{(j)}(\vec{A}) . \tag{2.139}$$

Thus, the basis for $\binom{0}{2}$ tensors is given by the outer product of the basis one-forms, and any $\binom{0}{2}$ tensor $\mathbf{F}$ can be expanded as

$$\boldsymbol{F} = F_{ij}\tilde{\omega}^{(i)} \otimes \tilde{\omega}^{(j)} . \tag{2.140}$$

Following the same procedure we can construct the basis for any type of tensor. For example, it is easy to show that the basis for a $\binom{2}{0}$ tensor is

$$\boldsymbol{e}_{(i)(j)} = \vec{e}_{(i)} \otimes \vec{e}_{(j)} . \tag{2.141}$$

Indeed, given a $\binom{2}{0}$ tensor

$$\mathbf{T} = T^{ij}\boldsymbol{e}_{(i)(j)} , \tag{2.142}$$

whose components are (see Eq. 2.128)

$$T^{ij} = \boldsymbol{T}(\tilde{\omega}^{(i)}, \tilde{\omega}^{(j)}) ; \tag{2.143}$$

when applied to any pair of one-forms $\tilde{\alpha}$ and $\tilde{\sigma}$, it gives

$$\boldsymbol{T}(\tilde{\alpha}, \tilde{\sigma}) = T^{ij}\boldsymbol{e}_{(i)(j)}(\tilde{\alpha}, \tilde{\sigma}) ; \tag{2.144}$$

on the other hand, we know that

$$\boldsymbol{T}(\tilde{\alpha}, \tilde{\sigma}) = T(\alpha_i\tilde{\omega}^{(i)}, \sigma_j\tilde{\omega}^{(j)}) = \alpha_i\sigma_j\boldsymbol{T}(\tilde{\omega}^{(i)}, \tilde{\omega}^{(j)}) = \alpha_i\sigma_j T^{ij} , \tag{2.145}$$

where we have used the linearity of tensors with respect to their arguments. According to Eq. 2.122, $\alpha_i = \vec{e}_{(i)}(\tilde{\alpha})$ and $\sigma_j = \vec{e}_{(j)}(\tilde{\sigma})$, and Eq. 2.145 becomes

$$\boldsymbol{T}(\tilde{\alpha}, \tilde{\sigma}) = T^{ij}\vec{e}_{(i)}(\tilde{\alpha})\vec{e}_{(j)}(\tilde{\sigma}) , \tag{2.146}$$

which, compared to Eq. 2.144, shows that

$$\boldsymbol{e}_{(i)(j)}(\tilde{\alpha}, \tilde{\sigma}) = \vec{e}_{(i)}(\tilde{\alpha})\vec{e}_{(j)}(\tilde{\sigma}) ; \tag{2.147}$$

this equation holds for any pair of one-forms. This allows us to write the basis for $\binom{0}{2}$ tensors as the outer product of the basis vectors as in Eq. 2.141, and consequently

$$\boldsymbol{T} = T^{ij}\vec{e}_{(i)} \otimes \vec{e}_{(j)} . \tag{2.148}$$

Box 2-L

Exercise

Prove that the $\binom{1}{1}$ tensor $\vec{V} \otimes \tilde{\sigma}$ has components $V^i\sigma_j$ and find the basis for $\binom{1}{1}$ tensors.

Transformation rules for tensors

We shall now find the transformation rules for tensor components and tensor bases, under a coordinate transformation $x^i = x^i(x^{j'})$. Let us start with a $\binom{0}{2}$ tensor

$$\boldsymbol{F} = F_{ij}\tilde{\omega}^{(i)} \otimes \tilde{\omega}^{(j)} . \tag{2.149}$$

If we change coordinates, we shall have a new set of coordinate basis one-forms $\{\tilde{\omega}^{(i')}\}$ which are related to the old ones by the equations

$$\tilde{\omega}^{(i)} = \Lambda^i{}_{j'}\tilde{\omega}^{(j')} \quad ; \quad \tilde{\omega}^{(j')} = \Lambda^{j'}{}_i\tilde{\omega}^{(i)} . \tag{2.150}$$

In the new basis the tensor is

$$\boldsymbol{F} = F_{i'j'}\tilde{\omega}^{(i')} \otimes \tilde{\omega}^{(j')} . \tag{2.151}$$

By equating Eqs. 2.149 and 2.151 we find

$$F_{i'j'}\tilde{\omega}^{(i')} \otimes \tilde{\omega}^{(j')} = F_{ij}\tilde{\omega}^{(i)} \otimes \tilde{\omega}^{(j)} . \tag{2.152}$$

By replacing $\tilde{\omega}^{(i)}$ and $\tilde{\omega}^{(j)}$ using the first of Eqs. 2.150

$$F_{i'j'}\tilde{\omega}^{(i')} \otimes \tilde{\omega}^{(j')} = F_{ij}\Lambda^i{}_{i'}\tilde{\omega}^{(i')} \otimes \Lambda^j{}_{j'}\tilde{\omega}^{(j')} = F_{ij}\Lambda^i{}_{i'}\Lambda^j{}_{j'}\tilde{\omega}^{(i')} \otimes \tilde{\omega}^{(j')} , \tag{2.153}$$

from which it follows[3]

$$F_{i'j'} = F_{ij}\Lambda^i{}_{i'}\Lambda^j{}_{j'} = F_{ij}\frac{\partial x^i}{\partial x^{i'}}\frac{\partial x^j}{\partial x^{j'}} . \tag{2.154}$$

In a similar way, by using the transformation rules for vector bases

$$\vec{e}_{(i)} = \Lambda^{j'}{}_i\, \vec{e}_{(j')} \quad ; \quad \vec{e}_{(j')} = \Lambda^i{}_{j'}\, \vec{e}_{(i)} \tag{2.155}$$

it is easy to show that

$$T^{i'j'} = T^{ij}\Lambda^{i'}{}_i\Lambda^{j'}{}_j, \tag{2.156}$$

and

$$T^{i'}{}_{j'} = T^i{}_j\Lambda^{i'}{}_i\Lambda^j{}_{j'} . \tag{2.157}$$

In general, for $\binom{N}{N'}$ tensors

$$T^{i'_1\cdots i'_N}{}_{j'_1\cdots j'_{N'}} = \Lambda^{i'_1}{}_{i_1}\ldots\Lambda^{i'_N}{}_{i_N}\Lambda^{j_1}{}_{j'_1}\ldots\Lambda^{j_{N'}}{}_{j'_{N'}}T^{i_1\cdots i_N}{}_{j_1\cdots j_{N'}} . \tag{2.158}$$

The following point should be stressed: the notion of tensor that we have introduced is *independent* of which coordinates, i.e. of which coordinate basis, we use. Indeed, the number that a $\binom{N}{N'}$ tensor associates to N one-forms and N' vectors does not depend on the particular basis we choose. This is the reason why, for example, we can equate Eqs. 2.149 and 2.151. Therefore, like vectors and one-forms, also *tensors are geometrical objects.*

[3]Note that this is the same transformation law for $\binom{0}{2}$ tensors given in Eq. 1.11.

Box 2-M

Definition of tensors from their transformation properties

Given a set of quantities $T^{i_1 \cdots i_N}{}_{j_1 \cdots j_{N'}}$ (with all indexes running from 1 to n), if for a general coordinate transformation $x^k \to x^{i'}(x^k)$ they transform as

$$T^{i'_1 \cdots i'_N}{}_{j'_1 \cdots j'_{N'}} = \Lambda^{i'_1}{}_{i_1} \ldots \Lambda^{i'_N}{}_{i_N} \Lambda^{j_1}{}_{j'_1} \ldots \Lambda^{j_{N'}}{}_{j'_{N'}} T^{i_1 \cdots i_N}{}_{j_1 \cdots j_{N'}} \tag{2.159}$$

(see Eq. 2.158), they are the components of a $\binom{N}{N'}$ tensor

$$\boldsymbol{T} = T^{i_1 \cdots i_N}{}_{j_1 \cdots j_{N'}} \vec{e}_{(i_1)} \otimes \cdots \otimes \vec{e}_{(i_N)} \otimes \tilde{\omega}^{(j_1)} \otimes \cdots \otimes \tilde{\omega}^{(j_{N'})} . \tag{2.160}$$

Indeed, replacing the transformation rules of basis vectors and one-forms (Eqs. 2.69, 2.99) and Eq. 2.159 in Eq. 2.160, the latter becomes

$$\boldsymbol{T} = T^{i'_1 \cdots i'_N}{}_{j'_1 \cdots j'_{N'}} \vec{e}_{(i'_1)} \otimes \cdots \otimes \vec{e}_{(i'_N)} \otimes \tilde{\omega}^{(j'_1)} \otimes \cdots \otimes \tilde{\omega}^{(j'_{N'})} \tag{2.161}$$

where we have used the inverse matrix relation, Eq. 2.24. Therefore $\mathbf{T}$ in different coordinate frames has the same form (Eqs. 2.160 and 2.161), and the applications from N one-forms and N' vectors to $\mathbb{R}$ are the same.

In some textbooks, tensors are defined from the transformation property of their components, Eq. 2.159. In the case of vectors (i.e. $\binom{1}{0}$ tensors) this is the "traditional definition" we discussed in Sec. 2.2.1.

Operations with tensors

The following operations with tensors are defined.

- *Multiplication by a real number*

  Given a tensor $\boldsymbol{T}$ of type $\binom{N}{N'}$ and a real number a, the quantity

$$\boldsymbol{W} = a\boldsymbol{T} \tag{2.162}$$

  is a tensor of the same type, with components

$$W^{i\ldots}{}_{j\ldots} = aT^{i\ldots}{}_{j\ldots} . \tag{2.163}$$

- *Sum of tensors*

  Given two tensors $\boldsymbol{T}, \boldsymbol{G}$ of the same type $\binom{N}{N'}$ the quantity

$$\boldsymbol{W} = \boldsymbol{T} + \boldsymbol{G} \tag{2.164}$$

  is a tensor of the same type, with components

$$W^{i\ldots}{}_{j\ldots} = T^{i\ldots}{}_{j\ldots} + G^{i\ldots}{}_{j\ldots} . \tag{2.165}$$

- *Outer product*

  Given two tensors $\boldsymbol{T}, \boldsymbol{G}$ of types $\binom{N_1}{N'_1}$ and $\binom{N_2}{N'_2}$, respectively, the quantity

$$\boldsymbol{W} = \boldsymbol{T} \otimes \boldsymbol{G} \tag{2.166}$$

is a tensor of type $\binom{N_1+N_2}{N_1'+N_2'}$, with components

$$W^{i\dots k\dots}{}_{j\dots l\dots} = T^{i\dots}{}_{j\dots}G^{k\dots}{}_{l\dots}\,. \tag{2.167}$$

For instance, if both $\boldsymbol{T}, \boldsymbol{G}$ are of type $\binom{0}{2}$,

$$W_{ijkl} = T_{ij}G_{kl}\,. \tag{2.168}$$

- *Contraction*

  Given a tensor $\boldsymbol{T}$ of type $\binom{N}{N'}$, with components $\{T^{i_1 i_2 \dots i_N}{}_{j_1 j_2 \dots j_{N'}}\}$ in a given frame, we define a new tensor $\boldsymbol{W}$ of type $\binom{N-1}{N'-1}$, the components of which are obtained by summing over one contravariant (i.e. upper) and one covariant (i.e. lower) index of $\boldsymbol{T}$, i.e.

$$W^{\dots i_{m-1}\; i_{m+1}\dots}{}_{\dots j_{n-1}\; j_{n+1}\dots} = \sum_k T^{\dots i_{m-1}\; k\; i_{m+1}\dots}{}_{\dots j_{n-1}\; k\; \beta_{n+1}\dots} \equiv T^{\dots i_{m-1}\; k\; i_{m+1}\dots}{}_{\dots j_{n-1}\; k\; \beta_{n+1}\dots}\,. \tag{2.169}$$

  This operation is called *contraction.*

  For instance, if $\boldsymbol{T}$ is of type $\binom{2}{3}$ and we choose to contract the first contravariant index with the second covariant index

$$W^j{}_{kl} = T^{ij}{}_{kil} = T^{1j}{}_{k1l} + T^{2j}{}_{k2l} + T^{3j}{}_{k3l} + \dots \tag{2.170}$$

  and $\boldsymbol{W}$ is a $\binom{1}{2}$ tensor.

The operations we have defined are called *tensor operations*, and equations involving tensor operations are *tensor equations.*

Since a tensor $\boldsymbol{W}$ has been defined as an application from vectors and one-forms to $\mathbb{R}$, it is defined on the product of a certain number of copies of the tangent and of the cotangent spaces *in a point* $\mathbf{p}$, $\mathbf{T_p}$, $\mathbf{T_p^*}$, called **tensor space**

$$\mathbf{T_p}^{NN'} = \overbrace{\mathbf{T_p}\otimes\dots\otimes\mathbf{T_p}}^{N} \otimes \overbrace{\mathbf{T_p^*}\otimes\dots\otimes\mathbf{T_p^*}}^{N'}\,. \tag{2.171}$$

It is possible to show that the operations of sum and multiplication by a real number defined above satisfy the properties 2.40 - 2.44; therefore the tensor space is also a vector space. Finally, we can define *tensor fields* as follows. Given an open set $\mathbf{S}\subset\mathbf{M}$, a tensor field is a mapping

$$\begin{aligned} \boldsymbol{W} : \mathbf{S} &\rightarrow \mathbf{T_S}^{NN'} \\ \mathbf{p} &\mapsto \boldsymbol{W}_{\mathbf{p}}\,, \end{aligned}$$

where $\mathbf{T_S}^{NN'} \equiv \bigcup_{\mathbf{p}\in\mathbf{S}} \mathbf{T_p}^{NN'}$, which associates to every point $\mathbf{p}\in\mathbf{S}$, a tensor $\boldsymbol{W}_{\mathbf{p}}$ defined on the tensor space in $\mathbf{p}$, $\mathbf{T_p}^{NN'}$.

2.4.2 Symmetries of a tensor

A $\binom{0}{2}$ tensor $\boldsymbol{F}$ is *symmetric* if

$$\boldsymbol{F}(\vec{A},\vec{B}) = \boldsymbol{F}(\vec{B},\vec{A}) \quad \forall \vec{A},\vec{B}\,. \tag{2.172}$$

As a consequence of Eq. 2.133 we see that if the tensor is symmetric

$$F_{ij}A^iB^j = F_{ij}B^iA^j\,, \tag{2.173}$$

and, by relabeling the indices on the right-hand side

$$F_{ij}A^iB^j = F_{ji}B^jA^i\,, \tag{2.174}$$

i.e.

$$F_{ij} = F_{ji}\,. \tag{2.175}$$

Thus, if a $\binom{0}{2}$ tensor is symmetric the matrix representing its components is symmetric.

Given any $\binom{0}{2}$ tensor $\boldsymbol{F}$ we can always construct from it a symmetric tensor $\boldsymbol{F}^{(s)}$

$$\boldsymbol{F}^{(s)}(\vec{A},\vec{B}) = \frac{1}{2}[\boldsymbol{F}(\vec{A},\vec{B}) + \boldsymbol{F}(\vec{B},\vec{A})]\,. \tag{2.176}$$

In fact $\forall \vec{A},\vec{B}$

$$\frac{1}{2}[\boldsymbol{F}(\vec{A},\vec{B}) + \boldsymbol{F}(\vec{B},\vec{A})] = \frac{1}{2}[\boldsymbol{F}(\vec{B},\vec{A}) + \boldsymbol{F}(\vec{A},\vec{B})]\,. \tag{2.177}$$

Moreover

$$\begin{aligned}
\boldsymbol{F}^{(s)}(\vec{A},\vec{B}) &= F^{(s)}_{ij}A^iB^j = \frac{1}{2}[F_{ij}A^iB^j + F_{ij}B^iA^j] = \frac{1}{2}[F_{ij}A^iB^j + F_{ji}B^jA^i] \\
&= \frac{1}{2}[F_{ij} + F_{ji}]A^iB^j\,,
\end{aligned}$$

and consequently the components of the symmetric tensor are

$$F^{(s)}_{ij} = \frac{1}{2}[F_{ij} + F_{ji}]\,. \tag{2.178}$$

These can be indicated as

$$F_{(ij)} = \frac{1}{2}[F_{ij} + F_{ji}]\,. \tag{2.179}$$

A $\binom{0}{2}$ tensor $\boldsymbol{F}$ is **antisymmetric** if

$$\boldsymbol{F}(\vec{A},\vec{B}) = -\boldsymbol{F}(\vec{B},\vec{A}) \quad \forall \vec{A},\vec{B}\,, \qquad \text{i.e.} \qquad F_{ij} = -F_{ji}\,. \tag{2.180}$$

Again from any $\binom{0}{2}$ tensor we can construct an antisymmetric tensor $\boldsymbol{F}^{(a)}$ defined as

$$\boldsymbol{F}^{(a)}(\vec{A},\vec{B}) = \frac{1}{2}[\boldsymbol{F}(\vec{A},\vec{B}) - \boldsymbol{F}(\vec{B},\vec{A})]\,. \tag{2.181}$$

Proceeding as before, we find that their components are

$$F^{(a)}_{ij} = \frac{1}{2}[F_{ij} - F_{ji}]\,, \tag{2.182}$$

also indicated as

$$F_{[ij]} = \frac{1}{2}[F_{ij} - F_{ji}]\,. \tag{2.183}$$

Any tensor $\binom{0}{2}$ can be written as the sum of its symmetric and antisymmetric parts

$$\boldsymbol{h}[\vec{A},\vec{B}] = \frac{1}{2}[\boldsymbol{h}(\vec{A},\vec{B}) + \boldsymbol{h}(\vec{B},\vec{A})] + \frac{1}{2}[\boldsymbol{h}(\vec{A},\vec{B}) - \boldsymbol{h}(\vec{B},\vec{A})]\,. \tag{2.184}$$

These definitions can be extended to tensors of any rank: an $\binom{N}{N'}$ tensor is symmetric/antisymmetric in its k-th and l-th contravariant indices if

$$T^{i_1\cdots i_{k-1}\, m\, i_{k+1}\ldots i_{l-1}\, n\, i_{l+1}\cdots i_N}{}_{j_1\cdots j_{N'}} = \pm T^{i_1\cdots i_{k-1}\, n\, i_{k+1}\ldots i_{l-1}\, m\, i_{l+1}\cdots i_N}{}_{j_1\cdots j_{N'}}\,; \tag{2.185}$$

the same definition applies for the covariant indices.

Box 2-N

Summary on tensors

Tensors are linear, real-valued functions of vectors and one-forms: a $\binom{N}{N'}$ tensor $\boldsymbol{T}$ maps N one-forms $\tilde{q}^{(1)},\ldots,\tilde{q}^{(N)}$ and N' vectors $\vec{V}_{(1)},\ldots,\vec{V}_{(N')}$ to the real number $\boldsymbol{T}(\tilde{q}^{(1)},\ldots,\tilde{q}^{(N)},\vec{V}_{(1)},\ldots,\vec{V}_{(N')})$, given by

$$\boldsymbol{T}(\tilde{q}^{(1)},\ldots,\tilde{q}^{(N)},\vec{V}_{(1)},\ldots,\vec{V}_{(N')}) = T^{i_1\cdots i_N}{}_{j_1\cdots j_{N'}}\, q^{(1)}_{i_1}\cdots q^{(N)}_{i_N}\, V^{j_1}_{(1)}\cdots V^{j_{N'}}_{(N')}\,. \tag{2.186}$$

$V^i_{(k)}$ and $q^{(l)}_i$ are the components of the vector $\vec{V}_{(k)}$ and of the one-form $\tilde{q}^{(l)}$ in a chosen basis, respectively. The tensor components $T^{i_1\cdots i_N}{}_{j_1\cdots j_{N'}}$ are obtained by applying the tensor $\boldsymbol{T}$ to the one-forms and vectors of the chosen basis:

$$T^{i_1\cdots i_N}{}_{j_1\cdots j_{N'}} = \boldsymbol{T}(\tilde{\omega}^{(1)},\ldots,\tilde{\omega}^{(N)},\vec{e}_{(1)},\ldots,\vec{e}_{(N')})\,. \tag{2.187}$$

Note that tensors are "attached" to the point $\mathbf{p}$, since they belong to the tensor space $\mathbf{T}^{NN'}_{\mathbf{p}}$, which is the outer product of N copies of the cotangent space $\mathbf{T}^*_{\mathbf{p}}$ and N' copies of the tangent space $\mathbf{T}_{\mathbf{p}}$ to the manifold $\mathbf{M}$ in $\mathbf{p}$.

Coordinate basis for tensors

Given a coordinate system $(x^1,\ldots,x^n)$, the coordinate basis for tensors is the outer product of N coordinate basis one-forms $\tilde{\omega}^{(i)} = dx^i$, and of N' coordinate basis vectors $\vec{e}_{(j)} = \frac{\partial}{\partial x^j}$:

$$\overbrace{\tilde{\omega}^{(1)}\otimes\cdots\otimes\tilde{\omega}^{(N)}}^{N}\otimes\overbrace{\vec{e}_{(1)}\otimes\cdots\otimes\vec{e}_{(N')}}^{N'}\,. \tag{2.188}$$

In this basis, any tensor $\boldsymbol{T}$ can be written as

$$\boldsymbol{T} = T^{i_1\cdots i_N}{}_{j_1\cdots j_{N'}}\,\tilde{\omega}^{(1)}\otimes\cdots\otimes\tilde{\omega}^{(N)}\otimes\vec{e}_{(1)}\otimes\cdots\otimes\vec{e}_{(N')}\,. \tag{2.189}$$

Transformation rules for tensors

Given the coordinate transformation $x^i = x^i(x^{j'})$, $i,j' = 1,\ldots,n$, the components of tensors transform as follows

$$T^{i'_1\cdots i'_N}{}_{j'_1\cdots j'_{N'}} = \Lambda^{i'_1}{}_{i_1}\ldots\Lambda^{i'_N}{}_{i_N}\Lambda^{j_1}{}_{j'_1}\ldots\Lambda^{j_{N'}}{}_{j'_{N'}}T^{i_1\cdots i_N}{}_{j_1\cdots j_{N'}}\,, \tag{2.190}$$

where $\Lambda^i{}_{j'} = \frac{\partial x^i}{\partial x^{j'}}$, and $\Lambda^{j'}{}_i = \frac{\partial x^{j'}}{\partial x^i}$ is the inverse matrix of $\Lambda^i{}_{j'}$. Vice versa, if some quantities $T^{i_1\cdots i_N}{}_{j_1\cdots j_{N'}}$ transform as in Eq. 2.190 for a general coordinate transformation, they are the components of a $\binom{N}{N'}$ tensor (see Box 2-M).

2.5 THE METRIC TENSOR AND ITS PROPERTIES

In Chapter 1 we have seen that the metric tensor has a central role in the relativistic theory of gravity. In this section we shall discuss its geometrical meaning.

Given a vector space $\mathbf{T_P}$, a **scalar product** is a mapping

$$\begin{aligned} \mathbf{T_P} \times \mathbf{T_P} &\rightarrow \mathbb{R} \\ (\vec{U},\vec{V}) &\mapsto \vec{U}\cdot\vec{V} \end{aligned} \tag{2.191}$$

which satisfies the following properties: $\forall\,\vec{U},\vec{V},\vec{W} \in \mathbf{T_P}$, $\forall\, a \in \mathbb{R}$,

$$\vec{U}\cdot\vec{V} = \vec{V}\cdot\vec{U} \tag{2.192}$$

$$(a\vec{U})\cdot\vec{V} = a(\vec{U}\cdot\vec{V}) \tag{2.193}$$

$$(\vec{U}+\vec{V})\cdot\vec{W} = \vec{U}\cdot\vec{W} + \vec{V}\cdot\vec{W}\,. \tag{2.194}$$

The scalar product allows to define the measure of a vector $\vec{V}$, i.e. the *norm* $\|\vec{V}\|$, given by

$$\|\vec{V}\|^2 \equiv \vec{V}\cdot\vec{V}\,. \tag{2.195}$$

The scalar product 2.191 is *positive definite* if $\vec{V}\cdot\vec{V} \geq 0\ \forall\vec{V}$, and $\vec{V}\cdot\vec{V} = 0$ if and only if $\vec{V} = \vec{0}$. It is *non-degenerate* if $\vec{V}\cdot\vec{W} = 0\ \forall\vec{W}$ implies $\vec{V} = \vec{0}$. A positive definite scalar product is always non-degenerate, but the converse is not necessarily true.

A differentiable manifold endowed with a positive-definite scalar product is called *Riemannian.* If the scalar product is non-degenerate and not positive definite, the manifold is called *pseudo-Riemannian.*

The **metric tensor $\boldsymbol{g}$** is a symmetric $\binom{0}{2}$ tensor which defines the scalar product between vectors on a manifold

$$\boldsymbol{g}(\vec{A},\vec{B}) \equiv \vec{A}\cdot\vec{B}\,, \qquad \forall \vec{A},\vec{B}\,. \tag{2.196}$$

A differentiable manifold on which a metric tensor is defined is called **metric space**.

The components of the metric tensor are (see Eq. 2.132)

$$g_{ij} = \boldsymbol{g}(\vec{e}_{(i)},\vec{e}_{(j)}) = \vec{e}_{(i)}\cdot\vec{e}_{(j)}\,. \tag{2.197}$$

Thus, the scalar product of two vectors $\vec{A} = A^i\vec{e}_{(i)}$, $\vec{B} = B^i\vec{e}_{(i)}$ is

$$\boldsymbol{g}(\vec{A},\vec{B}) = \boldsymbol{g}(A^i\vec{e}_{(i)},B^j\vec{e}_{(j)}) = A^iB^j\boldsymbol{g}(\vec{e}_{(i)},\vec{e}_{(j)}) = A^iB^jg_{ij}\,, \tag{2.198}$$

where we have used the linearity of tensors.

It is easy to show that in a Riemannian manifold the n eigenvalues of the (symmetric) matrix g_{ij} are all positive, while in a pseudo-Riemannian manifold some of them are positive, some others are negative. However, since the scalar product is non-degenerate in both cases, all eigenvalues are non-vanishing. For example, the Euclidean space is a Riemannian manifold with metric components δ_{ij}.

A particular case of pseudo-Riemannian space is the *Lorentzian space*, in which $n-1$ eigenvalues of the matrix $g_{\alpha\beta}$ are positive, and the remaining one is negative (or vice versa: $n-1$ are negative, one is positive). For instance, Minkowski's spacetime is Lorentzian, with metric $\eta_{\alpha\beta} = \mathrm{diag}(-1,1,1,1)$. The four-dimensional spacetime of General Relativity is a Lorentzian space. It should be noted that in some textbooks the metric tensor is defined with the opposite *signature*, i.e. $\eta_{\alpha\beta} = \mathrm{diag}(1,-1,-1,-1)$. This choice is just a matter of convention, and does not affect any observable quantity. In this book we shall use the signature $(-1,1,1,1)$[4].

We remark that the metric tensor allows to compute the scalar product in a general differentiable manifold and that, since it is a geometrical object, the scalar product of two vectors does not depend on the choice of the coordinate system.

The $\binom{0}{2}$ tensor $\boldsymbol{g}$ can be expanded as

$$\boldsymbol{g} = g_{ij}\tilde{\omega}^{(i)} \otimes \tilde{\omega}^{(j)} . \tag{2.199}$$

In a coordinate basis, $\tilde{\omega}^{(i)} = dx^i$, thus

$$\boldsymbol{g} = g_{ij} dx^i \otimes dx^j . \tag{2.200}$$

Hereafter, in order to follow the standard notation of textbooks, we shall call $\boldsymbol{g} \equiv ds^2$, we shall omit the symbol "$\otimes$", referring to the metric tensor as

$$ds^2 \equiv \boldsymbol{g} = g_{ij} dx^i dx^j . \tag{2.201}$$

We remind that the one-forms $\{dx^i\}$ can be seen as the coordinate changes for a displacement in an unspecified direction (at first order in the displacement); once we specify the direction assigning a vector $\vec{V}$

$$dx^i(\vec{V}) = V^i \tag{2.202}$$

(see Eq. 2.102). In the same way, the metric tensor can be seen as the square of the norm – i.e., the length squared – of a displacement in an unspecified direction, at first order in the displacement; once we specify a vector $\vec{V}$,

$$\boldsymbol{g}(\vec{V},\vec{V}) = \vec{V}\cdot\vec{V} = \|\vec{V}\|^2 . \tag{2.203}$$

The metric tensor allows to compute the distance between two points and the angle among vectors

Let us consider a four-dimensional spacetime described by the coordinates (x^0, x^1, x^2, x^3), and two infinitely close points $\mathbf{p}(x^0, x^1, x^2, x^3)$ and $\mathbf{p}'(x^0+\delta x^0, x^1+\delta x^1, x^2+\delta x^2, x^3+\delta x^3)$, where

$$\delta x^\alpha = dx^\alpha(\vec{\delta x}) .$$

Hereafter we shall write the components of the infinitesimal displacement as dx^α, leaving implicit the dependence on $\vec{\delta x}$. The displacement vector between $\mathbf{p}$ and $\mathbf{p}'$ is

$$\vec{\delta x} = dx^0\vec{e}_{(0)} + dx^1\vec{e}_{(1)} + dx^2\vec{e}_{(2)} + dx^3\vec{e}_{(3)} = dx^\alpha\vec{e}_{(\alpha)} . \tag{2.204}$$

The norm of $\vec{\delta x}$ measures the square of the spacetime distance between $\mathbf{p}$ and $\mathbf{p}'$

$$ds^2 = \boldsymbol{g}(dx^\alpha\vec{e}_{(\alpha)}, dx^\beta\vec{e}_{(\beta)}) = dx^\alpha dx^\beta g(\vec{e}_{(\alpha)}, \vec{e}_{(\beta)}) = g_{\alpha\beta}dx^\alpha dx^\beta \tag{2.205}$$

with $\alpha, \beta = 0, \ldots, 3$. For example, if the space is Minkowski's spacetime $g_{\alpha\beta} = \eta_{\alpha\beta} = \mathrm{diag}(-1,1,1,1)$, Eq. 2.205 gives

$$ds^2 = -(dx^0)^2 + (dx^1)^2 + (dx^2)^2 + (dx^3)^3 . \tag{2.206}$$

[4] Note that for coordinates and components in Lorentzian spaces we use Greek indices (see Notation for details).

If we change coordinates to $(x^{0'}, x^{1'}, x^{2'}, x^{3'})$, the distance between $\mathbf{p}$ and $\mathbf{p}'$ does not change, since ds^2 is a scalar quantity. Its expression in terms of the components of the metric tensor in the new frame, $g_{\alpha'\beta'}$, is

$$ds^2 = g(dx^{\alpha'}\vec{e}_{(\alpha')}, dx^{\beta'}\vec{e}_{(\beta')}) = g_{\alpha'\beta'}dx^{\alpha'}dx^{\beta'} . \tag{2.207}$$

Thus if we know the components of the metric tensor in any reference frame, we can compute the distance between two points infinitey close

$$ds = \sqrt{|ds^2|} = \sqrt{|g_{\mu\nu}dx^\mu dx^\nu|} . \tag{2.208}$$

Note that we consider the absolute value of ds^2 because, when the manifold is not Riemannian (as in the case of spacetime), the metric tensor is not positive-definite.

In order to compute *finite* distances, we need to proceed as follows. Let us consider a curve

$$\begin{array}{lcl} [a,b] \subset \mathbb{R} & \to & \mathcal{C} \\ \lambda & \mapsto & \mathbf{p}(\lambda) \end{array} \tag{2.209}$$

which, in a given coordinate system $\{x^\mu\}$, corresponds to the real functions

$$\lambda \mapsto \{x^\mu(\lambda)\} . \tag{2.210}$$

We define the *proper length* of the path $\mathcal{C}$ as (see Eq. 2.208)

$$\Delta s = \int_a^b ds = \int_a^b d\lambda \sqrt{\left| g_{\mu\nu}\frac{dx^\mu}{d\lambda}\frac{dx^\nu}{d\lambda} \right|} . \tag{2.211}$$

In other words, given the curve $\{x^\mu(\lambda)\}$, and the tangent vector

$$t^\mu = \frac{dx^\mu}{d\lambda} , \tag{2.212}$$

the measure element on the curve $ds/d\lambda$ (which, integrated in $d\lambda$, gives the proper length of the path) is

$$\frac{ds}{d\lambda} = \sqrt{|\boldsymbol{g}(\vec{t},\vec{t})|} = \sqrt{|g_{\mu\nu}t^\mu t^\nu|} = \sqrt{\left| g_{\mu\nu}\frac{dx^\mu}{d\lambda}\frac{dx^\nu}{d\lambda} \right|} . \tag{2.213}$$

Note that if we change coordinate system, $\{x^\mu\} \to \{x^{\alpha\prime}\}$, the quantity 2.213 does not change. Furthermore, if we change the parameter to the curve,

$$\lambda \to \lambda' = \lambda'(\lambda) , \tag{2.214}$$

the new measure element is

$$\frac{ds}{d\lambda'} = \sqrt{\left| g_{\mu\nu}\frac{dx^\mu}{d\lambda'}\frac{dx^\nu}{d\lambda'} \right|} = \sqrt{\left| g_{\mu\nu}\frac{dx^\mu}{d\lambda}\frac{d\lambda}{d\lambda'}\frac{dx^\nu}{d\lambda}\frac{d\lambda}{d\lambda'} \right|} = \frac{ds}{d\lambda}\left|\frac{d\lambda}{d\lambda'}\right| \tag{2.215}$$

and

$$\Delta s = \int_a^b d\lambda \frac{ds}{d\lambda} = \int_a^b \left(d\lambda' \left|\frac{d\lambda}{d\lambda'}\right| \right) \frac{ds}{d\lambda} = \int_{a'}^{b'} d\lambda' \frac{ds}{d\lambda'} . \tag{2.216}$$

We remark that Δs depends neither on the coordinate system, nor on the curve parametrization: it is characteristic of the path, not of the curve.

Finally, the metric tensor also allows to compute the angle between two vectors. The scalar product between two vectors $\vec{V}_1$ and $\vec{V}_2$ is, by definition,

$$\boldsymbol{g}(\vec{V}_1, \vec{V}_2) = \|\vec{V}_1\| \|\vec{V}_2\| \cos\theta \,, \tag{2.217}$$

where θ is the angle between them. Therefore,

$$\cos\theta = \frac{\boldsymbol{g}(\vec{V}_1, \vec{V}_2)}{\|\vec{V}_1\| \|\vec{V}_2\|} \,. \tag{2.218}$$

To summarize, the metric tensor allows to compute the length of vectors (i.e., their norm), the length of a curve, and the angle between vectors. These properties justify why it is called *metric* tensor: it allows to perform *metric* operations on a manifold, which are the building blocks of the measurement process in any physical system.

Box 2-O

Example: Minkowski coordinates and coordinate basis vectors

The metric of the four-dimensional Minkowski spacetime, in Minkowskian coordinates $\{\xi^\alpha\} = (ct, x, y, z)$, is

$$g_{\alpha\beta} = \begin{pmatrix} -1 & 0 & 0 & 0 \\ 0 & +1 & 0 & 0 \\ 0 & 0 & +1 & 0 \\ 0 & 0 & 0 & +1 \end{pmatrix} \equiv \eta_{\alpha\beta} \,, \tag{2.219}$$

i.e.

$$ds^2 = g_{\alpha\beta} d\xi^\alpha d\xi^\beta = -c^2 dt^2 + dx^2 + dy^2 + dz^2 \,. \tag{2.220}$$

This implies that the coordinate basis vectors

$$\begin{aligned} \vec{e}_{(0)} &= \vec{e}_{(t)} \to_O (1,0,0,0) \\ \vec{e}_{(1)} &= \vec{e}_{(x)} \to_O (0,1,0,0) \\ \vec{e}_{(2)} &= \vec{e}_{(y)} \to_O (0,0,1,0) \\ \vec{e}_{(3)} &= \vec{e}_{(z)} \to_O (0,0,0,1) \end{aligned}$$

are mutually orthogonal, i.e.

$$\vec{e}_{(\alpha)} \cdot \vec{e}_{(\beta)} = g_{\alpha\beta} = 0 \qquad \text{if} \qquad \alpha \neq \beta \,. \tag{2.221}$$

In addition, since

$$g_{11} = g_{22} = g_{33} = 1, \qquad \text{and} \qquad g_{00} = -1 \,, \tag{2.222}$$

the basis vectors are unit vectors. Since

$$\vec{e}_{(0)} \cdot \vec{e}_{(0)} = -1 \;, \quad \vec{e}_{(k)} \cdot \vec{e}_{(k)} = 1 \quad (k = 1, \ldots, 3) \;, \tag{2.223}$$

$\vec{e}_{(0)}$ is a timelike vector and $\vec{e}_{(i)}$ $(i = 1, 2, 3)$ are spacelike vectors.
Hereafter, $\eta_{\alpha\beta}$ will indicate the components of the metric tensor of Minkowski's spacetime when expressed in Minkowskian coordinates.

Box 2-P

Example: metric tensor and coordinate transformations. I

Let us consider the metric of the three-dimensional Minkowski spacetime, in Minkowskian coordinates $\xi^\alpha = (ct, x, y) \equiv (x^0, x^1, x^2)$

$$ds^2 = g_{\alpha\beta} dx^\alpha dx^\beta , \quad \alpha, \beta = 0, \ldots, 2 \qquad \text{and} \qquad g_{\alpha\beta} \equiv \eta_{\alpha\beta} = \text{diag}(-1, 1, 1) . \tag{2.224}$$

The vectors of the coordinate basis have components $\vec{e}_{(0)} \to_O (1, 0, 0)$, $\vec{e}_{(1)} \to_O (0, 1, 0)$, $\vec{e}_{(2)} \to_O (0, 0, 1)$. We now change to polar coordinates $(x^{0'}, x^{1'}, x^{2'}) = (x^{0'}, r, \theta)$, as in Box 2-G, and the new coordinate basis vectors are (see Eq. 2.73)

$$\begin{cases} x^0 = x^{0'} \\ x^1 = r\cos\theta \\ x^2 = r\sin\theta \end{cases} , \qquad \begin{cases} \vec{e}_{(0')} = \vec{e}_{(0)} \\ \vec{e}_{(1')} = \vec{e}_{(r)} = \cos\theta \vec{e}_{(1)} + \sin\theta \vec{e}_{(2)} \\ \vec{e}_{(2')} = \vec{e}_{(\theta)} = -r\sin\theta \vec{e}_{(1)} + r\cos\theta \vec{e}_{(2)} \end{cases} . \tag{2.225}$$

The components of the metric tensor in the new frame can be computed as the scalar product of the basis vectors of this frame:

$$\begin{aligned} g_{0'0'} &= \vec{e}_{(0')} \cdot \vec{e}_{(0')} = \vec{e}_{(0)} \cdot \vec{e}_{(0)} = -1 , \qquad g_{0'i'} = 0 \quad i' = 1, 2 \\ g_{1'1'} &= \vec{e}_{(1')} \cdot \vec{e}_{(1')} = 1 , \qquad g_{2'2'} = \vec{e}_{(2')} \cdot \vec{e}_{(2')} = r^2 \sin^2\theta + r^2 \cos^2\theta = r^2 , \\ g_{1'2'} &= -r\cos\theta\sin\theta + r\cos\theta\sin\theta = 0 , \end{aligned}$$

$$ds^2 = g_{\alpha'\beta'} dx^{\alpha'} dx^{\beta'} = -c^2 dt^2 + dr^2 + r^2 d\theta^2 , \qquad g_{\alpha'\beta'} = \begin{pmatrix} -1 & 0 & 0 \\ 0 & +1 & 0 \\ 0 & 0 & r^2 \end{pmatrix} . \tag{2.226}$$

We note that *the basis vectors are not necessarily unit vectors, even when the basis is a coordinate basis.*

Indeed, in the present example $\vec{e}_{(2')} \cdot \vec{e}_{(2')} = r^2 \neq 1$. An alternative way to determine the components of the metric tensor is to use the transformation law

$$g_{\mu'\nu'} = \Lambda^\alpha{}_{\mu'} \Lambda^\beta{}_{\nu'} \eta_{\alpha\beta} , \qquad \text{where} \qquad \Lambda^\alpha{}_{\mu'} = \frac{\partial x^\alpha}{\partial x^{\mu'}} . \tag{2.227}$$

$$\begin{aligned} g_{0'0'} &= \Lambda^\alpha{}_{0'} \Lambda^\beta{}_{0'} \eta_{\alpha\beta} = \left(\frac{\partial x^0}{\partial x^{0'}} \right)^2 \eta_{00} = 1 \cdot (-1) = -1 \\ g_{0'i'} &= \Lambda^\alpha{}_{0'} \Lambda^\beta{}_{i'} \eta_{\alpha\beta} = \frac{\partial x^0}{\partial x^{0'}} \frac{\partial x^0}{\partial x^{i'}} \eta_{00} + \frac{\partial x^1}{\partial x^{0'}} \frac{\partial x^1}{\partial x^{i'}} \eta_{11} + \frac{\partial x^2}{\partial x^{0'}} \frac{\partial \xi^2}{\partial x^{i'}} \eta_{22} = 0 \ (i' = 1, 2) \\ g_{1'1'} &= \Lambda^\alpha{}_{1'} \Lambda^\beta{}_{1'} \eta_{\alpha\beta} = -\left(\frac{\partial x^0}{\partial x^{1'}} \right)^2 \eta_{00} + \left(\frac{\partial x^1}{\partial x^{1'}} \right)^2 \eta_{11} + \left(\frac{\partial x^2}{\partial x^{1'}} \right)^2 \eta_{22} \\ &= \cos^2\theta + \sin^2\theta = 1 \\ g_{2'2'} &= \Lambda^\alpha{}_{2'} \Lambda^\beta{}_{2'} \eta_{\alpha\beta} = -\left(\frac{\partial x^0}{\partial x^{2'}} \right)^2 \eta_{00} + \left(\frac{\partial x^1}{\partial x^{2'}} \right)^2 \eta_{11} + \left(\frac{\partial x^2}{\partial x^{2'}} \right)^2 \eta_{22} \\ &= (-r\sin\theta)^2 + (r\cos\theta)^2 = r^2 , \end{aligned}$$

i.e. the metric given in Eq. 2.226.

Box 2-Q

Example: metric tensor and coordinate transformations. II

The simplest way to transform the metric tensor under a coordinate transformation consists in deriving the transformation of the basis one-forms by differentiating the coordinate transformation, and then replacing them in the metric. Let us consider, for example, the three-dimensional Euclidean space in Cartesian coordinates $\{x^i\} = (x^1, x^2, x^3)$; the spacetime metric is

$$ds^2 = (dx^1)^2 + (dx^2)^2 + (dx^3)^2 = g_{ij}dx^i dx^j \tag{2.228}$$

where $g_{ij} = \delta_{ij}$. The spherical coordinates $\{x^{i'}\} = (r, \theta, \varphi)$ are defined by the coordinate transformation $x^i = x^i(x^{k'})$:

$$\begin{aligned} x^1 &= r\sin\theta\cos\varphi \\ x^2 &= r\sin\theta\sin\varphi \\ x^3 &= r\cos\theta\,. \end{aligned} \tag{2.229}$$

Differentiating Eq. 2.229 we find

$$\begin{aligned} dx^1 &= \sin\theta\cos\varphi dr + r\cos\theta\cos\varphi d\theta - r\sin\theta\sin\varphi d\varphi \\ dx^2 &= \sin\theta\sin\varphi dr + r\cos\theta\sin\varphi d\theta + r\sin\theta\cos\varphi d\varphi \\ dx^3 &= \cos\theta dr - r\sin\theta d\theta\,. \end{aligned} \tag{2.230}$$

By replacing these expressions in Eq. 2.228, after some lengthy but trivial computation we find

$$ds^2 = dr^2 + r^2 d\theta^2 + r^2\sin^2\theta d\varphi^2 = g_{ij}dx^i dx^j \tag{2.231}$$

where

$$g_{ij} = \begin{pmatrix} 1 & 0 & 0 \\ 0 & r^2 & 0 \\ 0 & 0 & r^2\sin^2\theta \end{pmatrix} \tag{2.232}$$

and the inverse metric is (see Eq. 2.238)

$$g^{ij} = \begin{pmatrix} 1 & 0 & 0 \\ 0 & \frac{1}{r^2} & 0 \\ 0 & 0 & \frac{1}{r^2\sin^2\theta} \end{pmatrix}. \tag{2.233}$$

The metric tensor maps vectors into one-forms

The metric tensor is a real, linear function of two vectors, i.e. it takes two vectors and associates a real number to them, which is their scalar product

$$\boldsymbol{g}(\vec{W}, \vec{V}) = \vec{W}\cdot\vec{V}\,. \tag{2.234}$$

Now suppose that we leave the first argument empty, $\boldsymbol{g}(\ \ ,\vec{V})$. What is this object? We know that if we fill the empty slot with a generic vector $\vec{A}$ we get a number; thus $\boldsymbol{g}(\ \ ,\vec{V})$ *must be a one-form.* In addition, it is a particular one-form, because it depends on $\vec{V}$: if we change $\vec{V}$, the one-form will be different. Let us indicate this one-form as

$$\tilde{V} = \boldsymbol{g}(\ ,\vec{V}) \,. \tag{2.235}$$

By definition, the components of $\tilde{V}$ are

$$V_i = \tilde{V}(\vec{e}_{(i)}) = \boldsymbol{g}(\vec{e}_{(i)},\vec{V}) = \boldsymbol{g}(\vec{e}_{(i)},V^j\vec{e}_{(j)}) = V^j\boldsymbol{g}(\vec{e}_{(i)},\vec{e}_{(j)}) = V^j g_{ij}\,, \tag{2.236}$$

hence

$$V_i = g_{ij}V^j \,. \tag{2.237}$$

Thus the tensor $\boldsymbol{g}$ associates to any vector $\vec{V}$ a one-form $\tilde{V}$, which we call *dual of* $\vec{V}$, with components given by Eq. 2.237. In addition, if we multiply Eq. 2.237 by g^{ki}, where g^{ki} is the matrix inverse to g_{ki}, i.e.

$$g^{ki}g_{ij} = \delta^k{}_j \,, \tag{2.238}$$

we find

$$g^{ki}V_i = g^{ki}g_{ij}V^j = \delta^k{}_jV^j = V^k \,, \tag{2.239}$$

i.e.

$$V^k = g^{ki}V_i \,. \tag{2.240}$$

Consequently the metric tensor also maps one-forms into vectors. In a similar way the metric tensor maps a $\binom{2}{0}$ in a $\binom{1}{1}$ tensor

$$A^i{}_j = g_{jk}A^{ik} \,, \tag{2.241}$$

or in a $\binom{0}{2}$ tensor

$$A_{ij} = g_{ik}g_{jm}A^{km} \,, \tag{2.242}$$

or vice versa

$$A^{ij} = g^{ik}g^{jm}A_{km} \,. \tag{2.243}$$

Note that, since $g^{ij} = g^{ik}g^{jm}g_{km}$, g^{ij} are the components of the metric tensor in the contravariant form.

These maps are called **index raising** and **lowering** and will be widely used in the rest of the book.

The properties of the metric tensor are summarized in Box 2-R.

Box 2-R

Summary on the metric tensor

The metric tensor

- allows to compute the inner product of two vectors

$$\boldsymbol{g}(\vec{A},\vec{B}) = \vec{A}\cdot\vec{B} = A^i B^j g_{ij}\,, \tag{2.244}$$

  and consequently the norm of a vector $\boldsymbol{g}(\vec{A},\vec{A}) = \vec{A}\cdot\vec{A} = ||A||^2$. The components of the metric tensor are

$$g_{ij} = \boldsymbol{g}(\vec{e}_{(i)},\vec{e}_{(j)}) = \vec{e}_{(i)}\cdot\vec{e}_{(j)}\,; \tag{2.245}$$

- allows to compute the distance between two infinitely close points

$$ds^2 = \boldsymbol{g}(dx^i\vec{e}_{(i)}, dx^j\vec{e}_{(j)}) = g_{ij}dx^i dx^j\,; \tag{2.246}$$

- allows to compute the angle between two vectors

$$\cos\theta = \frac{\boldsymbol{g}(\vec{V}_1,\vec{V}_2)}{||\vec{V}_1||\,||\vec{V}_2||}\,; \tag{2.247}$$

- maps vectors into one-forms and vice versa

$$V_i = g_{ij}V^j\,, \tag{2.248}$$
$$V^k = g^{ki}V_i\,, \tag{2.249}$$

  and, in general, tensors of one type into tensors of a different type, i.e. the metric tensor allows to lower contravariant or raise covariant indices; for example

$$\begin{aligned} A^i{}_j &= g_{jk}A^{ik}\,, \\ A_{ij} &= g_{ik}g_{jm}A^{km}\,, \\ A^{ij} &= g^{ik}g^{jm}A_{km}. \end{aligned} \tag{2.250}$$

CHAPTER 3

Affine connection and parallel transport

In Chapter 1 we showed that, as a consequence of the Equivalence Principle, there are two quantities which describe the effects of a gravitational field on moving bodies: the metric tensor and Christoffel's symbols. In Sec. 2.4 we discussed the geometrical properties of the metric tensor. In this chapter we shall show that Christoffel's symbols allow to compute the derivative of vectors, one-forms, and tensors of any rank, and that they coincide with the quantities introduced in Chapter 1 (see Eq. 1.38).

In this book we will mainly deal with the spacetime manifold. Thus, even though most of the equations we will derive hold in general manifolds, hereafter tensor components and coordinates will be indicated with Greek indices, unless otherwise specified.

3.1 THE COVARIANT DERIVATIVE OF VECTORS

Let us consider a vector (field) $\vec{V} = V^\alpha \vec{e}_{(\alpha)}$ on a manifold $\mathbf{M}$, which we assume to be the spacetime, and let $\{x^\alpha\}$ be a chosen coordinate system. We want to compute the derivative of $\vec{V}$ with respect to the coordinates. By applying Leibniz's rule we get

$$\frac{\partial \vec{V}}{\partial x^\beta} = \frac{\partial V^\alpha}{\partial x^\beta}\vec{e}_{(\alpha)} + V^\alpha \frac{\partial \vec{e}_{(\alpha)}}{\partial x^\beta}\,. \tag{3.1}$$

The first term on the right-hand side is a linear combination of the basis vectors. The second term involves the derivative of the basis vectors, for which we need to compute the quantities $\vec{e}_{(\alpha)}(\mathbf{p}') - \vec{e}_{(\alpha)}(\mathbf{p})$, i.e. to subtract vectors which are applied at different points of the manifold. Note that the vectors $\vec{e}_{(\alpha)}(\mathbf{p})$ and $\vec{e}_{(\alpha)}(\mathbf{p}')$ belong, respectively, to $\mathbf{T_p}$ and to $\mathbf{T_{p'}}$, and that $\mathbf{T_p} \neq \mathbf{T_{p'}}$, since $\mathbf{p}$ and $\mathbf{p}'$ are distinct. Thus, to define the derivative of a vector field on a manifold, we need to specify a *rule* to compare vectors belonging to different tangent spaces; this rule is called **connection**.

Let us start by considering Minkowski's spacetime, where it is possible to define a *global* Minkowskian coordinate system $\{\xi^\mu\} = (ct, x, y, z)$ which covers the entire spacetime; at any given point $\mathbf{p}$ of the manifold there exists the coordinate basis $\vec{e}_{M(\alpha)}(\mathbf{p})$ which belongs to the tangent space $\mathbf{T_p}$, which is the same for any $\mathbf{p}$. In this case a simple rule to compare vectors at different points is to impose that each basis vector at a point $\mathbf{p}$ is equal to the corresponding basis vector at any other point $\mathbf{p}'$, i.e.

$$\vec{e}_{M(\alpha)}(\mathbf{p}) = \vec{e}_{M(\alpha)}(\mathbf{p}')\,. \tag{3.2}$$

This rule is called the *affine connection* of Minkowski's spacetime. Note that with this choice the basis vectors of the Minkowskian frame are, by definition, constant

$$\frac{\partial \vec{e}_{M(\alpha)}}{\partial \xi^{\beta}} = \vec{0}\,. \tag{3.3}$$

Let us now consider a general spacetime, on which we choose a coordinate system $\{x^\alpha\}$ and the associated coordinate basis $\vec{e}_{(\alpha)}$. According to the Equivalence Principle, at any point **p** we can set up a LIF $\{\xi^{\mu'}\}$, and the associated coordinate basis vectors $\vec{e}_{M(\mu')}$. As explained in Box 3-A, the spacetime metric is flat up to terms of order ξ^2, i.e. $g_{\mu\nu} = \eta_{\mu\nu} + O(\xi^2)$, which implies that $g_{\mu\nu,\alpha} = O(\xi)$. Therefore, in the point **p** the metric coincides with Minkowski's metric, and its first derivatives vanish; however, *the second derivatives of the metric, evaluated in* **p**, *do not vanish, unless the spacetime is flat.* Similarly, the first derivatives of the basis vectors $\vec{e}_{M(\mu')}$ evaluated in **p** vanish

$$\frac{\partial \vec{e}_{M(\mu')}}{\partial \xi^{\beta'}} = \vec{0}\,, \tag{3.4}$$

and they vanish also in the entire LIF, to first order in the displacement from **p**.

We know that the basis vectors $\vec{e}_{(\alpha)}$ of the generic frame $\{x^\alpha\}$, and the basis vectors $\vec{e}_{M(\mu')}$ of the LIF $\{\xi^{\mu'}\}$, are related by the transformation

$$\vec{e}_{(\alpha)} = \Lambda^{\mu'}{}_{\alpha}\vec{e}_{M(\mu')} \qquad \longleftrightarrow \qquad \vec{e}_{M(\mu')} = \Lambda^{\gamma}{}_{\mu'}\vec{e}_{(\gamma)}\,, \tag{3.5}$$

where $\Lambda^{\mu'}{}_{\alpha} = \frac{\partial \xi^{\mu'}}{\partial x^\alpha}$. By differentiating Eq. 3.5, using Eqs. 3.4 and 3.5, we find that

$$\frac{\partial \vec{e}_{(\alpha)}}{\partial x^\beta} = \left(\frac{\partial}{\partial x^\beta}\Lambda^{\mu'}{}_{\alpha}\right)\vec{e}_{M(\mu')} = \left(\frac{\partial}{\partial x^\beta}\Lambda^{\mu'}{}_{\alpha}\right)\Lambda^{\gamma}{}_{\mu'}\,\vec{e}_{(\gamma)}\,. \tag{3.6}$$

Defining

$$\Gamma^{\gamma}_{\alpha\beta} = \left(\frac{\partial}{\partial x^\beta}\Lambda^{\mu'}{}_{\alpha}\right)\Lambda^{\gamma}{}_{\mu'}\,, \tag{3.7}$$

we finally find, to first order in the displacement from **p**,

$$\frac{\partial \vec{e}_{(\alpha)}}{\partial x^\beta} = \Gamma^{\gamma}_{\alpha\beta}\,\vec{e}_{(\gamma)}\,; \tag{3.8}$$

the coefficients $\Gamma^{\gamma}_{\alpha\beta}$ have three indices: α indicates which basis vector $\vec{e}_{(\alpha)}$ we are differentiating, β indicates the coordinate with respect to which the differentiation is performed, and γ is the dummy index of summation.

Equation 3.8 is the rule we were looking for, i.e. the **affine connection**: it allows us to compute the derivatives of the basis vectors, i.e. to subtract vectors belonging to different tangent spaces. Replacing $\Lambda^{\mu'}{}_{\alpha} = \frac{\partial \xi^{\mu'}}{\partial x^\alpha}$ and $\Lambda^{\alpha}{}_{\mu'} = \frac{\partial x^\alpha}{\partial \xi^{\mu'}}$ in Eq. 3.7 we find

$$\Gamma^{\gamma}_{\alpha\beta} = \left(\frac{\partial}{\partial x^\beta}\Lambda^{\mu'}{}_{\alpha}\right)\Lambda^{\gamma}{}_{\mu'} = \frac{\partial x^\gamma}{\partial \xi^{\mu'}}\frac{\partial^2 \xi^{\mu'}}{\partial x^\beta \partial x^\alpha}\,, \tag{3.9}$$

which coincides with the definition of the Christoffel symbols in Eq. 1.38 of Chapter 1. Therefore, the quantities $\Gamma^{\gamma}_{\alpha\beta}$ introduced in Eq. 3.7 are the *Christoffel symbols*.

Note that

- In Minkowski's spacetime and in Minkowskian coordinates, since the coordinate basis vectors $\vec{e}_{M(\alpha)}$ are constant, the Christoffel symbols $\Gamma^{\mu}_{\beta\alpha}$ vanish.

- In a LIF at **p**, the vectors of the coordinate basis are constant (up to first order in the displacement from **p**) and $\Gamma^{\mu}_{\beta\alpha}$ vanish.
- *Christoffel's symbols are not the components of a tensor*: since they all vanish in a locally inertial frame, if they were they should all vanish in any other frame, which is not the case.

Having defined the connection, we can now compute the derivative of the vector $\vec{V}$ with respect to the coordinates x^{β}; substituting Eq. 3.8 in Eq. 3.1, we find

$$\frac{\partial \vec{V}}{\partial x^{\beta}} = \frac{\partial V^{\alpha}}{\partial x^{\beta}} \vec{e}_{(\alpha)} + V^{\alpha} \Gamma^{\mu}_{\beta\alpha} \vec{e}_{(\mu)} , \tag{3.10}$$

and relabeling the dummy indices

$$\frac{\partial \vec{V}}{\partial x^{\beta}} = \left[\frac{\partial V^{\alpha}}{\partial x^{\beta}} + V^{\sigma} \Gamma^{\alpha}_{\beta\sigma} \right] \vec{e}_{(\alpha)} . \tag{3.11}$$

If we introduce the following notation

$$V^{\alpha}{}_{,\beta} \equiv \frac{\partial V^{\alpha}}{\partial x^{\beta}} , \qquad \text{and} \quad V^{\alpha}{}_{;\beta} \equiv V^{\alpha}{}_{,\beta} + V^{\mu} \Gamma^{\alpha}_{\beta\mu} , \tag{3.12}$$

Eq. 3.11 finally becomes

$$\frac{\partial \vec{V}}{\partial x^{\beta}} = V^{\alpha}{}_{;\beta} \, \vec{e}_{(\alpha)} . \tag{3.13}$$

Box 3-A

The spacetime metric in a LIF

Let us consider a LIF $\{\xi^\mu\}$ around a point **p** of the spacetime. For simplicity, we choose the origin of the coordinate frame in **p**, i.e. $\xi^\mu(\mathbf{p}) = (0,0,0,0)$.
The Taylor expansion of the metric around the origin is

$$g_{\mu\nu}(\xi) = g_{\mu\nu}(\mathbf{p}) + \frac{\partial g_{\mu\nu}}{\partial \xi^\alpha}|_{\mathbf{p}}\, \xi^\alpha + \frac{1}{2}\frac{\partial^2 g_{\mu\nu}}{\partial \xi^\alpha \partial \xi^\beta}|_{\mathbf{p}}\, \xi^\alpha \xi^\beta + O(\xi^3)\,. \tag{3.14}$$

By construction, $g_{\mu\nu}$ coincides with Minkowski's metric *in the point* **p**, i.e. $g_{\alpha\beta}(\mathbf{p}) = \eta_{\mu\nu}$. Moreover, at the point **p** the geodesic equations 1.39

$$\frac{d^2 x^\alpha}{d\tau^2} + \Gamma^\alpha_{\mu\nu}\left(\frac{dx^\mu}{d\tau}\frac{dx^\nu}{d\tau}\right) = 0 \tag{3.15}$$

reduce to Eq. 1.34

$$\frac{d^2 \xi^\alpha}{d\tau^2} = 0\,. \tag{3.16}$$

Therefore $\Gamma^\alpha_{\mu\nu}(\mathbf{p}) = 0$. Since Christoffel's symbols can be expressed as a linear combination of the derivatives of the metric tensor as shown in Eq. 1.41 (see also Sec. 3.6), it follows that

$$\frac{\partial g_{\mu\nu}}{\partial \xi^\alpha}|_{\mathbf{p}} = 0\,. \tag{3.17}$$

Therefore, Eq. 3.14 reduces to

$$g_{\mu\nu}(\xi) = \eta_{\mu\nu} + \frac{1}{2}\frac{\partial^2 g_{\mu\nu}}{\partial \xi^\alpha \partial \xi^\beta}|_{\mathbf{p}}\xi^\alpha \xi^\beta + O(\xi^3) = \eta_{\mu\nu} + O(\xi^2)\,. \tag{3.18}$$

We conclude that, *at each point of a LIF, the metric tensor differs from Minkowski's metric by terms quadratic in the distance of the point from the origin, i.e. by terms of order $O(\xi^2)$*.
This also implies that in a LIF the proper distance between two events 1, 2 coincides with the spacetime coordinate separation among them, modulo $O(\xi^2)$ corrections:

$$\begin{aligned} \Delta s &= \int_{\text{event 1}}^{\text{event 2}} \sqrt{|(\eta_{\mu\nu} + O(\xi^2))d\xi^\mu d\xi^\nu|} \\ &= \int_{\text{event 1}}^{\text{event 2}} \sqrt{|-d(\xi^0)^2 + d(\xi^1)^2 + d(\xi^2)^2 + d(\xi^3)^2|} + O(\xi^2)\,. \end{aligned} \tag{3.19}$$

$V^\alpha{}_{;\beta}$ are the components of a tensor field

Let us define the following quantity:

$$\nabla \vec{V} \equiv V^\alpha{}_{;\beta}\, \vec{e}_{(\alpha)} \otimes \tilde{\omega}^{(\beta)}\,. \tag{3.20}$$

We shall now show that $\nabla\vec{V}$ is a $\binom{1}{1}$ tensor with components $V^\alpha{}_{;\beta}$. A $\binom{1}{1}$ tensor, say $\mathbf{F} = F^\alpha{}_\beta\, \vec{e}_{(\alpha)} \otimes \tilde{\omega}^{(\beta)}$, maps vectors to vectors. Indeed, if we apply **F** to a generic vector $\vec{V}$, remembering that $\tilde{\omega}^{(\beta)}(\vec{V}) = V^\beta$, we get

$$F(\ ,\vec{V}) = F^\alpha{}_\beta \vec{e}_{(\alpha)}\ \tilde{\omega}^{(\beta)}(\vec{V}) = F^\alpha{}_\beta V^\beta\ \vec{e}_{(\alpha)}\,. \tag{3.21}$$

Thus, the result of this operation is a vector with components $F^\alpha = F^\alpha{}_\beta V^\beta$.

Let us now consider a curve on the manifold $x^\mu(\lambda)$, passing through the point **p**, with tangent vector $t^\mu = \frac{dx^\mu}{d\lambda}$, and let us apply $\nabla\vec{V}$ to $\vec{t}$

$$\nabla\vec{V}(\vec{t}) = V^\alpha{}_{;\beta}\ \vec{e}_{(\alpha)}\ \tilde{\omega}^{(\beta)}(\vec{t}) = V^\alpha{}_{;\beta}\ t^\beta\ \vec{e}_{(\alpha)}\,. \tag{3.22}$$

The quantities $V^\alpha{}_{;\beta}t^\beta$ are the components of the directional derivative of the vector $\vec{V}$ along the curve; indeed, using Eq. 3.13 we find

$$\frac{d\vec{V}}{d\lambda} = \frac{\partial\vec{V}}{\partial x^\beta}\frac{dx^\beta}{d\lambda} = V^\alpha{}_{;\beta}t^\beta\ \vec{e}_{(\alpha)}\,. \tag{3.23}$$

The directional derivative of a vector field along a curve

$$\frac{d\vec{V}}{d\lambda} = \lim_{\Delta\lambda\to 0}\frac{\vec{V}(\lambda+\Delta\lambda) - \vec{V}(\lambda)}{\Delta\lambda} \tag{3.24}$$

is a vector, because it is the difference of two vectors (which we can compute after having defined a connection) divided by the real number $\Delta\lambda$. Therefore

$$\nabla\vec{V}(\vec{t}) = V^\alpha{}_{;\beta}t^\beta\ \vec{e}_{(\alpha)} = \frac{d\vec{V}}{d\lambda}\,. \tag{3.25}$$

We also denote the covariant derivative of $\vec{V}$ along $\vec{t}$ as $\nabla_{\vec{t}}\vec{V} \equiv \nabla\vec{V}(\vec{t})$ and, when $\vec{t}$ is a basis vector, $\nabla_\mu\vec{V} \equiv \nabla_{\vec{e}_{(\mu)}}\vec{V}$.

Thus, $\nabla\vec{V}$ maps the vector $\vec{t}$ to the vector $\frac{d\vec{V}}{d\lambda}$, i.e. it is a $\binom{1}{1}$ tensor field, called **covariant derivative of the vector** $\vec{V}$. Its components are

$$(\nabla\vec{V})^\alpha{}_\beta \equiv \nabla_\beta V^\alpha \equiv V^\alpha{}_{;\beta} = V^\alpha{}_{,\beta} + V^\mu\Gamma^\alpha_{\beta\mu}\,. \tag{3.26}$$

Since in a LIF Christoffel's symbols vanish, it follows that

$$V^\alpha{}_{;\beta} = V^\alpha{}_{,\beta} \longrightarrow \frac{\partial\vec{V}}{\partial x^\beta} = V^\alpha{}_{,\beta}\ \vec{e}_{(\alpha)}\,. \tag{3.27}$$

Thus, in a locally inertial frame covariant and ordinary derivatives coincide.

Note that we denote the components of the covariant derivative of a vector $\vec{V}$ either as $\nabla_\alpha V^\mu$ or as $V^\mu{}_{;\alpha}$. These two notations are equivalent in every respect, and will be extended in the next sections to one-forms and tensors.

3.2 THE COVARIANT DERIVATIVE OF SCALARS AND ONE-FORMS

Let us consider a scalar field Φ. At any given point **p** of the manifold, $\Phi(\mathbf{p})$ is a real number, the value of which does not depend on the choice of the coordinate system. However, $\Phi(\mathbf{p})$ has a specific dependence on the chosen coordinates. Therefore $\Phi(\mathbf{p}) = \Phi(x^\mu)$ is a real function of the coordinates.

Since a scalar function does not depend on the basis vectors, the covariant derivative of a scalar field on a manifold coincides with the ordinary derivative:

$$\nabla_\mu\Phi \equiv \frac{\partial\Phi}{\partial x^\mu}\,. \tag{3.28}$$

We remind that, as shown at the end of Sec. 2.3, the differential of a function Φ is a one-form whose components are

$$d\Phi_\mu = \frac{\partial \Phi}{\partial x^\mu}, \tag{3.29}$$

and that we have adopted the convention to omit the tilde over the differential $d\Phi$. Thus, the covariant derivative of a scalar function, whose components are $\nabla_\mu \Phi \equiv d\Phi_\mu$, is a $\binom{0}{1}$ tensor.

In order to define the covariant derivative of a one-form field $\tilde{q} = q_\alpha \tilde{\omega}^{(\alpha)}$, we may proceed as in Sec. 3.1 assuming that, due to the Equivalence Principle, the first derivatives of the basis one-forms in a LIF vanish,

$$\frac{\partial \tilde{\omega}_M^{(\mu')}}{\partial \xi^{\beta'}} = \tilde{0}. \tag{3.30}$$

However, we shall follow a simpler derivation, based on Eq. 3.28 and on the fact that derivative operators have to satisfy Leibniz's rule.

The one-form field $\tilde{q}$ is, by definition, a linear, real valued function of vectors such that

$$\tilde{q}(\vec{V}) = q_\alpha V^\alpha. \tag{3.31}$$

The value of $q_\alpha V^\alpha$ depends only on the point where $\tilde{q}$ and $\vec{V}$ are applied; therefore, once the coordinate system is fixed, $q_\alpha V^\alpha$ is a real function of the coordinates, i.e. a scalar field. By replacing $\Phi = q_\alpha V^\alpha$ in Eq. 3.28 we find

$$\nabla_\mu \Phi \equiv \frac{\partial \Phi}{\partial x^\mu} = \frac{\partial q_\alpha}{\partial x^\mu} V^\alpha + q_\alpha \frac{\partial V^\alpha}{\partial x^\mu}. \tag{3.32}$$

Substituting $\frac{\partial V^\alpha}{\partial x^\mu}$ from Eq. 3.12, we get

$$\nabla_\mu \Phi = \frac{\partial q_\alpha}{\partial x^\mu} V^\alpha + q_\alpha [V^\alpha{}_{;\mu} - V^\beta \Gamma^\alpha_{\mu\beta}], \tag{3.33}$$

and relabeling the indices

$$\nabla_\mu \Phi = [\frac{\partial q_\alpha}{\partial x^\mu} - q_\sigma \Gamma^\sigma_{\mu\alpha}] V^\alpha + q_\sigma V^\sigma{}_{;\mu}. \tag{3.34}$$

Since $\nabla_\mu \Phi$ are the components of a $\binom{0}{1}$ tensor, the right-hand side of this expression must be a tensor of the same rank. The second term is the result of the contraction of a $\binom{0}{1}$ and a $\binom{1}{1}$ tensor; therefore it is a $\binom{0}{1}$ tensor. The first term is a $\binom{0}{1}$ tensor only if the terms in square brackets are the components of a $\binom{0}{2}$ tensor, $\nabla \tilde{q}$, which we call **covariant derivative of the one-form** $\tilde{q}$. The components of $\nabla \tilde{q}$ are

$$(\nabla \tilde{q})_{\alpha\mu} \equiv \nabla_\mu q_\alpha \equiv q_{\alpha;\mu} = q_{\alpha,\mu} - q_\sigma \Gamma^\sigma_{\mu\alpha}. \tag{3.35}$$

Note that, with this definition, Eq. 3.34 can be written as

$$\nabla_\mu \Phi = \nabla_\mu (q_\alpha V^\alpha) = q_{\alpha;\mu} V^\alpha + q_\alpha V^\alpha{}_{;\mu}, \tag{3.36}$$

i.e., the covariant derivative satisfies the standard Leibniz rule.

3.3 SYMMETRIES OF CHRISTOFFEL'S SYMBOLS

Consider an arbitrary scalar field Φ. In Sec. 3.2 we have shown that its covariant derivative is a one-form with components

$$\nabla_\alpha \Phi = \Phi_{,\alpha}.$$

The second covariant derivative $\nabla\nabla\Phi$ is a $\binom{0}{2}$ tensor, the components of which can be computed using Eq. 3.35

$$\nabla_\beta \nabla_\alpha \Phi \equiv \Phi_{,\beta;\alpha} = \Phi_{,\beta,\alpha} - \Phi_{,\mu}\Gamma^{\mu}_{\beta\alpha} \,. \tag{3.37}$$

In a LIF, since $\Gamma^{\mu}_{\beta\alpha} = 0$, covariant and ordinary derivative coincide, and Eq. 3.37 becomes

$$\nabla_\alpha \nabla_\beta \Phi = \Phi_{,\beta;\alpha} = \Phi_{,\beta,\alpha} \,, \tag{3.38}$$

and since ordinary partial derivatives commute, i.e. $\Phi_{,\beta,\alpha} = \Phi_{,\alpha,\beta}$, it follows that in a LIF

$$\nabla_\alpha \nabla_\beta \Phi = \nabla_\beta \nabla_\alpha \Phi \,. \tag{3.39}$$

This equation shows that the tensor $\nabla\nabla\Phi$ is symmetric, and since symmetry is a tensorial property, it must hold in any frame; therefore from Eqs. 3.39 and 3.37 we find

$$\Phi_{,\beta,\alpha} - \Phi_{,\mu}\Gamma^{\mu}_{\beta\alpha} = \Phi_{,\alpha,\beta} - \Phi_{,\mu}\Gamma^{\mu}_{\alpha\beta} \tag{3.40}$$

in any coordinate system. It follows that for any Φ

$$\Phi_{,\mu}\Gamma^{\mu}_{\beta\alpha} = \Phi_{,\mu}\Gamma^{\mu}_{\alpha\beta} \,, \tag{3.41}$$

and consequently

$$\Gamma^{\mu}_{\beta\alpha} = \Gamma^{\mu}_{\alpha\beta} \,, \tag{3.42}$$

i.e. *Christoffel's symbols are symmetric in the lower indices.*

3.4 TRANSFORMATION RULES FOR CHRISTOFFEL'S SYMBOLS

According to Eq. 1.38 (or to Eq. 3.9), Christoffel's symbols are

$$\Gamma^{\rho}_{\tau\sigma} = \frac{\partial x^\rho}{\partial \xi^\mu} \frac{\partial^2 \xi^\mu}{\partial x^\tau \partial x^\sigma} \,.$$

Let us consider a general coordinate transformation $x^\mu = x^\mu(x^{\alpha'})$. In the new frame Christoffel's symbols are

$$\begin{aligned}
\Gamma^{\lambda'}_{\mu'\nu'} &= \frac{\partial x^{\lambda'}}{\partial \xi^\alpha} \frac{\partial^2 \xi^\alpha}{\partial x^{\mu'} \partial x^{\nu'}} = \\
&= \frac{\partial x^{\lambda'}}{\partial x^\rho} \frac{\partial x^\rho}{\partial \xi^\alpha} \frac{\partial}{\partial x^{\mu'}} \left(\frac{\partial \xi^\alpha}{\partial x^\sigma} \frac{\partial x^\sigma}{\partial x^{\nu'}} \right) = \\
&= \frac{\partial x^{\lambda'}}{\partial x^\rho} \frac{\partial x^\rho}{\partial \xi^\alpha} \left[\frac{\partial^2 \xi^\alpha}{\partial x^\tau \partial x^\sigma} \frac{\partial x^\tau}{\partial x^{\mu'}} \frac{\partial x^\sigma}{\partial x^{\nu'}} + \frac{\partial \xi^\alpha}{\partial x^\sigma} \frac{\partial^2 x^\sigma}{\partial x^{\mu'} \partial x^{\nu'}} \right] = \\
&= \frac{\partial x^{\lambda'}}{\partial x^\rho} \frac{\partial x^\tau}{\partial x^{\mu'}} \frac{\partial x^\sigma}{\partial x^{\nu'}} \Gamma^{\rho}_{\tau\sigma} + \frac{\partial x^{\lambda'}}{\partial x^\rho} \frac{\partial^2 x^\rho}{\partial x^{\mu'} \partial x^{\nu'}} \,.
\end{aligned} \tag{3.43}$$

It should be noted that if Christoffel's symbols were the components of a tensor, only the first term in the last row of Eq. 3.43 would be present (see Eq. 2.158). Therefore, as pointed out in Sec. 3.1, Christoffel's symbols are not the components of a tensor.

However, the $\Gamma^{\gamma}_{\alpha\beta}$'s transform as tensors for *linear transformations*, i.e. for coordinate transformations of the form $x'^\mu = \Lambda^{\mu'}{}_\alpha x^\alpha$ where $\Lambda^{\mu'}{}_\alpha$ are constants. Indeed in this case $\frac{\partial^2 x^\rho}{\partial x^{\mu'} \partial x^{\nu'}} = 0$.

Box 3-B

Exercise

Compute Christoffel's symbols $\Gamma^{\mu}_{\alpha\beta}$ using the affine connection

$$\frac{\partial \vec{e}_{(\alpha)}}{\partial x^{\beta}} = \Gamma^{\mu}_{\alpha\beta}\vec{e}_{(\mu)}\,. \tag{3.44}$$

Let us consider for example a two-dimensional flat space in polar coordinates, i.e. $(x^{1'}, x^{2'}) \equiv (r, \theta)$; the basis vectors are related to the coordinate basis associated to Cartesian coordinates by Eqs. 2.73

$$\begin{aligned} \vec{e}_{(1')} &= \vec{e}_{(r)} = \cos\theta\,\vec{e}_{(1)} + \sin\theta\,\vec{e}_{(2)}\,, & (3.45)\\ \vec{e}_{(2')} &= \vec{e}_{(\theta)} = -r\sin\theta\,\vec{e}_{(1)} + r\cos\theta\,\vec{e}_{(2)}\,. \end{aligned}$$

We shall indicate $(\vec{e}_{(1)}, \vec{e}_{(2)})$ with $(\vec{e}_{(x)}, \vec{e}_{(y)})$, and $(\vec{e}_{(1')}, \vec{e}_{(2')})$ with $(\vec{e}_{(r)}, \vec{e}_{(\theta)})$. From Eqs. 3.45 we find

$$\frac{\partial \vec{e}_{(r)}}{\partial r} = \frac{\partial}{\partial r}(\cos\theta\,\vec{e}_{(x)} + \sin\theta\,\vec{e}_{(y)}) = 0\,, \tag{3.46}$$

and consequently, from Eq. 3.44

$$\Gamma^{\mu}_{rr}\vec{e}_{(\mu)} = \Gamma^{r}_{rr}\vec{e}_{(r)} + \Gamma^{\theta}_{rr}\vec{e}_{(\theta)} = 0 \longrightarrow \Gamma^{r}_{rr} = \Gamma^{\theta}_{rr} = 0\,. \tag{3.47}$$

Moreover

$$\begin{aligned} \frac{\partial \vec{e}_{(r)}}{\partial \theta} &= \frac{\partial}{\partial \theta}(\cos\theta\,\vec{e}_{(x)} + \sin\theta\,\vec{e}_{(y)}) = \\ &= -\sin\theta\,\vec{e}_{(x)} + \cos\theta\,\vec{e}_{(y)} = \frac{1}{r}\vec{e}_{(\theta)}\,; \end{aligned}$$

therefore

$$\frac{1}{r}\vec{e}_{(\theta)} = \Gamma^{\mu}_{r\theta}\vec{e}_{(\mu)} = \Gamma^{r}_{r\theta}\vec{e}_{(r)} + \Gamma^{\theta}_{r\theta}\vec{e}_{(\theta)} \longrightarrow \Gamma^{r}_{r\theta} = 0\ ,\quad \Gamma^{\theta}_{r\theta} = \frac{1}{r}\,. \tag{3.48}$$

Proceeding along these lines it is easy to show that

$$\Gamma^{r}_{\theta r} = 0\,,\ \Gamma^{\theta}_{\theta r} = \frac{1}{r}\,,\ \Gamma^{r}_{\theta\theta} = -r\,,\ \Gamma^{\theta}_{\theta\theta} = 0\,. \tag{3.49}$$

3.5 THE COVARIANT DERIVATIVE OF TENSORS

Following the approach adopted in Sec. 3.2 we shall now define the covariant derivative of tensors of any rank. Let us consider, as an example, the covariant derivative of a $\binom{0}{2}$ tensor. Given the tensor $\mathbf{T} = T_{\alpha\beta}\,\tilde{\omega}^{(\alpha)} \otimes \tilde{\omega}^{(\beta)}$ and a vector $\vec{V}$, let $\tilde{q}$ be the one-form obtained by contracting $\mathbf{T}$ with $\vec{V}$, i.e., in components

$$q_{\alpha} = T_{\alpha\beta}V^{\beta}\,. \tag{3.50}$$

According to Eq. 3.35 the covariant derivative of $\tilde{q}$ is a $\binom{0}{2}$ tensor with components

$$\nabla_{\mu}q_{\alpha} = q_{\alpha,\mu} - \Gamma^{\sigma}_{\mu\alpha}q_{\sigma}\,, \tag{3.51}$$

which, substituting Eq. 3.50, yields

$$\nabla_\mu(T_{\alpha\beta}V^\beta) = (T_{\alpha\beta}V^\beta)_{,\mu} - T_{\sigma\beta}V^\beta\Gamma^\sigma_{\mu\alpha}\,. \tag{3.52}$$

By expanding this equation and replacing $V^\beta{}_{,\mu} = V^\beta{}_{;\mu} - \Gamma^\beta_{\delta\mu}V^\delta$ (see Eq. 3.26) we find

$$\begin{aligned}\nabla_\mu(T_{\alpha\beta}V^\beta) &= T_{\alpha\beta,\mu}V^\beta + T_{\alpha\beta}V^\beta{}_{,\mu} - \Gamma^\sigma_{\mu\alpha}T_{\sigma\beta}V^\beta \\ &= T_{\alpha\beta,\mu}V^\beta + T_{\alpha\beta}V^\beta{}_{;\mu} - \Gamma^\beta_{\delta\mu}V^\delta T_{\alpha\beta} - \Gamma^\sigma_{\mu\alpha}T_{\sigma\beta}V^\beta\,,\end{aligned}$$

which, relabeling the indices, becomes

$$\nabla_\mu(T_{\alpha\beta}V^\beta) = \left[T_{\alpha\beta,\mu} - \Gamma^\sigma_{\beta\mu}T_{\alpha\sigma} - \Gamma^\sigma_{\mu\alpha}T_{\sigma\beta}\right]V^\beta + T_{\alpha\beta}V^\beta{}_{;\mu}\,. \tag{3.53}$$

Since $\nabla_\mu(T_{\alpha\beta}V^\beta)$ are the components of a $\binom{0}{2}$ tensor (the covariant derivative of a one-form), the right-hand side of this expression must be a tensor of the same rank. The second term is indeed a $\binom{0}{2}$ tensor, since it is the contraction of a $\binom{0}{2}$ $(T_{\alpha\beta})$ and a $\binom{1}{1}$ tensor $(V^\beta{}_{;\mu})$. The first term is a $\binom{0}{2}$ tensor only if the terms in square brackets are the components of a $\binom{0}{3}$ tensor, $\nabla\mathbf{T}$, which we call **covariant derivative of the $\binom{0}{2}$ tensor T**. The components of $\nabla\mathbf{T}$ are

$$(\nabla T_{\alpha\beta})_\mu \equiv \nabla_\mu T_{\alpha\beta} \equiv T_{\alpha\beta;\mu} = T_{\alpha\beta,\mu} - \Gamma^\sigma_{\beta\mu}T_{\alpha\sigma} - \Gamma^\sigma_{\mu\alpha}T_{\sigma\beta}\,. \tag{3.54}$$

Box 3-C

Exercise

Given a $\binom{2}{0}$ tensor **A**, show that the covariant derivative $\nabla\mathbf{A}$ is the $\binom{2}{1}$ tensor with components

$$(\nabla A^{\mu\nu})_\beta \equiv \nabla_\beta A^{\mu\nu} \equiv A^{\mu\nu}{}_{;\beta} = A^{\mu\nu}{}_{,\beta} + A^{\alpha\nu}\Gamma^\mu_{\alpha\beta} + A^{\mu\alpha}\Gamma^\nu_{\alpha\beta}\,. \tag{3.55}$$

Given a $\binom{1}{1}$ tensor **B**, show that the covariant derivative $\nabla\mathbf{B}$ is the $\binom{1}{2}$ tensor with components

$$(\nabla B^\mu{}_\nu)_\beta \equiv \nabla_\beta B^\mu{}_\nu \equiv B^\mu{}_{\nu;\beta} = B^\mu{}_{\nu,\beta} + B^\alpha{}_\nu\Gamma^\mu_{\beta\alpha} - B^\mu{}_\alpha\Gamma^\alpha_{\beta\nu}\,. \tag{3.56}$$

The same procedure can be used to define the covariant derivative of $\binom{N}{N'}$ tensors:

$$\begin{aligned}\nabla_\mu T^{\alpha_1\alpha_2\cdots\alpha_N}{}_{\beta_1\beta_2\cdots\beta_{N'}} &= T^{\alpha_1\alpha_2\cdots\alpha_N}{}_{\beta_1\beta_2\cdots\beta_{N'},\mu} \\ &+ \Gamma^{\alpha_1}_{\mu\nu}T^{\nu\alpha_2\cdots\alpha_N}{}_{\beta_1\beta_2\cdots\beta_{N'}} + \Gamma^{\alpha_2}_{\mu\nu}T^{\alpha_1\nu\cdots\alpha_N}{}_{\beta_1\beta_2\cdots\beta_{N'}} + \cdots + \Gamma^{\alpha_N}_{\mu\nu}T^{\alpha_1\alpha_2\cdots\nu}{}_{\beta_1\beta_2\cdots\beta_{N'}} \\ &- \Gamma^\nu_{\mu\beta_1}T^{\alpha_1\alpha_2\cdots\alpha_N}{}_{\nu\beta_2\cdots\beta_{N'}} - \Gamma^\nu_{\mu\beta_2}T^{\alpha_1\alpha_2\cdots\alpha_N}{}_{\beta_1\nu\cdots\beta_{N'}} - \cdots - \Gamma^\nu_{\mu\beta_{N'}}T^{\alpha_1\alpha_2\cdots\alpha_N}{}_{\beta_1\beta_2\cdots\nu}\,.\end{aligned} \tag{3.57}$$

Box 3-D

The covariant derivative of the metric tensor

We shall now show that *the covariant derivative of the metric tensor vanishes in any coordinate frame.* According to Eq. 3.54

$$g_{\alpha\beta;\mu} = g_{\alpha\beta,\mu} - \Gamma^{\nu}_{\alpha\mu} g_{\nu\beta} - \Gamma^{\nu}_{\beta\mu} g_{\alpha\nu} \,. \tag{3.58}$$

At any spacetime point $\mathbf{p}$ we can set up a LIF such that, in that point, $g_{\alpha\beta} = \eta_{\alpha\beta}$, $g_{\alpha\beta,\mu} = 0$, and Christoffel's symbols vanish (see Sec. 3.1); consequently

$$g_{\alpha\beta;\mu} = \eta_{\alpha\beta;\mu} = 0 \,. \tag{3.59}$$

Note that $g_{\alpha\beta;\mu}$ are the components of a $\binom{0}{3}$ tensor. If all components of a tensor are zero in a given coordinate frame, they are zero in *any* frame. Therefore

$$g_{\alpha\beta;\mu} = 0 \tag{3.60}$$

in all frames.

3.6 CHRISTOFFEL'S SYMBOLS IN TERMS OF THE METRIC TENSOR

As shown in Box 3-D, in any coordinate frame

$$g_{\alpha\beta;\mu} = g_{\alpha\beta,\mu} - \Gamma^{\nu}_{\alpha\mu} g_{\nu\beta} - \Gamma^{\nu}_{\beta\mu} g_{\alpha\nu} = 0 \,, \tag{3.61}$$

therefore

$$g_{\alpha\beta,\mu} = \Gamma^{\nu}_{\alpha\mu} g_{\nu\beta} + \Gamma^{\nu}_{\beta\mu} g_{\alpha\nu} \,. \tag{3.62}$$

Relabeling the indices, we can write

$$g_{\alpha\mu,\beta} = \Gamma^{\nu}_{\alpha\beta} g_{\nu\mu} + \Gamma^{\nu}_{\mu\beta} g_{\alpha\nu} \,, \tag{3.63}$$

$$-g_{\beta\mu,\alpha} = -\Gamma^{\nu}_{\beta\alpha} g_{\nu\mu} - \Gamma^{\nu}_{\mu\alpha} g_{\beta\nu} \,. \tag{3.64}$$

The sum of the three equations above yields

$$g_{\alpha\beta,\mu} + g_{\alpha\mu,\beta} - g_{\beta\mu,\alpha} = (\Gamma^{\nu}_{\alpha\mu} - \Gamma^{\nu}_{\mu\alpha}) g_{\nu\beta} + (\Gamma^{\nu}_{\beta\mu} + \Gamma^{\nu}_{\mu\beta}) g_{\alpha\nu} + (\Gamma^{\nu}_{\alpha\beta} - \Gamma^{\nu}_{\beta\alpha}) g_{\nu\mu} \,, \tag{3.65}$$

where we have used the symmetry of the metric tensor. Since $\Gamma^{\alpha}_{\beta\gamma}$ are symmetric in β and γ, it follows that

$$g_{\alpha\beta,\mu} + g_{\alpha\mu,\beta} - g_{\beta\mu,\alpha} = 2\Gamma^{\nu}_{\beta\mu} g_{\alpha\nu} \,. \tag{3.66}$$

If we multiply by $g^{\alpha\gamma}$ and remember that (see Eq. 2.238)

$$g^{\alpha\gamma} g_{\alpha\nu} = \delta^{\gamma}_{\nu} \,, \tag{3.67}$$

we finally find

$$\Gamma^{\gamma}_{\beta\mu} = \frac{1}{2} g^{\gamma\alpha} (g_{\alpha\beta,\mu} + g_{\alpha\mu,\beta} - g_{\beta\mu,\alpha}) \,. \tag{3.68}$$

Thus, as anticipated in Chapter 1, it is possible to express Christoffel's symbols as a linear combination of the first derivatives of the metric tensor.

The connection given in Eq. 3.68 is called the **Levi-Civita connection**. In the following we shall compute Christoffel's symbols using Eq. 3.68, for some simple metric spaces.

Christoffel's symbols on the two-sphere in polar coordinates

Let us consider the two-dimensional manifold $\mathbf{S^2}$, i.e. the two-sphere which we have introduced in Box 2-C. The metric tensor on $\mathbf{S^2}$ can be found by noting that the two-sphere has been defined as a sub-manifold of the three-dimensional Euclidean space, corresponding (in terms of the polar coordinate r, see Box 2-Q) to $r = a$, where the constant a is the radius of the sphere. The line element of the two-sphere can be obtained by replacing $r = a$ in the line element of the three-dimensional Eucliedan space in polar coordinates, $ds^2 = dr^2 + r^2(d\theta^2 + \sin^2\theta d\varphi^2)$ (Eq. 2.231) [1]. Thus,

$$ds^2 = a^2\left(d\theta^2 + \sin^2\theta d\varphi^2\right) = g_{\mu\nu}dx^\mu dx^\nu\,, \quad g_{\mu\nu} = \begin{pmatrix} a^2 & 0 \\ 0 & a^2\sin^2\theta \end{pmatrix}. \tag{3.69}$$

In the following we shall consider a sphere of unit radius, i.e. $a = 1$. The inverse metric $g^{\mu\nu}$ is

$$g^{\mu\nu} = \begin{pmatrix} 1 & 0 \\ 0 & \frac{1}{\sin^2\theta} \end{pmatrix}. \tag{3.70}$$

Note that the only non-vanishing derivative of $g_{\mu\nu}$ is

$$g_{\varphi\varphi,\theta} = 2\sin\theta\cos\theta\,. \tag{3.71}$$

Using Eq. 3.68 we find

$$\begin{aligned} \Gamma^\theta_{\beta\mu} &= \frac{1}{2}g^{\theta\alpha}(g_{\alpha\beta,\mu} + g_{\alpha\mu,\beta} - g_{\beta\mu,\alpha}) = \frac{1}{2}g^{\theta\theta}(g_{\theta\beta,\mu} + g_{\theta\mu,\beta} - g_{\beta\mu,\theta})\,, \\ \Gamma^\varphi_{\beta\mu} &= \frac{1}{2}g^{\varphi\alpha}(g_{\alpha\beta,\mu} + g_{\alpha\mu,\beta} - g_{\beta\mu,\alpha}) = \frac{1}{2}g^{\varphi\varphi}(g_{\varphi\beta,\mu} + g_{\varphi\mu,\beta})\,, \end{aligned} \tag{3.72}$$

where we have used the fact that the metric is diagonal and that its φ-derivatives vanish. Thus, using Eq. 3.71 we find

$$\begin{aligned} &\Gamma^\theta_{\theta\theta} = \frac{1}{2}g^{\theta\theta}g_{\theta\theta,\theta} = 0\,, \quad \Gamma^\theta_{\theta\varphi} = \Gamma^\theta_{\varphi\theta} = \frac{1}{2}g^{\theta\theta}(g_{\theta\theta,\varphi} + g_{\theta\varphi,\theta} - g_{\theta\varphi,\theta}) = 0\,, \\ &\Gamma^\theta_{\varphi\varphi} = \frac{1}{2}g^{\theta\theta}(2g_{\theta\varphi,\varphi} - g_{\varphi\varphi,\theta}) = -\frac{1}{2}g^{\theta\theta}g_{\varphi\varphi,\theta} = -\sin\theta\cos\theta\,, \end{aligned} \tag{3.73}$$

and

$$\begin{aligned} &\Gamma^\varphi_{\theta\theta} = \frac{1}{2}g^{\varphi\varphi}(2g_{\varphi\theta,\theta}) = 0\,, \qquad \Gamma^\varphi_{\varphi\varphi} = \frac{1}{2}g^{\varphi\varphi}(2g_{\varphi\varphi,\varphi}) = 0\,, \\ &\Gamma^\varphi_{\theta\varphi} = \Gamma^\varphi_{\varphi\theta} = \frac{1}{2}g^{\varphi\varphi}(g_{\varphi\theta,\varphi} + g_{\varphi\varphi,\theta}) = \frac{1}{2}g^{\varphi\varphi}g_{\varphi\varphi,\theta} = \cot\theta\,. \end{aligned} \tag{3.74}$$

Therefore, the only non-vanishing Christoffel symbols on the two-sphere are

$$\Gamma^\theta_{\varphi\varphi} = -\sin\theta\cos\theta\,, \qquad \Gamma^\varphi_{\theta\varphi} = \Gamma^\varphi_{\varphi\theta} = \cot\theta\,. \tag{3.75}$$

Christoffel's symbols on the three-dimensional Euclidean space in spherical coordinates

Let us consider the three-dimensional Euclidean space in spherical coordinates $\{x^\mu\} = (r, \theta, \varphi)$. The spacetime metric is (see Box 2-Q)

$$ds^2 = dr^2 + r^2d\theta^2 + r^2\sin^2\theta d\varphi^2 = g_{\mu\nu}dx^\mu dx^\nu\,, \quad g_{\mu\nu} = \mathrm{diag}(1, r^2, r^2\sin^2\theta)\,. \tag{3.76}$$

[1] With this procedure, a manifold $\mathcal{M}$ with a metric $\boldsymbol{g}$ *induces* a metric on a submanifold $\mathcal{N} \subset \mathcal{M}$. A more rigorous definition of this procedure is given in Box 7-A.

The inverse metric is $g^{\mu\nu} = \mathrm{diag}(1, r^{-2}, r^{-2}\sin^{-2}\theta)$. The non-vanishing derivatives of the metric tensor are

$$g_{\theta\theta,r} = 2r\,, \quad g_{\varphi\varphi,r} = 2r^2\sin^2\theta\cos\theta\,, \quad g_{\varphi\varphi,\theta} = 2r^2\sin\theta\cos\theta\,. \tag{3.77}$$

Using Eq. 3.68 we find

$$\Gamma^r_{\beta\mu} = \frac{1}{2}g^{r\alpha}(g_{\alpha\beta,\mu} + g_{\alpha\mu,\beta} - g_{\beta\mu,\alpha}) = \frac{1}{2}g^{rr}(g_{r\beta,\mu} + g_{r\mu,\beta} - g_{\beta\mu,r})\,, \tag{3.78}$$

$$\Gamma^\theta_{\beta\mu} = \frac{1}{2}g^{\theta\alpha}(g_{\alpha\beta,\mu} + g_{\alpha\mu,\beta} - g_{\beta\mu,\alpha}) = \frac{1}{2}g^{\theta\theta}(g_{\theta\beta,\mu} + g_{\theta\mu,\beta} - g_{\beta\mu,\theta})\,, \tag{3.79}$$

$$\Gamma^\varphi_{\beta\mu} = \frac{1}{2}g^{\varphi\alpha}(g_{\alpha\beta,\mu} + g_{\alpha\mu,\beta} - g_{\beta\mu,\alpha}) = \frac{1}{2}g^{\varphi\varphi}(g_{\varphi\beta,\mu} + g_{\varphi\mu,\beta} - g_{\mu\beta,\varphi})\,, \tag{3.80}$$

and thus the only non-vanishing Christoffel symbols are those with the indexes (θ, θ, r), (φ, φ, r), and $(\varphi, \varphi, \theta)$:

$$\begin{aligned}
\Gamma^r_{\theta\theta} &= -\frac{1}{2}g^{rr}g_{\theta\theta,r} = -r\,, \qquad \Gamma^\theta_{r\theta} = \Gamma^\theta_{\theta r} = \frac{1}{2}g^{\theta\theta}g_{\theta\theta,r} = \frac{1}{r}\,,\\
\Gamma^r_{\varphi\varphi} &= -\frac{1}{2}g^{rr}g_{\varphi\varphi,r} = -r\sin^2\theta\,, \qquad \Gamma^\varphi_{r\varphi} = \Gamma^\varphi_{\varphi r} = \frac{1}{2}g^{\varphi\varphi}g_{\varphi\varphi,r} = \frac{1}{r}\,,\\
\Gamma^\theta_{\varphi\varphi} &= -\frac{1}{2}g^{\theta\theta}g_{\varphi\varphi,\theta} = -\sin\theta\cos\theta\,, \qquad \Gamma^\varphi_{\theta\varphi} = \Gamma^\varphi_{\varphi\theta} = \frac{1}{2}g^{\varphi\varphi}g_{\varphi\varphi,\theta} = \cot\theta\,.
\end{aligned} \tag{3.81}$$

Box 3-E

Laplacian operator in spherical coordinates

Let us consider the three-dimensional Euclidean space. The Laplacian operator $\nabla^2 \equiv \nabla_i \nabla^i$, which maps functions to functions, is easily defined in the Cartesian coordinate frame $\{x^i\} = (x^1, x^2, x^3)$ where the spacetime metric is $ds^2 = (dx^1)^2 + (dx^2)^2 + (dx^3)^2 = g_{ij} dx^i dx^j$, with $g_{ij} = \delta_{ij}$:

$$\nabla^2 f \equiv g^{ij} f_{,ij} = \frac{\partial f}{\partial (x^1)^2} + \frac{\partial f}{\partial (x^2)^2} + \frac{\partial f}{\partial (x^3)^2} \tag{3.82}$$

where f is any regular function on the manifold.
We want to write the Laplacian operator in a different coordinate frame: the spherical coordinates $\{x^{i'}\} = (r, \theta, \varphi)$, defined in Box 2-Q:

$$x^1 = r \sin\theta \cos\varphi\,, \quad x^2 = r \sin\theta \sin\varphi\,, \quad x^3 = r\cos\theta\,. \tag{3.83}$$

In the new frame the spacetime metric is $ds^2 = dr^2 + r^2 d\theta^2 + r^2 \sin^2\theta d\varphi^2 = g_{i'j'} dx^{i'} dx^{j'}$ where $g_{i'j'} = \mathrm{diag}(1, r^2, r^2 \sin^2\theta)$, $g^{i'j'} = \mathrm{diag}(1, r^{-2}, r^{-2}\sin^{-2}\theta)$, and the non-vanishing Christoffel symbols are given in Eq. 3.81. First of all we write the Laplacian operator, defined in Eq. 3.82, in tensorial form. Since in the Cartesian frame of Euclidean space ordinary and covariant derivative coincide, in that frame

$$\nabla^2 f \equiv g^{ij} f_{;ij}\,. \tag{3.84}$$

Being Eq. 3.84 a tensor equation it holds, with the same form, in any reference frame; in spherical coordinates

$$\nabla^2 f = g^{i'j'} f_{;i'j'} = g^{i'j'} (f_{,i'})_{;j'} = g^{i'j'} \left[f_{,i'j'} - \Gamma^{k'}_{i'j'} f_{,k'} \right]. \tag{3.85}$$

Replacing the explicit expressions in Eq. 3.81, we find

$$\begin{aligned}
\nabla^2 f &= f_{,rr} + \frac{1}{r^2} f_{,\theta\theta} + \frac{1}{r^2 \sin^2\theta} f_{,\varphi\varphi} - \left[\Gamma^r_{\theta\theta} g^{\theta\theta} + \Gamma^r_{\varphi\varphi} g^{\varphi\varphi} \right] f_{,r} - \Gamma^\theta_{\varphi\varphi} g^{\varphi\varphi} f_{,\theta} \\
&= f_{,rr} + \frac{2}{r} f_{,r} + \frac{1}{r^2} \left(f_{,\theta\theta} + \cot\theta f_{,\theta} + \frac{1}{\sin^2\theta} f_{,\varphi\varphi} \right) \\
&= \left[\frac{1}{r^2} \frac{\partial}{\partial r} \left(r^2 \frac{\partial}{\partial r} \right) + \frac{1}{r^2 \sin\theta} \frac{\partial}{\partial\theta} \left(\sin\theta \frac{\partial}{\partial\theta} \right) + \frac{1}{r^2 \sin^2\theta} \frac{\partial^2}{\partial\varphi^2} \right] f\,.
\end{aligned} \tag{3.86}$$

This is the well-known form of the Laplacian operator in spherical coordinates. It can also be written as

$$\nabla^2 = \frac{1}{r^2} \frac{\partial}{\partial r} r^2 \frac{\partial}{\partial r} + \frac{\mathbb{L}}{r^2} \tag{3.87}$$

where $\mathbb{L}$ is an operator acting on the angular variables defined as

$$\mathbb{L} = \frac{1}{r \sin\theta} \frac{\partial}{\partial\theta} \left(\sin\theta \frac{\partial}{\partial\theta} \right) + \frac{1}{\sin^2\theta} \frac{\partial^2}{\partial\varphi^2}\,. \tag{3.88}$$

The eigenfunctions of the $\mathbb{L}$ operator are the *spherical harmonics* $Y^{lm}(\theta, \varphi)$ ($l = 0, 1, \ldots,$ $m = -l, \ldots, l$), a set of complex functions (see Box 15-A and [67]) defined by

$$\mathbb{L} Y^{lm} = -l(l+1) Y^{lm}\,. \tag{3.89}$$

3.7 PARALLEL TRANSPORT

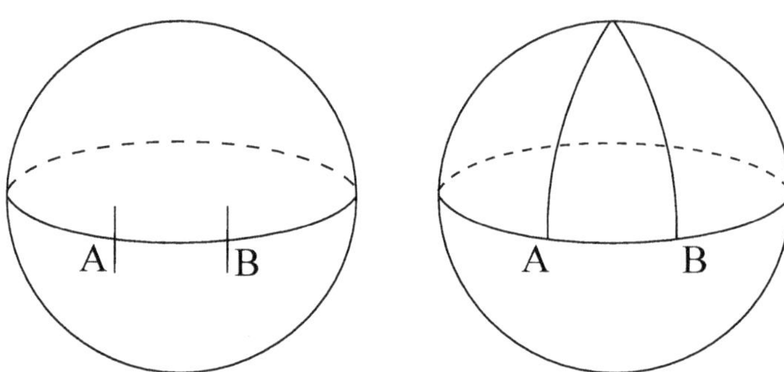

Figure 3.1: Parallel segments on a sphere do not remain parallel when prolonged.

In Chapter 1 we discussed and compared the intrinsic geometry of cylinders and spheres, and we noticed that while it is flat for cylinders, it is curved for spheres. This means, for example, that two lines which start parallel do not remain parallel when prolonged. Consider the two small segments in A and B of Fig. 3.1: they are perpendicular to the equator, i.e. parallel, but if prolonged they do not remain parallel.

It is also interesting to see what happens when we "parallely transport" a vector along a path. *Parallel transport means that for each infinitesimal displacement along the path, the displaced vector must be parallel to the original one, and must have the same length.* If the path belongs to a flat space (Fig. 3.2, left panel), when the vector returns to A it coincides with the original vector in A. If, instead, the path belongs to a sphere (Fig. 3.2, right panel), when the vector goes back to A it is rotated by 90° (remember that the vector is always tangent to the sphere, because it belongs to the tangent space). This result, which will be derived more rigorously in Sec. 3.7.1, is a consequence of the *curvature* of the sphere: *on a curved manifold it is impossible to define a globally parallel vector field. The parallel transport of a vector depends on the path along which it is transported.*

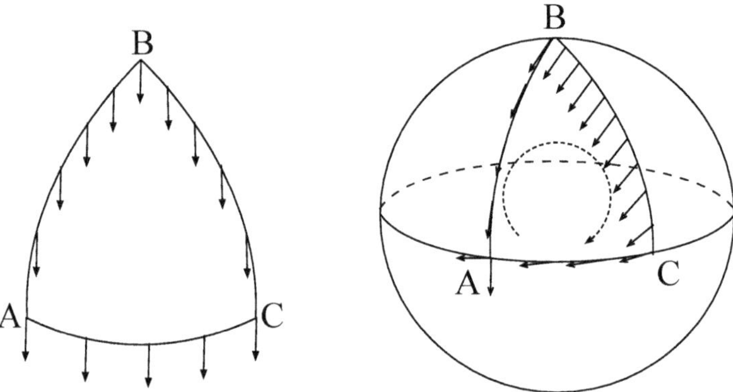

Figure 3.2: Parallel transport of a vector along a closed path, in flat space (left panel) and on a sphere (right panel).

Let us now compute how a vector changes when it is parallely transported along a path

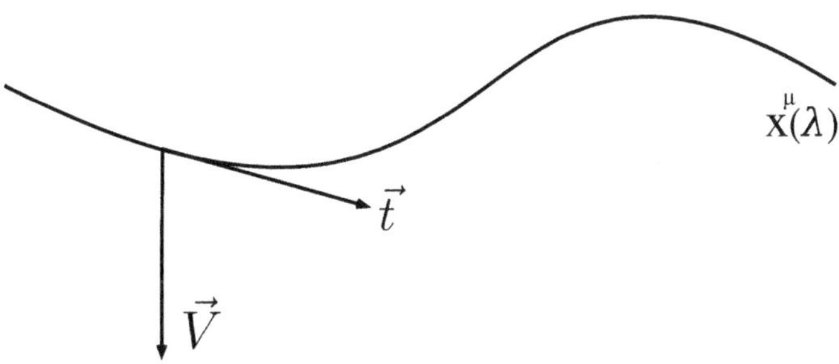

Figure 3.3: Parallel transport along a path.

in a general space. Consider a curve $x^\mu(\lambda)$ defined on the path, and a vector field $\vec{V}$ defined at every point of the curve. Let $t^\alpha = \frac{dx^\alpha}{d\lambda}$ be the vector tangent to the curve. At each point of the curve we can set up a LIF $\{\xi^\alpha\}$. In this frame, if we move $\vec{V}$ along the curve, parallel to itself and keeping its length unchanged, the vector components do not change

$$\frac{dV^\alpha}{d\lambda} = 0\,. \tag{3.90}$$

Therefore,

$$\frac{dV^\alpha}{d\lambda} = \frac{\partial V^\alpha}{\partial \xi^\beta}\frac{d\xi^\beta}{d\lambda} = V^\alpha{}_{,\beta}t^\beta = 0\,. \tag{3.91}$$

Since we are in a LIF, ordinary and covariant derivatives coincide, and Eq. 3.91 can be written as

$$V^\alpha{}_{;\beta}t^\beta = 0\,. \tag{3.92}$$

This is a tensor equation; consequently, if it is true in a LIF, it is true in any other frame. Equation 3.92 is the definition of the **parallel transport** of a vector $\vec{V}$ along a curve identified by its tangent vector $\vec{t}$. In a frame-independent form, Eq. 3.92 reads

$$\nabla_{\vec{t}}\,\vec{V} = 0\,, \tag{3.93}$$

i.e., the covariant derivative of the vector $\vec{V}$ along the direction of the vector $\vec{t}$ is zero. In a generic reference frame with coordinates $\{x^\alpha\}$,

$$\left(\nabla_{\vec{t}}\,\vec{V}\right)^\alpha \equiv V^\alpha{}_{;\beta}t^\beta = \left[\frac{\partial V^\alpha}{\partial x^\beta} + \Gamma^\alpha_{\beta\nu}V^\nu\right]\frac{dx^\beta}{d\lambda} = \frac{dV^\alpha}{d\lambda} + \Gamma^\alpha_{\beta\nu}V^\nu t^\beta = 0\,. \tag{3.94}$$

Summarizing, while in flat space the components of a vector parallely transported along a curve are constant, in curved space they change by an amount which can be found by solving the following equation

$$\frac{dV^\alpha}{d\lambda} = -\Gamma^\alpha_{\beta\nu}V^\nu t^\beta\,. \tag{3.95}$$

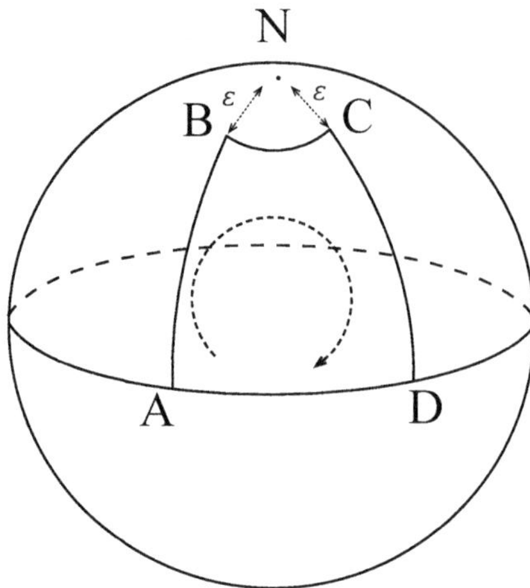

Figure 3.4: A closed path on a two-sphere, avoiding the North pole where the polar map is not defined.

3.7.1 Parallel transport of a vector along a closed path on a two-sphere

Let us consider the closed path on a two-sphere indicated in Fig. 3.4. In polar coordinates $\{x^\mu\} = (\theta, \varphi)$, the points indicated in the figure have coordinates

$$A = \left(\frac{\pi}{2}, 0\right) , \quad B = (\varepsilon, 0) , \quad C = \left(\varepsilon, \frac{\pi}{2}\right) , \quad D = \left(\frac{\pi}{2}, \frac{\pi}{2}\right) , \tag{3.96}$$

where ε is a small parameter. Note that we exclude the North pole N, since in that point the polar map is not defined (see Box 2-C). Thus, since in these coordinates we cannot choose the path $ANDA$ (which is equivalent to that considered in Fig. 3.2), we choose the path $ABCDA$, which (for $\varepsilon \ll 1$) almost touches the North pole and, in the $\varepsilon \to 0$ limit, reduces to $ANDA$.

Let us consider a vector $\vec{V}$ initially in A, where it has components

$$V^\mu(A) = (1, 0) . \tag{3.97}$$

We shall now show that if we parallely transport $\vec{V}$ along this path, when it returns to A it will be rotated by 90° (in the $\varepsilon \to 0$ limit), as in Fig. 3.2. The path is composed of four branches of coordinate lines:

- AB and CD, with tangent vector $\vec{e}_{(\theta)}$ with components $e^\alpha{}_{(\theta)} = (1, 0)$, and on which $\varphi = const.$
- BC and DA, with tangent vector $\vec{e}_{(\varphi)}$ with components $e^\alpha{}_{(\varphi)} = (0, 1)$, and on which $\theta = const.$

The equations of parallel transport are (see Eq. 3.95)

$$\frac{dV^\alpha}{d\lambda} = -\Gamma^\alpha_{\beta\nu} V^\nu t^\beta \tag{3.98}$$

where $\vec{t}$ is the tangent vector to the curves composing the path which, in our case, are the coordinate lines with the tangent vectors indicated above; thus, on AB and CD the parameter of the curves can be chosen to be $\lambda = \theta$, whereas $\lambda = \varphi$ on BC and DA. Recalling that the non-vanishing Christoffel symbols on the two-sphere are (see Sec. 3.6)

$$\Gamma^{\theta}_{\varphi\varphi} = -\sin\theta\cos\theta\,, \qquad \Gamma^{\varphi}_{\theta\varphi} = \Gamma^{\varphi}_{\varphi\theta} = \cot\theta\,, \tag{3.99}$$

Eq. 3.98 gives

- on AB and CD, parametrized with θ

$$\frac{dV^{\mu}}{d\theta} = -\Gamma^{\mu}_{\theta\nu}\, V^{\nu} \quad \rightarrow \quad \begin{cases} \dfrac{dV^{\theta}}{d\theta} = 0 \\[2ex] \dfrac{dV^{\varphi}}{d\theta} = -\cot\theta\; V^{\varphi}\,; \end{cases} \tag{3.100}$$

- on BC and DA, parametrized with φ

$$\frac{dV^{\mu}}{d\varphi} = -\Gamma^{\mu}_{\varphi\nu}\, V^{\nu} \quad \rightarrow \quad \begin{cases} \dfrac{dV^{\theta}}{d\varphi} = \sin\theta\cos\theta\; V^{\varphi} \\[2ex] \dfrac{dV^{\varphi}}{d\varphi} = -\cot\theta\; V^{\theta}\,. \end{cases} \tag{3.101}$$

These equations have to be solved with the appropriate boundary conditions as follows.

1. From $A = (\pi/2, 0)$ to $B = (\varepsilon, 0)$, with $V^{\mu}(A) = (1, 0)$.
   From the first of Eqs. 3.100 it follows that along AB V^{θ} is constant, therefore in B $V^{\theta}(B) = 1$. The equation for V^{φ} is

$$\begin{cases} \dfrac{dV^{\varphi}}{d\theta} = -\cot\theta V^{\varphi} \\[2ex] V^{\varphi}\left(\theta = \frac{\pi}{2}\right) = 0\,. \end{cases} \tag{3.102}$$

   This is a Cauchy problem, which admits a unique solution, and precisely $V^{\varphi}(\theta) \equiv 0$[2]. Thus, $V^{\mu}(B) = (1, 0)$.

2. From $B = (\varepsilon, 0)$ to $C = (\varepsilon, \pi/2)$, with $V^{\mu}(B) = (1, 0)$.
   On this path $\theta = \varepsilon$, and assuming that $\varepsilon \ll 1$, Eqs. 3.101 become

$$\begin{cases} \dfrac{dV^{\theta}}{d\varphi} = \sin\varepsilon\cos\varepsilon V^{\varphi} = \varepsilon V^{\varphi} + \mathcal{O}\left(\varepsilon^{3}\right) \\[2ex] \dfrac{dV^{\varphi}}{d\varphi} = -\cot\varepsilon V^{\theta} = -\dfrac{1}{\varepsilon} V^{\theta} + \mathcal{O}\left(\varepsilon\right)\,. \end{cases} \tag{3.103}$$

   By differentiating the first equation with respect to φ and using the second equation, we get

$$\frac{d^{2}V^{\theta}}{d\varphi^{2}} = -V^{\theta} + \mathcal{O}\left(\varepsilon^{2}\right)\,, \tag{3.104}$$

[2]Note that this is a general result: if the derivative of a function is proportional to the function itself (in this case V^{φ}), and if it vanishes at the initial point (in our case $\theta = \pi/2$), the function remains constant in all points of the integration domain, and equal to the initial value.

whereas from the first equation, we directly find

$$V^\varphi = \frac{1}{\varepsilon}\frac{dV^\theta}{d\varphi} + \mathcal{O}(\varepsilon) \,. \tag{3.105}$$

The general solution of Eq. 3.104 is $V^\theta = C_1 \cos\varphi + C_2 \sin\varphi + \mathcal{O}(\varepsilon^2)$, with C_1 and C_2 constants; the boundary condition $V^\theta(\varphi = 0) = 1$ implies that $C_1 = 1$ and $C_2 = 0$. Therefore, using Eq. 3.105, along BC the solution is

$$\begin{cases} V^\theta = \cos\varphi + \mathcal{O}(\varepsilon^2) \\ V^\varphi = -\frac{1}{\varepsilon}\sin\varphi + \mathcal{O}(\varepsilon) \,, \end{cases} \tag{3.106}$$

and in $C = (\varepsilon, \pi/2)$

$$V^\mu(C) = \left(\mathcal{O}(\varepsilon^2), -\frac{1}{\varepsilon} + \mathcal{O}(\varepsilon)\right) . \tag{3.107}$$

3. From $C = (\varepsilon, \pi/2)$ to $D = (\pi/2, \pi/2)$, with $V^\mu(C)$ given by 3.107.
Again we have to use Eqs. 3.100; therefore along this path V^θ is constant, i.e. $V^\theta(\theta) = \mathcal{O}(\varepsilon^2)$, whereas V^φ satisfies the equations

$$\begin{cases} \dfrac{dV^\varphi}{d\theta} = -\cot\theta V^\varphi \\ \\ V^\varphi(\theta = \epsilon) = -\frac{1}{\varepsilon} + \mathcal{O}(\varepsilon) \,. \end{cases} \tag{3.108}$$

The solution of this system is

$$V^\varphi = -\frac{1}{\sin\theta} + \mathcal{O}(\varepsilon) \,; \tag{3.109}$$

therefore in D, where $\theta = \pi/2$,

$$V^\mu(D) = \left(\mathcal{O}(\varepsilon^2), -1 + \mathcal{O}(\varepsilon)\right) . \tag{3.110}$$

4. From $D = (\pi/2, \pi/2)$ to $A = (\pi/2, 0)$, with $V^\mu(D)$ given by 3.110.
From Eqs. 3.101, given $\theta = \pi/2$, we get

$$\begin{cases} \dfrac{dV^\theta}{d\varphi} = 0 \\ \\ \dfrac{dV^\varphi}{d\varphi} = 0 \,. \end{cases} \tag{3.111}$$

Consequently, along DA the components of $\vec{V}$ remain constant, and in A

$$V^\mu(A) = \left(\mathcal{O}(\varepsilon^2), -1 + \mathcal{O}(\varepsilon)\right) . \tag{3.112}$$

In the limit $\varepsilon \to 0$, the path reduces to $ANDA$ and

$$V^\mu(A) = (0, -1) \,. \tag{3.113}$$

Since at the beginning of the journey the vector in A was $V^\mu = (1, 0)$, it follows that the transported vector has been rotated by 90° clockwise.

3.8 GEODESIC EQUATION

In Chapter 1 we introduced timelike geodesics as the curves which describe the motion of massive, *free* particles, i.e. those which move under the exclusive action of the gravitational field. We showed that they are the solution of the **geodesic equation**

$$\frac{d^2x^\alpha}{d\tau^2}+\Gamma^\alpha_{\mu\beta}\frac{dx^\mu}{d\tau}\frac{dx^\beta}{d\tau}=0\,, \tag{3.114}$$

where τ is the particle proper time (see Box 3-F). We shall now derive this equation, extending it to null geodesics, using the parallel transport introduced in the previous section.

Box 3-F

The proper time

In Special Relativity, the worldlines of massive particles are curves in the four-dimensional flat spacetime,

$$\tau \mapsto (x^0(\tau), x^1(\tau), x^2(\tau), x^3(\tau)) \tag{3.115}$$

where $x^0 = ct$, and t is the coordinate time; τ/c is the **proper time** of the particle, i.e. the time measured by an observer comoving with the particle. Thus, in the comoving frame $x^0 = \tau$, the space coordinates x^i are constant and the squared distance beween two infinitely close events on the worldline is

$$ds^2 = \eta_{\mu\nu}dx^\mu dx^\nu = -(dx^0)^2 = -d\tau^2\,; \tag{3.116}$$

thus in this frame

$$d\tau = \sqrt{-ds^2}\,. \tag{3.117}$$

Note that the proper time is a scalar quantity and Eq. 3.117 is a tensor equation; as such, it holds in any other frame.
Let us now suppose that the particle moves in a curved spacetime. According to the Equivalence Principle, in a LIF the laws of Special Relativity apply; therefore Eq. 3.117 holds in the LIF comoving with the particle and, being a tensor equation, in any other frame. Thus, in a general frame,

$$d\tau = \sqrt{-g_{\mu\nu}dx^\mu dx^\nu}\,. \tag{3.118}$$

In particular, in a LIF locally comoving with the particle the time coordinate coincides with the particle proper time, as in flat spacetime.
Note that τ has the dimensions of a length and the four-velocity $u^\mu = dx^\mu/d\tau$ is dimensionless. If one uses units in which $c = 1$, τ coincides with the proper time.

Let us consider a free particle which moves along the worldline $x^\mu(\lambda)$ with four-velocity $\vec{u}$, which is the tangent vector to the worldline, i.e. $u^\mu = dx^\mu/d\lambda$. By the Equivalence Principle, at any point of the worldline we can define a LIF, $\{\xi^{\alpha'}\}$, where the particle four-acceleration is zero, i.e.

$$\frac{du^{\mu'}}{d\lambda}=\frac{\partial u^{\mu'}}{\partial \xi^{\alpha'}}\frac{d\xi^{\alpha'}}{d\lambda}=u^{\alpha'}u^{\mu'}{}_{,\alpha'}=0\,. \tag{3.119}$$

In a locally inertial frame ordinary and covariant derivative coincide, thus

$$u^{\alpha'}u^{\mu'}{}_{;\alpha'}=0\,. \tag{3.120}$$

This is a tensor equation, and must hold in any coordinate frame; therefore, in a generic frame the equation of motion of the free particle is

$$u^\alpha u^\mu{}_{;\alpha} = 0\,. \tag{3.121}$$

If we expand the covariant derivative

$$u^\alpha u^\mu{}_{;\alpha} = u^\alpha u^\mu{}_{,\alpha} + u^\alpha \Gamma^\mu{}_{\alpha\beta} u^\beta\,, \tag{3.122}$$

and replace $u^\mu = dx^\mu/d\lambda$, Eq. 3.121 becomes

$$\frac{d^2x^\mu}{d\lambda^2} + \Gamma^\mu_{\alpha\beta}\frac{dx^\alpha}{d\lambda}\frac{dx^\beta}{d\lambda} = 0\,, \tag{3.123}$$

which is the geodesic equation, Eq. 3.114. Thus, the geodesic equation can also be written in the form 3.121, which describes the parallel transport of the tangent vector $\vec{u}$ along the geodesic. This means that if we parallely transport the tangent vector from a point $\mathbf{p}$ to a point $\mathbf{p}'$ along the geodesic line, the transported vector in $\mathbf{p}'$ will be tangent to the curve. Consequently, a curve $\mathcal{C}$ with tangent vector $\vec{u}$ is a geodesic if

$$\nabla_{\vec{u}}\, \vec{u} = 0\,. \tag{3.124}$$

Therefore, *geodesics are those curves which parallel-transport their own tangent vectors.*

The parameter along the geodesic is not unique. Let s be a new parameter; since $\frac{d}{d\lambda} = \frac{d}{ds}\frac{ds}{d\lambda}$, Eq. 3.123 becomes

$$\frac{d^2x^\alpha}{ds^2} + \Gamma^\alpha_{\mu\nu}\left[\frac{dx^\mu}{ds}\frac{dx^\nu}{ds}\right] = -\left[\frac{d^2s}{d\lambda^2}\Big/\left(\frac{ds}{d\lambda}\right)^2\right]\frac{dx^\alpha}{ds}\,. \tag{3.125}$$

This equation reduces to Eq. 3.123 only if s is related to λ by a linear transformation

$$s = a\lambda + b, \qquad a, b = \text{const}\,, \tag{3.126}$$

in which case the right-hand side of Eq. 3.125 vanishes. The parameters for which the geodesic equation takes the form 3.123 are called **affine parameters** and are all related by a linear transformation.

Eq. 3.121 describes timelike, spacelike, and null geodesics. If the geodesic is timelike, i.e. $\vec{u}\cdot\vec{u} < 0$, it represents the wordline of a massive particle; in this case, by performing the linear transformation in Eq. 3.126 it is possible to change the affine parameter in such a way that the new parameter is the particle proper time (see Box 2-F) multiplied by c, i.e. the parameter τ appearing in Eq. 3.114. With this choice, $\vec{u}\cdot\vec{u} = -1$. If the geodesic is a null curve, i.e. $\vec{u}\cdot\vec{u} = 0$, it represents the wordline of a massless particle; in this case it is not possible to choose the proper time as affine parameter, because it cannot be defined; the geodesic can be parametrized with a different affine parameter, for instance with the arc length. If the geodesic is spacelike, i.e. $\vec{u}\cdot\vec{u} > 0$, it does not represent the worldline of a physical particle.

3.9 FERMI COORDINATES

In this section we shall show how to construct a coordinate frame adapted to an observer. This frame is especially useful to describe physical experiments.

An observer in General Relativity is characterized by a timelike curve, $\tau \mapsto \gamma(\tau)$, where τ is the proper time (times c). The tangent vector to the curve is the four-velocity of the

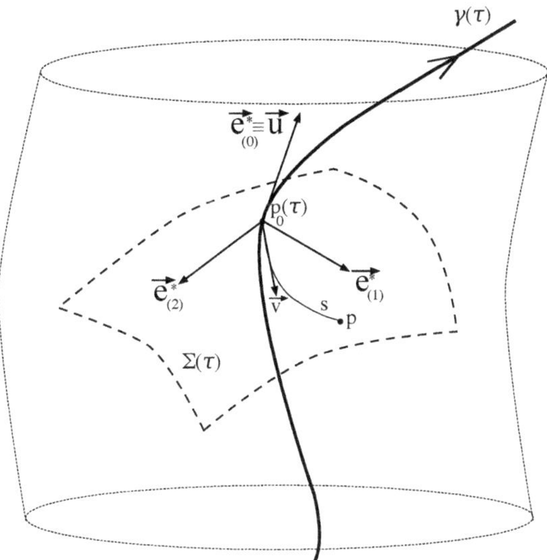

Figure 3.5: Fermi coordinates adapted to an observer. For simplicity, we do not show one of the three space dimensions.

observer $\vec{u}$, and $\vec{u}\cdot\vec{u} = -1$. The observer can either be freely-falling (i.e. with vanishing acceleration $a^\mu = u^\nu u^\mu{}_{;\nu} = 0$) or accelerated ($a^\mu \neq 0$).

It is possible to define, in a region close to the curve (a "worldtube" surrounding γ), coordinates $\{x^{*\mu}\}$ adapted to the observer, called the *Fermi coordinates*, as follows (see Fig. 3.5):

- At any point of the curve $\mathbf{p}_0(\tau) \in \gamma$, we can define a basis $\{\vec{e}^{\,*}_{(\mu)}\}$ of the vector space $\mathbf{T}_{\mathbf{p}_0}$ such that $\vec{e}^{\,*}_{(0)} \equiv \vec{u}$, and $\vec{e}^{\,*}_{(i)}$ $(i = 1, \ldots, 3)$ are orthogonal to $\vec{e}^{\,*}_{(0)}$ and to each other. In addition, we impose that $\vec{e}^{\,*}_{(i)}$ have norm one. Thus, the basis is orthonormal ($\vec{e}^{\,*}_{(\mu)} \cdot \vec{e}^{\,*}_{(\nu)} = \eta_{\mu\nu}$), and the timelike basis vector coincides with the four-velocity of the observer.
- At each value of τ, the three spacelike vectors $\vec{e}^{\,*}_{(i)}$ locally define a three-dimensional spacelike surface Σ, to which they are tangent in $\mathbf{p}_0(\tau)$.
- Given a point $\mathbf{p}$ in the worldtube, such that $\mathbf{p} \in \Sigma(\tau)$, let us consider a spacelike geodesic from $\mathbf{p}_0(\tau)$ to $\mathbf{p}$ on the surface $\Sigma(\tau)$. Let $\vec{v}$ be the tangent vector in $\mathbf{p}_0(\tau)$ to this geodesic with unit length. Since the geodesic belongs to $\Sigma(\tau)$, $\vec{v}$ is a linear combination of the three spacelike vectors $\vec{e}^{\,*}_{(i)}$ $(i = 1, \ldots, 3)$, i.e.

$$\vec{v} = v^i \vec{e}^{\,*}_{(i)} \,. \tag{3.127}$$

- Let s be the affine parameter of the geodesic from $\mathbf{p}_0(\tau)$ to $\mathbf{p}$. We define the Fermi coordinates of the point $\mathbf{p}$ as follows:

$$x^{*0} = \tau \,, \qquad x^{*i} = s v^i \,. \tag{3.128}$$

  It can be shown that s is also the proper length of the path (as defined in Eq. 2.211) of this spacelike geodesic between $\mathbf{p}_0$ and $\mathbf{p}$.

- Repeating the same procedure for all points of the observer worldline, we can define, in the entire worldtube, the Fermi coordinate system $\{x^{*\mu}\}$ adapted to the observer worldline.

It can be shown that (with an appropriate choice of the three-dimensional basis $\{\vec{e}^{\,*}_{(i)}\}$ for each surface $\Sigma(\tau)$) the spacetime metric in Fermi coordinates is

$$ds^2 = -(1 + 2a_\mu x^{*\mu})(dx^{*0})^2 + \delta_{ij}dx^{*i}dx^{*j} + O(x^{*2})\,, \tag{3.129}$$

where a^ν are the components of the observer's four-acceleration. If the observer is freely-falling ($\vec{a} = 0$), $ds^2 = \eta_{\mu\nu} + O(x^{*2})$, and this coordinate system coincides with the LIF centered in $\mathbf{p}_0$. Indeed, in a LIF the metric coincides with Minkowski's metric modulo *quadratic* terms in the coordinates. If, instead, the observer is accelerated, the deviation of the metric 3.129 from Minkowski's metric is *linear* in the coordinates $x^{*\mu}$: $g_{00} - \eta_{00} = -2a_\mu x^{*\mu}$. The linear term is precisely due to the acceleration of the observer.

Remarkably, the metric 3.129 does not depend on the spacetime curvature (up to $O(x^{*2})$). Indeed, introducing the Fermi coordinates for an accelerated observer in Special Relativity, i.e. $\xi^\alpha \to x^{*\mu}(\xi^\alpha)$, leads to the metric $ds^2 = -(1+2a_\mu x^{*\mu})(dx^{*0})^2 + \delta_{ij}dx^{*i}dx^{*j}$, which differs from Eq. 3.129 by the $O(x^{*2})$ terms only. We conclude that due to the Equivalence Principle, according to which the laws of physics in a LIF take, locally, the form prescribed by Special Relativity for inertial frames, the laws of physics in a Fermi frame take, locally, the form prescribed by Special Relativity for *accelerated* frames.

3.10 NON-COORDINATE BASES

In Boxes 2-G and 2-P we showed that changing from Minkowskian coordinates $\{\xi^{\alpha'}\} \equiv (ct, x, y)$ to polar coordinates $\{x^\alpha\} \equiv (ct, r, \theta)$, the coordinate basis for vectors [3]

$$\{\vec{e}_{(\alpha')}\} = \begin{cases} \vec{e}_{(0')} & \to \quad (1,0,0) \\ \vec{e}_{(1')} & \to \quad (0,1,0) \\ \vec{e}_{(2')} & \to \quad (0,0,1) \end{cases} \tag{3.130}$$

transforms to $\{\vec{e}_{(\alpha)}\}$, where (see Eq. 2.73)

$$\begin{cases} \vec{e}_{(0)} = \vec{e}_{(0')} \\ \vec{e}_{(1)} = \vec{e}_{(r)} = \cos\theta\vec{e}_{(1')} + \sin\theta\vec{e}_{(2')} \\ \vec{e}_{(2)} = \vec{e}_{(\theta)} = -r\sin\theta\vec{e}_{(1')} + r\cos\theta\vec{e}_{(2')}\,. \end{cases} \tag{3.131}$$

These expressions follow from the transformation laws

$$\vec{e}_{(\alpha)} = \Lambda^{\mu'}{}_{\alpha}\, \vec{e}_{(\mu')}\,, \tag{3.132}$$

where $\Lambda^{\mu'}{}_{\alpha} = \frac{\partial x^{\mu'}}{\partial x^\alpha}$ and $\Lambda^{\alpha}{}_{\mu'} = \frac{\partial x^\alpha}{\partial x^{\mu'}}$. We now want to choose a different basis for vectors. For example, while the vectors $\{\vec{e}_{(\alpha)}\}$ are not normalized because

$$\vec{e}_{(\alpha)} \cdot \vec{e}_{(\beta)} = g_{\alpha\beta} = \begin{pmatrix} -1 & 0 & 0 \\ 0 & 1 & 0 \\ 0 & 0 & r^2 \end{pmatrix} \neq \eta_{\alpha\beta}\,, \tag{3.133}$$

[3] Note that, at variance with Boxes 2-G and 2-P, we indicate with a prime the Minkowskian coordinates.

we may choose a basis of normalized vectors given by

$$\begin{cases} \vec{e}_{(\hat{0})} = \vec{e}_{(0)} \\ \vec{e}_{(\hat{r})} = \vec{e}_{(r)} \\ \vec{e}_{(\hat{\theta})} = \frac{1}{r}\vec{e}_{(\theta)}\,, \end{cases} \tag{3.134}$$

such that

$$\vec{e}_{(\hat{\alpha})} \cdot \vec{e}_{(\hat{\beta})} = \eta_{\hat{\alpha}\hat{\beta}}\,. \tag{3.135}$$

In this basis the transformation 3.131 becomes

$$\begin{aligned} \vec{e}_{(\hat{r})} &= \cos\theta \vec{e}_{(1')} + \sin\theta \vec{e}_{(2')} \\ \vec{e}_{(\hat{\theta})} &= -\sin\theta \vec{e}_{(1')} + \cos\theta \vec{e}_{(2')}\,. \end{aligned} \tag{3.136}$$

In the previous sections we have always introduced the basis vectors as the coordinate basis associated to a coordinate frame, but they can also be defined independently of the choice of the coordinate frame, as in Eq. 3.134. In this case, it is natural to ask whether there exists a set of coordinates $\{x^{\hat{\alpha}}\}$ such that

$$e_{(\hat{\alpha})} = \Lambda^{\beta}{}_{\hat{\alpha}}\, \vec{e}_{(\beta)} = \frac{\partial x^{\beta}}{\partial x^{\hat{\alpha}}}\, \vec{e}_{(\beta)}\,, \tag{3.137}$$

so that the basis $\{\vec{e}_{(\hat{\alpha})}\}$ is a coordinate basis. The same question can be formulated for the basis one-forms, i.e. whether there exist coordinates $\{x^{\hat{\alpha}}\}$ such that

$$\tilde{\omega}^{(\hat{\alpha})} = \Lambda^{\hat{\alpha}}{}_{\beta}\, \tilde{\omega}^{(\beta)} = \frac{\partial x^{\hat{\alpha}}}{\partial x^{\beta}}\, \tilde{\omega}^{(\beta)} \tag{3.138}$$

and

$$\Lambda^{\hat{\alpha}}{}_{\beta} = \frac{\partial x^{\hat{\alpha}}}{\partial x^{\beta}}\,. \tag{3.139}$$

For instance, in the considered example

$$\Lambda^{\beta}{}_{\hat{\alpha}} = \begin{pmatrix} 1 & 0 & 0 \\ 0 & 1 & 0 \\ 0 & 0 & \frac{1}{r} \end{pmatrix} \qquad \text{and} \qquad \Lambda^{\hat{\alpha}}{}_{\beta} = \begin{pmatrix} 1 & 0 & 0 \\ 0 & 1 & 0 \\ 0 & 0 & r \end{pmatrix}\,. \tag{3.140}$$

If Eq. 3.139 is true, the following condition must be satisfied

$$\frac{\partial}{\partial x^{\gamma}}\Lambda^{\hat{\alpha}}{}_{\beta} = \frac{\partial^2 x^{\hat{\alpha}}}{\partial x^{\gamma}\partial x^{\beta}} = \frac{\partial^2 x^{\hat{\alpha}}}{\partial x^{\beta}\partial x^{\gamma}} = \frac{\partial}{\partial x^{\beta}}\Lambda^{\hat{\alpha}}{}_{\gamma}\,. \tag{3.141}$$

This is an "integrability condition" that all the components of $\Lambda^{\hat{\alpha}}{}_{\gamma}$ must satisfy in order for the coordinates $\{x^{\hat{\alpha}}\}$ to exist. In the considered example Eq. 3.141 gives

$$\frac{\partial}{\partial\theta}\Lambda^{\hat{2}}{}_{1} = \frac{\partial}{\partial r}\Lambda^{\hat{2}}{}_{2} \qquad \to \qquad 0 = 1 \tag{3.142}$$

which is certainly not true. We conclude that the normalized vector basis in Eq. 3.134 and the dual basis for one forms $\{\tilde{\omega}^{(\hat{\alpha})}\}$ *are not* coordinate bases, since we cannot associate a coordinate transformation to them.

CHAPTER 4

The curvature tensor

In this chapter we shall introduce the curvature tensor, showing that it describes the curvature of the spacetime. This derivation is based on the parallel transport of a vector along a closed loop. We shall discuss the main properties of the curvature tensor and show how the latter is related to the equation of geodesic deviation.

4.1 PARALLEL TRANSPORT ALONG A LOOP

Let us consider an infinitesimal closed loop, having as sides the coordinate lines $x^1 = a$, $x^1 = a + \delta a$, $x^2 = b$, $x^2 = b + \delta b$ (see Fig. 4.1).

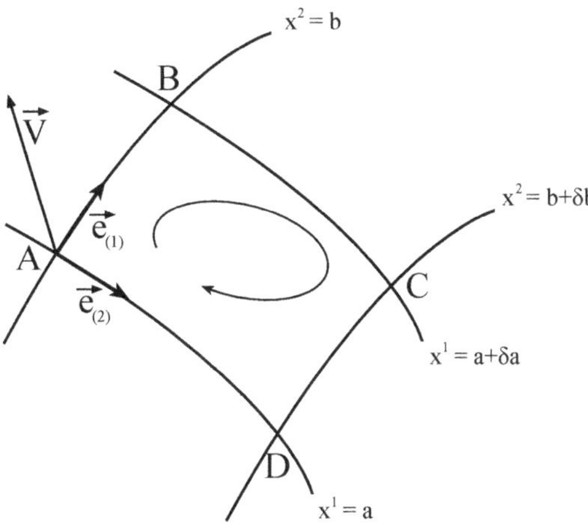

Figure 4.1: A generic vector $\vec{V}$ is parallely transported along an infinitesimal, closed loop whose boundaries are portions of coordinate lines.

We wish to compute how the components of a generic vector $\vec{V}$ change when the vector is parallely transported along the loop. The loop can be divided into four parts, so that the total variation of the vector components is

$$\delta V^\alpha = \delta V^\alpha_{AB} + \delta V^\alpha_{BC} + \delta V^\alpha_{CD} + \delta V^\alpha_{DA} \,. \tag{4.1}$$

The vector tangent to the curves AB and CD and that tangent to the curves BC and DA are, respectively, the basis vectors $\vec{e}_{(1)}$ and $\vec{e}_{(2)}$; since their only non-vanishing components are $e^1_{(1)} = 1$ and $e^2_{(2)} = 1$, the equation of parallel transport, Eq. 3.92, gives

$$\text{on AB and CD :} \qquad \nabla_{\vec{e}_{(1)}}\vec{V} = e^{\mu}_{(1)}V^{\alpha}{}_{;\mu} = 0 \quad \rightarrow \quad \frac{\partial V^{\alpha}}{\partial x^1} + \Gamma^{\alpha}_{\beta 1}V^{\beta} = 0\,, \tag{4.2}$$

$$\text{on BC and DA :} \qquad \nabla_{\vec{e}_{(2)}}\vec{V} = e^{\mu}_{(2)}V^{\alpha}{}_{;\mu} = 0 \quad \rightarrow \quad \frac{\partial V^{\alpha}}{\partial x^2} + \Gamma^{\alpha}_{\beta 2}V^{\beta} = 0\,. \tag{4.3}$$

Eq. 4.2 has to be integrated along the line $x^2 = b$ from A to B, and $x^2 = b + \delta b$ from C to D; Eq. 4.3 has to be integrated along $x^1 = a + \delta a$ from B to C, and on $x^1 = a$ from D to A. The integrals to evaluate thus are

$$\delta V^{\alpha}_{AB} = -\int_{A(x^2=b)}^{B}\Gamma^{\alpha}_{\beta 1}V^{\beta}dx^1\,, \qquad \delta V^{\alpha}_{BC} = -\int_{B(x^1=a+\delta a)}^{C}\Gamma^{\alpha}_{\beta 2}V^{\beta}dx^2\,, \tag{4.4}$$

$$\delta V^{\alpha}_{CD} = -\int_{C(x^2=b+\delta b)}^{D}\Gamma^{\alpha}_{\beta 1}V^{\beta}dx^1\,, \quad \delta V^{\alpha}_{DA} = -\int_{D(x^1=a)}^{A}\Gamma^{\alpha}_{\beta 2}V^{\beta}dx^2\,. \tag{4.5}$$

The integral in δV^{α}_{BC} is a function of $x^1 = a + \delta a$, with $\delta a \ll a$; therefore it can be Taylor-expanded as

$$\delta V^{\alpha}_{BC} = -\left[\int_{B(x^1=a)}^{C}\Gamma^{\alpha}_{\beta 2}V^{\beta}dx^2 + \frac{\partial}{\partial x^1}\left(\int_{B}^{C}\Gamma^{\alpha}_{\beta 2}V^{\beta}dx^2\right)\bigg|_{(x^1=a)}\delta a\right]. \tag{4.6}$$

The integral in δV^{α}_{CD} can be expanded in a similar way

$$\delta V^{\alpha}_{CD} = -\left[\int_{C(x^2=b)}^{D}\Gamma^{\alpha}_{\beta 1}V^{\beta}dx^1 + \frac{\partial}{\partial x^2}\left(\int_{C}^{D}\Gamma^{\alpha}_{\beta 1}V^{\beta}dx^1\right)\bigg|_{(x^2=b)}\delta b\right]. \tag{4.7}$$

Thus, finally

$$\begin{aligned}\delta V^{\alpha} \simeq & -\int_{A(x^2=b)}^{B}\Gamma^{\alpha}_{\beta 1}V^{\beta}dx^1 - \int_{D(x^1=a)}^{A}\Gamma^{\alpha}_{\beta 2}V^{\beta}dx^2 \\ & -\left[\int_{B(x^1=a)}^{C}\Gamma^{\alpha}_{\beta 2}V^{\beta}dx^2 + \frac{\partial}{\partial x^1}\left(\int_{B}^{C}\Gamma^{\alpha}_{\beta 2}V^{\beta}dx^2\right)\bigg|_{(x^1=a)}\delta a\right] \\ & -\left[\int_{C(x^2=b)}^{D}\Gamma^{\alpha}_{\beta 1}V^{\beta}dx^1 + \frac{\partial}{\partial x^2}\left(\int_{C}^{D}\Gamma^{\alpha}_{\beta 1}V^{\beta}dx^1\right)\bigg|_{(x^2=b)}\delta b\right].\end{aligned} \tag{4.8}$$

If we now replace the coordinates of the points A, B, C, D in the previous equation, i.e.

$$A = (a, b), \quad C = (a + \delta a, b + \delta b), \quad B = (a + \delta a, b), \quad \text{and} \quad D = (a, b + \delta b)\,, \tag{4.9}$$

only two terms survive in Eq. 4.8, and since the operators of integration and differentiation commute, the final expression of δV^{α} can be written as

$$\begin{aligned}\delta V^{\alpha} & \simeq -\delta a\int_{b}^{b+\delta b}\frac{\partial}{\partial x^1}\left(\Gamma^{\alpha}_{\beta 2}V^{\beta}\right)dx^2 + \delta b\int_{a}^{a+\delta a}\frac{\partial}{\partial x^2}\left(\Gamma^{\alpha}_{\beta 1}V^{\beta}\right)dx^1 \\ & \simeq \delta a\delta b\left[-\frac{\partial}{\partial x^1}\left(\Gamma^{\alpha}_{\beta 2}V^{\beta}\right) + \frac{\partial}{\partial x^2}\left(\Gamma^{\alpha}_{\beta 1}V^{\beta}\right)\right].\end{aligned} \tag{4.10}$$

By expanding the derivatives and making use of Eqs. 4.2 and 4.3, Eq. 4.10 can be further simplified to give

$$\begin{aligned}\delta V^{\alpha} &= \delta a \delta b \left[\frac{\partial \Gamma^{\alpha}_{\beta 1}}{\partial x^2} V^{\beta} + \Gamma^{\alpha}_{\sigma 1}\frac{\partial V^{\sigma}}{\partial x^2} - \frac{\partial \Gamma^{\alpha}_{\beta 2}}{\partial x^1} V^{\beta} - \Gamma^{\alpha}_{\sigma 2}\frac{\partial V^{\sigma}}{\partial x^1}\right] \\ &= \delta a \delta b \left[\frac{\partial \Gamma^{\alpha}_{\beta 1}}{\partial x^2} - \frac{\partial \Gamma^{\alpha}_{\beta 2}}{\partial x^1} - \Gamma^{\alpha}_{\sigma 1}\Gamma^{\sigma}_{\beta 2} + \Gamma^{\alpha}_{\sigma 2}\Gamma^{\sigma}_{\beta 1}\right] V^{\beta} .\end{aligned} \tag{4.11}$$

Note that δa and δb are the non-vanishing components of the displacement vectors $\vec{\delta x}_{(1)}$ and $\vec{\delta x}_{(2)}$ along the direction of the basis vectors $\vec{e}_{(1)}$ and $\vec{e}_{(2)}$, i.e.

$$\delta x^{\mu}_{(1)} = (0, \delta a, 0, 0) = \delta a \ \delta^{\mu}_{1} , \tag{4.12}$$

$$\delta x^{\mu}_{(2)} = (0, 0, \delta b, 0) = \delta b \ \delta^{\mu}_{2} ; \tag{4.13}$$

therefore Eq. 4.11 can be written as follows

$$\delta V^{\alpha} = \delta x^{\nu}_{(1)} \ \delta x^{\mu}_{(2)} \ \left[\frac{\partial \Gamma^{\alpha}_{\beta\nu}}{\partial x^{\mu}} - \frac{\partial \Gamma^{\alpha}_{\beta\mu}}{\partial x^{\nu}} - \Gamma^{\alpha}_{\sigma\nu}\Gamma^{\sigma}_{\beta\mu} + \Gamma^{\alpha}_{\sigma\mu}\Gamma^{\sigma}_{\beta\nu}\right] V^{\beta} . \tag{4.14}$$

If the loop is defined through different coordinate lines, the same proof leads to the same final expression, Eq. 4.14. A similar, but more involved, calculation shows that Eq. 4.14 also holds when the sides of the loop are not coordinate lines, i.e. when $\vec{\delta x}_{(1)}$ and $\vec{\delta x}_{(2)}$ are general vectors.

We shall now show that the terms in square brackets in Eq. 4.14 are the components of a tensor. According to the definition given in Chapter 2, a $\binom{1}{3}$ tensor $\mathbf{T}$ is a linear function of one one-form and three vectors which, when applied to these arguments, produces a scalar, i.e.

$$T(\tilde{q}, \vec{A}, \vec{B}, \vec{C}) = T^{\alpha}{}_{\beta\rho\delta} \ q_{\alpha} A^{\beta} B^{\rho} C^{\delta} , \tag{4.15}$$

where $T^{\alpha}{}_{\beta\rho\delta}$ are the components of $\mathbf{T}$. To begin with, we note that V^{α}, δV^{α}, $\delta x^{\mu}_{(1)}$, and $\delta x^{\mu}_{(2)}$ are the components of generic vectors. Let us consider a generic one-form field with components q_{α}. The quantity $\delta V^{\alpha} q_{\alpha}$ is a scalar, because it is the contraction of a vector and a one-form, and

$$\delta V^{\alpha} q_{\alpha} = \left[\frac{\partial \Gamma^{\alpha}_{\beta\nu}}{\partial x^{\mu}} - \frac{\partial \Gamma^{\alpha}_{\beta\mu}}{\partial x^{\nu}} - \Gamma^{\alpha}_{\sigma\nu}\Gamma^{\sigma}_{\beta\mu} + \Gamma^{\alpha}_{\sigma\mu}\Gamma^{\sigma}_{\beta\nu}\right] q_{\alpha} V^{\beta} \delta x^{\nu}_{(1)} \ \delta x^{\mu}_{(2)} . \tag{4.16}$$

Therefore, the contraction of the components of the vectors $\delta x^{\nu}_{(1)}$, $\delta x^{\mu}_{(2)}$, and V^{β}, and of the one-form q_{α} with the quantity in brackets is a scalar. In addition, Eq. 4.16 is linear in $\vec{V}$, $\tilde{q}$, $\vec{\delta x}_{(1)}$, and $\vec{\delta x}_{(2)}$. Indeed, if we consider for example a displacement $\vec{\delta x}_{(1a)} + \vec{\delta x}_{(1b)}$ along $\vec{e}_{(1)}$, it is immediate to check that

$$\delta V^{\alpha} q_{\alpha} = \delta x^{\nu}_{(1a)} \delta x^{\mu}_{(2)} \left[...\right] q_{\alpha} V^{\beta} + \delta x^{\nu}_{(1b)} \delta x^{\mu}_{(2)} \left[...\right] q_{\alpha} V^{\beta} , \tag{4.17}$$

and similarly for the other quantities. Thus, we can conclude that the terms in square brackets in Eq. 4.16 are the components of a $\binom{1}{3}$ tensor, the **Riemann tensor**

$$R^{\alpha}{}_{\beta\mu\nu} \equiv \Gamma^{\alpha}_{\beta\nu,\mu} - \Gamma^{\alpha}_{\beta\mu,\nu} - \Gamma^{\alpha}_{\sigma\nu}\Gamma^{\sigma}_{\beta\mu} + \Gamma^{\alpha}_{\sigma\mu}\Gamma^{\sigma}_{\beta\nu} . \tag{4.18}$$

Note that the Riemann tensor is antisymmetric in ν and μ; indeed, if we interchange $\vec{\delta x}_{(1)}$ and $\vec{\delta x}_{(2)}$ in Eq. 4.14, δV^{α} changes sign, because the loop goes in the opposite direction.

This shows that the sign of Eq. 4.18 can be chosen arbitrarily, and for this reason the definitions of the Riemann tensor given in textbooks may differ for an overall sign.

The Riemann tensor depends on the affine connection and on its first derivatives, i.e. on the first and second derivatives of the metric tensor. If we choose a LIF the $\Gamma^{\alpha}_{\beta\mu}$'s vanish, but their derivatives do not; the Riemann tensor is generically nonzero in a LIF.

We shall call a spacetime *flat* if the parallel transport of any vector field around any loop vanishes, otherwise we shall call it *curved.* Therefore, in a flat spacetime the Riemann tensor vanishes everywhere and (since it is a tensor) in any frame; in a curved spacetime, instead, the Riemann tensor is non-zero at some point of the manifold, and this is also true in any frame. It can be shown (but the rigorous derivation goes beyond the scope of this book) that a global, Minkowskian coordinate system can be defined if and only if the Riemann tensor vanishes on the entire spacetime[1].

In conclusion, *if the spacetime is flat the Riemann tensor is zero, if it is curved the Riemann tensor is non-zero, and this is true in any coordinate frame.* This justifies why the Riemann tensor is also called the **curvature tensor**.

Furthermore, Eq. 4.14 shows that if $R^{\alpha}{}_{\beta\gamma\delta} = 0$, i.e. the spacetime is flat, a vector $\vec{V}$ parallely transported along *any* closed loop does not change, i.e. $\delta V^{\alpha} = 0$. Conversely, if the Riemann tensor does not vanish, i.e. the spacetime is curved, there will exist at least one loop such that $\delta V^{\alpha} \neq 0$.

It should be stressed that

- The Riemann tensor is linear in the second derivatives of $g_{\mu\nu}$, and non-linear in the first derivatives. Note that it has the same differential structure of the Gaussian curvature introduced in Chapter 1, Eq. 1.12.
- In a LIF $\Gamma^{\alpha}_{\nu\sigma} = 0$, therefore the non-linear part of the Riemann tensor vanishes as well.

By contracting the Riemann tensor with the metric we can construct a $\binom{0}{2}$ tensor, called the **Ricci tensor**

$$R_{\mu\nu} = g^{\alpha\beta} R_{\alpha\mu\beta\nu} = R^{\alpha}{}_{\mu\alpha\nu} \,. \tag{4.19}$$

By further contracting the Ricci tensor with the metric, we can also define the **scalar (or Ricci) curvature**

$$R = g^{\alpha\beta} R_{\alpha\beta} = R^{\alpha}{}_{\alpha} \,. \tag{4.20}$$

As the Riemann tensor, both the Ricci tensor and the Ricci scalar are linear in the second derivatives of $g_{\mu\nu}$, and non-linear in the first derivatives.

In a LIF, the components of the Riemann tensor have a very simple form since the non-linear part of the tensor vanishes, i.e.

$$R^{\mu}{}_{\nu\alpha\beta} = \Gamma^{\mu}_{\nu\beta,\alpha} - \Gamma^{\mu}_{\nu\alpha,\beta} \,, \tag{4.21}$$

and by replacing the expression of the Christoffel symbols given in Eq. 3.68, we find

$$R^{\alpha}{}_{\beta\mu\nu} = \frac{1}{2} g^{\alpha\sigma} \left[g_{\sigma\nu,\beta\mu} - g_{\sigma\mu,\beta\nu} + g_{\beta\mu,\sigma\nu} - g_{\beta\nu,\sigma\mu} \right] , \tag{4.22}$$

or, lowering the index α,

$$R_{\alpha\beta\mu\nu} = g_{\alpha\lambda} R^{\lambda}{}_{\beta\mu\nu} = \frac{1}{2} \left[g_{\alpha\nu,\beta\mu} - g_{\alpha\mu,\beta\nu} + g_{\beta\mu,\alpha\nu} - g_{\beta\nu,\alpha\mu} \right] . \tag{4.23}$$

---

[1]This can be understood by noting that in a LIF $g_{\mu\nu} = \eta_{\mu\nu} + O(x^2)$, and the $O(x^2)$ terms depend on the second derivatives of the metric; it can be shown that the $O(x^2)$ terms can be expressed in terms of the Riemann tensor components. If these terms vanish everywhere, the LIF can be extended to the entire spacetime.

Consequently, in a LIF the components of the Ricci tensor are

$$R_{\beta\nu} = R^{\alpha}{}_{\beta\alpha\nu} = \frac{1}{2} g^{\alpha\sigma} \left[g_{\sigma\nu,\beta\alpha} - g_{\sigma\alpha,\beta\nu} + g_{\beta\alpha,\sigma\nu} - g_{\beta\nu,\sigma\alpha} \right] . \tag{4.24}$$

4.2 SYMMETRIES OF THE RIEMANN TENSOR

Using Eq. 4.23 it is easy to show that the curvature tensor is antisymmetric in the first and in the second pair of indices, i.e.

$$R_{\alpha\beta\mu\nu} = -R_{\beta\alpha\mu\nu} , \quad \text{and} \quad R_{\alpha\beta\mu\nu} = -R_{\alpha\beta\nu\mu} , \tag{4.25}$$

whereas it is symmetric under exchange of the two pairs of indices

$$R_{\alpha\beta\mu\nu} = R_{\mu\nu\alpha\beta} . \tag{4.26}$$

In addition it satisfies the *Ricci identities*

$$R_{\alpha\beta\mu\nu} + R_{\alpha\nu\beta\mu} + R_{\alpha\mu\nu\beta} = 0 . \tag{4.27}$$

Since $R_{\alpha\beta\mu\nu}$ is a tensor, these symmetry properties, which can easily be derived in a LIF using Eq. 4.23, hold in any reference frame. The symmetries of the Riemann tensor reduce the number of its independent components from $4^4 = 256$ to only 20.

As a consequence of Eq. 4.26, the Ricci tensor $R_{\mu\nu} = g^{\alpha\beta} R_{\mu\alpha\nu\beta}$ is symmetric under exchange of its indices:

$$R_{\mu\nu} = R_{\nu\mu} , \tag{4.28}$$

and therefore it has only 10 independent components.

4.3 THE RIEMANN TENSOR GIVES THE COMMUTATOR OF COVARIANT DERIVATIVES

Let us consider the second covariant derivatives of a vector field $\vec{V}$

$$\nabla_\alpha \nabla_\beta V^\mu = \nabla_\alpha (V^\mu{}_{;\beta}) = (V^\mu{}_{;\beta})_{,\alpha} + \Gamma^\mu_{\sigma\alpha} V^\sigma{}_{;\beta} - \Gamma^\sigma_{\beta\alpha} V^\mu{}_{;\sigma} . \tag{4.29}$$

In a LIF $\Gamma^\mu_{\sigma\alpha} = 0$, and Eq. 4.29 becomes

$$\nabla_\alpha \nabla_\beta V^\mu = (V^\mu{}_{;\beta})_{,\alpha} = V^\mu{}_{,\beta,\alpha} + \Gamma^\mu_{\nu\beta,\alpha} V^\nu . \tag{4.30}$$

By interchanging α and β

$$\nabla_\beta \nabla_\alpha V^\mu = (V^\mu{}_{;\alpha})_{,\beta} = V^\mu{}_{,\alpha,\beta} + \Gamma^\mu_{\nu\alpha,\beta} V^\nu . \tag{4.31}$$

The commutator of the covariant derivatives is then

$$[\nabla_\alpha, \nabla_\beta] V^\mu \equiv \nabla_\alpha \nabla_\beta V^\mu - \nabla_\beta \nabla_\alpha V^\mu = \left(\Gamma^\mu_{\nu\beta,\alpha} - \Gamma^\mu_{\nu\alpha,\beta} \right) V^\nu . \tag{4.32}$$

The term in round brackets in this equation coincides with the expression of the Riemann tensor in the LIF (see Eq. 4.21), therefore

$$[\nabla_\alpha, \nabla_\beta] V^\mu = R^\mu{}_{\nu\alpha\beta} V^\nu . \tag{4.33}$$

This is a tensor equation and since it is valid in a given reference frame, it will be valid in *any* frame. Eq. 4.33 shows that *in curved spacetime the covariant derivatives do not commute* and therefore the order in which they appear is important.

4.4 THE BIANCHI IDENTITIES

We shall now derive the *Bianchi identities*, a set of differential equations satisfied by the Riemann tensor. Since in a LIF the expression of the tensor is much simpler, we shall use Eq. 4.23 to prove the identities in this frame, and then generalize them to any frame.

Let us differentiate Eq. 4.23 with respect to x^λ

$$R_{\alpha\beta\mu\nu,\lambda} = \frac{1}{2}\left[g_{\alpha\nu,\beta\mu\lambda} - g_{\alpha\mu,\beta\nu\lambda} + g_{\beta\mu,\alpha\nu\lambda} - g_{\beta\nu,\alpha\mu\lambda}\right] . \tag{4.34}$$

Using Eq. 4.34 and the symmetry of $g_{\alpha\beta}$, it is easy to show that

$$R_{\alpha\beta\mu\nu,\lambda} + R_{\alpha\beta\lambda\mu,\nu} + R_{\alpha\beta\nu\lambda,\mu} = 0 . \tag{4.35}$$

In a LIF ordinary and covariant derivatives coincide, therefore Eq. 4.35 can also be written as

$$R_{\alpha\beta\mu\nu;\lambda} + R_{\alpha\beta\lambda\mu;\nu} + R_{\alpha\beta\nu\lambda;\mu} = 0 . \tag{4.36}$$

These equations have been derived in a LIF, and since they are tensor equations, they will be valid in any frame.

Eqs. 4.36 are the Bianchi identities which, as we shall later see, play an important role in the derivation of Einstein's equations.

4.5 THE EQUATION OF GEODESIC DEVIATION

If the spacetime is curved, the Equivalence Principle establishes that we can always choose a LIF where the affine connection vanishes. In this frame a free particle (i.e. one subjected to no other force than gravity) follows the geodesic equation with zero acceleration.

Conversely, if the spacetime is flat we can always choose a coordinate frame in which the affine connection does not vanish as, for instance, in Minkowski's spacetime when spherical coordinates are chosen. In this case a free particle would follow a geodesic with non-zero acceleration, and we would not be able to establish whether this acceleration is due to a true gravitational field or to the fact that the reference frame is not inertial. In other words, the local curvature of spacetime cannot be measured by observing the motion of a single particle.

We shall now show that the local curvature of spacetime can be measured by comparing the motion of two close particles, i.e. by comparing the behaviour of close geodesics.

Let us consider a two-parameter family of geodesics $x^\mu(\tau, p)$, where τ is the affine parameter and p labels different curves of the family. Along each geodesic $p = const$. We also consider the family of non-geodesic curves which join points of the family $x^\mu(\tau, p)$ having the same τ; these curves are parametrized by p and are indicated with $\tau = const$ in Fig. 4.2.

Two particles move along two infinitely close geodesics $x^\mu(\tau, p)$ and $x^\mu(\tau, p) + \delta x^\mu = x^\mu(\tau, p + \delta p)$. Here $\vec{\delta x}$ is the infinitesimal displacement vector between points having the same τ on the two geodesics. Let $\vec{t}$ be the tangent vector to the geodesics, and $\vec{b}$ the tangent vector to the $p = const$ curves, as indicated in Fig. 4.2, i.e.

$$t^\alpha = \frac{\partial x^\alpha}{\partial \tau} \qquad \text{and} \qquad b^\alpha = \frac{\partial x^\alpha}{\partial p} . \tag{4.37}$$

The displacement vector is $\delta x^\alpha = b^\alpha \delta p$. For simplicity, we choose the parameter p such that the tangent vector b^α and the displacement vector δx^α coincide. With this definition, $\delta x^\alpha = \frac{\partial x^\alpha}{\partial p}$, and

$$\frac{\partial t^\alpha}{\partial p} = \frac{\partial \delta x^\alpha}{\partial \tau} . \tag{4.38}$$

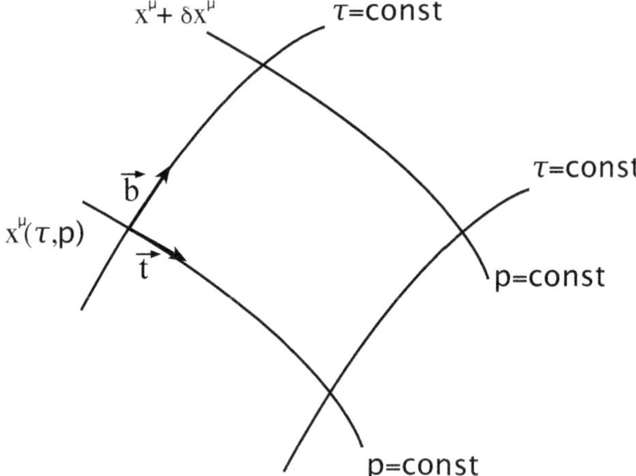

Figure 4.2: Worldlines of two infinitely close particles and their tangent vectors.

We now compute the covariant derivative of the vector $\vec{t}$ along the curve τ = const with tangent vector $\vec{\delta x}$, i.e. $\nabla_{\vec{\delta x}}\, \vec{t}$. The components of this vector are

$$\left(\nabla_{\vec{\delta x}}\, \vec{t}\right)^{\alpha} = \delta x^{\mu} t^{\alpha}{}_{;\mu} = \frac{\partial x^{\mu}}{\partial p}\left[\frac{\partial t^{\alpha}}{\partial x^{\mu}} + \Gamma^{\alpha}_{\mu\nu} t^{\nu}\right] = \frac{\partial t^{\alpha}}{\partial p} + \Gamma^{\alpha}_{\mu\nu} t^{\nu} \delta x^{\mu}\,. \tag{4.39}$$

Similarly, the covariant derivative of the vector $\vec{\delta x}$ along the curve p = const, i.e. along the geodesic, has components

$$\left(\nabla_{\vec{t}}\, \vec{\delta x}\right)^{\alpha} = t^{\mu} \delta x^{\alpha}{}_{;\mu} = \frac{\partial \delta x^{\alpha}}{\partial \tau} + \Gamma^{\alpha}_{\mu\nu} \delta x^{\nu} t^{\mu}\,. \tag{4.40}$$

From Eq. 4.38 and from the symmetry of $\Gamma^{\alpha}_{\mu\nu}$ in its lower indices it follows that

$$\nabla_{\vec{t}}\, \vec{\delta x} = \nabla_{\vec{\delta x}}\, \vec{t}. \tag{4.41}$$

The quantities $\left(\nabla_{\vec{t}}\, \vec{\delta x}\right)^{\alpha}$ and $\left(\nabla_{\vec{\delta x}}\, \vec{t}\right)^{\alpha}$ involve only the affine connection, therefore they do not give significant information on the gravitational field.

We shall now compute the second covariant derivative of the vector $\vec{\delta x}$ along the curve $p = const$, i.e $\nabla_{\vec{t}}\left(\nabla_{\vec{t}}\, \vec{\delta x}\right)$. This quantity is the relative acceleration of the two nearby particles as they move along the geodesics. By defining the operator

$$\frac{D}{d\tau}\delta x^{\alpha} \equiv \left(\nabla_{\vec{t}}\, \vec{\delta x}\right)^{\alpha}, \tag{4.42}$$

the quantity we need to compute is

$$\frac{D^2 \delta x^{\alpha}}{d\tau^2} = \left(\nabla_{\vec{t}}\left(\nabla_{\vec{t}}\, \vec{\delta x}\right)\right)^{\alpha}. \tag{4.43}$$

This is the **geodesic deviation**. In order to compute this quantity, let us expand the following commutator

$$\left[\nabla_{\vec{t}}, \nabla_{\vec{\delta x}}\right] \vec{t} = \nabla_{\vec{t}}\left(\nabla_{\vec{\delta x}}\, \vec{t}\right) - \nabla_{\vec{\delta x}}\left(\nabla_{\vec{t}}\, \vec{t}\right), \tag{4.44}$$

whose components are

$$\begin{aligned}
\left([\nabla_{\vec{t}}, \nabla_{\vec{\delta x}}]\ \vec{t}\right)^{\alpha} &= t^{\mu}\,(\delta x^{\nu}\ t^{\alpha}{}_{;\nu})_{;\mu} - \delta x^{\mu}\,(t^{\nu}\ t^{\alpha}{}_{;\nu})_{;\mu} \\
&= t^{\mu}\ \delta x^{\nu}{}_{;\mu} t^{\alpha}{}_{;\nu} + t^{\mu}\ \delta x^{\nu}\ t^{\alpha}{}_{;\nu;\mu} - \delta x^{\mu}\ t^{\nu}{}_{;\mu}\ t^{\alpha}{}_{;\nu} - \delta x^{\mu}\ t^{\nu} t^{\alpha}{}_{;\nu;\mu} \\
&= (t^{\mu}\ \delta x^{\nu}{}_{;\mu} - \delta x^{\mu}\ t^{\nu}{}_{;\mu})\, t^{\alpha}{}_{;\nu} + (t^{\alpha}{}_{;\nu;\mu} - t^{\alpha}{}_{;\mu;\nu})\, t^{\mu}\ \delta x^{\nu}\,. \qquad (4.45)
\end{aligned}$$

Since from Eq. 4.41

$$t^{\mu}\ \delta x^{\nu}{}_{;\mu} = \delta x^{\mu}\ t^{\nu}{}_{;\mu}\,, \qquad (4.46)$$

Eq. 4.45 gives

$$\left([\nabla_{\vec{t}}, \nabla_{\vec{\delta x}}]\ \vec{t}\right)^{\alpha} = (t^{\alpha}{}_{;\nu;\mu} - t^{\alpha}{}_{;\mu;\nu})\, t^{\mu}\delta x^{\nu}\,. \qquad (4.47)$$

We now remind that, according to Eq. 4.33, the commutator of covariant derivatives is

$$(t^{\alpha}{}_{;\nu;\mu} - t^{\alpha}{}_{;\mu;\nu}) = R^{\alpha}{}_{\beta\mu\nu}t^{\beta}\,, \qquad (4.48)$$

therefore Eq. 4.45 becomes

$$\left([\nabla_{\vec{t}}, \nabla_{\vec{\delta x}}]\ \vec{t}\right)^{\alpha} = R^{\alpha}{}_{\beta\mu\nu}t^{\beta}t^{\mu}\delta x^{\nu}\,. \qquad (4.49)$$

Moreover, since t^{μ} is the geodesic tangent vector, when it is parallely transported along the geodesic it satisfies the equation (see Sec. 3.8)

$$\nabla_{\vec{t}}\,\vec{t} = 0\,; \qquad (4.50)$$

as a consequence $\nabla_{\vec{\delta x}}\left(\nabla_{\vec{t}}\,\vec{t}\right) = 0$ and the commutator in Eq. 4.44 can be rewritten as

$$\left([\nabla_{\vec{t}}, \nabla_{\vec{\delta x}}]\ \vec{t}\right)^{\alpha} = \left(\nabla_{\vec{t}}(\nabla_{\vec{\delta x}}\ \vec{t})\right)^{\alpha} = \left(\nabla_{\vec{t}}\left(\nabla_{\vec{t}}\ \vec{\delta x}\right)\right)^{\alpha} = R^{\alpha}{}_{\beta\mu\nu}t^{\beta}t^{\mu}\delta x^{\nu}\,, \qquad (4.51)$$

where we have used Eq. 4.41. By direct substitution of this expression in Eq. 4.43 we finally find

$$\frac{D^2\delta x^{\alpha}}{d\tau^2} = R^{\alpha}{}_{\beta\mu\nu}\ t^{\beta}\ t^{\mu}\ \delta x^{\nu}\,. \qquad (4.52)$$

This is the **equation of geodesic deviation**, which shows that the relative acceleration of nearby particles moving along geodesics depends on the spacetime curvature. Since the Riemann tensor is zero if and only if the gravitational field is either zero or constant and uniform, the equation of geodesic deviation really contains the information on the gravitational field in a given spacetime.

Box 4-A

Summary on the curvature tensor

- The Riemann tensor, or curvature tensor, is a $\binom{1}{3}$ tensor describing the intrinsic curvature of the spacetime; it vanishes in a flat spacetime, while it is non-vanishing in a curved spacetime.
- Its components are

$$R^{\alpha}{}_{\beta\mu\nu} = \Gamma^{\alpha}_{\beta\nu,\mu} - \Gamma^{\alpha}_{\beta\mu,\nu} - \Gamma^{\alpha}_{\sigma\nu}\Gamma^{\sigma}_{\beta\mu} + \Gamma^{\alpha}_{\sigma\mu}\Gamma^{\sigma}_{\beta\nu} \,. \tag{4.53}$$

- It is antisymmetric in the first and in the second pair of indices; it is symmetric under exchange of the first and second pairs of indices.
- A contraction of the first and third index of the Riemann tensor yields the Ricci tensor, which is a symmetric $\binom{0}{2}$ tensor whose components are

$$R_{\mu\nu} = R^{\alpha}{}_{\mu\alpha\nu} = \Gamma^{\alpha}_{\mu\nu,\alpha} - \Gamma^{\alpha}_{\mu\alpha,\nu} - \Gamma^{\alpha}_{\sigma\nu}\Gamma^{\sigma}_{\mu\alpha} + \Gamma^{\alpha}_{\sigma\alpha}\Gamma^{\sigma}_{\mu\nu} \,. \tag{4.54}$$

- A contraction of the Ricci tensor yields the Ricci curvature, or scalar curvature, $R = g^{\mu\nu}R_{\mu\nu}$.
- A vector $\vec{V}$ parallely transported along an infinitesimal closed loop with sides $\vec{\delta x}_{(1)}$, $\vec{\delta x}_{(2)}$, undergoes a shift $\delta\vec{V}$ with components

$$\delta V^{\alpha} = \pm R^{\alpha}{}_{\beta\mu\nu} V^{\beta} \delta x^{\mu}_{(2)} \delta x^{\nu}_{(1)} \tag{4.55}$$

where the sign depends on the direction of the motion on the loop.

- The commutator of the covariant derivatives of a vector $\vec{V}$ is

$$[\nabla_{\mu}, \nabla_{\nu}]\, V^{\alpha} = R^{\alpha}{}_{\beta\mu\nu} V^{\beta} \,. \tag{4.56}$$

- The Riemann tensor satisfies the Bianchi identities

$$R_{\alpha\beta\mu\nu;\lambda} + R_{\alpha\beta\lambda\mu;\nu} + R_{\alpha\beta\nu\lambda;\mu} = 0 \,. \tag{4.57}$$

- The separation $\vec{\delta x}$ between two infinitely close geodesics with tangent vector $\vec{t}$ satisfies the equation of geodesic deviation

$$\frac{D^2 \delta x^{\alpha}}{d\tau^2} = R^{\alpha}{}_{\beta\mu\nu}\, t^{\beta}\, t^{\mu}\, \delta x^{\nu} \,. \tag{4.58}$$

CHAPTER 5

The stress-energy tensor

In the previous chapter we showed that there exists a tensor which allows to understand whether the spacetime is curved or flat, i.e. if we are in the presence of a non-constant or non-uniform gravitational field. In order to describe the dynamics of the gravitational field, we also need to understand how to include the contribution of matter and fields, which are the sources of such field. In this chapter we shall show that the distribution of matter and energy can be described – both in Special Relativity and in General Relativity – in terms of a *rank-two tensor field*, the *stress-energy tensor*, which will be indicated as $T_{\mu\nu}$. The relevance of this tensor for the theory of gravity will be clear in the next chapter, where we shall show that it is the source of the dynamical equations of the gravitational field.

We shall firstly introduce $T_{\mu\nu}$ in flat spacetime, using as an example the simplest physical system, a gas of non-interacting particles. Generalizing the concept of energy-momentum four-vector of a particle in Special Relativity, we shall show that the quantity we construct is a tensor, and that it satisfies a divergence-free equation which, in flat spacetime, leads to energy and momentum conservation laws; the meaning of the components of this tensor will be illustrated and discussed. In order to generalize these results to curved spacetimes, we shall state a fundamental principle of General Relativity, the *Principle of General Covariance*. In this way we show that the tensorial properties that $T_{\mu\nu}$ satisfies in Special Relativity, i.e. the symmetries and the divergence-free equation, hold in any frame.

Although derived for a simple physical system, the gas of non-interacting particles, these properties hold for any system which can be described by a stress-energy tensor as, for instance, a fluid.

The discussion on the stress-energy tensor will be further expanded in the following chapters: in Chapter 7 we shall provide a more formal derivation of this tensor for a generic physical system described by a Lagrangian; in Chapter 8 we shall show that conservation laws can be associated to $T_{\mu\nu}$ when the spacetime admits some symmetry; finally, the stress-energy tensor of a fluid will be derived in Chapter 16, where we will show how the laws of thermodynamics have to be modified in General Relativity.

5.1 THE STRESS-ENERGY TENSOR IN FLAT SPACETIME

In Special Relativity, the motion of a particle of mass m and three-velocity $\mathbf{v}$ with components $v^i = \frac{d\xi^i}{dt}$, $i = 1, \ldots, 3$ is described by the energy-momentum four-vector

$$p^\alpha = mcu^\alpha, \qquad \alpha = 0, \ldots, 3, \tag{5.1}$$

where $u^\alpha = \frac{d\xi^\alpha}{d\tau}$ are the components of the particle four-velocity $\vec{u}$, and τ (which has the dimensions of a length, see Box 3-F) is related to the particle proper time τ/c. With this notation and with our adopted convention for the metric signature, the square of the norm of the four-velocity is $u_\alpha u^\alpha = -1$. Also remember that $\{\xi^\alpha\}$ are Minkowskian coordinates. By defining $\xi^0 = ct$ and

$$\gamma = \frac{d\xi^0}{d\tau}\,, \tag{5.2}$$

we obtain

$$\begin{aligned} u^0 &= \gamma \\ u^i &= \frac{d\xi^i}{d\tau} = \frac{d\xi^i}{dt}\frac{dt}{d\tau} = v^i\frac{\gamma}{c} \\ \eta_{\alpha\beta}u^\alpha u^\beta &= -\gamma^2\left(1-\frac{v^2}{c^2}\right) = -1 \quad \rightarrow \quad \gamma = \left(1-\frac{v^2}{c^2}\right)^{-1/2}. \end{aligned} \tag{5.3}$$

Thus, the energy-momentum vector can be written as

$$p^\mu = m\gamma(c, \mathbf{v})\,. \tag{5.4}$$

The time component of this four-vector is proportional to the particle energy

$$p^0 = \frac{E}{c}, \quad \text{where} \quad E = mc^2\gamma\,, \tag{5.5}$$

whereas the space components are the components of the three-dimensional relativistic momentum $\mathbf{p} = m\gamma\mathbf{v}$.

When dealing with a continuous or discrete distribution of matter and energy, there are other quantities which we would like to define and measure, such as the mass and the energy which are contained into a certain volume, or the energy and momentum that flow across the surfaces enclosing this volume. All this information is contained in the stress-energy tensor which we are going to define.

Let us consider, as an example, the simple case of a system composed by a collection of *non-interacting* particles, each of which follows the worldline $\{\xi_n^\alpha(t)\}$, and let $\vec{p}_n$ be the energy-momentum vector of the n-th particle. We define the **energy density** of the system as

$$T^{00} \equiv \sum_n cp_n^0(t)\delta^3(\boldsymbol{\xi} - \boldsymbol{\xi}_n(t)) = \sum_n E_n\delta^3(\boldsymbol{\xi} - \boldsymbol{\xi}_n(t))\,, \tag{5.6}$$

where the function

$$\delta^3(\boldsymbol{\xi} - \boldsymbol{\xi}_n(t)) = \delta(\xi^1 - \xi_n^1(t))\delta(\xi^2 - \xi_n^2(t))\delta(\xi^3 - \xi_n^3(t)) \tag{5.7}$$

is the three-dimensional Dirac δ-function, whose main properties are summarized in Box 5-A. Note that, since $\delta^3(\boldsymbol{\xi} - \boldsymbol{\xi}_n(t))$ has the dimensions of an inverse cubic length, T^{00} has the dimensions of an energy divided by a volume, i.e. an energy density. In analogy with Eq. 5.6, we introduce the **density of momentum** $\frac{1}{c}T^{0i}$, where T^{0i} is defined as

$$T^{0i} \equiv \sum_n cp_n^i(t)\delta^3(\boldsymbol{\xi} - \boldsymbol{\xi}_n(t))\,, \qquad i = 1,2,3 \tag{5.8}$$

and the **momentum current** as

$$T^{ki} \equiv \sum_n p_n^k(t)\frac{d\xi_n^i(t)}{dt}\delta^3(\boldsymbol{\xi} - \boldsymbol{\xi}_n(t)), \qquad i,k = 1,2,3\,. \tag{5.9}$$

The definitions 5.6, 5.8 and 5.9 can be unified into a single formula

$$T^{\alpha\beta} = \sum_n p_n^\alpha\frac{d\xi_n^\beta(t)}{dt}\delta^3(\boldsymbol{\xi} - \boldsymbol{\xi}_n(t)), \qquad \alpha,\beta = 0,\ldots,3\,. \tag{5.10}$$

Furthermore, since

$$p_n^\alpha = \frac{E_n}{c^2}\frac{d\xi_n^\alpha(t)}{dt}\,, \tag{5.11}$$

Eq. 5.10 can also be written as

$$T^{\alpha\beta} = c^2\sum_n \frac{p_n^\alpha p_n^\beta}{E_n}\delta^3(\boldsymbol{\xi} - \boldsymbol{\xi}_n(t))\,, \tag{5.12}$$

which clearly shows that $T^{\alpha\beta}$ is *symmetric* in its two indices,

$$T^{\alpha\beta} = T^{\beta\alpha}. \tag{5.13}$$

Finally, an alternative way of writing Eq. 5.10 is

$$T^{\alpha\beta} = mc^2\sum_n \int u_n^\alpha u_n^\beta \delta^4(\vec{\xi} - \vec{\xi}_n(\tau_n))d\tau_n\,, \tag{5.14}$$

where

$$\delta^4(\vec{\xi} - \vec{\xi}_n) = \delta(\xi^0 - \xi_n^0)\delta(\xi^1 - \xi_n^1)\delta(\xi^2 - \xi_n^2)\delta(\xi^3 - \xi_n^3)\,. \tag{5.15}$$

Indeed, using the properties of the δ-function (see Box 5-A) it is easy to see that

$$\begin{aligned}
T^{\alpha\beta} &= mc^2\sum_n \int u_n^\alpha\ u_n^\beta\ \delta^4(\vec{\xi} - \vec{\xi}_n(\tau_n))\ d\tau_n \\
&= c\sum_n \int \left[p_n^\alpha\ \frac{d\xi_n^\beta}{d\tau_n}\ \delta^3(\boldsymbol{\xi} - \boldsymbol{\xi}_n(\tau_n))\right]\ \delta(\xi^0 - \xi_n^0(\tau_n))\ \frac{d\tau_n}{d\xi_n^0}\ d\xi_n^0 \\
&= c\sum_n \left[p_n^\alpha\ \frac{d\xi_n^\beta}{d\xi_n^0}\ \delta^3(\boldsymbol{\xi} - \boldsymbol{\xi}_n(\tau_n))\right]_{\xi_n^0(\tau_n)=\xi^0} \\
&= c\sum_n p_n^\alpha\ \frac{d\xi_n^\beta}{d\xi^0}\ \delta^3(\boldsymbol{\xi} - \boldsymbol{\xi}_n(\xi^0)) = \sum_n p_n^\alpha\ \frac{d\xi_n^\beta}{dt}\ \delta^3(\boldsymbol{\xi} - \boldsymbol{\xi}_n(\xi^0))\,,
\end{aligned} \tag{5.16}$$

which coincides with Eq. 5.12.

In the next sections, we shall show that:

- $T^{\alpha\beta}$ are the components of a tensor;
- this tensor satisfies a divergence-free equation;
- this equation can be generalized to the case of curved spacetimes, i.e. in the presence of a gravitational field.

Box 5-A

Properties of the δ-function

In one dimension, the Dirac δ-function is defined by the property that, for any smooth function $f(x)$,

$$\int dx\ f(x)\delta(x-x_0)=f(x_0)\,, \tag{5.17}$$

if x_0 is part of the integration domain, otherwise the above integral vanishes. Similarly, for the three-dimensional δ-function $\delta^3(\boldsymbol{\xi}-\boldsymbol{\xi}_n)$ given in Eq. 5.7, the following property holds

$$\int d^3\xi\, f(\boldsymbol{\xi})\delta^3(\boldsymbol{\xi}-\boldsymbol{\xi}_n)=f(\boldsymbol{\xi}_n)\,. \tag{5.18}$$

According to these definitions, the one-dimensional δ-function has the dimensions of the inverse of a length and $\delta^3(\boldsymbol{\xi}-\boldsymbol{\xi}_n)$ has dimensions $(\text{length})^{-3}$.
More generally, in D dimensions $\vec{x}=(x^1,\ldots,x^D)$

$$\int d^Dx\, f(\vec{x})\delta^D(\vec{x}-\vec{x}_n)=f(\vec{x}_n)\,, \tag{5.19}$$

where $\delta^D(\vec{x}-\vec{x}_n)=\delta(x^1-x^1_n(t))\delta(x^2-x^2_n(t))\cdots\delta(x^D-x^D_n(t))$.
It is easy to prove the following useful properties of the δ-function

$$\begin{aligned}
&\delta(x)=\delta(-x)\,, && \delta(const\,x)=\tfrac{1}{|const|}\delta(x)\,,\\
&\delta[g(x)]=\textstyle\sum_j \frac{1}{|g'(x^j)|}\delta(x-x^j)\,, && x\delta(x)=0\,,\\
&\textstyle\int dx f(x)\delta'(x-x_0)=-f'(x_0)\,,
\end{aligned} \tag{5.20}$$

where the prime indicates differentiation with respect to the argument of the function.

5.2 IS $T^{\alpha\beta}$ A TENSOR?

Let us consider the stress-energy tensor given in Eq. 5.14 and a generic coordinate transformation $\xi^\alpha=\xi^\alpha(x^{\alpha'})$. The four-momentum and the four-velocity transform as $p^\alpha=\Lambda^\alpha{}_{\gamma'}\,p^{\gamma'}$ and $u^\alpha=\Lambda^\alpha{}_{\gamma'}\,u^{\gamma'}$, where $\Lambda^\alpha{}_{\gamma'}=\frac{\partial\xi^\alpha}{\partial x^{\gamma'}}$. Since, as discussed in Box 5-B, under the transformation $\xi^\alpha=\xi^\alpha(x^{\alpha'})$ the $\delta^4(\vec{\xi}-\vec{\xi}_n)$ transforms as

$$\delta^4(\vec{\xi}-\vec{\xi}_n)=\frac{\delta^4(\vec{x}'-\vec{x}'_n)}{\sqrt{-g'}}, \tag{5.21}$$

Eq. 5.14 gives

$$T^{\alpha\beta}=mc^2\sum_n\int\Lambda^\alpha{}_{\mu'}\,\Lambda^\beta{}_{\nu'}\,u^{\mu'}_n\,u^{\nu'}_n\,\frac{\delta^4(\vec{x}'-\vec{x}'_n)}{\sqrt{-g'}}\,d\tau_n. \tag{5.22}$$

This shows that the quantity

$$T^{\alpha\beta} = mc^2 \sum_n \int u_n^\alpha u_n^\beta \, \frac{\delta^4(\vec{x} - \vec{x}_n)}{\sqrt{-g}} \, d\tau_n \,, \tag{5.23}$$

transforms as a tensor under a generic coordinate transformation, i.e.

$$T^{\alpha\beta} = \Lambda^\alpha{}_{\mu'} \, \Lambda^\beta{}_{\nu'} \, T^{\mu'\nu'} \,, \tag{5.24}$$

and therefore $T^{\alpha\beta}$ given in Eq. 5.23 *is a tensor* (see Box 2-M). In flat spacetime in Minkowskian coordinates and in a LIF, $\sqrt{-g} = 1$ and we recover the definition 5.14.

Therefore, we define the tensor $T^{\alpha\beta}$ given in Eq. 5.23 as the **stress-energy tensor** of a cloud of non-interacting particles. This definition is valid *both in flat and in curved spacetime.*

In general, different kinds of matter and/or fields may be present: fluids, electromagnetic fields, etc. In all cases it is possible to define a rank-two tensor, the *stress-energy tensor,* whose components have the same meaning as those discussed in Box 5-C for a cloud of non-interacting particles. In Chapter 7, we shall show that the stress-energy tensor can be derived by writing the action of the considered field, and by varying this action with respect to $g_{\mu\nu}$. For example, the stress energy tensor of a scalar field Ψ reads

$$T_{\mu\nu} = \partial_\mu \Psi \partial_\nu \Psi - \frac{1}{2} g_{\mu\nu} \partial_\sigma \Psi \partial^\sigma \Psi - g_{\mu\nu} V(\Psi) \,, \tag{5.25}$$

where $V(\Psi)$ is the scalar self-potential, which reduces to $V(\Psi) = \frac{m^2 c^2}{2\hbar^2} \Psi^2$ for a free scalar field with mass m.

Likewise, the electromagnetic stress-energy tensor in SI units reads

$$T_{\mu\nu} = -\frac{1}{\mu_0} \left(F_{\sigma(\mu} F_{\nu)\rho} g^{\sigma\rho} - \frac{1}{4} g_{\mu\nu} F_{\sigma\rho} F^{\sigma\rho} \right) , \tag{5.26}$$

where $F_{\mu\nu} = \nabla_\mu A_\nu - \nabla_\nu A_\mu$ is the electromagnetic tensor, A_μ is the four-potential, and μ_0 is the vacuum permeability (the overall coefficient depends on the system of units chosen for the electromagnetic quantities). As an exercise, the reader can easily show that it is superfluous to use covariant derivatives in the definition of $F_{\mu\nu}$, since $\nabla_\mu A_\nu - \nabla_\nu A_\mu = \partial_\mu A_\nu - \partial_\nu A_\mu$.

Finally, for a perfect fluid,

$$T_{\mu\nu} = (\epsilon + p) \, u_\mu u_\nu + p g_{\mu\nu} \,, \tag{5.27}$$

where ϵ and p are the energy density and the pressure of the fluid measured in a LIF co-moving with the fluid, and u^μ is the (dimensionless) four-velocity of a fluid element. This case will be discussed in detail in Chapter 16.

Box 5-B

Transformation rules for $\delta^4(\vec{\xi}-\vec{\xi}_n(\tau_n))$

In a four-dimensional spacetime the volume element which is invariant under a generic coordinate transformation $x^\alpha = x^\alpha(x^{\alpha'})$ is $\sqrt{-g}\, d^4x$, i.e.

$$\sqrt{-g}\; d^4x = \sqrt{-g'}\; d^4x' \,. \tag{5.28}$$

Indeed,

$$d^4x = |J|\; d^4x' \,, \tag{5.29}$$

where $J = \det\left(\frac{\partial x^\alpha}{\partial x^{\beta'}}\right)$ is the Jacobian associated to the coordinate transformation. Furthermore, since

$$g_{\alpha'\beta'} = \frac{\partial x^\mu}{\partial x^{\alpha'}}\frac{\partial x^\nu}{\partial x^{\beta'}} g_{\mu\nu} \,, \tag{5.30}$$

the determinant of both members gives

$$g' = J^2 g \qquad \text{and} \quad \text{therefore} \qquad |J| = \frac{\sqrt{-g'}}{\sqrt{-g}} \,, \tag{5.31}$$

which, replaced in Eq. 5.29, gives Eq. 5.28. Thus, if $\{\xi^\alpha\}$ is a Minkowskian frame, and $\{x^{\alpha'}\}$ is a generic frame, Eq. 5.29 gives

$$d^4\xi = \sqrt{-g'}\; d^4x' \,. \tag{5.32}$$

Let us now consider the δ-function in Minkowski's spacetime; by definition (see Eq. 5.19), for every function $f(\vec{\xi})$, we have

$$\int d^4\xi \; f(\vec{\xi})\delta^4(\vec{\xi}-\vec{\xi}_n) = f(\vec{\xi}_n) \,, \tag{5.33}$$

and, in a generic frame $\{x^{\alpha'}\}$,

$$\int d^4x' \; f(\vec{x}')\delta^4(\vec{x}'-\vec{x}'_n) = f(\vec{x}'_n) \,. \tag{5.34}$$

Let us now perform a coordinate transformation $\xi^\alpha \to x^{\alpha'}$, with $x^{\alpha'} = x^{\alpha'}(\xi^\mu)$, and define $\tilde{f}(\vec{\xi}) = f(\vec{x}'(\vec{\xi}))$; multiplying and dividing Eq. 5.34 by $\sqrt{-g'}$, we find

$$\int \sqrt{-g'}\; d^4x' \; f(\vec{x}')\frac{\delta^4(\vec{x}'-\vec{x}'_n)}{\sqrt{-g'}} = \int \tilde{f}(\vec{\xi})\frac{\delta^4(\vec{x}'-\vec{x}'_n)}{\sqrt{-g'}}\; d^4\xi = \tilde{f}(\vec{\xi}_n) \,, \tag{5.35}$$

which is valid for every function f. Comparing Eqs. 5.33 and 5.35 we finally find

$$\delta^4(\vec{\xi}-\vec{\xi}_n) = \frac{\delta^4(\vec{x}'-\vec{x}'_n)}{\sqrt{-g'}} \,. \tag{5.36}$$

Note that the Dirac δ-function does not transform as a scalar.

Box 5-C

Physical meaning of the stress-energy tensor components

In order to understand the physical meaning of the components of the stress-energy tensor, we shall consider again, as an example, the case of a gas of non-interacting particles. However, the results hold for any physical system. In flat space (and in Minkowskian coordinates) the components of the stress-energy tensor can be interpreted as follows.

- T^{00} = energy density.
  Indeed (see Eq. 5.6)
  $$T^{00} = \sum_n E_n \delta^3(\boldsymbol{\xi} - \boldsymbol{\xi}_n(t)) . \tag{5.37}$$
  In the non-relativistic limit $v \ll c$, $E_n \simeq m_n c^2$, and the energy density becomes $T^{00} = \rho c^2$, where ρ is the *matter density* :
  $$\rho = \sum_n m_n \delta^3(\boldsymbol{\xi} - \boldsymbol{\xi}_n(t)) . \tag{5.38}$$
- $\frac{1}{c}T^{0i}$ = density of momentum; cT^{0i} = energy current.
  Indeed (see Eq. 5.8)
  $$\frac{1}{c}T^{0i} = \sum_n p^i_n \delta^3(\boldsymbol{\xi} - \boldsymbol{\xi}_n(t)) , \tag{5.39}$$
  and since $p^i = m\gamma v^i = Ev^i/c^2$, we can also write
  $$cT^{0i} = \sum_n E_n v^i \delta^3(\boldsymbol{\xi} - \boldsymbol{\xi}_n(t)) , \tag{5.40}$$
  which is the energy flowing per unit time across the unit surface orthogonal to the axis ξ^i.
- T^{ij} = momentum current.
  Indeed (see Eq. 5.9)
  $$T^{ij} \equiv \sum_n p^i_n v^j_n \delta^3(\boldsymbol{\xi} - \boldsymbol{\xi}_n(t)) \tag{5.41}$$
  is the (i-th component of the) momentum flowing per unit time across the unit surface orthogonal to the axis x^j.

In curved spacetime, we can always define a LIF, where – due to the Equivalence Principle – the components of the stress-energy tensor have the same interpretation as in flat space. In this frame, then, T^{00} is the energy density, $\frac{1}{c}T^{01}$ is the momentum density, cT^{0i} is the energy current, and T^{ij} is the momentum current.

In a general coordinate frame, *the energy density as measured by an observer with four-velocity* $\vec{u}$, $\epsilon^{(\vec{u})}$, is
$$\epsilon^{(\vec{u})} = T_{\mu\nu} u^\mu u^\nu . \tag{5.42}$$
Indeed, $\epsilon^{(\vec{u})}$ is a scalar quantity (the contraction of a $\binom{0}{2}$ tensor with two vectors) and has the same form in any coordinate frame. In a LIF where the observer is at rest, its four-velocity has components $u^\mu = (1, 0, 0, 0)$, and Eq. 5.42 gives $\epsilon^{(\vec{u})} = T_{00}$, which is the energy density in this frame.

5.3 DOES $T^{\alpha\beta}$ SATISFY A CONSERVATION LAW?

Let us consider, as an example, the expression of the stress-energy tensor for a system of non-interacting particles (Eq. 5.10) in flat spacetime, and take the space-divergence [1] of the (αi)-components, i.e.

$$\frac{\partial T^{\alpha i}}{\partial \xi^i} = \sum_n p_n^\alpha(t) \, \frac{d\xi_n^i(t)}{dt} \, \frac{\partial}{\partial \xi^i} \, \delta^3(\boldsymbol{\xi} - \boldsymbol{\xi}_n(t)) \,, \tag{5.43}$$

where $\alpha = 0, \ldots, 3$ and $i = 1, \ldots, 3$. Since

$$\frac{\partial}{\partial \xi^i} \delta^3(\boldsymbol{\xi} - \boldsymbol{\xi}_n(t)) = -\frac{\partial}{\partial \xi_n^i} \delta^3(\boldsymbol{\xi} - \boldsymbol{\xi}_n(t)) \,, \tag{5.44}$$

Eq. 5.43 becomes

$$\begin{aligned} \frac{\partial T^{\alpha i}}{\partial \xi^i} &= -\sum_n p_n^\alpha(t) \, \frac{d\xi_n^i(t)}{dt} \, \frac{\partial}{\partial \xi_n^i} \, \delta^3(\boldsymbol{\xi} - \boldsymbol{\xi}_n(t)) \\ &= -\sum_n p_n^\alpha(t) \, \frac{\partial}{\partial t} \, \delta^3(\boldsymbol{\xi} - \boldsymbol{\xi}_n(t)) \,. \end{aligned} \tag{5.45}$$

Let us now differentiate the component $T^{\alpha 0}$ with respect to $\xi^0 = ct$:

$$\frac{\partial T^{\alpha 0}}{\partial \xi^0} = \sum_n \frac{dp_n^\alpha(t)}{dt} \delta^3(\boldsymbol{\xi} - \boldsymbol{\xi}_n(t)) + \sum_n p_n^\alpha(t) \, \frac{\partial}{\partial t} \delta^3(\boldsymbol{\xi} - \boldsymbol{\xi}_n(t)) \,. \tag{5.46}$$

Since

$$\frac{dp_n^\alpha(t)}{dt} = \frac{dp_n^\alpha(\tau)}{d\tau} \frac{d\tau}{dt} = \frac{d\tau}{dt} f_n^\alpha \,, \tag{5.47}$$

where f_n^α is the four-force, the first term in Eq. 5.46 is a density of force.

If the system of particles is isolated, the total four-force is zero and the first term in Eq. 5.46 vanishes. Therefore, adding Eq. 5.45 and Eq. 5.46 we find

$$\frac{\partial T^{\alpha\beta}}{\partial \xi^\beta} = 0 \,, \qquad \alpha, \beta = 0, \ldots, 3 \,. \tag{5.48}$$

This equation states that the ordinary four-divergence of the stress-energy tensor of an isolated system vanishes in flat spacetime. *Although this result has been derived for a gas of non-interacting particles, it can be shown that it holds for any distribution of matter and fields.*

We shall now show that this equation leads to the conservation of the energy and momentum of the considered system.

Let us consider the $\alpha = 0$ components of Eq. 5.48, and move the terms with space derivatives to the right-hand side,

$$\frac{\partial T^{00}}{\partial \xi^0} = -\frac{\partial T^{0k}}{\partial \xi^k} \,, \qquad k = 1, \ldots, 3 \,. \tag{5.49}$$

We can integrate this equation over a volume V which extends over all space; the integration is performed at a fixed time, i.e. on a hypersurface $\xi^0 = const$

$$\frac{\partial}{\partial \xi^0} \int_V T^{00} d^3\xi = -\int_V \frac{\partial T^{0k}}{\partial \xi^k} \, d^3\xi \,. \tag{5.50}$$

---

[1] We remind that $\frac{\partial T^{\alpha i}}{\partial \xi^i} = \frac{\partial T^{\alpha 1}}{\partial \xi^1} + \frac{\partial T^{\alpha 2}}{\partial \xi^2} + \frac{\partial T^{\alpha 3}}{\partial \xi^3}$.

Box 5-D

Gauss' theorem

Given a three-vector $\boldsymbol{A}$, the integral over a volume V of the three-divergence of $\boldsymbol{A}$ is equal to the flux of $\boldsymbol{A}$ across the boundary of V, ∂V, i.e.

$$\int_V \boldsymbol{\nabla} \cdot \boldsymbol{A}\, d^3\xi = \int_{\partial V} \boldsymbol{A} \cdot \boldsymbol{n}\, dS\,, \tag{5.51}$$

where $\mathbf{n}$ is the unit vector orthogonal to the surface element dS.
Since $\boldsymbol{\nabla} \cdot \boldsymbol{A} = \frac{\partial A^i}{\partial \xi^i}$, $i = 1, \ldots, 3$ and

$$\boldsymbol{A} \cdot \boldsymbol{n}\; dS = \delta_{ik} n^i A^k\; dS = A^k n_k dS\,, \tag{5.52}$$

by putting $dS_k = n_k dS$, Eq. 5.51 becomes

$$\int_V \frac{\partial A^i}{\partial \xi^i}\, d^3\xi = \int_{\partial V} A^k dS_k\,. \tag{5.53}$$

Using Gauss' theorem (see Box 5-D), we can write Eq. 5.50 as

$$\frac{\partial}{\partial \xi^0} \int_V T^{00} d^3\xi = -\int_{\partial V} T^{0k} dS_k\,. \tag{5.54}$$

Note that cT^{0k} is the energy which flows across the unit surface orthogonal to ξ^k (see Box 5-C); if we assume that the system is isolated (i.e. no energy and momentum flows across the surface ∂V which encloses the whole space) then cT^{0k} vanishes on ∂V, and Eq. 5.54 gives

$$\frac{\partial}{\partial \xi^0} \int_V T^{00} d^3\xi = 0 \quad \rightarrow \quad \int_V T^{00} d^3\xi = const\,, \tag{5.55}$$

which, since T^{00} is the energy density of the system, expresses the energy conservation law of an isolated system.

A similar procedure can be used to find the conservation law of the total momentum; by putting $\alpha = i = 1, \ldots, 3$ in Eq. 5.48 we find

$$\frac{\partial}{\partial \xi^0} \int_V T^{i0} d^3\xi = -\int_V \frac{\partial T^{ik}}{\partial \xi^k}\, d^3\xi = -\int_{\partial V} T^{ik} dS_k\,, \tag{5.56}$$

and, assuming that the momentum currents T^{ik} vanish at infinity,

$$\frac{\partial}{\partial \xi^0} \int_V T^{i0} d^3\xi = 0 \quad \rightarrow \quad \int_V T^{i0} d^3\xi = const\,. \tag{5.57}$$

Since $\frac{1}{c} T^{i0}$ is the density of momentum (see Box 5-C), Eq. 5.57 shows that the total momentum of an isolated system is conserved. In conclusion, in Special Relativity we can define the four-vector

$$P^\alpha = \int_V T^{\alpha 0} dV, \qquad \alpha = 0, \ldots, 3\,, \tag{5.58}$$

which, as a consequence of the divergence-free equation Eq. 5.48 satisfied by the stress-energy tensor, is conserved for isolated systems. The components P^0 and P^i are, respectively, the energy and the total momentum of the system.

The above derivation has been carried out in the framework of Special Relativity. We now need to show how this can be generalized to curved spacetimes. To this purpose, in Box 5-E we introduce the Principle of General Covariance which, together with the Equivalence Principle, lays at the foundations of the theory of General Relativity.

Box 5-E

Principle of General Covariance

When a physical law is preserved in form under an arbitrary coordinate transformation, we say that it is **generally covariant**.
The **principle of General Covariance** states that *a physical law is true if:*

1. *it is true in the absence of gravity, i.e. if it reduces to the laws of Special Relativity when* $g_{\mu\nu} \rightarrow \eta_{\mu\nu}$ *and* $\Gamma^{\alpha}_{\mu\nu}$ *vanish;*

2. *it is generally covariant.* This implies that all equations must be expressed in a tensorial form. Indeed, let us consider a law of physics written in a tensorial form, e.g.

$$A_{\mu\nu\rho} = B_{\mu\nu\rho} \,. \tag{5.59}$$

Upon a coordinate transformation, the above equation becomes

$$A_{\mu'\nu'\rho'}\Lambda^{\mu'}_{\mu}\Lambda^{\nu'}_{\nu}\Lambda^{\rho'}_{\rho} = B_{\mu'\nu'\rho'}\Lambda^{\mu'}_{\mu}\Lambda^{\nu'}_{\nu}\Lambda^{\rho'}_{\rho} \,, \tag{5.60}$$

which can be rearranged as

$$(A_{\mu'\nu'\rho'} - B_{\mu'\nu'\rho'})\Lambda^{\mu'}_{\mu}\Lambda^{\nu'}_{\nu}\Lambda^{\rho'}_{\rho} \quad \rightarrow \quad A_{\mu'\nu'\rho'} = B_{\mu'\nu'\rho'} \,. \tag{5.61}$$

Therefore, the law 5.59 has the same form in any coordinate frame. This proof can be repeated for tensorial equations of any rank.

Thus, according to the Principle of General Covariance, if a tensor equation is true in the absence of gravity, then it is true in the presence of an *arbitrary* gravitational field.

Since the divergence-free equation 5.48, $T^{\alpha\beta}{}_{,\beta} = 0$, is valid in Special Relativity, i.e. in the absence of gravity, according to the Equivalence Principle it holds in a LIF in curved spacetime. In this frame, the covariant and ordinary derivatives coincide, therefore Eq. 5.48 can be written in the alternative form

$$T^{\alpha\beta}{}_{;\beta} = 0. \tag{5.62}$$

Since Eq. 5.62 is a tensor equation, according to the Principle of General Covariance it holds not only in the LIF, but also in any other frame. Thus, the generalization of Eq. 5.48 in curved spacetime is Eq. 5.62, which establishes that the *covariant* divergence of the stress-energy tensor vanishes.

5.4 IS $T^{\alpha\beta}{}_{;\beta} = 0$ A CONSERVATION LAW?

To answer this question we need to compute the covariant divergence of $T^{\alpha\beta}$. From the expression of the affine connection in terms of the metric we find

$$\Gamma^{\mu}_{\lambda\mu} = \frac{1}{2} g^{\mu\rho} \left(\frac{\partial g_{\rho\lambda}}{\partial x^{\mu}} + \frac{\partial g_{\rho\mu}}{\partial x^{\lambda}} - \frac{\partial g_{\lambda\mu}}{\partial x^{\rho}} \right) . \tag{5.63}$$

The first and the third terms cancel each other out due to the symmetry of $g_{\alpha\beta}$; therefore

$$\Gamma^{\mu}_{\lambda\mu} = \frac{1}{2} g^{\mu\rho} \frac{\partial g_{\rho\mu}}{\partial x^{\lambda}} . \tag{5.64}$$

Given an arbitrary square matrix M, the following equality holds

$$\mathrm{Tr} \left[M^{-1}(x) \frac{\partial}{\partial x^{\lambda}} M(x) \right] = \frac{\partial}{\partial x^{\lambda}} \ln \left(|\det M(x)| \right) . \tag{5.65}$$

Using this equation, the right-hand side of Eq. 5.64 can be written as

$$\Gamma^{\mu}_{\lambda\mu} = \frac{1}{2} \frac{\partial}{\partial x^{\lambda}} \ln(-g) = \frac{1}{\sqrt{-g}} \frac{\partial}{\partial x^{\lambda}} \sqrt{-g} , \tag{5.66}$$

where we have defined g as the determinant of the matrix with components $g_{\mu\nu}$, and taken into account the fact that $g < 0$. Thus for example, given a vector $\vec{V}$

$$V^{\lambda}{}_{;\lambda} = V^{\lambda}{}_{,\lambda} + \Gamma^{\lambda}_{\alpha\lambda} V^{\alpha} = \frac{1}{\sqrt{-g}} \frac{\partial}{\partial x^{\lambda}} \left(\sqrt{-g} V^{\lambda} \right) . \tag{5.67}$$

Likewise, for a tensor $\boldsymbol{F}$

$$F^{\mu\nu}{}_{;\mu} = \frac{1}{\sqrt{-g}} \frac{\partial}{\partial x^{\mu}} (\sqrt{-g} F^{\mu\nu}) + \Gamma^{\nu}_{\lambda\mu} F^{\mu\lambda}. \tag{5.68}$$

In particular, if $F^{\mu\nu}$ is antisymmetric, the last term in Eq. 5.68 vanishes due to the symmetry of $\Gamma^{\nu}_{\lambda\mu}$ in the lower indices, and

$$F^{\mu\nu}{}_{;\mu} = \frac{1}{\sqrt{-g}} \frac{\partial}{\partial x^{\mu}} (\sqrt{-g} F^{\mu\nu}). \tag{5.69}$$

Using Eq. 5.68, Eq. 5.62 gives

$$\frac{\partial}{\partial x^{\mu}} (\sqrt{-g} T^{\mu\nu}) = -\sqrt{-g} \Gamma^{\nu}_{\lambda\mu} T^{\mu\lambda} , \tag{5.70}$$

which shows that Eq. 5.62 cannot be reduced to the vanishing of an ordinary four-divergence. Consequently, when integrated over the whole spacetime, as we did in Sec. 5.3, this equation does not lead to conserved quantities.

In analogy with what one does in Special Relativity, we may still define a four-vector

$$P^{\alpha} = \int_V \sqrt{-g} T^{\alpha 0} dV, \qquad \alpha = 0, \ldots, 3 , \tag{5.71}$$

but this is not a conserved quantity. The physical reason for this failure is that in General Relativity, the conservation of energy and momentum must also include the contribution of the energy and momentum carried by the gravitational field itself. How to include these contributions will be explained in Chapter 13.

CHAPTER 6

The Einstein equations

We now have all the tools needed to derive the equations governing the dynamics of the gravitational field. We expect such equations to be more complicated than the linear equations of the electromagnetic field. For example, electromagnetic waves are produced as a consequence of the motion of charged particles, but the energy and the momentum they carry *are not* a source of the electromagnetic field itself, and their contribution does not appear on the right-hand side of Maxwell's equations. For the gravitational interaction the situation is different. The famous equation

$$E = mc^2 \tag{6.1}$$

establishes that mass and energy can transform one into another: they are different manifestations of the same physical quantity. It follows that if the mass is a source of the gravitational field, so must be the energy, and consequently both mass and energy should contribute to the right-hand side of the field equations. This implies that the equations we are looking for have to be *non-linear*. For instance, a system of arbitrarily moving masses radiates gravitational waves which carry energy, which, in turn, is source of the gravitational field [1]. In addition, the Principle of General Covariance (see Box 5-E) states that the laws of physics must be generally covariant, i.e. their form must be invariant under a general coordinate transformation; this implies that the equations governing the gravitational field must be tensor equations.

As discussed in Chapter 5, the stress-energy tensor describes the matter and energy distribution, and since the latter are the source of the gravitational field we expect the stress-energy tensor to appear on the right-hand side of the tensor equations we are looking for.

Finally, since Newtonian gravity works remarkably well for non-relativistic systems, or in general when the gravitational field is weak, in formulating the new theory we shall require that in the weak-field limit the new equations reduce to Poisson's equation for the Newtonian potential Φ, namely

$$\nabla^2\Phi = 4\pi G\rho\,, \tag{6.2}$$

where ρ is the matter density, and ∇^2 is Laplace's operator in flat space which, in Cartesian coordinates, is (see Box 3-E)

$$\nabla^2 = \frac{\partial^2}{\partial x^2} + \frac{\partial^2}{\partial y^2} + \frac{\partial^2}{\partial z^2}\,. \tag{6.3}$$

Following these prescriptions, we shall start by showing how the new equations of gravity look like in the weak-field, stationary limit.

[1] John Archibal Wheeler had aptly put this property as *"Spacetime tells matter how to move, matter tells spacetime how to curve."*

6.1 GEODESIC EQUATIONS IN THE WEAK-FIELD, STATIONARY LIMIT

Consider a non-relativistic particle moving in a *weak and stationary* gravitational field. Let τ/c be the proper time of the particle. Since $v \ll c$, it follows that

$$\frac{dx^i}{dt} \ll c \quad \rightarrow \quad \frac{dx^i}{d\tau} \ll c\frac{dt}{d\tau} = \frac{dx^0}{d\tau}\,; \tag{6.4}$$

consequently, in this limit the geodesic equations become

$$\frac{d^2x^\mu}{d\tau^2} + \Gamma^\mu_{\alpha\beta}\frac{dx^\alpha}{d\tau}\frac{dx^\beta}{d\tau} = 0 \quad \rightarrow \quad \frac{d^2x^\mu}{d\tau^2} + \Gamma^\mu_{00}\left(c\frac{dt}{d\tau}\right)^2 = 0\,. \tag{6.5}$$

From the expressions of the affine connection in terms of $g_{\mu\nu}$ given in Eq. 3.68, we easily find

$$\Gamma^\mu_{00} = \frac{1}{2}g^{\mu\sigma}\left(2g_{0\sigma,0} - g_{00,\sigma}\right)\,. \tag{6.6}$$

In addition, since the field is stationary $g_{0\sigma,0} = 0$, and the above expression simplifies to

$$\Gamma^\mu_{00} = -\frac{1}{2}g^{\mu\sigma}\frac{\partial g_{00}}{\partial x^\sigma}\,. \tag{6.7}$$

Since we have assumed that the gravitational field is weak, we can choose a coordinate system such that

$$g_{\mu\nu} = \eta_{\mu\nu} + h_{\mu\nu}, \qquad |h_{\mu\nu}| \ll 1\,, \tag{6.8}$$

where $h_{\mu\nu}$ is a small perturbation of the flat metric. In other words, we are assuming that the field is so weak that the metric is nearly flat. Any quantity depending on the metric can be expanded in powers of $h_{\mu\nu}$; in such expansion, we shall denote the n-th order in $h_{\mu\nu}$ as $O(h^n)$. In the following we shall retain only first-order terms in $h_{\mu\nu}$, i.e. we shall neglect terms of order $O(h^2)$. On this assumption the inverse metric is

$$g^{\mu\nu} = \eta^{\mu\nu} - h^{\mu\nu} + O(h^2)\,. \tag{6.9}$$

Indeed, with this definition,

$$g^{\mu\nu}g_{\nu\alpha} = (\eta^{\mu\nu} - h^{\mu\nu})(\eta_{\nu\alpha} + h_{\nu\alpha}) = \delta^\mu_\alpha + O(h^2)\,. \tag{6.10}$$

Consequently, we shall raise and lower indices of quantities of order $O(h)$ with the flat metric $\eta^{\mu\nu}$. Indeed

$$h^\lambda{}_\nu = g^{\lambda\rho}h_{\rho\nu} = (\eta^{\mu\nu} - h^{\mu\nu})h_{\rho\nu} = \eta^{\lambda\rho}h_{\rho\nu} + O(h^2)\,. \tag{6.11}$$

Thus, neglecting $O(h^2)$ terms, Eq. 6.7 gives

$$\Gamma^\mu_{00} = -\frac{1}{2}\eta^{\mu\sigma}\frac{\partial h_{00}}{\partial x^\sigma}\,, \tag{6.12}$$

and the geodesic equations become

$$\frac{d^2x^\mu}{d\tau^2} = \frac{1}{2}\eta^{\mu\alpha}\frac{\partial h_{00}}{\partial x^\alpha}\left(c\frac{dt}{d\tau}\right)^2\,, \tag{6.13}$$

or, splitting the time- and the space-components,

$$\frac{d^2ct}{d\tau^2} = -\frac{1}{2}\frac{\partial h_{00}}{\partial ct}\left(c\frac{dt}{d\tau}\right)^2 \tag{6.14}$$

$$\frac{d^2\mathbf{x}}{d\tau^2} = \frac{1}{2}\nabla h_{00}\left(c\frac{dt}{d\tau}\right)^2\,, \tag{6.15}$$

where

$$\boldsymbol{\nabla} \equiv \left(\frac{\partial}{\partial x}, \frac{\partial}{\partial y}, \frac{\partial}{\partial z}\right) \tag{6.16}$$

is the gradient operator in Cartesian coordinates. Since we have assumed that the field is stationary ($\frac{\partial h_{00}}{\partial t} = 0$), the right-hand side of Eq. 6.14 vanishes and $c\,dt/d\tau = const.$ If we rescale the time coordinate in such a way that $c\frac{dt}{d\tau} = 1$, Eq. 6.15 becomes

$$\frac{d^2\mathbf{x}}{d\tau^2} = \frac{1}{2}\boldsymbol{\nabla} h_{00}\,. \tag{6.17}$$

Remeber that the corresponding Newtonian equation is

$$\frac{d^2\mathbf{x}}{dt^2} = -\boldsymbol{\nabla}\Phi\,, \tag{6.18}$$

where Φ is the gravitational potential, solution of the Poisson equation 6.2. By comparing Eqs. 6.17 and 6.18, and since $\tau = ct$, we find that the requirement that the geodesic equations reduce to the Newtonian equations yields:

$$h_{00} = -2\frac{\Phi}{c^2} + const\,, \quad \text{and} \quad g_{00} = -(1 + 2\frac{\Phi}{c^2})\,. \tag{6.19}$$

Thus, Eq. 6.19 establishes a relation between the 00-component of the metric tensor and the Newtonian potential.

This remarkable property gives us a hint on the form that the full field equations should have. Indeed, by applying the Laplace operator to the expression of g_{00} given in Eq. 6.19 and using the Poisson equation 6.2, we find

$$\nabla^2 g_{00} = -\frac{2}{c^2}\nabla^2\Phi = -\frac{8\pi G}{c^2}\rho\,. \tag{6.20}$$

Moreover, since we are considering non-relativistic particles, the matter distribution is described by the 00-component of the stress-energy tensor, i.e. $T^{00} = T_{00} \sim \rho c^2$ (see Box 5-C); consequently Eq. 6.20 becomes

$$\nabla^2 g_{00} = -\frac{8\pi G}{c^4}T_{00}\,. \tag{6.21}$$

As discussed in the previous section, the Principle of General Covariance imposes that the equations of the gravitational field must be written in a tensorial form. Thus, Eq. 6.21 has to be the weak-field, stationary limit of the 00-component of a *tensor equation.* This suggests that we should construct a rank-two tensor $G_{\mu\nu}$ starting from $g_{\mu\nu}$ and its derivatives, such that the field equations are

$$G_{\mu\nu} = \frac{8\pi G}{c^4}T_{\mu\nu}\,, \tag{6.22}$$

and that, in the weak-field limit and for a stationary field, the 00-component of these equations reduces to Eq. 6.21 which implies that, in this limit and neglecting $O(h^2)$ terms,

$$G_{00} = -\nabla^2 g_{00}\,. \tag{6.23}$$

Equation 6.22 has to apply to the general case of a gravitational field generated by an arbitrary distribution of energy and matter. Furthermore, since it is a tensor equation, it holds in any reference frame.

6.2 EINSTEIN'S FIELD EQUATIONS

We shall here construct the rank-two tensor $G_{\mu\nu}$ appearing in Eq. 6.22. To this aim we shall discuss and implement some reasonable properties of this tensor; we will also make use of the weak-field, stationary limit given in Eq. 6.23. This heuristic derivation follows the original derivation by Einstein in 1918 [44]. Einstein's equations were also obtained, independently, by Hilbert (see footnote 3) using the variational approach which will be discussed in Chapter 7.

Let us first discuss the differential structure of $G_{\mu\nu}$ in terms of derivatives of $g_{\mu\nu}$. Since in the weak-field, stationary limit $G_{00} \to -\nabla^2 g_{00}$, we expect $G_{\mu\nu}$ to depend at most on the second derivatives of $g_{\mu\nu}$. A further motivation to exclude a dependence on higher-order derivatives comes from the theory of partial differential equations, which shows that field equations of third or higher order generally lead to instabilities (indeed, all known theories of fundamental interactions are described by field equations of order no higher than two). This point will be further discussed in Box 7-D.

The tensor $G_{\mu\nu}$ is the geometric object which describes the gravitational field, therefore it should depend only on the metric tensor and its derivatives. Moreover, it is reasonable to assume that it does not contain any other dimensionful quantity. Since $g_{\mu\nu}$ is dimensionless and $G_{\mu\nu}$ has the dimensions of the inverse of a squared length (see Eq. 6.23), $G_{\mu\nu}$ has to be *linear* in the second derivatives of the metric, and *quadratic* in the first derivatives. Indeed, suppose that $G_{\mu\nu}$ contains terms of the following schematic type

$$\frac{\partial g_{\mu\nu}}{\partial x_\rho}\frac{\partial g_{\alpha\beta}}{\partial x_\sigma}\frac{\partial g_{\gamma\delta}}{\partial x_\tau}, \qquad \frac{\partial^2 g_{\mu\nu}}{\partial x_\sigma^2}, \qquad \frac{\partial g_{\mu\nu}}{\partial x_\rho}\frac{\partial g_{\alpha\beta}}{\partial x_\sigma}, \qquad \frac{\partial g_{\mu\nu}}{\partial x_\rho}, \qquad g_{\mu\nu}\,. \tag{6.24}$$

In order to be dimensionally homogeneous, each term should be multiplied by a constant having the dimensions of a suitable power of a length, e.g.

$$\frac{\partial g_{\mu\nu}}{\partial x_\rho}\frac{\partial g_{\alpha\beta}}{\partial x_\sigma}\frac{\partial g_{\gamma\delta}}{\partial x_\tau}l, \qquad \frac{\partial^2 g_{\mu\nu}}{\partial x_\sigma^2}, \qquad \frac{\partial g_{\mu\nu}}{\partial x_\rho}\frac{\partial g_{\alpha\beta}}{\partial x_\sigma}, \qquad \frac{\partial g_{\mu\nu}}{\partial x_\rho}\frac{1}{l}, \qquad g_{\mu\nu}\frac{1}{l^2}\,; \tag{6.25}$$

since we require that $G_{\mu\nu}$ does not depend on any dimensionful constant, only terms like $\frac{\partial^2 g_{\mu\nu}}{\partial x_\sigma^2}$ and $\frac{\partial g_{\mu\nu}}{\partial x_\rho}\frac{\partial g_{\alpha\beta}}{\partial x_\sigma}$ are allowed.

Let us summarize the assumptions that we need to make on $G_{\mu\nu}$ on the basis of Eqs. 6.22 and 6.23, and of the discussion above:

1. $G_{\mu\nu}$ must be a tensor and, as $T_{\mu\nu}$, must be symmetric; moreover, in the weak-field, stationary limit it must reduce to Eq. 6.23, i.e.

$$G_{00} \sim -\nabla^2 g_{00}\,. \tag{6.26}$$

2. It must be linear in the second derivatives of $g_{\mu\nu}$, it can contain terms quadratic in the first derivatives, and no linear terms in $g_{\mu\nu}$ should be included.

3. Since $T_{\mu\nu}$ satisfies the divergenceless equation $T^{\mu\nu}{}_{;\mu} = 0$, $G_{\mu\nu}$ must satisfy the same equation

$$G^{\mu\nu}{}_{;\nu} = 0\,. \tag{6.27}$$

As discussed below, there exists a theorem, due to Lovelock, which guarantees that, under the above assumptions, $G_{\mu\nu}$ is in fact unique.

In Chapter 4 we introduced the Riemann tensor and we showed that it carries the information on the curvature of the spacetime. It has the differential structure which we require for the tensor $G_{\mu\nu}$, i.e. it is linear in the second derivatives of $g_{\mu\nu}$, is quadratic in the first derivatives, and contains no linear terms in $g_{\mu\nu}$. Therefore, the Riemann tensor is a good starting candidate for the left-hand side of the gravitational field equations. However, it is a tensor of rank four, whereas the stress-energy tensor is of rank two.

As discussed in Chapter 4, by contracting the Riemann tensor with the metric we can construct the Ricci tensor (Eq. 4.19)

$$R_{\mu\nu} = g^{\alpha\beta} R_{\alpha\mu\beta\nu} = R^{\alpha}{}_{\mu\alpha\nu} \,, \tag{6.28}$$

which is a symmetric tensor, and the scalar curvature (Eq. 4.20)

$$R = g^{\alpha\beta} R_{\alpha\beta} = R^{\alpha}{}_{\alpha} \,. \tag{6.29}$$

It can be shown, by using the symmetries of the Riemann tensor, that $R_{\mu\nu}$ and R are the only second-rank tensors and scalars that can be constructed by contraction of $R_{\alpha\mu\beta\nu}$ with the metric tensor. Moreover, both $R_{\mu\nu}$ and R have the same differential structure of $R_{\alpha\mu\beta\nu}$. This suggests to write $G_{\mu\nu}$ as a linear combination of $R_{\mu\nu}$ and R

$$G_{\mu\nu} = A R_{\mu\nu} + B g_{\mu\nu} R \,, \tag{6.30}$$

where A and B are constants to be determined. The tensor $G_{\mu\nu}$ is symmetric, as required by the condition 1 in the above list, and has the differential structure imposed by condition 2. Furthermore, condition 3 requires that

$$G^{\mu\nu}{}_{;\nu} = A \left(R^{\mu\nu} + \frac{B}{A} g^{\mu\nu} R \right)_{;\nu} = 0 \,. \tag{6.31}$$

In order to find whether Eq. 6.31 can be satisfied, we shall use the Bianchi identities (see Sec. 4.4)

$$R_{\lambda\mu\nu\beta;\eta} + R_{\lambda\mu\eta\nu;\beta} + R_{\lambda\mu\beta\eta;\nu} = 0 \,. \tag{6.32}$$

By contracting these equations with $g^{\lambda\nu}$, and recalling that the covariant derivative of the metric tensor vanishes, we find

$$\begin{aligned} g^{\lambda\nu} \left(R_{\lambda\mu\nu\beta;\eta} + R_{\lambda\mu\eta\nu;\beta} + R_{\lambda\mu\beta\eta;\nu} \right) &= g^{\lambda\nu} \left(R_{\lambda\mu\nu\beta;\eta} - R_{\lambda\mu\nu\eta;\beta} \right) + g^{\lambda\nu} R_{\lambda\mu\beta\eta;\nu} \\ &= R_{\mu\beta;\eta} - R_{\mu\eta;\beta} + g^{\lambda\nu} R_{\lambda\mu\beta\eta;\nu} = 0 \,. \end{aligned} \tag{6.33}$$

Contracting once more,

$$g^{\mu\beta} \left(R_{\mu\beta;\eta} - R_{\mu\eta;\beta} \right) + g^{\lambda\nu} g^{\mu\beta} R_{\lambda\mu\beta\eta;\nu} = R_{;\eta} - R^{\beta}{}_{\eta;\beta} - R^{\nu}{}_{\eta;\nu} = 0 \,. \tag{6.34}$$

Contracting with $g^{\eta\alpha}$, the last expression can be rewritten in the following form

$$R^{\beta\alpha}{}_{;\beta} - \frac{1}{2} g^{\eta\alpha} R_{;\eta} = 0 \,,$$

which, relabeling the indices, becomes

$$\left(R^{\alpha\beta} - \frac{1}{2} g^{\alpha\beta} R \right)_{;\beta} = 0 \,. \tag{6.35}$$

Therefore, the Bianchi identities imply that, if

$$\frac{B}{A} = -\frac{1}{2} \,, \tag{6.36}$$

the equation

$$G^{\mu\nu}{}_{;\nu} = 0 \tag{6.37}$$

is *identically* satisfied. Thus,

$$G_{\mu\nu} = A\left(R_{\mu\nu} - \frac{1}{2}g_{\mu\nu}R\right). \tag{6.38}$$

We remark that Eq. 6.37 is an identity: any $g_{\mu\nu}$ has to satisfy this equation, even if it does not satisfy Einstein's equations (in the language of particle physics, it can be said that Eq. 6.37 is satisfied "off shell").

We still have to find A. To this aim we will use the condition 1 of the above list, which imposes that in the limit of weak, stationary field

$$G_{00} = A\left(R_{00} - \frac{1}{2}g_{00}R\right) \sim -\nabla^2 g_{00}. \tag{6.39}$$

Since the field is weak, as in Sec. 6.1 we shall assume that $g_{\mu\nu} = \eta_{\mu\nu} + h_{\mu\nu}$, with $|h_{\mu\nu}| \ll 1$ and that $g^{\mu\nu} = \eta^{\mu\nu} - h^{\mu\nu} + O(h^2)$. Under these conditions, the Christoffel symbols become

$$\Gamma^{\alpha}_{\mu\nu} = \eta^{\alpha\sigma}\left(h_{\sigma\mu,\nu} + h_{\sigma\nu,\mu} - h_{\mu\nu,\sigma}\right) + O(h^2). \tag{6.40}$$

The expression of the Ricci tensor is (Eq. 4.54)

$$R_{\mu\nu} = R^{\alpha}{}_{\mu\alpha\nu} = \Gamma^{\alpha}_{\mu\nu,\alpha} - \Gamma^{\alpha}_{\mu\alpha,\nu} - \Gamma^{\alpha}_{\sigma\nu}\Gamma^{\sigma}_{\mu\alpha} + \Gamma^{\alpha}_{\sigma\alpha}\Gamma^{\sigma}_{\mu\nu}, \tag{6.41}$$

and from Eq. 6.40 it follows that the terms which contain products of Γ's in Eq. 6.41 are of order $O(h^2)$, and can be neglected. Thus, in the weak field limit only the terms linear in the second derivatives of the metric tensor survive, and the Ricci tensor can be written as

$$R_{\mu\nu} = \frac{1}{2}\eta^{\alpha\rho}\left(h_{\mu\alpha,\rho\nu} + h_{\rho\nu,\mu\alpha} - h_{\mu\nu,\rho\alpha} - h_{\rho\alpha,\mu\nu}\right) + O(h^2). \tag{6.42}$$

The 00 component of $R_{\mu\nu}$ therefore is

$$R_{00} = \frac{1}{2}\eta^{\alpha\rho}\left(2h_{0\alpha,0\rho} - h_{00,\rho\alpha} - h_{\rho\alpha,00}\right) + O(h^2). \tag{6.43}$$

To hereafter we shall omit the term $O(h^2)$ for simplicity. If the field is stationary, the time derivatives of the metric tensor vanish, and Eq. 6.43 becomes

$$R_{00} = -\frac{1}{2}\eta^{ij}h_{00,ij} = -\frac{1}{2}\nabla^2 g_{00} \qquad i,j = 1,2,3. \tag{6.44}$$

In order to compute G_{00} we still need to compute R. In the weak-field limit [2]

$$|T_{ij}| \ll |T_{00}|. \tag{6.47}$$

We shall now compute the *trace* of $T_{\mu\nu}$, which is found by contracting the stress-energy tensor with the metric tensor, i.e.

$$T = g^{\mu\nu}T_{\mu\nu} \simeq \eta^{\mu\nu}T_{\mu\nu} = -T_{00} + \delta^{ij}T_{ij} \simeq -T_{00}. \tag{6.48}$$

---

[2] Consider for example the system on non-interacting, massive particles discussed in Chapter 5. Let ρ be the mass density,

$$\rho = \sum_n m_n \delta^3(\mathbf{r} - \mathbf{r}_n), \tag{6.45}$$

where $\mathbf{r}_n$ indicates the position of the n-th particle; in the weak-field limit the stress-energy tensor given in Eq. 5.10 can be written as

$$T^{\mu\nu} = \rho c^2 \frac{dx^\mu}{d\tau}\frac{dx^\nu}{d\tau}. \tag{6.46}$$

Since $\frac{dx^i}{d\tau} \ll \frac{dx^0}{d\tau}$ with $i = 1,2,3$ (see Eq. 6.4), the dominant term in $T^{\mu\nu}$ is T^{00}.

In this equation we have assumed that the stress-energy tensor, which is the source of the gravitational field, is of order h, consistently with the weak-field assumption. This will be discussed in more detail in Chapter 12. By taking the trace of the equation $G_{\mu\nu} = \frac{8\pi G}{c^4} T_{\mu\nu}$, we get

$$g^{\mu\nu} A \left(R_{\mu\nu} - \frac{1}{2} g_{\mu\nu} R \right) = \frac{8\pi G}{c^4} T \quad \rightarrow \quad -AR = \frac{8\pi G}{c^4} T \,. \tag{6.49}$$

In Eq. 6.49 we have used the property

$$g^{\mu\nu} g_{\mu\nu} = 4 \,, \tag{6.50}$$

which can easily be proved in a LIF, where $g_{\mu\nu} = \eta_{\mu\nu}$. Since $g^{\mu\nu} g_{\mu\nu}$ is a scalar quantity its value is the same in any frame. Alternatively, Eq. 6.50 can be proved by taking the trace of Eq. 2.238, $g^{\mu\nu} g_{\nu\alpha} = \delta^{\mu}{}_{\alpha}$.

Using Eq. 6.48, Eq. 6.49 yields

$$-AR = -\frac{8\pi G}{c^4} T_{00} \,, \tag{6.51}$$

and since the right-hand side of this equation is $-G_{00}$ we find

$$-AR = -A \left(R_{00} - \frac{1}{2} \eta_{00} R \right) \quad \rightarrow \quad R = 2R_{00} \,. \tag{6.52}$$

Using this relation and Eq. 6.44 we can finally compute G_{00}

$$G_{00} = A \left(R_{00} - \frac{1}{2} \eta_{00} R \right) = 2AR_{00} \quad \rightarrow \quad G_{00} = -A \nabla^2 g_{00} \,. \tag{6.53}$$

Comparing this equation with Eq. 6.26, we find that the relativistic field equations reduce, in the weak-field, stationary limit, to the Newtonian equations if

$$A = 1 \,. \tag{6.54}$$

Thus, in conclusion, **Einstein's equations** are[3]

$$G_{\mu\nu} = \frac{8\pi G}{c^4} T_{\mu\nu} \,, \tag{6.55}$$

where

$$G_{\mu\nu} = R_{\mu\nu} - \frac{1}{2} g_{\mu\nu} R \tag{6.56}$$

is called the **Einstein tensor**.

Einstein's equations can also be written in an alternative form. By taking the trace of Eq. 6.55

$$R = -\frac{8\pi G}{c^4} T \,, \tag{6.57}$$

and replacing this expression in the Einstein tensor, Eq. 6.55 becomes

$$R_{\mu\nu} = \frac{8\pi G}{c^4} \left(T_{\mu\nu} - \frac{1}{2} g_{\mu\nu} T \right) \,. \tag{6.58}$$

[3] Although we call these equations the Einstein equations, they were derived independently by Hilbert in the same year using a variational approach (see Chapter 7). However, Einstein showed the implications of these equations for the theory of the solar system, and in particular he showed that the precession of the perihelion of Mercury has a relativistic origin (see Chapter 11). This led to the acceptance of the theory, and since then the equations have been called the Einstein equations.

In vacuum $T_{\mu\nu} = 0$, and Eqs. 6.55 and 6.58 provide two equivalent forms of the Einstein equations

$$G_{\mu\nu} = 0\,, \qquad R_{\mu\nu} = 0\,. \tag{6.59}$$

These equations are equivalent because in vacuum $R = 0$ (see Eq. 6.57), and the Einstein and the Ricci tensors coincide. Therefore, in vacuum the Ricci scalar, the Ricci tensor, and the Einstein tensor all vanish, but the Riemann tensor does not, unless the gravitational field vanishes or is constant and uniform.

The above heuristic derivation of Einstein's equations might seem "ad hoc" and one might wonder whether there are other geometrical quantities that can be used in place of the Einstein tensor on the left-hand side of Eq. 6.55. A remarkable theorem, due to Lovelock (1972), proves that the above expression of the Einstein tensor is unique. Lovelock's theorem can be stated as follows

> *In four spacetime dimensions the only divergence-free, symmetric, rank-2 tensor constructed solely from the metric $g_{\mu\nu}$ and its derivatives up to second differential order, and preserving coordinate invariance, is the Einstein tensor plus a term proportional to $g_{\mu\nu}$.*

In other words, Lovelock's theorem shows that General Relativity emerges as the *unique* theory of gravity under the above assumptions[4].

Box 6-A

The cosmological constant

As proved by Lovelock's theorem, one may add to the Einstein tensor given in Eq. 6.56 a term proportional to $g_{\mu\nu}$, such that Einstein's equations would become

$$R_{\mu\nu} - \frac{1}{2} g_{\mu\nu} R + \Lambda g_{\mu\nu} = \frac{8\pi G}{c^4} T_{\mu\nu}\,, \tag{6.60}$$

where Λ is a constant. With this term $G_{\mu\nu}$ violates the condition 2 in the list of constraints that the Einstein tensor must satisfy; in addition Eq. 6.60 does not reduce to Newton's equations in the weak-field, stationary limit, as required by Eq. 6.26, unless Λ is extremely small. Such term is related to the *cosmological constant*, introduced by Einstein himself, and plays a crucial role in cosmology. In 1998 two independent experiments, using the observations of distant supernovae, discovered that the Universe is expanding at an increasing rate. This result (that was awarded the Nobel Prize in Physics in 2011) can be explained by a positive cosmological constant. The current measured value is $\Lambda \approx 1.11 \times 10^{-56}\,\text{cm}^{-2}$. The value of the cosmological constant is so small that it plays a role only on cosmological scales, whereas it can be safely neglected at astrophysical scales and in the study of compact objects. We shall therefore neglect such a term in the rest of the book. For an introduction on the cosmological implications of Einstein's equations, we refer to other monographs, e.g. [29].

---

[4] Although beyond the scope of this book, we would like to mention that Lovelock's theorem also provides a basis for possible modifications and extensions of general relativity, which can be defined by relaxing some of the assumptions of the theorem (for an overview, see [21]).

6.3 GAUGE INVARIANCE OF EINSTEIN'S EQUATIONS

Since the tensor $G_{\mu\nu}$ is symmetric, it has ten independent components. Therefore, Einstein's equations provide ten partial differential equations for the ten independent components of $g_{\mu\nu}$. However, these equations are not independent; indeed, as shown in the previous section, the Bianchi identities imply that $G^{\mu\nu}{}_{;\nu} = 0$, providing four additional conditions that the Einstein tensor must satisfy. Thus, the number of independent equations reduces to six. At first, this might seem a problem, since we now have six independent equations and ten unknown functions. However, it turns out that four of the components of the metric tensor are not physical, due to the invariance of the theory under coordinate transformations. In order to show this crucial point, let $g_{\mu\nu}$ be a solution of Einstein's equations. If we make a generic coordinate transformation, $x^{\mu'} = x^{\mu'}(x^\alpha)$, the "transformed" tensor $g'_{\mu\nu} = g_{\mu'\nu'}$ must still be a solution, as established by the Principle of General Covariance. This also means that $g_{\mu\nu}$ and $g'_{\mu\nu}$ represent the same physical solution (i.e., the same geometry) seen in different reference frames. The coordinate transformation involves four arbitrary functions $x^{\mu'}(x^\alpha)$; therefore the four non-physical degrees of freedom arise from the freedom of choosing the coordinate system, and disappear when we fix it. For example, we may choose a frame where four out of the ten components of $g_{\mu\nu}$ are identically zero; or, we may choose a frame in which four relations among the ten components are assigned, making only six of them really independent. In other words, Einstein's equations do not determine the solution $g_{\mu\nu}$ in a unique way, but only up to an arbitrary coordinate transformation.

A similar situation arises in the case of Maxwell's equations which, in terms of the vector potential A^μ, are

$$\Box_F A_\alpha - \frac{\partial^2 A^\beta}{\partial x^\alpha \partial x^\beta} = -\frac{4\pi}{c} J_\alpha \,, \tag{6.61}$$

where

$$\Box_F = -\frac{\partial^2}{c^2 \partial t^2} + \nabla^2 = \eta^{\alpha\beta} \frac{\partial}{\partial x^\alpha} \frac{\partial}{\partial x^\beta} \tag{6.62}$$

is the *d'Alembertian operator in flat spacetime.* These are four equations for the four components of A^μ. However, these equations are not independent, because the conservation law

$$J^\mu{}_{,\mu} = 0 \tag{6.63}$$

implies the constraint

$$\frac{\partial}{\partial x^\mu} \left(\Box_F A^\mu - \eta^{\mu\alpha} \frac{\partial^2 A^\beta}{\partial x^\alpha \partial x^\beta} \right) = 0 \,. \tag{6.64}$$

This equation, which is an identity valid "off-shell", plays the same role as the Bianchi identities in our context. Since the number of independent Maxwell equations is three, the vector potential A_α cannot be determined uniquely. This implies the existence of a non-physical degree of freedom, which corresponds to the gauge invariance of Maxwell's equations. Indeed, if A_α is a solution then, for an arbitrary scalar function Φ,

$$A'_\alpha = A_\alpha + \frac{\partial \Phi}{\partial x^\alpha} \,, \tag{6.65}$$

is also a solution. This can be directly checked upon substitution in Maxwell's equations,

$$\Box_F A'_\alpha - \frac{\partial}{\partial x^\alpha} \Box_F \Phi - \frac{\partial^2 A'^\beta}{\partial x^\alpha \partial x^\beta} + \eta^{\beta\delta} \frac{\partial^2}{\partial x^\alpha \partial x^\beta} \frac{\partial \Phi}{\partial x^\delta} = -\frac{4\pi}{c} J_\alpha \,, \tag{6.66}$$

and since the second and the last term on the left-hand side cancel, we obtain

$$\Box_F A'_\alpha - \frac{\partial^2 A'^\beta}{\partial x^\alpha \partial x^\beta} = -\frac{4\pi}{c} J_\alpha \,. \tag{6.67}$$

Since Φ is arbitrary, we can choose it to simplify the final form of the equations. For example by imposing the so-called *Lorenz gauge*:

$$\frac{\partial}{\partial x^\beta} A'^\beta = 0\,. \tag{6.68}$$

With this gauge choice, Eq. 6.67 becomes

$$\Box_F A'_\alpha = -\frac{4\pi}{c} J_\alpha\,. \tag{6.69}$$

To summarize, in the electromagnetic case the extra degree of freedom of A_μ is due to the fact that the vector potential is defined up to an arbitrary scalar function Φ. In our case the *four* extra degrees of freedom of the metric tensor are due to the fact that $g_{\mu\nu}$ is defined up to a coordinate transformation that involves four scalar functions. This gauge freedom is particularly useful when one is looking for solutions of Einstein's equations, or in the study of gravitational waves, as we shall discuss in Chapter 12.

6.4 THE HARMONIC GAUGE

As discussed above for Maxwell's equations, the gauge invariance of Einstein's equations can be used to simplify them. A notable gauge choice is defined by the condition

$$\Gamma^\lambda \equiv g^{\mu\nu}\Gamma^\lambda_{\mu\nu} = 0\,. \tag{6.70}$$

This is called **harmonic gauge**. As we shall see in Chapter 12, this gauge is of particular interest in the study of gravitational waves, because it simplifies the equations which govern their propagation in a way similar to that of Maxwell's equations in the Lorenz gauge. We shall now show that it is always possible to choose this gauge. Given a generic coordinate transformation, the affine connection $\Gamma^\alpha_{\beta\gamma}$ transforms as (see Eq. 3.43)

$$\Gamma^{\lambda'}_{\mu'\nu'} = \frac{\partial x^{\lambda'}}{\partial x^\rho}\frac{\partial x^\tau}{\partial x^{\mu'}}\frac{\partial x^\sigma}{\partial x^{\nu'}}\Gamma^\rho_{\tau\sigma} + \frac{\partial x^{\lambda'}}{\partial x^\sigma}\frac{\partial^2 x^\sigma}{\partial x^{\mu'}\partial x^{\nu'}}\,. \tag{6.71}$$

When contracted with $g^{\mu'\nu'}$, this equation gives

$$\Gamma^{\lambda'} = \frac{\partial x^{\lambda'}}{\partial x^\rho}\Gamma^\rho + g^{\mu'\nu'}\frac{\partial x^{\lambda'}}{\partial x^\sigma}\frac{\partial^2 x^\sigma}{\partial x^{\mu'}\partial x^{\nu'}}\,. \tag{6.72}$$

The last term can be written in the following form

$$\begin{aligned} g^{\mu'\nu'}\frac{\partial x^{\lambda'}}{\partial x^\sigma}\frac{\partial}{\partial x^{\mu'}}\left(\frac{\partial x^\sigma}{\partial x^{\nu'}}\right) &= g^{\mu'\nu'}\left[\frac{\partial}{\partial x^{\mu'}}\left(\frac{\partial x^{\lambda'}}{\partial x^\sigma}\frac{\partial x^\sigma}{\partial x^{\nu'}}\right) - \frac{\partial x^\sigma}{\partial x^{\nu'}}\frac{\partial^2 x^{\lambda'}}{\partial x^{\mu'}\partial x^\sigma}\right] \\ &= g^{\mu'\nu'}\left[\frac{\partial}{\partial x^{\mu'}}\delta^{\lambda'}{}_{\nu'} - \frac{\partial x^\sigma}{\partial x^{\nu'}}\frac{\partial x^\rho}{\partial x^{\mu'}}\frac{\partial^2 x^{\lambda'}}{\partial x^\rho \partial x^\sigma}\right], \end{aligned} \tag{6.73}$$

from which we find

$$\Gamma^{\lambda'} = \frac{\partial x^{\lambda'}}{\partial x^\rho}\Gamma^\rho - g^{\rho\sigma}\frac{\partial^2 x^{\lambda'}}{\partial x^\rho \partial x^\sigma}\,. \tag{6.74}$$

If Γ^ρ is non-zero, we can always find a frame where $\Gamma^{\lambda'} = 0$ and reduce to the harmonic gauge, because Eq. 6.74 (with $\Gamma^{\lambda'} = 0$) can be seen as a system of partial differential equations for the functions $x^{\mu'}(x^\alpha)$ [5].

[5] The existence and unicity of the solutions of this system can be proved, but this is beyond the scope of this book.

The condition $\Gamma^\rho = 0$ can be rewritten in a more elegant form if we use the expression of the affine connection in terms of the metric tensor,

$$\Gamma^\rho = \frac{1}{2} g^{\mu\nu} g^{\rho\beta} \left[\frac{\partial g_{\beta\mu}}{\partial x^\nu} + \frac{\partial g_{\beta\nu}}{\partial x^\mu} - \frac{\partial g_{\mu\nu}}{\partial x^\beta} \right] . \tag{6.75}$$

Since

$$g^{\rho\beta} \frac{\partial g_{\beta\mu}}{\partial x^\nu} = -g_{\beta\mu} \frac{\partial g^{\rho\beta}}{\partial x^\nu} , \tag{6.76}$$

and

$$\frac{1}{2} g^{\mu\nu} \frac{\partial g_{\mu\nu}}{\partial x^\beta} = \frac{1}{\sqrt{-g}} \frac{\partial}{\partial x^\beta} \sqrt{-g} , \tag{6.77}$$

it follows that

$$\Gamma^\rho = \frac{1}{2} g^{\mu\nu} \left[-g_{\beta\mu} \left(\frac{\partial g^{\rho\beta}}{\partial x^\nu} \right) - g_{\beta\nu} \left(\frac{\partial g^{\rho\beta}}{\partial x^\mu} \right) \right] - \frac{g^{\rho\beta}}{\sqrt{-g}} \frac{\partial}{\partial x^\beta} \sqrt{-g} . \tag{6.78}$$

The term in brackets is symmetric in μ and ν, therefore

$$\Gamma^\rho = -\frac{1}{2} \left[2 g^{\mu\nu} g_{\beta\mu} \frac{\partial g^{\rho\beta}}{\partial x^\nu} \right] - \frac{g^{\rho\beta}}{\sqrt{-g}} \frac{\partial}{\partial x^\beta} \sqrt{-g} \tag{6.79}$$

and, since $g^{\mu\nu} g_{\beta\mu} = \delta^\nu{}_\beta$, we get

$$\Gamma^\rho = -\frac{\partial g^{\rho\beta}}{\partial x^\beta} - \frac{g^{\rho\beta}}{\sqrt{-g}} \frac{\partial}{\partial x^\beta} \sqrt{-g} = -\frac{1}{\sqrt{-g}} \frac{\partial}{\partial x^\beta} \left(\sqrt{-g} g^{\rho\beta} \right) , \tag{6.80}$$

from which we find that

$$\Gamma^\rho = 0 \qquad \rightarrow \qquad \frac{\partial}{\partial x^\beta} \left(\sqrt{-g} g^{\rho\beta} \right) = 0 . \tag{6.81}$$

The reason why this gauge is called "harmonic" is the following. A function Φ is harmonic if

$$\Box \Phi = 0, \tag{6.82}$$

where $\Box$ is the *d'Alembertian operator in curved spacetime* defined as

$$\Box \Phi = g^{\rho\beta} \nabla_\rho \nabla_\beta \Phi , \tag{6.83}$$

and we remind that ∇_ρ is the covariant derivative. Note that

$$\begin{aligned} g^{\rho\beta} \nabla_\rho \nabla_\beta \Phi = g^{\rho\beta} \left(\frac{\partial \Phi_{;\beta}}{\partial x^\rho} - \Gamma^\alpha_{\beta\rho} \Phi_{;\alpha} \right) = \\ g^{\rho\beta} \left(\frac{\partial^2 \Phi}{\partial x^\rho \partial x^\beta} - \Gamma^\alpha_{\beta\rho} \frac{\partial \Phi}{\partial x^\alpha} \right) = g^{\rho\beta} \frac{\partial^2 \Phi}{\partial x^\rho \partial x^\beta} - \Gamma^\alpha \frac{\partial \Phi}{\partial x^\alpha} . \end{aligned} \tag{6.84}$$

Therefore, in the harmonic gauge ($\Gamma^\rho = 0$) Eq. 6.82 reduces to

$$\Box \Phi = g^{\rho\beta} \frac{\partial^2 \Phi}{\partial x^\beta \partial x^\rho} = 0 . \tag{6.85}$$

CHAPTER 7

Einstein's equations and variational principles

In this chapter we shall show that Einstein's equations can be derived from a variational principle, i.e. by defining an action for the gravitational field and by requiring this action to be stationary. We shall firstly recall how the variational approach can be applied in Special Relativity to derive Euler-Lagrange's equations for a given field. Using this approach, we shall then derive Einstein's equations in vacuum, and generalize the entire procedure to generic fields in the presence of gravity.

7.1 EULER-LAGRANGE'S EQUATIONS IN SPECIAL RELATIVITY

Let us consider a tensor field $\mathbf{\Phi}(x)$ in Special Relativity, where x denotes the spacetime point of coordinates $\{x^\mu\}$. The **action** for this field is a functional of $\mathbf{\Phi}(x)$ and of its first derivative, written as an integral of a Lagrangian density over a four-dimensional volume

$$S = \int d^4x \; \mathcal{L}\left(\mathbf{\Phi}, \partial_\mu \mathbf{\Phi}\right) . \tag{7.1}$$

Henceforth, we simplify the notation by defining $\partial_\mu \equiv \frac{\partial}{\partial x^\mu}$.

Let us consider a generic variation of the tensor field, $\delta\mathbf{\Phi}$, which we assume to vanish on the boundary of the integration volume or asymptotically, if the volume is infinite. The variation of the action, at first order in $\delta\mathbf{\Phi}$, is

$$\begin{aligned} \delta S &= \int d^4x \left(\frac{\partial \mathcal{L}}{\partial \mathbf{\Phi}} \delta\mathbf{\Phi} + \frac{\partial \mathcal{L}}{\partial\left(\partial_\mu \mathbf{\Phi}\right)} \delta \partial_\mu \mathbf{\Phi} \right) \\ &= \int d^4x \left(\frac{\partial \mathcal{L}}{\partial \mathbf{\Phi}} \delta\mathbf{\Phi} + \frac{\partial \mathcal{L}}{\partial\left(\partial_\mu \mathbf{\Phi}\right)} \partial_\mu (\delta \mathbf{\Phi}) \right) , \end{aligned} \tag{7.2}$$

where in the last step we have used the fact that the operations of variation and differentiation commute. Note that in this notation a sum over the indices of the tensor components (in any given frame) is assumed; for instance, if $\mathbf{\Phi}$ is a $\binom{0}{2}$ tensor with components – in a given frame – $\Phi_{\mu\nu}$, then

$$\frac{\partial \mathcal{L}}{\partial \mathbf{\Phi}} \delta\mathbf{\Phi} \equiv \frac{\partial \mathcal{L}}{\partial \Phi_{\mu\nu}} \delta\Phi_{\mu\nu} . \tag{7.3}$$

The last term of Eq. 7.2 can be integrated by parts

$$\int d^4x \frac{\partial \mathcal{L}}{\partial\left(\partial_\mu \mathbf{\Phi}\right)} \partial_\mu (\delta\mathbf{\Phi}) = \int d^4x \; \partial_\mu \left[\frac{\partial \mathcal{L}}{\partial\left(\partial_\mu \mathbf{\Phi}\right)} \delta\mathbf{\Phi} \right] - \int d^4x \; \partial_\mu \left[\frac{\partial \mathcal{L}}{\partial\left(\partial_\mu \mathbf{\Phi}\right)} \right] \delta\mathbf{\Phi} . \tag{7.4}$$

By Gauss' theorem, the volume integral of the four-divergence of $\frac{\partial \mathcal{L}}{\partial(\partial_\mu \mathbf{\Phi})}\delta\mathbf{\Phi}$ is equal to the integral of this quantity over the boundary of the volume[1]. Since $\delta\mathbf{\Phi}$ vanishes on the boundary, the first integral on the right-hand side of Eq. 7.4 vanishes and Eq. 7.2 becomes

$$\delta S = \int d^4x \left(\frac{\partial \mathcal{L}}{\partial \mathbf{\Phi}} - \partial_\mu \frac{\partial \mathcal{L}}{\partial\left(\partial_\mu \mathbf{\Phi}\right)} \right) \delta\mathbf{\Phi} \,. \tag{7.5}$$

The equations of motion for the field $\mathbf{\Phi}$ are then found by imposing the stationarity of δS

$$\delta S = 0, \qquad \forall\ \delta\mathbf{\Phi} \text{ vanishing on the boundary} \,. \tag{7.6}$$

Since the integral 7.5 has to vanish *for any* $\delta\mathbf{\Phi}(x)$ vanishing on the boundary, it follows that

$$\frac{\partial \mathcal{L}}{\partial \mathbf{\Phi}} - \partial_\mu \frac{\partial \mathcal{L}}{\partial\left(\partial_\mu \mathbf{\Phi}\right)} = 0 \,, \tag{7.7}$$

which are the **Euler-Lagrange equations** for the field $\mathbf{\Phi}$ in flat spacetime.

If the Lagrangian density depends on a collection of N tensor fields $\{\mathbf{\Phi}_A\}_{A=1,\ldots,N}$, a straightforward generalization of the above derivation yields the Euler-Lagrange equations for the tensor fields

$$\frac{\partial \mathcal{L}}{\partial \mathbf{\Phi}_A} - \partial_\mu \frac{\partial \mathcal{L}}{\partial\left(\partial_\mu \mathbf{\Phi}_A\right)} = 0 \qquad A = 1, \ldots, N \,. \tag{7.8}$$

Likewise, it can be shown that Euler-Lagrange's equations remain valid also if the Lagrangian density depends explicitly on the spacetime coordinates, i.e. if $\mathcal{L}\left(\mathbf{\Phi}_A, \partial_\mu \mathbf{\Phi}_A, x^\mu\right)$.

7.2 EULER-LAGRANGE'S EQUATIONS IN CURVED SPACETIME

Having derived Euler-Lagrange's equations in Special Relativity, we now move to curved spacetime. In this case, in addition to the fields $\mathbf{\Phi}_A$ $(A = 1, \ldots, N)$, the dynamics of the system also depend on the metric tensor $\mathbf{g}$, which describes the gravitational field.

Due to the strong Equivalence Principle (see Sec 1.3), in a locally inertial frame the dynamics of any field $\mathbf{\Phi}_A$ except gravity is described by the action 7.1. Therefore, according to the Principle of General Covariance, in a general frame the action (which is a scalar quantity) retains the same form as in the locally inertial frame, provided $\eta_{\mu\nu} \to g_{\mu\nu}$, partial derivatives ∂_μ are replaced by covariant derivatives ∇_μ, and the integration volume element d^4x is replaced by the *invariant volume element* $\sqrt{-g}\, d^4x$ introduced in Box 5-B. We recall that g is the determinant of the metric, and that $g < 0$ due to the Lorentzian signature. With these replacements, we shall now show that the results of the previous section remain valid.

We shall write the total action, describing both the gravitational field and a generic collection of fields $\{\mathbf{\Phi}_A\}$, as

$$S = S^{\text{EH}} + S^{\text{fields}} \,. \tag{7.9}$$

The first term, known as the Einstein-Hilbert action, describes the dynamics of the gravitational field in vacuum and depends only on $g_{\mu\nu}$ and its derivatives. In the next section we shall show that Einstein's equations in vacuum are obtained by imposing the stationarity of S^{EH} with respect to variations of $g_{\mu\nu}$. For the present discussion, it is sufficient to assume that S^{EH} does not depend on the fields $\{\mathbf{\Phi}_A\}$.

We write the second term in the above equation as

$$S^{\text{fields}} = \int d^4x \,\sqrt{-g}\,\mathcal{L}^{\text{fields}}\left(\mathbf{\Phi}_1, \nabla_\mu \mathbf{\Phi}_1, \ldots, \mathbf{\Phi}_N, \nabla_\mu \mathbf{\Phi}_N, \mathbf{g}\right) \,, \tag{7.10}$$

[1]The proof of Gauss' theorem in four-dimensional Minkowski's spacetime is the same as in Euclidean three-dimensional space (see Box 5-D).

where now the Lagrangian density $\mathcal{L}^{\text{fields}}$ depends explicitly on $\mathbf{g}$ because we have replaced $\eta_{\mu\nu}$ by $g_{\mu\nu}$ and ∂_μ by ∇_μ.

As in Special Relativity, the equations for a given field $\mathbf{\Phi}$ are found by varying the action with respect to that field, and since S^{EH} does not depend on $\mathbf{\Phi}$, we find

$$\begin{aligned} \delta S &\equiv \delta S^{\text{fields}} = \int d^4x \, \sqrt{-g} \left(\frac{\partial \mathcal{L}^{\text{fields}}}{\partial \mathbf{\Phi}} \delta \mathbf{\Phi} + \frac{\partial \mathcal{L}^{\text{fields}}}{\partial \left(\nabla_\mu \mathbf{\Phi} \right)} \delta \nabla_\mu \mathbf{\Phi} \right) \\ &= \int d^4x \, \sqrt{-g} \left(\frac{\partial \mathcal{L}^{\text{fields}}}{\partial \mathbf{\Phi}} \delta \mathbf{\Phi} + \frac{\partial \mathcal{L}^{\text{fields}}}{\partial \left(\nabla_\mu \mathbf{\Phi} \right)} \nabla_\mu \delta \mathbf{\Phi} \right) , \end{aligned} \tag{7.11}$$

where we have used the property $\delta \nabla_\mu = \nabla_\mu \delta$. Again, the last term in Eq. 7.11 can be integrated by parts

$$\begin{aligned} \int d^4x \, \sqrt{-g} \, \frac{\partial \mathcal{L}^{\text{fields}}}{\partial \left(\nabla_\mu \mathbf{\Phi} \right)} \nabla_\mu \delta \mathbf{\Phi} &= \int d^4x \, \sqrt{-g} \, \nabla_\mu \left[\frac{\partial \mathcal{L}^{\text{fields}}}{\partial \left(\nabla_\mu \mathbf{\Phi} \right)} \delta \mathbf{\Phi} \right] \\ &- \int d^4x \, \sqrt{-g} \, \nabla_\mu \left[\frac{\partial \mathcal{L}^{\text{fields}}}{\partial \left(\nabla_\mu \mathbf{\Phi} \right)} \right] \delta \mathbf{\Phi} . \end{aligned} \tag{7.12}$$

In order to show that the first integral on the right-hand side vanishes, we shall use Gauss' theorem generalized to curved spacetime as discussed in Box 7-A (see Eq. 7.19):

$$\int_\Omega d^4x \, \sqrt{-g} \, \nabla_\mu \left[\frac{\partial \mathcal{L}^{\text{fields}}}{\partial \left(\nabla_\mu \mathbf{\Phi} \right)} \delta \mathbf{\Phi} \right] = \int_{\partial \Omega} \frac{\partial \mathcal{L}^{\text{fields}}}{\partial \left(\nabla_\mu \mathbf{\Phi} \right)} \delta \mathbf{\Phi} dS_\mu = 0 ,$$

where in the last step we have assumed $\delta \mathbf{\Phi} = 0$ on the volume boundary $\partial \Omega$. Thus, Eq. 7.12 reduces to

$$\int d^4x \, \sqrt{-g} \, \frac{\partial \mathcal{L}^{\text{fields}}}{\partial \left(\nabla_\mu \mathbf{\Phi} \right)} \nabla_\mu \delta \mathbf{\Phi} = - \int d^4x \, \sqrt{-g} \, \nabla_\mu \left[\frac{\partial \mathcal{L}^{\text{fields}}}{\partial \left(\nabla_\mu \mathbf{\Phi} \right)} \right] \delta \mathbf{\Phi} ,$$

and Eq. 7.11 can be written as

$$\delta S \equiv \delta S^{\text{fields}} = \int d^4x \, \sqrt{-g} \left(\frac{\partial \mathcal{L}}{\partial \mathbf{\Phi}} - \nabla_\mu \frac{\partial \mathcal{L}}{\partial \left(\nabla_\mu \mathbf{\Phi} \right)} \right) \delta \mathbf{\Phi} . \tag{7.13}$$

Finally, by imposing

$$\delta S = 0, \qquad \forall \; \delta \mathbf{\Phi} \text{ vanishing on the boundary} , \tag{7.14}$$

we find the Euler-Lagrange equations for the field $\mathbf{\Phi}$, generalized to curved spacetime:

$$\frac{\partial \mathcal{L}}{\partial \mathbf{\Phi}} - \nabla_\mu \frac{\partial \mathcal{L}}{\partial \left(\nabla_\mu \mathbf{\Phi} \right)} = 0 . \tag{7.15}$$

In other words, we have proved that in curved spacetime Euler-Lagrange's equations have the same form as in flat spacetime (Eq. 7.7) with the replacement $\partial_\mu \to \nabla_\mu$ and $\eta_{\mu\nu} \to g_{\mu\nu}$.

Box 7-A

Gauss' theorem in curved space

Given a manifold $\mathcal{M}$ described by coordinates $\{x^\mu\}$, and a metric $g_{\mu\nu}$ on $\mathcal{M}$, let us consider a submanifold $\mathcal{N} \subset \mathcal{M}$ described by coordinates $\{y^i\}$, such that $x^\mu = x^\mu(y^i)$ on $\mathcal{N}$. We define the metric *induced* on $\mathcal{N}$ from $\mathcal{M}$ as

$$\gamma_{ij} \equiv \frac{\partial x^\mu}{\partial y^i}\frac{\partial x^\nu}{\partial y^j} g_{\mu\nu} \,. \tag{7.16}$$

With the above definition, Gauss' theorem in curved space can be stated as follows [a]. Let Ω be an open set of an n-dimensional manifold $\mathcal{M}$, described by the coordinates $\{x^\mu\}_{\mu=0,\ldots,n-1}$, and $g_{\mu\nu}$ the metric on Ω (note that an open set of a manifold has always the same dimensionality of the manifold itself; for this reason, we call Ω an "n-dimensional volume"). Let $\partial\Omega$ be the $(n-1)$-dimensional boundary of Ω, described by the coordinates $\{y^j\}_{j=0,\ldots,n-2}$ with normal vector n^μ (such that $|n_\mu n^\mu| = 1$ for timelike or spacelike vectors); let γ_{ij} be the metric induced on $\partial\Omega$ from $g_{\mu\nu}$. Then, for any vector field V^μ defined in Ω,

$$\int_\Omega d^n x \, \sqrt{-g} \, \nabla_\mu V^\mu = \int_{\partial\Omega} d^{n-1}y \, \sqrt{-\gamma} \, V^\mu n_\mu \,. \tag{7.17}$$

If we define the covariant surface element as

$$dS_\mu \equiv \sqrt{-\gamma} \, n_\mu d^{n-1} y \,, \tag{7.18}$$

Gauss' theorem can also be written as

$$\int_\Omega d^n x \, \sqrt{-g} \, \nabla_\mu V^\mu = \int_{\partial\Omega} V^\mu dS_\mu \,. \tag{7.19}$$

In particular, if one considers an infinite volume and if V^μ vanishes asymptotically, then the volume integral of $\nabla_\mu V^\mu$ vanishes.

[a] For the derivation of Gauss' theorem in curved space see e.g. [114].

7.3 EINSTEIN'S EQUATIONS IN VACUUM

The action for the gravitational field was proposed by David Hilbert in 1915 [63], and is called the **Einstein-Hilbert** action,

$$S^{\mathrm{EH}} = \frac{c^3}{16\pi G}\int d^4x \, \sqrt{-g} \, R \,, \tag{7.20}$$

where R is the Ricci scalar, $R = g^{\mu\nu}R_{\mu\nu}$. Since R is a combination of the metric tensor and of its derivatives, the action of the gravitational field depends on $g_{\mu\nu}$ and on its derivatives, up to second order.

Einstein's equations in vacuum can be derived from the Euler-Lagrange equations arising from the above action. However, it is simpler to proceed in a slightly different way. Varying the action 7.20 with respect to the metric tensor yields

$$\delta S^{\mathrm{EH}} = \frac{c^3}{16\pi G}\int d^4x \, \delta(\sqrt{-g} \, R) = \frac{c^3}{16\pi G}\int d^4x \, \left[\delta(\sqrt{-g}) \, R + \sqrt{-g} \, \delta R\right] \,. \tag{7.21}$$

The variation $\delta(\sqrt{-g})$ can be written as (see Box 7-B)

$$\delta(\sqrt{-g}) = -\frac{1}{2}\sqrt{-g}\; g_{\mu\nu}\delta g^{\mu\nu}\,. \tag{7.22}$$

In addition

$$\delta R = \delta(g^{\mu\nu}R_{\mu\nu}) = \delta g^{\mu\nu}R_{\mu\nu} + g^{\mu\nu}\delta R_{\mu\nu}\,. \tag{7.23}$$

Using these results, Eq. 7.21 becomes

$$\delta S^{\rm EH} = \frac{c^3}{16\pi G}\left\{\int\left(R_{\mu\nu} - \frac{1}{2}\, g_{\mu\nu}R\right)\delta g^{\mu\nu}\,\sqrt{-g}\; d^4x \;+\; \int g^{\mu\nu}\delta R_{\mu\nu}\,\sqrt{-g}\; d^4x\right\}. \tag{7.24}$$

In order to evaluate the second integral in Eq. 7.24, we need to compute the variation of the Ricci tensor. Using the definition given in Eq. 4.54 and Leibniz's chain rule, we find

$$\delta R_{\mu\nu} \;=\; \delta\Gamma^{\lambda}_{\mu\nu,\lambda} - \delta\Gamma^{\lambda}_{\mu\lambda,\nu} + \delta\Gamma^{\alpha}_{\mu\nu}\Gamma^{\lambda}_{\alpha\lambda} - \delta\Gamma^{\lambda}_{\alpha\nu}\Gamma^{\alpha}_{\mu\lambda} + \Gamma^{\alpha}_{\mu\nu}\delta\Gamma^{\lambda}_{\alpha\lambda} - \Gamma^{\lambda}_{\alpha\nu}\delta\Gamma^{\alpha}_{\mu\lambda}\,. \tag{7.25}$$

To evaluate $\delta\Gamma^{\lambda}_{\mu\nu}$ we define

$$\Gamma_{\mu\nu\,\delta} \equiv \Gamma^{\lambda}_{\mu\nu}\; g_{\lambda\delta} = \frac{1}{2}\left(g_{\mu\delta,\nu} + g_{\nu\delta,\mu} - g_{\mu\nu,\delta}\right)\,, \tag{7.26}$$

and, using the fact that $\delta g^{\lambda\delta} = -g^{\rho\lambda}g^{\sigma\delta}\delta g_{\rho\sigma}$ (see Eq. 7.45 in Box 7-B), we write $\delta\Gamma^{\lambda}_{\mu\nu}$ as follows

$$\begin{aligned}
\delta\Gamma^{\lambda}_{\mu\nu} = \delta\left(g^{\lambda\delta}\Gamma_{\mu\nu\,\delta}\right) \;&=\; \delta g^{\lambda\delta}\Gamma_{\mu\nu\,\delta} + g^{\lambda\delta}\delta\Gamma_{\mu\nu\,\delta}\\
&=\; -g^{\rho\lambda}g^{\sigma\delta}\delta g_{\rho\sigma}\Gamma_{\mu\nu\,\delta} + g^{\lambda\rho}\delta\Gamma_{\mu\nu\,\rho}\\
&=\; -g^{\lambda\rho}\delta g_{\rho\sigma}\Gamma^{\sigma}_{\mu\nu} + g^{\lambda\rho}\frac{1}{2}\left(\delta g_{\mu\rho,\nu} + \delta g_{\nu\rho,\mu} - \delta g_{\mu\nu,\rho}\right)\\
&=\; \frac{1}{2}g^{\lambda\rho}\left(\delta g_{\mu\rho,\nu} + \delta g_{\nu\rho,\mu} - \delta g_{\mu\nu,\rho} - 2\Gamma^{\sigma}_{\mu\nu}\delta g_{\rho\sigma}\right)\,.
\end{aligned} \tag{7.27}$$

The above equation can be rearranged as

$$\begin{aligned}
\delta\Gamma^{\lambda}_{\mu\nu} \;&=\; \frac{1}{2}g^{\lambda\rho}\left[\left(\delta g_{\mu\rho,\nu} - \Gamma^{\alpha}_{\mu\nu}\delta g_{\alpha\rho} - \Gamma^{\alpha}_{\nu\rho}\delta g_{\alpha\mu}\right) + \left(\delta g_{\nu\rho,\mu} - \Gamma^{\alpha}_{\nu\mu}\delta g_{\alpha\rho} - \Gamma^{\alpha}_{\rho\mu}\delta g_{\alpha\nu}\right)\right.\\
&\quad \left. - \left(\delta g_{\mu\nu,\rho} - \Gamma^{\alpha}_{\mu\rho}\delta g_{\alpha\nu} - \Gamma^{\alpha}_{\nu\rho}\delta g_{\alpha\mu}\right)\right]\\
&=\; \frac{1}{2}g^{\lambda\rho}\left(\delta g_{\mu\rho;\nu} + \delta g_{\nu\rho;\mu} - \delta g_{\mu\nu;\rho}\right)\,.
\end{aligned} \tag{7.28}$$

Since $\delta g_{\mu\nu}$ are the components of a tensor, from Eq. 7.28 it follows that $\delta\Gamma^{\lambda}_{\mu\nu}$ are the components of a tensor, despite the fact that the connection itself is *not* a tensor. Therefore, the quantity

$$(\delta\Gamma^{\lambda}_{\mu\nu})_{;\lambda} - (\delta\Gamma^{\lambda}_{\mu\lambda})_{;\nu} \tag{7.29}$$

is a tensor and can be evaluated with the usual rules of covariant differentiation:

$$\begin{aligned}
(\delta\Gamma^{\lambda}_{\mu\nu})_{;\lambda} - (\delta\Gamma^{\lambda}_{\mu\lambda})_{;\nu} \;&=\; \delta\Gamma^{\lambda}_{\mu\nu,\lambda} - \delta\Gamma^{\lambda}_{\mu\lambda,\nu}\\
&\quad + \delta\Gamma^{\alpha}_{\mu\nu}\Gamma^{\lambda}_{\alpha\lambda} - \delta\Gamma^{\lambda}_{\alpha\nu}\Gamma^{\alpha}_{\mu\lambda} + \delta\Gamma^{\lambda}_{\alpha\lambda}\Gamma^{\alpha}_{\mu\nu} - \Gamma^{\lambda}_{\alpha\nu}\delta\Gamma^{\alpha}_{\mu\lambda}\,.
\end{aligned} \tag{7.30}$$

By comparing Eq. 7.30 and Eq. 7.25 it follows that

$$\delta R_{\mu\nu} = (\delta\Gamma^{\lambda}_{\mu\nu})_{;\lambda} - (\delta\Gamma^{\lambda}_{\mu\lambda})_{;\nu}\,. \tag{7.31}$$

This equation is known as the *Palatini identity*[2]. Using this identity the term $g^{\mu\nu}\delta R_{\mu\nu}$ in Eq. 7.24 gives

$$\begin{aligned} g^{\mu\nu}\delta R_{\mu\nu} &= g^{\mu\nu}\left[(\delta\Gamma^{\lambda}_{\mu\nu})_{;\lambda} - (\delta\Gamma^{\lambda}_{\mu\lambda})_{;\nu}\right] = (g^{\mu\nu}\delta\Gamma^{\lambda}_{\mu\nu})_{;\lambda} - (g^{\mu\nu}\delta\Gamma^{\lambda}_{\mu\lambda})_{;\nu} \\ &= \left(g^{\mu\nu}\delta\Gamma^{\alpha}_{\mu\nu} - g^{\mu\alpha}\delta\Gamma^{\lambda}_{\mu\lambda}\right)_{;\alpha} . \end{aligned} \tag{7.33}$$

The right-hand side of this equation is the covariant divergence of a vector field; therefore, when it is integrated over the four-volume Ω, by Gauss' theorem it gives (see Box 7-A)

$$\int_{\Omega} g^{\mu\nu}\delta R_{\mu\nu}\sqrt{-g}\, d^4x = \int_{\partial\Omega}\left(g^{\mu\nu}\delta\Gamma^{\alpha}_{\mu\nu} - g^{\mu\alpha}\delta\Gamma^{\lambda}_{\mu\lambda}\right) dS_{\alpha} , \tag{7.34}$$

where $\partial\Omega$ is the three-dimensional boundary of Ω. Thus, finally, the variation $\delta S^{\rm EH}$ is

$$\delta S^{\rm EH} = \frac{c^3}{16\pi G}\left\{\int_{\Omega} d^4x\,\sqrt{-g}\left[R_{\mu\nu} - \frac{1}{2}g_{\mu\nu}R\right]\delta g^{\mu\nu} + \int_{\partial\Omega}\left(g^{\mu\nu}\delta\Gamma^{\alpha}_{\mu\nu} - g^{\mu\alpha}\delta\Gamma^{\lambda}_{\mu\lambda}\right) dS_{\alpha}\right\} . \tag{7.35}$$

The second integral is a surface term, which vanishes only if the variation of the Christoffel symbols are zero on the boundary, i.e. if $\delta g^{\mu\nu}{}_{,\alpha} = 0$ there. In the standard variational approach, only the variation of the field is required to vanish on the boundary of the integration volume; therefore $\delta g^{\mu\nu}{}_{,\alpha} = 0$ would be an extra condition to be imposed. However, it can be shown that this can be avoided by adding to the action a suitably defined boundary term, the variation of which cancels the surface term in Eq. 7.35 and does not contribute to the variation of the action in the four-volume Ω (see Box 7-C). Thus, the variation of the Einstein-Hilbert action can be written as

$$\delta S^{\rm EH} = \frac{c^3}{16\pi G}\int d^4x\,\sqrt{-g}\left[R_{\mu\nu} - \frac{1}{2}g_{\mu\nu}R\right]\delta g^{\mu\nu} , \tag{7.36}$$

and, by imposing

$$\delta S^{\rm EH} = 0, \qquad \forall\ \delta g^{\mu\nu} \text{ vanishing on the boundary} \tag{7.37}$$

we finally find Einstein's equations in vacuum

$$R_{\mu\nu} - \frac{1}{2}g_{\mu\nu}R = 0 , \tag{7.38}$$

or, in terms of the Einstein tensor $G_{\mu\nu} = R_{\mu\nu} - \frac{1}{2}g_{\mu\nu}R$,

$$G_{\mu\nu} = 0 .$$

[2]We invite the reader to prove, using a similar procedure, the following, more general version of Palatini's identity

$$\delta R^{\rho}{}_{\sigma\mu\nu} = \nabla_{\mu}(\delta\Gamma^{\rho}_{\nu\sigma}) - \nabla_{\nu}(\delta\Gamma^{\rho}_{\mu\sigma}) . \tag{7.32}$$

Box 7-B

Evaluation of $\delta(\sqrt{-g})$

The determinant g is a polynomial in $g_{\mu\nu}$, i.e. $g = g(g_{\mu\nu})$, therefore we can write

$$\delta g = \frac{\partial g}{\partial g_{\mu\nu}} \delta g_{\mu\nu} \,. \tag{7.39}$$

In particular, g is given by the following formula

$$g = \sum_{\nu} (-1)^{\mu+\nu} g_{\mu\nu} M_{\mu\nu} \quad \text{(no sum over } \mu\text{)} \tag{7.40}$$

where μ is fixed. $M_{\mu\nu}$ is the minor μ, ν, i.e. the determinant of the 3×3 matrix obtained by excluding the μ-th row and the ν-th column from the matrix $g_{\mu\nu}$. Note that since $g_{\mu\nu}$ is symmetric, $M_{\mu\nu} = M^T_{\mu\nu}$. Thus, by differentiating g with respect to $g_{\mu\nu}$ we find

$$\frac{\partial g}{\partial g_{\mu\nu}} = (-1)^{\mu+\nu} M_{\mu\nu} \quad \text{(no sum on } \mu \text{ and on } \nu\text{)} \,. \tag{7.41}$$

Since the components of the inverse matrix $g^{\mu\nu}$ are given by

$$g^{\mu\nu} = \frac{1}{g} M_{\mu\nu} (-1)^{\mu+\nu} \quad \text{(no sum on } \mu \text{ and on } \nu\text{)} \,, \tag{7.42}$$

Eq. 7.41 can be written as

$$\frac{\partial g}{\partial g_{\mu\nu}} = g g^{\mu\nu} \,. \tag{7.43}$$

Thus, the variation of the determinant reads

$$\delta g = g g^{\mu\nu} \delta g_{\mu\nu} \,. \tag{7.44}$$

Furthermore, since

$$\begin{aligned} \delta(g_{\mu\nu} g^{\mu\nu}) &= 0 \\ &= \delta g_{\mu\nu} g^{\mu\nu} + g_{\mu\nu} \delta g^{\mu\nu} \,, \end{aligned} \tag{7.45}$$

Eq. 7.44 reduces to

$$\delta g = -g g_{\mu\nu} \delta g^{\mu\nu} \,, \tag{7.46}$$

and we finally obtain

$$\delta(\sqrt{-g}) = -\frac{1}{2} \sqrt{-g} g_{\mu\nu} \delta g^{\mu\nu} \,. \tag{7.47}$$

Box 7-C

The boundary term of Einstein's equations

As shown in Sec. 7.3, the variation of Einstein-Hilbert's action is (see Eq. 7.24)

$$\delta S^{\rm EH} = \frac{c^3}{16\pi G}\left\{\int_\Omega d^4x\,\sqrt{-g}\left[R_{\mu\nu} - \frac{1}{2}g_{\mu\nu}R\right]\delta g^{\mu\nu} + \int_{\partial\Omega}\left(g^{\mu\nu}\delta\Gamma^\alpha_{\mu\nu} - g^{\mu\alpha}\delta\Gamma^\lambda_{\mu\lambda}\right)dS_\alpha\right\}. \tag{7.48}$$

In deriving Einstein's equations we have assumed that the surface term vanishes. As anticipated in Sec. 7.3, this requires that the variation of the connection, $\delta\Gamma^\alpha_{\mu\nu}$, vanishes on the boundary $\partial\Omega$ which, in turn, implies the vanishing of the variation on the *first derivatives* of the metric tensor. This additional boundary condition arises because the Ricci scalar R appearing in the Einstein-Hilbert action 7.20 contains the *second derivatives* of the metric tensor, whereas in other field theories the Lagrangian density depends only on the fields and on their first derivatives.

The additional requirement ($\delta\Gamma^\alpha_{\mu\nu} = 0$ on the boundary) can be avoided if the Einstein-Hilbert action is supplemented by an extra *boundary term*. The choice of this term is not unique; the standard choice is to modify the Einstein-Hilbert action as

$$S^{\rm EH} = \frac{c^3}{16\pi G}\int_\Omega d^4x\,\sqrt{-g}\,R + \frac{c^3}{8\pi G}\int_{\partial\Omega} d^3y\sqrt{\gamma}K\,, \tag{7.49}$$

where γ is the determinant of the metric $\gamma_{ab}(y)$ induced on the three-dimensional manifold $\partial\Omega$ (see Box 7-A), $K = \gamma_{ij}K^{ij}$ is the trace of the *extrinsic curvature* $K_{ij} \equiv -n_{\alpha;\beta}\frac{\partial x^\alpha}{\partial y^i}\frac{\partial x^\beta}{\partial y^j}$, and n^μ is the unit normal vector to the boundary hypersurface. The last term in Eq. 7.49 is called *Gibbons-Hawking-York boundary term* [121, 49]. The variation of this term cancels the surface term (i.e., the last integral) in Eq. 7.48, and thus the variation of the action, $\delta S^{\rm EH}$, reduces to Eq. 7.36.

In addition, it can be proved that – as a consequence of the Principle of General Covariance – the Gibbons-Hawking-York boundary term does not contribute to the variation of the action in the four-volume Ω and, hence, to the field equations. For this reason in this book we shall not include the Gibbons-Hawking-York boundary term in the action of the gravitational field (for further details, we refer to the original paper by Gibbons and Hawking [49]).

7.4 EINSTEIN'S EQUATIONS WITH SOURCES

The non-homogeneous form of Einstein's equations can be derived using the variational principle. Including the action of the source, a matter field or some other field (for instance electromagnetic, scalar, etc) in the total action, the equations are obtained by imposing the stationarity of the latter with respect to any variation of the metric tensor $\mathbf{g}$ such that

$$\delta S^{\rm EH} + \delta S^{\rm fields} = 0 \qquad \forall\ \delta g^{\mu\nu}\ \text{vanishing on the boundary}\,, \tag{7.50}$$

where $S^{\rm EH}$ and $S^{\rm fields}$ are given by Eqs. 7.20 and 7.10, respectively. The variation $\delta S^{\rm fields}$ with respect to the metric can easily be found using Eq. 7.22,

$$\begin{aligned}\delta S^{\rm fields} &= \int d^4x\, \delta\left[\sqrt{-g}\,\mathcal{L}^{\rm fields}\left(\mathbf{\Phi_1}, \nabla_\mu \mathbf{\Phi_1}, \ldots, \mathbf{\Phi_N}, \nabla_\mu \mathbf{\Phi_N}, \mathbf{g}\right)\right] \\ &= \int d^4x\, \sqrt{-g}\left[\frac{\partial \mathcal{L}^{\rm fields}}{\partial g^{\mu\nu}} - \frac{1}{2}\mathcal{L}^{\rm fields} g_{\mu\nu}\right]\delta g^{\mu\nu}\,. \end{aligned} \tag{7.51}$$

Therefore, if we define the *stress-energy tensor* as[3]

$$T_{\mu\nu} \equiv -2c\left[\frac{\partial \mathcal{L}^{\rm fields}}{\partial g^{\mu\nu}} - \frac{1}{2}\mathcal{L}^{\rm fields} g_{\mu\nu}\right], \tag{7.52}$$

the variation of the total action can be written, using Eq. 7.36, as

$$\delta S = \frac{c^3}{16\pi G}\int d^4x\,\sqrt{-g}\,\left[R_{\mu\nu} - \frac{1}{2}g_{\mu\nu}R - \frac{8\pi G}{c^4}T_{\mu\nu}\right]\delta g^{\mu\nu}\,. \tag{7.53}$$

By imposing the stationarity of this action for any $\delta g^{\mu\nu}$ vanishing on the boundary of the integration volume, the full Einstein equations follow:

$$G_{\mu\nu} = \frac{8\pi G}{c^4}T_{\mu\nu}\,. \tag{7.54}$$

Note that, although in Chapter 6 we derived Einstein's equations by *assuming* that $\nabla_\nu T^{\mu\nu} = 0$, one could also take the opposite, axiomatic approach. Namely, given the generic definition 7.52, and assuming Einstein's equations, one can take the covariant divergence of both sides of Eq. 7.54: the left-hand side of the resulting equation vanishes identically due to the Bianchi identities which, as shown in Chapter 6, imply $\nabla_\nu G^{\mu\nu} = 0$. As a consequence, the equation $\nabla_\nu T^{\mu\nu} = 0$ has to be satisfied by *any* stress-energy tensor satisfying Eq. 7.52. A further derivation of the equation $\nabla_\nu T^{\mu\nu} = 0$ will be discussed in Sec. 8.5.

7.4.1 The stress-energy tensor in some relevant cases

The definition of the stress-energy tensor given in Eq. 7.52 is very important, because it allows to compute the right-hand side of Einstein's equations for any given field from the corresponding Lagrangian in curved spacetime. The latter can be computed by applying the Principle of General Covariance to the corresponding Lagrangian in flat spacetime, i.e. we simply need to replace $\partial_\mu \to \nabla_\mu$ and $\eta_{\mu\nu} \to g_{\mu\nu}$. Using this powerful recipe, we can now compute the stress-energy tensor for the particular cases presented at the end of Sec. 5.2.

In Special Relativity, the Lagrangian density of a real scalar field Ψ, including self-interactions, reads

$$\mathcal{L}_{\rm scalar} = -\frac{1}{2c}\partial_\mu\Psi\partial^\mu\Psi - \frac{1}{c}V(\Psi)\,, \tag{7.55}$$

where $V(\Psi)$ is the scalar self-potential; in particular, for a free scalar field with mass m $V(\Psi) = \frac{m^2c^2}{2\hbar^2}\Psi^2$. The Lagrangian density of a scalar field in curved spacetime coincides with its flat-spacetime counterpart, since the covariant derivative of a scalar field coincides with its partial derivative. The two terms in Eq. 7.52 yield

$$-2c\frac{\partial \mathcal{L}^{\rm fields}}{\partial g^{\mu\nu}} = \partial_\mu\Psi\partial_\nu\Psi\,, \tag{7.56}$$

$$c g_{\mu\nu}\mathcal{L}^{\rm fields} = -\frac{1}{2}g_{\mu\nu}\partial_\rho\Psi\partial^\rho\Psi - g_{\mu\nu}V(\Psi)\,. \tag{7.57}$$

[3]Some textbooks define the Einstein-Hilbert action with an overall factor c^4 instead of c^3. With this choice the resulting quantity has not the dimensions of the action, but there would not appear unpleasant factors of c in the definition 7.52 of the stress-energy tensor.

Thus, from Eq. 7.52, the stress-energy tensor of a scalar field Ψ is

$$T_{\mu\nu} = \partial_\mu \Psi \partial_\nu \Psi - \frac{1}{2} g_{\mu\nu} \partial_\sigma \Psi \partial^\sigma \Psi - g_{\mu\nu} V(\Psi) \,. \tag{7.58}$$

It is easy to check that $\nabla_\mu T^{\mu\nu} = 0$ implies the *Klein-Gordon* field equation for the scalar field

$$\Box \Psi = dV/d\Psi \,. \tag{7.59}$$

We leave as an exercise to show that the above equation can equivalently be derived using the Euler-Lagrange equations in curved spacetime (Eq. 7.15) with $\mathbf{\Phi} = \Psi$ and $\mathcal{L} = \mathcal{L}_{\text{scalar}}$ given in Eq. 7.55.

The electromagnetic field is described by Maxwell's Lagrangian which in SI units reads (without currents) [4]

$$\mathcal{L}_{\text{EM}} = \frac{c}{4\mu_0} F_{\mu\nu} F^{\mu\nu} \tag{7.60}$$

where we recall that $F_{\mu\nu} = \nabla_\mu A_\nu - \nabla_\nu A_\mu$ is the electromagnetic tensor, A_μ is the four-potential, and μ_0 is the vacuum permeability. Euler-Lagrange's equations for A^μ reduce to Maxwell's equations in vacuum, $\partial_\mu F^{\mu\nu} = 0$ (the reader can check that this is equivalent to $\nabla_\mu F^{\mu\nu} = 0$). Furthermore, by writing

$$F_{\mu\nu} F^{\mu\nu} = g^{\rho\mu} g^{\sigma\nu} F_{\mu\nu} F_{\rho\sigma} \,, \tag{7.61}$$

using the definition 7.52 it is straightforward to obtain the electromagnetic stress-energy tensor

$$T_{\mu\nu} = -\frac{1}{\mu_0} \left(F_{\sigma(\mu} F_{\nu)\rho} g^{\sigma\rho} - \frac{1}{4} g_{\mu\nu} F_{\sigma\rho} F^{\sigma\rho} \right) . \tag{7.62}$$

We remark that the above expression is valid in any curved spacetime. It is easy to check that $\nabla_\mu T^{\mu\nu} = 0$ give Maxwell's equations in vacuum.

[4] In cgs-Gauss units, the dimensionful factor c/μ_0 disappears, and $\mathcal{L}_{\text{EM}} = -\frac{1}{16\pi} F_{\mu\nu} F^{\mu\nu}$.

Box 7-D

Uniqueness of the Einstein-Hilbert action

The Einstein-Hilbert action 7.20 might be derived from a more general action, as the *unique* term satisfying certain conditions which we shall specify. Let us construct a scalar quantity out of tensors that depend only on the metric. As discussed in Chapter 6, to linear order in the curvature the only possibility is the Ricci scalar, $R = R_{\mu\nu}g^{\mu\nu}$. If we include higher-order curvature terms, the action can be written as

$$S = \frac{c^3}{16\pi G}\int d^4x\,\sqrt{-g}\left(\Lambda + R + \alpha_1 R^2 + \alpha_2 R_{\mu\nu}R^{\mu\nu} + \alpha_3 K_1 + \alpha_4 K_2 + \mathcal{O}(R^3)\right)\,, \tag{7.63}$$

where $K_1 = R_{\mu\nu\rho\sigma}R^{\mu\nu\rho\sigma}$ and $K_2 = \varepsilon_{\mu\nu}{}^{\rho\sigma}R^{\mu\nu\alpha\beta}R_{\rho\sigma\alpha\beta}$ are the so-called Kretschmann scalar and Chern-Pontryagin scalar, respectively (see Sec. 9.3). In the action 7.63 we have included the cosmological constant (see Box 6-A), which gives a zeroth order term in the curvature, the Einstein-Hilbert term, all possible second order terms in the curvature, and we have omitted curvature terms of third or higher order. Thus, the general action contains an infinite series of scalar quantities of different order in the curvature, each multiplied by a dimensionful coupling constant α_i. We remind that in the derivation of Einstein's equations in Sec. 6.2 we assumed that the equations do not depend on any dimensionful quantity besides Newton's constant; the choice of the action 7.63 corresponds to dropping this assumption.

Clearly, there is an infinite number of scalar terms which can be constructed out of the contractions of the Riemann tensor. However, all terms in the action 7.63 which are of second or higher order in the curvature give rise to theories which are pathological. This is a consequence of a theorem due to Ostrogradsky [86], stating that Lagrangian densities that contain second- or higher-order derivatives in the field, and are non-linear in those terms, are generically unstable. Since terms like R^2, $R_{\mu\nu}R^{\mu\nu}$, etc., all contain second derivatives of the metric tensor *non-linearly*, they generically give rise to an unstable theory. An exception is given by the Gauss-Bonnet combination,

$$R^2 - 4R_{\mu\nu}R^{\mu\nu} + R_{\mu\nu\rho\sigma}R^{\mu\nu\rho\sigma}\,, \tag{7.64}$$

which, as the reader can check, does not contain non-linear terms in the second derivatives of $g_{\mu\nu}$, due to cancellations among the three contributions. However, it can be shown that the Gauss-Bonnet term is a *topological invariant*, i.e. it can be written in terms of a total derivative. Hence, it can be eliminated through integration by parts and it does not contribute to the field equations (modulo boundary terms). Therefore, Ostrogradsky's theorem guarantees that the only viable action constructed solely from the metric tensor is the Einstein-Hilbert one, plus at most a cosmological constant term.

7.5 EINSTEIN'S EQUATIONS IN THE PALATINI FORMALISM

The derivation of Einstein's vacuum equations from the Einstein-Hilbert action 7.20 implicitly assumes that the Levi-Civita connection (which enters in the definition of R, see Eq. 4.54) depends on the metric $g_{\mu\nu}$ through Eq. 3.68. We might wonder what would happen if both the metric $g_{\mu\nu}$ and the connection $\Gamma^{\gamma}_{\beta\mu}$ are treated as *independent* dynamical variables of the action. This approach is named after the Italian mathematician Attilio

Palatini, although it seems that the "Palatini formalism" was proposed by Einstein himself in 1925.

Let us consider the Einstein-Hilbert Lagrangian density,

$$\mathcal{L}^{\rm EH} = g^{\mu\nu} R_{\mu\nu} = g^{\mu\nu} \left(\Gamma^{\alpha}_{\mu\nu,\alpha} - \Gamma^{\alpha}_{\mu\alpha,\nu} - \Gamma^{\alpha}_{\sigma\nu}\Gamma^{\sigma}_{\mu\alpha} + \Gamma^{\alpha}_{\sigma\alpha}\Gamma^{\sigma}_{\mu\nu} \right) , \tag{7.65}$$

where we used Eq. 4.54. Assuming the connection and the metric as independent variables, we can write

$$\mathcal{L}^{\rm EH} = \mathcal{L}^{\rm EH}(g^{\mu\nu}, \Gamma^{\gamma}_{\beta\mu}, \Gamma^{\gamma}_{\beta\mu,\nu}) . \tag{7.66}$$

Note that in this form, the Lagrangian density does not depend on the derivatives of the metric tensor. Moreover, the Ricci tensor $R_{\mu\nu}$ can be written solely in terms of the connection and its derivatives, without using the metric explicitly (see Eq. 4.54). The total variation of the action contains two terms, namely

$$\delta S^{\rm EH} = \frac{\delta S^{\rm EH}}{\delta g_{\mu\nu}} \delta g_{\mu\nu} + \frac{\delta S^{\rm EH}}{\delta \Gamma^{\rho}_{\mu\nu}} \delta \Gamma^{\rho}_{\mu\nu} . \tag{7.67}$$

Since the metric and the connection are independent, $\delta S^{\rm EH} = 0$ implies that the two terms should vanish separately. Using Eq. 7.22 we can easily derive the variation of the action with respect to $g_{\mu\nu}$

$$\begin{aligned} \frac{\delta S^{\rm EH}}{\delta g_{\mu\nu}} \delta g^{\mu\nu} &= \frac{c^3}{16\pi G} \int d^4x \left(\sqrt{-g}\delta g^{\mu\nu} R_{\mu\nu} + \delta\sqrt{-g} g^{\mu\nu} R_{\mu\nu} \right) \\ &= \frac{c^3}{16\pi G} \int d^4x \sqrt{-g} \left(R_{\mu\nu} - \frac{1}{2} g_{\mu\nu} R \right) \delta g^{\mu\nu} = 0 , \end{aligned} \tag{7.68}$$

which gives Einstein's equations in vacuum. Note that, as mentioned above, the Ricci tensor depends only on Christoffel's symbols and their derivatives; therefore the variation of $R_{\mu\nu}$ with respect to variation of the metric tensor is zero.

Using the Palatini identity 7.31, the second term in Eq. 7.67 gives

$$\begin{aligned} \frac{\delta S^{\rm EH}}{\delta \Gamma^{\rho}_{\mu\nu}} \delta \Gamma^{\rho}_{\mu\nu} &= \frac{c^3}{16\pi G} \int d^4x \sqrt{-g} g^{\mu\nu} \delta R_{\mu\nu} \\ &= \frac{c^3}{16\pi G} \int d^4x \sqrt{-g} g^{\mu\nu} \left[(\delta\Gamma^{\lambda}_{\mu\nu})_{;\lambda} - (\delta\Gamma^{\lambda}_{\mu\lambda})_{;\nu} \right] = 0 . \end{aligned} \tag{7.69}$$

Integrating by parts and discarding surface terms with the usual argument, we obtain

$$\begin{aligned} \frac{\delta S^{\rm EH}}{\delta \Gamma^{\rho}_{\mu\nu}} \delta \Gamma^{\rho}_{\mu\nu} &= \frac{c^3}{16\pi G} \int d^4x \left[-\nabla_\lambda(\sqrt{-g} g^{\mu\nu}) \delta\Gamma^{\lambda}_{\mu\nu} + \nabla_\nu(\sqrt{-g} g^{\mu\nu}) \delta\Gamma^{\lambda}_{\mu\lambda} \right] , \\ &= \frac{c^3}{16\pi G} \int d^4x \left[-\nabla_\lambda(\sqrt{-g} g^{\mu\rho}) + \nabla_\nu(\sqrt{-g} g^{\mu\nu}) \delta^{\rho}_{\lambda} \right] \delta\Gamma^{\lambda}_{\mu\rho} = 0 . \end{aligned} \tag{7.70}$$

Since the variation of the action must be zero for all $\delta\Gamma^{\lambda}_{\mu\nu}$ vanishing at the boundary, the tensor in square brackets in the integrand of Eq. 7.70 has to vanish as well. However, since $\delta\Gamma^{\lambda}_{\mu\rho} = \delta\Gamma^{\lambda}_{\rho\mu}$, only the symmetric part of the tensor in square brackets should be considered (the antisymmetric part does not give any contribution when contracted with $\delta\Gamma^{\lambda}_{\mu\rho}$, and thus is not constrained by Eq. 7.70). Thus, we obtain (using the notation defined in Eq. 2.179)

$$\nabla_\nu (\sqrt{-g} g^{\nu(\mu}) \delta^{\rho)}{}_\lambda - \nabla_\lambda (\sqrt{-g} g^{\mu\rho}) = 0 . \tag{7.71}$$

By contracting the above equation with $g^{\lambda}{}_{\mu}$, we obtain $\nabla_\nu(\sqrt{-g}g^{\nu\rho}) = 0$; thus, the first term of the above equation vanishes and we get

$$\nabla_\lambda(\sqrt{-g}g^{\mu\rho}) = 0 \,. \tag{7.72}$$

A solution to this equation is

$$\nabla_\rho g_{\mu\nu} = 0 \,, \tag{7.73}$$

and it can be shown that this is the only solution. Thus, in the Palatini approach the vanishing of the covariant derivative of the metric tensor is not an *identity*: it is a *dynamical equation.* Eq. 7.73 is equivalent, as shown in Chapter 3, to the definition 3.68 of the Levi-Civita connection, which we repeat here for convenience,

$$\Gamma^{\gamma}_{\beta\mu} = \frac{1}{2} g^{\alpha\gamma}(g_{\alpha\beta,\mu} + g_{\alpha\mu,\beta} - g_{\beta\mu,\alpha}) \,. \tag{7.74}$$

In other words, by considering the metric and the connection as independent fields in the Einstein-Hilbert action, the field equations for the metric tensor give Einstein's equations, whereas the field equations for the independent connection *enforce* the fact that the latter has to be the Levi-Civita connection. Therefore, in the case of Einstein's equations without sources, the Palatini formalism is *dynamically equivalent* to the metric one.

The above derivation can be extended to include sources, provided the matter Lagrangian does not depend explicitly on the connection. This assumption is satisfied for scalar fields, Maxwell fields, and fluids, but it is violated, for instance, in the case of a fermionic (e.g., Dirac) field. For generic sources, the Palatini formulation is not equivalent to the metric one, and describes a gravitational theory which is different from General Relativity.

Finally, it is worth mentioning that a more general class of gravitational theories arises when the affine connection – in addition to be independent of the metric – is not assumed to be symmetric, $\Gamma^{\mu}_{\rho\sigma} \neq \Gamma^{\mu}_{\sigma\rho}$. These are called *metric-affine theories of gravity* [59].

CHAPTER 8

Symmetries

The solution of the equations describing a physical problem can be considerably simplified if the latter has some symmetry. For instance, it is easy to find the solution of the equations of Newtonian gravity for a spherically symmetric body, but it may be difficult to find the analytic solution for an arbitrary mass distribution.

Generally speaking, a physical quantity is symmetric if it is invariant with respect to some transformation. For instance, plane symmetry implies invariance with respect to translations on a plane, spherical symmetry implies invariance with respect to rotations on a sphere, and stationarity implies invariance with respect to time translations.

In General Relativity, a symmetry indicates invariance with respect to a **diffeomorphism** (see Sec. 2.1.4), which is a regular, invertible mapping of any point $\mathbf{p}$ to another point $\mathbf{p}'$ of the same manifold (the spacetime). In a given coordinate frame, a diffeomorphism is described by C^1 functions

$$x^\mu \mapsto x'^\alpha(x^\mu)\,. \tag{8.1}$$

Note that Eq. 8.1 is formally similar to a coordinate transformation, but it has a different meaning, because a coordinate transformation leaves the point of the manifold unchanged, and associates to it different coordinates (see Sec. 2.1.4 and Fig. 2.11). For this reason for coordinate transformations we use a different notation, i.e. $x^\mu \mapsto x^{\alpha'}(x^\mu)$.

For instance, translations in space or time and rotations are diffeomorphisms. A generic *infinitesimal diffeomorphism*

$$x^\mu \mapsto x^\mu + \delta x^\mu \tag{8.2}$$

can be seen as an *infinitesimal translation*. We stress that δx^μ is not constant and depends on the spacetime position.

A **spacetime isometry** is a diffeomorphism under which the metric tensor $\mathbf{g}$, i.e. the line element $ds^2 = g_{\mu\nu}dx^\mu dx^\nu$, is invariant. In many cases, an isometry defines a submanifold which is mapped into itself. For example, for spherically symmetric spacetimes the invariant submanifold is the two-sphere, for stationary spacetimes it is the time axis, etc. These definitions can be made more precise by introducing the notion of Killing vector fields.

8.1 KILLING VECTOR FIELDS

Consider a vector field $\vec{\xi}(x^\mu)$ defined at every point of a spacetime region. The vector $\vec{\xi}$ identifies a symmetry if an infinitesimal translation along $\vec{\xi}$ leaves the line element unchanged, i.e.

$$\delta(ds^2) = \delta(g_{\alpha\beta}dx^\alpha dx^\beta) = 0\,. \tag{8.3}$$

This implies that

$$\delta g_{\alpha\beta}dx^\alpha dx^\beta + g_{\alpha\beta}\left[\delta(dx^\alpha)dx^\beta + dx^\alpha\delta(dx^\beta)\right] = 0. \tag{8.4}$$

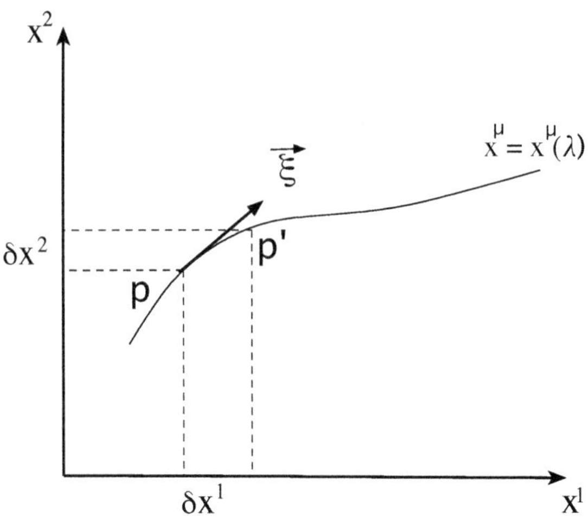

Figure 8.1: Infinitesimal translation along a curve. The Killing vector $\vec{\xi}$ is the tangent vector to the curve.

Let us consider a curve $x^\alpha(\lambda)$ to which $\vec{\xi}$ is tangent, i.e. $\xi^\alpha = \frac{dx^\alpha}{d\lambda}$. An infinitesimal translation in the direction of $\vec{\xi}$ is an infinitesimal translation along the curve from a point $\mathbf{p}$ to the point $\mathbf{p}'$ whose coordinates are, respectively,

$$\mathbf{p} = \{x^\alpha\} \qquad \text{and} \qquad \mathbf{p}' = \{x^\alpha + \delta x^\alpha\}. \tag{8.5}$$

For instance, in the case of the two-dimensional space in Fig. 8.1, $\vec{\xi}$ defines an infinitesimal translation along the curve from a point $\mathbf{p} = (x^1, x^2)$ to a point $\mathbf{p}' = (x^1 + \delta x^1, x^2 + \delta x^2)$, where

$$\delta x^1 = \frac{dx^1}{d\lambda}\delta\lambda = \xi^1 \delta\lambda \quad \text{and} \quad \delta x^2 = \frac{dx^2}{d\lambda}\delta\lambda = \xi^2 \delta\lambda\,. \tag{8.6}$$

In the general case of a four-dimensional spacetime, $\delta x^\alpha = \xi^\alpha \delta\lambda$, and the coordinates of the point $\mathbf{p}'$ can be written as

$$x'^\alpha = x^\alpha + \xi^\alpha\, \delta\lambda\,. \tag{8.7}$$

Under the diffeomorphism $\mathbf{p} \mapsto \mathbf{p}'$, the metric components change as follows

$$\begin{aligned} g_{\alpha\beta}(\mathbf{p}') &\simeq g_{\alpha\beta}(\mathbf{p}) + \frac{dg_{\alpha\beta}}{d\lambda}\delta\lambda + ... = g_{\alpha\beta}(\mathbf{p}) + \frac{\partial g_{\alpha\beta}}{\partial x^\mu}\frac{dx^\mu}{d\lambda}\delta\lambda + ... \\ &= g_{\alpha\beta}(\mathbf{p}) + g_{\alpha\beta,\mu}\xi^\mu \delta\lambda\,, \end{aligned} \tag{8.8}$$

hence

$$\delta g_{\alpha\beta} = g_{\alpha\beta,\mu}\xi^\mu \delta\lambda\,. \tag{8.9}$$

Moreover, since the operators δ and d commute, we find[1]

$$\delta(dx^\alpha) = d(\delta x^\alpha) = d(\xi^\alpha \delta\lambda) = d\xi^\alpha \delta\lambda = \xi^\alpha_{,\mu} dx^\mu \delta\lambda\,. \tag{8.10}$$

Thus, using Eqs. 8.9 and 8.10, Eq. 8.4 becomes

$$g_{\alpha\beta,\mu}\xi^\mu \delta\lambda dx^\alpha dx^\beta + g_{\alpha\beta}\left[\xi^\alpha_{,\mu} dx^\mu \delta\lambda dx^\beta + \xi^\beta_{,\gamma} dx^\gamma \delta\lambda dx^\alpha\right] = 0\,, \tag{8.11}$$

and, after relabeling the indices,

$$\left[g_{\alpha\beta,\mu}\xi^\mu + g_{\delta\beta}\xi^\delta_{,\alpha} + g_{\alpha\delta}\xi^\delta_{,\beta}\right] dx^\alpha dx^\beta \delta\lambda = 0\,. \tag{8.12}$$

In conclusion, a solution of Einstein's equations is invariant under infinitesimal translations along $\vec{\xi}$, if and only if

$$g_{\alpha\beta,\mu}\xi^\mu + g_{\delta\beta}\xi^\delta_{,\alpha} + g_{\alpha\delta}\xi^\delta_{,\beta} = 0\,. \tag{8.13}$$

The **Killing vector fields** of a given metric $g_{\alpha\beta}$ are the solution of Eq. 8.13. Sometimes, for brevity, the Killing vector fields are called simply *Killing vectors*.

Apparently, Eq. 8.13 is not covariant since it contains partial derivatives, but below we show that it is equivalent to the following covariant equation

$$\xi_{\alpha;\beta} + \xi_{\beta;\alpha} = 0\,, \tag{8.14}$$

which is called the **Killing equation**, after the mathematician Wilhelm Killing who found it. This equation implies that $\xi_{\alpha;\beta}$ is an antisymmetric tensor and can be written in a compact form as $\xi_{(\alpha;\beta)} = 0$ (see Sec. 2.4.2). The Killing equation 8.14 is therefore a system of ten independent differential equations for the four components of $\vec{\xi}$. As such, it might not admit a solution; in which case the spacetime has no symmetries.

Let us now show that Eq. 8.13 is equivalent to Eq. 8.14. Since

$$\xi_{\alpha;\beta} \;=\; (g_{\alpha\mu}\xi^\mu)_{;\beta} = g_{\alpha\mu}\xi^\mu_{;\beta} = g_{\alpha\mu}\left(\xi^\mu_{,\beta} + \Gamma^\mu_{\delta\beta}\xi^\delta\right), \tag{8.15}$$

we obtain

$$\begin{aligned}\xi_{\alpha;\beta} + \xi_{\beta;\alpha} &= g_{\alpha\mu}\left(\xi^\mu_{,\beta} + \Gamma^\mu_{\delta\beta}\xi^\delta\right) + g_{\beta\mu}\left(\xi^\mu_{,\alpha} + \Gamma^\mu_{\alpha\delta}\xi^\delta\right)\\ &= g_{\alpha\mu}\xi^\mu_{,\beta} + g_{\beta\mu}\xi^\mu_{,\alpha} + \left(g_{\alpha\mu}\Gamma^\mu_{\delta\beta} + g_{\beta\mu}\Gamma^\mu_{\alpha\delta}\right)\xi^\delta\,.\end{aligned} \tag{8.16}$$

Using Eq. 3.68, the term in parenthesis can be written as

$$\begin{aligned}&\frac{1}{2}\left[g_{\alpha\mu}g^{\mu\sigma}\left(g_{\delta\sigma,\beta} + g_{\sigma\beta,\delta} - g_{\delta\beta,\sigma}\right) + g_{\beta\mu}g^{\mu\sigma}\left(g_{\alpha\sigma,\delta} + g_{\sigma\delta,\alpha} - g_{\alpha\delta,\sigma}\right)\right]\\ &= \frac{1}{2}\left[\delta^\sigma_\alpha\left(g_{\delta\sigma,\beta} + g_{\sigma\beta,\delta} - g_{\delta\beta,\sigma}\right) + \delta^\sigma_\beta\left(g_{\alpha\sigma,\delta} + g_{\sigma\delta,\alpha} - g_{\alpha\delta,\sigma}\right)\right]\\ &= \frac{1}{2}\left[g_{\delta\alpha,\beta} + g_{\alpha\beta,\delta} - g_{\delta\beta,\alpha} + g_{\alpha\beta,\delta} + g_{\beta\delta,\alpha} - g_{\alpha\delta,\beta}\right] \qquad (8.17)\\ &= g_{\alpha\beta,\delta}\,, \qquad (8.18)\end{aligned}$$

and Eq. 8.16 becomes

$$\xi_{\alpha;\beta} + \xi_{\beta;\alpha} = g_{\alpha\mu}\xi^\mu_{,\beta} + g_{\beta\mu}\xi^\mu_{,\alpha} + g_{\alpha\beta,\delta}\xi^\delta \tag{8.19}$$

which coincides with Eq. 8.13.

---

[1] Note that since λ is a parameter, the differential operator d does not act on its variation, i.e. $d\delta\lambda = 0$.

Box 8-A

The Lie derivative

The variation of a $\binom{N}{N'}$ tensor **T** under an infinitesimal translation (i.e., an infinitesimal diffeomorphism) along the direction of a vector field $\vec{\xi}$ is called its **Lie derivative** along $\vec{\xi}$. Note that $\vec{\xi}$ does not necessarily have to be a Killing vector. The components of the Lie derivative of a tensor **T** along a vector field $\vec{\xi}$ are indicated as

$$\mathcal{L}_{\vec{\xi}} T^{\alpha_1 \cdots \alpha_N}{}_{\beta_1 \cdots \beta_{N'}} . \tag{8.20}$$

For instance, if **T** is a $\binom{0}{2}$ tensor, its Lie derivative along $\vec{\xi}$ can be obtained by repeating the derivation of Sec. 8.1, and is

$$\mathcal{L}_{\vec{\xi}} T_{\alpha\beta} = T_{\alpha\beta,\mu}\xi^\mu + T_{\delta\beta}\xi^\delta_{,\alpha} + T_{\alpha\delta}\xi^\delta_{,\beta} . \tag{8.21}$$

The Lie derivative along a vector field $\vec{\xi}$, $\mathcal{L}_{\vec{\xi}}$, should not be confused with the covariant directional derivative along $\vec{\xi}$, $\nabla_{\vec{\xi}}$ discussed in Chapter 3. For instance, for a $\binom{0}{2}$ tensor **T**, the components of the directional derivative along $\vec{\xi}$ are $\xi^\alpha T_{\mu\nu;\alpha}$, which is different from the expression in Eq. 8.21. Both of these derivatives give the modification of a tensor field, but with different prescriptions to compare tensors in different spacetime points. In the case of the directional derivative, the prescription is provided by the Levi-Civita connection, and hence by the metric: we require that the scalar product between two vectors is the same in different points (which is equivalent to requiring $g_{\mu\nu;\alpha} = 0$). In the case of the Lie derivative, instead, the tensors are transported by a diffeomorphism along the vector field $\vec{\xi}$. Note that the Lie derivative does not require the specification of the metric, only of the vector field $\vec{\xi}$. A more rigorous definition of the Lie derivative would require a discussion on the action of diffeomorphisms on tensor fields, which is beyond the scope of this book. We refer the interested reader to [57].

In the previous paragraphs we have shown that the Lie derivative of the metric tensor **g** along a vector field $\vec{\xi}$ is

$$\mathcal{L}_{\vec{\xi}} g_{\alpha\beta} = g_{\alpha\beta,\mu}\xi^\mu + g_{\delta\beta}\xi^\delta_{,\alpha} + g_{\alpha\delta}\xi^\delta_{,\beta} = \xi_{\alpha;\beta} + \xi_{\beta;\alpha} , \tag{8.22}$$

and that, if $\vec{\xi}$ is a Killing vector field, the Lie derivative of the metric tensor vanishes. Therefore, an alternative form of the Killing equation is

$$\mathcal{L}_{\vec{\xi}} g_{\alpha\beta} = 0 . \tag{8.23}$$

Moreover, since $g_{\alpha\beta}g^{\beta\gamma} = \delta^\gamma_\alpha$,

$$\mathcal{L}_{\vec{\xi}}(g_{\alpha\beta}g^{\beta\gamma}) = g_{\alpha\beta}\mathcal{L}_{\vec{\xi}} g^{\beta\gamma} + g^{\beta\gamma}\mathcal{L}_{\vec{\xi}} g_{\alpha\beta} = 0 , \tag{8.24}$$

and thus

$$\mathcal{L}_{\vec{\xi}} g^{\alpha\beta} = -\xi^{\alpha;\beta} - \xi^{\beta;\alpha} , \tag{8.25}$$

which also vanishes if $\vec{\xi}$ is a Killing vector. We leave as an exercise to prove that the Lie derivative of a vector field $\vec{V}$ along the direction of another vector field $\vec{\xi}$ is

$$\mathcal{L}_{\vec{\xi}} V^\alpha = [\vec{\xi}, \vec{V}]^\alpha = \xi^\mu V^\alpha{}_{,\mu} - V^\mu \xi^\alpha{}_{,\mu} . \tag{8.26}$$

8.2 KILLING VECTOR FIELDS AND THE CHOICE OF COORDINATE SYSTEMS

Given a vector field $\vec{\xi}$ (not necessarily a Killing vector field), the family of curves such that at any point **p** the vector $\vec{\xi}$ is tangent to the curve passing in **p** is called **congruence of worldlines**. They can be found by integrating the equations

$$\frac{dx^\mu}{d\lambda} = \xi^\mu(x^\alpha)\,. \tag{8.27}$$

If $\vec{\xi}$ is a Killing vector field, the line element (i.e. the metric tensor) is invariant under infinitesimal translations along any curve of the congruence; therefore, it is also invariant under a *finite transformation* $\mathbf{p} \mapsto \mathbf{p}'$ along any of these curves. The congruence of worldlines, then, defines a finite symmetry of the spacetime. For instance, if the spacetime is spherically symmetric, the curves are circumferences on spherical surfaces; if the spacetime is invariant for (space or time) translations the curves are straight lines, and so on. This finite transformation is a diffeomorphism, because it is a sequence of infinitesimal diffeomorphisms; we call it the *diffeomorphism generated by the vector field* $\vec{\xi}$.

The existence of Killing vector fields remarkably simplifies the problem of choosing a coordinate system appropriate to solve Einstein's equations. For instance, if we are looking for a solution which admits a timelike Killing vector field $\vec{\xi}$, it is convenient to choose, at each point of the manifold, the timelike basis vector $\vec{e}_{(0)}$ aligned with $\vec{\xi}$; with this choice, the time coordinate lines coincide with the worldlines to which $\vec{\xi}$ is tangent, i.e. with the congruence of worldlines of $\vec{\xi}$, and the components of $\vec{\xi}$ are

$$\xi^\alpha = (\xi^0, 0, 0, 0)\,. \tag{8.28}$$

If we parametrize these curves in such a way that ξ^0 is constant, $\xi^\mu{}_{,\alpha} = 0$ and the second and third terms in Eq. 8.13 identically vanish; therefore, Eq. 8.13 yields

$$\frac{\partial g_{\alpha\beta}}{\partial x^0} = 0\,. \tag{8.29}$$

This means that *if the spacetime admits a timelike Killing vector field, with an appropriate choice of the coordinate system the metric tensor can be made independent of time.* For simplicity, it is customary to choose the parametrization of the curve such that $\xi^0 = 1$; in this case, the timelike Killing vector field is $\xi^\mu = (1, 0, 0, 0)$.

A similar procedure can be used if the metric admits a spacelike Killing vector field; choosing, say, the vector $\vec{e}_{(1)}$ to be parallel to $\vec{\xi}$,

$$\xi^\alpha = (0, \xi^1, 0, 0)\,, \tag{8.30}$$

and choosing the parametrization such that ξ^1 is constant, i.e. $\xi^1{}_{,\alpha} = 0$, Eq. 8.13 shows that $\partial g_{\alpha\beta}/\partial x^1 = 0$, i.e. the metric is independent of x^1. By a suitable reparametrization of the corresponding congruence of coordinate lines, one can write $\xi^\alpha = (0, 1, 0, 0)$.

If the Killing vector field is null, it is necessary to define a coordinate basis in which at least one basis vector is null. Starting from generic coordinate basis vectors $(\vec{e}_{(0)}, \vec{e}_{(1)}, \vec{e}_{(2)}, \vec{e}_{(3)})$, it is always possible to construct a set of new basis vectors

$$\vec{e}_{(\alpha')} = \Lambda^\beta_{\alpha'}\vec{e}_{(\beta)}\,, \tag{8.31}$$

such that, for instance, the vector $\vec{e}_{(0')}$ is a null vector. Then, $\vec{e}_{(0')}$ can be chosen to be parallel to $\vec{\xi}$ at each point of the manifold, so that

$$\xi^\alpha = (\xi^0, 0, 0, 0) \tag{8.32}$$

with ξ^0 constant. The metric, then, is independent of the corresponding null coordinate x^0. By a suitable reparametrization of the corresponding coordinate lines, it is possible to choose $\xi^\mu = (1,0,0,0)$.

As an example, in the following we shall compute the Killing vector fields of the flat spacetime and of a spherical surface.

Killing vector fields of flat spacetime

The Killing vector fields of Minkowski's spacetime can be obtained very easily using Minkowskian coordinates. Since all Christoffel's symbols vanish, the Killing equation becomes

$$\xi_{\alpha,\beta} + \xi_{\beta,\alpha} = 0\,. \tag{8.33}$$

By combining the following equations

$$\xi_{\alpha,\beta\gamma} + \xi_{\beta,\alpha\gamma} = 0\,, \qquad \xi_{\beta,\gamma\alpha} + \xi_{\gamma,\beta\alpha} = 0\,, \qquad \xi_{\gamma,\alpha\beta} + \xi_{\alpha,\gamma\beta} = 0\,, \tag{8.34}$$

and by using Eq. 8.33, we find

$$\xi_{\alpha,\beta\gamma} = 0\,, \tag{8.35}$$

whose general solution is

$$\xi_\alpha = c_\alpha + \epsilon_{\alpha\gamma} x^\gamma\,, \tag{8.36}$$

where c_α, $\epsilon_{\alpha\beta}$ are constants. By substituting this expression into Eq. 8.33 we find

$$\epsilon_{\alpha\gamma} x^\gamma_{\ ,\beta} + \epsilon_{\beta\gamma} x^\gamma_{\ ,\alpha} = \epsilon_{\alpha\gamma}\delta^\gamma_\beta + \epsilon_{\beta\gamma}\delta^\gamma_\alpha = \epsilon_{\alpha\beta} + \epsilon_{\beta\alpha} = 0\,. \tag{8.37}$$

Thus, Eq. 8.36 is the solution to Eq. 8.33 if and only if

$$\epsilon_{\alpha\beta} = -\epsilon_{\beta\alpha}\,, \tag{8.38}$$

so that $\epsilon_{\alpha\beta}$ has six independent components. Therefore, flat spacetime admits ten linearly independent Killing vector fields. The general Killing vector field of the form 8.36 can be written as the linear combination of ten Killing vector fields $\xi^{(A)}_\alpha = \{\xi^{(1)}_\alpha, \xi^{(2)}_\alpha, \ldots, \xi^{(10)}_\alpha\}$ corresponding to ten independent choices of the constants c_α, $\epsilon_{\alpha\beta}$:

$$\xi^{(A)}_\alpha = c^{(A)}_\alpha + \epsilon^{(A)}_{\alpha\gamma} x^\gamma \qquad A = 1,\ldots,10\,. \tag{8.39}$$

For instance, we can choose

$$\begin{aligned}
c^{(1)}_\alpha &= (1,0,0,0) \qquad \epsilon^{(1)}_{\alpha\beta} = 0 \\
c^{(2)}_\alpha &= (0,1,0,0) \qquad \epsilon^{(2)}_{\alpha\beta} = 0 \\
c^{(3)}_\alpha &= (0,0,1,0) \qquad \epsilon^{(3)}_{\alpha\beta} = 0 \\
c^{(4)}_\alpha &= (0,0,0,1) \qquad \epsilon^{(4)}_{\alpha\beta} = 0 \\
c^{(5)}_\alpha &= 0 \quad \epsilon^{(5)}_{\alpha\beta} = \begin{pmatrix} 0 & 1 & 0 & 0 \\ -1 & 0 & 0 & 0 \\ 0 & 0 & 0 & 0 \\ 0 & 0 & 0 & 0 \end{pmatrix} \\
c^{(6)}_\alpha &= 0 \quad \epsilon^{(6)}_{\alpha\beta} = \begin{pmatrix} 0 & 0 & 1 & 0 \\ 0 & 0 & 0 & 0 \\ -1 & 0 & 0 & 0 \\ 0 & 0 & 0 & 0 \end{pmatrix} \\
c^{(7)}_\alpha &= 0 \quad \epsilon^{(7)}_{\alpha\beta} = \begin{pmatrix} 0 & 0 & 0 & 1 \\ 0 & 0 & 0 & 0 \\ 0 & 0 & 0 & 0 \\ -1 & 0 & 0 & 0 \end{pmatrix} \\
c^{(8)}_\alpha &= 0 \quad \epsilon^{(8)}_{\alpha\beta} = \begin{pmatrix} 0 & 0 & 0 & 0 \\ 0 & 0 & 1 & 0 \\ 0 & -1 & 0 & 0 \\ 0 & 0 & 0 & 0 \end{pmatrix} \\
c^{(9)}_\alpha &= 0 \quad \epsilon^{(9)}_{\alpha\beta} = \begin{pmatrix} 0 & 0 & 0 & 0 \\ 0 & 0 & 0 & 1 \\ 0 & 0 & 0 & 0 \\ 0 & -1 & 0 & 0 \end{pmatrix} \\
c^{(10)}_\alpha &= 0 \quad \epsilon^{(10)}_{\alpha\beta} = \begin{pmatrix} 0 & 0 & 0 & 0 \\ 0 & 0 & 0 & 0 \\ 0 & 0 & 0 & 1 \\ 0 & 0 & -1 & 0 \end{pmatrix} .
\end{aligned} \tag{8.40}$$

The symmetries generated by the Killing vector fields with $A = 1, \ldots, 4$ are spacetime translations; those generated by the Killing vector fields with $A = 5, 6, 7$ are Lorentz's boosts and those generated by the Killing vector fields with $A = 8, 9, 10$ are space rotations. The diffeomorphisms generated by these Killing vector fields form a group, called the *Poincaré group* $ISO(3,1)$.

Killing vector fields of a spherical surface

Let us consider a sphere of unit radius. Its metric is

$$ds^2 = d\theta^2 + \sin^2\theta d\varphi^2 = (dx^1)^2 + \sin^2 x^1 (dx^2)^2 . \tag{8.41}$$

Recalling Eq. 8.13,

$$g_{\alpha\beta,\mu}\xi^\mu + g_{\delta\beta}\xi^\delta_{,\alpha} + g_{\alpha\delta}\xi^\delta_{,\beta} = 0 , \tag{8.42}$$

we get

$$
\begin{aligned}
(\alpha,\beta) &= (1,1) \quad 2g_{\delta 1}\xi^{\delta}_{,1} = 0 \rightarrow \xi^1_{,1} = 0 \\
(\alpha,\beta) &= (1,2) \quad g_{\delta 2}\xi^{\delta}_{,1} + g_{1\delta}\xi^{\delta}_{,2} = 0 \rightarrow \xi^1_{,2} + \sin^2\theta\,\xi^2_{,1} = 0 \\
(\alpha,\beta) &= (2,2) \quad g_{22,\mu}\xi^{\mu} + 2g_{\delta 2}\xi^{\delta}_{,2} = 0 \rightarrow \cos\theta\,\xi^1 + \sin\theta\,\xi^2_{,2} = 0\,.
\end{aligned}
\tag{8.43}
$$

The general solution of Eqs. 8.43 is

$$
\xi^1 = c\sin(\varphi + a), \qquad \xi^2 = c\cos(\varphi + a)\cot\theta + b\,, \tag{8.44}
$$

with a, b, c arbitrary real constants[2]. Therefore a spherical surface admits three linearly independent Killing vector fields, associated to the choice of the integration constants (a, b, c). A set of three independent Killing vector fields for the sphere, $\vec{\xi}^{(A)}$ $(A = 1, 2, 3)$, is

$$
\begin{aligned}
\xi^{(1)\alpha} &= (0,1) \\
\xi^{(2)\alpha} &= (\sin\varphi, \cos\varphi\cot\theta) \\
\xi^{(3)\alpha} &= (\cos\varphi, -\sin\varphi\cot\theta)\,,
\end{aligned}
\tag{8.45}
$$

which correspond to the choices $(a, b, c) = (0, 1, 0)$, $(0, 0, 1)$, and $\left(\frac{\pi}{2}, 0, 1\right)$, respectively.

The diffeomorphisms generated by these Killing vector fields form a group called the *rotation group* $SO(3)$[3].

8.3 KILLING VECTOR FIELDS AND CONSERVATION LAWS

Killing vector fields allow to define conserved quantities, whose existence may be concealed by an unsuitable coordinate choice.

8.3.1 Conserved quantities in geodesic motion

Let us consider a particle moving along a geodesic of a spacetime which admits a Killing vector field $\vec{\xi}$. The geodesic equations written in terms of the particle four-velocity $u^\alpha = \frac{dx^\alpha}{d\lambda}$ (where λ is an affine parameter) read

$$
\frac{du^\alpha}{d\lambda} + \Gamma^\alpha_{\beta\nu}u^\beta u^\nu = 0\,. \tag{8.46}
$$

By contracting this equation with $\vec{\xi}$ we find

$$
\xi_\alpha\left[\frac{du^\alpha}{d\lambda} + \Gamma^\alpha_{\beta\nu}u^\beta u^\nu\right] = \frac{d(\xi_\alpha u^\alpha)}{d\lambda} - u^\alpha\frac{d\xi_\alpha}{d\lambda} + \Gamma^\alpha_{\beta\nu}u^\beta u^\nu\xi_\alpha\,. \tag{8.47}
$$

Since

$$
u^\beta\frac{d\xi_\beta}{d\lambda} = u^\beta\frac{\partial\xi_\beta}{\partial x^\nu}\frac{dx^\nu}{d\lambda} = u^\beta u^\nu\frac{\partial\xi_\beta}{\partial x^\nu}\,, \tag{8.48}
$$

Eq. 8.47 becomes

$$
\frac{d(\xi_\alpha u^\alpha)}{d\lambda} - u^\beta u^\nu\left[\frac{\partial\xi_\beta}{\partial x^\nu} - \Gamma^\alpha_{\beta\nu}\xi_\alpha\right] = 0\,, \tag{8.49}
$$

i.e.

$$
\frac{d(\xi_\alpha u^\alpha)}{d\lambda} - u^\beta u^\nu\xi_{\beta;\nu} = 0\,. \tag{8.50}
$$

---

[2]We leave as an exercise to prove, by direct substitution, that the vector fields 8.44 satisfy Eqs. 8.43.

[3]They satisfy the commutation relations $[\vec{\xi}^{(A)}, \vec{\xi}^{(B)}] = \epsilon^{ABC}\vec{\xi}^{(C)}$, where ϵ^{ABC} is the three-dimensional Levi-Civita symbol (see Box 8-C).

Since $\xi_{\beta;\nu}$ is antisymmetric in β and ν, while $u^\beta u^\nu$ is symmetric, the term $u^\beta u^\nu \xi_{\beta;\nu}$ identically vanishes, and Eq. 8.50 reduces to

$$\frac{d(\xi_\alpha u^\alpha)}{d\lambda} = 0 \qquad \to \qquad \xi_\alpha u^\alpha = const\,. \tag{8.51}$$

Thus, for every Killing vector field there exists an associated conserved quantity for the geodesic motion of particles.

Equation 8.51 can be written as follows:

$$g_{\alpha\mu}\xi^\mu u^\alpha = const\,. \tag{8.52}$$

The physical interpretation of the constants of motion associated to Killing vector fields will be discussed in Sec. 10.3.

8.3.2 Conserved quantities from the stress-energy tensor

In Chapter 7 we have shown that the stress-energy tensor satisfies the divergenceless equation

$$T^{\mu\nu}{}_{;\nu} = 0\,, \tag{8.53}$$

and that in curved spacetime this is not a genuine conservation law. If the spacetime admits a Killing vector field, then

$$(\xi_\mu T^{\mu\nu})_{;\nu} = \xi_{\mu;\nu}T^{\mu\nu} + \xi_\mu T^{\mu\nu}{}_{;\nu} = 0\,. \tag{8.54}$$

Indeed, the first term vanishes because $\xi_{\mu;\nu}$ is antisymmetric in μ and ν whereas $T^{\mu\nu}$ is symmetric, and the second term vanishes due to Eq. 8.53.

Since there is a contraction on the index μ, the quantity $(\xi_\mu T^{\mu\nu})$ is a vector, whose four-divergence vanishes according to Eq. 8.54. As discussed in Box 8-B, the vanishing of the covariant four-divergence of a vector leads to a conserved quantity which, in the present case, is (see Eq. 8.60)

$$\int_{x^0=const} \sqrt{-g}\,(\xi_\mu T^{\mu 0})\, dx^1 dx^2 dx^3 = const\,. \tag{8.55}$$

In classical mechanics energy is conserved when the Hamiltonian is independent of time; thus, the conservation of energy is associated to a symmetry with respect to time translations. In Sec. 8.2 we have shown that if a spacetime admits a timelike Killing vector field, with a suitable choice of coordinates the metric tensor can be made time independent; in this coordinate frame, the diffeomorphism along the Killing vector field is the time translation. Thus, in this case it is natural to interpret the quantity defined in Eq. 8.55 as a conserved energy. In a similar way, when the metric admits a spacelike Killing vector field, the associated conserved quantities are indicated as "momentum" or "angular momentum", depending on the coordinate choice (see Sec. 10.3).

Box 8-B

Divergence-free equations and conserved quantities

Let us consider a vector $\vec{A}$ which satisfies the divergence-free equation

$$A^{\nu}{}_{;\nu} = 0\,, \qquad \nu = 0, \ldots, 3\,, \tag{8.56}$$

which, according to Eq. 5.67, can be written as

$$A^{\nu}{}_{;\nu} = \frac{1}{\sqrt{-g}}\frac{\partial}{\partial x^{\nu}}\left(\sqrt{-g}A^{\nu}\right) = 0\,. \tag{8.57}$$

Splitting the $\nu = 0$ component from the space components, and integrating on a three-volume V which extends over all space, Eq. 8.57 gives

$$\frac{\partial}{\partial x^0}\int_V \sqrt{-g}\; A^0 d^3x = -\int_V \frac{\partial(\sqrt{-g}\; A^k)}{\partial x^k}\, d^3x\,, \qquad k = 1, \ldots, 3\,, \tag{8.58}$$

where the integration is performed at a fixed time, i.e. on a hypersurface $x^0 = const$. Since the integrand on the right-hand side of this equation is an *ordinary* three-dimensional divergence, and the integration element is just $d^3x = dx^1dx^2dx^3$, we can apply Gauss' theorem as in Euclidean space (see Box 5D, Eq. 5.53), and find

$$\int_V \frac{\partial(\sqrt{-g}\; A^k)}{\partial x^k}\, d^3x = \int_{\partial V} \sqrt{-g}\; A^k dS_k\,. \tag{8.59}$$

Since the volume V extends over all space, the boundary ∂V is at infinity and, assuming that A^i vanish at infinity, the right-hand side of Eq. 8.59 is zero. Consequently, the integral of $\sqrt{-g}A^0$ over the three-volume V is a conserved quantity

$$\int_V \sqrt{-g}\; A^0 d^3x = const\,. \tag{8.60}$$

8.4 HYPERSURFACE-ORTHOGONAL VECTOR FIELDS

Any vector field $\vec{V}$ of class C^1 identifies a congruence of worldlines, i.e. the set of curves to which the vector is tangent at any point (see Sec. 8.2). If there exists a family of hypersurfaces $\Sigma(x^\mu) = const$ such that, at each point, the worldlines of the congruence are perpendicular to that surface, $\vec{V}$ is said to be *hypersurface-orthogonal.* This is equivalent to require that $\vec{V}$ is, at any point $\mathbf{p}$, orthogonal to all vectors $\vec{t}$ tangent to the hypersurface in $\mathbf{p}$

$$\vec{t}\cdot\vec{V} = 0 \quad \rightarrow \quad t^\alpha V^\beta g_{\alpha\beta} = 0\,. \tag{8.61}$$

We shall now show that a hypersurface-orthogonal vector field is parallel to the gradient of Σ and that, vice versa, the gradient of Σ is hypersurface-orthogonal. We recall that, as described in Sec. 2.3, the gradient of a function $\Sigma(x^\mu)$ is a one-form,

$$d\Sigma \rightarrow_O \left(\frac{\partial\Sigma}{\partial x^0}, \frac{\partial\Sigma}{\partial x^1}, \cdots \frac{\partial\Sigma}{\partial x^n}\right) = \{\Sigma_{,\alpha}\}\,. \tag{8.62}$$

When we say that $\vec{V}$ is parallel to $d\Sigma$ we mean that the one-form dual to $\vec{V}$, i.e. $\tilde{V} \to \{g_{\alpha\beta}V^{\beta} \equiv V_{\alpha}\}$, satisfies the equation

$$V_{\alpha} = f\Sigma_{,\alpha} \,, \tag{8.63}$$

where f is a function of the coordinates $\{x^{\mu}\}$.

We shall first show that Eq. 8.63 implies Eq. 8.61. Since $\Sigma(x^{\mu}) = const$ on the hypersurface, given any curve $x^{\alpha}(s)$ lying on the hypersurface, and its tangent vector $t^{\alpha} = dx^{\alpha}/ds$, the directional derivative of $\Sigma(x^{\mu})$ along the curve vanishes

$$\frac{d\Sigma}{ds} = \frac{\partial\Sigma}{\partial x^{\alpha}}\frac{dx^{\alpha}}{ds} = \Sigma_{,\alpha}t^{\alpha} = 0\,. \tag{8.64}$$

Therefore Eq. 8.63 implies that for any vector $\vec{t}$ which is tangent to some curve on the surface

$$V_{\alpha}t^{\alpha} = f\Sigma_{,\alpha}t^{\alpha} = 0 \tag{8.65}$$

i.e. Eq. 8.61 is satisfied.

Let us now show the inverse, i.e. that if a vector field $\vec{V}$ satisfies Eq. 8.61 at any point **p** and for any $\vec{t}$ tangent to the surface in **p**, it also satisfies Eq. 8.63. The tangent space to that surface in **p** is a three-dimensional space, therefore there are three independent tangent vectors $\vec{t}_{(i)}$ $(i = 1, \ldots, 3)$, and

$$t^{\alpha}_{(i)}V_{\alpha} = 0\,. \tag{8.66}$$

These are three independent equations for the four unknowns V_{α}; their solution depends on one arbitrary constant. Since $t^{\alpha}_{(i)}\Sigma_{,\alpha} = 0$, V_{α} must coincide with $\Sigma_{,\alpha}$ modulo an overall constant, $V_{\alpha} = f\Sigma_{,\alpha}$. Repeating this procedure at any point **p** of the spacetime yields Eq. 8.63, with Σ function of the coordinates.

8.4.1 Frobenius' theorem

If the vector field $\vec{V}$ is hypersurface-orthogonal, it satisfies Eq. 8.63 and therefore

$$\begin{aligned}
V_{\alpha;\beta} - V_{\beta;\alpha} &= (f\Sigma_{,\alpha})_{;\beta} - (f\Sigma_{,\beta})_{;\alpha} \\
&= f\left(\Sigma_{,\alpha;\beta} - \Sigma_{,\beta;\alpha}\right) + \Sigma_{,\alpha}f_{;\beta} - \Sigma_{,\beta}f_{;\alpha} = \\
&= f\left(\Sigma_{,\alpha,\beta} - \Sigma_{,\beta,\alpha} - \Gamma^{\mu}_{\beta\alpha}\Sigma_{,\mu} + \Gamma^{\mu}_{\alpha\beta}\Sigma_{,\mu}\right) + \Sigma_{,\alpha}f_{,\beta} - \Sigma_{,\beta}f_{,\alpha} \\
&= V_{\alpha}\frac{f_{,\beta}}{f} - V_{\beta}\frac{f_{,\alpha}}{f}\,,
\end{aligned} \tag{8.67}$$

i.e.

$$V_{\alpha;\beta} - V_{\beta;\alpha} = V_{\alpha}\frac{f_{,\beta}}{f} - V_{\beta}\frac{f_{,\alpha}}{f}\,. \tag{8.68}$$

If we now define the following quantity, which is called *rotation* of the vector field $\vec{V}$

$$\omega^{\delta} = \frac{1}{2}\varepsilon^{\delta\alpha\beta\mu}V_{[\alpha;\beta]}V_{\mu}\,, \tag{8.69}$$

where $\varepsilon^{\delta\alpha\beta\mu}$ is the Levi-Civita tensor (see Box 8-C), from Eq. 8.68 it follows that

$$\omega^{\delta} = 0. \tag{8.70}$$

Thus, if the vector field $\vec{V}$ is hypersurface-orthogonal, its rotation vanishes identically. Actually, 8.70 is a necessary and sufficient condition for $\vec{V}$ to be hypersurface orthogonal; this result is known as *Frobenius' theorem.*

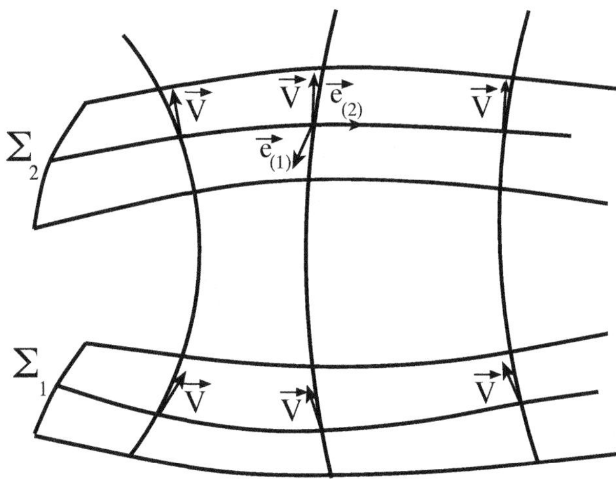

Figure 8.2: Hypersurface-orthogonal vector field.

8.4.2 Hypersurface-orthogonal vector fields and the choice of coordinate systems

The existence of a hypersurface-orthogonal vector field allows us to choose a coordinate frame such that the metric has a much simpler form. Let us consider, for the sake of simplicity, a three-dimensional spacetime described by the coordinates (x^0, x^1, x^2).

Let $\Sigma(x^\mu) = const$ be a family of surfaces to which the vector field $\vec{V}$ is orthogonal. In Fig. 8.2 we show two of such surfaces. As an example, let us assume that $\vec{V}$ is timelike, but a similar procedure can be used if $\vec{V}$ is spacelike. Since $\vec{V}$ is timelike, at each spacetime point **p** we can choose the basis vector $\vec{e}_{(0)}$ parallel to $\vec{V}$, and the remaining basis vectors as tangent to some curves lying on the surface passing through **p**, so that

$$g_{00} = g(\vec{e}_{(0)}, \vec{e}_{(0)}) = \vec{e}_{(0)} \cdot \vec{e}_{(0)} \neq 0\,, \tag{8.71}$$

$$g_{0i} = g(\vec{e}_{(0)}, \vec{e}_{(i)}) = 0, \qquad i = 1, 2\,. \tag{8.72}$$

Thus, with this choice, the metric becomes

$$ds^2 = g_{00}(dx^0)^2 + g_{ik}(dx^i)(dx^k), \qquad i, k = 1, 2\,. \tag{8.73}$$

The generalization of this example to the four-dimensional spacetime, in which case the surfaces $\Sigma(x^\mu) = const$ are hypersurfaces, is straightforward.

In general, given a timelike vector field $\vec{V}$, we can always choose a coordinate frame such that $\vec{e}_{(0)}$ is parallel to $\vec{V}$, so that in this frame

$$V^\alpha(x^\mu) = (V^0(x^\mu), 0, 0, 0)\,. \tag{8.74}$$

Such a coordinate system is called **comoving**. If, in addition, $\vec{V}$ is hypersurface-orthogonal, choosing the spacelike basis vectors as indicated above, $g_{0i} = 0$ and, as a consequence, the one-form associated to $\vec{V}$ also has the form

$$V_\alpha(x^\mu) = (V_0(x^\mu), 0, 0, 0)\,, \tag{8.75}$$

since $V_i = g_{i\mu}V^\mu = g_{i0}V^0 + g_{ik}V^k = 0\,,\ i = 1\ldots 3$.

Finally, let us assume that the spacetime admits a hypersurface-orthogonal timelike *Killing* vector field $\vec{V}$. In this case, as discussed in Sec. 8.2, if we choose the parameters of the curves such that V^0 is constant, then the metric 8.73 does not depend on time.

Box 8-C

The Levi-Civita completely antisymmetric tensor

We define the *Levi-Civita symbol* (also called *Levi-Civita tensor density*), $\epsilon_{\alpha\beta\gamma\delta}$ (with the indices running from 0 to 3), as an object whose components change sign under interchange of any pair of indices, and whose non-zero components are ± 1. Since it is completely antisymmetric, all the components with two equal indices are zero, and the only non-vanishing components are those for which all four indices are different. Conventionally, we set

$$\epsilon_{0123} = 1. \tag{8.76}$$

Under general coordinate transformations, $\epsilon_{\alpha\beta\gamma\delta}$ does not transform as a tensor; indeed, under the transformation $x^\alpha \to x^{\alpha'}$,

$$J\,\epsilon_{\alpha'\beta'\gamma'\delta'} = \frac{\partial x^\alpha}{\partial x^{\alpha'}}\frac{\partial x^\beta}{\partial x^{\beta'}}\frac{\partial x^\gamma}{\partial x^{\gamma'}}\frac{\partial x^\delta}{\partial x^{\delta'}}\,\epsilon_{\alpha\beta\gamma\delta}\,, \tag{8.77}$$

where the Jacobian J is defined as (see Chapter 5)

$$J \equiv \det\left(\frac{\partial x^\alpha}{\partial x^{\alpha'}}\right). \tag{8.78}$$

Indeed, the determinant of a square matrix $M = (M^\alpha{}_\beta)$ can always be written as $\det M = M^\alpha{}_0 M^\beta{}_1 M^\gamma{}_2 M^\delta{}_3 \epsilon_{\alpha\beta\gamma\delta}$.

We now define the *Levi-Civita tensor* as

$$\varepsilon_{\alpha\beta\gamma\delta} \equiv \sqrt{-g}\,\epsilon_{\alpha\beta\gamma\delta}\,. \tag{8.79}$$

Since, from Eq. 5.31, for a coordinate transformation $x^\alpha \to x^{\alpha'}$

$$|J| = \frac{\sqrt{-g'}}{\sqrt{-g}}\,, \tag{8.80}$$

then

$$\varepsilon_{\alpha'\beta'\gamma'\delta'} = \mathrm{sign}(J)\,\frac{\partial x^\alpha}{\partial x^{\alpha'}}\frac{\partial x^\beta}{\partial x^{\beta'}}\frac{\partial x^\gamma}{\partial x^{\gamma'}}\frac{\partial x^\delta}{\partial x^{\delta'}}\,\varepsilon_{\alpha\beta\gamma\delta}\,. \tag{8.81}$$

Strictly speaking, $\varepsilon_{\alpha\beta\gamma\delta}$ transforms as a tensor under a subset of the general coordinate transformations, i.e. those with $\mathrm{sign}(J) = +1$. Objects with this property are sometimes called *pseudo-tensors*, but we shall not follow this notation. In this book, "pseudo-tensors" are quantities which transform as tensors under linear coordinate transformations, see Sec. 13.6.1. *Warning:* do not confuse the Levi-Civita symbol, $\epsilon_{\alpha\beta\gamma\delta}$, with the Levi-Civita tensor, $\varepsilon_{\alpha\beta\gamma\delta}$. They only coincide if $g = -1$, as in flat space.

8.5 DIFFEOMORPHISM INVARIANCE OF GENERAL RELATIVITY

As discussed in Sec. 2.1.4 and at the beginning of this chapter, diffeomorphisms are different from coordinate transformations: the former are maps between different points of the manifold, while the latter associate different coordinates to the same point of the manifold, see e.g. Fig. 2.11. However, they have the same mathematical description, as a one-to-one set of C^1 functions:

$$x^\mu \mapsto x'^\alpha(x^\mu) \quad \text{for diffeomorphisms} \tag{8.82}$$

$$x^\mu \mapsto x^{\alpha'}(x^\mu) \quad \text{for coordinate transformations}\,. \tag{8.83}$$

We also know that, due to the Principle of General Covariance, Einstein's equations (both in vacuum and with sources) are invariant in form under general coordinate transformations, i.e. under a transformation 8.83. Consequently, since Eqs. 8.82 and 8.83 have the same form, Einstein's equations are *invariant under diffeomorphisms.*

In other words, the invariance of General Relativity under general coordinate transformations, i.e. the *gauge invariance* discussed in Sec 6.3, can be reformulated as the invariance of General Relativity under diffeomorphisms. The choice of the gauge (see e.g. Sec. 6.4) can be interpreted either as the choice of a coordinate transformation $x^{\alpha'}(x^\mu)$, or as the choice of a diffeomorphism $x'^\alpha(x^\mu)$.

Let us now consider the action of the gravitational field in the presence of sources (see Chapter 7),

$$S = S^{\rm EH} + S^{\rm fields}\,, \tag{8.84}$$

where the Einstein-Hilbert action $S^{\rm EH}$ depends on the metric and its derivatives, and $S^{\rm fields}$ depends on a set of fields, $\mathbf{\Phi}_{\boldsymbol{A}}$, on their covariant derivatives, and on the metric. Since Einstein's equations, both in vacuum and with sources, are invariant under diffeomorphisms, then both S and $S^{\rm EH}$ are diffeomorphism invariant. Thus, the action $S^{\rm fields}$ must also be diffeomorphism invariant. For an infinitesimal diffeomorphism along the direction of a vector field $\vec{\xi}$ (not necessarily a Killing vector),

$$\delta S^{\rm fields} = \frac{\delta S^{\rm fields}}{\delta g^{\mu\nu}}\delta g^{\mu\nu} + \frac{\delta S^{\rm fields}}{\delta \mathbf{\Phi}_{\boldsymbol{A}}}\delta\mathbf{\Phi}_{\boldsymbol{A}} = 0 \tag{8.85}$$

where $\delta g^{\mu\nu} = \mathcal{L}_{\vec{\xi}}\, g^{\mu\nu}$ and $\delta\mathbf{\Phi}_{\boldsymbol{A}} = \mathcal{L}_{\vec{\xi}}\,\mathbf{\Phi}_{\boldsymbol{A}}$ are the Lie derivatives (see Box 8-A) of the metric and of the fields $\mathbf{\Phi}_{\boldsymbol{A}}$, respectively.

If we restrict the action to the fields $\mathbf{\Phi}_{\boldsymbol{A}}$ which are solutions of their field equations, then (see Sec. 7.2)

$$\frac{\delta S^{\rm fields}}{\delta \mathbf{\Phi}_{\boldsymbol{A}}}\delta\mathbf{\Phi}_{\boldsymbol{A}} = \int d^4x\,\sqrt{-g}\left(\frac{\partial \mathcal{L}^{\rm fields}}{\partial \mathbf{\Phi}_{\boldsymbol{A}}} - \nabla_\mu \frac{\partial \mathcal{L}^{\rm fields}}{\partial(\nabla_\mu \mathbf{\Phi}_{\boldsymbol{A}})}\right)\delta\mathbf{\Phi}_{\boldsymbol{A}} = 0\,. \tag{8.86}$$

Thus, the second term on the right-hand side of Eq. 8.85 vanishes and, as a consequence, also the first term must be zero. Using Eqs. 7.47 and 7.52, this gives

$$\begin{aligned}
\int d^4x\,\frac{\partial(\sqrt{-g}\mathcal{L}^{\rm fields})}{\partial g^{\mu\nu}}\mathcal{L}_{\vec{\xi}}\, g^{\mu\nu} &= \int d^4x\,\sqrt{-g}\left(\frac{\partial\mathcal{L}^{\rm fields}}{\partial g^{\mu\nu}} - \frac{1}{2}g_{\mu\nu}\mathcal{L}^{\rm fields}\right)\mathcal{L}_{\vec{\xi}}\, g^{\mu\nu} \qquad (8.87)\\
&= -\frac{1}{2c}\int d^4x\,\sqrt{-g}\,T_{\mu\nu}\mathcal{L}_{\vec{\xi}}\, g^{\mu\nu}\\
&= -\frac{1}{c}\int d^4x\,\sqrt{-g}\,T_{\mu\nu}\xi^{\mu;\nu} = 0\,,
\end{aligned}$$

where in the last step we have used $\mathcal{L}_{\vec{\xi}}\, g^{\mu\nu} = -\xi^{\mu;\nu} - \xi^{\nu;\mu}$ (see Box 8-A) and the symmetry of the tensor $T_{\mu\nu}$, i.e. $T_{\mu\nu}(\xi^{\mu;\nu} + \xi^{\nu;\mu}) = 2T_{\mu\nu}\xi^{\mu;\nu}$. By integrating by parts and applying Gauss' theorem, we find

$$\int d^4x \sqrt{-g}\, T_{\mu\nu}{}^{;\nu}\xi^{\mu} = 0$$

for any vector field $\vec{\xi}$ vanishing at the boundary of the integration volume, and thus $T_{\mu\nu}{}^{;\nu} = 0$ or, in the usual form,

$$T^{\mu\nu}{}_{;\nu} = 0\,. \tag{8.88}$$

Thus, as a consequence of the diffeomorphism invariance of General Relativity, the divergenceless equation satisfied by the stress-energy tensor is valid for *any* metric (not necessarily solution to Einstein's equations), as long as the fields $\mathbf{\Phi_A}$ are the solutions of the field equations. Indeed, note that this derivation does not rely on the specific form of the Einstein-Hilbert action, but only on the diffeomorphism invariance of the latter.

CHAPTER 9

The Schwarzschild solution

The Schwarzschild solution was derived by Karl Schwarzschild in 1916. It describes the gravitational field exterior to a static, spherically symmetric, non-rotating body; in addition, it describes a non-rotating black hole. Even though it was the first exact solution of Einstein's equations ever found, a complete understanding of the Schwarzschild spacetime was achieved much later, in the mid-sixties of the 20th century.

It is interesting to know that Schwarzschild's paper was communicated to the Berlin Academy by Einstein on the 13th of January 1916, about two months after the publication of Einstein's seminal papers on the theory of General Relativity. In those years Schwarzschild was very ill, since he had contracted a fatal disease in 1915, while serving the German army at the eastern front. He died on the 11th of May 1916, and during his illness he wrote two papers on the newborn theory of General Relativity, one describing the solution which we are going to derive and discuss in this chapter, and the second describing the interior solution for a static, spherically symmetric star of constant density, which we shall discuss in Chapter 16.

9.1 STATIC AND SPHERICALLY SYMMETRIC SPACETIMES

We wish to find an exact solution of Einstein's equations in vacuum, which is *static* and *spherically symmetric*. This will be the relativistic generalization of the Newtonian potential

$$\Phi(r) = -\frac{GM}{r}\,, \tag{9.1}$$

and will describe the gravitational field in the exterior of a spherically symmetric, non-rotating body. Let us see how the symmetries of the problem can be used to simplify the expression of the metric tensor.

a) Symmetry with respect to time transformations

When dealing with symmetries with respect to transformations of the time coordinate, it is important to distinguish between stationary and static spacetimes.

A spacetime is **stationary** if it admits a *timelike Killing vector field* $\vec{k}$. As explained in Sec. 8.2, from the Killing equation it follows that, by a suitable choice of the coordinate system, the metric of a stationary spacetime can be made independent of time,

$$\frac{\partial g_{\alpha\beta}}{\partial x^0} = 0\,. \tag{9.2}$$

A spacetime is **static** if it admits a *hypersurface-orthogonal, timelike Killing vector field* $\vec{k}$.

In this case, as shown in Sec. 8.4, we can choose the vectors of the coordinate basis in such a way that, at each spacetime point, $\vec{e}_{(0)}$ coincides with $\vec{k}$, and the remaining basis vectors are tangent to the surfaces at which $\vec{k}$ is orthogonal, so that in this frame $k^\mu = (k^0(x^i), 0, 0, 0)$ (as usual, Greek indices run from 0 to 3, and Latin indices run from 1 to 3). Consequently, the line element takes the simple form

$$ds^2 = g_{00}(x^i)(dx^0)^2 + g_{kj}(x^i)dx^k dx^j\,, \tag{9.3}$$

where

$$g_{00} = \boldsymbol{g}(\vec{k}, \vec{k}) = \vec{k} \cdot \vec{k}\,. \tag{9.4}$$

Thus, if the spacetime is static the metric tensor is not only independent of time, but also invariant under time reversal $t \to -t$. Note that if terms like $dx^0 dx^i$ were present in the line element, this would not be true.

b) Spatial symmetry

Using the assumption of spherical symmetry, it is possible to express the spatial part of the metric in a simple form. To this aim, we "fill" the space with concentric spherical surfaces, which we describe with spherical coordinates $x^2 = \theta$ and $x^3 = \varphi$ (see Box 2-C). With this choice, the line element of a two-sphere of radius a is (see Eq. 3.69)

$$ds^2_{(2)} = g_{22}(dx^2)^2 + g_{33}(dx^3)^2 = a^2(d\theta^2 + \sin^2\theta d\varphi^2)\,. \tag{9.5}$$

The surface of this sphere is

$$A = \int \sqrt{g_{(2)}}\; d\theta d\varphi = \int_0^\pi a^2 \sin\theta d\theta \int_0^{2\pi} d\varphi = 4\pi a^2\,, \tag{9.6}$$

where $g_{(2)} = a^4 \sin^2\theta$ is the determinant of the metric 9.5. The length of its maximum circumference (we consider, without loss of generality, the circumference on the equatorial plane, $\theta = \pi/2$) is

$$C = \int_0^{2\pi} \sqrt{g_{\varphi\varphi}(\theta = \pi/2)d\varphi} = \int_0^{2\pi} a \sin(\pi/2)d\varphi = 2\pi a\,, \tag{9.7}$$

and the ratio between the two is $A/C = 2a$. The assumption of spherical symmetry implies that the function a does not depend on (θ, φ) but is an arbitrary function of the remaining coordinates x^0, x^1

$$ds^2_{(2)} = a^2(x^0, x^1)(d\theta^2 + \sin^2\theta d\varphi^2)\,. \tag{9.8}$$

Since in addition the spacetime is static, the metric tensor does not depend on time; therefore $a = a(x^1)$. We are now free to make a coordinate transformation and choose

$$r = a(x^1)\,. \tag{9.9}$$

Thus, we *define* the radial coordinate r as being half the ratio between the surface and the circumference of the two-sphere, as it is in flat space. However, we anticipate that the coordinate r *is not* the distance between the center of the sphere and the surface, as we shall later show.

Let us now consider two two-spheres with radii r and $r + dr$, respectively, as shown in Fig. 9.1. We choose the coordinates θ, φ on each sphere such that: i) the poles of the spheres are aligned along the same axis (the axis x^3 in Fig. 9.1), ii) the coordinate φ is measured having the axis x^1 as a reference.

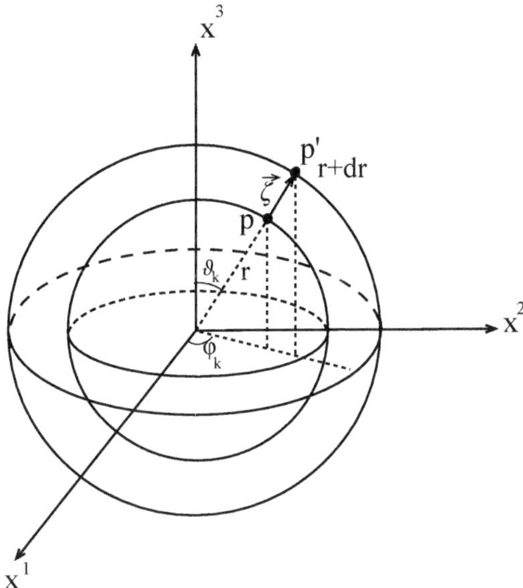

Figure 9.1: Choice of coordinates on the two-spheres filling the space. Poles are aligned along the axis x^3, whereas φ is measured as a counterclockwise rotation angle with respect to the axis x^1.

With this choice, the vector $\vec{\zeta}$ joining the points **p** and **p**$'$ which have the same values of $\theta = \theta_k$ and $\varphi = \varphi_k$ (see Fig. 9.1), is orthogonal to both spheres. If we now extend this choice to all the two-spheres filling the space, $\vec{\zeta}$ will be a vector field orthogonal, at each space point, to one of these two-spheres, i.e. $\vec{\zeta}$ is a hypersurface-orthogonal vector field. We can thus choose the three spacelike basis vectors as follows: i) $\vec{e}_{(r)}$ is parallel, at each space point, to $\vec{\zeta}$, and ii) $\vec{e}_{(\theta)}$ and $\vec{e}_{(\varphi)}$ are tangent to the coordinate lines $\varphi = const$, $\theta = const$, respectively, i.e. tangent to the two-spheres. With this choice, the basis vector $\vec{e}_{(r)}$ will be orthogonal to $\vec{e}_{(\theta)}$ and $\vec{e}_{(\varphi)}$. Consequently

$$g_{r\theta} = \vec{e}_{(r)} \cdot \vec{e}_{(\theta)} = 0, \qquad g_{r\varphi} = \vec{e}_{(r)} \cdot \vec{e}_{(\varphi)} = 0\,; \tag{9.10}$$

thus, the line element of the three-space becomes

$$ds^2_{(3)} = g_{rr}dr^2 + r^2(d\theta^2 + \sin^2\theta d\varphi^2), \tag{9.11}$$

and that of the four-dimensional spacetime given in Eq. 9.3 reduces to

$$ds^2 = g_{00}(dx^0)^2 + g_{rr}dr^2 + r^2(d\theta^2 + \sin^2\theta d\varphi^2)\,. \tag{9.12}$$

The metric components g_{00} and g_{rr} could, in principle, depend on (r, θ, φ). However, this is not the case for a spherically-symmetric metric. Indeed, if we choose a new set of global polar coordinates (θ', φ') to label the points on the concentric two-spheres which fill the space, neither $\vec{e}_{(0)}$, nor $\vec{e}_{(r)}$, will be affected by this change, being orthogonal to the new basis vectors $\vec{e}_{(\theta')}$ and $\vec{e}_{(\varphi')}$ tangent to the two-spheres; therefore $g_{00} = \vec{e}_{(0)} \cdot \vec{e}_{(0)}$ and $g_{rr} = \vec{e}_{(r)} \cdot \vec{e}_{(r)}$ cannot depend on the angular coordinates. Consequently

$$g_{00} = g_{00}(r), \qquad \text{and} \qquad g_{rr} = g_{rr}(r)\,. \tag{9.13}$$

Summarizing, the line element of a static and spherically-symmetric spacetime can always be written as

$$ds^2 = g_{00}(r)(dx^0)^2 + g_{rr}(r)dr^2 + r^2(d\theta^2 + \sin^2\theta d\varphi^2). \tag{9.14}$$

It is convenient to rewrite the metric in the following form

$$ds^2 = -e^{2\nu(r)}(dx^0)^2 + e^{2\lambda(r)}dr^2 + r^2(d\theta^2 + \sin^2\theta d\varphi^2), \tag{9.15}$$

where

$$g_{00}(r) = -e^{2\nu(r)}, \qquad \text{and} \qquad g_{rr}(r) = e^{2\lambda(r)}. \tag{9.16}$$

Let us now compute the proper distance between two points $P_1 = (x^0, r_1, \theta, \varphi)$, and $P_2 = (x^0, r_2, \theta, \varphi)$ (with the same x^0, θ and φ)

$$l = \int_{r_1}^{r_2} e^{\lambda} dr. \tag{9.17}$$

As anticipated above, the proper distance between two points having the same values of the time and of the angular coordinates (θ, φ) *does not* coincide with $(r_2 - r_1)$. This is due to the fact that, unless the spacetime is flat and $e^{\lambda} = 1$, the presence of the gravitational field distorts the spacetime and changes the metric relations among points with respect to those existing in flat spacetime.

9.2 THE SCHWARZSCHILD SOLUTION

We can now write the components of the Einstein tensor $G_{\mu\nu} = R_{\mu\nu} - \frac{1}{2}g_{\mu\nu}R$ in terms of the metric tensor 9.15. By replacing in the expression of the Ricci tensor, Eq. 4.54, the Christoffel symbols computed in Box 9-A (we leave the explicit computation as an exercise) we find

$$G_{00} = \frac{1}{r^2}e^{2\nu}\frac{d}{dr}\left[r(1 - e^{-2\lambda})\right], \tag{9.18}$$

$$G_{rr} = -\frac{1}{r^2}e^{2\lambda}\left(1 - e^{-2\lambda}\right) + \frac{2}{r}\nu_{,r}, \tag{9.19}$$

$$G_{\theta\theta} = r^2 e^{-2\lambda}\left[\nu_{,rr} + \nu_{,r}^2 + \frac{\nu_{,r}}{r} - \nu_{,r}\lambda_{,r} - \frac{\lambda_{,r}}{r}\right], \tag{9.20}$$

$$G_{\varphi\varphi} = \sin^2\theta\, G_{\theta\theta}. \tag{9.21}$$

The remaining components vanish identically. Since we are looking for a vacuum solution, the equations to solve are

$$G_{\mu\nu} = 0. \tag{9.22}$$

Equation 9.18 gives

$$e^{2\lambda} = \frac{1}{1 - \frac{K}{r}}, \tag{9.23}$$

where K is an integration constant. Equation 9.19 then gives

$$\nu_{,r} = \frac{1}{2}\frac{K}{r(r-K)}, \tag{9.24}$$

the solution of which is

$$\nu = \frac{1}{2}\log\left(1 - \frac{K}{r}\right) + \nu_0, \qquad \rightarrow \qquad e^{2\nu} = \left(1 - \frac{K}{r}\right)e^{2\nu_0}, \tag{9.25}$$

and ν_0 is an integration constant. We can rescale the time coordinate

$$x^0 \to e^{-\nu_0} x^0 , \qquad dx^0 \to e^{-\nu_0} dx^0 , \tag{9.26}$$

in such a way that

$$g_{00}(dx^0)^2 = -e^{2\nu}(dx^0)^2 \qquad \to \qquad -e^{2\nu}e^{-2\nu_0}(dx^0)^2 = -\left(1 - \frac{K}{r}\right)(dx^0)^2 . \tag{9.27}$$

Owing to the staticity of the spacetime, the rest of the metric is not affected by the rescaling 9.26.

The final form of the solution is [1]

$$ds^2 = -\left(1 - \frac{K}{r}\right)c^2dt^2 + \frac{dr^2}{1 - \frac{K}{r}} + r^2\left(d\theta^2 + \sin^2\theta d\varphi^2\right) . \tag{9.28}$$

This metric is said to be *asymptotically flat*, since at asymptotic infinity (when $r \to \infty$) it reduces to Minkowski's metric in spherical coordinates. It is easy to check that the solution 9.28 satisfies also the equation $G_{\theta\theta} = 0$ (see Eq. 9.20), which can be shown to be a consequence of $G_{00} = 0$ and $G_{rr} = 0$ (see Eqs. 9.18 and 9.19).

In order to understand what is the meaning of the integration constant K we go back to Sec. 6.1, where we showed that in the weak-field, stationary limit, the geodesic equations reduce to the Newtonian equations of motion provided

$$g_{00} = -\left(1 + \frac{2\Phi}{c^2}\right) , \tag{9.29}$$

where Φ is the Newtonian potential, solution of the Poisson equation 6.2. Since we are looking for a solution describing the gravitational field in the exterior of a spherical distribution of matter, we know from Newtonian gravity that $\Phi = -\frac{GM}{r}$ and hence in the weak-field, stationary limit

$$g_{00} = -\left(1 - \frac{2GM}{c^2r}\right) ; \tag{9.30}$$

from Eq. 9.28 we see that

$$g_{00} = -e^{2\nu} = -\left(1 - \frac{K}{r}\right) . \tag{9.31}$$

Comparing Eq. 9.30 with Eq. 9.31 we find

$$K = \frac{2GM}{c^2} . \tag{9.32}$$

The static, spherically symmetric solution of Einstein's equations is then

$$ds^2 = -\left(1 - \frac{2GM}{c^2r}\right)c^2dt^2 + \frac{dr^2}{1 - \frac{2GM}{c^2r}} + r^2\left(d\theta^2 + \sin^2\theta d\varphi^2\right) . \tag{9.33}$$

The metric 9.33 is called the **Schwarzschild solution**. We shall indicate the coordinates (t, r, θ, φ) as the *Schwarzschild coordinates*.

[1]If $r > K$, then $1 - \frac{K}{r} > 0$, therefore $g_{00} < 0$, $g_{11} > 0$ and the exponentials $e^{2\nu}$, $e^{2\lambda}$ are both positive, but – as we shall discuss in Sec. 9.4 – it may occur that $1 - \frac{K}{r} < 0$; in this case the exponentials are negative and their arguments are imaginary functions. However, the replacement $g_{00} = -e^{2\nu}$, $g_{11} = e^{2\lambda}$ is only used in the course of this derivation. Hereafter, we shall write the metric components in the form 9.28 (or in analogue forms, e.g. Eq. 9.33 below).

The constant K is equal to the physical mass of the solution multiplied by the factor $\frac{2G}{c^2}$ and it has dimensions of a length. The quantity

$$R_S = \frac{2GM}{c^2} \tag{9.34}$$

is called **Schwarzschild radius** and plays a crucial role in the interpretation of the metric 9.33, as we shall discuss in the next section.

In geometrized units $G = c = 1$ (see Box 9-B), the Schwarzschild solution reads

$$ds^2 = -\left(1 - \frac{2m}{r}\right) dt^2 + \frac{dr^2}{1 - \frac{2m}{r}} + r^2 \left(d\theta^2 + \sin^2\theta d\varphi^2\right) , \tag{9.35}$$

where m is related to the physical mass M by

$$m = \frac{GM}{c^2} , \tag{9.36}$$

and is called the *geometrical mass.*

Being static and spherically symmetric, the Schwarzschild metric admits four Killing vector fields (see Sec. 8.2): one timelike vector field, associated to time translations, and the three Killing vector fields of a spherical surface, given in Eq. 8.45:

$$\begin{aligned} \vec{k} &\equiv \frac{\partial}{\partial t} , \\ \vec{m}^{(1)} &\equiv \frac{\partial}{\partial \varphi} , \\ \vec{m}^{(2)} &\equiv \sin\varphi \frac{\partial}{\partial \theta} + \cos\varphi \cot\theta \frac{\partial}{\partial \varphi} , \\ \vec{m}^{(3)} &\equiv \cos\varphi \frac{\partial}{\partial \theta} - \sin\varphi \cot\theta \frac{\partial}{\partial \varphi} . \end{aligned} \tag{9.37}$$

In the following we shall use the Killing vectors $\vec{k}$ and $\vec{m} \equiv \vec{m}^{(1)}$. In coordinates (t, r, θ, φ), their components are

$$\begin{aligned} k^\mu &= (1, 0, 0, 0) \\ m^\mu &= (0, 0, 0, 1) . \end{aligned} \tag{9.38}$$

Box 9-A

Christoffel's symbols of the Schwarzschild spacetime

The Christoffel symbols of the metric 9.15,

$$ds^2 = -e^{2\nu(r)}(dx^0)^2 + e^{2\lambda(r)}dr^2 + r^2(d\theta^2 + \sin^2\theta d\varphi^2)\,, \tag{9.39}$$

can be easily computed using Eq. 3.68,

$$\Gamma^{\gamma}_{\beta\mu} = \frac{1}{2}g^{\gamma\alpha}(g_{\alpha\beta,\mu} + g_{\alpha\mu,\beta} - g_{\beta\mu,\alpha})\,. \tag{9.40}$$

The non-vanishing ones are (we leave the explicit computation to the reader as an exercise):

$$\begin{array}{lll} \Gamma^r_{00} = e^{2(\nu-\lambda)}\nu_{,r} & \Gamma^0_{0r} = \Gamma^0_{r0} = \nu_{,r} & \Gamma^r_{rr} = \lambda_{,r} \\ \Gamma^r_{\theta\theta} = -re^{-2\lambda} & \Gamma^\theta_{r\theta} = \Gamma^\theta_{\theta r} = \frac{1}{r} & \Gamma^\varphi_{r\varphi} = \Gamma^\varphi_{\varphi r} = \frac{1}{r} \\ \Gamma^r_{\varphi\varphi} = -r\sin^2\theta e^{-2\lambda} & \Gamma^\theta_{\varphi\varphi} = -\sin\theta\cos\theta & \Gamma^\varphi_{\theta\varphi} = \Gamma^\varphi_{\varphi\theta} = \cot\theta\,. \end{array} \tag{9.41}$$

By replacing $e^{2\nu} = e^{-2\lambda} = 1 - \frac{2m}{r}$ in Eq. 9.41, we find that the non-vanishing Christoffel symbols of the Schwarzschild spacetime are:

$$\begin{array}{lll} \Gamma^0_{0r} = \Gamma^0_{r0} = \frac{m}{r^2}\left(1 - \frac{2m}{r}\right)^{-1} & \Gamma^r_{00} = \frac{m}{r^2}\left(1 - \frac{2m}{r}\right) & \Gamma^r_{rr} = -\frac{m}{r^2}\left(1 - \frac{2m}{r}\right)^{-1} \\ \Gamma^r_{\theta\theta} = -(r-2m) & \Gamma^r_{\varphi\varphi} = -(r-2m)\sin^2\theta & \Gamma^\varphi_{r\varphi} = \Gamma^\varphi_{\varphi r} = \frac{1}{r} \\ \Gamma^\theta_{r\theta} = \Gamma^\theta_{\theta r} = \frac{1}{r} & \Gamma^\theta_{\varphi\varphi} = -\cos\theta\sin\theta & \Gamma^\varphi_{r\varphi} = \Gamma^\varphi_{\varphi r} = \cot\theta\,. \end{array} \tag{9.42}$$

Box 9-B

Geometrized units

It is easy to check that $R_S = \frac{2GM}{c^2}$ has the dimension of a length. Indeed the constants G and c, whose values are (see Table A)

$$G = 6.674 \times 10^{-8} \, \frac{\mathrm{cm}^3}{\mathrm{g\,s}^2} \,, \qquad c = 2.998 \times 10^{10} \ \mathrm{cm/s} \,,$$

have dimensions

$$[G] = \frac{(\mathrm{length})^3}{(\mathrm{mass})\,(\mathrm{time})^2} \,, \qquad [c] = \frac{(\mathrm{length})}{(\mathrm{time})} \,;$$

therefore

$$\left[\frac{G}{c^2}\right] = \frac{(\mathrm{length})}{(\mathrm{mass})} \,, \quad \text{with} \quad \frac{G}{c^2} = 7.425 \times 10^{-29} \mathrm{cm} \cdot \mathrm{g}^{-1} \,. \tag{9.43}$$

It is often convenient to use *geometrized units* putting

$$G = c = 1 \,. \tag{9.44}$$

In these units, masses, lengths, and time have the same dimensions. To recover a quantity in physical units, it is necessary to multiply it by suitable powers of G and c. For example, the quantities

$$M \,, \quad \frac{GM}{c^2} \,, \quad \frac{GM}{c^3} \,, \tag{9.45}$$

have the same expression in geometrized units, but their physical dimensions are those of mass, length, and time, respectively. Likewise the quantity

$$m = \frac{GM}{c^2} \tag{9.46}$$

has the dimension of a length, and it is usually referred to as the **geometrical mass** or the **gravitational radius**. As an example, the mass of the Sun, $M_\odot = 1.989 \times 10^{33}$ g, in geometrized units is

$$m_\odot = \frac{GM_\odot}{c^2} = (7.425 \times 10^{-29}) \cdot (1.989 \times 10^{33}) \ \mathrm{cm} = 1.477 \ \mathrm{km} \,. \tag{9.47}$$

Two useful conversion factors are the following:

$$m_\odot = 1.477 \, \mathrm{km} = 4.926 \times 10^{-6} \, \mathrm{s} \,. \tag{9.48}$$

While geometrized units $G = c = 1$ are convenient in General Relativity, there exist other possible choices, such as those in which $\hbar = c = 1$, mostly used in Quantum Field Theory. We note that in the $G = c = 1$ units the mass has the dimension of a length, and Planck's constant has the dimension of a length squared; its value is $\hbar = l_{\mathrm{P}}^2$, where $l_{\mathrm{P}} = 1.616 \times 10^{-33}$ cm is called *Planck length*. In the $\hbar = c = 1$ units, instead, the mass has the dimension of an inverse length, while Newton's constant has the dimension of a length squared (with $G = l_{\mathrm{P}}^2$). In physical units, $l_{\mathrm{P}}^2 = \hbar G / c^3$.

9.3 SINGULARITIES OF THE SCHWARZSCHILD SOLUTION

For the sake of simplicity, to hereafter we shall use the geometrized units introduced in Box 9-B and we will refer to the Schwarzschild solution in the form 9.35. An inspection of the metric shows that

$$\begin{aligned} \text{when} \quad r \to 2m: &\quad g_{00} \to 0\,, \qquad g_{rr} \to \infty\,, \\ r \to 0: &\quad g_{00} \to \infty\,, \qquad g_{rr} \to 0\,. \end{aligned} \tag{9.49}$$

In both cases, the components of the metric tensor are singular. Similarly, the (non-vanishing) components of the Riemann tensor (which can be computed by replacing the Christoffel symbols of Box 9-A in Eq. 4.18),

$$\begin{aligned} R^t{}_{rtr} &= -2\frac{m}{r^3}\left(1-\frac{2m}{r}\right)^{-1} \\ R^t{}_{\theta t\theta} &= \frac{1}{\sin^2\theta}R^t{}_{\varphi t\varphi} = \frac{m}{r^5} \\ R^\theta{}_{\varphi\theta\varphi} &= 2\frac{m}{r^5}\sin^2\theta \\ R^r{}_{\theta r\theta} &= \frac{1}{\sin^2\theta}R^r{}_{\varphi r\varphi} = -\frac{m}{r^5}\,, \end{aligned} \tag{9.50}$$

are singular both in $r = 0$ and in $r = 2m$. However, the fact that the components of these tensors diverge at some point is not indicative of a physical pathology, since the components of a tensor depend on the coordinate system. Thus, in principle, there may exist a frame in which one or both singularities disappear. Indeed we shall show that the two singularities are of a very different nature.

In order to check whether a singularity is physical or due to an improper coordinate choice, we must compute quantities that do not change under coordinate transformations, i.e. *scalar* quantities. If at least one of the scalars constructed from the Riemann tensor – the *curvature invariants* – diverges at some point, we can consider that point as a genuine **curvature singularity**, because the spacetime curvature becomes infinite regardless of the choice of the coordinate frame [2].

The only scalar quantity which is linear in the Riemann tensor is the Ricci curvature $R = g^{\mu\nu}R_{\mu\nu}$ (see Eq. 4.20), but for the Schwarzschild solution it does not give useful information, since in vacuum $R_{\mu\nu} \equiv 0$, and thus $R = 0$. The scalars quadratic in the Riemann tensor are (see Box 7-D):

- $R_{\mu\nu}R^{\mu\nu}$ (which also vanishes in vacuum);
- the *Kretschmann scalar* $K_1 = R_{\mu\nu\alpha\beta}R^{\mu\nu\alpha\beta}$;
- the *Chern-Pontryagin scalar* $K_2 = \varepsilon_{\mu\nu}{}^{\rho\sigma}R^{\mu\nu\alpha\beta}R_{\rho\sigma\alpha\beta}$.

Further curvature invariants can be constructed at higher polynomial orders in the Riemann tensor, therefore the number of curvature invariants is infinite [3]. For the Schwarzschild solution the Chern-Pontryagin scalar vanishes identically, whereas the Kretschmann scalar is

$$K_1 = \frac{48m^2}{r^6}\,, \tag{9.51}$$

[2] Actually, the concept of spacetime singularity is subtler than it may appear from this remark. A more rigorous characterization of singularities in General Relativity will be discussed in Sec. 9.5.1.

[3] However, it can be shown that all higher-order scalar polynomials in the Riemann tensor are combinations of fourteen independent curvature invariants (see e.g. [117]).

which is regular at $r = 2m$, and diverges at $r \to 0$. The higher-order curvature invariants are also regular at $r = 2m$ and singular at $r \to 0$. We conclude (see Sec. 9.5.2) that $r = 0$ is a true curvature singularity, whereas $r = 2m$ is only a **coordinate singularity**, due to an inappropriate choice of coordinates. Coordinate singularities will be discussed in more detail in Sec. 9.5, where we shall show that the $r = 2m$ singularity can be removed by a suitable coordinate choice.

Although $r = 2m$ is not a curvature singularity, this surface has some peculiar properties which we shall discuss in the next section.

9.4 SPACELIKE, TIMELIKE, AND NULL HYPERSURFACES

In order to analyse the properties of the surface $r = 2m$, we shall introduce a classification of hypersurfaces in a curved spacetime. Consider a generic hypersurface

$$\Sigma(x^\mu) = 0\,, \tag{9.52}$$

and a point $\mathbf{p}$ on Σ. Let $\vec{n}$ be the normal vector to Σ in $\mathbf{p}$, dual to the gradient one-form

$$n_\alpha = \Sigma_{,\alpha}\,. \tag{9.53}$$

Let $x^\alpha(\lambda)$ be *any* curve *on* Σ passing through $\mathbf{p}$, and $\vec{t}$ the tangent vector to the curve, $t^\alpha = \frac{dx^\alpha}{d\lambda}$ in $\mathbf{p}$. The vectors $\vec{n}$ and $\vec{t}$ are orthogonal, indeed

$$n_\alpha t^\alpha = \frac{\partial \Sigma}{\partial x^\alpha}\frac{dx^\alpha}{d\lambda} = \frac{d\Sigma}{d\lambda} = 0\,. \tag{9.54}$$

At any point of the hypersurface we can introduce a locally inertial frame, and rotate its axes in such a way that the components of $\vec{n}$ are

$$n^\alpha = (n^0, n^1, 0, 0) \qquad \text{and} \qquad n_\alpha n^\alpha = -(n^0)^2 + (n^1)^2\,. \tag{9.55}$$

Since $\vec{t}$ is orthogonal to $\vec{n}$, it follows that

$$n_\alpha t^\alpha = -n^0 t^0 + n^1 t^1 = 0 \qquad \to \qquad \frac{t^0}{t^1} = \frac{n^1}{n^0}. \tag{9.56}$$

Consequently, in this frame the tangent vector has components

$$t^\alpha = \Lambda(n^1, n^0, a, b) \qquad \text{with a, b, and } \Lambda \text{ arbitrary constants}\,, \tag{9.57}$$

and its norm is

$$t_\alpha t^\alpha = \Lambda^2[-(n^1)^2 + (n^0)^2 + (a^2+b^2)] = \Lambda^2[-n_\alpha n^\alpha + (a^2+b^2)]\,. \tag{9.58}$$

We classify Σ as *spacelike, timelike, and null hypersurfaces* according to the following:

1) $n_\alpha n^\alpha < 0$, $\to$ n^α is a timelike vector $\to$ Σ **is spacelike**
2) $n_\alpha n^\alpha > 0$, $\to$ n^α is a spacelike vector $\to$ Σ **is timelike**
3) $n_\alpha n^\alpha = 0$, $\to$ n^α is a null vector $\to$ Σ **is null**.

It should be stressed that the spacelike, timelike, or null nature of a hypersurface is a *local* property.

We shall now discuss how the light cones through $\mathbf{p}$ are oriented with respect to Σ in the three cases.

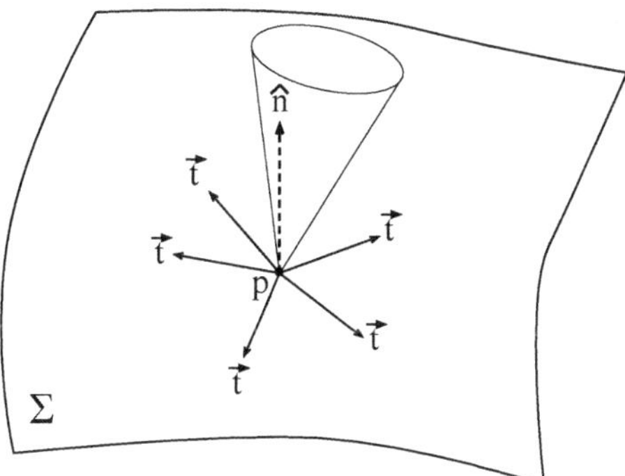

Figure 9.2: The light cone through a point **p** of a spacelike hypersurface Σ does not intersect the hypersurface. A massive/massless particle passing through **p** can cross Σ in only one direction (along the positive direction of the x^0 axis).

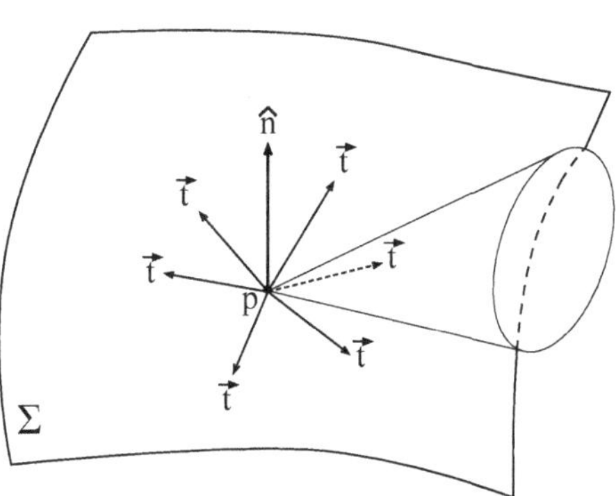

Figure 9.3: The light cone through a point **p** of a timelike hypersurface Σ intersects the hypersurface. A massive/massless particle passing through **p** can cross Σ inwards and outwards.

1. If $n_\alpha n^\alpha < 0$, from Eq. 9.58 it follows that $t_\alpha t^\alpha > 0$, i.e. $\vec{t}$ is spacelike. This is true for all tangent vectors to the curves laying on Σ and passing through **p**. Consequently, no tangent vector to Σ in **p** lies inside, or on the light cone through **p** (see Fig. 9.2). Particles passing through **p** must move inside the light cone if they are massive, or on the light cone if massless; since all particles must move in the positive time direction to preserve causality, *a spacelike hypersurface can be crossed only in one direction.*

2. If $n_\alpha n^\alpha > 0$, then $t_\alpha t^\alpha$ can be positive, negative, or null depending on the value of $a^2 + b^2$. Therefore tangent vectors can lie inside, outside, or on the light cone, which cuts the surface Σ as in Fig. 9.3 . Consequently a *timelike hypersurface can be crossed inwards and outwards.*

3. If $n_\alpha n^\alpha = 0$, then $t_\alpha t^\alpha$ is positive (t^α is spacelike), or null if $a = b = 0$. In this case there is only one tangent vector (and all its multiples) to Σ in **p** which lies on the light cone, as shown in Fig. 9.4. Consequently a *null hypersurface can be crossed only in*

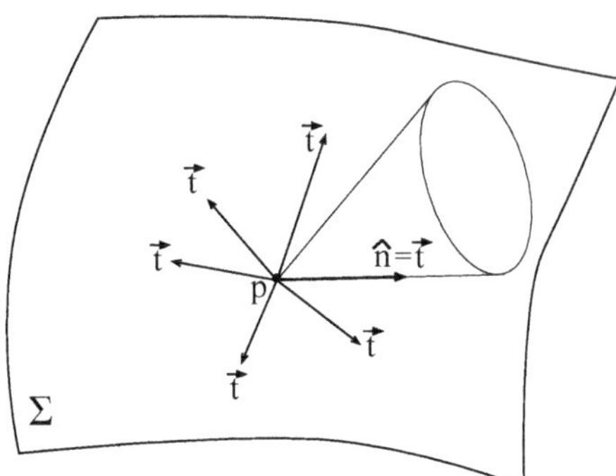

Figure 9.4: The light cone through a point **p** of a null hypersurface Σ is tangent to the hypersurface. A massive/massless particle passing through **p** can cross Σ only in one direction. Note that in this case, since $n^\mu n_\mu = 0$, the normal vector is also a tangent vector.

> *one direction.* In addition, while massive particles have to cross the surface, massless particles can either cross the null hypersurface, or move *on*.

For example, in Minkowski's spacetime $t = const$ is a spacelike hypersurface, and massive or null particles can cross it only in one direction without violating causality; $x^i = const$ (with $i = 1, \ldots, 3$) is a timelike hypersurface, and massive or null particles can cross it in either directions; $x^i - ct = const$ (with $i = 1, \ldots, 3$) is a null surface: massive particles can cross it only in one direction, whereas massless particles can also move on the null line along which the light cone is tangent to the surface.

9.4.1 Constant radius hypersurfaces in Schwarzschild's spacetime

Let us consider a generic hypersurface $r = const$ in the Schwarzschild geometry

$$\Sigma = r - const = 0\,. \tag{9.59}$$

The norm of the normal vector is

$$n_\alpha n^\alpha = g^{\alpha\beta} n_\alpha n_\beta = g^{\alpha\beta} \Sigma_{,\alpha} \Sigma_{,\beta} = g^{rr} \Sigma_{,r}^2 = \left(1 - \frac{2m}{r}\right), \tag{9.60}$$

therefore

$$\begin{array}{lllllll} r & > & 2m & \rightarrow & n_\alpha n^\alpha > 0, & \Sigma & \text{is timelike} \\ r & = & 2m & \rightarrow & n_\alpha n^\alpha = 0, & \Sigma & \text{is null} \\ r & < & 2m & \rightarrow & n_\alpha n^\alpha < 0, & \Sigma & \text{is spacelike}\,. \end{array}$$

It should be stressed that, for the Schwarzschild spacetime, the property of the hypersurfaces $\Sigma = r - const = 0$ of being timelike, spacelike, or null holds for any time and for all points on the hypersurfaces, due to the staticity and spherical symmetry of the metric.

Consider for example the hypersurfaces Σ_1 and Σ_2 shown in Fig. 9.5. Any signal which starts at some point of Σ_1 can be sent either toward the singularity at $r = 0$ or outwards, since Σ_1 is timelike. Conversely, a signal which starts at a point of Σ_2 in the interior of $r = 2m$ must necessarily move in one direction since Σ_2 is spacelike and, as we shall later

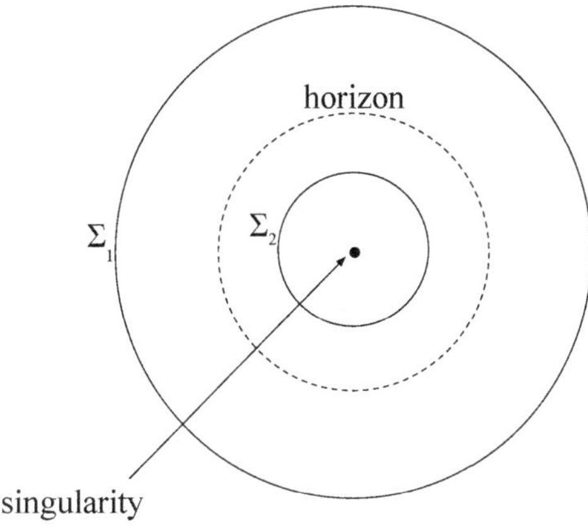

Figure 9.5: The timelike (Σ_1) and spacelike (Σ_2) hypersurfaces in Schwarzschild spacetime. The dashed circle represents the horizon $r = 2M$, and the point at the center represents the curvature singularity $r = 0$.

show, it is forced to go inwards, and be captured by the singularity in a finite amount of proper time. The hypersurface $r = 2m$ is null, and sets the transition from spacelike to timelike hypersurfaces. From Eqs. 9.4 and 9.35 it follows

$$g_{00} = \vec{k} \cdot \vec{k} = -\left(1 - \frac{2m}{r}\right) ; \tag{9.61}$$

therefore on $r = 2m$ the timelike Killing vector $\vec{k}$ becomes null and it is spacelike for $r < 2m$.

Since nothing, not even light, can escape $r = 2m$, this hypersurface is called the **event horizon**. The Schwarzschild radius 9.34 coincides with the location of the event horizon of the Schwarzschild solution.

Thus, the Schwarzschild solution describes not only the spacetime exterior to a static, spherical distribution of matter, but also a more mysterious object which is called **black hole**. *A black hole is a region of spacetime surrounded by an event horizon.* It owes this name to the fact that no signal sent by an observer inside the horizon toward the exterior can ever cross this surface and reach an observer outside. Strictly speaking, there is no way to directly see a black hole, since the latter appears as a "black region" in the spacetime (hence the name).

In the next chapters, we shall see that black holes are not only curious solutions of Einstein's equations in vacuum, but they actually exist for real in our universe, as shown by astrophysical and gravitational wave observations.

9.5 SINGULARITIES IN GENERAL RELATIVITY

9.5.1 Geodesic completeness

How can we define a singularity in General Relativity? This is not a trivial issue, since singularities (either curvature or coordinate ones) *do not belong* to the spacetime manifold. As discussed in Sec. 9.3, looking at the singular points of the metric components $g_{\mu\nu}$ can be misleading: changing coordinates some of them could be mapped to regular points, some others could be mapped to infinity.

We clarify the last remark with an example. Let us consider a two-dimensional manifold with metric

$$ds^2 = \frac{1}{(x^2+y^2)^2}\left(dx^2+dy^2\right) . \tag{9.62}$$

Since the metric components are singular at $(x, y) = (0, 0)$, one may think that this point (or, more precisely, this limit) is a singularity. However, the manifold with metric 9.62 is just the flat Euclidean space in a particular coordinate frame; indeed, changing to the frame (x', y') with

$$\begin{aligned} x' &= \frac{x}{x^2+y^2} , \\ y' &= \frac{y}{x^2+y^2} , \end{aligned} \tag{9.63}$$

the metric 9.62 becomes $ds^2 = (dx')^2 + (dy')^2$ (we leave the proof as an exercise). In the coordinates (x', y'), the point $(x, y) = (0, 0)$ is the limit $(x', y') \to (\infty, \infty)$ (the so-called "point at infinity"). Obviously, the point at infinity does not belong to the manifold (remember that a manifold is an open set, see Chapter 2), but this does not mean that the Euclidean space is singular. We can conclude that, although $(x, y) = (0, 0)$ does not belong to the manifold, it should not be considered a singularity because it is just the point at infinity in disguise.

This example suggests a way to characterize the singularities: they can be seen as a sort of "hole", or "edge" of the spacetime, which does not belong to the manifold, but which *can actually be reached by a physical object.* If, instead, the spacetime is ill-defined in a certain limit ($(x, y) \to (0, 0)$, i.e. $(x', y') \to (\infty, \infty)$ in the example above) which cannot be reached by physical objects, such limit should not be considered as a singularity.

We now introduce a formal, coordinate-invariant definition of singularities based on the following property: a spacetime is **geodesically complete** if every timelike and null geodesic can be extended to arbitrarily large values of the affine parameter. If the spacetime admits at least one incomplete (i.e., which cannot be extended) timelike or null geodesic, the spacetime is **geodesically incomplete**[4]. We say that the spacetime has a **singularity** if it is geodesically incomplete.

Coming back to the manifold with metric 9.62, no geodesic going towards $(x, y) = (0, 0)$ reaches that limit for a finite value of the affine parameter. For instance, the geodesic $(x'(\lambda), y'(\lambda)) = (\lambda, 0)$ in the coordinates (x, y) has the form $(x(\lambda), y(\lambda)) = (\lambda^{-1}, 0)$, and $x \to 0$ as $\lambda \to \infty$.

We remark that this definition applies both to curvature and coordinate singularities. In the following we shall discuss the difference between these two classes of singularities, and how a coordinate singularity can be removed.

9.5.2 How to remove a coordinate singularity

Some singularities can be removed with the following procedure. Let $\mathcal{M}$ be a spacetime manifold with a (timelike or null) geodesic which cannot be extended beyond a finite value of the affine parameter. Let $g_{\mu\nu}$ be the components of the metric tensor in a given coordinate frame $\{x^\mu\}$, defined in a domain $U \subset \mathbb{R}^4$. We remark that, by definition, the metric

[4]We only consider timelike and null geodesics because unlike spacelike geodesics, which cannot be associated to the motion of physical objects, they describe the motion of massive and massless particles, i.e. of observers and signals.

components $g_{\mu\nu}$ and the components of the inverse metric, $g^{\mu\nu}$, are regular in U (the latter requirement is equivalent to $g = \det(g_{\mu\nu}) \neq 0$ in U). To remove the singularity we follow these steps:

- We choose a new coordinate frame $\{x^{\alpha'}\}$. The domain $V \subset \mathbb{R}^4$ of the new coordinates is the image of U through the coordinate transformation; we choose the transformation $x^\mu \to x^{\alpha'}$ such that the metric components $g_{\alpha'\beta'}$ are regular and invertible in a *larger* domain in $\mathbb{R}^4$, $V' \supset V$.
- Let $\mathcal{M}'$ be the manifold described by the coordinates $\{x^{\alpha'}\}$ defined in the larger domain V', endowed with the metric tensor $g_{\alpha'\beta'}$. We have

  $$\mathcal{M}' \supset \mathcal{M}\,. \tag{9.64}$$

  We say that the spacetime has been *extended* if we consider the larger manifold $\mathcal{M}'$ as the spacetime manifold [5].

The singularity corresponding to the incomplete geodesic has been *removed* if, in the new spacetime, that geodesic can be extended to arbitrarily large values of the affine parameter.

As anticipated above, only some singularities can be removed with this procedure. They are called **coordinate singularities**. Those which cannot be removed are true spacetime singularities, and are called **curvature singularities**. As discussed in Sec. 9.3, curvature invariants are regular on coordinate singularities, while they can diverge approaching curvature singularities.

If several coordinate singularities are present the above procedure can be repeated, further extending the spacetime manifold. Once all coordinate singularities are removed, we have the *maximal extension* of the spacetime: in this case, all (timelike or null) geodesics which cannot be extended to arbitrarily large values of the affine parameter correspond to true curvature singularities.

In the following we shall discuss a simple example of spacetime with a coordinate singularity: the Rindler spacetime, which presents interesting similarities with the Schwarzschild geometry. Subsequently, we shall discuss how to remove the $r = 2m$ singularity of the Schwarzschild spacetime.

[5] For simplicity, we have assumed that the manifold $\mathcal{M}$ is described by a single chart, mapping the entire $\mathcal{M}$ to U, and $\mathcal{M}'$ to V. If, instead, a system of charts is needed to describe the manifold $\mathcal{M}$ (see Sec. 2.1.3), then the manifold $\mathcal{M}'$ is described by the same charts, with U replaced by V.

Box 9-C

Some remarks on coordinate singularities

A metric space $\mathcal{M}$ (i.e. a manifold endowed with a metric tensor, see Sec. 2.5) is called *extendible* if it coincides with a subset of another metric space $\mathcal{M}'$, and the metric of $\mathcal{M}'$, restricted to this subset, coincides with the metric of $\mathcal{M}$. It has been argued (see e.g. [47, 114, 56]) that the spacetime describing the Universe should be *inextendible*. This assumption means that our spacetime is a maximal extension, and all singularities are curvature singularities. Many authors assume inextendibility in modelling the spacetime, although strictly speaking there is no actual proof of this conjecture.

Thus, when we refer to curvature singularities as "true" spacetime singularities, and to coordinate singularities as mere artifacts of an inadequate choice of the coordinate frame, we are implicitly assuming that the spacetime is inextendible.

9.5.3 Extension of the Rindler spacetime

The metric of the Rindler spacetime in two spacetime dimensions is

$$ds^2 = -x^2 dt^2 + dx^2, \qquad -\infty < t < \infty, \qquad 0 < x < \infty\,. \tag{9.65}$$

The metric 9.65 is singular at $x \to 0$. Indeed, the determinant g vanishes in this limit, and $g^{\mu\nu}$ diverges.

Let $x^\mu(\tau)$ be a timelike geodesic in this spacetime, with proper time τ (we remind that the proper time is an affine parameter for timelike geodesics, see Chapter 3) and four-velocity $u^\mu = \frac{dx^\mu}{d\tau}$. Since the metric is independent of time, it admits a timelike Killing vector $\vec{k}$ with components, in the frame (t, x), $k^\mu = (1, 0)$. According to Eq. 8.52, and calling $-E$ the constant of motion associated to the Killing vector $\vec{k}$ (see Sec. 10.3 for a physical interpreation of this constant),

$$k_\alpha u^\alpha = g_{\alpha\beta} k^\alpha u^\beta = -x^2 u^0 = const \equiv -E\,, \tag{9.66}$$

therefore

$$u^0 = \frac{dt}{d\tau} = \frac{E}{x^2}\,. \tag{9.67}$$

Since the norm of the vector tangent to a timelike geodesic parametrized with proper time is -1,

$$u^\mu u^\nu g_{\mu\nu} = -x^2 \left(\frac{dt}{d\tau}\right)^2 + \left(\frac{dx}{d\tau}\right)^2 = -1\,, \tag{9.68}$$

it follows that

$$\left(\frac{dx}{d\tau}\right)^2 = \frac{E^2}{x^2} - 1 \qquad \frac{dx}{d\tau} = \pm\sqrt{\frac{E^2}{x^2} - 1}\,, \tag{9.69}$$

hence, choosing the solution moving towards the singularity (i.e., $\frac{dx}{d\tau} < 0$) and setting $x(\tau = 0) = x_0$ and $\tau(x = 0) = \bar{\tau}$,

$$\bar{\tau} = \tau(0) - \tau(x_0) = -\int_{x_0}^{0} \frac{x dx}{\sqrt{E^2 - x^2}} = E - \sqrt{E^2 - x_0^2}\,. \tag{9.70}$$

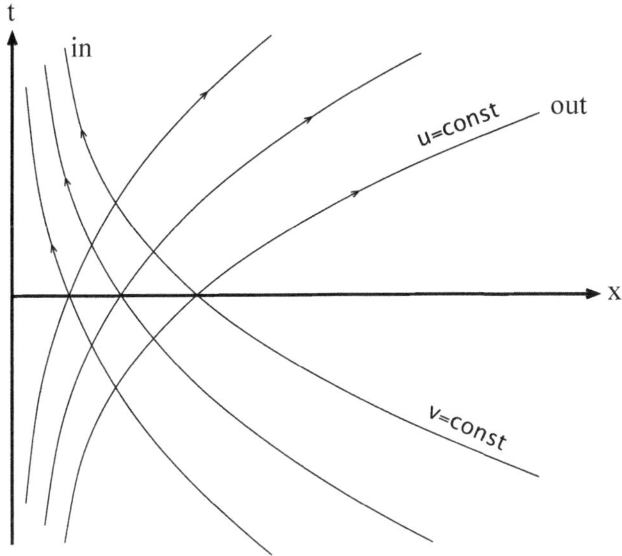

Figure 9.6: The Rindler spacetime in the coordinates (t, x). The logarithmic curves are, respectively, outgoing ($u = const$) and ingoing ($v = const$) null geodesics.

Thus, a particle starting its motion at some point x_0 reaches $x = 0$ in a finite interval $\bar{\tau}$ of the affine parameter: the Rindler spacetime is geodesically incomplete. However, as can easily be checked, the curvature scalars do not diverge at $x = 0$, therefore this could be a mere coordinate singularity, which might be removed with a coordinate transformation. Unfortunately, a systematic approach to the problem of finding the coordinates which allow to extend the spacetime does not exist. We shall describe a procedure which is based on the behaviour of null geodesics, and which – as can be seen *a posteriori* – in some cases allows to find the appropriate transformation to remove a coordinate singularity.

Let $x^\mu(\lambda)$ be a null geodesic, with affine parameter λ and tangent vector

$$u^\mu = \frac{dx^\mu}{d\lambda} \,. \tag{9.71}$$

Since the geodesic is null,

$$g_{\mu\nu} u^\mu u^\nu = -x^2 \left(\frac{dt}{d\lambda}\right)^2 + \left(\frac{dx}{d\lambda}\right)^2 = 0 \,, \tag{9.72}$$

hence

$$\left(\frac{dt}{dx}\right)^2 = \frac{1}{x^2} \,. \tag{9.73}$$

The solution of this equation

$$t = \pm \log x + const, \tag{9.74}$$

shows that there are two families of null geodesics belonging to the + and − sign, respectively. As can be seen from Fig. 9.6, the "+" sign identifies the outgoing geodesics, for which time increases as x increases, whereas the "−" sign identifies the ingoing geodesics, for which time increases as x decreases. Accordingly, we can define the *null outgoing and ingoing coordinates*

$$u = t - \log x \qquad \text{and} \qquad v = t + \log x \,. \tag{9.75}$$

They are constant along any outgoing or ingoing geodesic, respectively. Since $dudv = dt^2 - x^{-2}dx^2$ and $e^{v-u} = x^2$, the metric 9.65 becomes

$$ds^2 = -e^{v-u}dudv. \tag{9.76}$$

The coordinates u and v are defined in the range $(-\infty, +\infty)$, and cover the original region $x > 0$, $-\infty < t < +\infty$. Since the singular point $x = 0$ (with t finite) is mapped to the point at infinity $(u, v) \to (+\infty, -\infty)$, the coordinate frame (u, v) does not allow to extend the spacetime, i.e. to apply the procedure described in Sec. 9.5.2.

We shall now define a new coordinate system (U, V), such that along any null geodesic, one coordinate is an affine parameter and the other is constant. In this new frame it will be possible to extend the spacetime and remove the coordinate singularity $x = 0$.

Let us consider an *outgoing* null geodesic, $u = const$, and the timelike Killing vector field $\vec{k}$ admitted by Rindler's metric. From Eqs. 9.66 and 9.71 it follows that

$$k_\alpha u^\alpha = g_{\alpha\beta}k^\alpha u^\beta = -x^2u^0 = const \equiv -E, \qquad \to \qquad d\lambda = \frac{x^2}{E}dt \,. \tag{9.77}$$

Since along a $u = const$ geodesic $dt = \frac{1}{2}d(u+v) = \frac{1}{2}dv$, we get

$$d\lambda = \frac{x^2}{2E}dv = \frac{e^{v-u}}{2E}dv = Ce^v dv \tag{9.78}$$

where $C = e^{-u}/(2E)$ is constant. In the same way, if we consider an *ingoing* null geodesic $v = const$,

$$d\lambda = \frac{x^2}{2E}du = \frac{e^{v-u}}{2E}du = C'e^{-u}du \tag{9.79}$$

with $C' = e^v/(2E)$ constant. If we define

$$\begin{aligned} U(u) &= -e^{-u} \\ V(v) &= e^v \,, \end{aligned} \tag{9.80}$$

from Eqs. 9.78 and 9.79 it follows that along a null outgoing geodesic $d\lambda = CdV$, and along a null ingoing geodesic $d\lambda = C'dU$. This means that on null outgoing geodesics

$$\lambda = CV + const \,, \tag{9.81}$$

and on null ingoing geodesics

$$\lambda = C'U + const \,. \tag{9.82}$$

Eqs. 9.81, 9.82 show that V, U are linear functions of λ on null outgoing and ingoing geodesics, respectively. Thus, since linear transformations map affine parameters into affine parameters, V is an affine parameter for the outgoing geodesics $U = const$, and U is an affine parameter for the ingoing geodesics $V = const$, i.e. (U, V) are the coordinates we were looking for.

Since $dU = e^{-u}du$ and $dV = e^v dv$, in the new coordinates the line element 9.76 simply becomes

$$ds^2 = -dUdV \,. \tag{9.83}$$

This metric is clearly free of singularities.

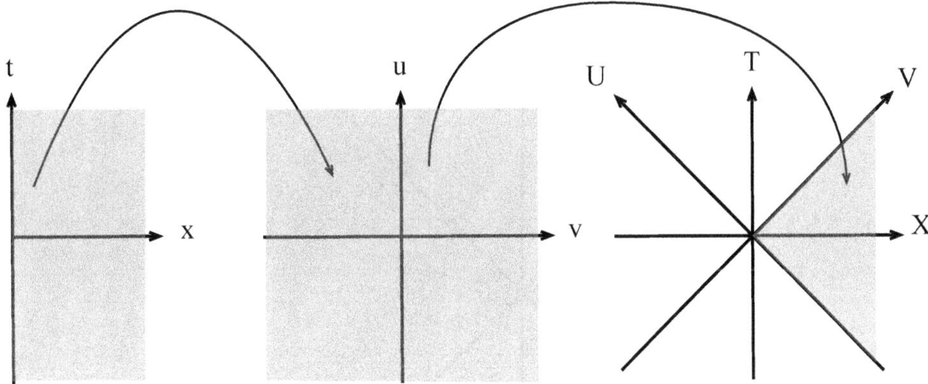

Figure 9.7: Different coordinate frames for Rindler's spacetime: (t, x) (with $x > 0$) is mapped to (u, v) (with (u, v) ranging in $(-\infty, +\infty)$), which is mapped to the region $U < 0$, $V > 0$ in the (U, V) plane. The (U, V) coordinates allow to extend the spacetime manifold..

At this point it is useful to remember that the Rindler metric 9.65 was defined in the region $0 < x < \infty$, $-\infty < t < \infty$ of the (t, x) coordinates; this region was mapped to $-\infty < (u, v) < +\infty$ in the (u, v) coordinates, which corresponds to the domain $U < 0$, $V > 0$ in the (U, V) plane (see Fig. 9.7). However, the line element 9.83 is perfectly well-behaved in the entire (U, V) plane, $-\infty < U < +\infty\,, -\infty < V < +\infty$, therefore we can extend the original (Rindler) spacetime manifold $\mathcal{M}$ to the entire (U, V) space, and by defining the new coordinates (T, X) through the transformation

$$U = T + X\,, \qquad V = T - X\,, \tag{9.84}$$

the line element 9.83 becomes

$$ds^2 = -dT^2 + dX^2\,. \tag{9.85}$$

This is the metric of the (two-dimensional) Minkowski spacetime in Cartesian coordinates, defined in the domain $-\infty < T < +\infty\,, -\infty < X < +\infty$, i.e. in the Minkowski manifold $\mathcal{M}' \supset \mathcal{M}$ corresponding to the entire $T - X$ plane. Indeed, the Rindler metric is just a boosted version of Minkowski's metric.

Since, as shown above, $x^2 = e^{v-u}$ and $U = -e^{-u}$, $V = e^v$, it follows that the $x \to 0$ singularity, in the (U, V) coordinates, is

$$x^2 = -UV \to 0 \qquad \Leftrightarrow \qquad U = 0 \text{ or } V = 0\,, \tag{9.86}$$

i.e. it corresponds to two semiaxes at the boundary of the $U < 0$, $V > 0$ region. By extending the spacetime to the entire (U, V) plane we include these two semiaxes. Therefore, with this procedure we have eliminated the coordinate singularity in $x = 0$ of Rindler's spacetime and we have extended the spacetime to a larger manifold.

The relation between the initial coordinates (t, x) and the final coordinates (T, X) in the region $U < 0$, $V > 0$ (we leave the derivation as an exercise) is

$$\begin{aligned} x &= (X^2 - T^2)^{\frac{1}{2}} \\ t &= \tanh^{-1}\left(\frac{T}{X}\right) = \frac{1}{2}\log\left(\frac{X+T}{X-T}\right). \end{aligned} \tag{9.87}$$

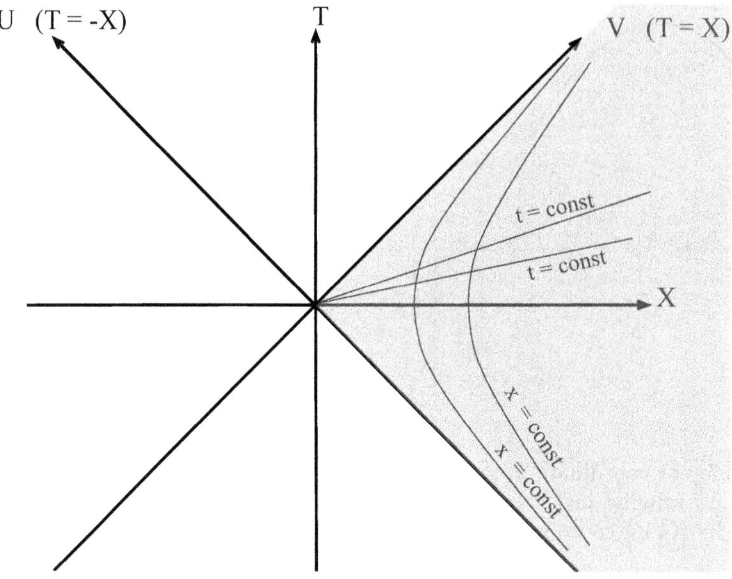

Figure 9.8: Rindler's spacetime in the (U, V) coordinates. The curves $t = const$ and $x = const$, i.e. the coordinate lines in the initial (t, x) frame, are hyperbolae and straight lines, respectively, in the (U, V) frame.

As discussed above, the singularity $x = 0$ corresponds to the lines $T = \pm X$. From the second of Eqs. 9.87

$$\begin{array}{lll} T = -X & \text{corresponds to} & t \to -\infty \\ T = X & \text{corresponds to} & t \to +\infty \, . \end{array} \tag{9.88}$$

The curves $x = const$ are the hyperbolae $X^2 - T^2 = const$, while the curves $t = const$ correspond to the straight lines

$$\frac{X + T}{X - T} = const \qquad \to \qquad T = const\, X \, . \tag{9.89}$$

An illustration of the original and extended spacetimes $\mathcal{M}$, $\mathcal{M}'$ is given in Fig. 9.8. The Rindler space corresponds to the shaded region in the figure.

9.5.4 Extension of the Schwarzschild spacetime

Let us now consider the Schwarzschild spacetime. In the coordinates (t, r, θ, φ) the metric is given by Eq. 9.35, i.e.

$$ds^2 = -\left(1 - \frac{2m}{r}\right) dt^2 + \frac{dr^2}{1 - \dfrac{2m}{r}} + r^2(d\theta^2 + \sin^2\theta d\varphi^2) \, . \tag{9.90}$$

Strictly speaking, since the metric 9.90 is not defined at $r = 0$ and $r = 2m$, it describes the union of two disconnected manifolds, $\mathcal{M}_1$ with $0 < r < 2M$ (the black hole interior) and $\mathcal{M}_2$ with $r > 2M$ (the black hole exterior). A timelike geodesic, i.e. the worldline of a point particle (or of an observer), falling toward the black hole cannot be extended across $r = 2M$, since this hypersurface does not belong to $\mathcal{M}_1 \cup \mathcal{M}_2$, and the geodesic terminates at a finite value of the affine parameter as[6] $r \to 2m$. However, since $r = 2m$ is not a curvature singularity, it can be removed with the procedure outline in Sec. 9.5.2.

Let us consider a null geodesic $x^\mu(\lambda)$ in the Schwarzschild spacetime 9.90, with $\theta = const$, $\varphi = const$. The tangent vector

$$u^\mu = \frac{dx^\mu}{d\lambda} = \left(\frac{dt}{d\lambda}, \frac{dr}{d\lambda}, 0, 0\right) \tag{9.91}$$

is a null vector, thus

$$g_{\mu\nu}u^\mu u^\nu = -\left(1-\frac{2m}{r}\right)\left(\frac{dt}{d\lambda}\right)^2 + \left(1-\frac{2m}{r}\right)^{-1}\left(\frac{dr}{d\lambda}\right)^2 = 0\,. \tag{9.92}$$

Hence

$$\left(\frac{dr}{dt}\right)^2 = \left(1-\frac{2m}{r}\right)^2 \quad \to \quad \frac{dt}{dr} = \pm\frac{r}{r-2m}\,, \tag{9.93}$$

the solution of which is

$$t = \pm r_* + const \tag{9.94}$$

where

$$r_* = r + 2m\log\left(\frac{r}{2m} - 1\right) \quad \text{if} \quad r > 2m \tag{9.95}$$

and

$$r_* = r + 2m\log\left(-\frac{r}{2m} + 1\right) \quad \text{if} \quad 0 < r < 2m\,. \tag{9.96}$$

The coordinate r_* is called "tortoise" coordinate[7]. As $r \to +\infty$, $r_* \sim r$, while as $r \to 2m$, $r_* \to -\infty$. In other words, this change of the radial variable "pushes" the horizon to $-\infty$. Thus, as in Rindler's spacetime, in the Schwarzschild spacetime there exist two congruences of null geodesics (with θ, φ constant) corresponding, respectively, to the $+$ and $-$ sign in Eq. 9.94, and it is natural to define the null coordinates

$$u \equiv t - r_*\,, \qquad v \equiv t + r_*\,. \tag{9.97}$$

Null outgoing geodesics correspond to $u = const$, and null ingoing geodesics to $v = const$. Note that there are *two* (u, v) maps, both defined in $-\infty < u < +\infty$, $-\infty < v < +\infty$: one, with the definition 9.95, corresponding to the manifold $\mathcal{M}_2$ $(r > 2m)$; the other, with the definition 9.96, corresponding to the manifold $\mathcal{M}_1$ $(0 < r < 2m)$.

Let us consider the manifold $\mathcal{M}_2$ (the black hole exterior) $r > 2m$. Since

$$\frac{dr_*}{dr} = \frac{1}{1-\frac{2m}{r}}\,, \tag{9.98}$$

the metric in the coordinates (u, v, θ, φ) is

$$\begin{aligned} ds^2 &= -\left(1-\frac{2m}{r}\right)(dt^2 - dr_*^2) + r^2(d\theta^2 + \sin^2\theta d\varphi^2) \\ &= -\left(1-\frac{2m}{r}\right)dudv + r^2(d\theta^2 + \sin^2\theta d\varphi^2)\,. \end{aligned} \tag{9.99}$$

[6] Actually, $r = 2m$ is not the unique coordinate singularity of the Schwarzschild spacetime. Other singular points of the metric 9.90 are $\theta = 0, \pi$, $\varphi = 0, 2\pi$. These coordinate singularities are also present in Minkowski space in polar coordinates, and can easily be removed with a space rotation (see Box 2-C). These coordinate singularities are therefore "trivial", and will not be discussed here. However, as we shall see in Chapter 18, in other spacetimes the coordinate singularities $\theta = 0, \pi$ can acquire a subtler meaning and have to be studied in detail.

[7] Like the famous Zeno's tortoise, the coordinate r_* "never" reaches the horizon $r = 2m$, but approaches it logarithmically.

Note that now r should not be considered as a coordinate: it is a function of the coordinates u and v, i.e. $r(u, v)$. The metric 9.99 is still singular at $r = 2m$. As in the Rindler case, we consider a new coordinate frame (U, V, θ, φ) such that – at least near the horizon – V is an affine parameter of the outgoing null geodesics, while U is an the affine parameter of ingoing null geodesics.

On null geodesics,

$$k_\alpha u^\alpha = g_{\alpha\beta} k^\alpha u^\beta = -\left(1 - \frac{2m}{r}\right)\frac{dt}{d\lambda} = const \equiv -E\,, \tag{9.100}$$

where $\vec{k}$ is the timelike Killing vector field admitted by the Schwarzschild spacetime whose components, in the (t, r, θ, φ) frame, are $k^\alpha = (1, 0, 0, 0)$. Moreover, on $\mathcal{M}_2$ from Eq. 9.95 we find

$$\frac{r_* - r}{2m} = \ln\left(\frac{r}{2m} - 1\right) \quad \to \quad 1 - \frac{2m}{r} = \frac{2m}{r} e^{\frac{r_* - r}{2m}} = \frac{2m}{r} e^{-\frac{r}{2m}} e^{\frac{v-u}{4m}}\,. \tag{9.101}$$

Therefore, on null outgoing geodesics, where $u = const$ and $dt = \frac{dv}{2}$,

$$d\lambda = \left(1 - \frac{2m}{r}\right)\frac{dv}{2E} = \frac{2m}{r} e^{-\frac{r}{2m}} e^{-\frac{u}{4m}} e^{\frac{v}{4m}} \frac{dv}{2E}\,. \tag{9.102}$$

Similarly, on null ingoing geodesics, $v = const$ and $dt = \frac{du}{2}$, therefore

$$d\lambda = \left(1 - \frac{2m}{r}\right)\frac{du}{2E} = \frac{2m}{r} e^{-\frac{r}{2m}} e^{\frac{v}{4m}} e^{-\frac{u}{4m}} \frac{du}{2E}\,. \tag{9.103}$$

We now define the coordinates

$$U \equiv -e^{-\frac{u}{4m}}\,, \qquad V \equiv e^{\frac{v}{4m}}\,. \tag{9.104}$$

Near the horizon , as $r \to 2m$, from Eq. 9.102 it follows that, on the null *outgoing* geodesics $U = const$, the affine parameter is given by $d\lambda \propto e^{\frac{v}{4m}} dv$, and consequently $d\lambda = C dV$ with C constant, i.e. V is an affine parameter for outgoing geodesics as in the Rindler case (see Eq. 9.81).

Similarly, from Eq. 9.103 it follows that, on the null *ingoing* geodesics $V = const$, the affine parameter is given by $d\lambda \propto e^{-\frac{u}{4m}} du$, and consequently $d\lambda = C' dU$ with C' constant, i.e. U is an affine parameter for ingoing geodesics, as in the Rindler case (see Eq. 9.82).

Thus, (U, V) are the coordinates we were looking for, but we should keep in mind that in the Schwarzschild case (U, V) are affine parameters along the null ingoing and outgoing null geodesics *only near the horizon*.

In the coordinates (U, V, θ, φ), called the **Kruskal coordinates**, the metric is

$$ds^2 = -\frac{32m^3}{r} e^{-\frac{r}{2m}} dU dV + r^2(d\theta^2 + \sin^2\theta d\varphi^2) \tag{9.105}$$

as can easily be shown by replacing Eqs. 9.101 and 9.104 in Eq. 9.99. The metric 9.105 is no longer singular on $r = 2m$.

Note that the coordinates (U, V) (Eq. 9.104) are defined in the quadrant $U < 0$, $V > 0$. Thus, the spacetime exterior to the black hole, i.e. the manifold $\mathcal{M}_2$, $r > 2m$ in the coordinates (t, r, θ, φ), has been mapped to the region $U < 0$, $V > 0$ in the Kruskal coordinates.

Since

$$UV = -e^{\frac{v-u}{4m}} = -e^{\frac{r_*}{2m}} = \left(1 - \frac{r}{2m}\right) e^{\frac{r}{2m}}\,, \tag{9.106}$$

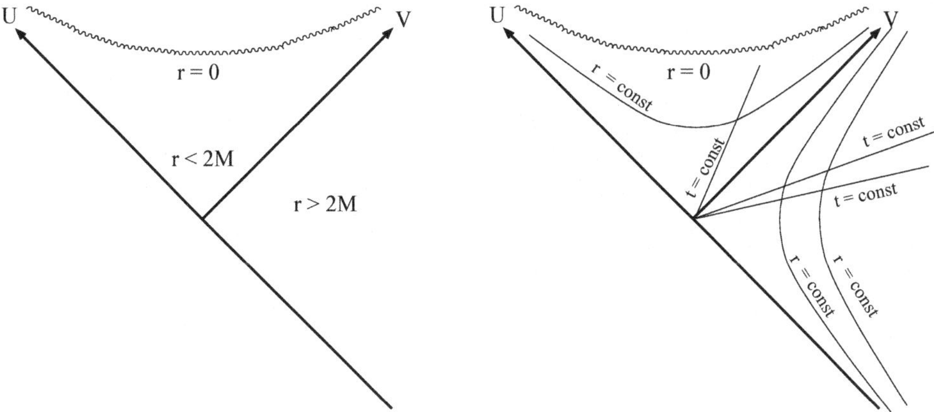

Figure 9.9: Interior and exterior of a Schwarzschild black hole in Kruskal coordinates. In the right panel we show the curves at t and r constant.

the limit $r \to 2m$ corresponds to $U \to 0$ or $V \to 0$.

Let us now consider the manifold $\mathcal{M}_1$ $(0 < r < 2m)$. The null coordinates are defined, as before, as $u = t - r_*$, $v = t + r_*$, but r_* is now given by Eq. 9.96. The tortoise coordinate is always negative, it tends to zero as $r \to 0$, and to $-\infty$ as $r \to 2M$. Since Eq. 9.98 still holds, the metric in the coordinates (u, v) is given by Eq. 9.99,

$$ds^2 = -\left(1 - \frac{2m}{r}\right) dudv + r^2(d\theta^2 + \sin^2\theta d\varphi^2)\,. \tag{9.107}$$

Since Eq. 9.96 gives

$$1 - \frac{2m}{r} = -\frac{2m}{r} e^{-\frac{r}{2m}} e^{\frac{v-u}{4m}}\,, \tag{9.108}$$

defining

$$U \equiv +e^{-\frac{u}{4m}}\,, \qquad V \equiv e^{\frac{v}{4m}} \tag{9.109}$$

we find the same expression for the metric in Kruskal coordinates given in Eq. 9.105. The coordinates (U, V) defined in Eq. 9.109 are defined in the domain $U > 0$, $V > 0$. They are still affine parameters, in the near-horizon limit, of null outgoing and ingoing geodesics, respectively (Eqs. 9.102 and 9.103 still hold, with the opposite sign). Note that the singularity $r = 0$ corresponds to $r_* = 0$, and thus $u = v$ (in $\mathcal{M}_1$) and $UV = e^{\frac{v-u}{4m}} = 1$, while $0 < r < 2M$ corresponds to $r_* < 0$ (see Eq. 9.96), and thus $v - u = 2r_* < 0$ and $UV < 1$.

Summarizing, the Kruskal coordinates (U, V, θ, φ) describe both the manifolds $\mathcal{M}_1$ and $\mathcal{M}_2$ (see Fig. 9.9). In these coordinates, the black hole exterior $r > 2m$ is mapped to the region $(U < 0,\ V > 0)$, while the interior $0 < r < 2m$ is mapped to $(U > 0,\ V > 0$ with $UV < 1)$; the coordinate singularity $r = 2m$ (and $t = +\infty$) corresponds to the semiaxis $(U = 0,\ V > 0)$; the curvature singularity $r = 0$ corresponds to the upper branch of the hyperbole $UV = 1$.

In the coordinate frame (U, V, θ, φ) the manifold $\mathcal{M}_1 \cup \mathcal{M}_2$ can be extended across the semiaxis $(U = 0,\ V > 0)$ separating $\mathcal{M}_1$ and $\mathcal{M}_2$, since the line metric 9.105 is not singular there. Thus, we consider a new manifold,

$$\mathcal{M} \supset \mathcal{M}_1 \cup \mathcal{M}_2\,, \tag{9.110}$$

defined by

$$V > 0\,, \qquad UV < 1\,. \tag{9.111}$$

Generally, when studying phenomena which occur near a Schwarzschild black hole such as the capture of particles, we implicitly consider the extended manifold $\mathcal{M}$: for instance, as we shall do in the next chapter, we assume that an object falling inside the black hole crosses the horizon $r = 2m$ which, therefore, has to belong to the manifold. The discussion in Sec. 9.4 about the $r = const$ hypersurfaces also assumes that the manifold is $\mathcal{M}$, since $r = 2m$ has been considered as a part of the manifold. However, as explained above in this section, the coordinates (t, r, θ, φ) do not cover the manifold $\mathcal{M}$. Strictly speaking, when we want to describe the black hole horizon, we should use a coordinate system in which the $r = 2m$ singularity is removed.

It is customary to define (as in Rindler's spacetime, see Eqs. 9.84) the coordinates T, X as

$$T = \frac{U+V}{2} \qquad X = \frac{U-V}{2}\,. \tag{9.112}$$

In terms of these coordinates, the metric 9.105 becomes

$$ds^2 = -\frac{32m^3}{r}e^{-\frac{r}{2m}}(-dT^2 + dX^2) + r^2(d\theta^2 + \sin^2\theta d\varphi^2)\,. \tag{9.113}$$

The relation between the coordinate frames (t, r, θ, φ) and (T, X, θ, φ) is similar to the expressions obtained in Rindler's spacetime (Eq. 9.87). Indeed,

$$X^2 - T^2 = -UV = \pm e^{\frac{v-u}{4m}} = \pm e^{\frac{r_*}{2m}} = \pm e^{\frac{r}{2m}}\left(\frac{r}{2m} - 1\right) \tag{9.114}$$

where the upper/lower sign refers to the first ($U < 0$) and second ($U > 0$) quadrants, respectively. We also note that

$$t = 2m \log e^{\frac{u+v}{4m}} = 2m \log\left(\pm\frac{V}{U}\right) = 2m \log\left|\frac{T+X}{T-X}\right|\,. \tag{9.115}$$

While in the case of Rindler's spacetime the coordinates (t, x) are defined in $U < 0$, $V > 0$, in Schwarzschild's spacetime the coordinates (t, r) are defined in $U \neq 0$, $V > 0$ (i.e., the exterior *and* the interior of the black hole), therefore the coordinate transformations are also defined in a larger domain.

The curves $t = const$ and $r = const$ are shown in the right panel of Fig. 9.9. Eq. 9.115 shows that the $t = const$ curves are straight lines in the $U - V$ (and $T - X$) plane; Eq. 9.114 that the $r = const$ curves are hyperbolae.

The manifold $\mathcal{M}$ can still be extended: timelike and null geodesics from $V = 0$ cannot be continued to large negative values of the affine parameter, unless we extend the manifold to $V \leq 0$. By including the region $-\infty < U < +\infty$, $-\infty < V < +\infty$ with $UV < 1$, we obtain the maximal extension of the Schwarzschild spacetime, shown in Fig. 9.10. The dashed line represents the worldline of an observer falling into the black hole, and the wave-like curves represent the curvature singularity $r = 0$.

In the Kruskal coordinates 9.105 the null worldlines with θ, φ constant are straight lines at 45°, i.e. $U = const.$ or $V = const.$; this can easily be seen from the metric 9.105: any worldline with tangent vector either $(1, 0, 0, 0)$ or $(0, 1, 0, 0)$ is null. Therefore, the light cones can be drawn as in Minkowski's spacetime, and the description of causal connections among events is easy and intuitive (see Fig. 9.10). In particular, we see that signals from region I can be sent only to region II; furthermore, there is a copy of region I, i.e. region IV, which is causally disconnected from I, but can receive signals only from region III and send signals

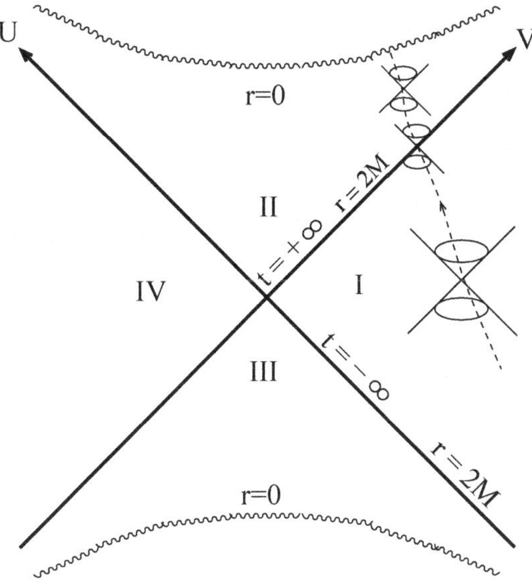

Figure 9.10: Maximal extension of the Schwarzschild spacetime in Kruskal coordinates. The dashed line represents the worldline of an observer falling into the black hole. Different regions are marked with I, II, III, and IV.

to region II only. Region III is often called *white hole*, since all signals starting in this region have to escape to regions I or IV across the horizon.

The incomplete (timelike and null) geodesics of the maximally extended manifold correspond to the true singularity $r = 0$, i.e., in Kruskal coordinates $UV = 1$. As we shall show in the next chapter, these geodesics reach the singularity at a finite value of the affine parameter, and cannot be extended through it; for instance, an observer that falls inside the black hole reaches the singularity in a finite amount of proper time.

The Kruskal coordinates, besides providing the maximal extension of the Schwarzschild spacetime, can be useful to clarify an important feature of the horizon. We have shown in Sec. 9.4 that, since $r = 2M$ is a null hypersurface, it can be crossed in one direction only; but which is this direction: inwards or outwards? The $r = 2M$ hypersurface *in the future of the events outside the black hole*, i.e. in the future of region I, is the semiaxis $U = 0$, $V > 0$, and can only be crossed inwards (see e.g. the worldline shown in Fig. 9.10). The $r = 2M$ hypersurface *in the past* of region I is a *different* hypersurface, the semiaxis $V = 0$, $U < 0$, which can only be crossed outwards. Similarly, the black hole interior in the future of region I is region II, where the spacelike $r = const$ hypersurfaces can only be crossed inwards, while the black hole interior in the past of region I is region III, where the spacelike $r = const$ hypersurfaces can only be crossed outwards.

It should be stressed that the maximal extension of the Schwarzschild spacetime has no meaning if we consider a black hole as an astrophysical object, formed in the gravitational collapse of a star. Indeed, it describes an *eternal black hole*, i.e. one which exists for $t \in (-\infty, +\infty)$, whereas the stellar collapse occurs at a finite value of t. In particular, region III cannot exist for an astrophysical black hole, because the semiaxis ($U < 0$, $V = 0$) corresponds to $t = -\infty$ when the black hole was not formed yet. The Kruskal coordinates are not appropriate to describe the stellar collapse, in which the horizon and the singularity appear at finite time. In Sec. 9.5.5 we shall show how to describe this process in a different coordinate system. Here we only note that we do not have to worry about the meaning of regions III and IV since they do not exist in astrophysical black holes, and we can leave the discussion on the existence of other universes (such as regions III and IV of the construction above) to science-fiction writers.

The final fate of an observer who reaches the singularity is unknown and this poses a problem for the predictability of the theory and for its self-consistence. On the other hand, such problem is not severe from an operational point of view, because no signal from the observer reaching the singularity can be sent outside the black hole: the consistency of the theory, in a certain sense, is preserved by the existence of the horizon. Roger Penrose has conjectured that there exists a fundamental principle, the *cosmic censorship hypothesis*, stating that all singularities in the Universe (with the exception of a possible initial singularity) are concealed behind a horizon, i.e. that *naked singularities* cannot exist in nature. There is no definitive proof of this conjecture, but there are indications supporting it, at least under some reasonable assumptions about the matter fields that can produce a singularity, e.g., during a gravitational collapse.

The presence of curvature singularities, although concealed within the event horizon of black holes, strongly suggests that Einstein's theory is not the last word on gravity. It is customary to consider General Relativity as an effective theory which is valid at curvature scales much smaller than Planckian curvature ($\sim 1/l_{\rm P}^2$, where $l_{\rm P}$ is the Planck length, see Box 9-B), above which quantum effects become relevant [8]. Near the singularity, the curvature is so large that higher-order curvature corrections to General Relativity might become dominant. There are several proposal for such corrections but, at the moment, none of them is supported by observations. Even if General Relativity will eventually have to be modified at the Planck scale, the corrections are expected to be negligible for the astrophysical objects and for the gravitational wave sources discussed in this book.

9.5.5 Eddington-Finkelstein coordinates

The Kruskal coordinates allow to define a maximal extension of the Schwarzschild spacetime covered by a unique coordinate choice. However, if we are only interested in removing the coordinate singularity $r = 2m$ between regions I and II, there is a simpler, and more practical choice: the Eddington-Finkelstein coordinates [9]

$$(v, r, \theta, \varphi) \qquad -\infty < v < +\infty \qquad 0 < r < +\infty\,. \tag{9.116}$$

Here, as before, $v = t + r_*$, and the tortoise coordinate is defined in Eqs. 9.95, 9.96, which can be unified in

$$r_* \equiv r + 2m \ln\left|\frac{r}{2m} - 1\right|\,. \tag{9.117}$$

Let us consider the Schwarzschild metric written as in Eq. 9.99

$$ds^2 = -\left(1 - \frac{2m}{r}\right)(dt^2 - dr_*^2) + r^2(d\theta^2 + \sin^2\theta d\varphi^2)\,. \tag{9.118}$$

It describes both the $\mathcal{M}_1$ and $\mathcal{M}_2$ manifolds. Since $dt = dv - dr_*$, from Eq. 9.98 we find

$$dt^2 - dr_*^2 = dv^2 - 2dvdr_* = dv^2 - 2\frac{dr_*}{dr}dvdr = dv^2 - 2\frac{dvdr}{1 - \frac{2m}{r}}\,, \tag{9.119}$$

---

[8] At present, we do not know any physical theory which describes all fundamental interactions including gravity at the Planck scale and beyond. Such theory of *quantum gravity*, which would generalize General Relativity and possibly unify it with the quantum field theories of the Standard Model, is one of the main challenges of modern physics.

[9] Strictly speaking, (v, r, θ, φ) are the *ingoing* Eddington-Finkelstein coordinates, while (u, r, θ, φ) are called *outgoing* Eddington-Finkelstein coordinates. We shall omit this specification because we only consider the ingoing coordinates, which allow to remove the coordinate singularity between the regions I and II.

therefore the metric in the Eddington-Finkelstein coordinates is

$$ds^2 = -\left(1 - \frac{2m}{r}\right) dv^2 + 2dvdr + r^2(d\theta^2 + \sin^2\theta d\varphi^2)\,. \tag{9.120}$$

This metric covers both the interior and the exterior of the black hole, i.e. the regions I and II of the Kruskal construction, and is regular and invertible on the horizon $r = 2m$ (the determinant does not vanish due to the off-diagonal components). Note that on the horizon v is finite because $t \to +\infty$ and $r_* \to -\infty$, while u diverges. All the computations and derivations involving the interior and the exterior of the black hole, such as the study of the $r = const$ surfaces of Sec. 9.4, can be rigorously performed in the Eddington-Finkelstein coordinates (note that $g^{rr} = 1 - 2m/r$ in these coordinates, too).

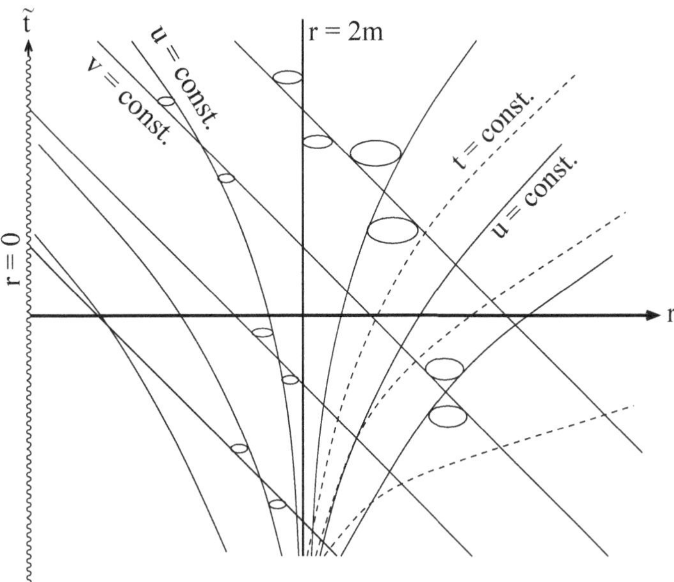

Figure 9.11: Finkelstein diagram of a Schwarzschild black hole.

The Finkelstein diagram

A useful way to visualize the Schwarzschild spacetime is the Finkelstein diagram, in which the axes are $(\tilde{t}, r)$ where

$$\tilde{t} \equiv v - r = t + 2m \ln\left|\frac{r}{2m} - 1\right|\,. \tag{9.121}$$

In this diagram the null lines $v = const$, corresponding to ingoing massless particles, are straight lines at 45°; the null lines $u = const$, corresponding to outgoing massless particles, are hyperbolic curves. These two sets of curves define the light cones centered in any spacetime point, and allow to establish the causal relations among different events. While the light cones in the (t, r) plane collapse to lines at the horizon, the light cones in the Finkelstein diagram remain regular.

Since $\tilde{t} \simeq t$ for $r \ll 2m$ and $r \gg 2m$, the coordinate $\tilde{t}$ coincides with t far away from the horizon, but they are very different near to the horizon (where, as we shall discuss in Chapter 10, the operational definition of the coordinate t is problematic). We also note that the coordinate $\tilde{t}$ cannot be considered a "time" inside the horizon, because the vector $\partial/\partial\tilde{t}$ is spacelike.

In Fig. 9.11 the coordinate lines $r = const$ are vertical straight lines, whereas $t = const$ are hyperbolic (dashed) curves; the $\tilde{t}$-axis represents the singularity, and for this reason it is drawn wave-like.

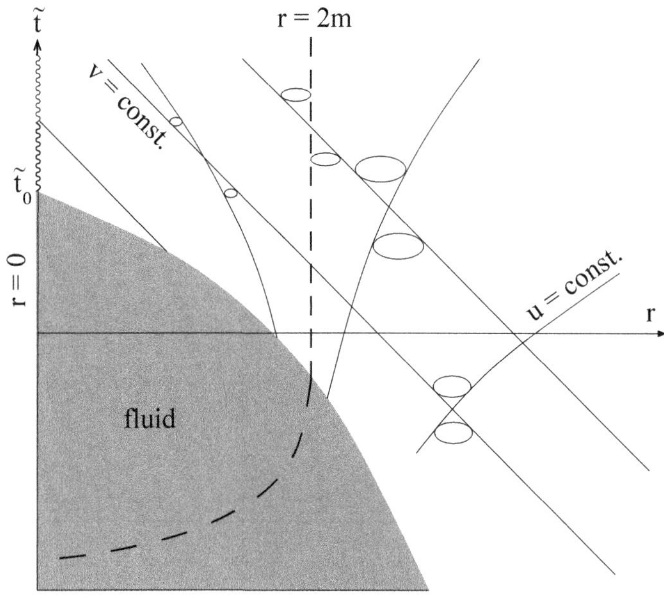

Figure 9.12: Finkelstein diagram of a stellar collapse originating a Schwarzschild black hole. The shaded area represents the fluid interior of the star. The curvature singularity (wave-like line) is formed at $\tilde{t} = \tilde{t}_0$. The horizon is represented by the dashed line.

As mentioned above, astrophysical black holes are the result of a gravitational collapse (see Chapter 16) occurring at some time and producing a singularity for some $t = t_0$. A qualitative view of the spacetime of a realistic black hole is shown in the Finkelstein diagram in Fig. 9.12, where the shaded area represents the interior of the star. The $r = 0$ axis is a curvature singularity for $\tilde{t} \geq \tilde{t}_0$, i.e. after the singularity forms; for $\tilde{t} < \tilde{t}_0$, the $r = 0$ axis is simply the (trivial) coordinate singularity at the origin of polar coordinates. The horizon, represented by the dashed line, is also formed during the collapse.

It is important to stress that, although we have discussed the entire Schwarzschild solution, only the $r > 2m$ region is directly relevant for astrophysical observations: relativity imposes that no signal can come from the interior of a black hole horizon.

Finally, it is worth mentioning that a useful way to represent the causal structure of a spacetime is through the so-called Penrose-Carter diagrams. We do not discuss this interesting topic in this book, and we refer the interested reader to more specialized work, e.g. [90].

9.6 THE BIRKHOFF THEOREM

In Sec. 9.2 we derived the Schwarzschild metric as the solution of Einstein's equations in vacuum, under the assumption of staticity and spherical symmetry. This solution describes the gravitational field external to a non-rotating, spherically symmetric body, the structure of which is time-independent. However, the Schwarzschild solution is more general, since, as shown by George Birkhoff in 1923, it is the only *spherically symmetric, asymptotically flat solution of Einstein's field equations in vacuum.* Thus, to prove Birkhoff's theorem we need to relax the assumption that the metric admits a timelike, hypersurface-orthogonal Killing vector field. We shall now generalize the results of Sec. 9.1, where we showed how to choose the coordinates by imposing the spherical symmetry, assuming that the metric depends on time. As in Sec. 9.1 we fill the three-dimensional space with two-spheres, with two-metric (see Eq. 9.8)

$$ds^2_{(2)} = a^2(x^0, x^1)(d\theta^2 + \sin^2\theta d\varphi^2)\,, \tag{9.122}$$

where $a^2(x^0, x^1)$ is an unspecified function. Contrary to what we did in Sec. 9.1, we shall now retain the dependence on x^0. The basis vectors $\vec{e}_{(\theta)}$ and $\vec{e}_{(\varphi)}$ are tangent, respectively, to the coordinate lines $\varphi = const$, $\theta = const$, which we choose on the two-spheres. Then, we align the poles of all spheres as explained in Sec. 9.1; in addition we choose the basis vector $\vec{e}_{(1)}$ parallel to the vector $\vec{\zeta}$ shown in Fig. 9.1, which joins points with the same values of θ and φ on different spheres. In this way $\vec{e}_{(1)}$ is orthogonal, at each space point, to both $\vec{e}_{(\theta)}$ and $\vec{e}_{(\varphi)}$, and the metric of the three-space can be written as

$$ds^2_{(3)} = g_{11}(x^0, x^1)(dx^1)^2 + a^2(x^0, x^1)(d\theta^2 + \sin^2\theta d\varphi^2)\,. \tag{9.123}$$

The metric of the four-dimensional spacetime therefore becomes

$$\begin{aligned} ds^2 &= g_{00}(x^0, x^1)(dx^0)^2 + g_{11}(x^0, x^1)(dx^1)^2 + 2g_{01}(x^0, x^1)dx^0dx^1 \\ &\quad + a^2(x^0, x^1)(d\theta^2 + \sin^2\theta d\varphi^2)\,. \end{aligned} \tag{9.124}$$

We now change coordinates from (x^0, x^1) to (x^0, r) where

$$r = a(x^0, x^1)$$

so that the metric becomes

$$ds^2 = g_{00}(x^0, r)(dx^0)^2 + g_{rr}(x^0, r)(dr^1)^2 + 2g_{0r}(x^0, r)dx^0dr + r^2(d\theta^2 + \sin^2\theta d\varphi^2)\,. \tag{9.125}$$

We wish to find a function $t(x^0, r)$ such that, if we choose it as a time coordinate, the cross term g_{tr} in the metric vanishes and the first three terms in Eq. 9.125 can be written as

$$g_{00}(x^0, r)(dx^0)^2 + g_{rr}(x^0, r)(dr^1)^2 + 2g_{0r}(x^0, r)dx^0dr = bdt^2 + cdr^2\,, \tag{9.126}$$

where b and c are functions of (x^0, r) to be determined. Given

$$dt = \frac{\partial t}{\partial x^0}dx^0 + \frac{\partial t}{\partial r}dr\,, \tag{9.127}$$

Eq. 9.126 becomes

$$\begin{aligned} & g_{00}(dx^0)^2 + g_{rr}dr^2 + 2g_{0r}dx^0dr \\ &= b\left(\frac{\partial t}{\partial x^0}\right)^2(dx^0)^2 + b\left(\frac{\partial t}{\partial r}\right)^2 dr^2 + 2b\frac{\partial t}{\partial x^0}\frac{\partial t}{\partial r}dtdr + cdr^2\,, \end{aligned} \tag{9.128}$$

which gives

$$\begin{aligned} b\left(\frac{\partial t}{\partial x^0}\right)^2 &= g_{00} \\ b\left(\frac{\partial t}{\partial r}\right)^2 + c &= g_{rr} \\ b\frac{\partial t}{\partial x^0}\frac{\partial t}{\partial r} &= g_{0r}\,. \end{aligned} \tag{9.129}$$

These are three equations for the three unknown functions $t(x^0, r)$, $b(x^0, r)$, and $c(x^0, r)$, which can in principle be solved. By inverting the function $t(x^0, r)$ with respect to x^0 and by replacing the result in b and c, these quantities become functions of (t, r) and the metric can be written as

$$ds^2 = b(t,r)dt^2 + c(t,r)dr^2 + r^2(d\theta^2 + \sin^2\theta d\varphi^2)\,. \tag{9.130}$$

As in Sec. 9.2, we replace $b(t, r)$ and $c(t, r)$ with the functions $\nu(t, r)$ and $\lambda(t, r)$, where

$$b(t,r) = -e^{2\nu(t,r)}\,, \qquad c = e^{2\lambda(t,r)}\,,$$

so that, finally, the metric of a time-dependent, spherically symmetric spacetime becomes

$$ds^2 = -e^{2\nu(t,r)}dt^2 + e^{2\lambda(t,r)}dr^2 + r^2(d\theta^2 + \sin^2\theta d\varphi^2)\,. \tag{9.131}$$

To prove Birkhoff's theorem we only need the components R_{tr} and $R_{\theta\theta}$ of the Ricci tensor, which can be computed by replacing the metric 9.131 in Eq. 4.54:

$$\begin{aligned} R_{tr} &= \frac{2}{r}\frac{\partial\lambda}{\partial t} = 0\,, \\ R_{\theta\theta} &= 1 - e^{-2\lambda}\left[1 + r\frac{\partial(\nu-\lambda)}{\partial r}\right] = 0\,. \end{aligned} \tag{9.132}$$

From the first equation in 9.132 it follows that λ must depend only on the radial coordinate r. Then, from the second equation it follows that

$$\frac{\partial\nu}{\partial r} = \frac{\partial\lambda}{\partial r} + \frac{e^{2\lambda(r)} - 1}{r}\,,$$

i.e. $\frac{\partial\nu}{\partial r}$ depends only on r. Consequently we can write

$$\nu = \nu(r) + f(t)\,, \tag{9.133}$$

and

$$ds^2 = -e^{2\nu(r)}e^{2f(t)}dt^2 + e^{2\lambda(r)}dr^2 + r^2(d\theta^2 + \sin^2\theta d\varphi^2)\,. \tag{9.134}$$

The term $e^{2f(t)}$ can be reabsorbed by a coordinate transformation such that

$$dt' = e^{f(t)}dt\,, \tag{9.135}$$

and the metric finally becomes

$$ds^2 = -e^{2\nu(r)}dt^2 + e^{2\lambda(r)}dr^2 + r^2(d\theta^2 + \sin^2\theta d\varphi^2)\,, \tag{9.136}$$

where the prime has been suppressed for simplicity. Thus, we have shown that even if we assume that the metric of a spherically symmetric spacetime depends on time, by suitable coordinate transformations it can be made time independent. Since, as we have shown in Sec. 9.1, the only asymptotically flat solution of the vacuum Einstein equations for the metric 9.136 is the Schwarzschild solution, we have proved the Birkhoff theorem stated at the beginning of this section.

An important consequence of this theorem is the following. The metric external to a spherically symmetric star is the Schwarzschild metric even when the star is collapsing, exploding, or radially pulsating. Thus, spherically symmetric systems can never emit gravitational waves. A similar situation occurs in electrodynamics: a spherically symmetric distribution of charges and currents does not radiate.

CHAPTER 10

Geodesic motion in Schwarzschild's spacetime

In this chapter we shall study the geodesics of the Schwarzschild spacetime for massive and massless particles. We shall also study the radial infall of a particle into a black hole, which will clarify the remarkable properties of the event horizon.

10.1 A VARIATIONAL PRINCIPLE FOR GEODESIC MOTION

The geodesic equations can be derived not only from the Equivalence Principle as shown in previous chapters, but also from a variational principle, as we shall now show. Let us define the Lagrangian of a free particle as

$$\mathcal{L}\left(x^\alpha,\dot{x}^\alpha\right)=\frac{1}{2}g_{\mu\nu}(x^\alpha)\frac{dx^\mu}{d\lambda}\frac{dx^\nu}{d\lambda}\equiv\frac{1}{2}g_{\mu\nu}(x^\alpha)\dot{x}^\mu\dot{x}^\nu\,, \tag{10.1}$$

in the space of the curves $\{x^\mu(\lambda),\,\lambda\in[\lambda_0,\lambda_1]\}$, and the action

$$S=\int\mathcal{L}\left(x^\alpha,\dot{x}^\alpha\right)d\lambda=\frac{1}{2}\int g_{\mu\nu}(x^\alpha)\dot{x}^\mu\dot{x}^\nu d\lambda\,, \tag{10.2}$$

where we have set

$$\dot{x}^\mu=\frac{dx^\mu}{d\lambda}\,. \tag{10.3}$$

Massive particles move along timelike geodesics, and we will refer to these curves also as the worldlines of possible "observers". For timelike geodesics the affine parameter λ can be chosen to be the proper time. Massless particles move along null geodesics; since in this case the proper time cannot be defined, λ can be any affine parameter, as for instance the proper length of the geodesics from an arbitrary origin λ_0.

The Euler-Lagrange equations are obtained by requiring the action S to be extremal under variation of the curves $x^\mu(\lambda)$

$$x^\mu(\lambda)\longrightarrow x^\mu(\lambda)+\delta x^\mu(\lambda)\,,\qquad\lambda\in[\lambda_0,\lambda_1]\,, \tag{10.4}$$

with

$$\delta x^\mu(\lambda_0)=\delta x^\mu(\lambda_1)=0\,, \tag{10.5}$$

i.e. by setting to zero the variation of S

$$\delta S=\int\left(\frac{\partial\mathcal{L}}{\partial x^\alpha}\delta x^\alpha+\frac{\partial\mathcal{L}}{\partial\dot{x}^\alpha}\delta(\dot{x}^\alpha)\right)d\lambda=0\,. \tag{10.6}$$

Since

$$\delta(\dot{x}^\alpha) = \delta\left(\frac{dx^\alpha}{d\lambda}\right) = \frac{d\delta x^\alpha}{d\lambda}\,, \tag{10.7}$$

the last term in Eq. 10.6 can be written as

$$\frac{\partial\mathcal{L}}{\partial\dot{x}^\alpha}\delta(\dot{x}^\alpha) = \frac{\partial\mathcal{L}}{\partial\dot{x}^\alpha}\frac{d\delta x^\alpha}{d\lambda} = \frac{d}{d\lambda}\left(\frac{\partial\mathcal{L}}{\partial\dot{x}^\alpha}\delta x^\alpha\right) - \frac{d}{d\lambda}\left(\frac{\partial\mathcal{L}}{\partial\dot{x}^\alpha}\right)\delta x^\alpha\,.$$

When integrated between λ_0 and λ_1 the first term on the right-hand side vanishes because of Eq. 10.5, therefore Eq. 10.6 yields

$$\delta S = \int\left[\frac{\partial\mathcal{L}}{\partial x^\alpha} - \frac{d}{d\lambda}\left(\frac{\partial\mathcal{L}}{\partial\dot{x}^\alpha}\right)\right]\delta x^\alpha d\lambda = 0\,, \tag{10.8}$$

which is satisfied for all δx^α vanishing in λ_0 and λ_1 if and only if

$$\frac{\partial\mathcal{L}}{\partial x^\alpha} - \frac{d}{d\lambda}\frac{\partial\mathcal{L}}{\partial(\dot{x}^\alpha)} = 0\,. \tag{10.9}$$

These are the Euler-Lagrange equations for this problem. We shall now show that these equations, when written for the action 10.1, are equivalent to the geodesic equation

$$\ddot{x}^\gamma + \Gamma^\gamma_{\mu\nu}\dot{x}^\mu\dot{x}^\nu = 0\,. \tag{10.10}$$

By substituting the Lagrangian 10.1 in Eqs. 10.9, and taking into account that $g_{\mu\nu} = g_{\mu\nu}(x^\alpha)$ and $\dot{x}^\mu = \dot{x}^\mu(\lambda)$, we find

$$\begin{aligned}\frac{\partial\mathcal{L}}{\partial x^\alpha} - \frac{d}{d\lambda}\frac{\partial\mathcal{L}}{\partial(\dot{x}^\alpha)} &= \\ &= \frac{1}{2}g_{\mu\nu,\alpha}\dot{x}^\mu\dot{x}^\nu - \frac{d}{d\lambda}\left[\frac{1}{2}g_{\mu\nu}(\delta^\mu_\alpha\dot{x}^\nu + \dot{x}^\mu\delta^\nu_\alpha)\right] \\ &= \frac{1}{2}\left[g_{\mu\nu,\alpha}\dot{x}^\mu\dot{x}^\nu - \frac{d}{d\lambda}(2g_{\alpha\nu}\dot{x}^\nu)\right] \\ &= \frac{1}{2}\left(g_{\mu\nu,\alpha}\dot{x}^\mu\dot{x}^\nu - 2g_{\alpha\nu,\beta}\dot{x}^\beta\dot{x}^\nu - 2g_{\alpha\nu}\ddot{x}^\nu\right) \\ &= \frac{1}{2}\left(-2g_{\alpha\nu}\ddot{x}^\nu + g_{\mu\nu,\alpha}\dot{x}^\mu\dot{x}^\nu - g_{\alpha\mu,\nu}\dot{x}^\nu\dot{x}^\mu - g_{\alpha\nu,\mu}\dot{x}^\mu\dot{x}^\nu\right) = 0\,.\end{aligned} \tag{10.11}$$

By contracting this equation with $-g^{\alpha\gamma}$ we find

$$\ddot{x}^\gamma + \frac{1}{2}g^{\alpha\gamma}\left[g_{\alpha\mu,\nu} + g_{\alpha\nu,\mu} - g_{\mu\nu,\alpha}\right]\dot{x}^\mu\dot{x}^\nu = 0\,, \tag{10.12}$$

which, using the definition of the Christoffel symbols (Eq. 3.68), coincides with Eq. 10.10. Thus, the Euler-Lagrange equations written for the Lagrangian 10.1 are equivalent to the geodesic equations 10.10; in different contexts it might be more convenient to use one or the other version of these equations.

10.2 EQUATIONS OF MOTION

The purpose of this section is to find the equations for the four components of a particle four-velocity $u^\mu = (\dot{t}, \dot{r}, \dot{\theta}, \dot{\varphi})$, which can be integrated to find the explicit form of $x^\mu(\lambda)$. The equation for $\dot{\theta}$ will be found using the θ-component of the Euler-Lagrange equations, whereas those for $(\dot{t}, \dot{r}, \dot{\varphi})$ will be derived using the constants of motion. To hereafter we put $G = c = 1$. For the Schwarzschild metric, the Lagrangian of a free particle 10.1 reads

$$\mathcal{L} = \frac{1}{2}\left[-\left(1-\frac{2m}{r}\right)\dot{t}^2 + \frac{\dot{r}^2}{\left(1-\frac{2m}{r}\right)} + r^2\dot{\theta}^2 + r^2\sin^2\theta\dot{\varphi}^2\right]. \tag{10.13}$$

The θ-component of the Euler-Lagrange equations (Eqs. 10.9) associated to this Lagrangian is

$$\frac{\partial\mathcal{L}}{\partial\theta} - \frac{d}{d\lambda}\frac{\partial\mathcal{L}}{\partial\dot{\theta}} = 0 \qquad \rightarrow \qquad \frac{d}{d\lambda}(r^2\dot{\theta}) = r^2\sin\theta\cos\theta\dot{\varphi}^2\,,$$

which gives

$$\ddot{\theta} = -\frac{2}{r}\dot{r}\dot{\theta} + \sin\theta\cos\theta\dot{\varphi}^2\,. \tag{10.14}$$

Due to spherical symmetry, the Schwarzschild spacetime is invariant under rotations of the axes. Using this freedom, we can choose these axes such that, for a given value of the affine parameter, say $\lambda = \lambda_0$, the particle is on the equatorial plane, $\theta = \frac{\pi}{2}$, and its three-velocity $(\dot{r}, \dot{\theta}, \dot{\varphi})$ lays on the same plane. Thus, without loss of generality, we set $\theta(\lambda = \lambda_0) = \frac{\pi}{2}$ and $\dot{\theta}(\lambda = \lambda_0) = 0$. The Cauchy problem

$$\begin{aligned}\ddot{\theta} &= -\frac{2}{r}\dot{r}\dot{\theta} + \sin\theta\cos\theta\dot{\varphi}^2\,,\\ \theta(\lambda = \lambda_0) &= \frac{\pi}{2}\,,\\ \dot{\theta}(\lambda = \lambda_0) &= 0\,,\end{aligned} \tag{10.15}$$

admits a unique solution. Since

$$\theta(\lambda) \equiv \frac{\pi}{2} \tag{10.16}$$

satisfies the differential equation and the initial conditions, it must be *the* solution. Thus, as in Newtonian theory, in General Relativity *the orbits around a spherically-symmetric object are planar*, and to hereafter we shall assume $\theta = \frac{\pi}{2}$ and $\dot{\theta} = 0$. In Chapter 19 we shall see that this is not the case if the object is spinning.

The equations for $\dot{t}$ and $\dot{\varphi}$ can be derived using the symmetries of the Schwarzschild metric. As explained in Sec. 8.3.1, if a given spacetime admits a Killing vector field $\vec{\xi}$, there exists an associated conserved quantity for geodesic motion (see Eq. 8.52)

$$g_{\alpha\mu}\xi^\mu u^\alpha = const\,. \tag{10.17}$$

The Schwarzschild metric admits a timelike and a spacelike Killing vector field, $k^\mu = (1,0,0,0)$ and $m^\mu = (0,0,0,1)$, respectively (see Eqs. 9.38). Therefore, there are two associated conserved quantities

$$g_{\alpha\mu}k^\mu u^\alpha = const_1 \qquad \rightarrow \qquad \left(1-\frac{2m}{r}\right)\dot{t} = E\,, \tag{10.18}$$

$$g_{\alpha\mu}m^\mu u^\alpha = const_2 \qquad \rightarrow \qquad r^2\sin^2\theta\dot{\varphi} = L\,, \tag{10.19}$$

where we have set

$$const_1 = -E \qquad\qquad const_2 = L\,. \tag{10.20}$$

Remembering that we have chosen the equatorial plane $\theta = \frac{\pi}{2}$ as the orbital plane, the equations for $\dot{t}$ and $\dot{\varphi}$ finally are

$$\dot{t} = \frac{E}{\left(1-\frac{2m}{r}\right)}\,, \qquad \text{and} \qquad \dot{\varphi} = \frac{L}{r^2}\,. \tag{10.21}$$

The interpretation of the constants E and L will be discussed in the next section.

It should be noted that the same equations can also be obtained using the t- and φ-components of the Euler-Lagrange equations which, since the Lagrangian is independent [1] of t and φ, yield

$$\frac{\partial \mathcal{L}}{\partial t} - \frac{d}{d\lambda}\frac{\partial \mathcal{L}}{\partial(\dot{t})} = 0 \quad \rightarrow \quad \left(1 - \frac{2m}{r}\right)\dot{t} = const_1\,, \tag{10.22}$$

$$\frac{\partial \mathcal{L}}{\partial \varphi} - \frac{d}{d\lambda}\frac{\partial \mathcal{L}}{\partial(\dot{\varphi})} = 0 \quad \rightarrow \quad r^2\sin^2\theta\dot{\varphi} = const_2\,,$$

which coincide with Eqs. 10.18 and 10.19.

The equation for $\dot{r}$ can be found from the r-component of the Euler-Lagrange equation or, more easily, using the constraint that the norm of the four-velocity is constant. Indeed, $u^\alpha u_\alpha = 0$ for massless particles, and – as long as the affine parameter is the proper time – $u^\alpha u_\alpha = -1$ for massive particles, in geometrical units. Therefore, for massive particles:

$$g_{\alpha\beta}u^\alpha u^\beta = -\left(1 - \frac{2m}{r}\right)\dot{t}^2 + \frac{\dot{r}^2}{\left(1 - \frac{2m}{r}\right)} + r^2\dot{\theta}^2 + r^2\sin^2\theta\dot{\varphi}^2 = -1\,; \tag{10.23}$$

by replacing the expressions for $\dot{t}$ and $\dot{\varphi}$ given in Eqs. 10.21, and setting $\theta = \pi/2$, this becomes

$$\dot{r}^2 + \left(1 - \frac{2m}{r}\right)\left(1 + \frac{L^2}{r^2}\right) = E^2\,. \tag{10.24}$$

For massless particles:

$$g_{\alpha\beta}u^\alpha u^\beta = -\left(1 - \frac{2m}{r}\right)\dot{t}^2 + \frac{\dot{r}^2}{\left(1 - \frac{2m}{r}\right)} + r^2\dot{\theta}^2 + r^2\sin^2\theta\dot{\varphi}^2 = 0\,, \tag{10.25}$$

which gives

$$\dot{r}^2 + \frac{L^2}{r^2}\left(1 - \frac{2m}{r}\right) = E^2\,. \tag{10.26}$$

Summarizing, the equations which describe geodesic motion in Schwarzschild's spacetime are:

- *for massive particles:*

$$\theta = \frac{\pi}{2}\,, \qquad \dot{t} = \frac{E}{\left(1 - \dfrac{2m}{r}\right)}\,, \tag{10.27}$$
$$\dot{\varphi} = \frac{L}{r^2}\,, \qquad \dot{r}^2 = E^2 - \left(1 - \frac{2m}{r}\right)\left(1 + \frac{L^2}{r^2}\right)\,;$$

- *for massless particles:*

$$\theta = \frac{\pi}{2}\,, \qquad \dot{t} = \frac{E}{\left(1 - \dfrac{2m}{r}\right)}\,, \tag{10.28}$$
$$\dot{\varphi} = \frac{L}{r^2}\,, \qquad \dot{r}^2 = E^2 - \frac{L^2}{r^2}\left(1 - \frac{2m}{r}\right)\,.$$

[1] It may be noted that, as a general property, if the Lagrangian does not depend explicitly on a given coordinate there exists an associated constant of motion; in this case, that coordinate is said to be cyclic.

10.3 THE CONSTANTS OF GEODESIC MOTION

As discussed in the previous section, geodesic motion in Schwarzschild's geometry is characterized by two constants of motion, E and L, associated to the timelike and spacelike Killing vector fields $\vec{k}$ and $\vec{m}$ defined in Eqs. 9.38. In order to understand the meaning of the constant E we need to define what is the energy of a particle, which is an *observer-dependent quantity*.

In Special Relativity, the energy of a particle with four-momentum $\vec{p}$, measured by an observer with four-velocity $\vec{U}$, is defined as

$$\mathcal{E}^{(U)} = -\eta_{\mu\nu}U^{\mu}p^{\nu} = -U^{\mu}p_{\mu}\,. \tag{10.29}$$

The energy defined by Eq. 10.29 is a scalar quantity, and does not depend on the choice of the coordinate frame. However, it depends on the choice of the observer $\vec{U}$. In particular, the *energy measured by a static observer* $U^{\mu}_{\mathrm{st}} = (1, 0, 0, 0)$ is

$$\mathcal{E}^{(U_{\mathrm{st}})} = -p_0 = +p^0\,. \tag{10.30}$$

In many textbooks, for brevity, one simply calls "energy" the energy measured by a static observer, given in Eq. 10.30. This quantity *does* depend on the coordinate choice, because different frames have different static observers. The energy measured by the static observer, $\mathcal{E}^{(U_{\mathrm{st}})}$, *has to be positive*; indeed, a negative energy would correspond to a particle moving backwards in time [2].

We shall now show that the energy measured by *any* observer has the same sign as the energy measured by the static observer. To this aim, let us consider a generic observer $U^{\mu} = (\gamma, \gamma V^i)$ with $\gamma = (1 - V^2)^{-1/2}$ and $V = \sqrt{V_i V^i}$ $(i = 1, \ldots, 3)$; the energy (s)he measures is

$$\mathcal{E}^{(U)} = -\gamma(p_0 + p_i V^i) = \gamma(p^0 - p_i V^i)\,. \tag{10.31}$$

Let $p = \sqrt{p_i p^i}$ $(i = 1, \ldots, 3)$. Since $p_i V^i \leq p\,V$, and since $V < 1$, it follows that $p_i V^i < p$. The particle can either be massive or massless, therefore the vector $\vec{p} = (p^0, p^i)$ is timelike or null, and $-(p^0)^2 + p^2 \leq 0$; consequently $p_i V^i < p^0$. A comparison of Eq. 10.31 and Eq. 10.30 shows that, by virtue of this inequality, $\mathcal{E}^{(U)}$ and $\mathcal{E}^{(U_{\mathrm{st}})}$ have the same sign. Since $\mathcal{E}^{(U_{\mathrm{st}})}$ is positive, it follows that the energy measured by *any* observer, $\mathcal{E}^{(U)}$, *has to be positive.*

In General Relativity, we can always consider a LIF in which the laws of Special Relativity hold, including the definition of energy given in Eq. 10.29. Since in the LIF $g_{\mu\nu} \equiv \eta_{\mu\nu}$, this equation can be written in a generally covariant form as

$$\mathcal{E}^{(U)} = -g_{\mu\nu}U^{\mu}p^{\nu} = -U^{\mu}p_{\mu}\,. \tag{10.32}$$

By the Principle of General Covariance Eq. 10.32 is valid in any coordinate frame, thus providing a covariant definition of a particle energy. Note that since $\mathcal{E}^{(U)} > 0$ in Special Relativity, the same property holds in General Relativity in a LIF, and consequently in any coordinate frame.

Let us now consider a static observer with $U^{\mu}_{st} = (1, 0, 0, 0)$, located at some point exterior to the horizon of a Schwarzschild spacetime. According to the definition 10.32, the energy of a particle measured by this observer is

$$\mathcal{E}^{(U_{st})} = -g_{00}U^0_{st}\,p^0 = \left(1 - \frac{2m}{r}\right)p^0\,. \tag{10.33}$$

[2] Moreover, in quantum field theory the existence of a negative energy particle implies that the vacuum state is unstable, because the creation of a cascade of particles with ever decreasing (negative) energy would be energetically favoured.

For a massive particle $p^0 = m_{\rm p}\dot{t}$, where $m_{\rm p}$ is the particle mass, and Eq. 10.33 becomes

$$\mathcal{E}^{(U_{st})} = \left(1 - \frac{2m}{r}\right) m_{\rm p}\dot{t}\,, \tag{10.34}$$

which, using the expression of $\dot{t}$ given in Eq. 10.21, gives

$$\mathcal{E}^{(U_{st})} = Em_{\rm p} \qquad \rightarrow \qquad E = \frac{\mathcal{E}^{(u_{st})}}{m_{\rm p}}\,. \tag{10.35}$$

Thus, the constant of motion E is *the energy of the particle per unit mass, measured by a static observer*. Note that in geometrical units E is a dimensionless quantity.

If the particle is massless we can always choose the affine parameter in such a way that $p^0 \equiv u^0 = \dot{t}$. Thus in this case

$$\mathcal{E}^{(U_{st})} = \left(1 - \frac{2m}{r}\right)\dot{t} = E\,, \tag{10.36}$$

and the constant E is the energy of the particle measured by a static observer.

Let us now consider the constant L. In order to understand its meaning, we need to remember the definition of angular momentum in Special Relativity.

In Special Relativity, the angular momentum of a particle with four-momentum $\vec{p}$ and position $\vec{x}$, measured by an observer with four-velocity $\vec{U}$, is defined as

$$l_\alpha = \epsilon_{\alpha\beta\gamma\delta} U^\beta x^\gamma p^\delta\,, \tag{10.37}$$

where $\epsilon_{\alpha\beta\gamma\delta}$ is the Levi-Civita completely antisymmetric symbol defined in Box 8-C (which, in Minkowski space, coincides with the Levi-Civita tensor). It is worth noting that Eq. 10.37 implies

$$l_\alpha U^\alpha = 0\,. \tag{10.38}$$

For a static observer $U^\beta = (1, 0, 0, 0)$, Eq. 10.38 implies that $l_0 = 0$ and, denoting the space indices as $i, j, k, \dots$, it yields

$$l_i = \epsilon_{ijk} x^j p^k \tag{10.39}$$

(where ϵ_{ijk} is the three-dimensional Levi-Civita symbol with $\epsilon_{123} = 1$), i.e., $\mathbf{l} = \mathbf{x} \wedge \mathbf{p}$. For a massive particle, $p^i = m_{\rm p}\dot{x}^i$, therefore $\mathbf{l} = m_{\rm p}\mathbf{x} \wedge \dot{\mathbf{x}}$; for a massless particle, we can always choose an affine parameter such that $p^i = \dot{x}^i$, and $\mathbf{l} = \mathbf{x} \wedge \dot{\mathbf{x}}$. Note that $\mathbf{l} = m_{\rm p}\mathbf{x} \wedge \dot{\mathbf{x}}$ is also the definition of angular momentum of massive particles in Newtonian physics; the only difference is that, in the relativistic case, the dot indicates differentiation with respect to the proper time, instead of the coordinate time.

In polar coordinates $x^i = (r \sin\theta \cos\varphi, r \sin\theta \sin\varphi, r \cos\theta)$, Eq. 10.39 with $i = 3$ gives the azimuthal angular momentum (we leave the explicit computation as an exercise)

$$l^3 = m_{\rm p}(x^1 \dot{x}^2 - x^2 \dot{x}^1) = m_{\rm p} r^2 \sin^2\theta \dot{\varphi}\,, \tag{10.40}$$

for a massive particle, and $l^3 = r^2 \sin^2\theta\dot{\varphi}$ for a massless particle. A comparison of Eq. 10.40 (which holds in Special Relativity) with Eq. 10.19, $L = r^2 \sin^2\theta\dot{\varphi}$, suggests that the constant of motion L which appears in Eq. 10.19 can be interpreted as *the azimuthal angular momentum measured by a static observer, in the case of massless particles, or the same divided by the particle mass (specific angular momentum), in the case of massive particles.*

A further motivation for this interpretation is provided in Box 10-A.

Box 10-A

Angular momentum as conjugate momentum of the rotation angle

In Special Relativity the Lagrangian of a free particle is (see Eq. 10.1)

$$\mathcal{L} = \frac{1}{2}\eta_{\mu\nu}\dot{x}^{\mu}\dot{x}^{\nu} = \frac{1}{2}(-\dot{t}^2 + \dot{r}^2 + r^2\dot{\theta}^2 + r^2\sin^2\theta\dot{\varphi}^2)\,. \qquad (10.41)$$

In any Lagrangian system, the *conjugate momentum* of a coordinate q^i is defined as

$$\hat{p}_i = \frac{\partial \mathcal{L}}{\partial \dot{q}^i}\,. \qquad (10.42)$$

Thus, the conjugate momentum of the polar angle φ (describing rotations around the axis x^3) is

$$\hat{p}_\varphi = r^2\sin^2\theta\dot{\varphi}\,; \qquad (10.43)$$

comparing this expression with Eq. 10.40 we find that $\hat{p}_\varphi = l^3/m_p$ for a massive particle, and $\hat{p}_\varphi = l^3$ for a massless particle [a].

Let us now consider the rotation angle around a generic axis $\hat{n}$. Due to spherical symmetry, it is always possible to define polar coordinates such that $\hat{n}$ is the azimuthal axis. The conjugate momentum of the rotation angle around the axis $\hat{n}$ is then the angular momentum (per unit mass in the case of massive particles) projected on the same axis $\hat{n}$.

Although the definition 10.37 cannot be extended to General Relativity, we can still call "orbital angular momentum of a particle along an axis" the conjugate momentum of the rotation angle along that axis (times the mass, if the particle is massive). With this definition, the constant of motion defined in Eq. 10.19,

$$L = \frac{\partial \mathcal{L}}{\partial \dot{\varphi}} = r^2\sin^2\theta\dot{\varphi} \qquad (10.44)$$

is, as argued above, just the angular momentum along the azimuthal axis per unit mass $L = l^3/m_p$ in the case of massive particles, and, with an appropriate choice of the affine parameter, the angular momentum along the azimuthal axis $L = l^3$ in the case of massless particles.

For a comprehensive review about the angular momentum in General Relativity, we refer the reader to [108].

[a] It is worth noting that the same result holds in Newtonian physics: the derivative of the point particle Lagrangian with respect to the polar angle gives the angular momentum along the azimuthal axis.

10.4 ORBITS OF MASSLESS PARTICLES

We shall first consider the orbits of massless particles. The equations of motion are given by Eqs. 10.28, which we write in the following form

$$\theta = \frac{\pi}{2}, \qquad \dot{t} = \frac{E}{\left(1 - \frac{2m}{r}\right)}, \qquad \dot{\varphi} = \frac{L}{r^2}, \tag{10.45}$$

$$\dot{r}^2 = E^2 - V(r), \qquad V(r) = \frac{L^2}{r^2}\left(1 - \frac{2m}{r}\right). \tag{10.46}$$

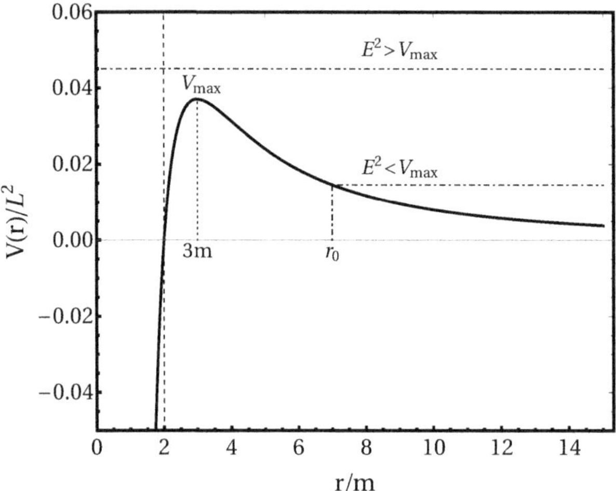

Figure 10.1: Effective potential $V(r)$ for massless particles (see Eq. 10.46).

Note that:

- For massless particles the angular momentum L acts as a scale factor for the *effective potential* $V(r)$;
- $V(r)$ tends to $-\infty$ as $r \to 0$, and approaches zero at $r \to \infty$;
- $V(r)$ has only one extremal point, a maximum, at $r_{max} = 3m$, where it takes the value

$$V_{max} = \frac{L^2}{27m^2}. \tag{10.47}$$

It is also useful to compute the radial acceleration, obtained by differentiating Eq. 10.46 with respect to the affine parameter λ

$$2\dot{r}\ddot{r} = -\frac{dV(r)}{dr}\dot{r} \qquad \to \qquad \ddot{r} = -\frac{1}{2}\frac{dV(r)}{dr}. \tag{10.48}$$

The plot of $V(r)$ is given in Fig. 10.1.

Let us assume that the particle, say a photon, starts its motion at a large value of r, moving towards the central body, i.e. with $\dot{r} < 0$. The photon orbit depends on the values of the constants of motion E and L. Since $\dot{r}^2 = E^2 - V(r)$, the value of $\dot{r}^2$ can be visualized in Fig. 10.1 as *the square of the distance between the straight, horizontal line* $E^2 = const$ *and the curve* $V(r)$.

Depending on the value of E we can have three possibilities:

1. $\mathbf{E^2 > V_{\max}}$. According to Eq. 10.46, $\dot{r}^2$ is always positive; the photon falls towards the central body with a radial velocity which *decreases (in absolute value)* as the particle approaches the body. If the body has a radius R_* larger than $r_{\max} = 3m$, the photon will end its motion impacting on its surface. If $R_* < 3m$, as it may occur for extremely compact stars or for black holes, the photon proceeds its motion with decreasing $|\dot{r}|$, reaches $r = 3m$, where $|\dot{r}|$ attains its minimum value $|\dot{r}|_{\min} = \sqrt{E^2 - V_{max}}$, and then falls toward the body with increasing radial velocity. Depending on the value of L, the photon can make several revolutions around the central body before falling in. This can be checked by integrating the equation for $\dot{r}$ and $\dot{\varphi}$ to find the trajectory $r(\varphi)$ in the equatorial plane.

2. $\mathbf{E^2 = V_{\max}}$. As in the previous case, $|\dot{r}|$ decreases as the photon approaches the central body: it stops on the surface if $R_* > 3m$, whereas if $R_* < 3m$ it proceeds with decreasing $|\dot{r}|$; in this case $|\dot{r}|$ vanishes when $r = 3m$. In addition since at that point $\frac{dV(r)}{dr} = 0$, the radial acceleration $\ddot{r}$ also vanishes, as shown by Eq. 10.48. The particle approaches $r = 3m$ asymptotically with ever decreasing velocity. If instead the particle is already at $r = 3m$ with $E^2 = V_{\max}$, then it remains at the same r at later times, i.e. its orbit is circular. However, since the potential is maximum, $r = r_{\max} = 3m$ is an unstable orbit. Indeed, if the position is perturbed, the particle will either

   - fall into the central body; this happens when the radial coordinate of the particle is displaced to $r = r_{\max} - \delta r$, since there the slope of the potential is positive, and the radial acceleration is negative (see Eq. 10.48); or
   - escape toward infinity; this happens when the radial coordinate is displaced to $r = r_{\max} + \delta r$, since there the radial acceleration is positive.

   Thus, for massless particles, there exists only one *circular, unstable* orbit, and for this orbit (see Eq.10.47)

   $$E^2 = \frac{L^2}{27m^2}. \tag{10.49}$$

   The quantity $r_{\max} = 3m$ is known as the radius of the **photon sphere** of a Schwarzschild black hole and plays a crucial role in defining the so-called *shadow* of a black hole (see Sec. 11.5). The corresponding circle on the equator is called **light ring**.

3. $\mathbf{E^2 < V_{\max}}$. Let r_0 be the abscissa of the point where $E^2 = V(r)$ (see Fig. 10.1). For $r > r_0$, $\dot{r}^2$ is positive and an incoming photon moves toward r_0 with decreasing (in modulus) radial velocity. When it reaches r_0, $\dot{r}^2$ vanishes. This is a *turning point*: the photon cannot penetrate the potential barrier, i.e. it cannot continue its motion for $r < r_0$, because $\dot{r}$ would become imaginary. At $r = r_0$ the radial acceleration is positive (the slope of $V(r)$ is negative, see Eq. 10.48), therefore the particle is forced to invert its radial velocity, and to escape toward infinity on an open trajectory.

Thus, contrary to Newtonian gravity, *General Relativity predicts that a light ray is bent by the gravitational field of a massive body.* In addition, if its energy satisfies the condition

$$E^2 < \frac{L^2}{27m^2}, \tag{10.50}$$

the light ray is deflected and returns to infinity on an open orbit.

10.5 ORBITS OF MASSIVE PARTICLES

Let us now consider the orbits of a massive particle. The equations of motion are

$$\theta = \frac{\pi}{2}, \qquad \dot{t} = \frac{E}{\left(1 - \frac{2m}{r}\right)}, \qquad \dot{\varphi} = \frac{L}{r^2}, \tag{10.51}$$

$$\dot{r}^2 = E^2 - V(r), \qquad V(r) = \left(1 - \frac{2m}{r}\right)\left(1 + \frac{L^2}{r^2}\right). \tag{10.52}$$

First of all we note that, contrary to the massless case, the potential does not scale with the angular momentum and that $V(r) \to 1$ when $r \to \infty$.

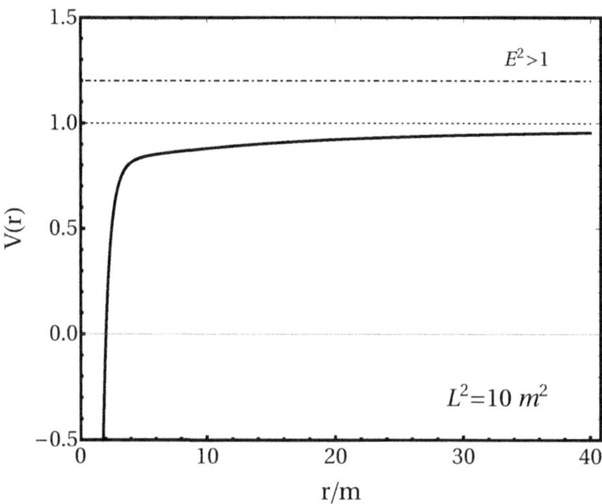

Figure 10.2: Effective potential for a massive particle, in a representative case in which $L^2 < 12m^2$.

To find whether the potential admits a minimum or a maximum we need to solve the equation

$$\frac{\partial V}{\partial r} = 2\frac{mr^2 - L^2 r + 3mL^2}{r^4} = 0,$$

which has two roots

$$r_\pm = \frac{L^2 \pm \sqrt{L^4 - 12m^2 L^2}}{2m}. \tag{10.53}$$

If $L^2 < 12m^2$ the roots $r_\pm$ are complex and the potential does not have extremal points, as shown in Fig. 10.2. In this case a massive particle arriving from infinity with $E^2 > 1$ and $\dot{r} \leq 0$ will be captured by the central body with increasing (in absolute value) radial velocity.

If $L^2 > 12m^2$ the roots $r_\pm$ are real, and $V(r)$ has a maximum in $r = r_-$ followed by a minimum in $r = r_+$. Depending on the value of L the maximum can be larger or smaller than unity, i.e.

$$\text{(a)} \quad L^2 > 16m^2 \quad \to \quad V_{\max} > 1, \tag{10.54}$$

$$\text{(b)} \quad 12m^2 < L^2 < 16m^2 \quad \to \quad V_{\max} < 1. \tag{10.55}$$

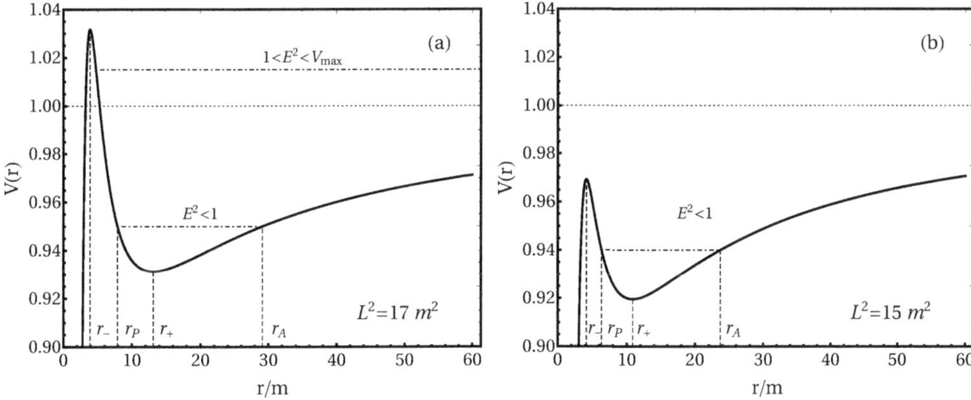

Figure 10.3: Effective potential for a massive particle with $L^2 > 12m^2$. Left and right panels refer to case (a) and (b) in the text, respectively.

The potential has the radial dependence shown in Fig. 10.3 for two values of L corresponding to case (a) (left panel) and case (b) (right panel).

We can therefore classify the geodesic motion of a massive particle in the following way:

- case (a)
  - A particle, which starts its motion from infinity with $E^2 > V_{\max}$ and $\dot{r} \leq 0$, falls towards the central body making a number of revolutions which depends on the value of L (see the discussion for $E^2 > V_{\max}$ in Sec. 10.4). The radial velocity $|\dot{r}|$ increases as the particle approaches r_+, decreases when $r_- < r < r_+$, and then increases again for $r < r_-$.
  - If $1 < E^2 < V_{\max}$ and $\dot{r} \leq 0$ the particle approaches the central body and reaches a turning point r_0 where $E^2 = V(r_0)$ and $\dot{r} = 0$. The particle cannot penetrate the barrier and since the radial acceleration given by Eq. 10.48, which holds also for massive particles, is positive in r_0, it inverts its radial velocity and escapes free at infinity on an open orbit (see the discussion for $E^2 < V_{\max}$ in Sec. 10.4).
  - A particle with $V_{\min} < E^2 < 1$ moves between two turning points, i.e. between the two values of r where $E^2 = V(r)$. The smaller value, r_{P}, is the *periastron*, the point of minimum distance from the central body; the largest, r_{A}, is the *apastron*, the point of maximum distance (see Fig. 10.3). The orbit is bound but, as we shall see in the next chapter, it is not closed, and therefore it is not an ellipse.
- case (b)
  - A particle with $V_{\min} < E^2 < V_{\max}$ moves between the two turning points as in the last point of case (a).
  - If $E^2 > 1$ and $\dot{r} \leq 0$, the particle approaches the central body and eventually falls in.
  - If $V_{\max} < E^2 < 1$, then we should distinguish between different cases: i) if $\dot{r} \leq 0$, the particle approaches the central body and eventually falls in; ii) if $\dot{r} > 0$, the particle first moves outwards, finds an inversion point, and eventually falls in. Note that in both cases the particle cannot come from infinity, since $E^2 < 1$.

Both in case (a) and in case (b), a particle with energy $E^2 = V(r_-) \equiv V_{\rm max}$ and $r = r_-$ moves on an *unstable* circular orbit (see the discussion for $E^2 = V_{\rm max}$ in Sec. 10.4), whereas if $E^2 = V(r_+) \equiv V_{\rm min}$ a particle at $r = r_+$ moves on a *stable* circular orbit.

From the expression of $r_\pm$ given in Eq. 10.53, we see that if $L^2 = 12m^2$ the two roots coincide,

$$r_- = r_+ = 6m\,. \tag{10.56}$$

Furthermore, r_+ is an increasing function of L and, as $L \to \infty$, $r_+ \to \infty$. This means that there cannot exist stable circular orbits with radius smaller than $6m$. For this reason, $r = 6m$ is called the radius of the **innermost stable circular orbit (or ISCO)** [3], and plays an important role in the dynamics of accretion disks around astrophysical black holes (see Sec. 11.5.1). In addition, it is easy to show that r_- is a decreasing function of L and, as $L \to \infty$,

$$r_- = \frac{L^2}{2m}\left(1 - \sqrt{1 - \frac{12m^2}{L^2}}\right) \to \frac{L^2}{2m}\left(1 - \left(1 - \frac{6m^2}{L^2} + ...\right)\right) = 3m + O\left(\frac{m}{L}\right)\,; \tag{10.57}$$

therefore, unstable circular orbits exist only in the range

$$3m < r_- < 6m\,. \tag{10.58}$$

In particular, the photon sphere ($r = 3m$) also represents the smallest, unstable circular orbit of a massive particle around a Schwarzschild black hole.

Let us consider a massive particle in a stable circular geodesic (which, as we have shown, exists for $r \geq 6m$). The *orbital frequency* is:

$$\omega = \frac{d\varphi}{dt} = \frac{\dot\varphi}{\dot t}\,; \tag{10.59}$$

this can be computed by replacing Eqs. 10.51 in Eq. 10.59, and imposing $V'(r) = 0$ and $E^2 = V(r)$. A simpler way to compute this quantity is by solving the Euler-Lagrange equations for r,

$$\frac{d}{d\tau}\frac{\partial\mathcal{L}}{\partial\dot r} = \frac{\partial\mathcal{L}}{\partial r}\,, \tag{10.60}$$

i.e. (see Eq. 10.13)

$$\frac{d}{d\tau}(g_{rr}\dot r) = \frac{1}{2}g_{\mu\nu,r}\dot x^\mu \dot x^\nu\,. \tag{10.61}$$

In the case of a circular geodesic, $\dot r = \ddot r = 0$, and Eq. 10.61 reduces to

$$g_{tt,r}\dot t^2 + g_{\varphi\varphi,r}\dot\varphi^2 = 0\,. \tag{10.62}$$

Replacing Eq. 10.59 yields

$$\omega^2 = -\frac{g_{tt,r}}{g_{\varphi\varphi,r}} \tag{10.63}$$

where we remember that $g_{tt} = -1 + 2m/r$ and (on the equatorial plane) $g_{\varphi\varphi} = r^2$. Thus, $\omega^2 = \frac{m}{r^3}$, i.e.

$$\omega = \pm\,\omega_K \qquad \text{where} \qquad \omega_K = \sqrt{\frac{m}{r^3}}\,. \tag{10.64}$$

[3] It can be shown that, on the ISCO, the second derivative of the potential is also vanishing (i.e., the ISCO is a marginally stable circular orbit). Thus, an alternative way to compute its location is by solving the three algebraic equations $V = E^2$ and $\partial V/\partial r = \partial^2 V/\partial r^2 = 0$ for the corresponding values of E, L, and r of the ISCO.

Remarkably, Eq. 10.64 coincides with Kepler's third law: although the orbital motion of a particle in General Relativity differs from that of Newtonian gravity, the expression of the orbital frequency of a massive particle on a circular orbit in the two theories is the same; ω_K is called *Keplerian frequency.*

The orbital frequency grows as the circular orbit approaches the object; its maximum value is reached at the ISCO, $r = 6m$, where $\omega_K = 6^{-3/2}/m$.

10.6 RADIAL CAPTURE OF A MASSIVE PARTICLE

Let us consider a massive particle falling radially into a Schwarzschild black hole. In this case $d\varphi/d\tau = 0$, therefore $L = 0$; moreover, since the particle is moving inwards, $\dot{r} < 0$. Eqs. 10.51 and 10.52 can be written as[4]

$$\frac{dt}{d\tau} = \frac{E}{1-\frac{2m}{r}} \qquad \frac{dr}{d\tau} = -\sqrt{E^2 - 1 + \frac{2m}{r}} \tag{10.65}$$

$$\frac{d\theta}{d\tau} = 0 \qquad \frac{d\varphi}{d\tau} = 0\,. \tag{10.66}$$

If the particle is at rest at infinity

$$\lim_{r\to\infty} \frac{dr}{d\tau} = 0\,, \tag{10.67}$$

from Eq. 10.65 it follows that

$$E = 1\,, \tag{10.68}$$

and the equations for t and r reduce to

$$\frac{dt}{d\tau} = \frac{1}{1-\frac{2m}{r}} \tag{10.69}$$

$$\frac{dr}{d\tau} = -\sqrt{\frac{2m}{r}}\,. \tag{10.70}$$

These equations can be integrated to find $\tau(r)$ and $t(r)$. By defining $r_0 \equiv r(\tau = 0)$, Eq. 10.70 gives

$$\tau(r) = -\int_{r_0}^{r} dr' \sqrt{\frac{r'}{2m}} = \frac{2}{3}\frac{1}{\sqrt{2m}}\left(r_0^{3/2} - r^{3/2}\right)\,. \tag{10.71}$$

Furthermore, to find $t(r)$, we combine Eqs. 10.69 and 10.70,

$$\frac{dt}{dr} = -\frac{1}{1-\frac{2m}{r}}\sqrt{\frac{r}{2m}}\,. \tag{10.72}$$

Setting $t = 0$ when $\tau = 0$, we find

$$t(r) = \int_0^t dt' = -\int_{r_0}^{r} \frac{dr'}{1-\frac{2m}{r'}}\sqrt{\frac{r'}{2m}}\,, \tag{10.73}$$

which gives

$$t(r) = \frac{2}{3}\frac{1}{\sqrt{2m}}\left[r_0^{3/2} - r^{3/2} + 6mr_0^{1/2} - 6mr^{1/2}\right] + 2m\ln\left[\frac{\sqrt{r_0}-\sqrt{2m}}{\sqrt{r_0}+\sqrt{2m}}\,\frac{\sqrt{r}+\sqrt{2m}}{\sqrt{r}-\sqrt{2m}}\right]\,. \tag{10.74}$$

[4]We remember that, since we are using geometrical units, τ is the proper time of the particle.

The inverse function of $t(r)$, i.e. $r(t)$, is not known analytically. In Fig. 10.4 we plot the functions $t(r)$ and $\tau(r)$.

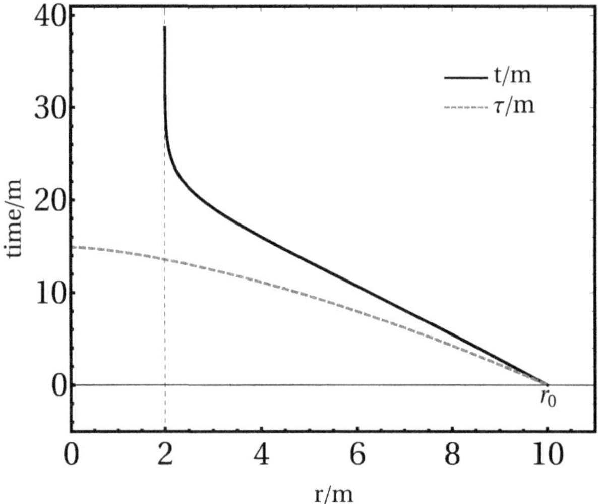

Figure 10.4: Proper time and coordinate time of a massive particle falling radially into a black hole, as functions of r.

Assuming for simplicity $r_0 \gg 2m$, the behaviour of $t(r)$ for $r \to 2m$ and $r \gg 2m$ is:

$$\text{for} \quad r \gg 2m \qquad t \simeq \frac{2}{3}\frac{1}{\sqrt{2m}}(r_0^{3/2} - r^{3/2}) = \tau\,, \tag{10.75}$$

$$\text{for} \quad r \simeq 2m \qquad t \simeq -2m\ln(\sqrt{r} - \sqrt{2m}) + const \;\to\; \infty\,.$$

Therefore $t(r)$ diverges for $r \to 2m$, while Eq. 10.71 shows that $\tau(r)$ is *regular* at $r = 2m$. The inverse functions $r(t)$ and $r(\tau)$ are plotted in Fig. 10.5. The function $r(\tau)$ is the radial coordinate of the particle as a function of the proper time, i.e. as seen by an observer moving with the particle, and Fig. 10.5 shows that for $r = 2m$ it has a regular behaviour: this observer crosses the horizon and reaches the singularity *in a finite amount of proper time.*

The function $r(t)$ is the radial coordinate as a function of the coordinate time, and approaches $r = 2m$ only asymptotically. This suggests that the coordinate t is not appropriate to describe the behaviour of a particle crossing the horizon. Indeed, as discussed in Sec. 9.5, the coordinates (t, r, θ, φ) are not defined on $r = 2m$. The motion of a body falling inside the black hole horizon should be described using, for instance, the Eddington-Finkelstein coordinates (v, r, θ, φ) (see Sec. 9.5.5), where $v = t + r_*$ and

$$r_* = r + 2m\log\left|\frac{r}{2m} - 1\right| \tag{10.76}$$

is the tortoise coordinate introduced in Eq. 9.117. As the particle falls radially into the black hole,

$$\frac{dv}{d\tau} = \frac{dt}{d\tau} + \frac{dr_*}{d\tau} = \frac{1}{1 - \frac{2m}{r}} - \sqrt{\frac{2m}{r}}\frac{1}{1 - \frac{2m}{r}} = \frac{1}{1 + \sqrt{\frac{2m}{r}}}\,, \tag{10.77}$$

where we have used $dr_*/dr = (1 - 2m/r)^{-1}$. Therefore, $\dot{v}$ does not diverge on the horizon:

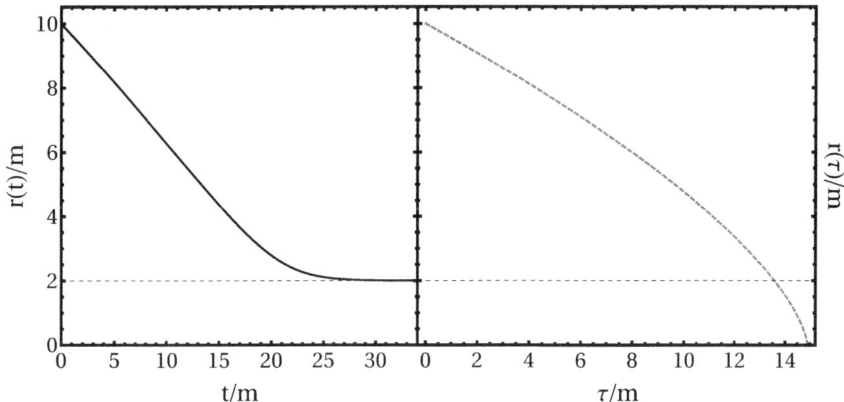

Figure 10.5: The radial coordinate of a massive particle falling radially into a Schwarzschild black hole is plotted as a function of the coordinate time (left panel) and of the proper time (right panel).

as r decreases, $\dot{v}$ also decreases, from $\dot{v} \to 1$ at infinity to $\dot{v} = 1/2$ at the horizon. Inside the horizon, $\dot{v}$ keeps decreasing, until $\dot{v} \to 0$ at the $r = 0$ curvature singularity. Thus, as the particle crosses the horizon and reaches the singularity, the coordinate v remains finite.

Let us now consider a spaceship which, while falling radially towards the black hole horizon, sends an SOS in the form of a sequence of equally spaced electromagnetic pulses; these signals are received by a static observer at radial infinity (we assume that the spaceship and the observer have the same $\varphi = const$), located at $r = r^{\mathrm{obs}}$. The SOS signal travels along null geodesics $t = t(\lambda)$, $r = r(\lambda)$, with θ, φ constants and $L = 0$; λ is the affine parameter along the null geodesics. Note that since the signals are sent toward the observer, they never cross the black hole horizon, and can be studied in the (t, r, θ, φ) coordinate frame. From Eqs. 10.28 we find

$$\theta = \frac{\pi}{2}, \quad \varphi = const, \quad \frac{dt}{d\lambda} = \frac{E}{1 - \frac{2m}{r}}, \quad \frac{dr}{d\lambda} = \pm E, \tag{10.78}$$

hence

$$\frac{dt}{dr} = \pm \frac{r}{r - 2m}, \tag{10.79}$$

and the solution is

$$t = \pm r_* + const\,. \tag{10.80}$$

This result has already been found in Sec. 9.5.4, where we showed that ingoing null geodesics are those for which $v = t + r_* = const$, while outgoing null geodesics, like those describing the signals emitted by the spaceship, are those for which $u = t - r_* = const$.

Let us consider two electromagnetic pulses sent from the spaceship, the first at $\tau = \tau_1$, the second at $\tau = \tau_2$ where τ is the proper time of the spaceship (see Fig. 10.6). The two pulses correspond to $u = u_1$ and $u = u_2$, respectively. The observer detects the pulses at two values of its own proper time which, since the observer is at rest at infinity, coincides with

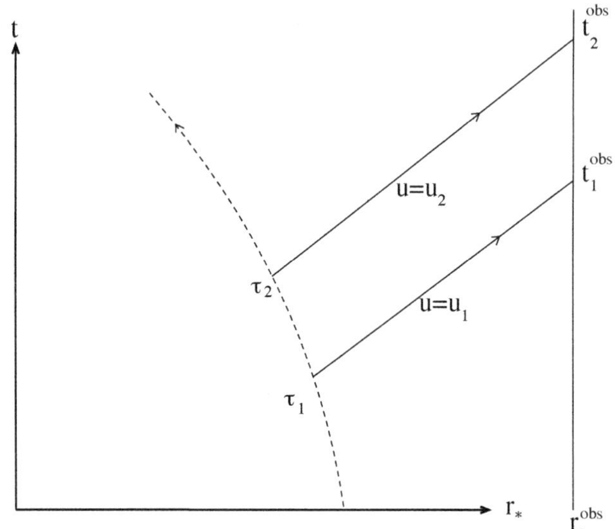

Figure 10.6: A spaceship falling radially into a black hole sends electromagnetic signals to a distant observer. The spaceship trajectory is indicated by the dashed curve.

the coordinate time, i.e. at $t = t_1^{obs}$ and $t = t_2^{obs}$. Thus, while the spaceship pilot measures a proper time interval between the pulses

$$\Delta\tau = \tau_2 - \tau_1 \,, \tag{10.81}$$

the observer at infinity measures a corresponding coordinate time interval

$$\Delta t^{obs} = t_2^{obs} - t_1^{obs} = (u_2 + r_*^{\rm obs}) - (u_1 + r_*^{\rm obs}) = u_2 - u_1 = \Delta u \,. \tag{10.82}$$

Since u is constant along each of the two null geodesics, u_1 and u_2 can also be evaluated in terms of points along the spaceship geodesic:

$$\begin{aligned} u_1 &= t(\tau_1) - r_*(\tau_1) \,, \\ u_2 &= t(\tau_2) - r_*(\tau_2) \,, \end{aligned} \tag{10.83}$$

thus $\Delta u = \Delta t - \Delta r_*$. Therefore

$$\frac{\Delta t^{obs}}{\Delta\tau} = \frac{\Delta u}{\Delta\tau} = \frac{\Delta t}{\Delta\tau} - \frac{\Delta r_*}{\Delta\tau} \,. \tag{10.84}$$

Assuming that the pulses are emitted at very short time intervals, in the limit $\Delta\tau \to 0$, this equation becomes

$$\frac{dt^{obs}}{d\tau} = \frac{du}{d\tau} = \frac{dt}{d\tau} - \frac{dr_*}{d\tau} = \frac{dt}{d\tau} - \frac{dr_*}{dr}\frac{dr}{d\tau} = \frac{1}{1-\frac{2m}{r}}\left(1 + \sqrt{\frac{2m}{r}}\right) \,. \tag{10.85}$$

Eq. 10.85 shows that as $r \to 2m$, $\dfrac{dt^{obs}}{d\tau} \to \infty$; this means that the static observer at infinity receives the signals from the spaceship separated by time intervals which increase, and finally diverge, as the spaceship approaches the horizon.

CHAPTER 11

Kinematical tests of General Relativity

In this chapter we shall describe some of the most interesting predictions of Einstein's theory of gravity related to the motion of particles and light in a static spacetime. Some of these predictions were derived almost immediately after the formulation of the theory in 1916 and are considered as the *classical tests of General Relativity*: the redshift of spectral lines, the bending of light propagating in the gravitational field of a massive body, and the precession of Mercury's perihelion. A fourth classical test is the Shapiro time delay, which was proposed much later and verified in the early 2000s. These four tests probe the so-called "weak-field" regime of the theory, in which the gravitational potential and the curvature of spacetime are weak. In 2019 the Event Horizon Telescope Collaboration obtained the first radio image of the "silhouette" of a black hole, which can be considered a further kinematical test of General Relativity (this time in the strong-field regime) and confirmed Einstein's theory once more. This chapter is devoted to discuss the above tests and their verification. A completely different class of tests concerns the highly-dynamical regime of the theory and the emission of radiation from non-stationary sources. These tests will be discussed in the next chapters. A further kinematical test, the precession of gyroscopes in a gravitational field, will be discussed in Sec. 17.4, after showing how the angular momentum of a body affects the metric in the far-field limit.

11.1 GRAVITATIONAL SHIFT OF SPECTRAL LINES

General Relativity predicts that the spectral lines of a source in the vicinity of a massive body are redshifted when detected far away from the body.

In order to explain this effect, we first need to clarify how to measure the proper time between two spacetime events. These time intervals are measured using clocks, which are instruments whose functioning is based on the repetition of a periodic phenomenon, such as atomic oscillations, the oscillations of a quartz crystal, or of a pendulum. In this section we shall use physical units, and we shall indicate the proper time as $T = \tau/c$ (see Box 3-F). The proper time interval between two infinitely close events can be expressed, in a coordinate frame $\{x^\mu\}$, as

$$dT = \frac{1}{c}\sqrt{-ds^2} \equiv \frac{1}{c}\sqrt{-g_{\mu\nu}(x^\mu)dx^\mu dx^\nu}\,. \tag{11.1}$$

From the above equation we find the *time dilation factor*

$$\frac{dT}{dt} = \frac{1}{c}\sqrt{-g_{\alpha\beta}(x^\mu)\frac{dx^\alpha}{dt}\frac{dx^\beta}{dt}}\,, \tag{11.2}$$

which gives the ratio between the interval of proper time between two infinitely close events and the corresponding interval of coordinate time, and depends both on the metric and on the clock velocity. The proper time T should not be confused with the *coordinate time* which is indicated with t.

If Eq. 11.1 refers to the time interval between two ticks of the clock, $g_{\mu\nu}$ must be evaluated at the clock position. If the clock is at rest with respect to the reference frame, $dx^i = 0$ ($i = 1, \ldots, 3$) along its worldline, then the proper time interval between two ticks is

$$dT = \frac{1}{c}\sqrt{-g_{00}(x^\mu)}dx^0 = \sqrt{-g_{00}(x^\mu)}dt\,, \tag{11.3}$$

and dt is the corresponding interval of coordinate time. Note that we are assuming that dT is very small, so that we can use the infinitesimal expression of the proper time without integrating over the clock worldline.

We shall now show that, when a signal propagates in a gravitational field, its frequency changes with respect to the emission frequency. Let us assume that the gravitational field is stationary, which implies that there exists a timelike Killing vector field and that, by a suitable coordinate choice [1], the metric can be made independent of time (see Sec. 8.2). In this case, the coordinate x^0 will be referred to as the *universal time*. Let S be a light source located, say, on the surface of a massive body, and O an observer located far away from the source, as shown in Fig. 11.1; source and observer are assumed to be at rest with respect to each other.

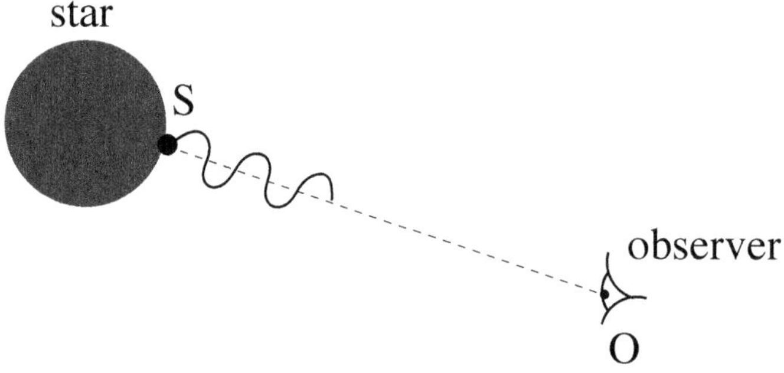

Figure 11.1: A light signal is sent from a source S located on the surface of a star, and detected far away from it. The source and the observer are at fixed relative position in a stationary spacetime.

The source S emits a wave crest which reaches O after an interval $\Delta t = \frac{1}{c}\Delta x^0$ of coordinate time. The interval Δx^0 that *each wave crest* takes to go from S to O only depends on the components of the metric. Indeed, since along a light ray

$$ds^2 = g_{00}(dx^0)^2 + 2g_{0i}dx^0dx^i + g_{ik}dx^idx^k = 0\,, \qquad i,k = 1,\ldots,3$$

then

$$\Delta x^0 = \int_{\text{light path}} dx^0 = \int_{\text{light path}} \frac{-g_{0i}dx^i \pm \sqrt{(g_{0i}dx^i)^2 - g_{00}g_{ik}dx^idx^k}}{g_{00}}\,. \tag{11.4}$$

[1] This choice is not univocal, because we can always shift the origin of time, and we can rescale x^0 by an arbitrary constant factor.

The physical solution is that corresponding to the $-$ sign (the solution with the $+$ sign corresponds to a signal emitted by O at some negative value of x^0, that reaches S at $x^0 = 0$). Even without explicitly computing Eq. 11.4 we see that, since the metric is independent of time, Δx^0 does not depend on time as well. Therefore, the time interval needed to a wave crest to go from S to O is the same for all crests. Consequently, if two wave crests are emitted with a coordinate time separation $\delta t_{\rm em}$ by S, they will be detected by the observer O with the *same coordinate time separation*, i.e. $\delta t_{\rm obs} = \delta t_{\rm em}$.

The period of the emitted wave, $\delta T_{\rm em}$, is the interval of proper time elapsed between the emission of two successive wave crests, i.e.

$$\delta T_{\rm em} = \sqrt{-g_{00}(x^\mu_{\rm em})}\ \delta t_{\rm em}\,,$$

and the emission frequency is

$$\nu_{\rm em} = \frac{1}{\delta T_{\rm em}} = \frac{1}{\sqrt{-g_{00}(x^\mu_{\rm em})}\ \delta t_{\rm em}}\,.$$

Similarly, the period measured by the observer is the interval of the observer proper time elapsed between the detection of two wave crests, i.e.

$$\delta T_{\rm obs} = \sqrt{-g_{00}(x^\mu_{\rm obs})}\ \delta t_{\rm obs}\,,$$

and the observed frequency is

$$\nu_{\rm obs} = \frac{1}{\delta T_{\rm obs}} = \frac{1}{\sqrt{-g_{00}(x^\mu_{\rm obs})}\ \delta t_{\rm obs}}\,.$$

Since $\delta t_{\rm em} = \delta t_{\rm obs}$ we finally find

$$\frac{\nu_{\rm obs}}{\nu_{\rm em}} = \frac{\lambda_{\rm em}}{\lambda_{\rm obs}} = \sqrt{\frac{g_{00}(x^\mu_{\rm em})}{g_{00}(x^\mu_{\rm obs})}}\,. \tag{11.5}$$

Thus, in general the observed frequency does not coincide with the emitted one, since the metric in the two points is different.

It should be stressed that Eq. 11.5 has been derived under the assumptions that i) the gravitational field is stationary, ii) the source and the observer are at fixed relative positions. If one or both assumptions are removed, the derivation of the shift formula must be modified and Eq. 11.5 is no longer appropriate to describe the phenomenon.

Box 11-A

“Weak” and “strong” gravitational fields

We wish to establish when the gravitational field generated by a massive body can be considered “weak” or “strong”. For example, let us consider the gravitational field in the vicinity of the Sun. The Sun mass and radius are (see Table A)

$$M_\odot = 1.989 \times 10^{33}\,\text{g}, \qquad R_\odot = 6.960 \times 10^{10}\,\text{cm}\,; \tag{11.6}$$

moreover, given $G = 6.674 \times 10^{-8}\,\text{cm}^3/(\text{g}\,\text{s}^2)$ and $c = 2.998 \times 10^{10}\,\text{cm/s}$,

$$\frac{GM_\odot}{c^2} = 1.477\,\text{km}\,, \quad \text{and} \qquad \frac{GM_\odot}{R_\odot c^2} \simeq 2.1 \times 10^{-6}\,. \tag{11.7}$$

The dimensionless quantity $\frac{GM}{Rc^2}$ is called *compactness*, and can be taken as a measure for the effects of General Relativity near the surface of an object with mass M and radius R. The compactness of the Sun is much smaller than unity therefore we can say that the gravitational field of the Sun is weak.

The Earth has mass $M_\oplus = 5.972 \times 10^{27}\,\text{g}$ and equatorial radius $R_\oplus = 6.378 \times 10^8\,\text{cm}$. Since

$$M_\odot/M_\oplus \simeq 3 \times 10^5, \qquad \text{and} \qquad R_\odot/R_\oplus \simeq 10^2,$$

$$\frac{GM_\odot}{R_\odot c^2}\Big/\frac{GM_\oplus}{R_\oplus c^2} \simeq 3 \times 10^3\,, \tag{11.8}$$

i.e. the compactness of the Sun is about 3000 times larger than that of the Earth. Conversely, if we consider a neutron star with typical mass and radius (see Chapter 16)

$$M_{\text{NS}} \simeq 1.4\,M_\odot\,, \qquad R_{\text{NS}} \simeq 10\,\text{km}\,, \tag{11.9}$$

the compactness is

$$\frac{GM_{\text{NS}}}{R_{\text{NS}}c^2} \simeq 0.21\,, \tag{11.10}$$

which is close to unity and much larger than that of the Sun. Thus, the effects of General Relativity will be much more important close to a neutron star than in the vicinity of the Sun. On the event horizon of a Schwarzschild black hole $R_{\text{S}} = 2GM_{\text{BH}}/c^2$ (see Eq. 9.34), thus

$$\frac{GM_{\text{BH}}}{R_{\text{S}}c^2} = \frac{1}{2}\,, \tag{11.11}$$

and in this case the compactness is even larger than that of a neutron star.

Note that in Newtonian gravity the gravitational potential on the surface of a spherical body is $\Phi = -GM/R$, therefore its compactness is the dimensionless quantity $|\Phi/c^2|$. Thus, if $|\Phi/c^2|$ is much smaller than one, as it is for the Earth and for the Sun, we can say that the gravitational field generated by that body is weak. More in general, if Φ is the solution of Poisson's equation for an arbitrary distribution of matter, the field is considered weak if $|\Phi/c^2| \ll 1$.

11.1.1 Redshift of spectral lines in the weak-field limit

Let us now consider Eq. 11.5 in the weak-field limit. In Sec. 6.1 we showed that if the gravitational field is stationary and weak, the 00-component of the metric tensor is related to the Newtonian potential Φ, solution of the Poisson equation $\nabla^2\Phi = 4\pi G\rho$, as follows

$$g_{00} \simeq -\left(1+\frac{2\Phi}{c^2}\right) .$$

Consequently, in this limit Eq. 11.5 gives

$$\frac{\nu_{obs}-\nu_{em}}{\nu_{em}} = \frac{\lambda_{em}-\lambda_{obs}}{\lambda_{obs}} = \sqrt{\frac{1+\frac{2\Phi_{em}}{c^2}}{1+\frac{2\Phi_{obs}}{c^2}}} - 1 \simeq \sqrt{\left(1+\frac{2\Phi_{em}}{c^2}\right)\left(1-\frac{2\Phi_{obs}}{c^2}\right)} - 1$$
$$\simeq \sqrt{1+\frac{2}{c^2}\left(\Phi_{em}-\Phi_{obs}\right)} - 1 \simeq \frac{1}{c^2}\left(\Phi_{em}-\Phi_{obs}\right) ,$$

and finally

$$\frac{\Delta\nu}{\nu} \simeq \frac{1}{c^2}\left(\Phi_{em}-\Phi_{obs}\right) . \tag{11.12}$$

To derive this equation we have assumed that the gravitational field is weak both near the source and near the observer, i.e. $\frac{|\Phi_{em}|}{c^2}, \frac{|\Phi_{obs}|}{c^2} \ll 1$. If we assume, for example, that the light source is on the surface of the Sun, and that the observer is on the Earth, since $|\Phi_{Earth}| \ll |\Phi_{Sun}|$ (see Box 11-A) we can also neglect the contribution of the gravitational field of the Earth. Thus the potential to be considered in Eq. 11.12 is that of the Sun, $\Phi = -\frac{GM_\odot}{r}$, the emission radius is $r_{em} = R_\odot$ and $r_{obs} = r_{Sun-Earth} \sim 149.6 \times 10^6$ km is the average distance between the center of the Sun and the surface of the Earth. With these assumptions Eq. 11.12 gives

$$\frac{\Delta\nu}{\nu} \simeq \frac{GM_\odot}{c^2}\left(-\frac{1}{R_\odot}+\frac{1}{r_{\mathrm{Sun-Earth}}}\right) \simeq -\frac{GM_\odot}{R_\odot c^2} \simeq -2.1\times 10^{-6} , \tag{11.13}$$

where we have used the fact that $r_{\mathrm{Sun-Earth}} \gg r_{em}$.

Note that:

- Eq. 11.13 shows that $\Delta\nu < 0$, which means that the spectral lines detected by an observer on the Earth have a frequency smaller than the emission frequency, i.e. they are *redshifted.*
- The relative redshift of spectral lines produced by the Sun is of the order of its compactness (see Box 11-A).

11.1.2 Redshift of spectral lines in a strong gravitational field

Let us now consider the case when the source emitting light is located in a strong gravitational field, for instance near a neutron star or a black hole. The spacetime exterior to both these objects[2] is described – assuming, for simplicity, spherical symmetry – by the Schwarzschild metric

$$ds^2 = -\left(1-\frac{2m}{r}\right)dt^2 + \frac{1}{1-\frac{2m}{r}}dr^2 + r^2\left(d\theta^2+\sin^2\theta d\varphi^2\right) ,$$

[2]In the case of black holes, "exterior" means outside the horizon.

where, as usual, the gravitational radius $m = GM/c^2$ is the mass of the object in geometrized units. If we assume that the observer is located very far from the source emitting light, i.e. $r_{obs} \gg r_{em}$, Eq. 11.5 gives

$$\frac{\nu_{obs}}{\nu_{em}} = \sqrt{\frac{-g_{00}(x^\mu_{em})}{-g_{00}(x^\mu_{obs})}} = \sqrt{\frac{1-\dfrac{2m}{r_{em}}}{1-\dfrac{2m}{r_{obs}}}} \simeq \sqrt{1-\frac{2m}{r_{em}}}\,. \tag{11.14}$$

If the light source is located on a neutron star surface $r_{em} = R_{NS} \sim 10$ km (see Box 11-A), and Eq. 11.14 gives

$$\frac{\nu_{obs}}{\nu_{em}} \simeq \sqrt{1-\frac{2GM_{NS}}{R_{NS}c^2}} \simeq \sqrt{1-2\times 0.21} = 0.76 \quad \rightarrow \quad \frac{\Delta\nu}{\nu} = \frac{\nu_{obs}-\nu_{em}}{\nu_{em}} \simeq -0.24\,. \tag{11.15}$$

This means that an observer located at large distance from the neutron star will see the emitted light reddened ($\Delta\nu < 0$) by quite a large amount, much larger than that produced by the Sun given in Eq. 11.13.

Let us now assume that the source emitting light is located outside the horizon of a black hole, and at rest with respect to it[3]. Eq. 11.14 shows that as the source approaches the horizon

$$\nu_{obs} \simeq \sqrt{1-\frac{2m}{r_{em}}}\,\nu_{em} \quad \rightarrow \quad 0\,, \tag{11.16}$$

i.e. the signal detected by the distant observer fades away. This is why the black hole horizon is also called a *surface of infinite redshift.* The case of a source in motion with respect to the black hole has been studied in Sec. 10.6, where we have shown that the expression of dt/dT is different, but it still diverges as the source crosses the horizon.

If the light signal travels from a point where the gravitational field is weaker towards one where the field is stronger, the observed frequency is larger than the emitted one, i.e. the light is *blue-shifted.*

The shift of spectral lines was measured for the first time by W.S. Adams in 1925. He measured the redshift of the light emitted by Sirius B, a very compact star (a white dwarf, see Chapter 16), providing one of the first experimental verifications of the theory of General Relativity. Since then, the gravitational redshift has been measured with higher and higher accuracy, including the redshift of Sirius B measured by the Hubble Space Telescope.

Box 11-B

General Relativity and the Global Positioning System

The Global Positioning System (GPS) which is used by smartphones, satellite-based navigation systems, etc., must account for the gravitational redshift produced by the Earth gravitational field. When the first GPS satellite was launched, it showed the predicted shift of 38 μarcsec/day. This tiny shift accumulates over time and, if not accounted for, is sufficient to substantially affect the GPS localization within hours (for a complete overview of general relativistic corrections to the GPS, we refer to [9]).

[3]Note that a body at rest near a black hole is *not* in geodesic motion. An energy source (e.g., the engines of a spaceship) is needed to prevent the body from falling into the horizon.

11.2 THE DEFLECTION OF LIGHT

According to Einstein's theory massless particles move in a gravitational field following null geodesics. As a consequence, photons are deflected by the gravitational field generated by a massive body, as discussed in Sec. 10.4, or by any other distribution of matter/energy. We shall now compute to what extent a beam of massless particles is deflected by a massive

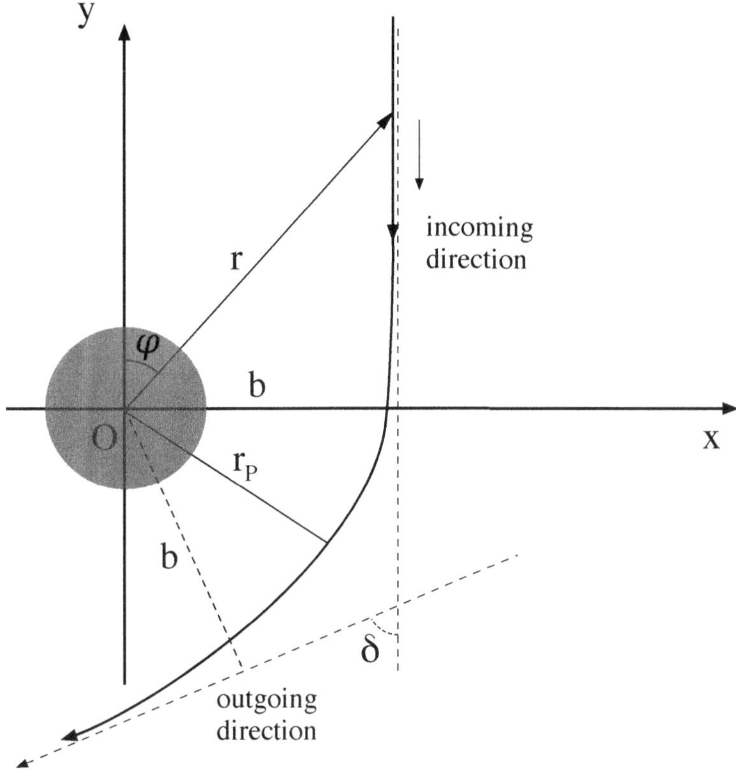

Figure 11.2: The trajectory of a massless particle deflected by a massive body.

body, under the assumption that the field generated by the body is weak (see Box 11-A). Referring to Fig. 11.2, we shall use the following notation:

- r is the radial coordinate of the particle in a frame centered in the center of mass of the body; the orbit is on the equatorial plane, and the position vector of the particle forms an angle φ with the y-axis.
- b is the *impact parameter*, i.e. the distance between the direction of the incoming particle (dashed vertical line) and the center of attraction,

$$b = \lim_{\varphi \to 0} r \sin \varphi \,. \tag{11.17}$$

- δ is the deflection angle which we wish to evaluate: this is the angle between the incoming direction and the outgoing direction of the deflected particle.

Note that, since the Schwarzschild metric is invariant under time reflection, the particle can go through the trajectory of the figure either in the direction indicated by the arrow, or in the opposite one. Thus, the trajectory must be symmetric with respect to the periastron, the location of which is indicated in the Fig. 11.2 by $r_{\rm P}$. When the particle starts its motion at radial infinity

$$\varphi^{in} = 0\,; \tag{11.18}$$

when it escapes at infinity along the trajectory

$$\varphi^{out} = \pi + \delta\,, \tag{11.19}$$

and δ is the deflection angle to be computed. To study the motion, we are going to use the geodesic equations for massless particles, i.e. Eqs. 10.45 and 10.46.

We shall now express the impact parameter b in terms of the constants of motion of the particle E and L, i.e. its energy (measured by a static observer) and angular momentum. When the particle arrives from infinity, r is large, $\varphi \simeq 0$, and Eq. 11.17 gives

$$b \simeq r\varphi \qquad \rightarrow \qquad \frac{d\varphi}{dr} \simeq -\frac{b}{r^2}\,. \tag{11.20}$$

The derivative $\frac{d\varphi}{dr}$ can also be computed by combining the equations for $\dot{\varphi}$ and $\dot{r}$ given in Eqs. 10.45 and 10.46

$$\frac{d\varphi}{dr} = \frac{d\varphi}{d\lambda}\frac{d\lambda}{dr} = \pm\frac{L}{r^2\sqrt{E^2 - V(r)}}\,, \tag{11.21}$$

which, taking the limit for $r \to \infty$, gives

$$\frac{d\varphi}{dr} \simeq \pm\frac{L}{r^2 E}\,; \tag{11.22}$$

hence, Eqs. 11.20 and 11.22 yield

$$b = \frac{L}{E}\,. \tag{11.23}$$

In Sec. 10.4 we showed that a massless particle is deflected if $E^2 < \frac{L^2}{27m^2}$ (see Eq. 10.50), and this imposes a constraint on b, i.e.

$$b \geq \sqrt{27}m \equiv b_{crit}; \tag{11.24}$$

if b is smaller than this critical value, the particle is captured by the central body. Note that, unless the central body is a black hole or a very compact star, it has a radius which is larger than b_{crit}, $R > \sqrt{27}m$. In this case, the actual constraint on the impact parameter b is

$$b \geq R\,. \tag{11.25}$$

To find the deflection angle it is convenient to introduce a new variable

$$u \equiv \frac{1}{r}\,, \tag{11.26}$$

which satisfies the initial condition

$$u(\varphi = 0) = 0\,. \tag{11.27}$$

Furthermore

$$u(\varphi = \pi + \delta) = 0\,, \tag{11.28}$$

since this value of φ corresponds to the particle escaping to infinity. In terms of the variable u, Eq. 10.45 for $\dot{\varphi}$ becomes

$$\dot{\varphi} = Lu^2\,,$$

and, indicating with a prime the differentiation with respect to φ,

$$\dot{r} = r'\dot{\varphi} = -\frac{1}{u^2}u'\dot{\varphi} = -Lu' \,. \tag{11.29}$$

By substituting this expression in the equation for $\dot{r}$ (Eq. 10.46), we find

$$L^2(u')^2 + u^2L^2 - 2mL^2u^3 = E^2 \,,$$

which, differentiating with respect to φ and dividing by $2L^2u'$, yields

$$u'' + u - 3mu^2 = 0 \,. \tag{11.30}$$

This equations has to be solved with the boundary conditions

$$u(\varphi = 0) = 0 \,, \qquad u'(\varphi = 0) = \frac{1}{b} \,. \tag{11.31}$$

The second condition is obtained using the relation

$$u(\varphi \simeq 0) = \frac{\varphi}{b} \,,$$

which comes from Eq. 11.17.

If the mass of the central body vanishes, Eq. 11.30 becomes

$$u'' + u = 0 \,, \tag{11.32}$$

the solution of which

$$u(\varphi) = \frac{1}{b}\sin\varphi \qquad \rightarrow \qquad b = r\sin\varphi \,, \tag{11.33}$$

describes the trajectory of a particle which is not deflected, and satisfies the boundary conditions 11.31.

Equations 11.30 and 11.32 differ by the term $3mu^2$. At this point, to make some progress analytically, we need to use the aforementioned weak-field assumption. We require that, for all values of r reached by the particle,

$$\frac{m}{r} \ll 1 \,,$$

where m is the geometrical mass of the central body. This condition is satisfied, for instance, for a photon which passes very near the surface of the Sun; in this case $r \geq R_\odot$, and (see Box 11-A)

$$\frac{m_\odot}{r} \leq \frac{m_\odot}{R_\odot} \sim 10^{-6} \,.$$

Under this assumption, the term $3mu^2$ in Eq. 11.30 is much smaller than the term u by a factor

$$\frac{3mu^2}{u} = \frac{3m}{r} \ll 1 \,. \tag{11.34}$$

Consequently, Eq. 11.30 can be solved by using a perturbative approach, proceeding as follows. We put

$$u = u^{(0)} + u^{(1)} \,, \tag{11.35}$$

where $u^{(0)}$ is the solution of Eq. 11.32, i.e.

$$u^{(0)} \equiv \frac{1}{b}\sin\varphi \,, \tag{11.36}$$

and we assume that

$$|u^{(1)}| \ll |u^{(0)}| . \tag{11.37}$$

By substituting Eq. 11.35 in Eq. 11.30 we find

$$(u^{(0)})'' + u^{(0)} - 3m(u^{(0)})^2 + (u^{(1)})'' + u^{(1)} - 3m(u^{(1)})^2 - 6mu^{(0)}u^{(1)} = 0 . \tag{11.38}$$

Since $u^{(0)}$ satisfies Eq. 11.32, Eq. 11.38 becomes

$$(u^{(1)})'' + u^{(1)} - 3m(u^{(0)})^2 - 3m(u^{(1)})^2 - 6mu^{(0)}u^{(1)} = 0 . \tag{11.39}$$

The terms $3m(u^{(1)})^2$ and $6mu^{(0)}u^{(1)}$ are of higher order with respect to $3m(u^{(0)})^2$, therefore the leading terms in equation (11.30) are

$$(u^{(1)})'' + u^{(1)} - 3m(u^{(0)})^2 = 0 . \tag{11.40}$$

By replacing the expression of $u^{(0)}$ given in Eq. 11.36 we finally find

$$(u^{(1)})'' + u^{(1)} = \frac{3m}{b^2} \sin^2 \varphi \quad \rightarrow \quad (u^{(1)})'' + u^{(1)} = \frac{3m}{2b^2} (1 - \cos 2\varphi) . \tag{11.41}$$

The solution of this equation which satisfies the boundary conditions 11.31 is

$$u^{(1)} = \frac{3m}{2b^2} \left(1 + \frac{1}{3} \cos 2\varphi - \frac{4}{3} \cos \varphi \right) , \tag{11.42}$$

as can be checked by direct substitution. Thus, the complete solution $u = u^{(0)} + u^{(1)}$ is

$$u = \frac{1}{b} \sin \varphi + \frac{3m}{2b^2} \left(1 + \frac{1}{3} \cos 2\varphi - \frac{4}{3} \cos \varphi \right) . \tag{11.43}$$

The deflection angle δ is then found by imposing the condition 11.28, with $\delta \ll 1$ (which is consistent with the weak-field approximation, since the deflection is small in this regime)

$$u(\pi + \delta) \simeq -\frac{\delta}{b} + \frac{3m}{2b^2} \frac{8}{3} = 0 , \tag{11.44}$$

which gives

$$\delta = \frac{4m}{b} . \tag{11.45}$$

For a light ray passing close to the surface of the Sun, $m = m_\odot$, and $b = R_\odot$, thus

$$\delta \sim 1.75 \, \text{arcsec} . \tag{11.46}$$

The first measurement of the deflection of light was done by Eddington, Dayson, and Davidson during a solar eclipse in 1919. They measured the apparent position of a star behind the Sun (see Fig. 11.3) during the eclipse, when some light coming from the star was able to reach the Earth because the luminosity of the Sun was obscured by the eclipse. Comparing the apparent position with the true position of the star, measured when the Earth was on the opposite side of its orbit around the Sun, they were able to infer the deflection angle δ. At that time it was measured with an accuracy of about 10% [4]. Nowadays, the bending of radio waves by quasars is measured with an accuracy of the order of 1%.

[4] Eddington's measurement was repeated on August 21st, 2017, during a total eclipse happening in the U.S.; using a modern apparatus, the deflection was measured with an accuracy of 3%.

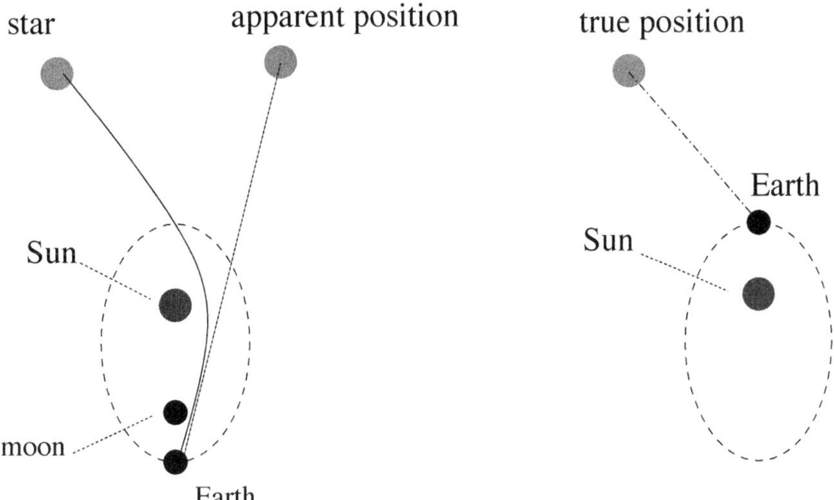

Figure 11.3: (Left) The apparent position of a star, indicated by the dotted segment, was measured by Eddington, Dayson, and Davidson during a solar eclipse. (Right) The true position of the star, indicated by the dot-dashed segment, was then measured when the Earth was on the opposite side of its orbit. The angle between the two segments is the deflection angle δ.

11.3 PERIASTRON PRECESSION

Planets orbiting around a central star, satellites orbiting planets, or stars orbiting a black hole move along timelike geodesics, governed by Eqs. 10.51 and 10.52. In Sec. 10.5 we showed that when the constants of motion E and L are in the range

$$\begin{aligned} & L^2 > 16m^2 \qquad && V_{min} < E^2 < 1 \,, \\ & 12m^2 < L^2 < 16m^2 \qquad && V_{min} < E^2 < V_{max} \,, \end{aligned} \tag{11.47}$$

the orbits of massive particles around a spherically symmetric, non-rotating body are bound; in this case the radial coordinate of the particle varies between an apastron and a periastron, as for elliptic orbits in Newtonian theory. However, we shall show that in General Relativity these orbits are not closed, since at each revolution the periastron advances by some amount. This explains the precession of Mercury's perihelion observed by Le Verrier in 1845 and discussed in Chapter 1.

As in Sec. 11.2, we shall indicate with a prime differentiation with respect to φ, and we shall use the variable $u = 1/r$. In terms of u and φ Eq. 10.52, i.e.

$$\dot{r}^2 - E^2 + V(r) = 0 \qquad \text{with} \qquad V(r) = 1 - \frac{2m}{r} + \frac{L^2}{r^2} - \frac{2mL^2}{r^3} \tag{11.48}$$

becomes

$$L^2(u')^2 - E^2 + 1 - 2mu + u^2L^2 - 2mL^2u^3 = 0 \,, \tag{11.49}$$

where we have used Eq. 11.29, $\dot{r} = -Lu'$.

Differentiating with respect to φ and dividing by $2L^2u'$, Eq. 11.49 yields

$$u'' + u - \frac{m}{L^2} - 3mu^2 = 0 \,. \tag{11.50}$$

We shall now assume

$$\frac{m}{r} \ll 1\,. \tag{11.51}$$

For instance, in the case of Mercury orbiting the Sun, given the Sun-Mercury average distance $r_{\rm S-M} \sim 5.8 \times 10^7$ km, Eq. 11.51 yields

$$\frac{m_\odot}{r_{\rm S-M}} = \frac{1.477}{5.8 \times 10^7} \sim 2.5 \times 10^{-8}\,,$$

and our assumption is fully justified.

The Newtonian equation

The Newtonian equation which corresponds to Eq. 11.50 can be derived classically from the conservation of mechanical energy. Since the (squared) velocity of the particle is $v^2 = \dot{r}^2 + r^2\dot{\varphi}^2$, we get

$$\frac{1}{2}m_p\left[\dot{r}^2 + r^2\dot{\varphi}^2\right] - \frac{m_p\,m}{r} = const \qquad \rightarrow \qquad \dot{r}^2 - \frac{2m}{r} + \frac{L^2}{r^2} = const\,,$$

where m_p is the particle mass and we are using $G = c = 1$. By expressing this equation in terms of u and differentiating with respect to φ, we find

$$u'' + u - \frac{m}{L^2} = 0\,, \tag{11.52}$$

which can be written as

$$\left(u - \frac{m}{L^2}\right)'' + \left(u - \frac{m}{L^2}\right) = 0\,.$$

The solution of this equation is

$$u - \frac{m}{L^2} = A\cos(\varphi - \varphi_0) \qquad \rightarrow \qquad u = \frac{m}{L^2}\left(1 + \frac{L^2 A}{m}\cos(\varphi - \varphi_0)\right)\,, \tag{11.53}$$

where φ_0 and A are integration constants. If we define

$$e = \frac{L^2 A}{m}\,, \tag{11.54}$$

Eq. 11.53 gives

$$u(\varphi) = \frac{m}{L^2}\left[(1 + e\cos(\varphi - \varphi_0)\right]\,, \tag{11.55}$$

or, written in terms of r,

$$r(\varphi) = \frac{L^2}{m}\frac{1}{1 + e\cos(\varphi - \varphi_0)}\,, \tag{11.56}$$

which describes an ellipse with eccentricity e. Setting $\varphi_0 = 0$, the periastron, i.e. the minimum distance the planet reaches in its motion around the central body (this is called perihelion if the central body is the Sun), occurs when $\varphi = 0$, i.e. at

$$r_{\rm P} = \frac{L^2}{m}\frac{1}{1 + e}\,. \tag{11.57}$$

The *apastron*, i.e. the maximum distance from the central body (aphelion in the case of the Sun), is

$$r_{\rm A} = \frac{L^2}{m}\frac{1}{1 - e}\,. \tag{11.58}$$

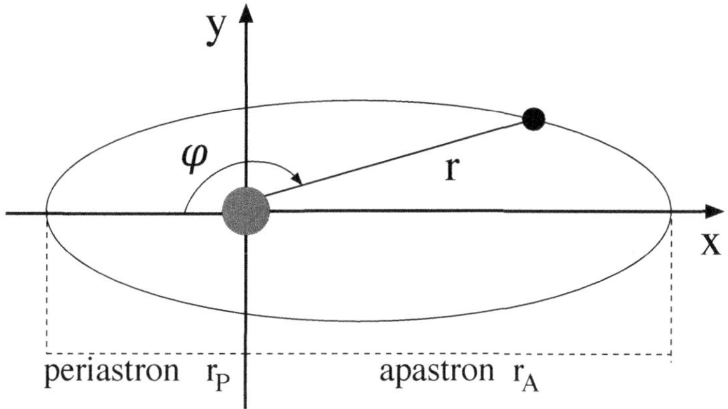

Figure 11.4: Bound orbits in Newtonian gravity are ellipses.

It is worth noting that, since

$$\frac{m}{L^2} = \frac{1}{r_P(1+e)} \qquad \to \qquad \frac{m^2}{L^2} = \frac{m}{r_P(1+e)}, \tag{11.59}$$

and since $m/r_P \ll 1$ in the weak-field regime, it follows that

$$\frac{m^2}{L^2} \ll 1. \tag{11.60}$$

This inequality will be useful in the following.

The relativistic equation

The Newtonian equation 11.52 differs from the relativistic equation 11.50 only by the term $3mu^2$, which is smaller than u by a factor

$$3mu = \frac{3m}{r} \ll 1.$$

Therefore, to solve Eq. 11.50 we shall adopt the same perturbative approach used in Sec. 11.2 to study the bending of light. We search for a solution in the form

$$u = u^{(0)} + u^{(1)}, \qquad |u^{(1)}| \ll |u^{(0)}|, \tag{11.61}$$

where $u^{(0)}$ is the solution of the Newtonian equation 11.52, given by Eq. 11.55, namely

$$u^{(0)} = \frac{m}{L^2}(1 + e\cos\varphi), \tag{11.62}$$

where we have put $\varphi_0 = 0$ without loss of generality. By substituting Eq. 11.61 in Eq. 11.50, and using the fact that $u^{(0)}$ satisfies the zeroth-order equation 11.55, we find

$$\left(u^{(1)}\right)'' + u^{(1)} - 3m(u^{(0)})^2 - 3m(u^{(1)})^2 - 6mu^{(0)}u^{(1)} = 0. \tag{11.63}$$

The terms $3m(u^{(1)})^2$ and $6mu^{(0)}u^{(1)}$ are of higher order with respect to $3m(u^{(0)})^2$ therefore, at leading order, Eq. 11.63 gives

$$\left(u^{(1)}\right)'' + u^{(1)} = 3m(u^{(0)})^2,$$

i.e.

$$\left(u^{(1)}\right)'' + u^{(1)} = 3\frac{m^3}{L^4}\left(1 + e\cos\varphi\right)^2 . \tag{11.64}$$

Expanding the right-hand side of this equation we find

$$\begin{aligned} \left(u^{(1)}\right)'' + u^{(1)} &= 3\frac{m^3}{L^4}\left[1 + e^2\cos^2\varphi + 2e\cos\varphi\right] \\ &= 3\frac{m^3}{L^4}\left[const + \frac{1}{2}e^2\cos 2\varphi + 2e\cos\varphi\right] . \end{aligned} \tag{11.65}$$

This is the equation of a harmonic oscillator with three forcing terms. They are all very small because, as shown in Eq. 11.60, $\frac{m^2}{L^2} \ll 1$. However, the term $2e\cos(\varphi)$ is in resonance with the free oscillations of the harmonic oscillator, i.e., they have the same frequency of the solution of the homogeneous equation $\left(u^{(1)}\right)'' + u^{(1)} = 0$; therefore, even if its amplitude is comparable to that of the other terms, it dominates the solution, and determines a secular perturbation of the planet motion which, after a long time, becomes relevant. For this reason, we will retain only this term and look for the solution of the resulting equation

$$\left(u^{(1)}\right)'' + u^{(1)} = 6e\frac{m^3}{L^4}\cos(\varphi) , \tag{11.66}$$

which is

$$u^{(1)} = \frac{3em^3}{L^4}\varphi\sin\varphi ,$$

as can be checked by direct substitution. Thus, the complete solution of the relativistic equation 11.50 is

$$u = u^{(0)} + u^{(1)} = \frac{m}{L^2}\left[1 + e\left(\cos\varphi + 3\frac{m^2}{L^2}\varphi\sin\varphi\right)\right] . \tag{11.67}$$

Since, as shown by Eq. 11.60, $m^2/L^2 \ll 1$, at first order in m^2/L^2

$$\cos\left(\frac{3m^2}{L^2}\varphi\right) \simeq 1 \qquad \text{and} \qquad \sin\left(\frac{3m^2}{L^2}\varphi\right) \simeq \frac{3m^2}{L^2}\varphi ;$$

consequently we can write

$$u \simeq \frac{m}{L^2}\left\{1 + e\cos\left[\varphi\left(1 - 3\frac{m^2}{L^2}\right)\right]\right\} . \tag{11.68}$$

A comparison with the corresponding Newtonian solution given by Eq. 11.62 shows that the term $\frac{3m^2}{L^2}\varphi$ determines a secular advance of the periastron. Indeed, when the argument of the sinusoidal function in Eq. 11.68 goes from zero to 2π, i.e. when the planet reaches again $r = r_P$, φ changes by an amount

$$\Delta\varphi = \frac{2\pi}{1 - \frac{3m^2}{L^2}} \simeq 2\pi\left(1 + \frac{3m^2}{L^2}\right) .$$

Consequently, in one period the periastron *advances* by

$$\delta\varphi = \frac{6\pi m^2}{L^2} . \tag{11.69}$$

Thus, as shown in Fig. 11.5, in General Relativity the orbit of a planet around a non-rotating, spherically symmetric body is open, and the periastron shifts by $\delta\varphi$ at each revolution.

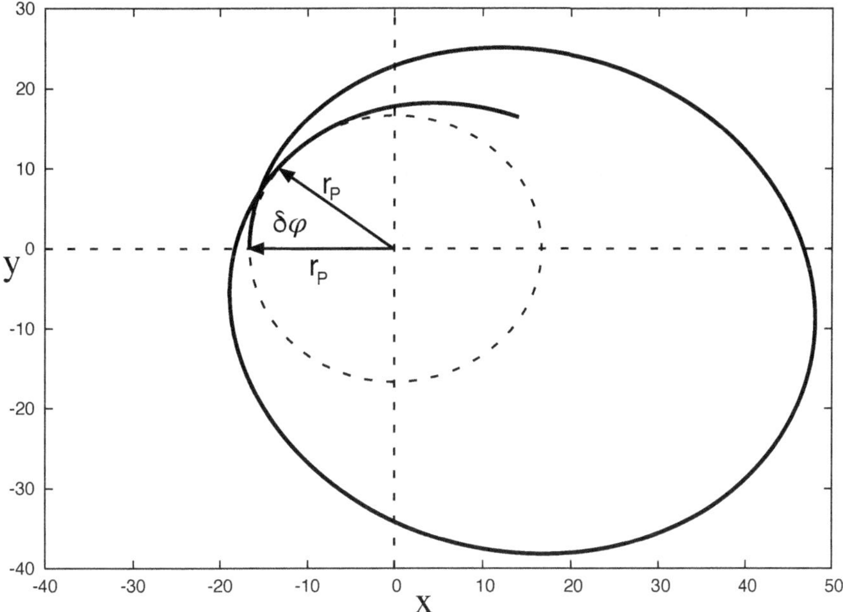

Figure 11.5: Bound orbit of a body in Schwarzschild geometry. At each revolution, the periastron precedes by an amount $\delta\varphi$.

For example, for Mercury Eq. 11.69 gives a precession of 42.98 arcsec/century. The observed value, after all effects which can be explained with Newtonian theory (precession of the equinoxes, perturbations of other planets on Mercury's orbit, etc.) have been subtracted [5], is

$$\delta\varphi_{\text{Mercury}} = 42.98 \pm 0.04 \text{ arcsec/century}, \tag{11.70}$$

in perfect agreement with the predictions of General Relativity. The other planets of the solar system experience a perihelion shift as well; however, since they are farther from the Sun, their constant of motion L is larger than that of Mercury (see e.g. Eq. 11.56). Let us consider, for example, the Earth. Since $r_{P_\oplus} \gg r_{P_{\text{Mercury}}}$, Eq. 11.57 gives

$$L_\oplus^2 \gg L_{\text{Mercury}}^2, \quad \rightarrow \quad \frac{m^2}{L_\oplus^2} \ll \frac{m^2}{L_{\text{Mercury}}^2}$$

and consequently $\delta\varphi_\oplus \ll \delta\varphi_{\text{Mercury}}$. Indeed, for the Earth $\delta\varphi_\oplus = 3.84$ arcsec/century.

The perihelion shift of Mercury is extremely small, and it was only using the data from centuries of observations that Le Verrier discovered this effect. However, other binary systems can have a much larger periastron precession. In the system J0737-3039, composed by two neutron stars orbiting at a distance of less than one million kilometers from each other, the periastron advances approximately by *16 degrees per year* [72]. Note, however, that in this case the derivation discussed in this chapter cannot be applied, because we have assumed that the body of mass m_p moves along a geodesic of the Schwarzschild spacetime, which implies that the mass m_p is so small, compared to the mass m generating the gravitational field, that it does not perturb it. When (as in the case of two neutron stars)

[5] For an accurate account of these effects, we refer to the monograph by Poisson and Will [93].

the two bodies have comparable masses, the spacetime is different from Schwarzschild [6], and the derivation of the periastron precession is more involved.

11.4 THE SHAPIRO TIME DELAY

Another genuinely relativistic effect involving light propagation in curved spacetime is the so-called *Shapiro time delay.* In 1964, Irwin Shapiro proposed that, due to gravitational time dilation, radar signals passing near a massive star would take longer to travel to a target and longer to return than they would take in flat spacetime.

To investigate this effect, let us consider a light ray propagating in the Schwarzschild spacetime. For null geodesics on the equatorial plane ($\theta = \pi/2$), the condition $ds^2 = 0$ yields

$$\left(1 - \frac{2m}{r}\right) dt^2 = \left(1 - \frac{2m}{r}\right)^{-1} dr^2 + r^2 d\varphi^2 . \tag{11.71}$$

For a given trajectory $r = r(\varphi)$, by replacing $d\varphi = \frac{d\varphi}{dr} dr$ in Eq. 11.71 and integrating, we obtain the function $t(r)$. As shown in Sec. 11.2, the trajectory of a light ray moving in the gravitational field of a massive body is deflected. To isolate the time delay from light bending effect, we shall neglect the effect of the gravitational field on the orbit, i.e. we shall consider the zeroth-order solution, $u = \sin\varphi/b$, where we remind that $u = 1/r$ and b is the impact parameter (see Fig. 11.6). In the weak-field regime, both the time delay and the light bending are small effects, therefore, to leading order, the time delay on the unperturbed orbit coincides with that of the deflected one.

Since $r = \frac{b}{\sin\varphi}$,

$$dr = -\frac{b}{\sin^2\varphi} \cos\varphi d\varphi \qquad \rightarrow \qquad d\varphi = -\frac{\tan\varphi}{r} dr , \tag{11.72}$$

and

$$\tan^2\varphi = \frac{\frac{b^2}{r^2}}{1 - \frac{b^2}{r^2}} = \frac{b^2}{r^2 - b^2} , \tag{11.73}$$

it follows that

$$r^2 d\varphi^2 = \tan^2\varphi dr^2 = \frac{b^2}{r^2 - b^2} dr^2 . \tag{11.74}$$

By replacing the latter expression into Eq. 11.71, and solving for dt, we get

$$dt = \pm dr \left(1 - \frac{2m}{r}\right)^{-1} \sqrt{\frac{r^3 - 2b^2 m}{r^3 - b^2 r}} = \pm \frac{r dr}{\sqrt{r^2 - b^2}} \left(1 - \frac{2m}{r}\right)^{-1} \sqrt{1 - \frac{2mb^2}{r^3}} . \tag{11.75}$$

Let us consider a photon moving from $r = r_i$ to $r = r_f$ (see Fig. 11.6). The total time delay reads

$$\Delta t = \int_{r_i}^{r_f} dr \frac{dt}{dr} . \tag{11.76}$$

Unfortunately, the above equation cannot be integrated analytically when dt is given by

[6] This is the *two-body problem*, which has a simple and elegant solution in Newtonian physics: the motion of two masses m_1, m_2 is equivalent to that of one mass $\mu = m_1 m_1/(m_1 + m_2)$ in the static gravitational field generated by a mass $m_1 + m_2$. In General Relativity, the two-body problem can been solved analytically in the small-velocity ($v \ll c$) and weak-field ($GM/(c^2 r) \ll 1$) regime using a complex mathematical framework, the post-Newtonian expansion (see, e.g., [93] for a monograph on the subject). The two-body problem will be further discussed in Chapter 14.

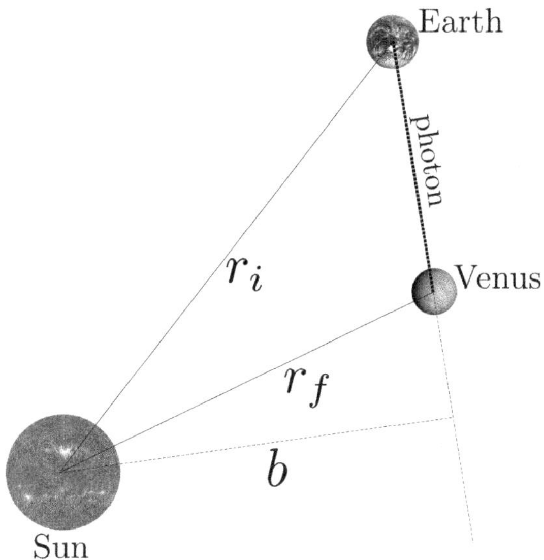

Figure 11.6: The Shapiro delay. A photon beam is sent from Earth (located at $r = r_i$ relative to the Sun) and travels on an approximately straight line (neglecting light bending) to Venus (located at $r = r_f$). The impact parameter of the photon is b. In the absence of gravitational time delay and relativistic corrections, the travel time would simply be $\Delta t = \sqrt{r_i^2 - b^2} - \sqrt{r_f^2 - b^2}$.

Eq. 11.75. However, if the photon travels sufficiently far away from the source with mass m we can expand Eq. 11.75 for $m \ll r$, $m \ll b$. This gives

$$dt \simeq \pm \frac{dr}{\sqrt{r^2 - b^2}} \left[r + \left(2 - \frac{b^2}{r^2} \right) m \right] , \tag{11.77}$$

which can be integrated to give

$$\Delta t \simeq \pm \left[\frac{(r - m)\sqrt{r^2 - b^2}}{r} + 2m \log \left(r + \sqrt{r^2 - b^2} \right) \right]_{r_i}^{r_f} . \tag{11.78}$$

The above expression gives the *coordinate* time delay of a photon traveling from r_i to r_f. As expected, when $m = 0$ one gets simply $\Delta t = \pm\sqrt{r^2 - b^2}$ (see Fig. 11.6).

A case of interest is when the photon starts from Earth at $r = r_i$ (e.g., it is sent by a radar) and reaches a nearby planet at $r = r_f$ in the gravitational field of the Sun ($m = m_\odot \approx 1.48\,\text{km}$). Notice that Δt is not the *proper* time delay on Earth. In geometrized units, the latter is defined as

$$d\tau^2 = -ds^2_{\text{Earth}} \,. \tag{11.79}$$

For simplicity, let us assume that the orbit of the Earth is circular, so that $dr = d\theta = 0$ and therefore

$$d\tau^2 = \left(1 - \frac{2m}{r_i} \right) dt^2 - r^2 d\varphi^2 , \tag{11.80}$$

where r_i is the orbital radius of the Earth around the Sun. The orbital frequency is the Keplerian frequency (see Eq. 10.64) $(d\varphi/dt)_{\rm Earth} = \omega_K = \sqrt{m/r^3}$. Therefore, the above equation can be written as

$$d\tau = dt\sqrt{1 - \frac{3m}{r_i}}\,, \tag{11.81}$$

and consequently $\Delta\tau = \sqrt{1 - 3m/r_i}\Delta t$. Note that the proper time delay coincides with the coordinate time delay when $r_i/m \gg 1$.

Finally, the proper time delay measured on Earth is

$$\Delta\tau \sim \sqrt{1 - \frac{3m}{r_i}}\left[\frac{(r-m)\sqrt{r^2-b^2}}{r} + 2m\log\left(r + \sqrt{r^2-b^2}\right)\right]_{r_i}^{r_f} . \tag{11.82}$$

The difference between the time delay that would be measured on Earth if the spacetime was flat, $\Delta\tau = [\sqrt{r^2-b^2}]_{r_i}^{r_f}$, and the above formula grows logarithmically when r increases; this mild growth can be used to magnify the effect of the gravitational delay. Shapiro proposed to send from Earth a radar signal which would be reflected off Venus and be detected back on Earth. The above formula predicts that the Sun induces a gravitational delay $\Delta\tau \sim \mathcal{O}(100\,\mu{\rm s})$, the exact value depending on the relative position of the two planets. The measured value agreed with the theoretical prediction within 20%. The best constraint to date on the Shapiro time delay comes from the measurement obtained by sending a radar signal from Earth to the Cassini satellite [22]; this confirmed the prediction of General Relativity with an accuracy of one part in 10^5.

11.5 THE SHADOW OF A BLACK HOLE

A "black hole" owns its name (allegedly due to John Archibald Wheeler, although he probably heard it from an unknown participant in a conference in 1967 [60]) to the fact that nothing – not even light – can escape from the region enclosed by the event horizon. An isolated black hole would therefore appear truly as a "hole" in the sky, since we observe objects by receiving the light they either emit or reflect. The boundary of this hole, i.e. the "silhouette" of a black hole, is called the **black-hole shadow**. It can be defined by studying the path of photons emitted by a light source at infinity and deflected by the black hole. This "ray tracing" requires solving the geodesic equations in the black hole spacetime. We consider here the simplest case in which the spacetime is described by the Schwarzschild solution.

The shadow of a black hole does not coincide with the event horizon, and it is actually larger. This follows from our previous discussion on null geodesics in the Schwarzschild metric. Let us consider a source located at infinity and behind the black hole (see Fig. 11.7, upper panel). As shown in Sec. 10.4, a photon coming from infinity will be deflected only if $E^2 < L^2/(27m^2)$ (see Eq. 10.50); otherwise it is captured by the black hole. As shown in Sec. 11.2, the impact parameter is $b = L/E$, thus the deflected photons are those with impact parameter $b > 3\sqrt{3}m$. Therefore, the size of the shadow of a Schwarzschild black hole is $3\sqrt{3}m$, slightly larger than the photon sphere. The critical photon with impact parameter $3\sqrt{3}m$ approaches the light ring orbit at $r = 3m$, making an infinite number of revolutions. Let us now consider a source located *at finite distance behind the black hole*, which emits photons in all directions, and an observer located at infinity along the direction $\hat{n}$. The observer receives the photons which are deflected by the black hole and propagate along the direction $\hat{n}$. This process is described by the time-reversal of the geodesics considered in the case of a source at infinity, as in the lower panel of Fig. 11.7. If the source is behind the black hole, and if it is located at $r > 3m$, then the only photons which can reach the

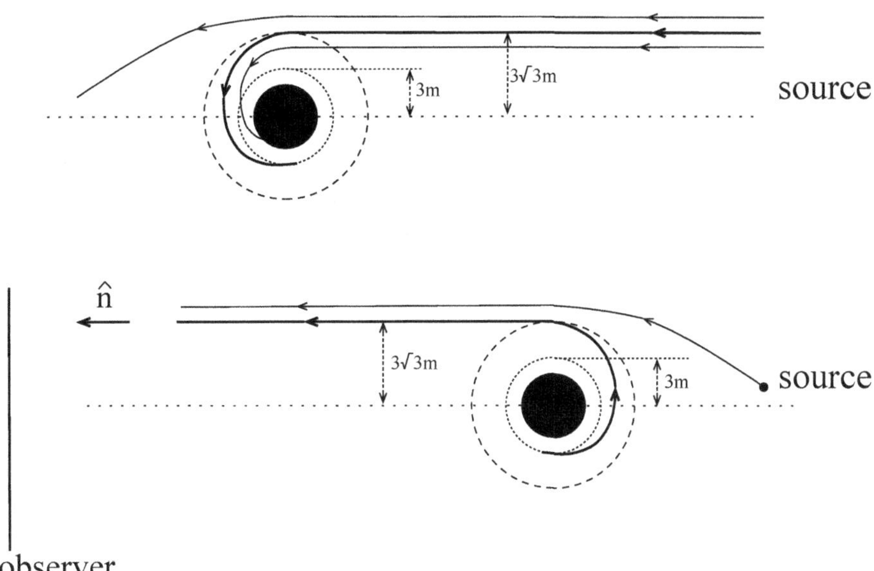

Figure 11.7: Deflection of photons coming from infinity to a Schwarzschild black hole (upper panel) and emitted to infinity from a source close to the black hole (lower panel). The lower panel is the time-reversal of the upper one. The observer is represented as a vertical screen (a "detector") orthogonal to $\hat{n}$.

observer are those with impact parameter $b > 3\sqrt{3}m$. Therefore, the observer sees a circular silhouette with radius $3\sqrt{3}m$, slightly larger than the photon sphere.

Let us study this problem in more detail. Consider a static source at $r = r_0$, as in Fig. 11.8, emitting photons isotropically. The orbit of a given photon is parametrized by the constants E and L. We define the dimensionless constant $K = L/(mE)$. For a given value of E, a light ray can escape to infinity if the angular momentum is larger than the critical value $L_{\text{crit}} \equiv K_{\text{crit}} mE$, where $K_{\text{crit}} = 3\sqrt{3}$.

To analyse the motion of the emitted photons, it is convenient to consider the local Fermi frame $\{x^{*\mu}\}$ adapted to the source (see Sec. 3.9). Note that the source is static but not in free fall, therefore this coordinate frame is *not* a LIF. We remind (see discussion at the end of Sec. 3.9) that the laws of physics in this frame have locally the form prescribed by Special Relativity *for accelerated frames*; thus, this is the appropriate frame to describe the local process of photon emission.

Let $\{\vec{e}_{(\mu)}\}$ be the vector basis associated to the coordinates $\{x^\mu\} = (t, r, \theta, \varphi)$, and $\{\vec{e}^{\,*}_{(\mu)}\}$ the vector basis associated to the local Fermi coordinates $\{x^{*\mu}\}$. This is, by construction, an orthonormal basis ($\vec{e}^{\,*}_{(\mu)} \cdot \vec{e}^{\,*}_{(\nu)} = \eta_{\mu\nu}$), whose timelike vector coincides with the four-velocity of the source, $\vec{e}^{\,*}_{(0)} = \vec{u}$, with $\vec{e}^{\,*}_{(0)} \cdot \vec{e}^{\,*}_{(0)} = -1$. Since the source is static, $u^\mu = (u^0, 0, 0, 0)$, hence $\vec{u} \parallel \vec{e}_{(0)}$, i.e. $\vec{e}^{\,*}_{(0)} = const\ \vec{e}_{(0)}$, with $const = (-\vec{e}_{(0)} \cdot \vec{e}_{(0)})^{-1/2}$. Thus

$$\vec{e}^{\,*}_{(0)} = \frac{\vec{e}_{(0)}}{\sqrt{-\vec{e}_{(0)} \cdot \vec{e}_{(0)}}} = \left(1 - \frac{2m}{r}\right)^{-1/2} \vec{e}_{(0)} \,. \tag{11.83}$$

Moreover, the three vectors $\vec{e}^{\,*}_{(i)}$ ($i = 1, \ldots, 3$) are orthogonal to $\vec{e}_{(0)}$. Since $\vec{e}_{(i)}$ are also orthogonal to $\vec{e}_{(0)}$, it follows that $\vec{e}^{\,*}_{(i)}$ are linear combinations of the vectors $\vec{e}_{(i)}$. Since

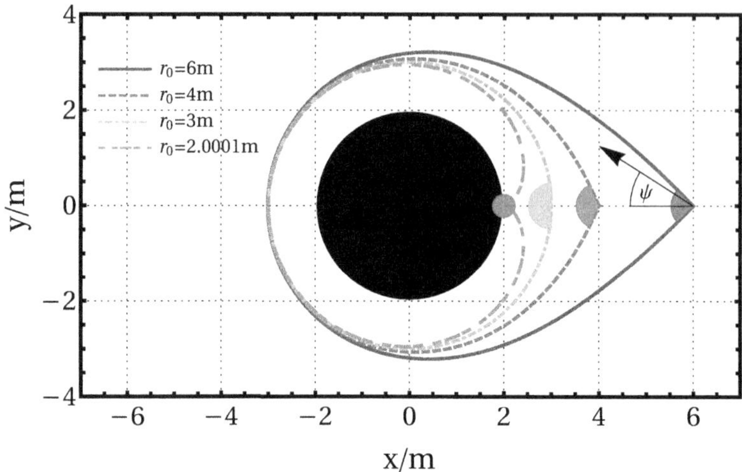

Figure 11.8: Critical escape trajectories of radiation in the Schwarzschild geometry, projected in the equatorial plane. A static source located at $r = r_0$ (the vertex of the conical sectors) emits photons isotropically, but those emitted within the conical sectors (i.e., with an angle $\psi \leq \psi_{\text{crit}}$) will not reach infinity.

$\vec{e}^{\,*}_{(i)} \cdot \vec{e}^{\,*}_{(j)} = \delta_{jk}$, we choose $\vec{e}^{\,*}_{(i)} = \vec{e}_{(i)}/\sqrt{\vec{e}_{(i)} \cdot \vec{e}_{(i)}}$, i.e.

$$
\begin{aligned}
\vec{e}^{\,*}_{(1)} &= \left(1 - \frac{2m}{r}\right)^{1/2} \vec{e}_{(1)} \,, \\
\vec{e}^{\,*}_{(2)} &= \frac{1}{r}\vec{e}_{(2)} \,, \\
\vec{e}^{\,*}_{(3)} &= \frac{1}{r\sin\theta}\vec{e}_{(3)} \,.
\end{aligned}
\tag{11.84}
$$

We can now decompose the four-momentum of the emitted photon in the local Fermi basis: $\vec{p} = p^{*\mu}\vec{e}^{\,*}_{(\mu)}$ where $p^{*0} = -\vec{p}\cdot\vec{e}^{\,*}_{(0)}$ and $p^{*i} = \vec{p}\cdot\vec{e}^{\,*}_{(i)}$ since $\vec{e}^{\,*}_{(\mu)} \cdot \vec{e}^{\,*}_{(\nu)} = \eta_{\mu\nu}$. Using Eq. 11.84 and the fact that $\vec{p}\cdot\vec{e}_{(\mu)} = \mathbf{g}(p^\alpha \vec{e}_{(\alpha)}, \vec{e}_{(\mu)}) = p^\alpha g_{\alpha\mu} = p_\mu$, we find

$$
\begin{aligned}
p^{*0} &= -p_0\left(1 - \frac{2m}{r}\right)^{-1/2}, \qquad & p^{*1} &= p_1\left(1 - \frac{2m}{r}\right)^{1/2} \\
p^{*2} &= \frac{p_2}{r}\,, & p^{*3} &= \frac{p_3}{r\sin\theta}\,.
\end{aligned}
\tag{11.85}
$$

If the photon is emitted at $r = r_0$ along the equatorial plane ($\theta = \pi/2$, $p^\theta = 0$), with constants of motion $E = -p_0$ and $L = p_3$, then $p^{*0} = (1 - 2m/r_0)^{-1/2} E$ and $p^{*3} = L/r_0$. The components of the *velocity* of the emitted photon measured in the local Fermi frame are then

$$
\begin{aligned}
v^{*r} &= \frac{p^{*1}}{p^{*0}} = \left(1 - \frac{2m}{r_0}\right)\frac{p_1}{E} = g^{11}(r_0)\frac{p_1}{E} = \frac{p^1}{E} \\
v^{*\varphi} &= \frac{p^{*3}}{p^{*0}} = \frac{L}{r_0 E}\left(1 - \frac{2m}{r_0}\right)^{1/2} .
\end{aligned}
\tag{11.86}
$$

Since $p^1 = \dot{r} = \pm\sqrt{E^2 - \frac{L^2}{r_0^2}\left(1 - \frac{2m}{r_0}\right)}$ (see Eq. 10.26) and $L = mKE$, Eqs. 11.86 can be written as

$$v^{*r} = \pm\sqrt{1 - \frac{K^2 m^2}{r_0^2}\left(1 - \frac{2m}{r_0}\right)}, \qquad v^{*\varphi} = \frac{mK}{r_0}\left(1 - \frac{2m}{r_0}\right)^{1/2}, \tag{11.87}$$

where the plus and minus signs in the above equation correspond to photons moving outwards and inwards, respectively.

Let us now compute the local emission angle ψ of a photon. For a generic emission point on the equatorial plane, ($x = r\cos\varphi, y = r\sin\varphi$) in Schwarzschild coordinates, the components of the velocity on the equatorial plane are $v^x = \dot{x} = \dot{r}\cos\varphi - r\sin\varphi\dot{\varphi}$ and $v^y = \dot{y} = \dot{r}\sin\varphi + r\cos\varphi\dot{\varphi}$, where a dot indicates the derivative with respect to the time coordinate t. As in Fig. 11.8, we can assume that the emission occurs on the x axis, i.e. at $\varphi = 0$, without loss of generality. Therefore $v^x = \dot{r} = v^{*r}$ and $v^y = r\dot{\varphi} = v^{*\varphi}$. Since for a massless particle $|\mathbf{v}| = 1$ (we are using geometrized units in which $c = 1$), we have $v^x = \cos\psi$ and $v^y = \sin\psi$ (see Fig. 11.8). Thus, we find a relation between the emission angle and the polar component of the velocity:

$$\sin\psi = v^y = r\dot{\varphi} = v^{*\varphi}. \tag{11.88}$$

The photon can escape to infinity for $K > K_{\text{crit}}$, i.e. for local emission angles larger than the *critical angle*

$$\sin\psi_{\text{crit}} = v^{*\varphi}|_{K=K_{\text{crit}}} = \frac{3m\sqrt{3(1-2m/r_0)}}{r_0}. \tag{11.89}$$

When $\psi < \psi_{\text{crit}}$, the photon is captured by the black hole (see Fig. 11.8). When the source is close to the horizon ($r_0 \approx 2m$), Eq. 11.89 reduces to

$$\psi_{\text{crit}} \simeq \left(\frac{3}{2}\right)^{3/2}\sqrt{\frac{r_0 - 2m}{m}}. \tag{11.90}$$

Therefore, as the source approaches the horizon not only are the emitted photons redshifted as discussed in Sec. 11.1, but their escape angle tends to zero. For emission angles smaller than ψ_{crit}, the light rays are absorbed by the black hole. The trajectory $r(\varphi)$ of these rays can be studied by solving the ordinary differential equation

$$\frac{dr}{d\varphi} = \frac{\dot{r}(t)}{\dot{\varphi}(t)} = \frac{rv_r^*}{v_\varphi^*} = \pm\frac{r^2\sqrt{1 - \frac{K^2m^2}{r^2}\left(1 - \frac{2m}{r}\right)}}{Km\sqrt{1 - \frac{2m}{r}}}, \tag{11.91}$$

supplemented by the initial condition $r(0) = r_0$ (the initial value of φ is arbitrary due to the spherical symmetry of the Schwarzschild metric). The above equation does not admit a solution in closed form and one has to solve it numerically. Some numerical solutions are shown in Fig. 11.8 for different emission points r_0. The figure shows some important features:

- The emission angle ψ of a photon which is free to escape (i.e., the escape angle) is larger than 90° when $r_0 > 3m$, whereas it is smaller than 90° when $r_0 < 3m$. In the former case the critical escape velocity can be either directed outward or inward, whereas in the latter case it is only directed outward relative to the black hole.

- In all cases, the critical trajectory ($K = K_{\text{crit}}$) corresponds to light rays that start at $r = r_0$ and *approach the photon sphere*, i.e. $r(\varphi) \to 3m$ either from above (when $r_0 > 3m$) or from below (when $r_0 < 3m$).

Suppose the source is at $r_0 > 3m$ on the right of the black hole, as in the bottom panel of Fig. 11.7, and the observer is on the left of the black hole, i.e. the source is *behind* the black hole. In this case, the observer would only receive those photons emitted with an angle $\psi > \psi_{\rm crit}$. Photons emitted precisely at the critical angle (with $\psi = \psi_{\rm crit}$) will orbit the black hole approaching the photon sphere ($r = 3m$) and making an infinite number of revolutions, whereas photons emitted with $\psi < \psi_{\rm crit}$ will be captured by the black hole. In other words, part of the light coming from *behind* the black hole (emitted at $r_0 > 3m$) is trapped by the photon sphere, whereas the other will escape to infinity, with impact parameter $b > 3\sqrt{3}m$.

It is interesting to note that, if $\psi > \psi_{\rm crit} + \epsilon$ (with $\epsilon \ll 1$), the photons can make a large number of revolutions before eventually escaping to infinity. Photons that escape to infinity after making multiple revolutions contribute to create *multiple images* of the same source, giving a peculiar structure to the region corresponding to $\psi \simeq \psi_{\rm crit}$.

11.5.1 The accretion disk of a black hole

The above discussion assumes that the black hole is isolated and only enlightened by a point-like isotropic source behind it. However, astrophysical black holes are usually not isolated but rather surrounded by an **accretion disk** of gas and plasma orbiting around them. As previously discussed, these orbits can be circular only up to the ISCO at $r = 6m$, whereas when $r < 6m$ matter will fall into the horizon. This process is called *accretion.* The surrounding matter accreting onto the hole heats up through viscous dissipation and converts gravitational energy into radiation.

In order to understand how efficient can this process be, we can make a back-of-the-envelope estimate. For clarity here we use physical units. Let us consider a body of mass m_p falling from infinity onto a body of mass $M \gg m_p$ and radius R. The gravitational potential energy released in this process is $\Delta E_{accr} = GMm_p/R$, most of which is emitted in electromagnetic radiation. If a steady flux of matter falls into the central body with accretion rate $\dot{m}_p$, the emitted energy flux (i.e., the energy emitted per unit of time) is $L_{\rm em} = GM\dot{m}_p/R$. Since the flux of accreting mass-energy is $L_{accr} = \dot{m}_p c^2$, this mass-energy is converted in electromagnetic radiation with an *efficiency* $\eta = L_{\rm em}/L_{accr} = GM/(Rc^2)$. For a black hole, we can consider R as the ISCO radius, therefore $\eta \simeq 0.1$. This is an enormous efficiency: as a comparison, the efficiency of conversion of mass to energy in thermonuclear reactions from hydrogen to helium is $\eta \simeq 0.007$.

As previously shown, an electromagnetic signal emitted by a particle freely falling onto a black hole is strongly redshifted when approaching the horizon. In addition, an element of fluid typically emits isotropically in all directions and not all photons will be able to reach infinity as the fluid element approaches the black hole, because some of them would be bent to the horizon. Fig. 11.8 and Eq. 11.89 show that, as the fluid falls toward the horizon ($r_0 \to 2m$), the escape angle of radiation decreases until it vanishes precisely when the emitting source is on the horizon. Therefore, the fraction of energy that is able to reach infinity vanishes as $r_0 \to 2m$ and the signal becomes more feeble due to both the gravitational redshift and the extreme bending of all photons not emitted perfectly in radial direction.

In other words, most of the radiation that illuminates a black hole comes from the accretion disk (i.e., from $r_0 \gtrsim 6m$). The small amount of radiation coming from matter falling into the horizon is highly redshifted and has a small escape angle, so it does not contribute much to illuminate the region close to the horizon.

The shadow of a black hole has been observed for the first time in 2019 by the Event Horizon Telescope Collaboration, that provided the first radio image of the supermassive ($M \approx 6.5 \times 10^9 M_\odot$) black hole in the galaxy M87 [111], which is shown in Fig. 11.9. This

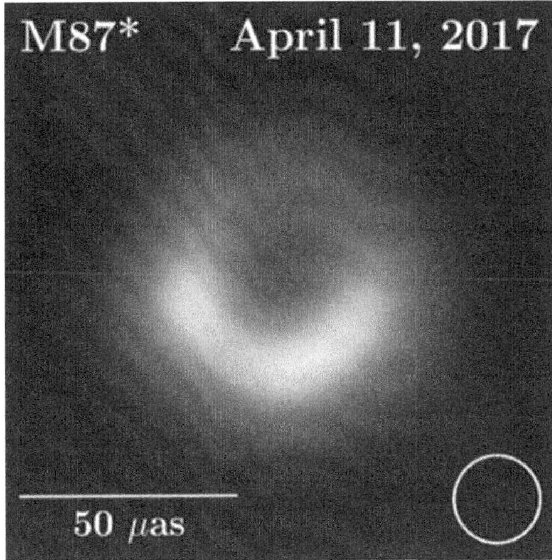

Figure 11.9: Shadow of the supermassive black hole in the galaxy M87, observed by the Event Horizon Telescope Collaboration. From: K. Akiyama *et al.*, *First M87 Event Horizon Telescope Results. I. The Shadow of the Supermassive Black Hole* [111].

black hole is likely rotating; the spin affects geodesic motion, and hence the shadow, as will be discussed in Box 19-B.

CHAPTER 12

Gravitational waves

One of the most interesting predictions of the theory of General Relativity is the existence of gravitational waves. The idea that a perturbation of the gravitational field should propagate as a wave is, in some sense, intuitive. For example, electromagnetic waves were introduced when the Coulomb theory of electrostatics was replaced by the theory of electrodynamics, and it was shown that they transport through the space the information about the dynamics of charged systems. In a similar way, when a mass-energy distribution changes in time the information about this dynamics should propagate in the form of waves. However, gravitational waves have a distinctive feature: due to the twofold nature of $g_{\mu\nu}$, which is both the metric tensor and the gravitational potential, gravitational waves are *metric waves*. Thus, when a gravitational wave propagates, the geometry and, consequently, the proper distance between spacetime points changes in time. In other words, gravitational waves do not propagate *in* the spacetime, they actually are waves *of* the spacetime.

Gravitational waves can be studied by solving Einstein's equations with different approaches: *"exact"*, *perturbative*, or *numerical*. In the *"exact"* approach we look for analytical solutions of Einstein's equations describing these waves; these can be found only by imposing some particular symmetry, for example plane, spherical, or cylindrical. Exact solutions can also describe the interaction of gravitational waves which, due to the non-linearity of the equations of gravity, is very different from the interaction of electromagnetic waves [52].

The *perturbative* approach considers gravitational waves emitted by a source as small perturbations of a background solution. It does not require a particular symmetry of the spacetime, but it requires the knowledge of the background solution.

Finally, when we cannot impose a particular symmetry to the spacetime and the metric cannot be considered as a small perturbation of a known background solution, we have to follow the *numerical* approach, i.e. numerical integration of the full, non-linear Einstein equations. Due to the non-linearity of these equations, this approach is very challenging. An entire branch of General Relativity, named *Numerical Relativity*, studies all related issues. The topic of Numerical Relativity deserves a monograph on its own and we refer the interested reader to e.g. [6] and references therein.

In the rest of this book we shall only focus on the perturbative approach, treating gravitational waves as small perturbations of a given background solution of Einstein's equations.

12.1 PERTURBATIVE APPROACH

Let $g^0_{\mu\nu}$ be a known solution of Einstein's equations; this will be referred to as the *background*. It can be, for instance, the metric of flat spacetime $\eta_{\mu\nu}$, or the metric describing a Schwarzschild black hole. Let us consider a small perturbation of $g^0_{\mu\nu}$ caused by some source described by a stress-energy tensor $\delta T_{\mu\nu}$. We write the metric tensor of the perturbed spacetime, $g_{\mu\nu}$, as

$$g_{\mu\nu} = g^0_{\mu\nu} + h_{\mu\nu}\,, \tag{12.1}$$

where $h_{\mu\nu}$ is the small perturbation, such that

$$|h_{\mu\nu}| \ll |g^0_{\mu\nu}|\,. \tag{12.2}$$

We remark that Eqs. 12.1 and 12.2 hold in a given reference frame. As we shall discuss in Sec. 12.2, although Einstein's equations are invariant under a general coordinate transformation, only a subset of these transformations is consistent with the perturbative framework defined by Eq. 12.1, i.e. those which preserve the condition 12.2.

Similarly to the case of flat background discussed in Sec. 6.1, the inverse metric is

$$g^{\mu\nu} = g^{0\,\mu\nu} - h^{\mu\nu} + O(h^2)\,, \tag{12.3}$$

where the indices of $h^{\mu\nu}$ have been raised with the *unperturbed* metric

$$h^{\mu\nu} \equiv g^{0\,\mu\alpha}g^{0\,\nu\beta}h_{\alpha\beta}\,. \tag{12.4}$$

It is easy to show that $g^{\mu\nu}$ defined in Eq. 12.3 is the inverse metric; indeed with this definition,

$$(g^{0\,\mu\nu} - h^{\mu\nu})(g^0_{\nu\alpha} + h_{\nu\alpha}) = \delta^\mu_\alpha + O(h^2)\,. \tag{12.5}$$

The equations governing the dynamics of $h_{\mu\nu}$ are obtained by linearizing Einstein's equations, which are conveniently written in the following form

$$R_{\mu\nu} = \frac{8\pi G}{c^4}\left(T_{\mu\nu} - \frac{1}{2}g_{\mu\nu}T\right)\,; \tag{12.6}$$

the stress-energy tensor $T_{\mu\nu}$ is the sum of two terms, one associated to the source that generates the background geometry $g^0_{\mu\nu}$, say $T^0_{\mu\nu}$, and one associated to the source of the perturbation $\delta T_{\mu\nu}$, which is of order h.

Let us first consider the left-hand side of Eq. 12.6. We remind that the Ricci tensor $R_{\mu\nu}$ is (see Box 4-A)

$$R_{\mu\nu} = \Gamma^\alpha_{\mu\nu,\alpha} - \Gamma^\alpha_{\mu\alpha,\nu} + \Gamma^\alpha_{\sigma\alpha}\Gamma^\sigma_{\mu\nu} - \Gamma^\alpha_{\sigma\nu}\Gamma^\sigma_{\mu\alpha}\,, \tag{12.7}$$

and that the affine connections $\Gamma^\gamma_{\beta\mu}$ are

$$\Gamma^\gamma_{\beta\mu} = \frac{1}{2}g^{\gamma\alpha}\left(g_{\alpha\beta,\mu} + g_{\alpha\mu,\beta} - g_{\beta\mu,\alpha}\right)\,. \tag{12.8}$$

The $\Gamma^\gamma_{\beta\mu}$ computed for the perturbed metric 12.1 are

$$\begin{aligned}
\Gamma^\gamma_{\beta\mu}(g_{\mu\nu}) &= \frac{1}{2}\left(g^{0\,\gamma\alpha} - h^{\gamma\alpha}\right)\left[\left(g^0_{\alpha\beta,\mu} + g^0_{\alpha\mu,\beta} - g^0_{\beta\mu,\alpha}\right) + \left(h_{\alpha\beta,\mu} + h_{\alpha\mu,\beta} - h_{\beta\mu,\alpha}\right)\right] \\
&= \frac{1}{2}g^{0\,\gamma\alpha}\left(g^0_{\alpha\beta,\mu} + g^0_{\alpha\mu,\beta} - g^0_{\beta\mu,\alpha}\right) + \frac{1}{2}g^{0\,\gamma\alpha}\left(h_{\alpha\beta,\mu} + h_{\alpha\mu,\beta} - h_{\beta\mu,\alpha}\right) \\
&- \frac{1}{2}h^{\gamma\alpha}\left(g^0_{\alpha\beta,\mu} + g^0_{\alpha\mu,\beta} - g^0_{\beta\mu,\alpha}\right) + O(h^2) \\
&\equiv \Gamma^\gamma_{\beta\mu}(g^0) + \delta\Gamma^\gamma_{\beta\mu}(g^0, h) + O(h^2)\,,
\end{aligned} \tag{12.9}$$

where $\delta\Gamma^\gamma_{\beta\mu}(g^0, h)$ are the first-order corrections, which are linear in $h_{\mu\nu}$,

$$\delta\Gamma^\gamma_{\beta\mu}(g^0, h) = \frac{1}{2}g^{0\,\gamma\alpha}\left(h_{\alpha\beta,\mu} + h_{\alpha\mu,\beta} - h_{\beta\mu,\alpha}\right) - \frac{1}{2}h^{\gamma\alpha}\left(g^0_{\alpha\beta,\mu} + g^0_{\alpha\mu,\beta} - g^0_{\beta\mu,\alpha}\right)\,. \tag{12.10}$$

Note that, as shown in Sec. 7.3, $\delta\Gamma^{\alpha}_{\mu\nu}$ are the components of a tensor, and can be written as (see Eq. 7.28)

$$\delta\Gamma^{\lambda}_{\mu\nu} = \frac{1}{2}g^{\lambda\rho}\left(h_{\mu\rho;\nu} + h_{\nu\rho;\mu} - h_{\mu\nu;\rho}\right) = \frac{1}{2}g^{0\,\lambda\rho}\left(h_{\mu\rho;\nu} + h_{\nu\rho;\mu} - h_{\mu\nu;\rho}\right) + O(h^2) \,. \tag{12.11}$$

When we substitute the expression of $\Gamma^{\gamma}_{\beta\mu}(g_{\mu\nu})$ given in Eq. 12.9 in the Ricci tensor we get

$$\begin{aligned} R_{\mu\nu}(g_{\mu\nu}) &= R^{0}_{\mu\nu}(g^0) + \frac{\partial}{\partial x^{\alpha}}\delta\Gamma^{\alpha}_{\mu\nu}(g^0,h) - \frac{\partial}{\partial x^{\nu}}\delta\Gamma^{\alpha}_{\mu\alpha}(g^0,h) \\ &+ \Gamma^{\alpha}_{\sigma\alpha}(g^0)\,\delta\Gamma^{\sigma}_{\mu\nu}(g^0,h) + \delta\Gamma^{\alpha}_{\sigma\alpha}(g^0,h)\,\Gamma^{\sigma}_{\mu\nu}(g^0) \\ &- \Gamma^{\alpha}_{\sigma\nu}(g^0)\,\delta\Gamma^{\sigma}_{\mu\alpha}(g^0,h) - \delta\Gamma^{\alpha}_{\sigma\nu}(g^0,h)\,\Gamma^{\sigma}_{\mu\alpha}(g^0) + O(h^2) \\ &= R^{0}_{\mu\nu}(g^0) + \delta\Gamma^{\alpha}_{\mu\nu;\alpha}(g^0,h) - \delta\Gamma^{\alpha}_{\mu\alpha;\nu}(g^0,h) + O(h^2) \\ &\equiv R^{0}_{\mu\nu}(g^0) + \delta R_{\mu\nu}(g^0,h) + O(h^2)\,, \end{aligned} \tag{12.12}$$

where the covariant derivative is performed with respect to the background metric $g^{0}_{\mu\nu}$. Note that the expression of $\delta R_{\mu\nu}$ in Eq. 12.12 can also be obtained using the Palatini identity 7.31.

Let us now consider the right-hand side of the Einstein equations 12.6, i.e. $\left(T_{\mu\nu} - \frac{1}{2}g_{\mu\nu}T\right)$. Since $T_{\mu\nu} = T^{0}_{\mu\nu} + \delta T_{\mu\nu}$, we get

$$\begin{aligned} T &= g^{\mu\nu}T_{\mu\nu} = \left(g^{0\mu\nu} - h^{\mu\nu}\right)\left(T^{0}_{\mu\nu} + \delta T_{\mu\nu}\right) + O(h^2) \\ &= g^{0\mu\nu}T^{0}_{\mu\nu} - h^{\mu\nu}T^{0}_{\mu\nu} + g^{0\mu\nu}\delta T_{\mu\nu} + O(h^2) \equiv T^0 + \delta T \,. \end{aligned} \tag{12.13}$$

Consequently

$$\begin{aligned} \left(T_{\mu\nu} - \frac{1}{2}g_{\mu\nu}T\right) &= T^{0}_{\mu\nu} + \delta T_{\mu\nu} - \frac{1}{2}\left(g^{0}_{\mu\nu} + h_{\mu\nu}\right)\left(T^0 + \delta T\right) \\ &= \left(T^{0}_{\mu\nu} - \frac{1}{2}g^{0}_{\mu\nu}T^0\right) + \left[\delta T_{\mu\nu} - \frac{1}{2}\left(g^{0}_{\mu\nu}\delta T + h_{\mu\nu}T^0\right)\right] + O(h^2)\,. \end{aligned} \tag{12.14}$$

Combining Eqs. 12.12 and 12.13, and reminding that $g^{0}_{\mu\nu}$ is, by assumption, a solution of Einstein's equations in vacuum, $R_{\mu\nu}(g^0) = \frac{8\pi G}{c^4}\left(T^{0}_{\mu\nu} - \frac{1}{2}g^{0}_{\mu\nu}T^0\right)$, the field equations for the perturbations $h_{\mu\nu}$ reduce to

$$\delta\Gamma^{\alpha}_{\mu\nu;\alpha}(g^0,h) - \delta\Gamma^{\alpha}_{\mu\alpha;\nu}(g^0,h) = \frac{8\pi G}{c^4}\left[\delta T_{\mu\nu} - \frac{1}{2}\left(g^{0}_{\mu\nu}\delta T + h_{\mu\nu}T^0\right)\right] + O(h^2)\,. \tag{12.15}$$

As expected, the above equations are linear in $h_{\mu\nu}$. Their solution describes the generation of gravitational waves and their propagation in the considered background. This approximation works sufficiently well in a variety of physical situations because gravitational waves are very weak. This point will be discussed in more detail in Chapter 13, when we will study the generation of gravitational waves.

12.2 GRAVITATIONAL WAVES AS PERTURBATIONS OF FLAT SPACETIME

Let us consider the flat spacetime described by the metric tensor $\eta_{\mu\nu}$ and a small perturbation $h_{\mu\nu}$ thereof, such that the resulting metric can be written as

$$g_{\mu\nu} = \eta_{\mu\nu} + h_{\mu\nu}\,, \qquad |h_{\mu\nu}| \ll 1\,. \tag{12.16}$$

In this case the equations derived in the previous section considerably simplify. The affine connections 12.9 computed for the metric 12.16 give

$$\Gamma^{\lambda}_{\mu\nu} = \frac{1}{2}\eta^{\lambda\rho}\left[h_{\rho\mu,\nu} + h_{\rho\nu,\mu} - h_{\mu\nu,\rho}\right] + O(h^2)\,. \tag{12.17}$$

Since the metric $g^0_{\mu\nu} \equiv \eta_{\mu\nu}$ is constant, $\Gamma^{\lambda}_{\mu\nu}(g^0) = 0$ and $\Gamma^{\lambda}_{\mu\nu} \equiv \delta\Gamma^{\lambda}_{\mu\nu}(h)$ is of order $O(h)$; moreover, since the background is flat $T^0_{\mu\nu} = 0$, and $T_{\mu\nu} = \delta T_{\mu\nu}$ is also of order $O(h)$.

In this case the left-hand side of Eq. 12.15 simply reduces to

$$\begin{aligned} &\frac{\partial\Gamma^{\alpha}_{\mu\nu}}{\partial x^{\alpha}} - \frac{\partial\Gamma^{\alpha}_{\mu\alpha}}{\partial x^{\nu}} + O(h^2) \\ = \;& \frac{1}{2}\left\{-\Box_F h_{\mu\nu} + \left[\frac{\partial^2}{\partial x^{\lambda}\partial x^{\mu}}h^{\lambda}_{\nu} + \frac{\partial^2}{\partial x^{\lambda}\partial x^{\nu}}h^{\lambda}_{\mu} - \frac{\partial^2}{\partial x^{\mu}\partial x^{\nu}}h^{\lambda}_{\lambda}\right]\right\} + O(h^2)\,. \end{aligned} \tag{12.18}$$

The operator $\Box_F$ is the d'Alembertian in flat spacetime,

$$\Box_F = \eta^{\alpha\beta}\frac{\partial}{\partial x^{\alpha}}\frac{\partial}{\partial x^{\beta}} = -\frac{1}{c^2}\frac{\partial^2}{\partial t^2} + \nabla^2\,. \tag{12.19}$$

Thus, Einstein's equations 12.6 for $h_{\mu\nu}$ become

$$\left\{\Box_F h_{\mu\nu} - \left[\frac{\partial^2}{\partial x^{\lambda}\partial x^{\mu}}h^{\lambda}_{\nu} + \frac{\partial^2}{\partial x^{\lambda}\partial x^{\nu}}h^{\lambda}_{\mu} - \frac{\partial^2}{\partial x^{\mu}\partial x^{\nu}}h^{\lambda}_{\lambda}\right]\right\} = -\frac{16\pi G}{c^4}\left(T_{\mu\nu} - \frac{1}{2}\eta_{\mu\nu}T\right)\,. \tag{12.20}$$

This equation can be considerably simplified by using the gauge freedom of Einstein's equations discussed in Chapter 6. Indeed, the solution of Eq. 12.20 is not uniquely determined. If we make a coordinate transformation, the transformed metric tensor is still a solution: it describes the same physical situation seen from a different frame. However, since we are working in the weak-field limit we are entitled to make only those transformations which preserve the condition $|h_{\mu\nu}| \ll |\eta_{\mu\nu}|$, i.e. $|h_{\mu\nu}| \ll 1$. To this purpose, let us consider an infinitesimal coordinate transformation

$$x^{\alpha\prime} = x^{\alpha} + \epsilon^{\alpha}(x)\,, \tag{12.21}$$

(note that the prime refers to the coordinate x^{α}, not to the index α) where ϵ^{α} is an arbitrary vector such that $\epsilon_{\alpha,\mu}$ are of the same order of $h_{\mu\nu}$ (see discussion below). Then,

$$\frac{\partial x^{\alpha\prime}}{\partial x^{\mu}} = \delta^{\alpha}_{\mu} + \frac{\partial\epsilon^{\alpha}}{\partial x^{\mu}}\,. \tag{12.22}$$

Furthermore, since [1]

$$\begin{aligned} g_{\mu\nu} &= g'_{\alpha\beta}\frac{\partial x^{\alpha\prime}}{\partial x^{\mu}}\frac{\partial x^{\beta\prime}}{\partial x^{\nu}} = \left(\eta_{\alpha\beta} + h'_{\alpha\beta}\right)\left(\delta^{\alpha}_{\mu} + \frac{\partial\epsilon^{\alpha}}{\partial x^{\mu}}\right)\left(\delta^{\beta}_{\nu} + \frac{\partial\epsilon^{\beta}}{\partial x^{\nu}}\right) \\ &= \eta_{\mu\nu} + h'_{\mu\nu} + \epsilon_{\nu,\mu} + \epsilon_{\mu,\nu} + O(h^2)\,, \end{aligned} \tag{12.23}$$

and $g_{\mu\nu} = \eta_{\mu\nu} + h_{\mu\nu}$, it follows that, up to $O(h^2)$,

$$h'_{\mu\nu} = h_{\mu\nu} - (\epsilon_{\nu,\mu} + \epsilon_{\mu,\nu})\,. \tag{12.24}$$

This equation can also be interpreted as the transformation of the metric perturbation for an infinitesimal diffeomorphism (see Box 12-A).

[1] Note that in this chapter we denote the transformed tensor as (say) $h'_{\mu\nu}$ rather than as $h_{\mu'\nu'}$, since this simplifies the discussion on the infinitesimal coordinate transformations.

From the above equation it follows that, if $h_{\mu\nu}$ satisfies the condition $|h_{\mu\nu}| \ll 1$, $h'_{\mu\nu}$ satisfies the same condition only if the derivatives $\epsilon_{\nu,\mu}$ are of order $h_{\mu\nu}$ as anticipated above. Notice that $h'_{\mu\nu}$ is symmetric, as it should be.

In order to simplify Eq. 12.20 it appears convenient to choose a coordinate system in which the harmonic gauge condition introduced in Sec. 6.4,

$$\Gamma^\lambda = g^{\mu\nu}\Gamma^\lambda_{\mu\nu} = 0\,, \tag{12.25}$$

is satisfied. Indeed, using Eq. 12.17,

$$\begin{aligned} g^{\mu\nu}\Gamma^\lambda_{\mu\nu} &= \frac{1}{2}\eta^{\mu\nu}\eta^{\lambda\rho}\left[h_{\rho\mu,\nu} + h_{\rho\nu,\mu} - h_{\mu\nu,\rho}\right] + O(h^2) \\ &= \frac{1}{2}\eta^{\lambda\rho}\left\{h^\nu{}_{\rho,\nu} + h^\mu{}_{\rho,\mu} - h^\nu{}_{\nu,\rho}\right\} + O(h^2)\,. \end{aligned} \tag{12.26}$$

Since the first two terms are equal to each other, we find

$$g^{\mu\nu}\Gamma^\lambda_{\mu\nu} = \eta^{\lambda\rho}\left(h^\mu{}_{\rho,\mu} - \frac{1}{2}h^\nu{}_{\nu,\rho}\right) + O(h^2)\,. \tag{12.27}$$

Thus, up to terms that are first order in $h_{\mu\nu}$, the harmonic gauge condition is equivalent to

$$\frac{\partial}{\partial x^\mu}h^\mu{}_\rho = \frac{1}{2}\frac{\partial}{\partial x^\rho}h\,, \tag{12.28}$$

where we have set

$$h = \eta^{\mu\nu}h_{\mu\nu} \equiv h^\nu{}_\nu\,. \tag{12.29}$$

Using this condition the term in square brackets in Eq. 12.20 vanishes, and the linearized Einstein equations reduce to a simple wave equation supplemented by the condition 12.28,

$$\begin{cases} \Box_F h_{\mu\nu} = -\dfrac{16\pi G}{c^4}\left(T_{\mu\nu} - \dfrac{1}{2}\eta_{\mu\nu}T\right) \\ \dfrac{\partial}{\partial x^\mu}h^\mu{}_\nu = \dfrac{1}{2}\dfrac{\partial}{\partial x^\nu}h\,. \end{cases} \tag{12.30}$$

If we introduce the tensor

$$\bar{h}_{\mu\nu} \equiv h_{\mu\nu} - \frac{1}{2}\eta_{\mu\nu}h\,, \tag{12.31}$$

Eqs. 12.30 become

$$\begin{cases} \Box_F \bar{h}_{\mu\nu} = -\dfrac{16\pi G}{c^4}T_{\mu\nu} \\ \dfrac{\partial}{\partial x^\mu}\bar{h}^\mu{}_\nu = 0\,, \end{cases} \tag{12.32}$$

and outside the source, where $T_{\mu\nu} = 0$,

$$\begin{cases} \Box_F \bar{h}_{\mu\nu} = 0 \\ \dfrac{\partial}{\partial x^\mu}\bar{h}^\mu{}_\nu = 0\,. \end{cases} \tag{12.33}$$

Since the first of Eqs. 12.33 is the standard D'Alembert equation, it shows that *a perturbation of a flat spacetime propagates as a wave travelling at the speed of light*, and that *Einstein's theory of gravity predicts the existence of gravitational waves.* The tensor $\bar{h}_{\mu\nu}$ is also called the "trace-reversed" perturbation tensor, since

$$\bar{h} = \eta^{\mu\nu}\bar{h}_{\mu\nu} = -h\,. \tag{12.34}$$

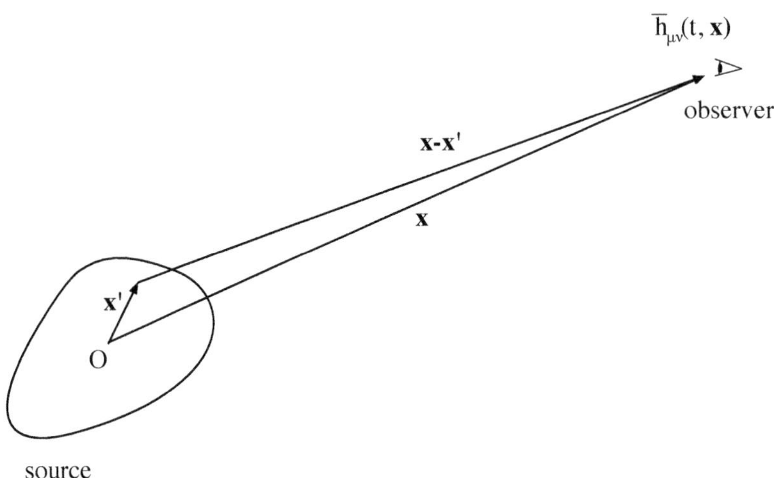

Figure 12.1: An observer located at a distance $\mathbf{x}$ from the source, at the time t receives a wave $\bar{h}_{\mu\nu}(t, \mathbf{x})$ which is the sum of the contributions emitted from each source element located in $\mathbf{x}'$ at the retarded time $t - \frac{|\mathbf{x}-\mathbf{x}'|}{c}$.

As in electrodynamics, the solution of Eqs. 12.32 can be written in terms of retarded potentials [67]

$$\bar{h}_{\mu\nu}(t, \mathbf{x}) = \frac{4G}{c^4} \int_V \frac{T_{\mu\nu}(t - \frac{|\mathbf{x}-\mathbf{x}'|}{c}, \mathbf{x}')}{|\mathbf{x} - \mathbf{x}'|} \, d^3x' \,, \tag{12.35}$$

where V is the three-dimensional source volume, $\mathbf{x}'$ is the distance of an element of the emitting source from the origin of a frame centered in some point of the source, $\mathbf{x}$ is the distance of the observer from the source (see Fig. 12.1).

Remarkably, the retarded potentials solution in Eq. 12.35 automatically satisfies the harmonic gauge condition

$$\frac{\partial}{\partial x^\mu} \bar{h}^{\mu\nu} = 0 \,. \tag{12.36}$$

Indeed, if we define the *Green function*

$$\mathcal{G}\left(\vec{x} - \vec{x}'\right) \equiv \frac{4G}{c^5} \frac{1}{|\mathbf{x} - \mathbf{x}'|} \delta \left[t' - \left(t - \frac{|\mathbf{x} - \mathbf{x}'|}{c} \right) \right] , \tag{12.37}$$

where $\vec{x} = (ct, \mathbf{x})$ and $\vec{x}\,' = (ct', \mathbf{x}')$, Eq. 12.35 can be written as a four-dimensional integral as follows

$$\bar{h}_{\mu\nu}(\vec{x}) = \int_\Omega T_{\mu\nu}(\vec{x}\,') \, \mathcal{G}\left(\vec{x} - \vec{x}'\right) \, d^4x' \,, \tag{12.38}$$

where $\Omega \equiv V \times I$, and I is a time interval to be taken such that $\mathcal{G}(\vec{x} - \vec{x}')$ vanishes at the extrema of I. This condition is satisfied if I is so large that, for all $\mathbf{x}' \in V$, the instant $t' = t - \frac{|\mathbf{x}-\mathbf{x}'|}{c}$ is contained in I; indeed, from Eq. 12.37, the Green function $\mathcal{G}$ is different from zero only for $t' = t - \frac{|\mathbf{x}-\mathbf{x}'|}{c}$.

Since $\mathcal{G}$ is a function of the *difference* $(\vec{x} - \vec{x}')$, then

$$\frac{\partial}{\partial x^\mu} \left[\mathcal{G}\left(\vec{x} - \vec{x}'\right)\right] = -\frac{\partial}{\partial x'^\mu} \left[\mathcal{G}\left(\vec{x} - \vec{x}'\right)\right] \,. \tag{12.39}$$

Consequently,

$$\frac{\partial}{\partial x^{\mu}}\bar{h}^{\mu\nu}(\vec{x}) = \int_{\Omega} T^{\mu\nu}(\vec{x}\,')\frac{\partial}{\partial x^{\mu}}\mathcal{G}\,(\vec{x}-\vec{x}')\; d^4x' = -\int_{\Omega} T^{\mu\nu}(\vec{x}\,')\frac{\partial}{\partial x'^{\mu}}\mathcal{G}\,(\vec{x}-\vec{x}')\; d^4x' \,. \tag{12.40}$$

The last term can be integrated by parts,

$$\int_{\Omega} T^{\mu\nu}(\vec{x}\,')\frac{\partial}{\partial x'^{\mu}}\mathcal{G}\,(\vec{x}-\vec{x}')\; d^4x' = \tag{12.41}$$
$$\int_{\Omega} d^4x'\; \frac{\partial}{\partial x'^{\mu}}\left[T^{\mu\nu}(\vec{x}\,')\mathcal{G}\,(\vec{x}-\vec{x}')\right] - \int_{\Omega} d^4x'\; \left[\frac{\partial}{\partial x'^{\mu}}T^{\mu\nu}(\vec{x}\,')\right]\mathcal{G}\,(\vec{x}-\vec{x}')\; d^4x' \,.$$

We now show that both integrals above vanish. By Gauss' theorem, the first integral over the 4-volume Ω on the right-hand side of Eq. 12.41 is equal to the flux of $[T^{\mu\nu}(\vec{x}\,')\mathcal{G}\,(\vec{x}-\vec{x}')]$ across the surface $\partial\Omega$ enclosing that volume, i.e.

$$\int_{\Omega} d^4x'\; \frac{\partial}{\partial x'^{\mu}}\left[T^{\mu\nu}(\vec{x}\,')\mathcal{G}\,(\vec{x}-\vec{x}')\right] = \int_{\partial\Omega}\left[T^{\mu\nu}(\vec{x}\,')\mathcal{G}\,(\vec{x}-\vec{x}')\right] dS_{\mu} \,, \tag{12.42}$$

where dS is the surface element on $\partial\Omega$, with normal vector n^{μ}, and $dS_{\mu} = n_{\mu}dS$. This integral vanishes since $T^{\mu\nu} = 0$ on the boundary of V and $\mathcal{G} = 0$ on the boundary of I. The second integral on the right-hand side of Eq. 12.41 also vanishes, since the stress-energy tensor satisfies the conservation law ${T^{\mu\nu}}_{,\nu} = 0$. Consequently, Eq. 12.41 vanishes and hence

$$\frac{\partial}{\partial x^{\mu}}\bar{h}^{\mu\nu}(\vec{x}) = 0 \,, \tag{12.43}$$

as previously stated.

Box 12-A

Transformation of metric perturbations under infinitesimal diffeomorphisms

As discussed in Sec. 8.5, the invariance of Einstein's equations with respect to general coordinate transformations is formally equivalent to the invariance of the same equations under diffeomorphisms, and a gauge transformation can be interpreted either as a coordinate transformation or as a diffeomorphism. Thus Eq. 12.21,

$$x^{\alpha'} = x^\alpha + \epsilon^\alpha(x) \tag{12.44}$$

with $\epsilon^\alpha(x)$ of order $O(h)$, can be interpreted as an *infinitesimal diffeomorphism of order* $O(h)$. Under such diffeomorphism, the change in the metric components is given by the Lie derivative (see Box 8-A)

$$\delta g_{\mu\nu} = \mathcal{L}_{\vec{\epsilon}}\, g_{\mu\nu} = \epsilon_{\mu;\nu} + \epsilon_{\nu;\mu}\,. \tag{12.45}$$

Therefore,

$$g_{\mu\nu} + \delta g_{\mu\nu} = (g^0_{\mu\nu} + h_{\mu\nu}) + \delta g_{\mu\nu} = g^0_{\mu\nu} + h'_{\mu\nu}\,, \tag{12.46}$$

consequently,

$$h'_{\mu\nu} = h_{\mu\nu} + \delta g_{\mu\nu} = h_{\mu\nu} + \epsilon_{\mu;\nu} + \epsilon_{\nu;\mu}\,, \tag{12.47}$$

which, if $g^0_{\mu\nu} = \eta_{\mu\nu}$, is equivalent up to $O(h^2)$ terms, and a sign difference, to Eq. 12.24, $h'_{\mu\nu} = h_{\mu\nu} - (\epsilon_{\nu,\mu} + \epsilon_{\mu,\nu})$.
The reason of this sign difference is the following. Eq. 12.44 corresponds to both a coordinate transformation and a diffeomorphism, but these affect scalar functions (and vector and tensor field as well) in a different way. Let us consider, for instance, a scalar function $f(x)$; it is transformed by a *coordinate transformation* 12.21 into a function f' such that $f'(x') = f(x)$. Therefore, $f'(x+\epsilon) = f(x)$ and thus $(f' - f)(x) = -\epsilon^\mu f_{,\mu}(x)$. Conversely, the function f is transformed by a *diffeomorphism* 12.21 into a function f' such that $f'(x) = f(x')$, and thus in that case $(f' - f)(x) = \epsilon^\mu f_{,\mu}(x)$. The same holds for vector and tensor fields.

12.3 HOW TO CHOOSE THE HARMONIC GAUGE

We shall now show that if the harmonic gauge condition is not satisfied in a reference frame, we can always find a new frame where it is, by making an infinitesimal coordinate transformation

$$x^{\lambda'} = x^\lambda + \epsilon^\lambda\,, \tag{12.48}$$

provided ϵ^λ satisfies the following equation

$$\Box_F \epsilon_\rho = \frac{\partial h^\beta{}_\rho}{\partial x^\beta} - \frac{1}{2}\frac{\partial h}{\partial x^\rho}\,. \tag{12.49}$$

Indeed, under a change of coordinates the quantity $\Gamma^\lambda = g^{\mu\nu}\Gamma^\lambda_{\mu\nu}$ transforms according to Eq. 6.74, i.e.

$$\Gamma^{\lambda'} = \frac{\partial x^{\lambda'}}{\partial x^\rho}\Gamma^\rho - g^{\rho\sigma}\frac{\partial^2 x^{\lambda'}}{\partial x^\rho \partial x^\sigma}\,. \tag{12.50}$$

Since from Eq. 12.48

$$\frac{\partial x^{\lambda'}}{\partial x^\sigma} = \delta^\lambda_\sigma + \frac{\partial \epsilon^\lambda}{\partial x^\sigma}\,, \tag{12.51}$$

the last term in Eq. 12.50 gives

$$g^{\rho\sigma}\frac{\partial^2 x^{\lambda'}}{\partial x^\rho \partial x^\sigma} = g^{\rho\sigma}\left[\frac{\partial}{\partial x^\rho}\left(\delta^\lambda_\sigma + \frac{\partial \epsilon^\lambda}{\partial x^\sigma}\right)\right] = \eta^{\rho\sigma}\left(\frac{\partial^2 \epsilon^\lambda}{\partial x^\rho \partial x^\sigma}\right) = \Box_F \epsilon^\lambda . \tag{12.52}$$

In addition, from Eq. 12.27

$$\Gamma^\rho = \eta^{\rho\nu}\left(h^\mu{}_{\nu,\mu} - \frac{1}{2}h_{,\nu}\right), \tag{12.53}$$

therefore in the new gauge the condition $\Gamma^{\lambda'} = 0$ becomes

$$\Gamma^{\lambda'} = \left(\delta^\lambda_\rho + \frac{\partial \epsilon^\lambda}{\partial x^\rho}\right)\eta^{\rho\nu}\left(\frac{\partial h^\mu{}_\nu}{\partial x^\mu} - \frac{1}{2}\frac{\partial h}{\partial x^\nu}\right) - \Box_F \epsilon^\lambda = 0 . \tag{12.54}$$

If we neglect second-order terms in h, Eq. 12.54 reduces to

$$\Gamma^{\lambda'} = \eta^{\lambda\nu}\left(\frac{\partial h^\mu{}_\nu}{\partial x^\mu} - \frac{1}{2}\frac{\partial h}{\partial x^\nu}\right) - \Box_F \epsilon^\lambda = 0 , \tag{12.55}$$

which, contracting with $\eta_{\lambda\alpha}$ and reminding that $\eta_{\lambda\alpha}\eta^{\lambda\nu} = \delta^\nu_\alpha$, finally gives

$$\Box_F \epsilon_\alpha = \frac{\partial h^\mu{}_\alpha}{\partial x^\mu} - \frac{1}{2}\frac{\partial h}{\partial x^\alpha} , \tag{12.56}$$

i.e. Eq. 12.49. This is an inhomogeneous wave equation, which can be solved with standard techniques to find the components ϵ_α. These identify the coordinate system in which the harmonic gauge condition is satisfied.

It is important to note that *the harmonic gauge condition 12.25 does not determine the gauge uniquely*. Indeed, suppose that we are in a frame where this condition is satisfied, i.e. $\Gamma^\rho = 0$. If we make an infinitesimal coordinate transformation 12.48 and impose that ϵ^μ satisfies the equation $\Box_F \epsilon^\mu = 0$, according to Eqs. 12.50 and 12.52 in the new frame $\Gamma^{\lambda\prime} = 0$, i.e. the harmonic gauge condition still holds. This gauge freedom will be used in Sec. 12.5.

12.4 PLANE GRAVITATIONAL WAVES

We have shown that the perturbations of flat spacetime satisfy the wave equation

$$\Box_F \bar{h}_{\mu\nu} = 0 \tag{12.57}$$

supplemented by the harmonic gauge condition

$$\frac{\partial}{\partial x^\mu}\bar{h}^{\mu\nu} = 0 . \tag{12.58}$$

The general solution of Eq. 12.57 is a linear superposition of monochromatic plane waves, each one of the form

$$\bar{h}_{\mu\nu} = \Re\left\{A_{\mu\nu}e^{ik_\alpha x^\alpha}\right\}, \tag{12.59}$$

where $A_{\mu\nu}$ is the polarization tensor, i.e. the wave amplitude, $\vec{k}$ is the wave four-vector, and $\Re(z)$ denotes the real part of any complex number z, which we shall often omit for simplicity.

By direct substitution of Eq. 12.59 into Eq. 12.57 we find

$$\begin{aligned}\Box_F \bar{h}_{\mu\nu} &= A_{\mu\nu}\eta^{\alpha\beta}\frac{\partial}{\partial x^\alpha}\frac{\partial}{\partial x^\beta}\left(e^{ik_\gamma x^\gamma}\right) = A_{\mu\nu}\eta^{\alpha\beta}\frac{\partial}{\partial x^\alpha}\left[ik_\gamma \delta^\gamma{}_\beta e^{ik_\gamma x^\gamma}\right] \\ &= A_{\mu\nu}\eta^{\alpha\beta}\frac{\partial}{\partial x^\alpha}\left[ik_\beta e^{ik_\gamma x^\gamma}\right] = -A_{\mu\nu}\eta^{\alpha\beta}k_\alpha k_\beta\, e^{ik_\gamma x^\gamma} = 0 .\end{aligned} \tag{12.60}$$

Thus, neglecting the trivial solution $A_{\mu\nu} = 0$, Eq. 12.59 is a solution to Eq. 12.57 if

$$\eta^{\alpha\beta} k_\alpha k_\beta = 0 \,, \tag{12.61}$$

that is, if $\vec{k}$ is a null vector.

The harmonic gauge condition 12.58 can be written as

$$\eta^{\mu\alpha} \frac{\partial}{\partial x^\mu} \bar{h}_{\alpha\nu} = 0 \,. \tag{12.62}$$

Using Eq. 12.59, this yields

$$\eta^{\mu\alpha} \frac{\partial}{\partial x^\mu} A_{\alpha\nu} e^{ik_\gamma x^\gamma} = 0 \quad \rightarrow \quad \eta^{\mu\alpha} A_{\alpha\nu} k_\mu = 0 \quad \rightarrow \quad k_\mu A^\mu{}_\nu = 0 \,. \tag{12.63}$$

This further condition imposes the orthogonality of the wave four-vector to the polarization tensor.

The perturbation $\bar{h}_{\mu\nu}$ in Eq. 12.59 is constant on those surfaces where

$$k_\alpha x^\alpha = const \qquad \rightarrow \qquad k_i x^i = const - ck_0 t \,. \tag{12.64}$$

At fixed time, the spatial surfaces where $\bar{h}_{\mu\nu}$ is constant, i.e. the *wavefronts* of the gravitational wave, are the planes $k_i x^i = const.$

It is conventional to refer to k^0 as $\frac{\omega}{c}$, where ω is the wave pulsation (we shall often refer to it simply as "frequency"). Consequently

$$\vec{k} = \left(\frac{\omega}{c}, \mathbf{k} \right) \,, \tag{12.65}$$

where $\mathbf{k}$ is the wave three-vector with components (k^1, k^2, k^3), and is orthogonal to the wavefront. It is related to the *wavelength* λ by $|\mathbf{k}| = 2\pi/\lambda$. Since $\vec{k}$ is a null vector

$$-(k^0)^2 + (k^1)^2 + (k^2)^2 + (k^3)^2 = -(k^0)^2 + |\mathbf{k}| = 0 \,, \tag{12.66}$$

from which it follows

$$\omega = ck_0 = c|\mathbf{k}| \,, \tag{12.67}$$

which gives the expected dispersion relation for a wave moving at the speed of light.

12.5 THE TT GAUGE

Let us now discuss how many of the ten components of $h_{\mu\nu}$ have a real physical meaning, i.e. what are the degrees of freedom of a gravitational wave. Consider a plane wave propagating in flat spacetime along the x-direction, using Cartesian coordinates $(x^1, x^2, x^3) = (x, y, z)$. Since $h_{\mu\nu}$ is independent of y and z, Eq. 12.57 becomes (as before we raise and lower indices with $\eta_{\mu\nu}$)

$$\left(-\frac{1}{c^2} \frac{\partial^2}{\partial t^2} + \frac{\partial^2}{\partial x^2} \right) \bar{h}^\mu{}_\nu = 0 \,. \tag{12.68}$$

This is the one-dimensional wave equation, the solution of which is an arbitrary function of $t \pm \frac{x}{c}$. Let us consider, for example, a progressive wave $\bar{h}^\mu{}_\nu\,[f(t,x)]$, where $f(t,x) = t - \frac{x}{c}$. Since

$$\begin{aligned} \frac{\partial}{\partial t} \bar{h}^\mu{}_\nu &= \frac{\partial \bar{h}^\mu{}_\nu}{\partial f} \frac{\partial f}{\partial t} = \frac{\partial \bar{h}^\mu{}_\nu}{\partial f} \,, \\ \frac{\partial}{\partial x} \bar{h}^\mu{}_\nu &= \frac{\partial \bar{h}^\mu{}_\nu}{\partial f} \frac{\partial f}{\partial x} = -\frac{1}{c} \frac{\partial \bar{h}^\mu{}_\nu}{\partial f} \,, \end{aligned} \tag{12.69}$$

the harmonic gauge condition $\frac{\partial}{\partial x^\mu}\bar{h}^\mu{}_\nu = 0$ gives

$$\frac{\partial}{\partial x^\mu}\bar{h}^\mu{}_\nu = \frac{1}{c}\frac{\partial \bar{h}^t{}_\nu}{\partial t} + \frac{\partial \bar{h}^x{}_\nu}{\partial x} = \frac{1}{c}\frac{\partial}{\partial f}\left[\bar{h}^t{}_\nu - \bar{h}^x{}_\nu\right] = 0\,. \tag{12.70}$$

The solution of this equation is $\bar{h}^t{}_\nu - \bar{h}^x{}_\nu = const$ and, since we are interested only in the time-dependent part of the solution, we can set the integration constants to zero[2]. Consequently

$$\bar{h}^t{}_t = \bar{h}^x{}_t, \qquad \bar{h}^t{}_x = \bar{h}^x{}_x, \qquad \bar{h}^t{}_y = \bar{h}^x{}_y, \qquad \bar{h}^t{}_z = \bar{h}^x{}_z\,. \tag{12.71}$$

We now remind that, as shown at the end of Sec. 12.3, we still have the freedom of making an infinitesimal coordinate transformation

$$x^{\mu\prime} = x^\mu + \epsilon^\mu \tag{12.72}$$

with $\epsilon_{\mu,\nu} = O(h)$. If $\Box_F \epsilon^\mu = 0$, the harmonic gauge condition is satisfied also in the new frame, and consequently the tensor $\bar{h}'_{\mu\nu}$ satisfies the wave equation

$$\Box_F \bar{h}'_{\mu\nu} = 0\,. \tag{12.73}$$

Using this gauge freedom and without loss of generality, we can choose the four functions ϵ^μ to set to zero the following four quantities

$$\bar{h}^t{}_x = \bar{h}^t{}_y = \bar{h}^t{}_z = \bar{h}^y{}_y + \bar{h}^z{}_z = 0\,. \tag{12.74}$$

From Eq. 12.71 it then follows that

$$\bar{h}^x{}_x = \bar{h}^x{}_y = \bar{h}^x{}_z = \bar{h}^t{}_t = 0\,. \tag{12.75}$$

The remaining non-vanishing components are $\bar{h}^z{}_y$ and $\bar{h}^y{}_y - \bar{h}^z{}_z$. These components cannot be set equal to zero, because we have exhausted our gauge freedom.

From Eqs. 12.74 and 12.75 it follows that, with the above choice,

$$\bar{h} = \bar{h}^\mu{}_\mu = \bar{h}^t{}_t + \bar{h}^x{}_x + \bar{h}^y{}_y + \bar{h}^z{}_z = 0\,. \tag{12.76}$$

Since $\bar{h} = -h$ (see Eq. 12.34) it follows that $h = 0$, consequently

$$\bar{h}_{\mu\nu} = h_{\mu\nu} - \frac{1}{2}\eta_{\mu\nu} h \equiv h_{\mu\nu}\,, \tag{12.77}$$

i.e. in this gauge $h_{\mu\nu}$ and $\bar{h}_{\mu\nu}$ coincide and are *traceless*. Thus, a plane gravitational wave propagating along the x-axis is characterized by two functions h_{xy} and $h_{yy} = -h_{zz}$, while the remaining components can be set to zero by choosing the gauge, as we have shown

$$h^{\mathrm{TT}}_{\mu\nu} = \begin{pmatrix} 0 & 0 & 0 & 0 \\ 0 & 0 & 0 & 0 \\ 0 & 0 & h_{yy} & h_{yz} \\ 0 & 0 & h_{yz} & -h_{yy} \end{pmatrix}. \tag{12.78}$$

[2]We remark that Eq. 12.57 is *linear* in the metric perturbation, like e.g. Maxwell's equations for the electromagnetic field, or Klein-Gordon's equation for the scalar field. Linear, differential equations satisfy the *superposition principle*: a linear combination of solutions is still a solution. Therefore, it is legitimate to study the time-dependent solutions of Eq. 12.57, and discard the solutions constant in time (the latter will be studied in Chapter 17). The superposition principle does not apply to the full non-linear Einstein equations; in that case, different solutions cannot be studied separately.

In conclusion, *a gravitational wave has only* **two physical degrees of freedom,** *which correspond to two polarization states.* The gauge in which this is clearly manifested is called the **TT gauge**, where 'TT' stands for "transverse traceless": it indicates that the metric tensor $h_{\mu\nu}$ is *traceless*, i.e. $h = 0$, and *transverse*, i.e. the components of $h_{\mu\nu}$ along the direction of propagation vanish (in this case, $h_{\mu x} = 0$). Note that the transversality condition, together with the harmonic gauge condition $\bar{h}^{\mu}{}_{\nu,\mu} = 0$, implies that in the TT gauge the components $h_{\mu 0}$ also vanish.

12.6 HOW DOES A GRAVITATIONAL WAVE AFFECT THE MOTION OF A SINGLE PARTICLE

Consider a particle at rest in flat spacetime before the passage of a gravitational wave. We take the x-axis coincident with the direction of propagation of an incoming wave in the TT gauge. The particle follows a geodesic of the curved spacetime generated by the wave

$$\frac{d^2x^\alpha}{d\tau^2} + \Gamma^\alpha_{\mu\nu}\frac{dx^\mu}{d\tau}\frac{dx^\nu}{d\tau} \equiv \frac{du^\alpha}{d\tau} + \Gamma^\alpha_{\mu\nu}u^\mu u^\nu = 0\,. \tag{12.79}$$

At $\tau = 0$ the particle is at rest ($u^\alpha = (1,0,0,0)$) and, from the above equation, the initial acceleration produced by the wave is

$$\left(\frac{du^\alpha}{d\tau}\right)_{(\tau=0)} = -\Gamma^\alpha_{00} = -\frac{1}{2}\eta^{\alpha\beta}\left[h^{\rm TT}_{\beta 0,0} + h^{\rm TT}_{0\beta,0} - h^{\rm TT}_{00,\beta}\right]\,. \tag{12.80}$$

Since in the TT gauge all time components of $h_{\mu\nu}$ are zero (see Eq. 12.78), Eq. 12.80 gives

$$\left(\frac{du^\alpha}{d\tau}\right)_{(\tau=0)} = 0. \tag{12.81}$$

Thus, u^α remains constant, which means that the particle is not accelerated by the wave and remains at a *fixed coordinate position.* We conclude that *the motion of a single particle is not affected by the passage of the gravitational wave.*

12.7 GEODESIC DEVIATION INDUCED BY A GRAVITATIONAL WAVE

We shall now study the relative motion of neighbouring particles induced by a gravitational wave. Let us consider two particles A and B, initially at rest, with coordinates x^μ_A, x^μ_B. We shall assume that a plane-fronted gravitational wave reaches them at some coordinate time $t = 0$, propagating along the x-axis. We shall use the TT gauge, so that the only non-vanishing components of the wave are those on the (y, z)-plane as in Eq. 12.78. In this frame, the metric is

$$ds^2 = g_{\mu\nu}dx^\mu dx^\nu = (\eta_{\mu\nu} + h^{\rm TT}_{\mu\nu})dx^\mu dx^\nu\,. \tag{12.82}$$

As shown in the previous section, if the two particles are initially at rest, they will remain at the same coordinate position even later, when the wave passes by: thus their *coordinate* separation

$$\delta x^\mu = x^\mu_B - x^\mu_A \tag{12.83}$$

remains constant. We may visualize the TT coordinate frame as a grid of massive particles initially at rest: the position of each particle corresponds to a given space point of the frame. At the passage of a gravitational wave, the grid stretches but the coordinate positions of the particles remain, by definition, constant.

However, the gravitational wave does affect the system. Indeed, *it changes the proper distance between particles*, and this is a scalar quantity invariant under coordinate transformations. For example if two particles are on the y-axis,

$$\begin{aligned}\Delta l &= \int ds = \int_{y_A}^{y_B} |g_{yy}|^{\frac{1}{2}} dy = \int_{y_A}^{y_B} \left[1 + h_{yy}^{\rm TT}(t - x/c)\right]^{\frac{1}{2}} dy \\ &\sim \left[1 + \frac{1}{2} h_{yy}^{\rm TT}(t - x/c)\right] \neq const\,. \end{aligned} \qquad (12.84)$$

The fact that the coordinate separation given in Eq. 12.83 remains constant, whereas the proper distance 12.84 changes, may seem in contradiction. However, we remind that in General Relativity non-tensorial quantities (such as the coordinate separation among two events in the TT frame) may not be appropriate to describe physical processes, because they are not covariant under general coordinate transformations; therefore, it may be difficult to discriminate between true, physical effects and artifacts of the choice of the coordinate frame. In order to study the physical effects produced by the wave, we should consider *tensorial relations*, which are covariant under coordinate transformations. To this purpose it is useful to study the geodesic deviation introduced in Sec. 4.5, which gives the relative acceleration between two particles, and is a tensorial quantity. It is worth reminding that the geodesic deviation

$$\frac{D^2 \delta x^\alpha}{d\tau^2} \equiv \left(\nabla_{\vec{t}}\left(\nabla_{\vec{t}}\, \vec{\delta x}\right)\right)^\alpha , \qquad (12.85)$$

where $\vec{t}$ is the tangent vector to one of the two neighbouring geodesics, is given in terms of the Riemann tensor through Eq. 4.52, which we rewrite here for convenience:

$$\frac{D^2 \delta x^\alpha}{d\tau^2} = R^\alpha{}_{\beta\mu\nu}\, t^\beta\, t^\mu\, \delta x^\nu . \qquad (12.86)$$

In the presence of a gravitational wave, the Riemann tensor does not vanish, whatever frame we use; therefore, the geodesic deviation $\frac{D^2\delta x^i}{d\tau^2} \neq 0$ in any frame, including the TT frame. The quantity $\frac{d^2\delta x^i}{d\tau^2}$, instead, is not a tensorial quantity, and – as noted above – it vanishes in the TT frame.

We shall integrate the equation of geodesic deviation in the LIF $\{\xi^\alpha\}$ centered on one of the two particles, say the particle A (see Sec. 3.9). Note that the coordinate separation in the LIF, $\delta\xi^\alpha$, is not a tensorial quantity; nevertheless it is appropriate to describe the relative motion. Indeed, $\{\xi^\alpha\}$ are the coordinates of a local observer who performs the measurement and, by the Equivalence Principle, the laws of physics in this frame are those of Special Relativity, therefore no spurious coordinate-dependent effects can affect the measurement.

As explained in Box 3-A, in the neighbourhood of A the metric differs from Minkowski's metric by terms of order $|\xi|^2$, i.e.

$$ds^2 = \eta_{\alpha\beta} d\xi^\alpha d\xi^\beta + O(|\xi|^2) . \qquad (12.87)$$

In this frame the particle A has space coordinates $\xi^i_A = 0$ ($i = 1, 2, 3$), and since it is at rest, its proper time coincides with the coordinate time

$$t_A = \tau/c , \qquad \frac{d\xi^\mu}{d\tau}_{|A} = (1, 0, 0, 0) ; \qquad (12.88)$$

in addition

$$g_{\mu\nu\,|A} = \eta_{\mu\nu} , \qquad g_{\mu\nu,\alpha\,|A} = 0 \quad (\text{i.e.}\ \Gamma^\alpha_{\mu\nu\,|A} = 0) , \qquad (12.89)$$

where the subscript $_{|A}$ means that the quantities are computed along the geodesic of the particle A. The vector t^α tangent to the geodesic of the particle A is $t^\mu = \frac{d\xi^\mu}{d\tau}_{|A} = (1, 0, 0, 0)$,

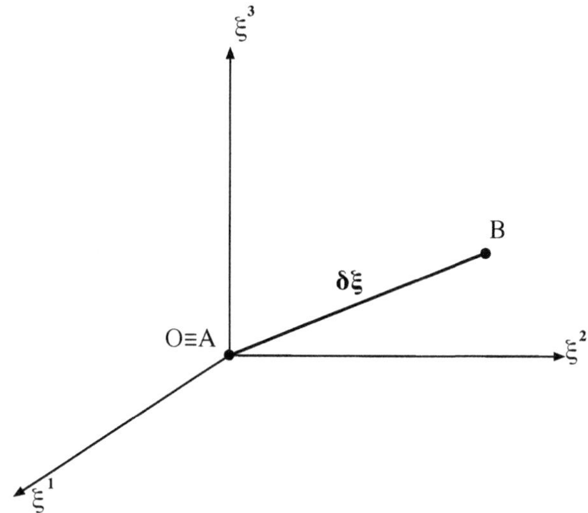

Figure 12.2: Two nearby particles in the LIF $\{\xi^\mu\}$ centered on the particle A.

and $\delta x^\gamma = \delta\xi^\gamma$ are the components of the separation vector between the particles A and B. Thus Eq. 12.86 becomes

$$\frac{D^2\delta\xi^\mu}{d\tau^2} = R^\mu{}_{\alpha\beta\gamma}\frac{d\xi^\alpha}{d\tau}_{|A}\frac{d\xi^\beta}{d\tau}_{|A}\delta\xi^\gamma \quad \to \quad \frac{1}{c^2}\frac{d^2\delta\xi^\mu}{dt^2} = R^\mu{}_{00\gamma}\delta\xi^\gamma\,. \tag{12.90}$$

In the LIF centered in the worldline of the particle A the Riemann tensor is (see Eq. 4.23)

$$R^{\rm LIF}_{\alpha\kappa\lambda\mu} = \frac{1}{2}\left(g_{\alpha\mu,\lambda\kappa} + g_{\kappa\lambda,\mu\alpha} - g_{\alpha\lambda,\mu\kappa} - g_{\kappa\mu,\lambda\alpha}\right)\,, \tag{12.91}$$

and, since $g_{\mu\nu} = \eta_{\mu\nu} + h^{\rm LIF}_{\mu\nu}$, it takes the form

$$R^{\rm LIF}_{\alpha\kappa\lambda\mu} = \frac{1}{2}\left(h^{\rm LIF}_{\alpha\mu,\lambda\kappa} + h^{\rm LIF}_{\kappa\lambda,\mu\alpha} - h^{\rm LIF}_{\alpha\lambda,\mu\kappa} - h^{\rm LIF}_{\kappa\mu,\lambda\alpha}\right)\,. \tag{12.92}$$

In the TT gauge, where $g_{\mu\nu} = \eta_{\mu\nu} + h^{\rm TT}_{\mu\nu}$ and $h^{\rm TT}_{\mu\nu} \neq h^{\rm LIF}_{\mu\nu}$, the Riemann tensor has formally the same expression as in the LIF, i.e.

$$R^{\rm TT}_{\alpha\kappa\lambda\mu} = \frac{1}{2}\left(h^{\rm TT}_{\alpha\mu,\lambda\kappa} + h^{\rm TT}_{\kappa\lambda,\mu\alpha} - h^{TT}_{\alpha\lambda,\mu\kappa} - h^{\rm TT}_{\kappa\mu,\lambda\alpha}\right) + O(h^2)\,; \tag{12.93}$$

however, it greatly simplifies because in this gauge the only non-vanishing components of $h^{\rm TT}_{\mu\nu}$ are those orthogonal to the direction of propagation of the wave. Thus, it is useful to find the relation between the Riemann tensor computed in the TT gauge and that computed in the LIF. As explained in Sec. 12.5, in order to choose a TT frame we need to make an infinitesimal coordinate transformation $x^\mu \to x^\mu + \epsilon^\mu$, with $\Box_F\epsilon^\mu = 0$, and impose the transverse-traceless conditions. Under this transformation $h_{\mu\nu}$ transforms as $h^{\rm LIF}_{\mu\nu} = h^{\rm TT}_{\mu\nu} + \epsilon_{\mu,\nu} + \epsilon_{\nu,\mu}$ which, substituted in Eq. 12.92, gives

$$\begin{aligned} R^{\rm TT}_{\alpha\kappa\lambda\mu} = R^{\rm LIF}_{\alpha\kappa\lambda\mu} - \frac{1}{2}\big(&\epsilon_{\alpha,\mu\kappa\lambda} + \epsilon_{\mu,\alpha\kappa\lambda} + \epsilon_{\kappa,\lambda\alpha\mu} + \epsilon_{\lambda,\kappa\alpha\mu} \\ &-\epsilon_{\alpha,\lambda\kappa\mu} - \epsilon_{\lambda,\alpha\kappa\mu} - \epsilon_{\kappa,\mu\alpha\lambda} - \epsilon_{\mu,\kappa\alpha\lambda}\big) = R^{\rm LIF}_{\alpha\kappa\lambda\mu}\,. \end{aligned} \tag{12.94}$$

Hence

$$R^{\rm TT}_{\alpha\kappa\lambda\mu} = R^{\rm LIF}_{\alpha\kappa\lambda\mu}\,, \tag{12.95}$$

which shows that the Riemann tensor is *invariant* (and not only covariant) under the infinitesimal coordinate transformation 12.72 between a generic LIF and the TT frame [3]. Due to this property, we can solve Eq. 12.90, which holds in the LIF centered in A, using the Riemann tensor 12.93 computed in the TT gauge

$$\frac{1}{c^2}\frac{d^2\delta\xi^\mu}{dt^2} = R^{\rm TT\,\mu}{}_{00\gamma}\delta\xi^\gamma\,. \tag{12.96}$$

We stress that in Eq. 12.96 the Riemann tensor is written in terms of the TT metric $h^{\rm TT}_{\mu\nu}$, while $\delta\xi^\alpha$ is the coordinate separation in the LIF.

Assuming that the wave travels along ξ^1, the only non-vanishing components of $h^{\rm TT}_{\mu\nu}$ are those with $\mu,\nu = 2,3$, and these depend on ξ^0 and ξ^1; for a progressive wave $h^{\rm TT}_{\mu\nu,0} = h^{\rm TT}_{\mu\nu,1}$, and using Eq. 12.93 it is easy to show that the non-vanishing components of the Riemann tensor are

$$R^{\rm TT}_{i00m} = \frac{1}{2}h^{\rm TT}_{im,00}\,, \qquad i,m = 2,3\,. \tag{12.97}$$

It follows that

$$R^{\rm TT\,\mu}{}_{00m} = \eta^{\mu i}R^{\rm TT}_{i00m} = \frac{1}{2}\,\eta^{\mu i}\,\frac{1}{c^2}\frac{\partial^2 h^{\rm TT}_{im}}{\partial t^2}\,, \qquad i,m = 2,3\,, \tag{12.98}$$

and Eq. 12.96 becomes

$$\frac{d^2\delta\xi^\mu}{dt^2} = \frac{1}{2}\,\eta^{\mu i}\,\frac{\partial^2 h^{\rm TT}_{im}}{\partial t^2}\delta\xi^m\,. \tag{12.99}$$

For $t \le 0$ the two particles are at rest relative to each other, and consequently

$$t \le 0 \qquad \delta\xi^j = \delta\xi^j_0\,. \tag{12.100}$$

Since $h^{\rm TT}_{\mu\nu}$ is a small perturbation, when the wave arrives the relative position of the particles changes only by infinitesimal quantities, therefore we put

$$t > 0 \qquad \delta\xi^j(t) = \delta\xi^j_0 + \delta\xi^j_1(t), \qquad |\delta\xi^j_1| \ll |\delta\xi^j_0|\,, \tag{12.101}$$

i.e. $\delta\xi^j_1(t)$ is a small perturbation of the initial separation $\delta\xi^j_0$. Substituting Eq. 12.101 in Eq. 12.99, remembering that $\delta\xi^j_0$ is constant, and retaining only terms of order $O(h)$, Eq. 12.99 becomes

$$\frac{d^2\delta\xi^j_1}{dt^2} = \frac{1}{2}\,\eta^{ji}\frac{\partial^2 h^{\rm TT}_{ik}}{\partial t^2}\delta\xi^k_0\,. \tag{12.102}$$

The solution of this equation (with the condition $\delta\xi^j = \frac{d}{dt}\delta\xi^j = 0$ for $t \le 0$) is

$$\delta\xi^j = \delta\xi^j_0 + \frac{1}{2}\,\eta^{ji}\,h^{\rm TT}_{ik}\,\delta\xi^k_0\,. \tag{12.103}$$

If the wave propagates along ξ^1 only the components $h_{22} = -h_{33}, h_{23} = h_{32}$ are different from zero, and Eq. 12.103 yields

$$\begin{aligned}
\delta\xi^1 &= \delta\xi^1_0 + \frac{1}{2}\,\eta^{11}h^{\rm TT}_{1k}\,\delta\xi^k_0 = \delta\xi^1_0 \\
\delta\xi^2 &= \delta\xi^2_0 + \frac{1}{2}\,\eta^{22}h^{\rm TT}_{2k}\,\delta\xi^k_0 = \delta\xi^2_0 + \frac{1}{2}\left(h^{\rm TT}_{22}\,\delta\xi^2_0 + h^{\rm TT}_{23}\,\delta\xi^3_0\right) \\
\delta\xi^3 &= \delta\xi^3_0 + \frac{1}{2}\,\eta^{33}h^{\rm TT}_{3k}\,\delta\xi^k_0 = \delta\xi^3_0 + \frac{1}{2}\left(h^{\rm TT}_{32}\,\delta\xi^2_0 + h^{\rm TT}_{33}\,\delta\xi^3_0\right)\,.
\end{aligned} \tag{12.104}$$

[3] We leave as an exercise for the reader to show that this property is more generic: the Riemann tensor linearized on a flat spacetime and written in any frame is invariant (rather than only covariant) under an infinitesimal coordinate transformation 12.72.

Thus, the particles are accelerated only in the plane orthogonal to the direction of propagation, and this clearly shows the transverse nature of gravitational waves.

Let us now study the effect of the polarization of the wave, and to simplify the notation let us put $(\xi^1, \xi^2, \xi^3) = (x, y, z)$. Let us consider a progressive plane wave (see Eq. 12.59) with frequency ω and period $P = 2\pi/\omega$. In the TT gauge, its non-vanishing components are (we omit in the following the superscript TT)

$$\begin{aligned} h_{yy} &= -h_{zz} = 2\Re\left\{A_+ e^{i\omega(t-\frac{x}{c})}\right\}, \\ h_{yz} &= h_{zy} = 2\Re\left\{A_\times e^{i\omega(t-\frac{x}{c})}\right\}, \end{aligned} \tag{12.105}$$

where A_+ and $A_\times$ are complex constants. Consider two particles, **1** and **2**, initially located at $(0, y_0, 0)$ and $(0, 0, z_0)$, respectively, as indicated in the first plot of the upper panel of Fig. 12.3. The particle A to which the LIF is attached is sitting at the origin and is not indicated in Fig. 12.3 and in the following figures. Thus the space components of the vectors which separate A from the two particles coincide with the particles coordinates (see Fig. 12.2). Let us first consider the polarization '+', i.e.

$$A_+ \neq 0 \qquad \text{and} \qquad A_\times = 0\,. \tag{12.106}$$

Assuming the initial phase is zero (i.e. A_+ is real[4]) Eq. 12.105 gives

$$h_{yy} = -h_{zz} = 2A_+ \cos\omega\left(t - \frac{x}{c}\right), \qquad h_{yz} = h_{zy} = 0\,, \tag{12.107}$$

and Eq. 12.104 written for the two particles for $t \geq 0$ gives

$$\begin{aligned} &\text{particle } \mathbf{1}: \quad z = 0, \quad y = y_0 + \frac{1}{2}h_{yy}\, y_0 = y_0\left[1 + A_+ \cos\omega\left(t - \frac{x}{c}\right)\right], \\ &\text{particle } \mathbf{2}: \quad y = 0, \quad z = z_0 + \frac{1}{2}h_{zz}\, z_0 = z_0\left[1 - A_+ \cos\omega\left(t - \frac{x}{c}\right)\right]. \end{aligned} \tag{12.108}$$

Since for $t \leq 0$ the particles are at $(y_0, 0)$ and $(0, z_0)$, at $t = 0$ $\omega(t - \frac{x}{c}) = \frac{\pi}{2}$. Therefore, at t=0

$$\begin{aligned} &\text{particle } \mathbf{1}: \quad z = 0, \qquad y = y_0\,, \\ &\text{particle } \mathbf{2}: \quad y = 0, \qquad z = z_0\,. \end{aligned} \tag{12.109}$$

After a quarter of a period, $t = P/4$, $\omega\left(t - \frac{x}{c}\right) = \pi$, and

$$\begin{aligned} &\text{particle } \mathbf{1}: \quad z = 0, \qquad y = y_0\,(1 - A_+)\,, \\ &\text{particle } \mathbf{2}: \quad y = 0, \qquad z = z_0\,(1 + A_+)\,. \end{aligned} \tag{12.110}$$

After half a period, $t = P/2$, $\omega\left(t - \frac{x}{c}\right) = \frac{3}{2}\pi$ and

$$\begin{aligned} &\text{particle } \mathbf{1}: \quad z = 0, \qquad y = y_0\,, \\ &\text{particle } \mathbf{2}: \quad y = 0, \qquad z = z_0\,. \end{aligned} \tag{12.111}$$

After three quarters of a period, $t = 3P/4$

$$\begin{aligned} &\text{particle } \mathbf{1}: \quad z = 0, \qquad y = y_0\,(1 + A_+)\,, \\ &\text{particle } \mathbf{2}: \quad y = 0, \qquad z = z_0\,(1 - A_+)\,. \end{aligned} \tag{12.112}$$

[4]The phase of the complex amplitude $A_+ = |A_+|e^{i\phi_0}$ is the initial phase of the wave, since $A_+ e^{i\omega(t-\frac{x}{c})} = |A_+|e^{i(\omega(t-\frac{x}{c})+\phi_0)}$.

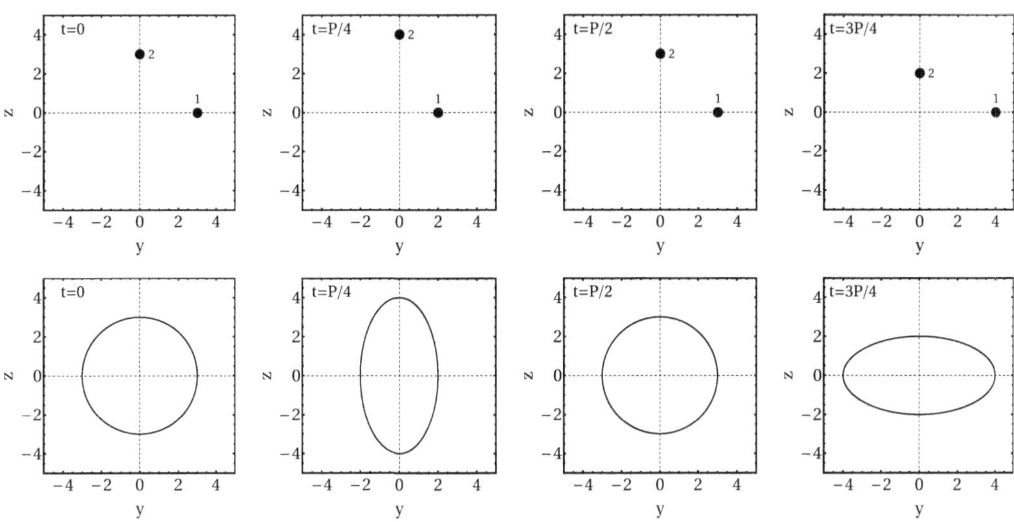

Figure 12.3: In the upper panel we show the displacement of two particles in the (y, z) plane due to the passage of a gravitational wave propagating along x with polarization '+'. Initially the particles **1** and **2** are located at $(3, 0)$ and at $(0, 3)$, respectively. We set the gravitational wave amplitude to $A_+ = 1$ (as discussed in the next chapters, this deformation is enormously exaggerated relative to the real effects of a gravitational wave detectable on Earth). The time snapshots correspond to $t = 0$, $t = P/4$, $t = P/2$, $t = 3P/2$ (left upper corner of each plot), where P is the wave period. In the lower panel the deformation of a small ring of particles in the (y, z) plane produced by the same wave is shown for the same values of time.

Similarly, if we consider a small ring of particles centered at the origin, the effect of the wave with polarization '+' is that of deforming the circle in an ellipse prolate (after half a period) and oblate (after three quarters of a period), as shown in the lower panel of Fig. 12.3. Note that in this figure the effect of the gravitational wave is enormously exaggerated: as we shall discuss in Chapter 14, for typical gravitational waves detectable on Earth, a ring of particles with size roughly 1 meter would be deformed by 10^{-21} m only!

Let us now study the effect of the polarization '×', i.e. $A_\times \neq 0$ and $A_+ = 0$. Consider four particles **1**, **2**, **3**, and **4**, initially located at $(0, y_0, z_0)$, $(0, -y_0, z_0)$, $(0, -y_0, -z_0)$, $(0, y_0, -z_0)$, respectively, as indicated in the first plot of the upper panel of Fig. 12.4.

Assuming the initial phase is zero (i.e. $A_\times$ is real) Eq. 12.105 gives

$$h_{yy} = h_{zz} = 0, \qquad h_{yz} = h_{zy} = 2A_\times \cos\omega\left(t - \frac{x}{c}\right) \tag{12.113}$$

and Eq. 12.104 written for each of the four particles for $t \geq 0$ gives

$$y = y_0 + \frac{1}{2}h_{yz}\, z_0 = y_0 + A_\times \cos\omega\left(t - \frac{x}{c}\right) z_0 \tag{12.114}$$

$$z = z_0 + \frac{1}{2}h_{zy}\, y_0 = z_0 + A_\times \cos\omega\left(t - \frac{x}{c}\right) y_0\,. \tag{12.115}$$

If, for simplicity, we assume that $y_0 = z_0 = r$, at $t = 0$ the positions of the four particles

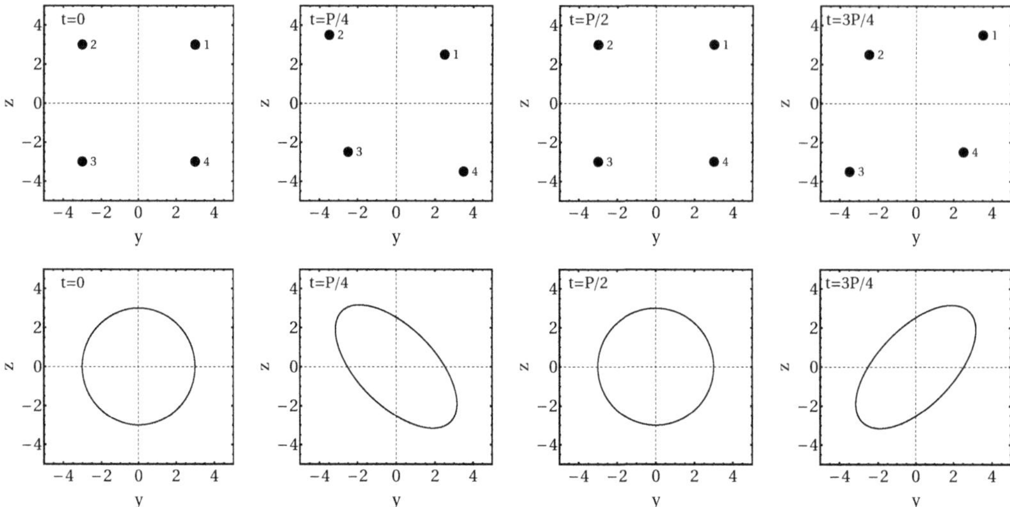

Figure 12.4: Upper panel: displacement produced by a wave propagating along x with polarization '$\times$', on four particles in the (y,z) plane. The particles are initially located at $(y_0, z_0) = (3,3)$, $(y_0, z_0) = (-3,3)$, $(y_0, z_0) = (-3,3)$, and $(y_0, z_0) = (3,-3)$, respectively, and the gravitational wave amplitude is $A_\times = 1$. As in Fig. 12.3 the time snapshots correspond to $t = 0$, $t = P/4$, $t = P/2$, $t = 3P/2$. Lower panel: deformation produced by the same wave on a small ring of particles in the (y,z) plane, for the same values of time.

are:

$$\begin{aligned} &\text{particle } \mathbf{1}: \quad y = r \qquad z = r\,, \\ &\text{particle } \mathbf{2}: \quad y = -r \qquad z = r\,, \\ &\text{particle } \mathbf{3}: \quad y = -r \qquad z = -r\,, \\ &\text{particle } \mathbf{4}: \quad y = r \qquad z = -r\,. \end{aligned} \tag{12.116}$$

At $t = P/4$, $\cos\omega(t - \frac{x}{c}) = -1$, and

$$\begin{aligned} &\text{particle } \mathbf{1}: \quad y = r\,(1 - A_\times) \qquad z = r\,(1 - A_\times)\,, \\ &\text{particle } \mathbf{2}: \quad y = -r\,(1 + A_\times) \qquad z = r\,(1 + A_\times)\,, \\ &\text{particle } \mathbf{3}: \quad y = -r\,(1 - A_\times) \qquad z = -r\,(1 - A_\times)\,, \\ &\text{particle } \mathbf{4}: \quad y = r\,(1 + A_\times) \qquad z = -r\,(1 + A_\times)\,. \end{aligned} \tag{12.117}$$

At $t = P/2$, $\cos\omega(t - \frac{x}{c}) = 0$ and the particles go back to the initial positions. At $t = 3P/4$, $\cos\omega(t - \frac{x}{c}) = 1$ and

$$\begin{aligned} &\text{particle } \mathbf{1}: \quad y = r\,(1 + A_\times) \qquad z = r\,(1 + A_\times)\,, \\ &\text{particle } \mathbf{2}: \quad y = -r\,(1 - A_\times) \qquad z = r\,(1 - A_\times)\,, \\ &\text{particle } \mathbf{3}: \quad y = -r\,(1 + A_\times) \qquad z = -r\,(1 + A_\times)\,, \\ &\text{particle } \mathbf{4}: \quad y = r\,(1 - A_\times) \qquad z = -r\,(1 - A_\times)\,. \end{aligned} \tag{12.118}$$

The snapshots of the particles position at these times are shown in the upper panel of Fig. 12.4.

It follows that a small ring of particles centered at the origin will again become an ellipse after a quarter of a period, but rotated at 45° with respect to the case previously analysed (see the lower panel of Fig. 12.3). In conclusion, we can define A_+ and $A_\times$ as the **polarization amplitudes** of the wave. The wave will be linearly polarized when only one of the two polarization amplitudes is different from zero. The effect produced by a general wave containing both polarizations will be a superposition of the effects shown in Fig. 12.3 and Fig. 12.4.

12.8 GRAVITATIONAL WAVES AND MICHELSON INTERFEROMETERS

The Michelson interferometer is a device consisting of two tubes ("arms") orthogonal to each other. A source sends a light beam to a beam splitter (e.g. a half-silvered mirror), and the two parts of the beam are reflected by mirrors put at the end of the arms (see Fig. 12.5). These beams go back and forth along the arms, and when they reach the screen (the detector) they produce the interference pattern. During the 19th century, this instrument played a fundamental role in the crisis of classical physics, since it was used to prove that the speed of light is a universal constant, eventually leading to the formulation of Special Relativity. After one century and a half, in 2015, a similar device was used in the LIGO experiment to detect – for the first time – the gravitational waves emitted by an astrophysical source [1]: the coalescence of a binary system composed by two black holes (see Chapter 14). In the following years, LIGO (and a similar interferometric detector, Virgo) detected several gravitational wave signals emitted by compact sources like neutron stars and black holes.

The LIGO and Virgo interferometers are of course much more sophisticated instruments than that used by Michelson in the 19th century: for instance the light beams are laser beams, they cross the arms back and forth hundreds of times before reaching the detector, where a photodetector replaces the screen; moreover in order to detect the incredibly small variation of the interference pattern induced by a gravitational wave, the interferometers must be accurately isolated from any source of noise. However, they work on the same basic principles as the Michelson instrument.

Let us assume, for instance, that the arms of the interferometer lie in the y and z directions, and that a gravitational wave propagates in the x direction, with polarization '+' in the plane yz (see Eq. 12.107). When the wave crosses the interferometer, the proper lengths of the two arms change, and the paths of the light rays change as well. The difference of the paths determines a shift in the interference pattern on the detector.

This description may appear too simplistic. One could remark, for instance, that the number of light wavelengths contained in an arm does not change when the gravitational wave passes through the interferometer, because both the arm and the wavelength are stretched by the same amount. Does the gravitational wave affect at all the interference pattern?

The answer to this question is "yes!", because the interference pattern is affected by the *time delay* in the light propagation produced by the gravitational wave. In order to estimate this delay, we describe the interferometer and the gravitational wave (with '+' polarization) in the TT gauge (see Eq. 12.82):

$$ds^2 = (\eta_{\mu\nu} + h^{TT}_{\mu\nu})dx^\mu dx^\nu = -c^2dt^2 + dx^2 + (1 + h_+)dy^2 + (1 - h_+)dz^2 \,. \tag{12.119}$$

Let l_0 be the proper length of the two arms (between the beam splitter and the mirrors), measured in the TT frame, before the arrival of the wave, and let ω be the frequency of the gravitational wave.

If we assume that the wavelength of the impinging wave, $\lambda_{GW} = 2\pi c/\omega$, is much larger than the arm-length l_0, i.e. that $\lambda_{GW} \gg l_0$, the gravitational perturbation h_+ can be

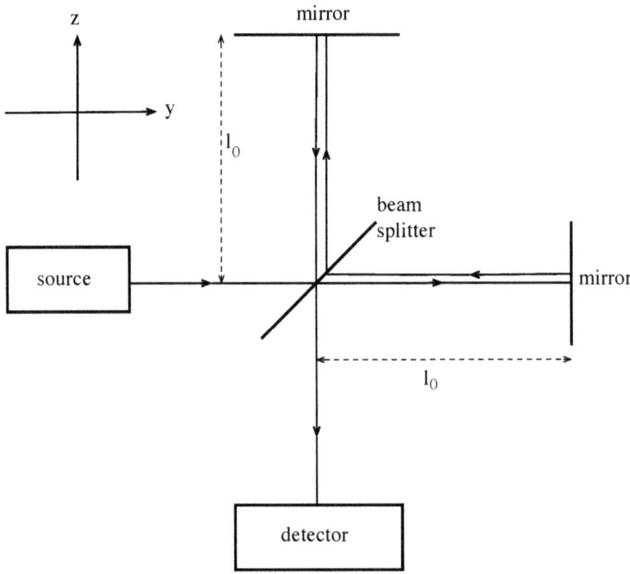

Figure 12.5: Schematic structure of a Michelson interferometer.

considered as constant as the light ray crosses the arm. A light ray moving in the y direction follows a null geodesic with $c^2dt^2 = (1 + h_+)dy^2$, thus $dt = c^{-1}(1 + h_+/2)\,dy + O(h^2)$ and the time to cross back and forth the y-arm is

$$t_{(y)} = \left(1 + \frac{h_+}{2}\right)\frac{2l_0}{c}\,. \tag{12.120}$$

A light ray moving in the z direction, instead, follows a null geodesic with $c^2dt^2 = (1 - h_+)dz^2$, therefore it crosses back and forth the z-arm in the time

$$t_{(z)} = \left(1 - \frac{h_+}{2}\right)\frac{2l_0}{c}\,. \tag{12.121}$$

Therefore, although – as discussed in Sec. 12.6 – the coordinate positions of the arm points in the TT gauge are not affected by the gravitational wave, the time needed to cross the arms *is* affected. When the rays join in the detector, there is a time delay

$$\Delta t = t_{(y)} - t_{(z)} = \frac{2l_0}{c}h_+ \tag{12.122}$$

between them, which produces a shift $\sim c\Delta t = 2l_0h_+$ in the interference fringes (this shift was measured on a screen in the original Michelson-Morley experiment, while in modern gravitational wave interferometers it is measured with a photodetector). If the amplitude of the wave is large enough, as we shall discuss in the next chapters, this shift can be directly measured.

Strictly speaking, the condition $l_0 \ll \lambda_{\rm GW}$ is only marginally satisfied by the interferometers LIGO and Virgo, for which $l_0 \lesssim \lambda_{\rm GW}$. In this case Eq. 12.122 gives an approximate estimate of the time delay. A more detailed analysis can be found in Chapter 9 of [82], where it is shown that the interferometer is most sensitive when

$$l_0 \simeq \frac{1}{4}\lambda_{\rm GW} \,. \tag{12.123}$$

For instance, for a signal with frequency $\sim$ 100 Hz, well within the sensitivity band of LIGO and Virgo, $\lambda_{\rm GW} \sim 3000$ km, and thus the optimal length is $l_0 \sim 750$ km. This is close to the effective length of the arms, i.e. to their actual length (4 km) multiplied by the times the light ray goes back and forth before reaching the photodetector (a few hundreds).

CHAPTER 13

Gravitational waves in the quadrupole approximation

In this chapter we will study the gravitational waves emitted by dynamical systems described by a stress-energy tensor $T^{\mu\nu}$, on the assumption that the gravitational field is weak and that the speed of the bodies is small compared to the speed of light. This assumption is called the *quadrupole approximation* since, as we shall show, in this regime the emitted radiation depends only on the quadrupole moment of the source. The quadrupole approximation allows to estimate the gravitational energy and the waveforms emitted by these systems. We remark that we shall use Latin indices for the space components of tensors, and Greek indices for the spacetime components. Furthermore, unless otherwise stated, in this Chapter we shall keep physical units, writing G and c explicitly.

13.1 THE WEAK-FIELD, SLOW-MOTION APPROXIMATION

In Sec. 12.2 we showed that in the weak-field limit, i.e. when the metric tensor can be written as $g_{\mu\nu} = \eta_{\mu\nu} + h_{\mu\nu}$ with $|h_{\mu\nu}| \ll 1$, and with a suitable choice of the gauge, Einstein's equations reduce to Eq. 12.32, which we here re-write for convenience

$$\Box_F \bar{h}_{\mu\nu} = -\frac{16\pi G}{c^4} T_{\mu\nu} \tag{13.1}$$

$$\frac{\partial}{\partial x^\mu} \bar{h}^\mu{}_\nu = 0 \,.$$

The second equation is the *harmonic gauge* condition.

We shall now solve these equations assuming that the region (of size ϵ) where the source is confined, namely

$$T_{\mu\nu} \begin{cases} \neq 0 & \text{if} \quad |x^i| < \epsilon \\ = 0 & \text{otherwise} \end{cases} , \tag{13.2}$$

is much smaller than the wavelength of the emitted radiation, $\lambda_{\rm GW} = \frac{2\pi c}{\omega}$. This implies that

$$\frac{2\pi c}{\omega} \gg \epsilon \quad \to \quad \epsilon\omega \ll c \quad \to \quad v_{\rm typical} \ll c \,. \tag{13.3}$$

In other words, the typical velocities of the system, $v_{\rm typical} \sim \epsilon\omega$, are much smaller than the speed of light; for this reason this is called the *slow-motion approximation.*

Let us consider the first of Eqs. 13.1,

$$\Box_F \bar{h}_{\mu\nu} = -\frac{16\pi G}{c^4} T_{\mu\nu} \,, \tag{13.4}$$

where we remind that

$$\bar{h}_{\mu\nu} = h_{\mu\nu} - \frac{1}{2}\eta_{\mu\nu} h \qquad \text{and} \qquad \Box_F = \eta^{\mu\nu}\frac{\partial}{\partial x^\mu}\frac{\partial}{\partial x^\nu} = -\frac{1}{c^2}\frac{\partial^2}{\partial t^2} + \nabla^2 . \tag{13.5}$$

As discussed in Sec. 12.2 the general solution of the inhomogeneous d'Alembertian equation 13.4 can be written in terms of *retarded potentials*,

$$\bar{h}_{\mu\nu}(t, \mathbf{x}) = \frac{4G}{c^4}\int_V \frac{T_{\mu\nu}(t - \frac{|\mathbf{x}-\mathbf{x}'|}{c}, \mathbf{x}')}{|\mathbf{x}-\mathbf{x}'|}\, d^3x' , \tag{13.6}$$

where $t - \frac{|\mathbf{x}-\mathbf{x}'|}{c}$ is the retarded time, and $\frac{|\mathbf{x}-\mathbf{x}'|}{c}$ is the time that the wave takes to travel from the source element at $\mathbf{x}'$ to the observer located at $\mathbf{x}$ (see Fig. 12.1). The above integral is performed over the entire three-dimensional source volume V.

By defining the Fourier transforms of both $\bar{h}_{\mu\nu}$ and $T_{\mu\nu}$,

$$\begin{aligned} T_{\mu\nu}(t, \mathbf{x}) &= \int_{-\infty}^{+\infty} T_{\mu\nu}(\omega, \mathbf{x}) e^{-i\omega t}\, d\omega , \\ \bar{h}_{\mu\nu}(t, \mathbf{x}) &= \int_{-\infty}^{+\infty} \bar{h}_{\mu\nu}(\omega, \mathbf{x}) e^{-i\omega t}\, d\omega , \end{aligned} \tag{13.7}$$

Eq. 13.6 gives

$$\bar{h}_{\mu\nu}(\omega, \mathbf{x}) = \frac{4G}{c^4}\int_V T_{\mu\nu}(\omega, \mathbf{x}')\frac{e^{i\omega\frac{|\mathbf{x}-\mathbf{x}'|}{c}}}{|\mathbf{x}-\mathbf{x}'|} d^3x' . \tag{13.8}$$

Indeed, multiplying both sides of Eq. 13.8 with $e^{-i\omega t}$ and integrating in $d\omega$ we obtain Eq. 13.6.

The frequencies ω contributing to the Fourier transform are the typical pulsations of the gravitational wave, which – in the slow-motion approximation – satisfy the condition $\lambda_{\rm GW} = \frac{2\pi c}{\omega} \gg \epsilon$. Since, for $\mathbf{x}'$ within the source, $|\mathbf{x}'| < \epsilon$, it follows that $\omega|\mathbf{x}'|/c \ll 2\pi$. Therefore[1],

$$\frac{e^{i\omega\frac{|\mathbf{x}-\mathbf{x}'|}{c}}}{|\mathbf{x}-\mathbf{x}'|} \simeq \frac{e^{i\omega\frac{|\mathbf{x}|}{c}}}{|\mathbf{x}|} = \frac{e^{i\omega\frac{r}{c}}}{r} , \tag{13.9}$$

where we have set $r = |\mathbf{x}|$. Thus, Eq. 13.8 can be approximated as

$$\bar{h}_{\mu\nu}(\omega, r) = \frac{4G}{c^4}\frac{e^{i\omega\frac{r}{c}}}{r}\int_V T_{\mu\nu}(\omega, \mathbf{x}')\, d^3x' . \tag{13.10}$$

The solution in the time domain is the inverse Fourier transform of the above expression,

$$\bar{h}_{\mu\nu}(t, r) = \frac{4G}{rc^4}\int_V T_{\mu\nu}\left(t - \frac{r}{c}, \mathbf{x}'\right)\, d^3x' . \tag{13.11}$$

This is the gravitational wave signal emitted by the source, to leading order in the weak-field, slow-motion approximation.

[1]Note that while $|\mathbf{x}-\mathbf{x}'| \simeq |\mathbf{x}|$ is always satisfied as long as $|\mathbf{x}'| \ll |\mathbf{x}|$, the approximation $e^{i\omega\frac{|\mathbf{x}-\mathbf{x}'|}{c}} \simeq e^{i\omega\frac{|\mathbf{x}|}{c}}$ requires the stronger condition $\omega|\mathbf{x}'|/c \ll 2\pi$. Indeed, given two real numbers a, A with $a \ll A$, we always have $A + a \simeq A$, but $\sin(A + a) \simeq \sin(A)$ only if the condition $a \ll 2\pi$ is also satisfied.

In the next section we shall further simplify the integral in Eq. 13.11. Meanwhile, we recall that, as shown in Sec. 12.2, the solution 13.6 (and thus the solution 13.11) automatically satisfies the harmonic gauge condition

$$\frac{\partial}{\partial x^\mu}\bar{h}^\mu{}_\nu = 0\,. \tag{13.12}$$

We also note that, in order to extract the physical components of the wave, we still have to transform to the TT gauge. This will be done explicitly in Sec. 13.3.

We remark that Eq. 13.11 has been derived on *two* very strong assumptions: weak field ($g_{\mu\nu} = \eta_{\mu\nu} + h_{\mu\nu}$, with $|h_{\mu\nu}| \ll \eta_{\mu\nu}|$) and slow motion ($v_{\text{typical}} \ll c$). For this reason that expression has to be considered as an estimate of the radiation emitted by the source; it fails in various cases of interest, namely when the sources move with relativistic velocities or in spacetime regions where the gravitational field is strong.

13.2 THE QUADRUPOLE FORMULA

In order to simplify the integral in Eq. 13.11 we shall use the conservation law satisfied by $T_{\mu\nu}$ in the weak-field approximation

$$\frac{\partial T^{\mu\nu}}{\partial x^\nu} = 0, \qquad \to \qquad \frac{1}{c}\frac{\partial T^{\mu 0}}{\partial t} = -\frac{\partial T^{\mu k}}{\partial x^k}, \qquad \mu = 0,\ldots,3\,, \quad k = 1,2,3\,. \tag{13.13}$$

Let us integrate this equation over the source volume V, for a generic index μ,

$$\frac{1}{c}\frac{\partial}{\partial t}\int_V T^{\mu 0} d^3x = -\int_V \frac{\partial T^{\mu k}}{\partial x^k} d^3x\,. \tag{13.14}$$

By Gauss' theorem (see Box 5-D), the integral over the volume is equal to the flux of $T^{\mu k}$ across the surface ∂V enclosing that volume; thus the right-hand side becomes

$$\int_V \frac{\partial T^{\mu k}}{\partial x^k} d^3x = \int_{\partial V} T^{\mu k} dS_k\,, \tag{13.15}$$

where dS is the surface element on the boundary ∂V with normal three-vector n^k, and $dS_k = n_k dS$.

By definition, $T^{\mu\nu} = 0$ on ∂V and consequently the surface integral vanishes; thus

$$\frac{1}{c}\frac{\partial}{\partial t}\int_V T^{\mu 0} d^3x = 0\,, \qquad \to \qquad \int_V T^{\mu 0} d^3x = \text{const}\,. \tag{13.16}$$

From Eq. 13.11 it follows that

$$\bar{h}^{\mu 0} = \text{const}\,. \tag{13.17}$$

Since we are interested in the time-dependent part of the field (see footnote 2 in Chapter 12) we shall put

$$\bar{h}^{\mu 0} = 0\,. \tag{13.18}$$

13.2.1 The tensor-virial theorem

We shall now prove the **tensor-virial theorem** which establishes that, in the weak-field limit,

$$\frac{1}{c^2}\frac{\partial^2}{\partial t^2}\int_V T^{00}\, x^k\, x^n\; d^3x = 2\int_V T^{kn}\, d^3x, \qquad k,n = 1,2,3\,. \tag{13.19}$$

Let us consider the space components of the conservation law 13.13

$$\frac{\partial T^{n0}}{\partial x^0} = -\frac{\partial T^{ni}}{\partial x^i}, \qquad i, n = 1, 2, 3 \,; \tag{13.20}$$

multiplying both members by x^k and integrating over the source volume we find

$$\begin{aligned} \frac{1}{c}\frac{\partial}{\partial t}\int_V T^{n0}\, x^k\, d^3x &= -\int_V \frac{\partial T^{ni}}{\partial x^i}\, x^k\, d^3x \\ &= -\left[\int_V \frac{\partial\left(T^{ni}\, x^k\right)}{\partial x^i}\, d^3x - \int_V T^{ni}\,\frac{\partial x^k}{\partial x^i}\, d^3x\right] \\ &= -\int_{\partial V}\left(T^{ni}\, x^k\right)\, dS_i + \int_V T^{nk}\, d^3x\,. \end{aligned} \tag{13.21}$$

As before $\int_{\partial V}\left(T^{ni}\, x^k\right)\, dS_i = 0$, therefore

$$\frac{1}{c}\frac{\partial}{\partial t}\int_V T^{n0}\, x^k\, d^3x = \int_V T^{nk}\, d^3x\,. \tag{13.22}$$

Since T^{nk} is symmetric we can rewrite this equation in the following form

$$\frac{1}{2c}\frac{\partial}{\partial t}\int_V \left(T^{n0}\, x^k + T^{k0}\, x^n\right)\, d^3x = \int_V T^{nk}\, d^3x\,. \tag{13.23}$$

Let us now consider the 0-component of the conservation law

$$\frac{1}{c}\frac{\partial T^{00}}{\partial t} + \frac{\partial T^{0i}}{\partial x^i} = 0\,; \tag{13.24}$$

multiplying by $x^k x^n$ and integrating over V we get

$$\begin{aligned} \frac{1}{c}\frac{\partial}{\partial t}\int_V T^{00}\, x^k\, x^n\, d^3x &= -\int_V \frac{\partial T^{0i}}{\partial x^i}\, x^k\, x^n\, d^3x \\ &= -\left[\int_V \frac{\partial\left(T^{0i}\, x^k\, x^n\right)}{\partial x^i}\, d^3x - \int_V \left(T^{0i}\,\frac{\partial x^k}{\partial x^i}\, x^n + T^{0i}\, x^k\,\frac{\partial x^n}{\partial x^i}\right)\, d^3x\right] \\ &= -\int_{\partial V}\left(T^{0i}\, x^k\, x^n\right)\, dS_i + \int_V \left(T^{0k}\, x^n + T^{0n}\, x^k\right)\, d^3x\,; \end{aligned} \tag{13.25}$$

the first integral vanishes and this equation becomes

$$\frac{1}{c}\frac{\partial}{\partial t}\int_V T^{00}\, x^k\, x^n\, d^3x = \int_V \left(T^{0k}\, x^n + T^{0n}\, x^k\right)\, d^3x\,. \tag{13.26}$$

If we now differentiate with respect to x^0

$$\frac{1}{c^2}\frac{\partial^2}{\partial t^2}\int_V T^{00}\, x^k\, x^n\, d^3x = \frac{1}{c}\frac{\partial}{\partial t}\int_V \left(T^{0k}\, x^n + T^{0n}\, x^k\right)\, d^3x\,, \tag{13.27}$$

and, using Eq. 13.23, we finally prove the tensor-virial theorem 13.19:

$$\frac{1}{c^2}\frac{\partial^2}{\partial t^2}\int_V T^{00}\, x^k\, x^n\, d^3x = 2\int_V T^{kn}\, d^3x\,. \tag{13.28}$$

13.2.2 The quadrupole moment tensor

In the weak-field limit, the metric perturbation $h_{\mu\nu}$ can be considered as a (tensor) field living in the Minkowski spacetime **M**. Indeed, in this approximation we raise and lower the indices of tensors, to order $O(h)$, using Minkowski's metric, and this applies also to $h_{\mu\nu}$. In addition, the stress-energy tensor satisfies the conservation law of Special Relativity, $T^{\mu\nu}{}_{,\nu} = 0$.

In the following we shall consider (at each value of the time coordinate t) the $t = const$ three-dimensional submanifold of **M**, i.e. the three-dimensional Euclidean space, whose metric – in Cartesian coordinates $\{x^i\}_{i=1,2,3}$ – is simply δ_{ij}; therefore, we shall not distinguish between covariant and contravariant indices. In this space, we introduce the **quadrupole moment tensor** of a dynamical system

$$q^{kn}(t) = \frac{1}{c^2} \int_V T^{00}(t, \mathbf{x})\, x^k\, x^n d^3x, \qquad k, n = 1, 2, 3\,; \tag{13.29}$$

note that q^{kn} *is a function of time only.*

The left-hand side of the tensor-virial theorem, Eq. 13.28, is the second time derivative of the quadrupole moment of the system. Therefore,

$$\int_V T^{kn}(t, \mathbf{x})\, d^3x = \frac{1}{2} \frac{d^2}{dt^2} q^{kn}(t)\,. \tag{13.30}$$

Using this equation and Eq. 13.18, the solution to Einstein's equations for the perturbation $h_{\mu\nu}$, given by Eq. 13.11, can be written as

$$\begin{aligned} \bar{h}^{\mu 0} &= 0\,, \\ \bar{h}^{ik}(t, r) &= \frac{2G}{c^4 r} \frac{d^2}{dt^2}\, q^{ik}\left(t - \frac{r}{c}\right), \end{aligned} \tag{13.31}$$

where $\mu = 0, \ldots, 3$ and $i, k = 1, 2, 3$. Eq. 13.31, which is often called **quadrupole formula**, describes the gravitational wave emitted by a gravitating system evolving in time. Note that any form of mass and energy can be a source of gravitational waves, as long as the second time derivative of the quadrupole moment of the system is non-vanishing. Furthermore, in the weak-field, slow-motion approximation the metric perturbation depends only on the T^{00} component of the stress energy tensor; the other components do not source gravitational waves in this approximation.

The factor $\frac{2G}{c^4}$ affects the intensity of the source term in Eq. 13.31 and is extremely small:

$$\frac{G}{c^4} \sim 8 \times 10^{-50}\, \frac{\mathrm{s}^2}{\mathrm{g\,cm}}\,. \tag{13.32}$$

This is the reason why gravitational waves are typically extremely weak.

It is important to remark that the quadrupole formula holds under the conditions of weak gravitational field (not only far away from the source, but also in the source itself) and of slow motion. The former condition implies that $T^{\mu\nu}{}_{,\nu} = 0$, i.e. that the motion of bodies is dominated by non-gravitational forces. The latter is equivalent to the requirement that the source is much smaller than the wavelength of the emitted radiation.

13.2.3 Absence of monopolar and dipolar gravitational waves

General Relativity predicts that gravitational waves do not include monopolar and dipolar contributions. Let us discuss this point in more detail by comparing the gravitational case with that of electrodynamics. According to Maxwell's theory, a system of accelerated charges emits *dipole* radiation [67], the flux of which depends on the *second time derivative* of the electric dipole moment

$$\vec{d}_{\rm EM} = \sum_i q_i \vec{r}_i\,, \tag{13.33}$$

where q_i and $\vec{r}_i$ are the charge and location of the i-th particle. If $d^2\vec{d}_{\rm EM}/dt^2 = \vec{0}$, the system does not emit electromagnetic waves.

For an isolated system of masses we can define a gravitational dipole moment

$$\vec{d}_{\rm G} = \sum_i m_i \vec{r}_i\,, \tag{13.34}$$

which satisfies the conservation law of the total momentum of isolated systems

$$\frac{d}{dt}\vec{d}_{\rm G} = \vec{0}\,. \tag{13.35}$$

Consequently *gravitational waves do not have a dipolar contribution.* Similarly, the conservation of the total energy implies the absence of a monopolar component (the same is true in electromagnetism due to the total charge conservation). Thus, the leading contribution is due to a time-dependent quadrupole moment.

It should be stressed that for a spherical or axisymmetric, stationary distribution of matter (or energy) the quadrupole moment is constant, even if the body is rotating. Thus, a spherical or axisymmetric star does not emit gravitational waves. The same is true for a star which collapses, or explodes, maintaining a spherically symmetric shape, or for a star that radially pulsates: they have a vanishing $\frac{d^2q^{ik}}{dt^2}$ and do not emit gravitational waves [2]. To produce gravitational waves the source must have a certain degree of asymmetry, as it occurs for instance in the non-radial pulsations of a star, in a non-spherical gravitational collapse, in the coalescence of massive bodies, in deformed rotating stars, etc.

13.3 HOW TO TRANSFORM TO THE TT GAUGE

The solution 13.31 describes a spherical wave far from the emitting source. Locally, it looks like a plane wave propagating along the direction of the unit vector $\mathbf{n}$ orthogonal to the wavefront

$$n^\alpha = (0, n^i), \qquad i = 1, 2, 3 \tag{13.36}$$

where

$$n^i = \frac{x^i}{r}\,. \tag{13.37}$$

As previously discussed, $\mathbf{n}$ is a vector in the three-dimensional Euclidean space, which is a $t = const$ submanifold of Minkowski's space, and n^i are its components in the Cartesian frame $\{x^i\}$.

In order to express this waveform in the TT gauge we shall make an infinitesimal coordinate transformation $x^{\mu\,\prime} = x^\mu + \epsilon^\mu$, with $\Box_F \epsilon^\mu = 0$, so that the harmonic gauge condition is preserved, as explained in Chapter 12. The conditions to impose on the perturbed metric are

$$\bar{h}'_{\alpha i}\, n^i = 0, \qquad \text{trasverse wave condition} \tag{13.38}$$

$$\bar{h}'_{\alpha\beta}\, \eta^{\alpha\beta} = 0, \qquad \text{traceless condition.} \tag{13.39}$$

[2] For spherically symmetric pulsating stars or for spherical collapse the absence of gravitational wave emission can also be seen as a consequence of Birkhoff's theorem (Sec. 9.6): a vacuum spherically symmetric spacetime must be static.

It should be mentioned that the transverse-wave condition implies $\bar{h}^{\mu 0} = 0$ $(\mu = 0, \ldots, 3)$ which we imposed in Eq. 13.18. Indeed, given the wave vector $k^\mu = k^0(1, n^i)$, we know from Eq. 12.63 that $k^\mu \bar{h}'_{\mu\nu} = 0$, i.e.

$$k^0 \bar{h}'_{0\nu} + k^0 n^i \bar{h}'_{i\nu} = 0\,. \tag{13.40}$$

The second term vanishes because of the transverse-wave condition, therefore

$$h'_{0\nu} = \bar{h}'_{0\nu} = 0\,. \tag{13.41}$$

We remind here that, as shown in Eq. 12.77, in the TT gauge $\bar{h}_{\mu\nu}$ and $h_{\mu\nu}$ coincide.

We shall now describe a procedure to project the wave in the TT gauge, which is equivalent to performing the coordinate transformation mentioned above. As a first step, we define the operator which projects a three-dimensional vector onto the plane orthogonal to the direction of **n**

$$P_{jk} \equiv \delta_{jk} - n_j n_k\,. \tag{13.42}$$

It is easy to verify that, for any vector V^j, $P_{jk}V^k$ is orthogonal to n^j, i.e. $(P_{jk}V^k)n^j = 0$, and that

$$P^j{}_k P^k{}_l V^l = P^j{}_l V^l\,. \tag{13.43}$$

Note that $P_{jk} = P_{kj}$, i.e. P_{jk} is *symmetric*. The projector is *transverse*, i.e.

$$n^j P_{jk} = 0\,. \tag{13.44}$$

As a second step, we define the **transverse-traceless projector**:

$$\mathcal{P}_{jkmn} \equiv P_{jm}P_{kn} - \frac{1}{2}P_{jk}P_{mn}\,, \tag{13.45}$$

which "extracts" the transverse-traceless part of a rank-two tensor on the three-dimensional Euclidean space. Indeed, using the definition 13.45, it is easy to check that this operator satisfies the following properties (we remind that Latin indices run from 1 to 3 and are raised and lowered by δ_{ij})

- $\mathcal{P}_{jklm} = \mathcal{P}_{lmjk}$
- $\mathcal{P}_{jklm} = \mathcal{P}_{kjml}$
- $\mathcal{P}_{jkmn}\mathcal{P}_{mnrs} = \mathcal{P}_{jkrs}$
- it is transverse:

$$n^j \mathcal{P}_{jkmn} = n^k \mathcal{P}_{jkmn} = n^m \mathcal{P}_{jkmn} = n^n \mathcal{P}_{jkmn} = 0 \tag{13.46}$$

- it is traceless:

$$\delta^{jk}\mathcal{P}_{jkmn} = \delta^{mn}\mathcal{P}_{jkmn} = 0\,. \tag{13.47}$$

Since h_{jk} and $\bar{h}_{jk}$ differ only by the trace, and since the projector $\mathcal{P}_{jklm}$ extracts the traceless part of a tensor (Eq. 13.47), the components of the perturbed metric tensor in the TT gauge can be obtained by applying the projector $\mathcal{P}_{jkmn}$ either to h_{jk} or to $\bar{h}_{jk}$, i.e.

$$h^{\mathbf{TT}}_{jk} = \mathcal{P}_{jkmn}h_{mn} = \mathcal{P}_{jkmn}\bar{h}_{mn}\,. \tag{13.48}$$

By applying $\mathcal{P}$ to $\bar{h}_{jk}$ defined in Eq. 13.31 we get

$$h^{\mathbf{TT}}_{\mu 0} = 0, \tag{13.49}$$

$$h^{\mathbf{TT}}_{jk}(t,r) = \frac{2G}{c^4 r}\frac{d^2}{dt^2}\,Q^{\mathbf{TT}}_{jk}\left(t - \frac{r}{c}\right),$$

where

$$Q^{\mathbf{TT}}_{jk} \equiv \mathcal{P}_{jkmn} q_{mn} \tag{13.50}$$

is the **transverse-traceless part of the quadrupole moment.**

As we shall see, to compute the luminosity of a gravitational wave source, i.e. the energy emitted per unit time, it is useful to define the **reduced quadrupole moment** Q_{jk}

$$Q_{jk} \equiv q_{jk} - \frac{1}{3}\delta_{jk} q_m{}^m \tag{13.51}$$

which is traceless by definition, i.e.

$$\delta^{jk} Q_{jk} = 0\,, \tag{13.52}$$

and from Eq. 13.47 it follows that

$$Q^{\mathbf{TT}}_{jk} = \mathcal{P}_{jkmn} q_{mn} = \mathcal{P}_{jkmn} Q_{mn}\,. \tag{13.53}$$

13.4 GRAVITATIONAL WAVES EMITTED BY A HARMONIC OSCILLATOR

Let us consider a harmonic oscillator composed of two equal masses m connected by a spring, oscillating at a frequency $\nu = \frac{\omega}{2\pi}$ with amplitude A. Let l_0 be the proper length of the spring when the system is at rest (see Fig. 13.1). Assuming that the masses move along

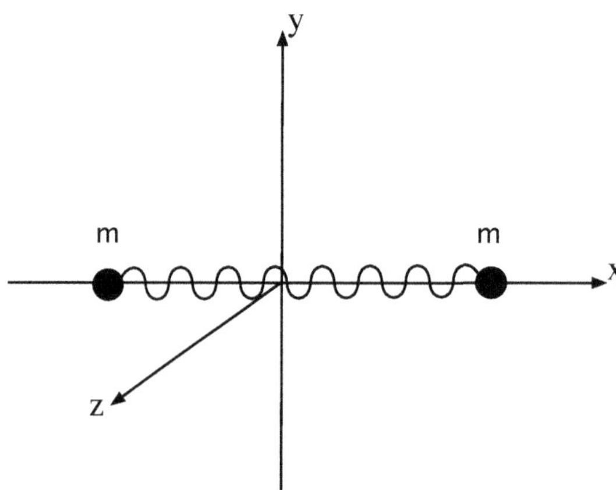

Figure 13.1: Two masses connected by a spring, oscillating along the x direction.

the x-axis, their coordinates at the time t are

$$\begin{cases} x_1 = -\frac{1}{2} l_0 - A\cos\omega t & y_1 = 0 \qquad z_1 = 0 \\ x_2 = +\frac{1}{2} l_0 + A\cos\omega t & y_2 = 0 \qquad z_2 = 0\,, \end{cases} \tag{13.54}$$

and the 00-component of the stress-energy tensor of the system is (see Eq. 5.6)

$$T^{00} = \sum_{n=1}^{2} c p^0 \; \delta(x - x_n)\; \delta(y)\; \delta(z)\,. \tag{13.55}$$

In the slow-motion approximation $v \ll c$, therefore $p^0 \simeq mc$ and T^{00} reduces to

$$T^{00} = mc^2 \sum_{n=1}^{2} \delta(x - x_n)\, \delta(y)\, \delta(z)\,. \tag{13.56}$$

The xx-component of the quadrupole moment $q^{ik}(t) = \frac{1}{c^2} \int_V T^{00}(t, \mathbf{x})\, x^i\, x^k d^3x$ reads

$$\begin{aligned} q^{xx} = q_{xx} &= m\left(\int_V \delta(x - x_1)\, x^2\, dx\, \delta(y)\, dy\, \delta(z)\, dz\right. \\ &\quad \left. + \int_V \delta(x - x_2)\, x^2\, dx\, \delta(y)\, dy\, \delta(z)\, dz\right) \\ &= m\left(x_1^2 + x_2^2\right) = m\left(\frac{1}{2} l_0^2 + 2A^2 \cos^2 \omega t + 2Al_0\, \cos \omega t\right) \\ &= m\left(const + A^2 \cos 2\omega t + 2Al_0\, \cos \omega t\right)\,. \end{aligned} \tag{13.57}$$

The zz-component of q^{ik} is

$$\begin{aligned} q^{zz} = \quad q_{zz} = \quad & m\left[\int_V \delta(x - x_1)\, dx\, \delta(y)\, dy\, \delta(z)\, z^2\, dz\right. \\ & \left. + \int_V \delta(x - x_2)\, dx\, \delta(y)\, dy\, \delta(z)\, z^2\, dz\right] = 0 \end{aligned} \tag{13.58}$$

because $\int_V z^2\, \delta(z)\, dz = 0$. Likewise, since the motion is confined to the x-axis, all remaining components of q^{ik} vanish.

Let us compute, as an example, the wave propagating along the z-direction. In this case $\mathbf{n} = \frac{\mathbf{x}}{r} = (0, 0, 1)$ and

$$P_{jk} = \delta_{jk} - n_j n_k = \begin{pmatrix} 1 & 0 & 0 \\ 0 & 1 & 0 \\ 0 & 0 & 0 \end{pmatrix}\,. \tag{13.59}$$

By applying the transverse-traceless projector $\mathcal{P}_{jkmn}$ constructed from P_{jk} to the reduced quadrupole moment tensor Q_{ij}, we find

$$\begin{aligned} Q^{\mathbf{TT}}{}_{xx} &= \left(P_{xm}P_{xn} - \frac{1}{2}P_{xx}P_{mn}\right) q_{mn} = \left(P_{xx}P_{xx} - \frac{1}{2}P_{xx}^2\right) q_{xx} = \frac{1}{2} q_{xx}, \\ Q^{\mathbf{TT}}{}_{yy} &= \left(P_{ym}P_{yn} - \frac{1}{2}P_{yy}P_{mn}\right) q_{mn} = -\frac{1}{2} P_{yy} P_{xx} q_{xx} = -\frac{1}{2} q_{xx}\,, \\ Q^{\mathbf{TT}}{}_{xy} &= \left(P_{xm}P_{yn} - \frac{1}{2}P_{xy}P_{mn}\right) q_{mn} = 0\,, \\ Q^{\mathbf{TT}}{}_{zz} &= \left(P_{zm}P_{zn} - \frac{1}{2}P_{zz}P_{mn}\right) q_{mn} = 0\,. \end{aligned} \tag{13.60}$$

In addition $Q^{\mathbf{TT}}{}_{zx} = Q^{\mathbf{TT}}{}_{zy} = 0$. Using these expressions, Eqs. 13.49 give

$$\begin{cases} h^{\mathbf{TT}}{}_{\mu 0} = 0 \\ h^{\mathbf{TT}}{}_{zi} = 0, \qquad h^{\mathbf{TT}}{}_{xy} = 0 \\ h^{\mathbf{TT}}{}_{xx}(t, z) = -h^{\mathbf{TT}}{}_{yy}(t, z) = \dfrac{G}{c^4 z} \dfrac{d^2}{dt^2}\, q_{xx}\left(t - \dfrac{z}{c}\right), \end{cases} \tag{13.61}$$

and, using Eq. 13.57,

$$\begin{aligned} h^{\mathrm{TT}}{}_{xx} &= -h^{\mathrm{TT}}{}_{yy} = \frac{G}{c^4 z}\left[\frac{d^2}{dt^2}\, q_{xx}\left(t-\frac{z}{c}\right)\right], \\ &= -\frac{2Gm}{c^4 z}\,\omega^2\left[2A^2\cos 2\omega\left(t-\frac{z}{c}\right)+Al_0\,\cos\omega\left(t-\frac{z}{c}\right)\right]. \end{aligned} \tag{13.62}$$

Thus, the radiation emitted by the harmonic oscillator along the z-axis is *linearly polarized.*

Let us assume that the amplitude of the oscillation is much smaller than the size of the system, $A \ll l_0$. In this case the first term on the right-hand side of the above equation is negligible compared to the second one:

$$h^{\mathrm{TT}}{}_{xx} \simeq -\frac{2Gm}{c^4 z}\,\omega^2 A l_0\,\cos\omega\left(t-\frac{z}{c}\right), \tag{13.63}$$

which shows that the wave is emitted at the *same frequency* of the harmonic oscillator. For example, if $m = 10^6$ g, $l_0 = 10^2$ cm, $A = 10^{-2}$ cm, and $\omega = 10^4$ rad/s, we get

$$h^{\mathrm{TT}}{}_{xx} \sim \frac{1.6\times 10^{-35}\,\mathrm{cm}}{z}. \tag{13.64}$$

As expected, the wave amplitude is extremely small.

Due to the symmetry of the system, the waveform emitted along the y direction has the same form with $y \leftrightarrow z$. We leave as an exercise to show that the oscillator does not emit gravitational waves along the x direction (choose $\mathbf{n} = (1, 0, 0)$ and use the same procedure).

13.5 GRAVITATIONAL WAVE EMITTED BY A BINARY SYSTEM IN CIRCULAR ORBIT

We shall now estimate the gravitational signal emitted by a binary system composed of two stars moving on a circular orbit around their common center of mass. For simplicity we shall assume that the two stars of mass m_1 and m_2 are point-like. Let l_0 be their orbital separation, M the *total mass*

$$M \equiv m_1 + m_2\,, \tag{13.65}$$

and μ the *reduced mass*

$$\mu \equiv \frac{m_1 m_2}{M}. \tag{13.66}$$

Let us place the origin of the coordinate frame at the center of mass of the system, as indicated in Fig. 13.2, so that

$$l_0 = r_1 + r_2, \qquad r_1 = \frac{m_2 l_0}{M}, \qquad r_2 = \frac{m_1 l_0}{M}. \tag{13.67}$$

The orbital frequency can be found by equating the gravitational to the centrifugal force

$$G\frac{m_1 m_2}{l_0^2} = m_1\omega_K^2\frac{m_2 l_0}{M}, \qquad G\frac{m_1 m_2}{l_0^2} = m_2\omega_K^2\frac{m_1 l_0}{M}, \tag{13.68}$$

from which we find the Keplerian frequency

$$\omega_K = \sqrt{\frac{GM}{l_0^3}}. \tag{13.69}$$

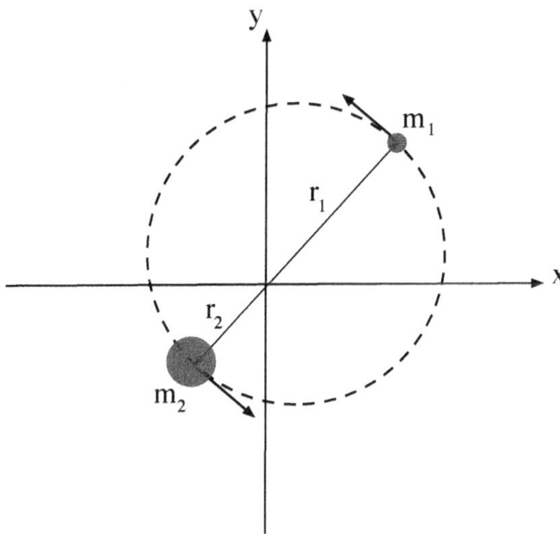

Figure 13.2: Two point masses in circular orbit around the common center of mass

Let (x_1, x_2) and (y_1, y_2) be the coordinates of the masses m_1 and m_2 on the orbital plane

$$\begin{aligned} x_1 &= \frac{m_2}{M} l_0 \cos\omega_K t \qquad x_2 = -\frac{m_1}{M} l_0 \cos\omega_K t \\ y_1 &= \frac{m_2}{M} l_0 \sin\omega_K t \qquad y_2 = -\frac{m_1}{M} l_0 \sin\omega_K t\,. \end{aligned} \tag{13.70}$$

The 00-component of the stress-energy tensor of the system reads

$$T^{00} = c^2 \sum_{n=1}^{2} m_n \; \delta(x - x_n)\; \delta(y - y_n)\; \delta(z)\,, \tag{13.71}$$

and the non-vanishing components of the quadrupole moment are

$$\begin{aligned} q_{xx} &= m_1 \int_V x^2 \delta(x - x_1)\, dx\, \delta(y - y_1)\, dy\, \delta(z)\, dz \\ &+ m_2 \int_V x^2 \delta(x - x_2)\, dx\, \delta(y - y_2)\, dy\, \delta(z)\, dz = m_1 x_1^2 + m_2 x_2^2 \\ &= \mu\, l_0^2 \cos^2 \omega_K t = \frac{\mu}{2}\, l_0^2 \cos 2\omega_K t + const\,, \end{aligned} \tag{13.72}$$

$$\begin{aligned} q_{yy} &= m_1 \int_V \delta(x - x_1)\, dx\, y^2\, \delta(y - y_1)\, dy\, \delta(z)\; dz \\ &+ m_2 \int_V \delta(x - x_2)\, dx\, y^2\, \delta(y - y_2)\, dy\, \delta(z)\, dz = m_1 y_1^2 + m_2 y_2^2 \\ &= \mu\, l_0^2 \sin^2 \omega_K t = \frac{\mu}{2}\, l_0^2 \cos 2\omega_K t + const_1\,, \end{aligned} \tag{13.73}$$

and

$$\begin{aligned} q_{xy} &= m_1 \int_V x\delta(x-x_1)\,dx\,y\,\delta(y-y_1)\,dy\,\delta(z)\,dz \\ &+ m_2 \int_V x\delta(x-x_2)\,dx\,y\,\delta(y-y_2)\,dy\,\delta(z)\,dz \\ &= m_1x_1y_1 + m_2x_2y_2 = \mu\, l_0^2 \cos\omega_K t\, \sin\omega_K t = \frac{\mu}{2}\, l_0^2 \sin 2\omega_K t\,, \end{aligned} \tag{13.74}$$

where we have used the trigonometric relations $\cos^2\alpha = (1+\cos 2\alpha)/2$, $\sin^2\alpha = (1 - \cos 2\alpha)/2$, and $m_1m_2 = \mu M$.

In summary, we find

$$q_{xx} = \frac{\mu}{2}\, l_0^2 \cos 2\omega_K t + const \tag{13.75}$$

$$q_{yy} = -\frac{\mu}{2}\, l_0^2 \cos 2\omega_K t + const_1 \tag{13.76}$$

$$q_{xy} = \frac{\mu}{2}\, l_0^2 \sin 2\omega_K t, \tag{13.77}$$

and

$$q^k{}_k = \eta^{kl}\, q_{kl} = q_{xx} + q_{yy} = const_2\,. \tag{13.78}$$

Therefore, the time-varying parts of q_{ij} and of $Q_{ij} = q_{ij} - \frac{1}{3}\,\delta_{ij}\, q^k{}_k$ coincide, and

$$\begin{aligned} Q_{xx} &= -Q_{yy} = \frac{\mu}{2}\, l_0^2 \cos 2\omega_K t \\ Q_{xy} &= \frac{\mu}{2}\, l_0^2 \sin 2\omega_K t\,. \end{aligned} \tag{13.79}$$

By defining a matrix A_{ij} as

$$A_{ij}(t) = \begin{pmatrix} \cos 2\omega_K t & \sin 2\omega_K t & 0 \\ \sin 2\omega_K t & -\cos 2\omega_K t & 0 \\ 0 & 0 & 0 \end{pmatrix} \tag{13.80}$$

we can write

$$Q_{ij} = \frac{\mu}{2}\, l_0^2\, A_{ij}\,. \tag{13.81}$$

Since according to Eq. 13.49 the gravitational perturbation is given by the second time derivative of the quadrupole moment, Eqs. 13.81 and 13.80 show that *a binary system in circular orbit emits waves at twice the orbital frequency.*

The wave emitted along a generic direction **n** in the TT gauge is (see Eq. 13.49)

$$h_{ij}^{\mathbf{TT}}(t,r) = \frac{2G}{rc^4}\frac{d^2}{dt^2}\left[Q_{ij}^{\mathbf{TT}}\left(t-\frac{r}{c}\right)\right], \tag{13.82}$$

where

$$Q_{ij}^{\mathbf{TT}}\left(t-\frac{r}{c}\right) = \mathcal{P}_{ijkl}Q_{kl}\left(t-\frac{r}{c}\right) = \mathcal{P}_{ijkl}q_{kl}\left(t-\frac{r}{c}\right). \tag{13.83}$$

Then, using Eqs. 13.80, 13.81, and the expression of the orbital (Keplerian) frequency 13.69, we find

$$h_{ij}^{\mathbf{TT}} = -\frac{2G}{rc^4}\frac{\mu}{2}\, l_0^2\, (2\omega_K)^2\, \mathcal{P}_{ijkl}A_{kl} = -\frac{4\,\mu\, M\, G^2}{r\, l_0\, c^4}\, \mathcal{P}_{ijkl}A_{kl}\,. \tag{13.84}$$

By defining

$$\mathcal{A} = \frac{4\mu M G^2}{l_0 c^4} \tag{13.85}$$

we can finally write the emitted wave as

$$h_{ij}^{\mathbf{TT}}(t,r) = -\frac{\mathcal{A}}{r}\,A_{ij}^{\mathbf{TT}}\left(t-\frac{r}{c}\right), \tag{13.86}$$

where

$$A_{ij}^{\mathbf{TT}}\left(t-\frac{r}{c}\right) = \mathcal{P}_{ijkl}A_{kl}\left(t-\frac{r}{c}\right) \tag{13.87}$$

depends on the orientation of the line of sight with respect to the orbital plane.

For example, if $\mathbf{n}=(0,0,1)$, $P_{ij}=\mathrm{diag}(1,1,0)$,

$$A_{ij}^{\mathbf{TT}}(t) = \begin{pmatrix} \cos 2\omega_K t & \sin 2\omega_K t & 0 \\ \sin 2\omega_K t & -\cos 2\omega_K t & 0 \\ 0 & 0 & 0 \end{pmatrix} \tag{13.88}$$

and

$$\begin{aligned} h^{\mathbf{TT}}{}_{xx} &= -h^{\mathbf{TT}}{}_{yy} = -\frac{\mathcal{A}}{z}\cos 2\omega_K\left(t-\frac{z}{c}\right) \\ h^{\mathbf{TT}}{}_{xy} &= -\frac{\mathcal{A}}{z}\sin 2\omega_K\left(t-\frac{z}{c}\right). \end{aligned} \tag{13.89}$$

Thus, the wave emitted in the direction orthogonal to the orbital plane has both polarizations and, since $h^{\mathbf{TT}}{}_{xx}$ and $h^{\mathbf{TT}}{}_{xy}$ are cosine and sine of the same phase, with the same amplitude, it is *circularly polarized.*

If $\mathbf{n}=(1,0,0)$, $P_{ij}=\mathrm{diag}(0,1,1)$, and

$$A_{ij}^{\mathbf{TT}} = \begin{pmatrix} 0 & 0 & 0 \\ 0 & -\frac{1}{2}\cos 2\omega_K t & 0 \\ 0 & 0 & \frac{1}{2}\cos 2\omega_K t \end{pmatrix}, \tag{13.90}$$

thus

$$h^{\mathbf{TT}}{}_{yy} = -h^{\mathbf{TT}}{}_{zz} = \frac{1}{2}\frac{\mathcal{A}}{x}\cos 2\omega_K\left(t-\frac{x}{c}\right), \tag{13.91}$$

i.e. the wave emitted along the x-axis is linearly polarized.

Similarly, if $\mathbf{n}=(0,1,0)$, $P_{ij}=\mathrm{diag}(1,0,1)$,

$$A_{ij}^{\mathbf{TT}} = \begin{pmatrix} \frac{1}{2}\cos 2\omega_K t & 0 & 0 \\ 0 & 0 & 0 \\ 0 & 0 & -\frac{1}{2}\cos 2\omega_K t \end{pmatrix} \tag{13.92}$$

and the wave is linearly polarized,

$$h^{\mathbf{TT}}{}_{xx} = -h^{\mathbf{TT}}{}_{zz} = -\frac{1}{2}\frac{\mathcal{A}}{y}\cos 2\omega_K\left(t-\frac{y}{c}\right). \tag{13.93}$$

Eq. 13.85 can be used to estimate the amplitude of the gravitational wave emitted by the binary system PSR 1913+16 discovered in 1975 [64], which consists of two neutron stars (these compact objects will be discussed in Chapter 16) orbiting at a relatively short distance from each other. The data obtained from pulsar[3] timing (i.e., from tracking the electromagnetic pulses emitted by one of the stars and inferring the properties of the system from this periodic signal) are

$$\begin{aligned} m_1 \sim m_2 &= 1.4M_\odot\,, & l_0 &= 1.9\times 10^{11}\ \mathrm{cm} \\ P = 7\mathrm{h}\ 45\mathrm{m}\ 7\mathrm{s} &= 2.8\times 10^4\,\mathrm{s}\,, & \nu_K &= \frac{\omega_K}{2\pi} = 3.58\times 10^{-5}\ \mathrm{Hz}\,, \end{aligned} \tag{13.94}$$

[3]Pulsars are rotating neutron stars with strong magnetic fields, which emit periodic beams of radio waves (see Chapter 16).

where P is the orbital period. The distance of the system from Earth is $r = 5$ kpc, where pc denotes the *parsec*, and since (see Table A)

$$1 \text{ pc} = 3.09 \times 10^{18} \text{ cm}, \quad \rightarrow \quad r = 1.5 \times 10^{22} \text{ cm}. \tag{13.95}$$

Note that the two stars have nearly equal masses, comparable to that of the Sun, and their orbital separation is about twice the radius of the Sun. For this system the emission frequency is

$$\nu_{\rm GW} = 2\nu_K = 7.16 \times 10^{-5} \text{ Hz}, \tag{13.96}$$

therefore the wavelength of the emitted radiation is

$$\lambda_{\rm GW} = \frac{c}{\nu_{\rm GW}} \sim 10^{14} \text{ cm} \qquad \text{i.e.} \qquad \lambda_{\rm GW} \gg l_0\,. \tag{13.97}$$

Thus, the slow-motion approximation, on which the quadrupole formalism is based, is certainly satisfied in this case.

The orbit of PSR 1913+16 is eccentric ($e \simeq 0.617$); however for simplicity we shall assume that it is circular and estimate the wave amplitude. By defining

$$h_0 = \frac{\mathcal{A}}{r}, \tag{13.98}$$

and using Eq. 13.85 we get

$$h_0 = \frac{4\mu M G^2}{r l_0 c^4} \sim 5 \times 10^{-23}\,. \tag{13.99}$$

A new binary pulsar has more recently been discovered, which has an even shorter orbital period and it is closer to Earth than PSR 1913+16: it is the double pulsar PSR J0737-3039 [27], whose orbital parameters are

$$\begin{aligned} m_1 &= 1.337 M_\odot\,, & m_2 &= 1.250 M_\odot \\ P &= 2.4\,\text{h} = 8.64 \times 10^3\,\text{s}\,, & e &= 0.08 \\ r &= 500 \text{ pc}\,, & l_0 &= 1.2 R_\odot. \end{aligned} \tag{13.100}$$

In this case the orbit is nearly circular and our previous formulas are more accurate. For this source we find

$$\mu = \frac{m_1 m_2}{m_1 + m_2} = 0.646 M_\odot \quad \rightarrow \quad h_0 = \frac{4\mu M G^2}{r l_0 c^4} \sim 1.1 \times 10^{-21}, \tag{13.101}$$

and waves are emitted at frequency

$$\nu_{GW} = 2\nu_K = 2.3 \times 10^{-4} \text{ Hz}\,. \tag{13.102}$$

In this section we have considered only circular orbits; the calculations can be generalized to the case of eccentric or open orbits by replacing the equation of motion of the two masses (Eq. 13.70) by those appropriate to the chosen orbit. By this procedure it is possible to show that when the orbits are elliptical, gravitational waves are emitted at frequencies which are a *multiple* of the orbital frequency ν_K, and that the number of equally spaced spectral lines increases with the eccentricity (for a detailed discussion see, e.g., [82]).

13.6 ENERGY CARRIED BY A GRAVITATIONAL WAVE

In order to evaluate how much energy and momentum are radiated in gravitational waves by a dynamical system, we would need a tensor that properly describes these quantities for the gravitational field. Unfortunately such a tensor does not exist and the reason will be explained in this section. However, it is possible to define a class of pseudo-tensors, i.e. quantities which behave like tensors only under linear coordinate transformations, which carry information on the energy and momentum of the gravitational field. Within this class, we shall consider the Landau-Lifshitz *stress-energy pseudo-tensor* [75], which will be useful for our current purpose.

13.6.1 The stress-energy pseudo-tensor of the gravitational field

As discussed in Chapter 5, the conservation laws of energy and momentum which, in Special Relativity, can be derived from the equation

$$T^{\mu\nu}{}_{,\nu} = 0\,, \tag{13.103}$$

cannot be extended to General Relativity. Indeed, the generalization of Eq. 13.103 in curved spacetime is

$$T^{\mu\nu}{}_{;\nu} = \frac{1}{\sqrt{-g}}\frac{\partial}{\partial x^{\nu}}(\sqrt{-g}T^{\mu\nu}) + \Gamma^{\mu}{}_{\lambda\alpha}T^{\lambda\alpha} = 0 \tag{13.104}$$

from which, for the reasons explained in Sec. 5.4, conserved quantities cannot be found.

The fact that the energy and momentum described by $T^{\mu\nu}$ alone are not conserved is not surprising; indeed $T^{\mu\nu}$ describes the energy and momentum of matter and fields, which are sources of the gravitational field, but it does not include the energy and momentum of the gravitational field itself. What we need is a conservation law which includes *both* contributions.

An equation which fulfills this requirement was found by Landau and Lifshitz [75], who showed that Einstein's equations, $G_{\mu\nu} = \frac{8\pi G}{c^4}T^{\mu\nu}$, can be written in the form

$$\frac{\partial \zeta^{\mu\nu\alpha}}{\partial x^{\alpha}} = (-g)(T^{\mu\nu} + t^{\mu\nu})\,, \tag{13.105}$$

where $\zeta^{\mu\nu\alpha}$ is antisymmetric in the last two indices, i.e. $\zeta^{\mu\nu\alpha} = -\zeta^{\mu\alpha\nu}$, $T^{\mu\nu}$ is the stress-energy tensor and $t^{\mu\nu}$ is a quantity called **stress-energy pseudo-tensor**. From the symmetry properties of $\zeta^{\mu\nu\alpha}$ it follows that $\zeta^{\mu\nu\alpha}{}_{,\alpha\nu} = 0$, and consequently

$$\frac{\partial}{\partial x^{\nu}}\left[(-g)\left(t^{\mu\nu} + T^{\mu\nu}\right)\right] = 0\,. \tag{13.106}$$

This equation has the form of the vanishing of the *ordinary four-divergence* of the quantity $(-g)\left(t^{\mu\nu} + T^{\mu\nu}\right)$, from which conserved quantities can be derived following the procedure described in Sec. 5.3. As we shall show, Eq. 13.106 allows to interpret the stress-energy pseudo-tensor $t^{\mu\nu}$ as the quantity which describes the energy and momentum of the gravitational field. We shall now derive Eq. 13.105 and define the relevant quantities which appear in that equation.

The Landau-Lifshitz stress-energy pseudo-tensor

Let us consider a LIF $\{\xi^{\mu}\}$ centered in a spacetime point $\mathbf{p}$. In $\mathbf{p}$ the metric coincides with Minkowski's metric and its first derivatives vanish ($\Gamma^{\mu}_{\alpha\beta}(\mathbf{p}) = 0$), whereas in a generic point of the LIF we can write (see Box 3-A)

$$g_{\mu\nu}(\xi) = \eta_{\mu\nu} + O(\xi^2)\,, \qquad \Gamma^{\alpha}_{\mu\nu}(\xi) = O(\xi)\,. \tag{13.107}$$

Note that, unless the spacetime is flat, the second derivatives of the metric tensor in **p** are generally non-vanishing, and the Riemann and the Ricci tensor do not vanish as well. Indeed, the Ricci tensor reduces to the expression given in Eq. 4.24, which we rewrite here for clarity:

$$R^{\mu\nu} = \frac{1}{2} g^{\mu\alpha} g^{\nu\beta} g^{\gamma\delta} \left(\frac{\partial^2 g_{\gamma\beta}}{\partial x^\alpha \partial x^\delta} + \frac{\partial^2 g_{\alpha\delta}}{\partial x^\gamma \partial x^\beta} - \frac{\partial^2 g_{\gamma\delta}}{\partial x^\alpha \partial x^\beta} - \frac{\partial^2 g_{\alpha\beta}}{\partial x^\gamma \partial x^\delta} \right) . \tag{13.108}$$

Here and in the following terms of order $O(\xi)$ are absent because we will always consider quantities evaluated in **p**.

We remark that if we consider any coordinate system $\{x^\mu\}$ related to the LIF by a *linear transformation*, i.e. $x^\mu = A^\mu{}_\nu \xi^\nu$ with $A^\mu{}_\nu$ constant, the metric tensor in the new frame is generally different from $\eta_{\mu\nu}$, but its first derivatives still vanish, and Eq. (13.108) holds in the new frame as well.

Einstein's equations can be cast in the form

$$T^{\mu\nu} = \frac{c^4}{8\pi G} \left(R^{\mu\nu} - \frac{1}{2} g^{\mu\nu} R \right) . \tag{13.109}$$

Replacing Eq. 13.108 in Eq. 13.109, after lengthy but simple calculations we obtain

$$T^{\mu\nu} = \frac{\partial}{\partial x^\alpha} \left\{ \frac{c^4}{16\pi G} \frac{1}{(-g)} \frac{\partial}{\partial x^\beta} \left[(-g) \left(g^{\mu\nu} g^{\alpha\beta} - g^{\mu\alpha} g^{\nu\beta} \right) \right] \right\} . \tag{13.110}$$

By defining

$$\zeta^{\mu\nu\alpha} = \frac{c^4}{16\pi G} \frac{\partial}{\partial x^\beta} \left[(-g) \left(g^{\mu\nu} g^{\alpha\beta} - g^{\mu\alpha} g^{\nu\beta} \right) \right] , \tag{13.111}$$

Eq. 13.110 can be written as

$$\frac{\partial \zeta^{\mu\nu\alpha}}{\partial x^\alpha} - (-g) T^{\mu\nu} = 0 , \tag{13.112}$$

where we have used the fact that, since $g_{\mu\nu,\alpha} = 0$, $\partial g / \partial x^\alpha = 0$. Note that with these definitions, the quantity $\zeta^{\mu\nu\alpha}$ is the ordinary four-divergence of

$$\lambda^{\mu\nu\alpha\beta} = \frac{c^4}{16\pi G} \left[(-g) \left(g^{\mu\nu} g^{\alpha\beta} - g^{\mu\alpha} g^{\nu\beta} \right) \right] , \tag{13.113}$$

and is anti-symmetric in its last two indices,

$$\zeta^{\mu\nu\alpha} = -\zeta^{\mu\alpha\nu} . \tag{13.114}$$

Moreover, Eq. 13.112 implies that the divergence $\zeta^{\mu\nu\alpha}{}_{,\alpha}$ is symmetric (note that this does not apply to the quantity $\zeta^{\mu\nu\alpha}$ itself, i.e. in general $\zeta^{\mu\nu\alpha} \neq \zeta^{\nu\mu\alpha}$).

Eq. 13.112 has been derived in a LIF; if we now consider a *general coordinate frame*[4], the difference in the two terms of Eq. 13.112 will be non-zero, and we call this quantity $(-g)t^{\mu\nu}$:

$$(-g) t^{\mu\nu} \equiv \frac{\partial \zeta^{\mu\nu\alpha}}{\partial x^\alpha} - (-g) T^{\mu\nu} . \tag{13.115}$$

Using Einstein's equations 13.109, $(-g)t^{\mu\nu}$ can be written as

$$(-g) t^{\mu\nu} = \frac{\partial \zeta^{\mu\nu\alpha}}{\partial x^\alpha} - (-g) \frac{c^4}{8\pi G} \left(R^{\mu\nu} - \frac{1}{2} g^{\mu\nu} R \right) . \tag{13.116}$$

[4]The definition 13.111 of $\zeta^{\mu\nu\alpha}$ is assumed to hold also in a general frame.

By replacing in the right-hand side of this equation the general definition of the Ricci tensor (see e.g. Box 4-A), after some lengthy but simple manipulation we find

$$\begin{aligned} t^{\mu\nu} &= \frac{c^4}{16\pi G}\left\{\left(2\Gamma^\delta{}_{\alpha\beta}\Gamma^\sigma{}_{\delta\sigma} - \Gamma^\delta{}_{\alpha\sigma}\Gamma^\sigma{}_{\beta\delta} - \Gamma^\delta{}_{\alpha\delta}\Gamma^\sigma{}_{\beta\sigma}\right)\left(g^{\mu\alpha}g^{\nu\beta} - g^{\mu\nu}g^{\alpha\beta}\right)\right. \\ &+ g^{\mu\alpha}g^{\beta\delta}\left(\Gamma^\nu{}_{\alpha\sigma}\Gamma^\sigma{}_{\beta\delta} + \Gamma^\nu{}_{\beta\delta}\Gamma^\sigma{}_{\alpha\sigma} - \Gamma^\nu{}_{\delta\sigma}\Gamma^\sigma{}_{\alpha\beta} - \Gamma^\nu{}_{\alpha\beta}\Gamma^\sigma{}_{\delta\sigma}\right) \\ &+ g^{\nu\alpha}g^{\beta\delta}\left(\Gamma^\mu{}_{\alpha\sigma}\Gamma^\sigma{}_{\beta\delta} + \Gamma^\mu{}_{\beta\delta}\Gamma^\sigma{}_{\alpha\sigma} - \Gamma^\mu{}_{\delta\sigma}\Gamma^\sigma{}_{\alpha\beta} - \Gamma^\mu{}_{\alpha\beta}\Gamma^\sigma{}_{\delta\sigma}\right) \\ &+ \left. g^{\alpha\beta}g^{\delta\sigma}\left(\Gamma^\mu{}_{\alpha\delta}\Gamma^\nu{}_{\beta\sigma} - \Gamma^\mu{}_{\alpha\beta}\Gamma^\nu{}_{\delta\sigma}\right)\right\}. \end{aligned} \quad (13.117)$$

This expression is quadratic in the Christoffel symbols (a very important property, which will be discussed below). This is a consequence of the way this quantity has been constructed. Indeed, Eq. 13.112 shows that in a LIF the difference between $\frac{\partial\zeta^{\mu\nu\alpha}}{\partial x^\alpha}$ and the expression of $T^{\mu\nu}$ obtained using Eq. 13.110 is zero; note that both terms in Eq. 13.112 depend only on second derivatives of $g_{\mu\nu}$, since in the LIF the first derivatives vanish. When we write Eq. 13.116 in a general frame, we include both in $\frac{\partial\zeta^{\mu\nu\alpha}}{\partial x^\alpha}$ and in $T^{\mu\nu}$ all non-linear terms containing products of first derivatives of $g_{\mu\nu}$, and these are the terms which survive in the expression of $t^{\mu\nu}$, since the second derivatives cancel by construction.

Since $\zeta^{\mu\nu\alpha}{}_{,\alpha}$ is symmetric, so is $t^{\mu\nu}$. Eq. 13.115 can be written in the form of Eq. 13.105,

$$\frac{\partial\zeta^{\mu\nu\alpha}}{\partial x^\alpha} = (-g)(T^{\mu\nu} + t^{\mu\nu}), \quad (13.118)$$

and can be considered as an alternative way to cast Einstein's equation. As discussed above, the symmetry property 13.114 implies $\zeta^{\mu\nu\alpha}{}_{,\alpha\nu} = 0$, from which the conservation law 13.106 follows.

Energy and momentum of the gravitational field

In the absence of gravitational field, i.e. in flat spacetime, Christoffel's symbols vanish (in Minkowskian coordinates), hence $t^{\mu\nu} = 0$ and Eq. 13.106 reduces to the conservation law for the stress-energy tensor of Special Relativity, $T^{\mu\nu}{}_{,\nu} = 0$. This suggests to interpret Eq. 13.106 as the generalization of that conservation law in General Relativity. Indeed, from Eq. 13.106 it is possible to define a conserved four-vector, as follows (see also Sec. 5.3 where a similar procedure was applied to the stress-energy tensor).

Let us consider an asymptotically flat spacetime, and a three-dimensional spacelike hypersurface with equation $x^0 = const$, having three-volume V and boundary ∂V; Eq. 13.106 integrated over V gives

$$\int_V (-g)(T^{0\mu}+t^{0\mu})_{,0}\, d^3x = -\int_V (-g)(T^{0i}+t^{0i})_{,i}\, d^3x = -\int_{\partial V} (-g)(T^{0i}+t^{0i})\, dS_i\,, \quad (13.119)$$

where $dS_i = n_i dS$, dS is the surface element on ∂V, and n^i is the normal three-vector to this surface; the last term has been obtained using Gauss' theorem. Thus, if we define

$$P^\mu = \int_V (-g)(T^{0\mu} + t^{0\mu})\, d^3x \quad (13.120)$$

from Eq. 13.119 it follows that

$$\frac{\partial}{\partial x^0}P^\mu = -\int_{\partial V} (-g)(T^{\mu i} + t^{\mu i})\, dS_i\,. \quad (13.121)$$

Eq. 13.121 shows that the time derivative of P^μ, which is defined in Eq. 13.120 as an integral over V, can be expressed as a flux across the boundary ∂V. If $T^{\mu i} + t^{\mu i}$ decays sufficiently fast approaching ∂V, then the time derivative of P^μ vanishes, i.e. $P^\mu = const.$ Thus, Eq. 13.121 provides four *conservation laws* for the four components of the vector $\vec{P}$, like those discussed in Sec. 5.3.

We can argue that (P^0, P^i) are the total energy and momentum enclosed in V; they are contributed by $T^{\mu\nu}$, which describes the energy and momentum of non-gravitational fields and matter, and by $t^{\mu\nu}$, which describes the energy and momentum of the gravitational field. This interpretation is also supported by the following remark. Many known examples of stress-energy tensors are bilinear in the first derivatives of the fields. This is the case, for instance, for the scalar field and for the electromagnetic field (see Sec. 5.2). Similarly, the stress-energy pseudo-tensor defined in Eq. 13.117 is bilinear in the Christoffel symbols, i.e. in the first derivatives of the spacetime metric.

However, there is a fundamental difference between $T^{\mu\nu}$ and the quantity $t^{\mu\nu}$ defined in Eq. 13.117: the latter *is not a tensor field*, because Christoffel's symbols are not tensors. $t^{\mu\nu}$ is instead a *pseudo-tensor*, which means that it transforms as a tensor only under *linear transformations*, i.e. coordinate transformations of the form $x^{\mu'} = A^{\mu'}{}_\alpha x^\alpha$ with $A^{\mu'}{}_\alpha$ constants. As shown in Sec. 3.4, under linear transformations Christoffel's symbols transform as tensors, i.e.

$$\Gamma^{\lambda'}_{\mu'\nu'} = \frac{\partial x^{\lambda'}}{\partial x^\rho}\frac{\partial x^\tau}{\partial x^{\mu'}}\frac{\partial x^\sigma}{\partial x^{\nu'}}\Gamma^\rho_{\tau\sigma}\,. \tag{13.122}$$

Replacing this equation (and the transformation law for the metric tensor) in Eq. 13.117, it is a simple exercise to show that under a linear transformation $t^{\mu\nu}$ transform as a tensor, i.e.

$$t^{\mu'\nu'} = \frac{\partial x^{\mu'}}{\partial x^\tau}\frac{\partial x^{\nu'}}{\partial x^\sigma}t^{\tau\sigma}\,. \tag{13.123}$$

The fact that a stress-energy *tensor* of the gravitational field does not exist can be understood on general grounds. The Equivalence Principle establishes that the effect of the gravitational field on a particle vanishes in a LIF centered on that particle; therefore the energy density of the gravitational field also vanishes in the LIF. But if a tensor field vanishes in one frame, it vanishes in all frames, and since we can set up a LIF at any spacetime point, it follows that the gravitational energy density cannot be described by a tensor field.

One could be tempted to conclude that the energy and momentum densities described by the stress-energy pseudo-tensor $t^{\mu\nu}$ have no physical reality, since we are defining quantities which can be set to zero at any point with a suitable coordinate transformation. However, this is not the case, because it is possible to define procedures of *integration* or *averaging* involving the stress-energy pseudo-tensor, which allow to construct well-defined gauge-invariant quantities.

Let us consider an asymptotically flat spacetime, and a general coordinate transformation. If we want to keep the spacetime asymptotically flat, the transformation should reduce, in the asymptotic region (and thus on the boundary ∂V of the integral in Eq. 13.121), to a Lorentz transformation $x^{\mu'} = \Lambda^{\mu'}{}_\nu x^\nu$. We have shown that for linear coordinate transformations $t^{\mu\nu}$ transforms as a tensor; therefore, since Lorentz transformations are linear, $t^{\mu\nu}$ transforms as a tensor on ∂V. Consequently, from Eq. 13.121 it follows that $P^\mu{}_{,0}$ transforms as (the time derivative of) a four-vector in Special Relativity. Integrating over time, we can conclude that the total energy-momentum P^μ defined in Eq. 13.120 transforms as a four-vector of Special Relativity, and thus, although $t^{\mu\nu}$ is not a tensor, the *global* notion of energy and momentum, which includes the contribution of the gravitational field, in an asymptotically flat spacetime is well defined. We shall further discuss these issues in Chapter 17, where we will study the far-field limit of stationary, isolated objects.

We have thus introduced a *global* notion of energy and momentum in General Relativity.

It is also possible to define a *local* notion of energy and momentum, in the case of a perturbed spacetime $g_{\mu\nu} = g^0_{\mu\nu} + h_{\mu\nu}$, on the assumption that the characteristic length-scale λ of the perturbation $h_{\mu\nu}$ (which, in case of gravitational waves, is the wavelength) is much smaller than the characteristic length-scale L of the background $g^0_{\mu\nu}$:

$$\lambda/L \ll 1\,. \tag{13.124}$$

In this regime (which is typical, for instance, of the propagation of gravitational waves far away from their source), it can be shown that the stress-energy pseudo-tensor, once *averaged over several wavelengths* λ, transforms as a tensor for coordinate transformations of order $O(h)$ (i.e. for coordinate transformations $x^\mu \to x^\mu + \epsilon^\mu$ with $\epsilon^\mu{}_{,\alpha} \sim O(h)$). We denote this average (also called *Brill-Hartle average*) with $\langle\cdot\rangle$. Therefore, provided the condition 13.124 is satisfied, the averaged stress-energy pseudo-tensor

$$\langle t^{\mu\nu}\rangle \tag{13.125}$$

describes the energy and momentum carried by the perturbation, and it can be shown that it transforms as a tensor. This result, obtained by Isaacson in 1968 [66], has been the definitive proof that gravitational waves are associated with a flux of energy and momentum. Before this was clarified, even the actual physical reality of gravitational waves was still a matter of debate.

13.6.2 Energy flux of a gravitational wave

Having defined the stress-energy pseudo-tensor of the gravitational field, we shall now show that it allows to compute the energy carried by a gravitational wave. Let us consider a source of gravitational waves and an associated three-dimensional coordinate frame (O, x, y, z), with the origin O within the source (for instance, at its center of mass). An observer is located at $P = (x_1, y_1, z_1)$, as shown in Fig. 13.3. The observer detects a wave coming along the direction identified by the unit three-vector $\mathbf{n} = \frac{\mathbf{r}}{r}$, where $\mathbf{r} = (x_1, y_1, z_1)$ and $r = |\mathbf{r}| = \sqrt{x_1^2 + y_1^2 + z_1^2}$ is the distance of the observer from the origin. Let us now consider a second frame $O'(x', y', z')$, with origin O' coincident with O, having the axis x' aligned with $\mathbf{n}$. The corresponding metric tensor is

$$g_{\mu'\nu'} = \eta_{\mu'\nu'} + h^{\mathbf{TT}}_{\mu'\nu'} = \begin{pmatrix} -1 & 0 & 0 & 0 \\ 0 & 1 & 0 & 0 \\ 0 & 0 & [1 + h_+(t, x')] & h_\times(t, x') \\ 0 & 0 & h_\times(t, x') & [1 - h_+(t, x')] \end{pmatrix}. \tag{13.126}$$

The observer wishes to measure the energy which flows per unit time across the unit surface orthogonal to x', i.e. $t^{0x'}$, therefore (s)he needs to compute the Christoffel symbols i.e. the derivatives of $h^{\mathbf{TT}}_{\mu'\nu'}$. According to Eq. 13.49, the metric perturbation (for instance, with '+' polarization) has the form $h_+(t, x') = \frac{const}{x'} f\left(t - \frac{x'}{c}\right)$, and since the only derivatives which matter are those with respect to time and x', we find

$$\frac{\partial h_+}{\partial t} \equiv \dot{h}_+ = \frac{const}{x'} \dot{f}, \tag{13.127}$$

$$\frac{\partial h_+}{\partial x'} \equiv h'_+ = -\frac{const}{x'^2} f + \frac{const}{x'} f' \simeq -\frac{1}{c}\frac{const}{x'} \dot{f} = -\frac{1}{c}\dot{h}_+\,, \tag{13.128}$$

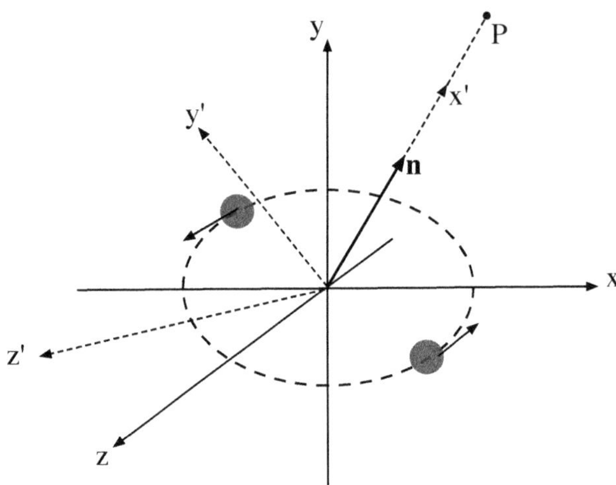

Figure 13.3: A binary system lies in the $z-x$ plane. An observer located at P detects the gravitational wave emitted by the system.

where only the dominant $1/x'$ term has been considered. Similarly, for the '×' polarization, $h'_\times = -\frac{1}{c}\dot{h}_\times$. Therefore, the non-vanishing Christoffel symbols are

$$\begin{aligned}
\Gamma^0{}_{y'y'} &= -\Gamma^0{}_{z'z'} = -\tfrac{1}{2c}\,\dot{h}_+ & \Gamma^{y'}{}_{0y'} &= -\Gamma^{z'}{}_{0z'} = \frac{1}{2c}\,\dot{h}_+ \\
\Gamma^{x'}{}_{y'y'} &= -\Gamma^{x'}{}_{z'z'} = \tfrac{1}{2c}\,\dot{h}_+ & \Gamma^{y'}{}_{y'x'} &= -\Gamma^{z'}{}_{z'x'} = -\frac{1}{2c}\,\dot{h}_+ \\
\Gamma^0{}_{y'z'} &= -\tfrac{1}{2c}\,\dot{h}_\times & \Gamma^{y'}{}_{0z'} &= -\Gamma^{z'}{}_{0y'} = \frac{1}{2c}\,\dot{h}_\times \\
\Gamma^x{}_{y'z'} &= \tfrac{1}{2c}\,\dot{h}_\times & \Gamma^{y'}{}_{xz'} &= -\Gamma^{z'}{}_{xy'} = -\frac{1}{2c}\,\dot{h}_\times\,.
\end{aligned} \tag{13.129}$$

The components of the stress-energy pseudo-tensor are quadratic in the Christoffel symbols (see Eq. 13.117); thus, in the TT gauge, they are quadratic in $\dot{h}_+$ and $\dot{h}_\times$.

In general, the energy flowing across a unit surface orthogonal to the direction x' per unit time is given by (c times) the component $0x'$ of the stress-energy tensor (see Box 5-C). Similarly, the energy flux of a gravitational wave propagating in the same direction is given by (c times) the component $0x'$ of the stress-energy pseudo-tensor averaged over several wavelengths, i.e.

$$\frac{dE_{\mathrm{GW}}}{dtdS} = \left\langle ct^{0x'}\right\rangle\,. \tag{13.130}$$

By replacing the Christoffel symbols 13.129 in Eq. 13.117, we find (by a simple but lengthy computation, which we leave to the reader as an exercise)

$$\frac{dE_{\mathrm{GW}}}{dtdS} = c\left\langle t^{0x'}\right\rangle = \frac{c^3}{16\pi G}\left\langle \left(\dot{h}_+(t,x')\right)^2 + \left(\dot{h}_\times(t,x')\right)^2\right\rangle\,. \tag{13.131}$$

The quantity in Eq. 13.131 is the energy per unit time which flows across a unit surface orthogonal to the direction x'. However, the direction x' is arbitrary: if the observer is at a different position and computes the energy flux, (s)he finds formally the same Eq. 13.131, with the polarizations h_+ and $h_\times$ referred to the TT gauge associated with the new direction. Thus, the right-hand side of Eq. 13.131 is a general expression for the gravitational wave flux per unit of surface orthogonal to the direction of propagation. In a general TT frame we can write Eq. 13.131 as

$$\frac{dE_{\rm GW}}{dtdS} = \frac{c^3}{32\pi G}\left\langle \sum_{jk}\left(\dot{h}^{\mathbf{TT}}_{jk}(t,r)\right)^2\right\rangle . \tag{13.132}$$

Since

$$\begin{cases} h^{\mathbf{TT}}{}_{\mu 0} = 0, & \mu = 0,\ldots,3 \\ h^{\mathbf{TT}}_{ik}(t,r) = \dfrac{2G}{c^4 r}\dfrac{d^2}{dt^2}\,Q^{\mathbf{TT}}_{ik}\left(t-\dfrac{r}{c}\right) \end{cases} \tag{13.133}$$

by direct substitution we find

$$\frac{dE_{\rm GW}}{dtdS} = \frac{G}{8\pi c^5\, r^2}\left\langle \sum_{jk}\left(\dddot{Q}^{\mathbf{TT}}_{jk}\left(t-\frac{r}{c}\right)\right)^2\right\rangle . \tag{13.134}$$

As explained in Sec. 13.3, $Q^{\mathbf{TT}}_{jk} \equiv \mathcal{P}_{jkmn}q_{mn}$ is the quadrupole tensor projected onto the TT gauge and

$$Q^{\mathbf{TT}}_{jk} = \mathcal{P}_{jkmn}q_{mn} = \mathcal{P}_{jkmn}Q_{mn}\,, \tag{13.135}$$

where we recall that $Q_{jk} \equiv q_{jk} - \frac{1}{3}\delta_{jk}q_m{}^m$ is the reduced quadrupole moment tensor, which is traceless by definition.

In order to obtain the **gravitational wave luminosity** $L_{\rm GW} = \frac{dE_{\rm GW}}{dt}$, i.e. the gravitational energy emitted by a source per unit time, it is more convenient to use the reduced quadrupole moment, therefore we shall write Eq. 13.134 in terms of Q_{jk} using Eq. 13.135, i.e.

$$\frac{dE_{\rm GW}}{dtdS} = \frac{G}{8\pi c^5\, r^2}\left\langle \sum_{jk}\left(\mathcal{P}_{jkmn}\dddot{Q}_{mn}\left(t-\frac{r}{c}\right)\right)^2\right\rangle . \tag{13.136}$$

The luminosity therefore reads

$$\begin{aligned} L_{\rm GW} &= \int \frac{dE_{\rm GW}}{dtdS}\,dS = \int \frac{dE_{\rm GW}}{dtdS}\,r^2 d\Omega \\ &= \frac{G}{2c^5}\frac{1}{4\pi}\int d\Omega \left\langle \sum_{jk}\left(\mathcal{P}_{jkmn}\dddot{Q}_{mn}\left(t-\frac{r}{c}\right)\right)^2\right\rangle , \end{aligned} \tag{13.137}$$

where $d\Omega = (d\cos\theta)d\varphi$ is the solid angle element. This integral can be computed by using the properties of $\mathcal{P}_{jkmn}$ described in Sec. 13.3:

$$\begin{aligned} &\sum_{jk}\left(\mathcal{P}_{jkmn}\dddot{Q}_{mn}\right)^2 = \sum_{jk}\mathcal{P}_{jkmn}\dddot{Q}_{mn}\mathcal{P}_{jkrs}\dddot{Q}_{rs} = \\ &= \left[\sum_{jk}\mathcal{P}_{mnjk}\mathcal{P}_{jkrs}\right]\dddot{Q}_{mn}\dddot{Q}_{rs} = \mathcal{P}_{mnrs}\,\dddot{Q}_{mn}\dddot{Q}_{rs} \\ &= \left[(\delta_{mr}-n_m n_r)(\delta_{ns}-n_n n_s) - \frac{1}{2}(\delta_{mn}-n_m n_n)(\delta_{rs}-n_r n_s)\right]\dddot{Q}_{mn}\dddot{Q}_{rs}\,. \end{aligned} \tag{13.138}$$

Since

- $\delta_{mn}\dddot{Q}_{mn} = \delta_{rs}\dddot{Q}_{rs} = 0$, because the trace of Q_{ij} vanishes by definition, and
- $n_m n_r \delta_{ns}\dddot{Q}_{mn}\dddot{Q}_{rs} = n_n n_s \delta_{mr}\dddot{Q}_{mn}\dddot{Q}_{rs}$, because Q_{ij} is symmetric,

after some manipulation we obtain

$$\sum_{jk}\left(\mathcal{P}_{jkmn}\dddot{Q}_{mn}\right)^2 = \dddot{Q}_{rn}\dddot{Q}_{rn} - 2n_m\dddot{Q}_{ms}\dddot{Q}_{sr}n_r + \frac{1}{2}n_m n_n n_r n_s \dddot{Q}_{mn}\dddot{Q}_{rs}\,. \tag{13.139}$$

By substituting this expression in Eq. 13.137, we find

$$L_{\rm GW} = \frac{G}{2c^5}\frac{1}{4\pi}\left\langle \dddot{Q}_{rn}\dddot{Q}_{rn}\int d\Omega - 2\dddot{Q}_{ms}\dddot{Q}_{sr}\int n_m n_r d\Omega + \frac{1}{2}\dddot{Q}_{mn}\dddot{Q}_{rs}\int n_m n_n n_r n_s d\Omega\right\rangle . \tag{13.140}$$

The integrals to be performed over the solid angle are:

$$\frac{1}{4\pi}\int n_m n_r d\Omega, \qquad \text{and} \qquad \frac{1}{4\pi}\int n_m n_n n_r n_s d\Omega\,.$$

Let us compute the first. In polar coordinates, the unit vector $\mathbf{n}$ is

$$n^i = (\sin\theta\cos\varphi, \sin\theta\sin\varphi, \cos\theta). \tag{13.141}$$

Thus, since the versor $\mathbf{n}$ is odd under a parity transformation, i.e. $\mathbf{n}\to-\mathbf{n}$ if $\theta\to\pi-\theta$, $\varphi\to\varphi+\pi$,

$$\frac{1}{4\pi}\int d\Omega\, n_m n_r = 0 \quad \text{when } m\neq r\,. \tag{13.142}$$

Furthermore, since there is no preferred direction in the integration (isotropy), it must be

$$\int d\Omega\; n_1^2 = \int d\Omega\; n_2^2 = \int d\Omega\; n_3^2 \quad\to\quad \frac{1}{4\pi}\int d\Omega\, n_m n_r = \text{const}\,\delta_{mr}\,. \tag{13.143}$$

For instance,

$$\frac{1}{4\pi}\int d\Omega (n_3)^2 = \frac{1}{4\pi}\int d\cos\theta d\varphi\cos^2\theta = \frac{1}{4\pi}\int_0^{2\pi} d\varphi\int_{-1}^{1} d\cos\theta\cos^2\theta = \frac{1}{3}\,, \tag{13.144}$$

and consequently

$$\frac{1}{4\pi}\int d\Omega\, n_m n_r = \frac{1}{3}\delta_{mr}\,. \tag{13.145}$$

The second integral in Eq. 13.141 can be computed in a similar way and gives

$$\frac{1}{4\pi}\int d\Omega n_m n_n n_r n_s = \frac{1}{15}\left(\delta_{mn}\delta_{rs} + \delta_{mr}\delta_{ns} + \delta_{ms}\delta_{nr}\right). \tag{13.146}$$

By substituting Eqs. 13.145 and 13.146 in Eq. 13.140, we find

$$\begin{aligned} L_{\rm GW} &= \frac{G}{2c^5}\left\langle \dddot{Q}_{rn}\dddot{Q}_{rn} - \frac{2}{3}\dddot{Q}_{ms}\dddot{Q}_{sr}\delta_{mr} + \frac{1}{30}\dddot{Q}_{mn}\dddot{Q}_{rs}\left(\delta_{mn}\delta_{rs} + \delta_{mr}\delta_{ns} + \delta_{ms}\delta_{nr}\right)\right\rangle \\ &= \frac{G}{2c^5}\left\langle \dddot{Q}_{rn}\dddot{Q}_{rn} - \frac{2}{3}\dddot{Q}_{rs}\dddot{Q}_{sr} + \frac{1}{30}\left(\dddot{Q}_{mn}\delta_{mn}\dddot{Q}_{rs}\delta_{rs} + \dddot{Q}_{rn}\dddot{Q}_{rn} + \dddot{Q}_{sn}\dddot{Q}_{ns}\right)\right\rangle \\ &= \frac{G}{2c^5}\left\langle \dddot{Q}_{rn}\dddot{Q}_{rn}\right\rangle\left[1 - \frac{2}{3} + \frac{2}{30}\right] = \frac{G}{5c^5}\left\langle \dddot{Q}_{rn}\dddot{Q}_{rn}\right\rangle , \end{aligned} \tag{13.147}$$

where we have used the property $Q_{mn}\delta_{mn} = Q_{rs}\delta_{rs} = 0$. In conclusion, the luminosity of a gravitational wave source is

$$L_{\text{GW}}(t) = \frac{G}{5c^5}\left\langle \dddot{Q}_{ij}\left(t - \frac{r}{c}\right)\dddot{Q}_{ij}\left(t - \frac{r}{c}\right)\right\rangle , \qquad i, j = 1, 2, 3 . \tag{13.148}$$

This expression is called the **luminosity quadrupole formula**. This landmark result was derived in 1918 by Einstein in the paper *Über Gravitationswellen* [44].

CHAPTER 14

Gravitational wave sources

In this chapter we shall apply the formalism introduced in Chapter 13 to study the gravitational wave emission of the most relevant astrophysical sources of gravitational waves: black holes and neutron stars. We will refer to these sources as *compact objects.*

General Relativity predicts that the emission of gravitational waves affects the orbital motion of compact binary systems: the orbit shrinks and the orbital period decreases in time. As will be shown in Sec. 14.1, the observation of this effect in binary pulsars provided the first indirect evidence of the existence of gravitational waves.

As the binary evolves and the two compact objects approach each other while revolving around their common center of mass, their "cosmic dance" becomes faster and faster and the emitted signal becomes stronger and stronger, until the two bodies merge, forming a single compact object. In 2015 the signal emitted by the *coalescence* of two black holes was detected for the first time by the interferometers of the LIGO experiment, and Sec. 14.2 will be devoted to this historical discovery.

Finally, in Sec. 14.3 a further source of gravitational waves will be described, rotating non-axisymmetric neutron stars, whose gravitational wave whispers – at the time of writing – have not been detected yet.

14.1 EVOLUTION OF A COMPACT BINARY SYSTEM

In this section we shall show how the orbital period P, the orbital distance l_0, and the Keplerian frequency ω_K of a binary system composed of two compact objects change in time, due to the emission of gravitational waves. To this purpose, we shall assume that the two bodies move on a circular orbit, and use the quadrupole approximation introduced in Chapter 13.

Using the reduced quadrupole moment of a binary system given in Eqs. 13.80 and 13.81, it is easy to show that

$$\begin{aligned}\sum_{k,n=1}^{3} \dddot{Q}_{kn}\dddot{Q}_{kn} &= \frac{\mu^2}{4} l_0^4 (2\omega_K)^6 (2\cos^2 2\omega_K t + 2\sin^2 2\omega_K t) \\ &= 32\mu^2 l_0^4 \omega_K^6 = 32\mu^2 G^3 \frac{M^3}{l_0^5}\,,\end{aligned} \tag{14.1}$$

where the total mass M and the reduced mass μ have been defined in Eqs. 13.65, 13.66, and $\omega_K = \sqrt{GM/l_0^3}$. Substituting Eq. 14.1 in Eq. 13.148, we find that the **gravitational wave luminosity of a compact binary system** is

$$L_{\mathrm{GW}} \equiv \frac{dE_{\mathrm{GW}}}{dt} = \frac{32}{5}\frac{G^4}{c^5}\frac{\mu^2 M^3}{l_0^5}\,. \tag{14.2}$$

This expression has to be considered as an average over several wavelengths, as discussed in Sec. 13.6.2. Since the emission frequency is twice the orbital frequency (see Eq. 13.96), one wavelength is emitted in half a period; thus averaging over several wavelengths corresponds to averaging in time over several periods. Therefore Eq. 14.2 applies when the orbital parameters do not change significantly over a large number of periods. This assumption is called *adiabatic approximation* and it is certainly satisfied for systems like PSR 1913+16 or PSR J0737-3039, described in Sec. 13.5, which are very far from merging [1]. In the adiabatic regime, the system has the time compensate the energy lost in gravitational waves by changing its orbital energy, in such a way that

$$\frac{dE_{\rm orb}}{dt} + L_{\rm GW} = 0\,. \tag{14.3}$$

For circular orbits, this equation is sufficient to compute the adiabatic evolution of the orbit. We remind that we are working in the weak-field, slow-motion approximation, in which the orbital motion can be described using the laws of Newtonian gravity.

The orbital energy is

$$E_{\rm orb} = E_K + U\,, \tag{14.4}$$

where the kinetic and the gravitational energy, E_K and U, respectively, are

$$\begin{aligned} E_K &= \frac{1}{2}m_1\omega_K^2 r_1^2 + \frac{1}{2}m_2\omega_K^2 r_2^2 = \frac{1}{2}\omega_K^2\left(\frac{m_1 m_2^2 l_0^2}{M^2} + \frac{m_2 m_1^2 l_0^2}{M^2}\right) \\ &= \frac{1}{2}\omega_K^2 \mu l_0^2 = \frac{1}{2}\frac{G\mu M}{l_0}\,, \end{aligned} \tag{14.5}$$

and

$$U = -\frac{Gm_1m_2}{l_0} = -\frac{G\mu M}{l_0}\,. \tag{14.6}$$

Therefore

$$E_{\rm orb} = -\frac{1}{2}\frac{G\mu M}{l_0}\,, \tag{14.7}$$

and its time derivative is

$$\frac{dE_{\rm orb}}{dt} = \frac{1}{2}\frac{G\mu M}{l_0}\left(\frac{1}{l_0}\frac{dl_0}{dt}\right) = -E_{\rm orb}\left(\frac{1}{l_0}\frac{dl_0}{dt}\right)\,. \tag{14.8}$$

Substituting Eq. 14.8 in Eq. 14.3, using Eqs. 14.7 and 14.2, we find

$$\frac{1}{l_0}\frac{dl_0}{dt} = \frac{L_{\rm GW}}{E_{\rm orb}} = -\frac{64}{5}\frac{G^3}{c^5}\mu M^2\frac{1}{l_0^4}\,. \tag{14.9}$$

Assuming that at some initial time $t = 0$ the orbital separation is $l_0(t = 0) = l_0^{\rm in}$, the integration of Eq. 14.9 yields

$$l_0^4(t) = (l_0^{\rm in})^4 - \frac{256}{5}\frac{G^3}{c^5}\mu M^2 t\,, \tag{14.10}$$

and defining

$$t_c \equiv \frac{5}{256}\frac{c^5}{G^3}\frac{\left(l_0^{\rm in}\right)^4}{\mu M^2}\,, \tag{14.11}$$

[1]In these systems, the change of the period during one orbit, i.e. over a time interval of the order of P, is $\delta P \sim (dP/dt)P$; since $dP/dt \ll 1$, as shown in Eq. 14.21, $\delta P \ll P$.

Eq. 14.10 becomes

$$l_0(t) = l_0^{\rm in}\left(1 - \frac{t}{t_c}\right)^{1/4}, \tag{14.12}$$

which shows that *the orbital separation decreases with time and becomes zero when* $t = t_c$. This last statement is a consequence of the assumption that the two compact bodies are point particles. Of course, stars and black holes have finite sizes, therefore they merge before $t = t_c$. In addition, in the last cycles of the inspiralling before merging both the slow-motion approximation and the weak-field assumption, on which the quadrupole formalism relies, fail and strong field effects have to be considered. However, these cycles are very fast compared to those spent by the system in the adiabatic regime and do not contribute significantly to the total time needed to merge. Therefore, t_c provides a reliable estimate of the **time of coalescence** of a compact binary with initial orbital separation $l_0^{\rm in}$.

Using Eq. 14.12 we can now compute how the Keplerian frequency and the orbital period change in time. From the definition given in Eq. 13.69

$$\omega_K = \frac{(GM)^{1/2}}{l_0^{3/2}} \qquad \rightarrow \qquad \omega_K(t) = \frac{(GM)^{1/2}}{(l_0^{\rm in})^{3/2}}\left(1 - \frac{t}{t_c}\right)^{-3/8}, \tag{14.13}$$

i.e.

$$\omega_K(t) = \omega_K^{\rm in}\left(1 - \frac{t}{t_c}\right)^{-3/8}, \quad \text{where} \quad \omega_K^{\rm in} \equiv \frac{(GM)^{1/2}}{(l_0^{\rm in})^{3/2}}. \tag{14.14}$$

Moreover, since $\omega_K = 2\pi P^{-1}$, we get

$$P(t) = P^{\rm in}\left(1 - \frac{t}{t_c}\right)^{3/8}. \tag{14.15}$$

From Eqs. 14.14 and 14.15 it follows that *as* $t \rightarrow t_c$ *the orbital frequency increases and the orbital period decreases.*

It is also possible to evaluate the rate at which the orbital period changes due to gravitational wave emission. Since

$$\omega_K^2 = GMl_0^{-3} \;\rightarrow\; P^2 = \frac{4\pi^2}{GM}l_0^3 \;\rightarrow\; 2\ln P = \ln\frac{4\pi^2}{GM} + 3\ln l_0 \;\rightarrow\; \frac{1}{P}\frac{dP}{dt} = \frac{3}{2}\frac{1}{l_0}\frac{dl_0}{dt}, \tag{14.16}$$

and using Eq. 14.9 we find

$$\frac{1}{P}\frac{dP}{dt} = \frac{3}{2}\frac{L_{\rm GW}}{E_{\rm orb}} = -\frac{96}{5}\frac{G^3}{c^5}\frac{\mu M^2}{l_0^4}. \tag{14.17}$$

Using the relation $l_0 = (GM)^{1/3}\left(\frac{P}{2\pi}\right)^{2/3}$, Eq. 14.17 becomes

$$\frac{1}{P}\frac{dP}{dt} = -\frac{96}{5}\frac{G^{5/3}}{c^5}\mu M^{2/3}\left(\frac{2\pi}{P}\right)^{8/3}. \tag{14.18}$$

For example if we consider PSR 1913+16, and neglect the eccentricity of its orbit, using the data given in Eq. 13.94, we find

$$\frac{dP}{dt} \simeq -2.0 \times 10^{-13}. \tag{14.19}$$

As mentioned in Sec. 13.5, the orbit PSR 1913+16 is strongly eccentric, with $e \simeq 0.62$. In this case the calculations above have to be repeated using the equations of motion appropriate for eccentric orbits. In addition, it can be shown that gravitational wave emission also affects the eccentricity of the orbit; in particular $de/dt < 0$, i.e. the orbit slowly circularizes during the inspiral. The final result for the period evolution reads

$$\frac{1}{P}\frac{dP}{dt} = -\frac{96}{5}\frac{G^{5/3}}{c^5}\mu M^{2/3}\left(\frac{2\pi}{P}\right)^{8/3}\frac{1}{(1-e^2)^{7/2}}\left(1+\frac{73}{24}e^2+\frac{37}{96}e^4\right), \tag{14.20}$$

which gives

$$\frac{dP}{dt} = -2.4\times 10^{-12}. \tag{14.21}$$

The interested reader can find the derivation of Eq. 14.20 in [82, 93]. The secular variation of the orbital period predicted by Eq. 14.21 has been confirmed by astrophysical observations. After the binary system PSR 1913+16 was discovered in 1975, the times $\{t_i\}_{i=0,\dots}$ of the periastron passages have been measured for decades, finding that – as expected – they are not separated by integer multiples of the initial period $P^{\rm in} = t_1 - t_0$. Instead, the time residuals $\Delta t_i = t_i - t_0 - iP^{\rm in}$ have a *quadratic* dependence on t_i, as shown in Fig. 14.1.

In order to understand this behaviour, let us compute the phase of the orbital motion $\phi(t) = \int_0^t \omega_K(t)dt$ (see Eq. 13.70) assuming that $\dot\omega_K$ is approximately constant

$$\phi(t) = \int_0^t \omega_K(t)dt \simeq \int_0^t \left(\omega_K^{\rm in} + \dot\omega_K t\right)dt = \omega_K^{\rm in}t + \frac{1}{2}\dot\omega_K t^2, \tag{14.22}$$

where we have set $t_0 = 0$. The times of the periastron passages are those for which $\phi(t_i) = 2\pi i$, i.e.

$$\omega_K^{\rm in}t_i + \frac{1}{2}\dot\omega_K t_i^2 = 2\pi i \quad\rightarrow\quad t_i + \frac{1}{2}\frac{\dot\omega_K}{\omega_K^{\rm in}}t_i^2 = \frac{2\pi i}{\omega_K^{\rm in}}. \tag{14.23}$$

Since

$$\frac{\dot\omega_K}{\omega_K^{\rm in}} = -\frac{\dot P}{P^{\rm in}}, \tag{14.24}$$

the time residuals are

$$\Delta t_i = t_i - iP^{\rm in} = t_i - \frac{2\pi i}{\omega_K^{\rm in}} = -\frac{1}{2}\frac{\dot\omega_K}{\omega_K^{\rm in}}t_i^2 = \frac{1}{2}\frac{\dot P}{P^{\rm in}}t_i^2, \tag{14.25}$$

i.e., they are expected to be fitted by a parabola with coefficient $\frac{\dot P}{P^{\rm in}} < 0$. In Fig. 14.1 Eq. 14.25 is plotted as a continuous line.

The measured values of the time residuals Δt_i for PSR 1913+16, indicated by dots in Fig. 14.1, confirm the prediction of General Relativity. It should be mentioned that, in addition to the secular changes of the orbital period induced by gravitational wave emission, the observed variation of P in PSR 1913+16 is also affected by kinematic effects due to the center-of-mass acceleration along the line of sight, and to the variation of the orbital inclination. They both produce a Doppler shift of the pulse frequency of the pulsar, which affects the measurement of P and have to be properly subtracted in order to isolate the effect genuinely due to General Relativity [79]. After this subtraction, the ratio between the observed value of $\dot P$ and the theoretical value predicted by General Relativity is [118]

$$\frac{\dot P_{\rm obs}}{\dot P_{GR}} = 1.0013 \pm 0.021. \tag{14.26}$$

This result provided the first indirect evidence of the existence of gravitational waves and for this discovery Hulse and Taylor were awarded the Nobel Prize in Physics in 1993.

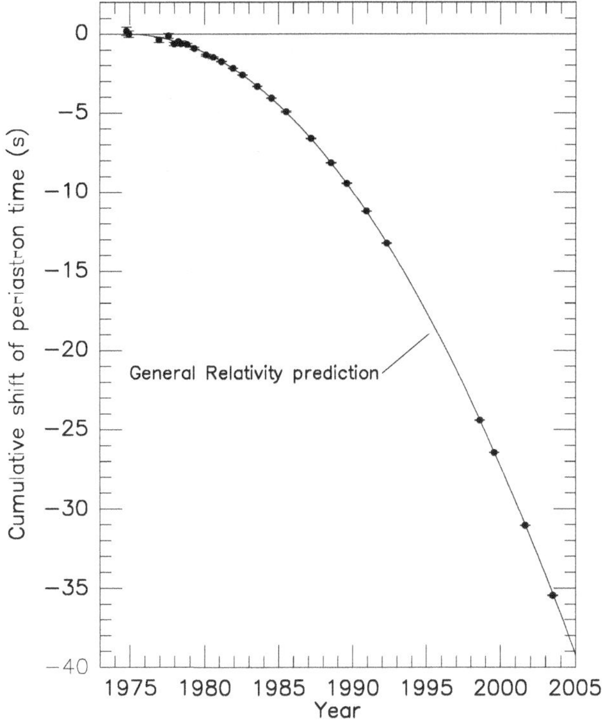

Figure 14.1: Shift in the time of periastron passage of the binary pulsar PSR 1913+16. The parabolic behaviour agrees with the predictions of General Relativity based on the assumption that the system loses energy due to gravitational wave emission, as shown in Eq. 14.25. From: *J.M. Weisberg & J.H. Taylor, Relativistic binary pulsar B1913+16: Thirty years of observations and analysis* [118].

For the double pulsar PSR J0737-3039 [81] discovered in 2003 (see Sec. 13.5), using the data given in Eq. 13.100, Eq. 14.20 gives

$$\frac{dP}{dt} = -1.2 \times 10^{-12}, \tag{14.27}$$

which is also in agreement with the observations, with an accuracy similar to that of Eq. 14.26.

14.2 GRAVITATIONAL WAVES FROM INSPIRALLING COMPACT OBJECTS

Equations 14.14 and 14.12 show how the Keplerian frequency ω_K and the orbital distance l_0 of a binary in the adiabatic regime evolve as the two compact objects inspiral around the common center of mass; during this phase – the *inspiralling* or simply the **inspiral** – the orbit goes through a sequence of stationary circular orbits. We shall now compute how the wave frequency and amplitude change with time. In Sec. 13.5 we showed that the signal emitted at a time t by a binary system moving on a circular orbit, and located at distance r from the observer, is

$$h_{ij}^{\mathbf{TT}}(t,r) = -h_0 \; A_{ij}^{\mathbf{TT}}\left(t - \frac{r}{c}\right) , \tag{14.28}$$

where h_0 is the instantaneous wave amplitude

$$h_0 = \frac{4\mu M G^2}{l_0 c^4 r} , \tag{14.29}$$

and $A_{ij}^{\mathbf{TT}}(t - \frac{r}{c}) = \mathcal{P}_{ijkl} A_{kl}(t - \frac{r}{c})$, where

$$A_{ij}(t) = \begin{pmatrix} \cos 2\omega_K t & \sin 2\omega_K t & 0 \\ \sin 2\omega_K t & -\cos 2\omega_K t & 0 \\ 0 & 0 & 0 \end{pmatrix} . \tag{14.30}$$

The wave is monochromatic, and is emitted at twice the Keplerian frequency, i.e.

$$\nu_{\text{GW}} = 2\nu_K = \frac{\omega_K}{\pi} . \tag{14.31}$$

From Eq. 14.14 it follows that, as the two bodies approach each other, the wave frequency changes as

$$\nu_{\text{GW}}(t) = \nu_{\text{GW}}^{\text{in}} \left(1 - \frac{t}{t_c}\right)^{-3/8} , \qquad \nu_{\text{GW}}^{\text{in}} \equiv \frac{1}{\pi}\sqrt{\frac{GM}{(l_0^{\text{in}})^3}} . \tag{14.32}$$

In addition, since the orbital distance decreases according to Eq. 14.12, from Eq. 14.29 it follows that the wave amplitude changes with time as

$$\begin{aligned} h_0(t) &= \frac{4\mu M G^2}{l_0(t) c^4 r} = \frac{4\mu M G^2}{r c^4} \frac{\omega_K^{2/3}(t)}{G^{1/3} M^{1/3}} \\ &= \frac{4\pi^{2/3} G^{5/3} \mathcal{M}^{5/3}}{r c^4} \nu_{\text{GW}}^{2/3}(t) , \end{aligned} \tag{14.33}$$

where we have defined

$$\mathcal{M} = \mu^{3/5} M^{2/5} = \frac{(m_1 m_2)^{3/5}}{M^{1/5}} . \tag{14.34}$$

Eqs. 14.32 and 14.33 show that both the frequency and the amplitude of the signal emitted by a binary system during the inspiral increase with time. This feature is typical of the chirp of a singing bird, and for this reason this part of the signal is named *chirp*, and $\mathcal{M}$ is called the **chirp mass.**

The phase Φ of the gravitational signal is obtained by integrating the instantaneous phase of the polarization tensor 14.30

$$\Phi(t) = \int_0^t 2\omega_K(t) dt + \Phi_{\text{in}} = \int_0^t 2\pi\nu_{\text{GW}}(t) dt + \Phi_{\text{in}}, \qquad \text{where} \quad \Phi_{\text{in}} = \Phi(t=0) . \tag{14.35}$$

To compute this integral it is convenient to use the following relations. First of all we write Eq. 14.32 in the form

$$\nu_{\text{GW}}(t) = \frac{\nu_{\text{GW}}^{\text{in}} t_c^{3/8}}{(t_c - t)^{3/8}} . \tag{14.36}$$

Using Eq. 14.11 and given $256^{3/8} = 8$, it is easy to show that

$$\nu_{\text{GW}}^{\text{in}} t_c^{3/8} = \frac{5^{3/8}}{8\pi} \left(\frac{c^3}{G\mathcal{M}}\right)^{5/8} , \tag{14.37}$$

so that $\nu_{GW}(t)$ can be written as

$$\nu_{GW}(t) = \frac{5^{3/8}}{8\pi}\left(\frac{c^3}{G\mathcal{M}}\right)^{5/8}\left[\frac{1}{t_c - t}\right]^{3/8}. \tag{14.38}$$

The integrated phase therefore is

$$\Phi(t) = -2\left[\frac{c^3\,(t_c - t)}{5G\mathcal{M}}\right]^{5/8} + \Phi_c\,, \tag{14.39}$$

where $\Phi_c = \Phi_{\rm in} - 2\left(\frac{c^3 t_c}{5G\mathcal{M}}\right)^{5/8}$ is the phase at $t = t_c$. Eq. 14.39 shows that, within our approximations, *the phase of the gravitational signal depends on the masses m_1 and m_2 only through the combination given by the chirp mass $\mathcal{M}$.* An alternative expression of the phase 14.39 in terms of the wave frequency is

$$\Phi(t) = -\frac{1}{16}\left(\frac{c^3}{G}\right)^{5/3}(\pi\mathcal{M}\nu_{GW})^{-5/3} + \Phi_c\,. \tag{14.40}$$

In conclusion, the signal emitted during the inspiral is

$$h^{\mathbf{TT}}_{ij}(t,r) = -h_0\left(t - \frac{r}{c}\right)\mathcal{P}_{ijkl}A_{kl}\left(t - \frac{r}{c}\right), \tag{14.41}$$

where

$$h_0(t) = \frac{4\pi^{2/3}G^{5/3}}{c^4}\frac{\mathcal{M}}{r}(\mathcal{M}\nu_{GW}(t))^{2/3} \tag{14.42}$$

is the wave amplitude, and

$$A_{ij}(t) = \begin{pmatrix} \cos\Phi(t) & \sin\Phi(t) & 0 \\ \sin\Phi(t) & -\cos\Phi(t) & 0 \\ 0 & 0 & 0 \end{pmatrix} \tag{14.43}$$

is the polarization tensor.

14.2.1 September 14th, 2015: the detection of gravitational waves

On September 14th, 2015 the interferometric antennas of the experiment LIGO, located in Livingston (Louisiana, USA) and in Hanford (Washington, USA) detected, for the first time, the gravitational wave signal emitted in the coalescence of two black holes, which is shown in Fig. 14.2. In the upper panels the output of the two detectors is shown as a function of time; the middle panels show the signal extracted from the raw data using suitable filtering techniques; the bottom panels show how the frequency of the signal grows with time. This signal was named "GW150914".

The analysis of the data showed that the detected wave was emitted by two black holes with masses

$$m_1 = 35.6^{+4.8}_{-3.0}M_\odot\,, \qquad m_2 = 30.6^{+3.0}_{-4.4}M_\odot\,. \tag{14.44}$$

As explained in previous sections, due to the emission of gravitational waves the two black holes orbit around each other in an ever-decreasing orbit, until they "merge" and form a single black hole. This process is the aforementioned *coalescence.* The remnant black hole of GW150914 has mass and dimensionless angular momentum (spinning black holes will be discussed in Chapter 18)

$$M_{\rm fin} = 63.1^{+3.3}_{-3.0}M_\odot\,, \qquad \frac{cJ_{\rm fin}}{GM^2_{\rm fin}} = 0.69^{+0.05}_{-0.04}\,. \tag{14.45}$$

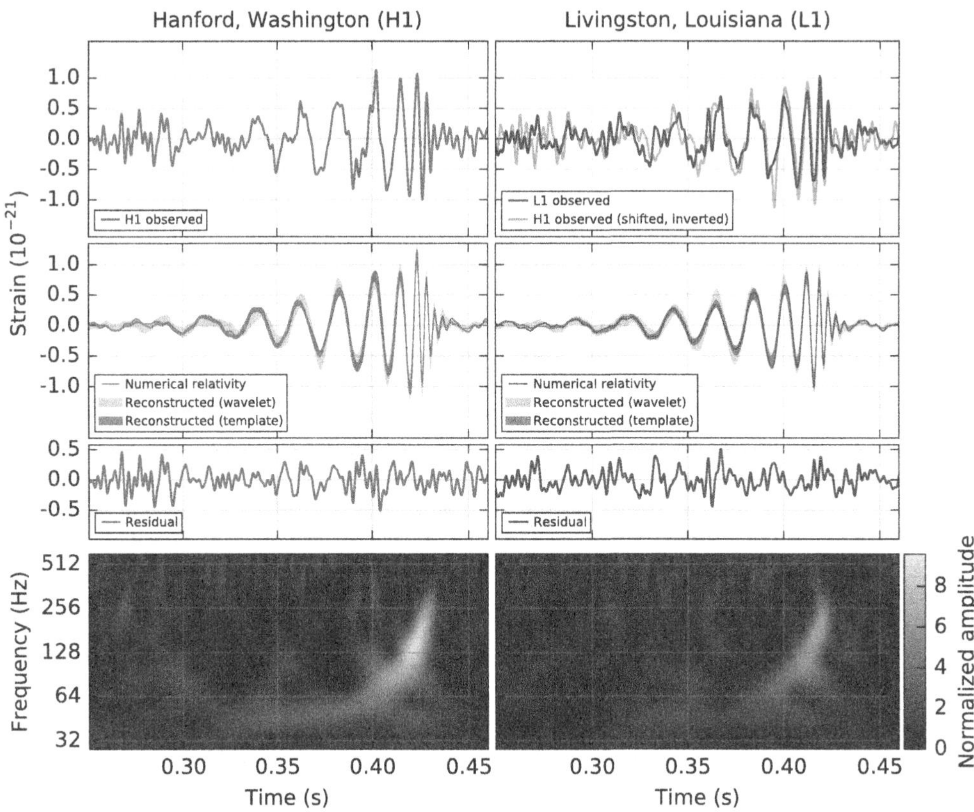

Figure 14.2: The signal of the first gravitational wave event, GW150914, see text for details. From: B.P. Abbott *et al.*, *Observation of gravitational waves from a binary black hole merger* [1].

A comparison of the total mass of the binary before merging with the mass of the final black hole, shows that $\sim 3.1 M_\odot$ have been lost. This corresponds to a huge amount of energy being radiated in gravitational waves:

$$E_{\rm GW} \sim 3.1 M_\odot c^2 \quad \to \quad 3.1 \times 1.989 \times 10^{33}\,{\rm g} \times (2.998 \times 10^{10} {\rm cm/s})^2 = 5.5 \times 10^{54}\,{\rm erg}\,. \tag{14.46}$$

As a comparison, it may be recalled that the total energy emitted in neutrinos and electromagnetic radiation in the most energetic supernova explosions is of the order $\sim 10^{53}$ erg.

We shall now discuss how the parameters of the source have been extracted from the detected signal.

14.2.2 The chirp mass and the luminosity distance

The phase and the amplitude of the gravitational wave signal emitted during the inspiralling of two compact bodies, the chirp, are given in Eqs. 14.39 and 14.42, respectively. From the detected phase it is possible to measure the chirp mass $\mathcal{M}$ as follows. If we define the constant

$$K = \frac{5^{3/8}}{8\pi} \left(\frac{c^3}{G\mathcal{M}} \right)^{5/8}, \tag{14.47}$$

Eq. 14.38 can be written as

$$\nu_{\rm GW}(t) = \frac{K}{(t_c - t)^{3/8}} \tag{14.48}$$

and

$$\dot{\nu}_{\rm GW} = \frac{3}{8}\frac{K}{(t_c - t)^{11/8}} \quad \to \quad \dot{\nu}_{\rm GW}\nu_{\rm GW}^{-11/3} = \frac{3}{8}K^{-8/3}. \tag{14.49}$$

Using the definition of K the last equation gives

$$\left[\frac{8\pi}{5^{3/8}}\left(\frac{G\mathcal{M}}{c^3}\right)^{5/8}\right]^{8/3} = \frac{8}{3}\dot{\nu}_{\rm GW}\nu_{\rm GW}^{-11/3}, \tag{14.50}$$

which can be solved for $\mathcal{M}$, and gives

$$\mathcal{M} = \frac{c^3}{G}\left(\frac{5}{96}\frac{1}{\pi^{8/3}}\dot{\nu}_{\rm GW}\nu_{\rm GW}^{-11/3}\right)^{3/5}. \tag{14.51}$$

Thus, the chirp mass can be obtained by measuring the wave frequency and its time derivative from the inspiralling part of the signal.

When the source is located at cosmological distance, we can no longer assume that the gravitational wave propagates in Minkowski's spacetime: we need to take into account the effects of the cosmological expansion, which are briefly discussed in Sec. 14.4. In this case, due to the expansion of the Universe, the frequency $\nu_{\rm GW}^{\rm obs}$ measured at the detector differs from the frequency $\nu_{\rm GW}$ emitted by the source by (see Eq. 14.127)

$$\nu_{\rm GW}^{\rm obs} = \frac{\nu_{\rm GW}}{1+z}, \tag{14.52}$$

where z is the cosmological redshift. Moreover, the observed time separation between two pulses is rescaled according to Eq. 14.128, $dt_{\rm obs} = (1+z)dt_{\rm em}$.

Therefore, the quantity which is actually measured is the so-called *redshifted chirp mass*

$$\mathcal{M}^{\rm obs} = \frac{c^3}{G}\left[\frac{5}{96}\frac{1}{\pi^{8/3}}\left(\frac{d}{dt_{\rm obs}}\nu_{\rm GW}^{\rm obs}\right)(\nu_{\rm GW}^{\rm obs})^{-11/3}\right]^{3/5}, \tag{14.53}$$

and since

$$\left[\left(\frac{d}{dt_{\rm obs}}\nu_{\rm GW}^{\rm obs}\right)(\nu_{\rm GW}^{\rm obs})^{-11/3}\right]^{3/5} = [\nu_{\rm GW}^{-11/3}\dot{\nu}_{\rm GW}]^{3/5} \times (1+z), \tag{14.54}$$

we get

$$\mathcal{M}^{\rm obs} = \frac{c^3}{G}\left(\frac{5}{96}\frac{1}{\pi^{8/3}}\dot{\nu}_{\rm GW}\nu_{\rm GW}^{-11/3}\right)^{3/5} \times (1+z). \tag{14.55}$$

The redshifted chirp mass is related to the true chirp mass $\mathcal{M}$, given in Eq. 14.51, by

$$\mathcal{M}^{\rm obs} = \mathcal{M}\,(1+z). \tag{14.56}$$

Let us now see which information can be extracted from the amplitude of the chirp. As discussed in Sec. 14.4, the amplitude of a gravitational wave propagating in the curved background of an expanding universe has the same form as that of a wave propagating in Minkowski's space, with the radial distance r replaced by the proper distance D at the time t. Thus, for the wave emitted by a compact binary inspiral, Eq. 14.42 becomes:

$$h_0(t) = \frac{4\pi^{2/3}G^{5/3}}{c^4} \times \frac{\mathcal{M}}{D(t)} \times (\mathcal{M}\nu_{\rm GW})^{2/3}. \tag{14.57}$$

Expressing Eq. 14.57 in terms of the quantities measured by the detector at the time $t_{\rm obs}$, i.e. $\nu_{\rm GW}^{\rm obs}$ and $\mathcal{M}^{\rm obs}$, we obtain

$$h_0(t_{\rm obs}) = \frac{4\pi^{2/3}G^{5/3}}{c^4} \times \frac{\mathcal{M}^{\rm obs}}{D_L(t_{\rm obs})} \times (\mathcal{M}^{\rm obs}\nu_{\rm GW}^{\rm obs})^{2/3}, \tag{14.58}$$

where $D_L = (1+z)D$ is the *luminosity distance* of the source, defined in Sec. 14.4.3 (see Eq. 14.144).

Summarizing, by detecting the gravitational wave signal emitted by a compact binary inspiral, we can determine the redshifted chirp mass $\mathcal{M}^{\rm obs}$ from the time evolution of the phase, and the luminosity distance from the amplitude h_0, by inverting Eq. 14.58 [2]. In this way, the luminosity distance of GW150914 has been found to be

$$D_L = 430^{+150}_{-170}\ \text{Mpc}. \tag{14.59}$$

This means that (since 1 pc $\simeq$ 3.262 ly) the coalescence occurred approximately 1 Gyr before its detection: the gravitational wave had to travel for a cosmological time before reaching the Earth and leaving its tiny imprint on the instruments!

In order to find the true chirp mass $\mathcal{M} = \mathcal{M}^{\rm obs}/(1+z)$, we need to know the source redshift z; this can be found if the sky position of the source is localized with sufficient precision, so that the hosting galaxy can be identified. This is possible if an electromagnetic counterpart of the gravitational event can be detected. However, coalescing black holes are not expected to have an electromagnetic counterpart because, according to what is presently known about their evolutionary path, they should not be surrounded by accretion disks (see Sec. 11.5.1) sufficiently massive to produce a significant electromagnetic emission at merger. In addition, the localization of a gravitational wave event is done by triangulation of the signal detected by multiple interferometers. Using only the two LIGO detectors, the position of GW150914 was localized in a region of sky of 180 deg^2, too large to be spanned by the existing electromagnetic detectors. To restrict this area more detectors are needed, as we will see in Sec. 14.2.6.

In the absence of an independent measure of the source redshift, this important parameter can be inferred in the following way. Given a cosmological model, the luminosity distance can be expressed in terms of the redshift and of the cosmological parameters. In the Friedmann-Lemaître-Robertson-Walker model, in the case of small redshift, its expression is given by Eq. 14.151, which we anticipate here:

$$D_L(z) = \frac{c}{H_0}\left[z + \frac{1}{2}(1-q_0)z^2 + \dots\right], \tag{14.60}$$

where H_0 is the Hubble constant, and q_0 is the deceleration parameter (see Sec. 14.4.2). The LIGO-Virgo collaboration used the values of the cosmological parameters measured by the Planck mission [92]. The value of the luminosity distance was found from the measured value of the wave amplitude h_0 and Eq. 14.58; then, using Eq. 14.60 it was possible to infer the redshift of the source, finding $z = 0.09^{+0.03}_{-0.03}$. Knowing the redshift, it was finally possible to obtain the true value of the chirp mass of GW150914:

$$\mathcal{M} = 28.6^{+1.6}_{-1.5}\,M_\odot\,. \tag{14.61}$$

[2] It should be mentioned that in order to measure D_L we should also know the orientation of the source with respect to the line of sight. This introduces further parameters (the orientation angles) in the waveform which have not been considered in our simplified analysis. The degeneracy among these parameters can be resolved by detecting the signal with multiple detectors.

14.2.3 A lower bound for the total mass of the system

From the estimated value of the chirp mass it is possible to set a lower bound for the total mass of the binary as follows. In Fig. 14.3 we plot the total mass of the system, $M = m_1+m_2$,

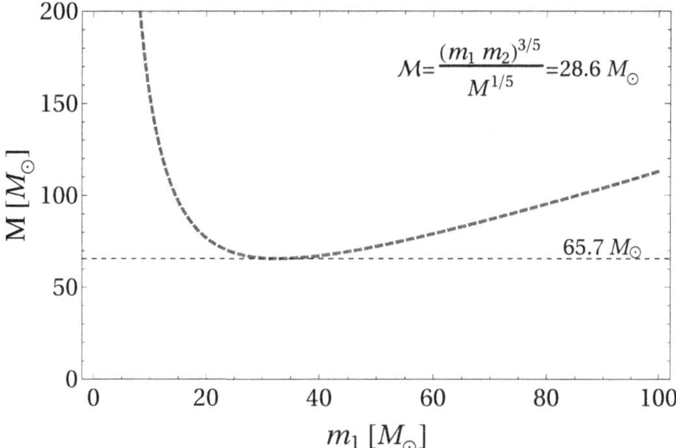

Figure 14.3: The total mass of the coalescing system, $M = m_1 + m_2$, as a function of the mass m_1, assuming the chirp mass of GW150914, $\mathcal{M} = 28.6\, M_\odot$.

as a function of the mass of one of the two bodies, say m_1, using the equation

$$\frac{(m_1 m_2)^{3/5}}{M^{1/5}} = \mathcal{M} = 28.6\, M_\odot\,. \tag{14.62}$$

We see that, to be compatible with the observed chirp mass, the total mass of the system must be larger than $M = 65.7\, M_\odot$. From this value we understand that the coalescing bodies cannot be neutron stars, since the maximum mass observed for neutron stars is $\sim 2\, M_\odot$ (see Chapter 16).

Further information can be obtained by evaluating the distance between the two objects just before merging. We know that the wave frequency is related to the orbital distance l_0 and to the total mass $M = m_1 + m_2$ by

$$\nu_{\rm GW}(t) = 2\nu_{\rm orb} = \frac{1}{\pi}\sqrt{\frac{GM}{l_0^3(t)}}\,. \tag{14.63}$$

The data show that the frequency of GW150914 spanned the range from 35 to 150 Hz in a time interval of about 0.2 s; therefore, when $\nu_{\rm GW} = 150\,\text{Hz}$, the distance between the two objects was approximately[3]

$$l_0 \sim \left(\frac{GM}{(\nu_{\rm GW}\pi)^2}\right)^{1/3} \sim 3.4\times 10^2\,\text{km}\,, \tag{14.64}$$

where we assumed $M \sim 65.7\, M_\odot$. Considering how large is the mass of the two bodies, this distance is extremely small, and this indicates that they must be extremely compact. To the best of our knowledge, only black holes can be so compact and massive. As a comparison, the Schwarzschild radius of a black hole with mass $65.7\, M_\odot$ is 1.94×10^2 km.

[3]We are assuming that the two bodies are non-spinning; this is justified by the data which are compatible with vanishing spins (spinning black holes will be discussed in Chapter 18).

At this point we have exploited all information which can be extracted from the inspiral (at leading order in the weak-field, slow-motion, and adiabatic approximation):

- the chirp mass from the wave phase;
- the luminosity distance from the wave amplitude;
- a lower bound on the total mass, which excludes the possibility that the two bodies are neutron stars;
- an approximate value of the orbital distance just before merging, which indicates that the two bodies are extremely compact.

Let us now see what else can be inferred from the latest stages of the coalescence.

14.2.4 The final stages of the inspiral

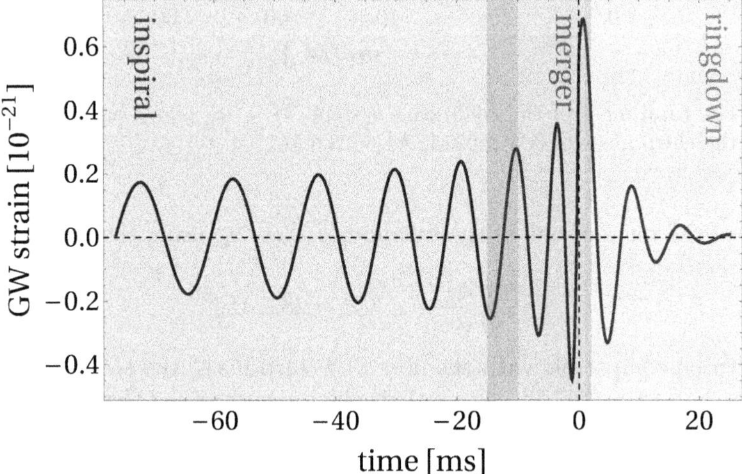

Figure 14.4: Representative example of the signal emitted by coalescing black holes. The signal within the leftmost shaded band is the chirp emitted during the inspiral, when the two bodies approach each other; the signal within the central shaded band is emitted during the merger, whereas the last part, the ringing tail within the rightmost shaded band, is emitted by the final distorted black hole which oscillates according to its characteristic quasi-normal modes. Note that the separation between inspiral, merger, and ringdown is only approximate, since in the last stages of the coalescence the system is highly non-linear.

In Fig. 14.4 we show the plot of a typical waveform emitted in the coalescence of two black holes. The first part is the chirp, emitted during the adiabatic inspiralling and is described by Eqs. 14.41, 14.42, and 14.43; these equations describe the signal with a good approximation, provided the orbital velocities are much smaller than the speed of light. When this condition is no longer verified the quadrupole approach, which we have used to derive the waveform, becomes inaccurate; this occurs approximately when the orbital distance between the two bodies approaches the innermost, stable circular orbit, the ISCO, of a black hole with mass $m_1 + m_2$; assuming that the two bodies have zero spins, this distance is (see Sec. 10.5):

$$l_{\rm ISCO} = \frac{6G(m_1 + m_2)}{c^2} \,. \tag{14.65}$$

The wave frequency which corresponds to the orbital distance $l_{\rm ISCO}$ is

$$\nu_{\rm GW}^{\rm ISCO} = \frac{\omega_K}{\pi} = \frac{1}{\pi}\sqrt{\frac{G(m_1+m_2)}{{l_{\rm ISCO}}^3}} = \frac{c^3}{\pi G\sqrt{6^3}}\frac{1}{(m_1+m_2)}\,. \tag{14.66}$$

Note that $\nu_{\rm GW}^{\rm ISCO}$ scales as the inverse of the total mass of the system.

The waveform derived within the quadrupole approximation corresponds to the lowest-order term in a **post-Newtonian expansion** of the equations of motion in the parameter v/c, where v is the orbital velocity of the coalescing bodies. At distances comparable to $l_{\rm ISCO}$ the waveform must be corrected including more terms with higher power in (v/c). Customarily, the parameter

$$x = (v/c)^2 = \frac{1}{c^2}[G(m_1+m_2)\pi\nu_{\rm GW}]^{2/3} \tag{14.67}$$

indicates the order of the correction to the quadrupole waveform. The first correction to the phase of the signal given in Eq. 14.40 is of order $O(x)$, and depends on the mass ratio of the binary, m_1/m_2. This information can be extracted by an accurate data analysis and, knowing the chirp mass, allows to estimate the individual values of m_1 and m_2.

The next corrections to the phase account for the effect of the spins of the binary components; these terms are of order $O(x^{3/2})$ and $O(x^2)$. Therefore spin corrections start to be significant when the velocities become large, the two bodies are very close to merging, and a large signal amplitude is needed to extract this contribution from the detector noise. For instance, even though GW150914 was a very loud signal, the individual spins of the coalescing bodies have only been weakly constrained, and appear to be compatible with zero.

If the coalescing bodies are neutron stars, phase corrections of order $O(x^5)$ carry information on the deformations these bodies undergo due to mutual tidal interactions. For a detailed discussion of the effects of tidal interactions and other post-Newtonian corrections on the waveform, we refer the interested reader to the monograph [93].

14.2.5 Merger and ringdown: identifying the nature of coalescing compact objects

When the two bodies are very close to merging, strongly non-linear effects take over and even the post-Newtonian approach becomes inaccurate. The part of the signal corresponding to the **merger**, the central shaded band in Fig. 14.4, has to be found by solving numerically Einstein's equations in the fully non-linear regime. These studies started in the late 1990s with the Grand Challenge project, which aimed at simulating the collision of two black holes. This goal was finally achieved in 2006, when the first waveforms were obtained [96, 28, 14]. After this breakthrough, a bank of templates was set up, which has been instrumental to extract the gravitational wave signal from the detectors' noise. Indeed from these waveforms, which include initially eccentric orbits and the spins of the bodies, one can find some very useful fitting formulae which allow to estimate the masses and the spins of the two coalescing bodies (to be compared with those obtained from the analysis of the inspiral part of the signal), and the mass and spin of the final black hole.

The last part of the signal, the rightmost shaded band in Fig. 14.4, is the *ringing tail* or **ringdown**, and is emitted by the distorted black hole which forms as a result of the merger. As we shall discuss in Chapter 15, the gravitational signal emitted by a perturbed black hole will, after a transient, decay as a superposition of its quasi-normal modes (QNMs) of oscillations, which are described by damped sinusoids

$$h(t) = \sum_i A^{(i)}\sin(\omega_R^{(i)}t + \Phi^{(i)})e^{-t/\tau^{(i)}}\,, \tag{14.68}$$

where $\omega_R^{(i)}$ and $\omega_I^{(i)} \equiv -1/\tau^{(i)}$ are the real and imaginary part of the eigenfrequency of the i-th mode, and $A^{(i)}$ and $\Phi^{(i)}$ are the corresponding amplitude and phase. The imaginary part of the frequency is the inverse of the damping time; indeed the oscillations are damped because the black hole loses energy emitting gravitational waves.

Remarkably, the QNM frequencies depend only on the mass and on the angular momentum of the black hole. This result is known as the *no-hair theorem* (see Chapter 18). If the black hole is non-rotating, the fundamental ($i = 0$) QNM frequency, in geometrized units, is given by Eq. 15.111, which we anticipate here

$$M\omega^{(0)} \simeq 0.3736 - i\,0.0890\,. \tag{14.69}$$

For a black hole with mass $M = n\,M_\odot$, this equation yields the frequency and damping time

$$\nu_0 \simeq \frac{12}{n}\,\text{kHz}, \qquad \tau_0 \simeq n\,5.5 \times 10^{-5}\,\text{s}\,, \tag{14.70}$$

which shows that the larger the black hole mass, the smaller the frequency and the longer the damping time. As an example, if the black hole has a mass of 10 $M_\odot$, the frequency and the damping time of the fundamental QNM are $\nu_0 = 1.2$ kHz, $\tau = 5.5 \times 10^{-4}$ s; if, instead, it is a supermassive black hole with $M = 10^6\ M_\odot$, like those sitting at the center of most galaxies (see Chapter 18), they are $\nu_0 = 1.2 \times 10^{-2}$ Hz, $\tau = 55$ s.

Note that, in general, the remnant black hole produced in a coalescence is spinning, and calculating its QNM eigenfrequencies is more complicated than what is discussed in Chapter 15 (see Box 15-D). For more details, we refer the interested reader to the monograph by Chandrasekhar [34].

In conclusion, by extracting more than one QNM eigenfrequency from the last part of a sufficiently strong signal emitted by coalescing compact binaries, it is possible not only to infer the mass and angular momentum of the final object, but also to test whether the QNM spectrum is consistent with the predictions of General Relativity. In other words, the ringdown will allow us to perform a *spectroscopy of black holes.*

From the data of GW150914 only the eigenfrequency of the lowest QNM has been measured; the mass and the angular momentum have then been deduced, and their values agree with those estimated from the merging part of the signal, within the experimental errors.

To summarize, General Relativity provides a complete description of the waveforms emitted in a binary coalescence. When compared to the data, these waveforms allowed to interpret the detected signal GW150914 as due to the coalescence of two black holes with masses given in Eq. 14.44 which, after merging, gave birth to a rotating black hole, with mass and angular momentum given in Eq. 14.45.

14.2.6 More signals from coalescences

The analysis of the data collected by the LIGO interferometers during two observational runs between 2015 and 2017 has disclosed ten more gravitational events [4], whose parameters are shown in Table 14.1. The data are tabulated as follows. In column 1 there is the name of the event; in columns 2, 3, and 4 the masses of the two coalescing bodies and the chirp mass; in columns 5 and 6 the mass and the dimensionless angular momentum of the black hole which forms as a result of the merging; in columns 7, 8, and 9 the source luminosity distance, the redshift, and the sky localization area.

[4] During the third observational run of the LIGO and Virgo interferometers in 2019, binary black hole mergers have been detected on a weekly basis. At the time of writing, the data analysis of this run is still ongoing.

Table 14.1: Source parameters of the events detected in the first two observational runs of the LIGO-Virgo collaboration between the years 2015 and 2017 [3].

Event	$\frac{m_1}{M_\odot}$	$\frac{m_2}{M_\odot}$	$\frac{\mathcal{M}}{M_\odot}$	$\frac{M_{\rm fin}}{M_\odot}$	$\frac{cJ_{\rm fin}}{GM_{\rm fin}^2}$	$\frac{D_L}{\rm Mpc}$	z	$\frac{\Delta\Omega}{\rm deg^2}$
GW150914	$35.6^{4.8}_{-3.0}$	$30.6^{+3.0}_{-4.4}$	$28.6^{+1.6}_{-1.5}$	$63.1^{+3.3}_{-3.0}$	$0.69^{+0.05}_{-0.04}$	430^{+150}_{-170}	$0.09^{+0.03}_{-0.03}$	180
GW151012	$23.3^{+14.0}_{-5.5}$	$13.6^{+4.1}_{-4.8}$	$15.2^{+2.0}_{-1.1}$	$35.7^{+9.9}_{-3.8}$	$0.67^{+0.13}_{-0.11}$	1060^{+540}_{-480}	$0.21^{+0.09}_{-0.09}$	1555
GW151226	$13.7^{+8.8}_{-3.2}$	$7.7^{+2.2}_{-2.6}$	$8.9^{+0.3}_{-0.3}$	$20.5^{+6.4}_{-1.5}$	$0.74^{+0.07}_{-0.05}$	440^{+180}_{-190}	$0.09^{+0.04}_{-0.04}$	1033
GW170104	$31.0^{+7.2}_{-5.6}$	$20.1^{+4.9}_{-4.5}$	$21.5^{+2.1}_{-1.7}$	$49.1^{+5.2}_{-3.9}$	$0.66^{+0.08}_{-0.10}$	960^{+430}_{-410}	$0.19^{+0.07}_{-0.08}$	924
GW170608	$10.9^{+5.3}_{-1.7}$	$7.6^{+1.3}_{-2.1}$	$7.9^{+0.2}_{-0.2}$	$17.8^{+3.2}_{-0.7}$	$0.69^{+0.04}_{-0.04}$	320^{+120}_{-110}	$0.07^{+0.02}_{-0.02}$	396
GW170729	$50.6^{+16.6}_{-10.2}$	$34.3^{+9.1}_{-10.1}$	$35.7^{+6.5}_{-4.7}$	$80.3^{+14.6}_{-10.2}$	$0.81^{+0.07}_{-0.13}$	2750^{+1350}_{-1320}	$0.48^{+0.19}_{-0.20}$	1033
GW170809	$35.2^{+8.3}_{-6.0}$	$23.8^{+5.2}_{-5.1}$	$25.0^{+2.1}_{-1.6}$	$56.4^{+5.2}_{-3.7}$	$0.70^{+0.08}_{-0.09}$	990^{+320}_{-380}	$0.20^{+0.05}_{-0.07}$	340
GW170814	$30.7^{+5.7}_{-3.0}$	$25.3^{+2.9}_{-4.1}$	$24.2^{+1.4}_{-1.1}$	$53.4^{+3.2}_{-2.4}$	$0.72^{+0.07}_{-0.05}$	580^{+160}_{-210}	$0.12^{+0.03}_{-0.04}$	87
GW170817	$1.46^{+0.12}_{-0.10}$	$1.27^{+0.09}_{-0.09}$	$1.186^{+0.001}_{-0.001}$	≤ 2.8	≤ 0.89	40^{+10}_{-10}	$0.01^{+0.00}_{-0.00}$	16
GW170818	$35.5^{+7.5}_{-4.7}$	$26.8^{+4.3}_{-5.2}$	$26.7^{+2.1}_{-1.7}$	$59.8^{+4.8}_{-3.8}$	$0.67^{+0.07}_{-0.08}$	1020^{+430}_{-360}	$0.20^{+0.07}_{-0.07}$	39
GW170823	$39.6^{+10.0}_{-6.6}$	$29.4^{+6.3}_{-7.1}$	$29.3^{+4.2}_{-3.2}$	$65.6^{+9.4}_{-6.6}$	$0.71^{+0.08}_{-0.10}$	1850^{+840}_{-840}	$0.34^{+0.13}_{-0.14}$	1651

Among these events there are three, GW170814, GW170817, and GW170818, which have been localized with much higher precision with respect to the others (see last column of Table 14.1). The reason is that these events have been detected in coincidence with a third interferometer, Virgo, located in Cascina, near Pisa in Italy. The sky position of a source is primarily determined through the difference in the arrival time of the signals at the detectors, their phase differences, and amplitude ratios. With a network of detectors the position can be inferred by triangulation, thus reducing the uncertainty on the source localization. All events in Table 14.1 have been identified as due to the coalescence of two black holes, except one, GW170817. The table shows that the mass of the individual black holes ranges between $7.6\,M_\odot$ and $50.6\,M_\odot$. GW170729 is not only the heaviest black hole merger, but also the most distant observed in the first two observing runs. It is interesting to note that – owing to the high orbital angular momentum acquired by the systems before the merger – the dimensionless spins of the black hole which forms after the merger cluster in the range between $\in [0.66, 0.81]$, i.e. black holes formed in these processes are rapidly rotating. Indeed, as we shall show in Chapter 18, the maximum dimensionless spin of a rotating black hole is $cJ/(GM^2) = 1$.

It should be recalled that, when the coalescing bodies reach the ISCO, which can be considered the end point of the inspiral, the orbital distance and the wave frequency are approximately given by Eqs. 14.65 and 14.66. These equations show that the *larger is the total mass of the system, the larger is the orbital distance at the ISCO, and the smaller is the corresponding wave frequency.* This means that although more massive systems emit signals with larger amplitude, they span a smaller region of the detector bandwidth, and consequently stay in the bandwidth for a shorter time.

These properties are shown in Fig. 14.5, where we compare the noise spectral density [5] of the LIGO detector, also called the *sensitivity curve* (thick black curve), with some representative signals from coalescences of black hole binaries. The sensitivity curve has typically a

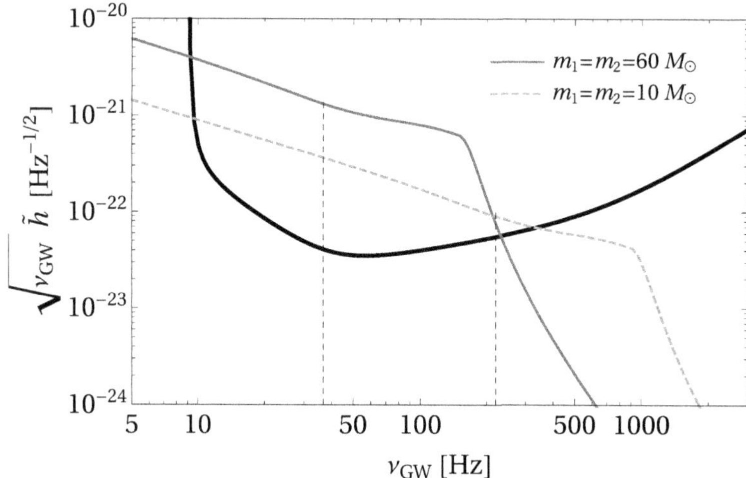

Figure 14.5: The strain amplitude of the gravitational wave signal emitted in the coalescence of two black holes, $\sqrt{\nu_{GW}}\tilde{h}$, where $\tilde{h}$ is the Fourier transform of the waveform, is plotted as a function of the frequency and compared to the noise spectral density of the advanced LIGO interferometer (thick black curve). Roughly speaking, a signal is detectable when the strain amplitude of the signal lies well above the sensitivity curve. We show two signals for $m_1 = m_2 = 10M_\odot$ and $m_1 = m_2 = 60M_\odot$, respectively, for a source located at the distance of 100 Mpc. Note that the amplitude decreases sharply after some frequency which approximately corresponds to the merger of each binary. The dashed vertical lines correspond to the ISCO frequency given in Eq. 14.66. See text for more details.

"bucket" shape: at low frequencies the sensitivity deteriorates (i.e., the curve grows) because of seismic and environmental noise, whereas at high frequencies it deteriorates because of quantum noise of the laser beam of the interferometer (see Sec. 12.8). The central plateau identifies the frequency range where the detector is more sensitive. The quantity that needs to be compared with the sensitivity curve is the *strain amplitude* of the signal, given by the Fourier transform of the gravitational waveform $\tilde{h}$, multiplied by the square root of the wave frequency, i.e. $\sqrt{\nu_{GW}}\tilde{h}$. Roughly speaking, a signal is detectable when the strain amplitude $\sqrt{\nu_{GW}}\tilde{h}$ lies well above the sensitivity curve (in reality the detection requires much more sophisticated filtering techniques, see [82]).

The signals shown in Fig. 14.5 are obtained using analytical waveforms that approximate the whole inspiral, merger, and ringdown signal. The leftmost part of the signal (linear in the log-log scale) is associated with the inspiral; the small bump corresponds to the merging, after which the signal dies off quickly during the ringdown.

As previously discussed, as the total mass of the system increases the ISCO frequency decreases, and the overall amplitude of the signal increases (see Eq. 14.33). Since each detector is sensitive only in a certain frequency range (e.g., the frequency band for LIGO and Virgo is roughly $\nu_{GW} \in (10, 1000)$ Hz, see Fig. 14.5), the peak frequency lies in the detector band only when the total mass is in a certain range. This explains why LIGO and

[5] The noise spectral density of a detector, $S_n(f)$, is the noise power per unit frequency. This quantity has units of Hz$^{-1/2}$.

Virgo are most sensitive to the coalescence of black hole binaries with total mass ranging from (approximately) 10 $M_\odot$ to a few hundred $M_\odot$.

Several sources of noise determine the precise shape of the sensitivity curve for each detector, but a discussion of their characterization is beyond the scope of this book. For a laser interferometer the frequency band depends on the length of the arms, as discussed in Sec. 12.8: the range of frequencies where the detector is most sensitive approximately scales with the inverse of the interferometer arm length (see Eq. 12.123). Kilometer-size interferometers like LIGO and Virgo can detect black hole binaries with total mass in the above mentioned range. Supermassive black hole binaries (total mass about $10^6 M_\odot$) emit signals in the range $\nu_{\rm GW} \in (10^{-3}, 10^{-1})$ Hz. Their detection requires 10^5 km-size interferometers, which can not be built on the Earth. The Laser Interferometer Space Antenna (LISA) is a joint ESA-NASA space mission which will be launched around the year 2034, with the goal of detecting gravitational waves from supermassive black holes for the first time.

As for 2018, the only event which is due to the coalescence of two neutron stars is GW170817[6]; this is also the closest event, detected at a distance of 40 Mpc. The inspiral part of the signal emitted in these processes is a chirp with the same features discussed for black hole mergers: as shown by Eqs. 14.41, 14.42, and 14.43 it is a sinusoid with increasing amplitude and frequency, and the phase depends only on the chirp mass. This is because neutron stars are extremely compact and when they are far apart they can be treated as point particles. Since the masses are low compared to those of black holes, as explained above the signal stays in the detector bandwidth for a much longer time (~ 100 s, whereas it was ~ 0.2 s for GW150914), and this allows a better estimate of the binary parameters. The part of the signal emitted by GW170817 in the latest phases before the merger deviates from that emitted by black holes because the two stars are deformed by the mutual tidal interaction. These deformations are encoded in some parameters, called *Love's numbers*, which measure the quadrupole moment of a star induced by the tidal field of its companion. The Love numbers depend on the internal composition of the star, and give important information on the behaviour of matter in the extreme conditions typical of a neutron star core. In that region, densities can exceed the nuclear saturation density, and the equation of state of matter is largely unknown (see Sec. 16.3). Many theoretical models have been proposed, which cannot be validated by terrestrial experiments, since such extreme conditions cannot be reached in a laboratory. The extraction of Love's numbers from the gravitational wave emitted by coalescing neutron stars is expected to provide information on this fundamental issue, which cannot be obtained otherwise. A first estimate of the quadrupolar tidal Love numbers has been possible in GW170817 [2], but more events will be needed to confirm and refine these results.

It is important to stress that an electromagnetic counterpart, comprising a gamma-ray burst and optical and radio observations, was detected in coincidence with the detection of GW170817; this allowed obtaining an independent measure of the redshift z and the best source localization to date, marking the dawn of *multi-messenger astronomy.* Thanks to the electromagnetic counterpart, it has also been possible to measure the speed of the gravitational waves emitted by GW170817: the gamma-ray burst was detected only 1.7 s after the gravitational wave signal at merger, showing that the speed of gravitational waves coincides with the speed of light – as predicted by General Relativity – within one part in 10^{15}.

The merging part and the ringing tail of the signal emitted in binary neutron star coalescences can be significantly different from that of black holes. The presence of matter

[6]A more recent gravitational wave event from the third observational run, GW190425, has been reported [4]. Although information for this event is much more limited than for GW170817, it is consistent with another binary neutron star coalescence. If confirmed, it would also be the heaviest neutron star binary known to date.

produces a rich phenomenology which depends essentially on the equation of state, i.e. on the relation between the pressure and the energy density of the fluid inside the stars. For instance, depending on the compressibility of matter, the object which forms after the merging can bounce several times, oscillate, and then collapse to form either a black hole, or a neutron star which, depending on the mass, will later collapse to a black hole; alternatively, a black hole can form immediately after the bounce. These different behaviours are reflected in different waveforms, which have to be computed by integrating numerically Einstein's equations with a matter source, choosing an equation of state motivated by nuclear physics studies.

Finally, the last part of the signal is the ringing tail, emitted when the final object which forms after the merging oscillates (either a black hole or a neutron star), releasing its residual energy in gravitational waves, at the frequencies and damping times of its proper QNMs.

14.3 GRAVITATIONAL WAVES FROM ROTATING COMPACT STARS

In this section we shall show that a rotating star emits gravitational waves only if its shape deviates from axial symmetry. Contrary to binary coalescences, rotating stars are *continuous gravitational wave sources*: they emit signals lasting much longer than the observation time, with an approximately constant frequency, which depends on the star rotation rate.

To study the features of these signals, we shall use the quadrupole formalism introduced in Chapter 13; we shall model the star as an ellipsoid with uniform density ρ, rigidly rotating around an axis with angular velocity Ω.

14.3.1 Stars rigidly rotating around a principal axis

The quadrupole moment of the star

$$q_{ij} = \int_V \rho\, x_i x_j \, d^3x \,, \qquad i,j = 1,\ldots,3 \tag{14.71}$$

is related to the *inertia tensor*

$$I_{ij} = \int_V \rho \left(r^2 \delta_{ij} - x_i x_j\right) \, d^3x \tag{14.72}$$

by the equation

$$q_{ij} = -I_{ij} + \delta_{ij}\, q^k_{\ k} \,. \tag{14.73}$$

Consequently, the reduced quadrupole moment can be written as

$$Q_{ij} = q_{ij} - \frac{1}{3}\delta_{ij}\, q^k_{\ k} = -\left(I_{ij} - \frac{1}{3}\delta_{ij}\, I^k_{\ k}\right) \,. \tag{14.74}$$

Let us first consider a non-rotating, homogeneous ellipsoid, with semiaxes a, b, c, volume $V = \frac{4}{3}\pi abc$, and equation:

$$\left(\frac{x^1}{a}\right)^2 + \left(\frac{x^2}{b}\right)^2 + \left(\frac{x^3}{c}\right)^2 = 1 \,. \tag{14.75}$$

In this frame, x^i are the principal axes of the ellipsoid. The inertia tensor of the ellipsoid is

$$I_{ij} = \frac{M}{5}\begin{pmatrix} b^2+c^2 & 0 & 0 \\ 0 & c^2+a^2 & 0 \\ 0 & 0 & a^2+b^2 \end{pmatrix} = \begin{pmatrix} I_1 & 0 & 0 \\ 0 & I_2 & 0 \\ 0 & 0 & I_3 \end{pmatrix}, \tag{14.76}$$

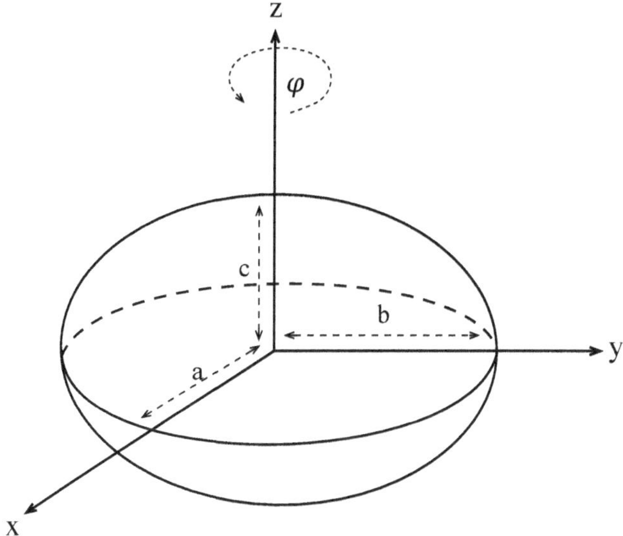

Figure 14.6: Ellipsoidal star rotating around the principal axis z.

where I_1, I_2, I_3 are the principal moments of inertia.

Let us now consider a homogeneous ellipsoid which rotates around one of its principal axes, for instance x^3, with angular velocity $(0, 0, \Omega)$, as shown in Fig. 14.6. Let $x^i = (x, y, z)$ be the coordinates of the inertial frame, and $x^{i'} = (x', y', z')$ those of the co-rotating frame. Then,

$$x_i = R_{ij} x_{j'}, \tag{14.77}$$

where R_{ij} is the rotation matrix

$$R_{ij} = \begin{pmatrix} \cos\varphi & -\sin\varphi & 0 \\ \sin\varphi & \cos\varphi & 0 \\ 0 & 0 & 1 \end{pmatrix}, \qquad \text{with} \quad \varphi = \Omega t\,. \tag{14.78}$$

For instance, a point at rest in the co-rotating frame, with coordinates $x^{i'} = (1, 0, 0)$, has coordinates $x^i = (\cos\Omega t, \sin\Omega t, 0)$ in the inertial frame, i.e. it rotates around the z-axis with angular velocity Ω. Since in the co-rotating frame $\{x^{i'}\}$

$$I'_{ij} = \begin{pmatrix} I_1 & 0 & 0 \\ 0 & I_2 & 0 \\ 0 & 0 & I_3 \end{pmatrix}, \tag{14.79}$$

in the inertial frame $\{x^i\}$ the inertia tensor is

$$
\begin{aligned}
I_{ij} &= R_{ik}R_{jl}I'_{kl} = (RI'R^T)_{ij} \\
&= \begin{pmatrix} I_1\cos^2\varphi + I_2\sin^2\varphi & -(I_2-I_1)\sin\varphi\cos\varphi & 0 \\ -(I_2-I_1)\sin\varphi\cos\varphi & I_1\sin^2\varphi + I_2\cos^2\varphi & 0 \\ 0 & 0 & I_3 \end{pmatrix}.
\end{aligned}
$$

It is easy to check that $I^k_{\ k} = I_1+I_2+I_3 = const$. The reduced quadrupole moment therefore is

$$Q_{ij} = -\left(I_{ij} - \frac{1}{3}\delta_{ij}I^k_{\ k}\right) = -I_{ij} + const\,. \tag{14.80}$$

Using the relations $\cos 2\varphi = 2\cos^2\varphi - 1$, and $\sin 2\varphi = 2\sin\varphi\cos\varphi$, Q_{ij} can be written as

$$Q_{ij} = \frac{I_2 - I_1}{2}\begin{pmatrix} \cos 2\varphi & \sin 2\varphi & 0 \\ \sin 2\varphi & -\cos 2\varphi & 0 \\ 0 & 0 & 0 \end{pmatrix} + const\,. \tag{14.81}$$

Since

$$I_1 = \frac{M}{5}(b^2+c^2), \qquad \text{and} \qquad I_2 = \frac{M}{5}(c^2+a^2)\,, \tag{14.82}$$

from Eq. 14.81 it follows that if a, b are equal, i.e. if the ellipsoid is axisymmetric, the quadrupole moment is constant; thus *axisymmetric stars do not emit gravitational waves.* This is a generic result: an axisymmetric object rigidly rotating around its symmetry axis does not radiate gravitational waves.

If the rotating star is a *triaxial ellipsoid*, i.e. the semiaxes are all different, $a \neq b$ and the star emits gravitational waves; however for compact stars the difference $a - b$ is expected to be extremely small. It is convenient to express the reduced quadrupole moment in terms of a dimensionless parameter ϵ, the *oblateness*, which expresses the deviation from axisymmetry

$$\epsilon \equiv \frac{a-b}{(a+b)/2}\,. \tag{14.83}$$

It is easy to show that

$$\frac{I_2 - I_1}{I_3} = \epsilon + O(\epsilon^3)\,. \tag{14.84}$$

Indeed, given $a - b = \epsilon(a+b)/2$,

$$\frac{I_2 - I_1}{I_3} = \frac{a^2-b^2}{a^2+b^2} = \frac{1}{2}\epsilon\frac{(a+b)^2}{a^2+b^2}\,. \tag{14.85}$$

On the other hand,

$$(a-b)^2 = O(\epsilon^2) = a^2 + b^2 - 2ab\,, \tag{14.86}$$

therefore $2ab = a^2 + b^2 + O(\epsilon^2)$ and

$$\frac{I_2 - I_1}{I_3} = \frac{1}{2}\epsilon\frac{a^2+b^2+2ab}{a^2+b^2} = \epsilon + O(\epsilon^3)\,. \tag{14.87}$$

Consequently, neglecting terms of order $O(\epsilon^3)$, Q_{ij} becomes

$$Q_{ij} = \frac{\epsilon\, I_3}{2}\begin{pmatrix} \cos 2\varphi & \sin 2\varphi & 0 \\ \sin 2\varphi & -\cos 2\varphi & 0 \\ 0 & 0 & 0 \end{pmatrix} + const\,. \tag{14.88}$$

According to Eq. 13.49, the emitted signal has the form

$$h_{ij}^{\mathbf{TT}}(t,r) = \frac{2G}{rc^4}\,\mathcal{P}_{ijlm}\,\frac{d^2}{dt^2}\,Q_{lm}\left(t-\frac{r}{c}\right), \tag{14.89}$$

and using Eq. 14.88 this becomes

$$h_{ij}^{\mathbf{TT}} = h_0 \mathcal{P}_{ijlm}\mathcal{A}_{lm}\,, \tag{14.90}$$

with

$$\mathcal{A}_{lm} = \begin{pmatrix} -\cos 2\varphi_{ret} & -\sin 2\varphi_{ret} & 0 \\ -\sin 2\varphi_{ret} & \cos 2\varphi_{ret} & 0 \\ 0 & 0 & 0 \end{pmatrix}, \qquad \varphi_{ret} = \Omega\left(t-\frac{r}{c}\right), \tag{14.91}$$

and

$$h_0 = \frac{4G\,\Omega^2}{c^4 r} I_3\epsilon = \frac{16\pi^2 G}{c^4 P^2 r} I_3\epsilon\,, \tag{14.92}$$

where $P = 2\pi/\Omega$ is the rotation period of the star. The term $\mathcal{P}_{ijlm}\mathcal{A}_{lm}$ in Eq. 14.91 depends on the direction of the observer relative to the star axes. Eq. 14.91 shows that *a triaxial star rotating around a principal axis emits gravitational waves at twice the rotation frequency*

$$\nu_{\mathrm{GW}} = 2\nu_{\mathrm{rot}} = \frac{2}{P}\,. \tag{14.93}$$

Rapidly rotating neutron stars have rotation periods of the order of a few milliseconds; typical values of neutron stars moment of inertia are of the order $\sim 10^{45}\,\mathrm{g\,cm^2}$; for galactic sources the distance from Earth is of a few kpc. If we assume an oblateness as small as $\sim 10^{-6}$, and consider $I_3 \simeq I$ as the average moment of inertia of the star, we find a typical amplitude of the order of

$$\frac{16\pi^2 G}{c^4\, r\, P^2}\, I\,\epsilon \sim \frac{16\pi^2 G}{c^4}\,(\mathrm{ms})^{-2}(\mathrm{kpc})^{-1}(10^{45}\mathrm{g\,cm^2})(10^{-6}) = 4.2\times 10^{-24}\,. \tag{14.94}$$

We can thus normalize the amplitude emitted by a generic triaxial neutron star with respect to this value as follows

$$h_0 = 4.2\times 10^{-24}\left[\frac{\mathrm{ms}}{P}\right]^2\left[\frac{\mathrm{kpc}}{r}\right]\left[\frac{I}{10^{45}\mathrm{g\,cm^2}}\right]\left[\frac{\epsilon}{10^{-6}}\right]. \tag{14.95}$$

We remark that Eq. 14.95 gives only an estimate of the gravitational wave amplitude for these sources, since neutron stars are not homogeneous and the density profile $\rho(r)$ depends on the equation of state of matter in their interior which, as we shall see in Chapter 16, is largely unknown. However, the moment of inertia can be estimated once an equation of state is chosen among those proposed in the literature.

The rotation period and the distance can be measured for many galactic neutron stars. Conversely the oblateness ϵ is unknown. Astrophysical observations allow to set an upper limit on this parameter. It is known that the rotation frequency of many observed pulsars decreases with time. The rotational energy decreases mainly because these stars have a time-varying magnetic dipole moment, and therefore radiate electromagnetic waves. A further braking mechanism is provided by gravitational wave emission, although the energy dissipated in this channel is expected to be subleading with respect to the electromagnetic emission. An upper limit on ϵ can then be obtained assuming that the amount of energy lost in gravitational waves is smaller than – or equal to, in the limiting case – the loss of rotational energy. To this purpose, we shall evaluate the gravitational luminosity 13.148 in terms of the source quadrupole moment 14.88. We leave as an exercise to show that

$$L_{\rm GW} = \frac{32G}{5c^5}\left\langle \epsilon^2\Omega^6 I^2\right\rangle . \tag{14.96}$$

In the following we shall assume (as in the case of binary neutron stars) that the source evolves *adiabatically*, i.e. that the parameters of the source (in this case, ϵ, I, and Ω) do not change significantly on the timescale of the gravitational wave emission (which is, in this case, the rotation period). This approximation is satisfied by all known rotating neutron stars (for instance, for the Crab pulsar $\dot{P} \sim 10^{-12}$). Therefore, since averaging over several wavelengths of the gravitational wave is equivalent to averaging over several periods, Eq. 14.96 reduces to

$$L_{\rm GW} = \frac{32G}{5c^5}\epsilon^2\Omega^6 I^2 . \tag{14.97}$$

The rotational energy and its time derivative are, in the Newtonian limit,

$$E_{\rm rot} = \frac{1}{2}I\Omega^2 , \qquad \dot{E}_{\rm rot} = I\Omega\dot{\Omega} . \tag{14.98}$$

By imposing $L_{\rm GW} \leq -\dot{E}_{\rm rot}$, we find

$$\frac{32G}{5c^5}(2\pi\nu)^6\epsilon^2 I^2 \leq I(2\pi)^2\nu|\dot{\nu}| \tag{14.99}$$

(note that $|\dot{\nu}| = -\dot{\nu}$), and consequently

$$\epsilon \leq \epsilon_{\rm sd} \equiv \left(\frac{5c^5|\dot{\nu}|}{512\pi^4 G\nu^5 I}\right)^{1/2} . \tag{14.100}$$

$\epsilon_{\rm sd}$ is called the *spin-down limit* for the oblateness. For instance, the Crab pulsar has rotation frequency $\nu = 30$ Hz, spin-down frequency $\dot{\nu} = -3.7 \times 10^{-10}$ Hz/s, and distance from Earth $r = 2\,$kpc; if we assume a fiducial moment of inertia of $I = 10^{45}\,{\rm g\,cm}^2$, Eq. 14.100 gives

$$\epsilon_{\rm sd} = 7.5 \times 10^{-4} . \tag{14.101}$$

In Table 14.2 we show the values of the frequency of the emitted gravitational signal and of the spin-down limit for the oblateness, for some known pulsars.

Table 14.2: Upper limits for the oblateness of an ensemble of known pulsars, obtained from spin-down measurements [51].

name	$\nu_{\rm GW}$ (Hz)	$\epsilon_{\rm sd}$
Vela	22	1.8×10^{-3}
Crab	60	7.5×10^{-4}
Geminga	8.4	2.3×10^{-3}
PSR B 1509-68	13.2	1.4×10^{-2}
PSR B 1706-44	20	1.9×10^{-3}
PSR B 1957+20	1242	1.6×10^{-9}
PSR J 0437-4715	348	2.9×10^{-8}

Independent constraints on the oblateness of the Crab pulsar have been set by the analysis of the data collected by the LIGO and Virgo interferometers. At the time of writing, these instruments did not detect gravitational waves from spinning neutron stars; however, their sensitivity was such that a gravitational signal emitted by these stars would had been detected if its amplitude exceeded $h_0 \sim 10^{-25}$. Since no signal has been detected, it follows that the amplitude of the wave emitted by the pulsar must be smaller than this value. Therefore, using this constraint and substituting the values of the distance of the Crab from Earth ($r \sim 2\,\mathrm{kpc}$), its rotation period ($P \sim 33\,\mathrm{ms}$), and the fiducial value of the moment of inertia ($I = 10^{45}\,\mathrm{g\,cm^2}$) in Eq. 14.95, the upper limit for the Crab oblateness is

$$\epsilon \leq 5 \times 10^{-5} \,. \tag{14.102}$$

Similarly, for the Vela pulsar ($r \sim 0.2\,\mathrm{kpc}$, $P \sim 89\,\mathrm{ms}$), the upper limit is $\epsilon \leq 4 \times 10^{-5}$. Remarkably, these bounds are *more restrictive* than the spin-down limits obtained from astrophysical observations. Note that the above constraints are derived under simplified assumptions and for a fiducial value of the moment of inertia. A more rigorous analysis for the Vela pulsar using the recent data from advanced LIGO and Virgo provides a much more stringent bound on the oblateness, $\epsilon \leq 3 \times 10^{-8}$ [83].

14.3.2 Wobbling stars

Let us now assume that the star rotates about an axis which forms an angle θ, called *"wobble angle"*, with one of the principal axes, say, the semi-axis c. In this case, the angular velocity

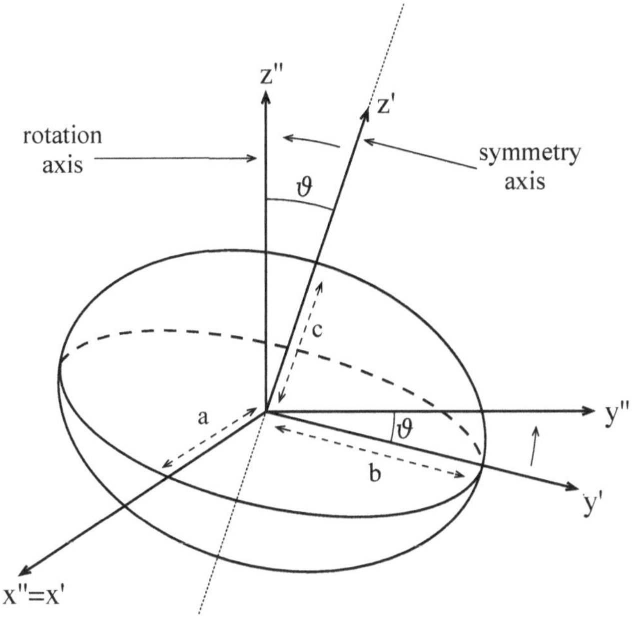

Figure 14.7: Rotation from O' to O''.

precedes around this axis. For simplicity, let us assume that the semi-axis c is a symmetry axis of the ellipsoid, i.e.

$$a = b \qquad \rightarrow \qquad I_1 = I_2 \,, \tag{14.103}$$

and that the wobble angle θ is small, $\theta \ll 1$. Let $x^{i'} = (x', y', z')$ be the coordinates of the co-rotating frame O' and $x^i = (x, y, z)$ those of the inertial frame O, related by $x_i = R_{ij}x'_j$, where R_{ij} is the rotation matrix. The transformation from O' to O is the composition of two rotations:

- A rotation of O' around the x-axis by a small angle θ (constant), as shown in Fig. 14.7; in the new frame O'' the rotation axis coincides with the z''-axis. The corresponding rotation matrix is

$$R_x = \begin{pmatrix} 1 & 0 & 0 \\ 0 & \cos\theta & \sin\theta \\ 0 & -\sin\theta & \cos\theta \end{pmatrix} = \begin{pmatrix} 1 & 0 & 0 \\ 0 & 1 & \theta \\ 0 & -\theta & 1 \end{pmatrix} + O(\theta^2)\,. \tag{14.104}$$

- A time-dependent rotation around the z''-axis, by an angle $\varphi = \Omega t$, as shown in Fig. 14.8; the corresponding rotation matrix is

$$R_z = \begin{pmatrix} \cos\varphi & -\sin\varphi & 0 \\ \sin\varphi & \cos\varphi & 0 \\ 0 & 0 & 1 \end{pmatrix}\,. \tag{14.105}$$

  After this rotation, the symmetry axis of the ellipsoid precedes around the z-axis, with angular velocity Ω.

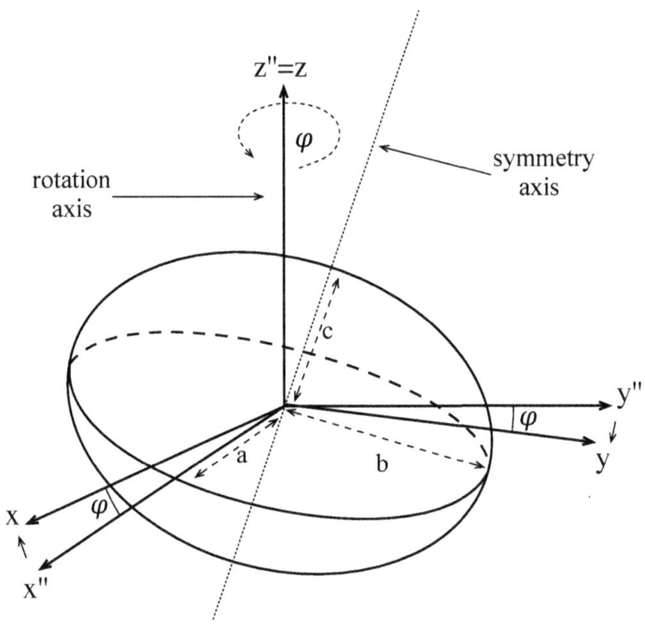

Figure 14.8: Rotation from O'' to O.

The rotation matrix from O' to O therefore is

$$\begin{aligned} R &= R_z R_x = \begin{pmatrix} \cos\varphi & -\sin\varphi & 0 \\ \sin\varphi & \cos\varphi & 0 \\ 0 & 0 & 1 \end{pmatrix} \begin{pmatrix} 1 & 0 & 0 \\ 0 & 1 & \theta \\ 0 & -\theta & 1 \end{pmatrix} + O(\theta^2) \\ &= \begin{pmatrix} \cos\varphi & -\sin\varphi & -\theta\sin\varphi \\ \sin\varphi & \cos\varphi & \theta\cos\varphi \\ 0 & -\theta & 1 \end{pmatrix} + O(\theta^2) . \end{aligned} \tag{14.106}$$

Since in the co-rotating frame O'

$$I'_{ij} = \begin{pmatrix} I_1 & 0 & 0 \\ 0 & I_1 & 0 \\ 0 & 0 & I_3 \end{pmatrix} , \tag{14.107}$$

in the inertial frame O the inertia tensor is

$$I_{ij} = R_{ik} R_{jl} I'_{kl} = (R I' R^T)_{ij} \tag{14.108}$$

$$= \begin{pmatrix} I_1 & 0 & (I_1 - I_3)\theta\ \sin\varphi \\ 0 & I_1 & -(I_1 - I_3)\theta\ \cos\varphi \\ (I_1 - I_3)\theta\ \sin\varphi & -(I_1 - I_3)\theta\ \cos\varphi & I_3 \end{pmatrix} + O(\theta^2) . \tag{14.109}$$

Then, neglecting terms of order $O(\theta^2)$, the quadrupole moment is

$$Q_{ij} = -I_{ij} + const = (I_1 - I_3)\ \theta \begin{pmatrix} 0 & 0 & -\sin\varphi \\ 0 & 0 & \cos\varphi \\ -\sin\varphi & \cos\varphi & 0 \end{pmatrix} + const , \tag{14.110}$$

and using Eq. 13.49 the waveform is

$$h^{\mathbf{TT}}_{ij} = h_0 \mathcal{P}_{ijlm} A_{lm} , \tag{14.111}$$

where

$$A_{lm} = \begin{pmatrix} 0 & 0 & \sin\varphi_{ret} \\ 0 & 0 & -\cos\varphi_{ret} \\ \sin\varphi_{ret} & -\cos\varphi_{ret} & 0 \end{pmatrix} , \qquad \varphi_{ret} = \Omega\left(t - \frac{r}{c}\right) , \tag{14.112}$$

and

$$h_0 = \frac{2G\ \Omega^2}{c^4 r} (I_1 - I_3)\ \theta = \frac{8\pi^2 G}{c^4\ r\ P^2} (I_1 - I_3)\ \theta . \tag{14.113}$$

Thus, when an axisymmetric star rotates around an axis which does not coincide with a principal axis, gravitational waves are emitted at the rotation frequency

$$\nu^{\rm GW} = \nu_{\rm rot} . \tag{14.114}$$

Like the oblateness, the wobble angle is an unknown parameter. Comparing the amplitude of the wave emitted by a star rotating around a symmetry axis given in Eq. 14.92, and that of a wobbling star given in Eq. 14.113 we see that, apart from a factor 2, the difference is that the moment of inertia I_3 and the oblateness ϵ in Eq. 14.92 are replaced by the difference $(I_1 - I_3)$ and by the wobble angle θ in Eq. 14.113.

14.4 COSMOLOGICAL PARAMETERS

In this section we briefly summarize some basic concepts of cosmology. We refer the interested reader to any textbook dedicated to this subject, e.g. [71], [102].

Cosmological observations show that our Universe is approximately homogeneous and isotropic on spatial scales of the order of $\sim 100\,\mathrm{Mpc}$ (we remind – see Table A – that $1\,\mathrm{pc} = 3.086 \times 10^{18}\,\mathrm{cm}$, i.e. 1 pc = 3.262 ly). This implies that at each space point the geometry must be the same (homogeneity), and that the space is spherically symmetric (isotropy) about every point, i.e. no direction is preferred. The line element of the cosmological geometry which satisfies these properties is the *Friedmann-Lemaître-Robertson-Walker* (FLRW) metric

$$ds^2 = -c^2dt^2 + a^2(t)\left[\frac{dr^2}{1-kr^2} + r^2d\theta^2 + r^2\sin^2\theta d\phi^2\right], \tag{14.115}$$

where $a(t)$ is the *scale factor*, a function of the coordinate time t, and the constant k takes the values $k = 0, \pm1$, i.e. it is dimensionless. If $k = 1$, the $t = const$ hypersurfaces have positive curvature, in which case the Universe is closed, while if $k = -1$ they have negative curvature and the Universe is open. If $k = 0$, the $t = const$ hypersurfaces have vanishing curvature, i.e. they are flat. Indeed, in this case the part of the metric in square brackets can be reduced to

$$dx^2 + dy^2 + dz^2 \tag{14.116}$$

by the standard transformation from polar to Cartesian coordinates, and if we set $t = const$ and choose new coordinates $x' = a(t)x$, $y' = a(t)y$, $z' = a(t)z$, the line element of the three-dimensional space becomes $ds_3^2 = dx'^2 + dy'^2 + dz'^2$.

Note that if $k = \pm1$, r must be dimensionless, while the scale factor a has the dimension of a length. However, the metric 14.115 is invariant under a rescaling $r \to Lr$, $a \to a/L$, $k \to k/L^2$, where L has the dimension of a length; therefore, we can choose a more "physical" radial coordinate, with the dimension of a length. With this choice the line element 14.115 has the same form, but $k = \pm1/L^2$, and the scale factor a is dimensionless. The scalar curvature of the three-dimensional submanifold $t = const$ is $R^{(3)} = 6k/a^2$; thus the length-scale of this curvature, $l_{\mathrm{curv}} = 1/\sqrt{|R^{(3)}|}$, is

$$l_{\mathrm{curv}} \sim a/\sqrt{|k|}, \tag{14.117}$$

which shows that, for $k = \pm1$, $l_{\mathrm{curv}} \sim a$, while for $k = \pm1/L^2$, $l_{\mathrm{curv}} \sim aL$.

According to the latest results by the Planck mission (released in 2018 [92]), we live in a FLRW Universe with negligible curvature, therefore for practical purposes it is nowadays often assumed $k = 0$.

In the FLRW model, galaxies are considered as particles of a cosmological, pressureless fluid, each located at the space coordinates x^i, $i = 1\ldots3$. Each particle has vanishing three-velocity, otherwise they would establish a preferred direction, contradicting the isotropy assumption. We remark that this should be considered an average description, accurate at length-scales much larger than the size of a single galaxy. The time coordinate t is called *cosmic time*, while the space coordinates (r, θ, ϕ) are called *comoving coordinates*, since in this frame each galaxy has the same space coordinates (r, θ, ϕ) for all times. Thus, whereas the coordinate distance between two galaxies is constant, their proper spatial distance

$$D(t) = \int_{\mathrm{galaxy\ 1}}^{\mathrm{galaxy\ 2}} \sqrt{g_{ik}dx^idx^k}, \qquad i,k = 1,\ldots,3 \tag{14.118}$$

is not, and can increase, decrease, or remain constant depending on the behaviour of the scale factor in time.

14.4.1 The cosmological redshift

Let us assume that a source in a distant galaxy located at $r = R$, emits a light signal at some coordinate time $t_{\rm em}$ with frequency $\nu_{\rm em}$, and that the pulse is emitted in the radial direction. An observer in a galaxy in $r = 0$ receives this signal at a coordinate time $t_{\rm obs}$, with frequency $\nu_{\rm obs}$.

Let us compute whether the observed frequency differs from $\nu_{\rm em}$. Since both the galaxy and the observer are at rest, and since in the FLRW metric 14.115 $g_{00} = -c^2$, the proper times of the source and of the observer coincide with the corresponding coordinate times (see Sec. 11.1). Therefore, the coordinate time separation of the wave crests as they are emitted is the inverse of the emission frequency, $\delta t_{\rm em} = 1/\nu_{\rm em}$, and the coordinate time separation of the observed wave crests is the inverse of the observed frequency, $\delta t_{\rm obs} = 1/\nu_{\rm obs}$.

The light pulse emitted at time $t_{\rm em}$ at $r = R$ travels along a null geodesic of the FLRW spacetime, and setting $ds^2 = 0$ in Eq. 14.115 we find

$$\int_{t_{\rm em}}^{t_{\rm obs}} \frac{cdt}{a(t)} = \int_0^R \frac{dr}{\sqrt{1-kr^2}}\,. \tag{14.119}$$

Note that the right-hand side of this equation is the same at all times, since in the comoving frame the coordinates of the source and the observer do not change with time. The next wave crest, emitted at $t_{\rm em} + \delta t_{\rm em}$, travels along the same radial path and is received at time $t_{\rm obs} + \delta t_{\rm obs}$; thus in this case

$$\int_{t_{\rm em}+\delta t_{\rm em}}^{t_{\rm obs}+\delta t_{\rm obs}} \frac{cdt}{a(t)} = \int_0^R \frac{dr}{\sqrt{1-kr^2}}\,. \tag{14.120}$$

It follows that

$$\int_{t_{\rm em}}^{t_{\rm obs}} \frac{cdt}{a(t)} = \int_{t_{\rm em}+\delta t_{\rm em}}^{t_{\rm obs}+\delta t_{\rm obs}} \frac{cdt}{a(t)}\,. \tag{14.121}$$

The last integral in this equality can be written as

$$\int_{t_{\rm em}+\delta t_{\rm em}}^{t_{\rm obs}+\delta t_{\rm obs}} = \int_{t_{\rm em}}^{t_{\rm obs}} - \int_{t_{\rm em}}^{t_{\rm em}+\delta t_{\rm em}} + \int_{t_{\rm obs}}^{t_{\rm obs}+\delta t_{\rm obs}}\,, \tag{14.122}$$

therefore Eq. 14.121 reduces to

$$-\int_{t_{\rm em}}^{t_{\rm em}+\delta t_{\rm em}} \frac{cdt}{a(t)} + \int_{t_{\rm obs}}^{t_{\rm obs}+\delta t_{\rm obs}} \frac{cdt}{a(t)} = 0 \tag{14.123}$$

and, assuming $\delta t_{\rm em}$ and $\delta t_{\rm obs}$ to be small, we get

$$-\frac{\delta t_{\rm em}}{a(t_{\rm em})} + \frac{\delta t_{\rm obs}}{a(t_{\rm obs})} = 0\,. \tag{14.124}$$

Since $\delta t_{\rm em} = 1/\nu_{\rm em}$ and $\delta t_{\rm obs} = 1/\nu_{\rm obs}$,

$$\frac{\nu_{\rm obs}}{\nu_{\rm em}} = \frac{a(t_{\rm em})}{a(t_{\rm obs})}\,. \tag{14.125}$$

Cosmological observations show that our Universe is expanding, therefore $a(t)$ is an increasing function of the cosmic time t. Consequently, a signal emitted at frequency $\nu_{\rm em}$ by a source in a distant galaxy will be detected by our instruments with a frequency $\nu_{\rm obs} < \nu_{\rm em}$, i.e. will be *redshifted*. We define the *cosmological redshift* z as

$$z = \frac{\nu_{\rm em} - \nu_{\rm obs}}{\nu_{\rm obs}} = \frac{\lambda_{\rm obs} - \lambda_{\rm em}}{\lambda_{\rm em}}\,, \tag{14.126}$$

hence

$$1+z=\frac{\nu_{\rm em}}{\nu_{\rm obs}}=\frac{\lambda_{\rm obs}}{\lambda_{\rm em}}=\frac{a(t_{\rm obs})}{a(t_{\rm em})}\,. \tag{14.127}$$

From Eqs. 14.124 and 14.127 it also follows that if two light (or gravitational wave) pulses are emitted with time separation $\delta t_{\rm em}$, they will be observed with time separation

$$\delta t_{\rm obs}=(1+z)\delta t_{\rm em}\,. \tag{14.128}$$

14.4.2 The Hubble constant

Let us Taylor-expand the ratio $a(t_{\rm em})/a(t_{\rm obs})$ around the present time $t_{\rm obs}$

$$\frac{a(t_{\rm em})}{a(t_{\rm obs})}=1+H_0(t_{\rm em}-t_{\rm obs})-\frac{1}{2}q_0H_0^2(t_{\rm em}-t_{\rm obs})^2+\ldots \tag{14.129}$$

where

$$H_0\equiv\frac{\dot a(t_{\rm obs})}{a(t_{\rm obs})} \tag{14.130}$$

is called the *Hubble constant*, and

$$q_0=-\frac{1}{H_0^2}\frac{\ddot a\,(t_{\rm obs})}{a(t_{\rm obs})}=\frac{a(t_{\rm obs})\,\ddot a\,(t_{\rm obs})}{\dot a^2(t_{\rm obs})} \tag{14.131}$$

is called the *deceleration parameter.* In the previous equation a dot indicates differentiation with respect to the cosmic time. It should be stressed that, although the name of q_0 seems to indicate a deceleration in the expansion of the Universe, observations show that the expansion is actually accelerating, i.e. $q_0<0$.

Let us now consider, as in Sec. 14.4.1, a source located at $r=R$ emitting at the time $t_{\rm em}$ a signal which is observed in $r=0$ at the time $t_{\rm obs}$. The proper distance between the source and the observer, measured at the time $t_{\rm obs}$, is obtained writing Eq. 14.118 for the FLRW metric 14.115

$$D=a(t_{\rm obs})\int_0^R\frac{dr}{\sqrt{1-kr^2}}\,. \tag{14.132}$$

If the source is at a (relatively) small distance from us, $a(t)$ does not change significantly during the time interval $(t_{\rm obs}-t)$ needed for the light signal to reach us. In the same approximation the rescaled radial distance, aR, is much smaller than the cosmological curvature $l_{\rm curv}\sim a/\sqrt{|k|}$ (see Eq. 14.117), and thus $|k|R^2\ll 1$. Therefore, Eq. 14.132 reduces to

$$D=a(t_{\rm obs})R \tag{14.133}$$

(note that Eq. 14.133 is exactly satisfied in the case of flat spatial hypersurfaces, $k=0$). Light moves on null geodesics, therefore setting $ds^2=0$ in the metric 14.115 and neglecting kr^2 we find

$$-c^2dt^2+a(t)^2dr^2=0\quad\rightarrow\quad c(t_{\rm obs}-t_{\rm em})\simeq a(t_{\rm em})R\,, \tag{14.134}$$

and consequently from Eq. 14.127

$$z=\frac{a(t_{\rm obs})-a(t_{\rm em})}{a(t_{\rm em})}\simeq\frac{\dot a(t_{\rm obs})}{a(t_{\rm em})}(t_{\rm obs}-t_{\rm em})\simeq\dot a(t_{\rm obs})\frac{R}{c}\simeq\frac{\dot a(t_{\rm obs})}{a(t_{\rm obs})}\frac{D}{c}\,, \tag{14.135}$$

where we have used Eq. 14.133 and the fact that $(t_{\rm obs}-t_{\rm em})$ is small since the source is close to the observer. From the Doppler shift formula, we know that if a light source travels at a speed $V \ll c$ with respect to the observer, the radiation emitted with a wavelength $\lambda_{\rm em}$ will be received with a shift

$$\frac{\lambda_{\rm obs} - \lambda_{\rm em}}{\lambda_{\rm em}} = \frac{V}{c} \,. \tag{14.136}$$

Thus, using Eq. 14.126 and Eq. 14.135 we find

$$V = z\,c = \frac{\dot{a}(t_{\rm obs})}{a(t_{\rm obs})} D \quad \rightarrow \quad V = H_0 D \,. \tag{14.137}$$

This equation is the *Hubble law*, which shows that, since our Universe is expanding, the velocity at which galaxies in our local Universe (up to a few hundred Mpc away, see Eq. 14.151) are receding from us is proportional to their proper distance. The values of the Hubble constant and of the deceleration parameter are [92] [7]

$$H_0 = (67.36 \pm 0.54)\ \text{km s}^{-1}\text{Mpc}^{-1} \,, \tag{14.138}$$

and

$$q_0 = -0.527 \pm 0.011 \,. \tag{14.139}$$

The inverse of the Hubble constant is the *Hubble time* t_H which provides a rough estimate of the age of the Universe

$$t_H = \frac{1}{H_0} \simeq 14.5\ \text{Gyr} \,. \tag{14.140}$$

14.4.3 Luminosity distance

Let us consider again an electromagnetic signal emitted at time $t_{\rm em}$ in $r = R$ and observed at time $t_{\rm obs}$ in $r = 0$, where we remind that (t, r) are coordinates in the FLRW frame defined in Eq. 14.115. If the emitted power, i.e. the luminosity of the source, is

$$L = \frac{d\mathcal{E}_{\rm em}}{dt_{\rm em}} \,, \tag{14.141}$$

then the power transmitted at the observer is

$$\frac{d\mathcal{E}_{\rm obs}}{dt_{\rm obs}} = \frac{1}{(1+z)^2} L \,; \tag{14.142}$$

indeed, two wave crests emitted with time separation $\delta t_{\rm em}$ arrive with time separation $\delta t_{\rm obs} = (1+z)\delta t_{\rm em}$ (see Eq. 14.128); moreover, the energy transmitted between the two wave crests is also redshifted, i.e. $\delta\mathcal{E}_{\rm obs} = \delta\mathcal{E}_{\rm em}/(1+z)$, because the signal corresponds to a certain number of photons, and each of them is emitted with energy $\hbar\nu_{\rm em}$ and detected with energy $\hbar\nu_{\rm obs} = \hbar\nu_{\rm em}/(1+z)$.

Since the wavefront, at the time $t_{\rm obs}$, is a sphere of radius R, whose proper area is (see Eq. 14.115) $4\pi a(t_{\rm obs})^2 R^2$, the *observed energy flux* $\mathcal{F}$ is

$$\mathcal{F} = \frac{d\mathcal{E}_{\rm em}}{dt_{\rm em}} \frac{1}{4\pi a(t_{\rm obs})^2 R^2} = \frac{L}{4\pi a(t_{\rm obs})^2 R^2 (1+z)^2} = \frac{L}{4\pi D_L^2} \tag{14.143}$$

[7] At the time of writing, there exists a statistically significant tension between the value of the Hubble constant measured by the Planck mission (Eq. 14.138), based on measurements of the cosmic microwave background, and independent measurements based on calibrated distance ladder techniques using standard candles, which give a value of H_0 approximately 8% larger.

where we have defined the *luminosity distance*

$$D_L = a(t_{\rm obs})R(1+z) = D(1+z)\,. \tag{14.144}$$

If we know *a priori* the intrinsic luminosity L of a source, then the source is called *standard candle.* For instance, a certain class of stars, the Cepheids, have a variability which is related to their luminosity. Similarly, a class of supernovae called "type Ia" have an intrinsic luminosity which is related to the behaviour of the electromagnetic emission with time. Our theoretical understanding of these (and other) sources is good enough to know their luminosity L. Therefore, since the flux $\mathcal{F}$ can be measured, Eq. 14.143 provides an operational way to measure the luminosity distances of these sources:

$$D_L = \sqrt{\frac{L}{4\pi\mathcal{F}}}\,. \tag{14.145}$$

The luminosity distance can also be written in terms of the redshift z as follows. From the Taylor expansion 14.129, which is performed about the present time (assuming $H_0(t_{\rm em} - t_{\rm obs}) \ll 1$, i.e. that $t_{\rm em} - t_{\rm obs}$ is much smaller than the age of the Universe), we have

$$\frac{a(t_{\rm obs})}{a(t_{\rm em})} = 1 - H_0(t_{\rm em} - t_{\rm obs}) + \frac{1}{2}q_0H_0^2(t_{\rm em} - t_{\rm obs})^2 \tag{14.146}$$

and since $a(t_{\rm obs})/a(t_{\rm em}) = (1+z)$ (Eq. 14.127),

$$z = H_0(t_{\rm obs} - t_{\rm em}) + \frac{q_0}{2}H_0^2(t_{\rm obs} - t_{\rm em})^2 + \dots \tag{14.147}$$

from which $(t_{\rm obs} - t_{\rm em})$ can be found in terms of z

$$(t_{\rm obs} - t_{\rm em}) = \frac{1}{H_0}\left[z - \frac{q_0}{2}z^2 + \dots\right]. \tag{14.148}$$

Let us now consider Eq. 14.119. The expansion of its left-hand side yields (using Eqs. 14.130 and 14.147)

$$\begin{aligned}
\int_{t_{\rm em}}^{t_{\rm obs}} \frac{cdt}{a(t)} &= \int_{t_{\rm em}}^{t_{\rm obs}} \frac{cdt}{a(t_{\rm obs}) + \dot{a}(t_{\rm obs})(t - t_{\rm obs}) + \dots} \\
&= \frac{c}{a(t_{\rm obs})}\int_{t_{\rm em}}^{t_{\rm obs}} [1 - H_0(t - t_{\rm obs}) + \dots]\,dt \\
&= \frac{c}{a(t_{\rm obs})}\left[(t_{\rm obs} - t_{\rm em}) - \frac{1}{2}H_0(t_{\rm obs} - t_{\rm em})^2 + \dots\right] \\
&= \frac{c}{a(t_{\rm obs})H_0}\left[z - \frac{1}{2}(1+q_0)z^2 + \dots\right].
\end{aligned} \tag{14.149}$$

The expansion of the right-hand side of Eq. 14.119 yields [8]

$$\int_0^R \frac{dr}{\sqrt{1-kr^2}} = R + \dots\,. \tag{14.150}$$

[8] If $k \neq 0$, the expansion in Eq. 14.150 is performed assuming that the distance aR is much smaller than the curvature of the Universe 14.117. If, instead, $k = 0$, the higher-order terms in the expansion 14.150 identically vanish.

Therefore,

$$D_L(z) = (1+z)a(t_{\rm obs})R = (1+z)\frac{c}{H_0}\left[z - \frac{1}{2}(1+q_0)z^2 + \dots\right] \tag{14.151}$$
$$= \frac{c}{H_0}\left[z + \frac{1}{2}(1-q_0)z^2 + \dots\right].$$

The linear term in z in the above equation corresponds to the Hubble law, Eq. 14.137. When the redshift is comparable to 1 (or larger), the quantities $H_0(t_{\rm obs} - t_{\rm em})$ and kR^2 are not negligible, and Eq. 14.151 is no longer applicable. In this case the luminosity distance can be written as an integral over the scale factor, which has to be solved numerically for the chosen cosmological model.

14.4.4 Standard sirens: coalescing binaries as standard candles

In Sec. 14.2 we have shown that the gravitational wave signal emitted by an inspiralling binary system has amplitude given by

$$h_0(t) = \frac{4\pi^{2/3}G^{5/3}}{c^4} \times \frac{\mathcal{M}}{R} \times (\mathcal{M}\nu_{\rm GW}(t))^{2/3}. \tag{14.152}$$

This signal has been computed by solving Einstein's equations perturbed around Minkowski's spacetime, i.e. assuming (in the TT-gauge) $g_{\mu\nu} = \eta_{\mu\nu} + h^{\mathbf{TT}}_{\mu\nu}$. If the gravitational wave propagates across cosmological distances (i.e., if we cannot neglect the cosmological redshift of the source), it should be considered as a perturbation of a curved background

$$g_{\mu\nu} = g^0_{\mu\nu} + h^{\mathbf{TT}}_{\mu\nu} \tag{14.153}$$

where $g^0_{\mu\nu}$ is the FLRW spacetime 14.115.

In order to understand which are the corrections due to the expansion of the universe to Eq. 14.152, we shall proceed in two steps. To begin with, let us consider a space region far enough from the source for the metric perturbation to propagate as a spherical wave, but close enough that the wave reaches this region at a time t' such that $a(t) \simeq a(t')$. The spacetime in this region can then be considered as flat, and the Minkowskian coordinates are obtained by a simple rescaling by the (approximately) constant factor $a(t)$. Therefore, in this region the gravitational wave amplitude becomes

$$h_0(t') = \frac{4\pi^{2/3}G^{5/3}}{c^4} \times \frac{\mathcal{M}}{a(t')R} \times (\mathcal{M}\nu_{\rm GW})^{2/3}. \tag{14.154}$$

We now want to determine how the gravitational wave propagates along cosmological distances, from the time t' to the observation time $t_{\rm obs}$. To this aim, as mentioned above, we should consider Einstein's equations linearized around the curved FLRW background (the procedure was briefly discussed in Sec. 12.1; see also Eq. 15.37). The resulting equations are much more involved than the simple wave equation studied in Chapter 13. However, if the wavelength of the propagating radiation is much smaller than the curvature radius of the background (which is always the case in a cosmological background), these equations can be solved using the approximation of geometric optics [65, 112]. The result is that the cosmological background affects in the same way the propagation of both gravitational and electromagnetic waves:

- The frequency of the wave is redshifted, as discussed in Sec. 14.4.1,

$$\nu^{\rm obs}_{\rm GW} = \frac{\nu^{\rm em}_{\rm GW}}{1+z}. \tag{14.155}$$

- The amplitude of the wave falls off with the *proper curvature radius of the spherical wavefront*[9]. Since the sphere at comoving radius R (and proper time $t_{\rm obs}$) has surface area $4\pi R^2 a(t_{\rm obs})^2$ (see Eq. 14.115), its radius is $a(t_{\rm obs})R$.

Thus the amplitude in Eq. 14.152, after propagation on a cosmological background, is rescaled as

$$h_0(t_{\rm obs}) = \frac{4\pi^{2/3}G^{5/3}}{c^4} \times \frac{\mathcal{M}}{a(t_{\rm obs})R} \times (\mathcal{M}\nu_{\rm GW}^{\rm em})^{2/3} . \tag{14.156}$$

If we assume that the Universe is spatially flat ($k = 0$), $a(t_{\rm obs})R$ coincides with the proper distance D between source and observer at time $t_{\rm obs}$ given in Eq. 14.133.

Thus, as discussed in Sec. 14.2.2, the amplitude 14.156, expressed in terms of the observed quantities ($\nu_{\rm GW}^{\rm obs}$ and $\mathcal{M}^{\rm obs} = \mathcal{M}(1+z)$), reads

$$h_0(t_{\rm obs}) = \frac{4\pi^{2/3}G^{5/3}}{c^4} \times \frac{\mathcal{M}^{\rm obs}}{D_L(t_{\rm obs})} \times (\mathcal{M}^{\rm obs}\nu_{\rm GW}^{\rm obs})^{2/3} , \tag{14.157}$$

where $D_L(t_{\rm obs}) = (1+z)a(t_{\rm obs})R$ is the luminosity distance of the source (Eq. 14.144).

Similarly to the standard candles discussed in Sec. 14.4.3, compact binary inspirals allow to directly measure the luminosity distance D_L of the source. While for ordinary standard candles the luminosity distance can be measured from Eq. 14.145 once the luminosity of the source is known and the flux is measured, for compact binary inspirals the luminosity distance can be measured from Eq. 14.157 once $h_0(t_{\rm obs})$, $\mathcal{M}^{\rm obs}$, and $\nu_{\rm GW}^{\rm obs}$ are measured from a gravitational wave event. This latter measurement of the luminosity distance is also more direct than for ordinary standard candles, since it comes entirely from observations and does not require any assumption on the intrinsic luminosity of the source. Compact binary inspirals are also called *standard sirens* [104].

Detecting the gravitational wave signal from compact binaries, in coincidence with an electromagnetic counterpart (see the discussion on neutron star coalescing binaries in Sec. 14.2.6), allows to measure the luminosity distance of the sources and, using Eq. 14.151, to obtain a direct measure of the Hubble constant.

[9]This is a general property in geometric optics, also satisfied by light rays: the propagation equations require that the amplitude falls off as the rays diverge. Note that this guarantees that the energy flux (which is proportional to the squared amplitude of the wave), integrated over the wavefront, is constant.

CHAPTER 15

Gravitational waves from oscillating black holes

In this chapter we discuss the dynamics of small metric perturbations of a Schwarzschild black hole. To leading order in their amplitude, these perturbations are governed by Einstein's equations linearized on the Schwarzschild background. We shall follow the approach pioneered by Regge and Wheeler [99] in which the metric perturbations are decomposed in a complete basis of spherical harmonics according to their parity. We warn the reader that our treatment in this chapter will be basic and limited. Some of the computations we are going to present are considerably more involved than in the previous chapters, therefore in some cases we shall provide the final result and leave its derivation as an exercise. A thorough discussion on the theory of black hole perturbations can be found in Chandrasekhar's monograph [34].

Our goal is to compute how Eqs. 12.32 and 12.33 should be modified if, instead of considering the perturbation of a flat spacetime, we consider the perturbation of a curved background, in particular the Schwarzschild solution for a non-rotating black hole.

Strictly speaking, the linearized Einstein equations on a curved background have already been derived in Sec. 12.1. In vacuum (neglecting terms $O(h^2)$) they reduce to (see Eq. 12.12)

$$\delta R_{\mu\nu} = \delta\Gamma^{\alpha}_{\mu\nu;\alpha} - \delta\Gamma^{\alpha}_{\mu\alpha;\nu} = 0 \,. \tag{15.1}$$

However, the form 15.1 is impractical for explicit integration, and the physical content of the equations is not manifest. We shall show how to transform Eq. 15.1 – in the case of the Schwarzschild background – in two simple wave equations with potential barriers, which can be solved with standard numerical techniques.

In this chapter we shall use geometrized units $G = c = 1$. We shall denote the mass of the black hole in these units by M, while the symbol m will denote the azimuthal number.

15.1 A TOY MODEL: SCALAR PERTURBATIONS

Before discussing the perturbations of a Schwarzschild black hole, it is instructive to consider a much simpler problem, namely the dynamics of a scalar field ψ on the Schwarzschild background. We shall assume that ψ is a "test" scalar field, i.e. its amplitude is sufficiently small, so that we can neglect its backreaction on the metric. Indeed, the stress-energy tensor of a scalar field given in Eq. 5.25 is quadratic in ψ; therefore a background solution of Einstein's equation in vacuum (e.g., the Schwarzschild metric) is not modified by the perturbation to the leading order in the field's amplitude.

To first order in the field, we wish to solve the *Klein-Gordon equation* (see Eq. 7.59) on a Schwarzschild background. For simplicity, we consider a free scalar field, i.e. we neglect the potential term in Eq. 7.59. Then, the Klein-Gordon equation reduces to

$$\Box\psi \equiv \nabla_\mu \nabla^\mu \psi = 0\,, \tag{15.2}$$

where $\psi = \psi(t, r, \theta, \varphi)$ is a small scalar field and ∇_μ is, as usual, the covariant derivative operator (we remind that $\Box$ is the d'Alembertian operator in curved spacetime). Note that while the background is static and spherically symmetric, the perturbation is a generic function of all spacetime coordinates.

The differential operator can be written as shown in Eq. 5.67 with $V^\mu = \nabla^\mu \psi = \partial^\mu \psi$, namely

$$\nabla_\mu \nabla^\mu \psi = \frac{1}{\sqrt{-g}} \partial_\mu \left(\sqrt{-g}\, \partial^\mu \psi \right) = 0\,, \tag{15.3}$$

where $\sqrt{-g} = r^2 \sin\theta$ is the determinant of the Schwarzschild metric. Since the determinant does not depend on t and φ, and the metric is diagonal, Eq. 15.3 reduces to

$$g^{tt} \psi_{,tt} + \frac{1}{\sqrt{-g}} \left(\sqrt{-g}\, g^{rr} \psi_{,r} \right)_{,r} + \frac{1}{\sqrt{-g}} \left(\sqrt{-g}\, g^{\theta\theta} \psi_{,\theta} \right)_{,\theta} + g^{\varphi\varphi} \psi_{,\varphi\varphi} = 0\,. \tag{15.4}$$

Here and in the following we define

$$f = 1 - \frac{2M}{r}\,, \tag{15.5}$$

therefore the Schwarzschild metric components read

$$g_{tt} = -f\,, \qquad g_{rr} = f^{-1}\,, \qquad g_{\theta\theta} = r^2\,, \qquad g_{\varphi\varphi} = r^2 \sin^2\theta\,. \tag{15.6}$$

Replacing in Eq. 15.4, we get

$$\begin{aligned} & -f^{-1} \psi_{,tt} + \frac{1}{r^2 \sin\theta} \left(r^2 \sin\theta\, f \psi_{,r} \right)_{,r} + \frac{1}{r^2 \sin\theta} \left(\sin\theta\, \psi_{,\theta} \right)_{,\theta} + \frac{1}{r^2 \sin^2\theta} \psi_{,\varphi\varphi} \\ = & -f^{-1} \frac{\partial^2 \psi}{\partial t^2} + f \frac{\partial^2 \psi}{\partial r^2} + \left(f_{,r} + \frac{2f}{r} \right) \frac{\partial \psi}{\partial r} + \frac{1}{r^2} \left[\frac{\partial^2 \psi}{\partial \theta^2} + \cot\theta \frac{\partial \psi}{\partial \theta} + \frac{1}{\sin^2\theta} \frac{\partial^2 \psi}{\partial \varphi^2} \right] = 0\,. \end{aligned} \tag{15.7}$$

Box 15-A

Scalar spherical harmonics

The scalar spherical harmonics are a set of functions $\{Y^{lm}\}$ on the two-sphere $\mathbf{S^2}$. Since the two-sphere can be described by the polar coordinates (θ, φ), spherical harmonics can be defined as functions of these coordinates. The integer label l is called *harmonic index*, whereas m is called *azimuthal number*. Regularity at the boundaries implies that $l = 0, 1, \dots$ and $m = -l, -l+1, \dots, l-1, l$.

We have introduced the spherical harmonics in Box 3-E, as the eigenfunctions of the angular part of the Laplacian operator:

$$\mathbb{L}Y^{lm} \equiv \frac{\partial^2 Y^{lm}}{\partial \theta^2} + \cot\theta \frac{\partial Y^{lm}}{\partial \theta} + \frac{1}{\sin^2\theta}\frac{\partial^2 Y^{lm}}{\partial \varphi^2} = -l(l+1)Y^{lm}\,. \tag{15.8}$$

Due to the orthogonality relations

$$\int Y^{lm\,*} Y^{l'm'} \, d\Omega = \delta_{ll'}\delta_{mm'} \tag{15.9}$$

(where * denotes complex conjugation, and $d\Omega = \sin\theta d\theta d\varphi$ is the solid angle integration element) the scalar spherical harmonics form a *complete basis* of functions on the two-sphere, i.e. any function $\psi(\theta, \varphi)$ on $\mathbf{S^2}$ can be written as an expansion in spherical harmonics:

$$\psi(\theta, \varphi) = \sum_{l=0}^{\infty}\sum_{m=-l}^{l} R_{lm} Y^{lm}(\theta, \varphi) \tag{15.10}$$

where the coefficients R_{lm} can be obtained using Eq. 15.9:

$$R_{lm} = \int d\Omega \psi(\theta, \varphi) Y^{lm\,*}(\theta, \varphi)\,. \tag{15.11}$$

The explicit form of the scalar spherical harmonics is

$$Y^{lm}(\theta, \varphi) = \sqrt{\frac{2l+1}{4\pi}\frac{(l-m)!}{(l+m)!}} P^{lm}(\cos\theta) e^{im\varphi} \tag{15.12}$$

where $P^{lm}(x)$ are the *associated Legendre polynomials*

$$P^{lm}(x) = (-1)^m (1-x^2)^{m/2} \frac{d^m}{dx^m} P^l(x) \tag{15.13}$$

and $P^l(x)$ are the *Legendre polynomials* given by Rodrigues' formula

$$P^l(x) = \frac{1}{2^l l!}\frac{d^l}{dx^l}((x^2-1)^l)\,. \tag{15.14}$$

The lowest-order spherical harmonics with $m \geq 0$ are:

$$\begin{array}{lll} Y^{00} = \frac{1}{\sqrt{4\pi}} & Y^{10} = \sqrt{\frac{3}{4\pi}}\cos\theta & Y^{11} = -\sqrt{\frac{3}{8\pi}}\sin\theta e^{i\varphi} \\ Y^{20} = \sqrt{\frac{5}{16\pi}}(3\cos^2\theta - 1) & Y^{21} = -\sqrt{\frac{15}{8\pi}}\sin\theta\cos\theta e^{i\varphi} & Y^{22} = \sqrt{\frac{15}{8\pi}}\sin^2\theta e^{2i\varphi} \end{array} \tag{15.15}$$

while those with $m < 0$ can be obtained from the relation $Y^{l-m} = (-1)^m Y^{lm\,*}$.
For further details on the spherical harmonics, see e.g. [67].

Since the background is spherically symmetric, to simplify Eq. 15.7 it is very convenient to perform a *spherical harmonic decomposition* of the scalar field, i.e. to expand the field $\psi(t,r,\theta\,\varphi)$ in a basis of complex functions of the angular variables, the *spherical harmonics* $\{Y^{lm}(\theta,\varphi)\}$ (see Box 15-A):

$$\psi(t,r,\theta,\varphi) = \sum_{lm} R_{lm}(t,r) Y^{lm}(\theta,\varphi)\,. \tag{15.16}$$

Indeed, any spherically symmetric manifold can be considered as "filled" with two-spheres (e.g. Sec. 9.1). The metric of a spherically symmetric spacetime can be written as (see Sec. 9.6)

$$ds^2 = g_{AB}(x^0,x^1)dx^A dx^B + a(x^0,x^1)(d\theta^2 + \sin^2\theta d\varphi^2)\,, \tag{15.17}$$

where $A = 0,1$ and we can choose $x^0 = t$ and $x^1 = r$. In other words, the surfaces with t and r constant are two-spheres, and the four-dimensional manifold $\mathbf{M}$ can be written as the product of a two-dimensional manifold $\mathbf{M_2}$ and the two-sphere $\mathbf{S^2}$: $\mathbf{M} = \mathbf{M_2} \times \mathbf{S^2}$.

Any scalar function defined on a spherically symmetric manifold can be expanded in spherical harmonics as in Eq. 15.16; the coefficient functions $R_{lm}(t,r)$ can be obtained as shown in Eq. 15.11 of Box 15-A:

$$R_{lm}(t,r) = \int d\Omega \psi(t,r,\theta,\varphi) Y^{lm\,*}(\theta,\varphi)\,. \tag{15.18}$$

Note that, while in Eq. 15.11 the coefficients of the expansion are constants, here R_{lm} are functions of t and r.

The differential operator appearing in the square brakets of Eq. 15.7 is the angular part of the Laplacian operator in polar coordinates, $I\!\!L$, defined in Eqs. 3.88 and 15.8. Since the spherical harmonics are eigenfunctions of this operator,

$$I\!\!L Y^{lm} = -l(l+1) Y^{lm}\,, \tag{15.19}$$

by replacing the expansion 15.16 into Eq. 15.7 we get

$$\sum_{lm}\left[-f^{-1}\frac{\partial^2 R_{lm}}{\partial t^2} + f\frac{\partial^2 R_{lm}}{\partial r^2} + \left(f_{,r} + \frac{2f}{r}\right)\frac{\partial R_{lm}}{\partial r} - \frac{l(l+1)}{r^2} R_{lm}\right] Y^{lm} = 0\,. \tag{15.20}$$

We can multiply the above equation by $Y^{l'm'\,*}$ and integrate over the solid angle. Using the orthogonality condition 15.9, we obtain an infinite set of *decoupled* equations for the functions $R_{l'm'}(t,r)$ only:

$$f\frac{\partial^2 R_{l'm'}}{\partial r^2} - f^{-1}\frac{\partial^2 R_{l'm'}}{\partial t^2} + \left(f_{,r} + \frac{2f}{r}\right)\frac{\partial R_{l'm'}}{\partial r} - \frac{l(l+1)}{r^2} R_{l'm'} = 0\,. \tag{15.21}$$

Since l', m' are free indices, we can redefine them as l and m for simplicity. Remarkably, while Eqs. 15.7 are partial differential equations in (t,r,θ,ϕ), Eqs. 15.21 are partial differential equations in t and r only. By performing a Fourier transform (as we shall discuss below), they become *ordinary differential equations* in r, which are much easier to solve than partial differential equations.

It is also convenient to redefine the scalar field such that

$$R_{lm}(t,r) = \frac{\psi_{lm}(t,r)}{r}\,. \tag{15.22}$$

Thus,

$$R_{lm\,,r} = \frac{\psi_{lm\,,r}}{r} - \frac{\psi_{lm}}{r^2}\,, \qquad R_{lm\,,rr} + \frac{2R_{lm\,,r}}{r} = \frac{\psi_{lm\,,rr}}{r} \tag{15.23}$$

and

$$f R_{lm,rr} + \left(f_{,r} + \frac{2f}{r} \right) R_{lm,r} = f \frac{\psi_{lm,rr}}{r} + f_{,r} \frac{\psi_{lm,r}}{r} - f_{,r} \frac{\psi_{lm}}{r^2} \,; \tag{15.24}$$

therefore, after multiplying by $rf(r)$ and noting that $f_{,r} = 2M/r^2$, Eq. 15.21 takes the form

$$f^2 \frac{\partial^2 \psi_{lm}}{\partial r^2} - \frac{\partial^2 \psi_{lm}}{\partial t^2} + f f_{,r} \frac{\partial \psi_{lm}}{\partial r} - f \left(\frac{l(l+1)}{r^2} + \frac{2M}{r^3} \right) \psi_{lm} = 0 \,. \tag{15.25}$$

Note that the azimuthal number m does not appear explicitly in this equation. This is a general property of linear perturbations of spherically symmetric spacetimes.

It is also convenient to introduce the tortoise coordinate (defined in Eq. 9.117),

$$r_* = r + 2M \ln \left| \frac{r}{2M} - 1 \right| \,. \tag{15.26}$$

Since $dr/dr_* = f$, we obtain

$$\frac{\partial^2}{\partial r_*^2} = f \frac{\partial}{\partial r} f \frac{\partial}{\partial r} = f^2 \frac{\partial^2}{\partial r^2} + f_{,r} f \frac{\partial}{\partial r} \,, \tag{15.27}$$

and Eq. 15.25 can be written in the form of a one-dimensional wave equation

$$\frac{\partial^2 \psi_{lm}(t,r)}{\partial r_*^2} - \frac{\partial^2 \psi_{lm}(t,r)}{\partial t^2} - V_l^{\rm scalar}(r) \psi_{lm}(t,r) = 0 \,, \tag{15.28}$$

where

$$V_l^{\rm scalar}(r) = \left(1 - \frac{2M}{r} \right) \left(\frac{l(l+1)}{r^2} + \frac{2M}{r^3} \right) \tag{15.29}$$

is the effective potential, which is due to the spacetime curvature. Indeed, in the flat spacetime limit ($M = 0$) the potential displays only the classical centrifugal term $V_l^{\rm scalar}(r) = l(l+1)/r^2$ and is a monotonically decreasing function. Thus, the scalar field propagates in the Schwarzschild background as a wave scattered by the effective potential.

This is plotted in Fig. 15.1 as a function of the tortoise coordinate, expressing r in terms of r_*. Note that $V_l^{\rm scalar}(r) \to 0$ both near the horizon, $r \to 2M$ ($r_* \to -\infty$), and at infinity, $r \to \infty$ ($r_* \to \infty$). Furthermore, it can be shown that the scalar potential has a maximum outside the horizon at

$$r_{\rm max} = \frac{3l(l+1) + \sqrt{l(l+1)(9l(l+1)+14)+9} - 3}{2l(l+1)} M \,. \tag{15.30}$$

It may be noted that, in the $l \to \infty$ limit, the location of the maximum coincides with the location of the photon sphere, i.e. the light ring $r = 3M$ (see Sec. 10.4).

Finally, since the background is static, it is often convenient to work in the Fourier domain. We define the Fourier transform $\tilde{\psi}_{lm}(\omega, r)$ of the function $\psi_{lm}(t,r)$ as

$$\psi_{lm}(t,r) = \int_{-\infty}^{+\infty} d\omega \, \tilde{\psi}_{lm}(\omega, r) e^{-i\omega t} \,, \tag{15.31}$$

and, from Eq. 15.28, one can immediately obtain the equation in the frequency domain

$$\frac{\partial^2 \tilde{\psi}_{lm}(\omega, r)}{\partial r_*^2} + \left[\omega^2 - V_l^{\rm scalar}(r) \right] \tilde{\psi}_{lm}(\omega, r) = 0 \,. \tag{15.32}$$

This is an ordinary differential equation for the radial wavefunction $\tilde{\psi}_{lm}(\omega, r)$. Note that

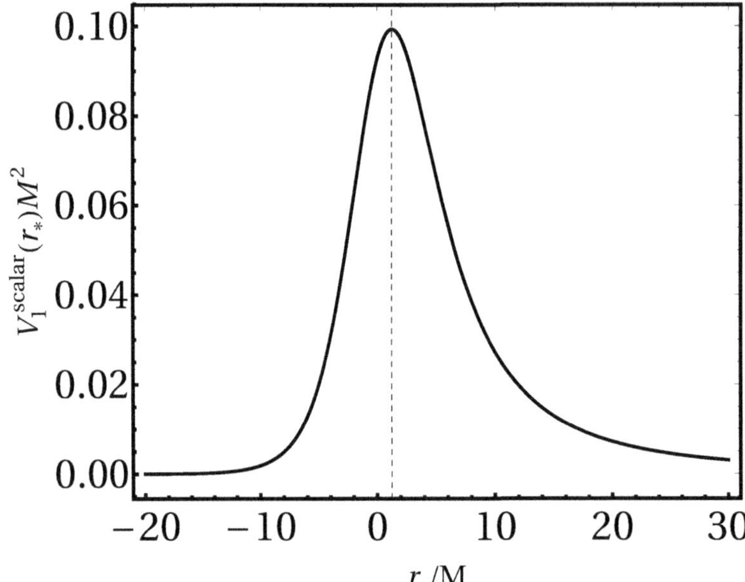

Figure 15.1: Effective potential $V_l^{\rm scalar}$ for scalar perturbations on a Schwarzschild black hole in tortoise coordinates and for $l = 1$.

Eq. 15.32 has the same form of the time-independent Schrödinger equation for a quantum mechanical problem in one dimension, with the identification $\omega^2 \equiv E$, where E is the energy of the quantum-mechanical eigenstate. Therefore, the above equation can be solved and analysed with the standard methods developed for quantum mechanics. In particular, the scattering of a scalar wave by a Schwarzschild black hole can be simply analysed as the scattering of a wave packet off the effective potential given in Eq. 15.29.

To summarize, to leading order in the scalar field, the Klein-Gordon equation on a Schwarzschild spacetime reduces to the Schrödinger-like equation 15.32 with an effective potential barrier given by Eq. 15.29.

15.2 PERTURBATIONS OF THE SCHWARZSCHILD SPACETIME

Let us now consider the more interesting case of spacetime perturbations. As we shall see, the procedure to analyse this problem is similar to that discussed above for a test scalar field, but it is more complicated due to the tensorial nature of the metric.

15.2.1 Linearized Einstein's equations in vacuum

Following what we did in Chapter 12, we expand the metric as

$$g_{\mu\nu} = g^0_{\mu\nu} + h_{\mu\nu} \,, \tag{15.33}$$

where $g^0_{\mu\nu}$ is the background metric and $h_{\mu\nu}$ is a small perturbation, $|h_{\mu\nu}| \ll |g^0_{\mu\nu}|$.

Einstein's equations in vacuum reduce to $R_{\mu\nu}(g) = R^0_{\mu\nu} + \delta R_{\mu\nu}(h) = 0$, and the Ricci tensor linearized on a generic, vacuum background is given in Eq. 12.12. Since $R^0_{\mu\nu}(g^0) = 0$ by definition, the linearized Einstein equations reduce to

$$\delta R_{\mu\nu} = \delta\Gamma^\alpha{}_{\mu\nu;\alpha} - \delta\Gamma^\alpha{}_{\mu\alpha;\nu} = 0 \,, \tag{15.34}$$

where (see Eq. 12.11)

$$\delta\Gamma^{\mu}{}_{\alpha\beta} = \frac{1}{2} g^{0\,\mu\nu}(h_{\nu\alpha;\beta} + h_{\nu\beta;\alpha} - h_{\alpha\beta;\nu}) \,. \tag{15.35}$$

Note that covariant derivatives in Eqs. 15.34 and 15.35 are performed with respect to the background metric $g^0_{\mu\nu}$.

By replacing Eq. 15.35 in Eq. 15.34 we find

$$\begin{aligned} \delta R_{\mu\nu} &= \frac{1}{2} g^{0\,\alpha\rho}\,(h_{\nu\rho;\mu\alpha} + h_{\mu\rho;\nu\alpha} - h_{\mu\nu;\rho\alpha} - h_{\alpha\rho;\mu\nu} - h_{\mu\rho;\alpha\nu} + h_{\mu\alpha;\rho\nu}) \quad (15.36) \\ &= \frac{1}{2}\left(h^{\rho}{}_{\nu;\mu\rho} + h^{\rho}{}_{\mu;\nu\rho} - h^{\rho}{}_{\rho;\mu\nu} - g^{0\,\sigma\rho} h_{\mu\nu;\sigma\rho}\right) \,. \end{aligned}$$

Thus, the perturbation of a spacetime which is solution of Einstein's equations in vacuum must satisfy the partial differential equation

$$2\delta R_{\mu\nu} = h^{\rho}{}_{\nu;\mu\rho} + h^{\rho}{}_{\mu;\nu\rho} - h^{\rho}{}_{\rho;\mu\nu} - g^{0\,\sigma\rho} h_{\mu\nu;\sigma\rho} = 0 \,. \tag{15.37}$$

This equation holds for *any* vacuum, background spacetime, and in the following we shall consider the case in which the background is the Schwarzschild metric.

We remark that in the perturbative approach we deal with two different manifolds: the physical manifold $\mathbf{M}$, with metric $g_{\mu\nu} = g^0_{\mu\nu} + h_{\mu\nu}$, and the background manifold $\mathbf{M^0}$, with metric $g^0_{\mu\nu}$. The quantity $h_{\mu\nu}$ can be interpreted either as the metric perturbation on the manifold $\mathbf{M}$, or as a tensor field on the manifold $\mathbf{M^0}$, which satisfies the dynamical equations 15.37.

With the latter interpretation (which we shall follow in the rest of this section) Eq. 15.37 is a tensor equation with respect to the manifold $\mathbf{M^0}$, and thus it is generally covariant. To hereafter, covariant derivatives and index raising/lowering will always be performed with respect to the background metric $g^0_{\mu\nu}$.

15.2.2 Harmonic decomposition

Since the Schwarzschild spacetime is spherically symmetric, Eq. 15.37 can be greatly simplified by performing a decomposition in spherical harmonics. However, due to the tensorial nature of $h_{\mu\nu}$ the harmonic decomposition is more involved than in the scalar case.

As discussed in Sec. 15.1, the Schwarzschild manifold (and, more generally, any spherically symmetric spacetime) can be decomposed as the product of a two-dimensional manifold $\mathbf{M_2}$ described by the coordinates (t, r), and the two-sphere $\mathbf{S^2}$, $\mathbf{M_2} \times \mathbf{S^2}$. With this decomposition, the coordinates of the Schwarzschild spacetime are split as (we follow the notation of [46]):

$$x^{\mu} = (z^A, y^a) \tag{15.38}$$

where $A = 0, 1$, $a = 2, 3$ and

$$z^A = t,\, r, \quad y^a = \theta, \varphi \,. \tag{15.39}$$

The tangent space at any point $\mathbf{p} \in \mathbf{M}$ is the outer product (see Sec. 2.4.1) of the tangent spaces to $\mathbf{M_2}$ and to $\mathbf{S^2}$, $\mathbf{T_{M_2}} \otimes \mathbf{T_{S^2}}$. Therefore, any tensor can be decomposed in a tensor in $\mathbf{M_2}$ and a tensor in $\mathbf{S^2}$. With this decomposition, the components of a vector field V^{μ} are split as

$$V^{\mu} = (V^A, V^a) \,, \tag{15.40}$$

while the components of a rank-two symmetric tensor field $X_{\mu\nu}$ are split as

$$X_{\mu\nu} = \begin{pmatrix} X_{AB} & X_{Aa} \\ X_{aA} & X_{ab} \end{pmatrix} . \tag{15.41}$$

We decompose the metric perturbation $h_{\mu\nu}$ as in Eq. 15.41. We then expand the components h_{AB}, h_{Aa}, h_{ab} in scalar, vector, and rank-two tensor harmonics (see Box 15-B), as follows:

$$\begin{aligned} h_{AB}(t,r,\theta,\varphi) &= \sum_{lm} \bar{h}_{AB\,lm}(t,r)Y^{lm}(\theta,\varphi)\,, \\ h_{aA}(t,r,\theta,\varphi) &= \sum_{lm} \left[\bar{h}^{\rm pol}_{A\,lm}(t,r)Y^{lm}_a(\theta,\varphi) + \bar{h}^{\rm ax}_{A\,lm}(t,r)S^{lm}_a(\theta,\varphi)\right]\,, \\ h_{ab}(t,r,\theta,\varphi) &= \sum_{lm} \left[r^2\left(K_{lm}(t,r)\gamma_{ab}Y^{lm}(\theta,\varphi) + G_{lm}(t,r)Z^{lm}_{ab}(\theta,\varphi)\right)\right. \\ &\quad \left. +2\bar{h}_{lm}(t,r)S^{lm}_{ab}(\theta,\varphi)\right]\,, \end{aligned} \tag{15.42}$$

where γ_{ab} is the metric of the two-sphere (Eq. 15.46), and $\bar{h}_{ABlm}(t,r)$, $\bar{h}^{\rm pol}_{A\,lm}(t,r)$, $\bar{h}^{\rm ax}_{A\,lm}(t,r)$, $K_{lm}(t,r)$, $G_{lm}(t,r)$, and $\bar{h}_{lm}(t,r)$ are functions of (t,r) to be determined by solving Einstein's equations, i.e. Eq. 15.37. To hereafter, sums over l,m will be omitted. Eqs. 15.42 are the generalisation of the expansion of a scalar field in spherical harmonics given in Eq. 15.16 to the case of a symmetric, rank-two tensor field. We remark that X_{AB}, X_{Aa}, and X_{ab} transform as scalar, vector, and tensor quantities for coordinate transformations on $\mathbf{S^2}$ (i.e., for rotations), hence the decomposition.

Remarkably, it is possible to prove[1] that as the scalar spherical harmonics $\{Y^{lm}\}$ form a complete basis for scalar functions on $\mathbf{S^2}$, the polar and axial vector harmonics, $\{Y^{lm}_a, S^{lm}_a\}$, form a complete basis for vectors on $\mathbf{S^2}$. Similarly, the polar and axial rank-two tensor harmonics, $\{Z^{lm}_{ab}, S^{lm}_{ab}\}$, form a complete basis for symmetric, traceless, rank-two tensors on $\mathbf{S^2}$. We also note that any rank-two tensor can be decomposed in the sum of a traceless tensor and of a tensor proportional to the metric (the so-called "trace part" of the tensor),

$$X_{ab} = X^{\rm traceless}_{ab} + \frac{1}{4}(\gamma^{cd}X_{cd})\gamma_{ab}\,. \tag{15.43}$$

Since $\gamma^{cd}X_{cd} = X$ is a scalar (with respect to coordinate transformations on $\mathbf{S^2}$), it can be decomposed in scalar spherical harmonics. Therefore, any symmetric rank-two tensor can be decomposed in a *complete basis* of scalar, vector, and tensor spherical harmonics, as done in Eq. 15.42 for the symmetric tensor $h_{\mu\nu}$. The perturbation functions $\bar{h}_{AB,lm}(t,r)$, $h^{\rm pol}_{A\,lm}(t,r)$, $\bar{h}^{\rm ax}_{A\,lm}(t,r)$, $K_{lm}(t,r)$, $G_{lm}(t,r)$, and $\bar{h}_{lm}(t,r)$ can be obtained in terms of the components $h_{\mu\nu}(t,r,\theta,\varphi)$ using the orthogonality properties discussed in Box 15-C.

The scalar spherical harmonics transform under parity inversion ($\theta\to\pi-\theta$, $\varphi\to\varphi+\pi$) as[2]

$$Y^{lm}(\pi-\theta,\varphi+\pi) = (-1)^l Y^{lm}(\theta,\varphi)\,. \tag{15.44}$$

The harmonics Y^{lm}_a and Z^{lm}_{ab} transform in the same way. Conversely, the harmonics S^{lm}_a transform as

$$S^{lm}_a(\pi-\theta,\varphi+\pi) = (-1)^{l+1}S^{lm}_a(\theta,\varphi)\,, \tag{15.45}$$

and the same holds for S^{lm}_{ab}. The first group of harmonics is called *polar* (or *even* or *electric*), and the second group is called *axial* (or *odd* or *magnetic*). Therefore, the perturbations in the decomposition 15.42 split in two decoupled sets. Correspondingly, we shall call $\bar{h}_{AB\,lm}(t,r)$, $\bar{h}^{\rm pol}_{A\,lm}(t,r)$, $K_{lm}(t,r)$, $G_{lm}(t,r)$ *polar functions*, and $\bar{h}^{\rm ax}_{A\,lm}(t,r)$, $\bar{h}_{lm}(t,r)$ *axial functions.*

[1] The proof requires an extensive use of the tools of group theory, and is beyond the scope of this book.
[2] Eq. 15.44 can be obtained by applying parity inversion to Eqs. 15.12, 15.13, and 15.14.

Box 15-B

Scalar, vector, and tensor spherical harmonics

Let γ_{ab} be the metric of the two-sphere $\mathbf{S^2}$:

$$\gamma_{ab} = \text{diag}(1, \sin^2\theta). \tag{15.46}$$

We shall denote as ":" the covariant derivative on the two-sphere. Since the non-vanishing Christoffel's symbols on $\mathbf{S^2}$ are (see Sec. 3.6) $\Gamma^\theta_{\varphi\varphi} = -\sin\theta\cos\theta$, $\Gamma^\varphi_{\theta\varphi} = \Gamma^\varphi_{\varphi\theta} = \cot\theta$, the Laplacian operator on a scalar function $f(\theta,\varphi)$ yields

$$\gamma^{ab} f_{:ab} = \gamma^{ab} f_{,ab} + \gamma^{ab}\Gamma^c_{ab} f_{,c} = f_{,\theta\theta} + \cot\theta f_{,\theta} + \sin^{-2}\theta f_{,\varphi\varphi}\,. \tag{15.47}$$

Thus the eigenvalue equation defining the spherical harmonics, Eq. 15.8, can be written as

$$\gamma^{ab} Y^{lm}_{:ab} = -l(l+1)Y^{lm}\,. \tag{15.48}$$

We also introduce the Levi-Civita tensor on $\mathbf{S^2}$. It is defined as (see Box 8-C) $\varepsilon_{ab} = \sqrt{\gamma}\epsilon_{ab}$ where $\gamma = \det(\gamma_{ab}) = \sin^2\theta$ and ϵ_{ab} is the two-dimensional Levi-Civita symbol ($\epsilon_{ab} = -\epsilon_{ba}$, $\epsilon_{12} = 1$). Therefore, $\varepsilon_{\theta\varphi} = -\varepsilon_{\varphi\theta} = \sin\theta$.
We now introduce a set of vectors and of rank-two tensors on $\mathbf{S^2}$, called *spherical vector harmonics* and *spherical tensor harmonics*, respectively.

- The *polar vector harmonics* are:

$$Y^{lm}_a \equiv Y^{lm}_{:a} = \left(Y^{lm}_{,\theta}, Y^{lm}_{,\varphi}\right)\,. \tag{15.49}$$

- The *axial vector harmonics* are:

$$S^{lm}_a \equiv -\varepsilon_a{}^b Y^{lm}_{:b} = \left(-\frac{1}{\sin\theta}Y^{lm}_{,\varphi}, \sin\theta Y^{lm}_{,\theta}\right)\,. \tag{15.50}$$

- The *polar rank-two tensor harmonics* are:

$$Z^{lm}_{ab} \equiv Y^{lm}_{:ab} + \frac{l(l+1)}{2}\gamma_{ab}Y^{lm} = \frac{1}{2}\begin{pmatrix} W^{lm} & X^{lm} \\ X^{lm} & -\sin^2\theta W^{lm} \end{pmatrix}, \tag{15.51}$$

  where $X^{lm} \equiv 2\left(Y^{lm}_{,\theta\varphi} - \cot\theta Y^{lm}_{,\varphi}\right)$ and $W^{lm} \equiv Y^{lm}_{,\theta\theta} - \cot\theta Y^{lm}_{,\theta} - \sin^{-2}\theta Y^{lm}_{,\varphi\varphi}$.

- The *axial rank-two tensor harmonics* are:

$$S^{lm}_{ab} \equiv \frac{1}{2}(S^{lm}_{a:b} + S^{lm}_{b:a}) = \frac{1}{2}\begin{pmatrix} -\sin^{-1}\theta X^{lm} & \sin\theta W^{lm} \\ \sin\theta W^{lm} & \sin\theta X^{lm} \end{pmatrix}. \tag{15.52}$$

The indexes l and m are integers such that $|m| \leq l = 0, 1, 2, ...$ We note that the tensor harmonics are symmetric, since all terms in $Y^{lm}_{:ab} = Y^{lm}_{,ab} + \Gamma^c_{ab}Y^{lm}_{,c}$ are symmetric in the indexes (a,b). Moreover, they are traceless:

$$\begin{aligned} \gamma^{ab}Z^{lm}_{ab} &= \gamma^{ab}Y^{lm}_{:ab} + \frac{l(l+1)}{2}\gamma^{ab}\gamma_{ab}Y^{lm} = 0 \\ \gamma^{ab}S^{lm}_{ab} &= \gamma^{ab}S^{lm}_{a:b} = -\gamma^{ab}\varepsilon_a{}^c Y^{lm}_{:cb} = -\varepsilon^{bc}Y^{lm}_{:cb} = 0\,. \end{aligned} \tag{15.53}$$

Box 15-C

Orthogonality properties of scalar, vector, and tensor spherical harmonics

We define the scalar product on the two-sphere as

$$\langle f,g\rangle \equiv \int d\Omega f^* g = \int d\theta d\varphi \sin\theta f^* g\,. \tag{15.54}$$

With this notation the orthogonality property for the scalar spherical harmonics (Eq. 15.9) reads

$$\langle Y^{lm}, Y^{l'm'}\rangle = \delta^{ll'}\delta^{mm'}\,. \tag{15.55}$$

Using this and Eq. 15.48, we obtain the orthogonality property of the polar vector harmonics:

$$\begin{aligned}
\gamma^{ab}\langle Y^{lm}_a, Y^{l'm'}_b\rangle &= \int d\theta d\varphi \sin\theta \left(Y^{lm\,*}_{,\theta} Y^{l'm'}_{,\theta} + \frac{1}{\sin^2\theta} Y^{lm\,*}_{,\varphi} Y^{l'm'}_{,\varphi}\right) \\
&= -\int d\theta d\varphi \sin\theta Y^{lm\,*}\left(Y^{l'm'}_{,\theta\theta} + \cot\theta Y^{l'm'}_{,\theta} + \frac{1}{\sin^2\theta} Y^{l'm'}_{,\varphi\varphi}\right) \\
&= l'(l'+1)\int d\theta d\varphi \sin\theta Y^{lm\,*} Y^{l'm'} = l(l+1)\delta^{ll'}\delta^{mm'}\,.
\end{aligned} \tag{15.56}$$

Similarly, it is possible to derive the orthogonality property of axial vector harmonics

$$\gamma^{ab}\langle S^{lm}_a, S^{l'm'}_b\rangle = l(l+1)\delta^{ll'}\delta^{mm'}\,, \tag{15.57}$$

and it can be shown that

$$\begin{aligned}
&\frac{1}{2}\langle Z^{lm}_{ab}, Z^{l'm'}_{ab}\rangle = \frac{1}{2}\langle S^{lm}_{ab}, S^{l'm'}_{ab}\rangle \\
&= \int d\theta d\varphi \sin\theta \left(W^{*lm}_{ab} W^{l'm'}_{ab} + \frac{1}{\sin^2\theta} X^{*lm}_{ab} X^{l'm'}_{ab}\right) \\
&= (l-1)l(l+1)(l+2)\delta^{ll'}\delta^{mm'}\,.
\end{aligned} \tag{15.58}$$

15.2.3 The Regge-Wheeler gauge

As discussed in Sec. 6.3, Einstein's equations do not determine the spacetime metric in a unique way, but only up to an arbitrary coordinate transformation (which is formally equivalent to a diffeomorphism, see Secs. 2.1.4 and 8.5). In the case of perturbations of a flat spacetime, a coordinate transformation of order $O(h)$ is equivalent to an infinitesimal diffeomorphism $x^\alpha \to x^\alpha + \epsilon^\alpha(x)$, which yields a transformation of the metric components (see Box 12-A)

$$h_{\mu\nu} \equiv \delta g_{\mu\nu} = \mathcal{L}_{\vec{\epsilon}}\, g_{\mu\nu} = \epsilon_{\mu;\nu} + \epsilon_{\nu;\mu}\,. \tag{15.59}$$

Therefore (see Eq. 12.47),

$$h'_{\mu\nu} = h_{\mu\nu} + \epsilon_{\mu;\nu} + \epsilon_{\nu;\mu}\,. \tag{15.60}$$

We have thus the freedom to choose four functions $\epsilon^\alpha(x)$ to simplify the perturbation equations.

It can be shown that is always possible to find four functions $\epsilon^\alpha(x)$ such that the transformation 15.60 enforces the so-called **Regge-Wheeler gauge** [99], defined by[3]

$$\bar{h}^{\rm pol}_{0\,lm}(t,r) = \bar{h}^{\rm pol}_{1\,lm}(t,r) = G_{lm}(t,r) = \bar{h}_{lm}(t,r) = 0\,. \tag{15.61}$$

We are therefore left with two axial functions (which for ease of notation we can simply redefine as $\bar{h}^{\rm ax}_{A\,lm} \equiv h_{A\,lm}$ without ambiguity) and four polar functions:

$$h_{\mu\nu}(t,r,\theta,\varphi) = \begin{pmatrix} \bar{h}_{AB\,lm}(t,r)Y^{lm}(\theta,\varphi) & h_{A\,lm}(t,r)S^{lm}_a(\theta,\varphi) \\ h_{A\,lm}(t,r)S^{lm}_a(\theta,\varphi) & r^2K_{lm}(t,r)\gamma_{ab}Y^{lm}(\theta,\varphi) \end{pmatrix}, \tag{15.62}$$

where

$$\begin{aligned} \bar{h}_{AB\,lm}(t,r) &= \begin{pmatrix} f(r)H_{0\,lm}(t,r) & H_{1\,lm}(t,r) \\ H_{1\,lm}(t,r) & f(r)^{-1}H_{2\,lm}(t,r) \end{pmatrix}, \\ h_{A\,lm}(t,r) &= (h_{0\,lm}(t,r), h_{1\,lm}(t,r))\,. \end{aligned} \tag{15.63}$$

We can thus decompose the metric perturbation in a polar and an axial part,

$$h_{\mu\nu}(t,r,\theta,\varphi) = h^{\rm pol}_{\mu\nu}(t,r,\theta,\varphi) + h^{\rm ax}_{\mu\nu}(t,r,\theta,\varphi) \tag{15.64}$$

whose harmonic expansions in the Regge-Wheeler gauge are

$$h^{\rm pol}_{\mu\nu}(t,r,\theta,\varphi) = \tag{15.65}$$
$$\begin{pmatrix} fH_{0\,lm}(t,r) & H_{1\,lm}(t,r) & 0 & 0 \\ * & f^{-1}H_{2\,lm}(t,r) & 0 & 0 \\ * & * & r^2K_{lm}(t,r) & 0 \\ * & * & * & r^2\sin^2\theta K_{lm}(t,r) \end{pmatrix} Y^{lm},$$

$$h^{\rm ax}_{\mu\nu}(t,r,\theta,\varphi) = \begin{pmatrix} 0 & 0 & -h_{0\,lm}(t,r)\csc\theta Y^{lm}_{,\varphi} & h_{0\,lm}(t,r)\sin\theta Y^{lm}_{,\theta} \\ * & 0 & -h^{lm}_1(t,r)\csc\theta Y^{lm}_{,\varphi} & h_{1\,lm}(t,r)\sin\theta Y^{lm}_{,\theta} \\ * & * & 0 & 0 \\ * & * & * & 0 \end{pmatrix}, \tag{15.66}$$

where asterisks indicate symmetric components and we remind that $f = 1 - 2M/r$.

Since the background is static, it is convenient to perform a Fourier transform of all perturbation functions, as done in the scalar case. The expansion of the metric perturbations is then

$$h_{\mu\nu}(t,r,\theta,\varphi) = \int_{-\infty}^{+\infty} d\omega \begin{pmatrix} \tilde{h}_{AB\,lm}(\omega,r)Y^{lm}(\theta,\varphi) & \tilde{h}_{A\,lm}(\omega,r)S^{lm}_a(\theta,\varphi) \\ \tilde{h}_{A\,lm}(\omega,r)S^{lm}_a(\theta,\varphi) & r^2\tilde{K}_{lm}(\omega,r)\gamma_{ab}Y^{lm}(\theta,\varphi) \end{pmatrix} e^{-i\omega t} \tag{15.67}$$

where

$$\begin{aligned} \tilde{h}_{AB\,lm}(\omega,r) &= \begin{pmatrix} f(r)\tilde{H}_{0\,lm}(\omega,r) & \tilde{H}_{1\,lm}(\omega,r) \\ \tilde{H}_{1\,lm}(\omega,r) & f(r)^{-1}\tilde{H}_{2\,lm}(\omega,r) \end{pmatrix} \\ \tilde{h}_{A\,lm}(\omega,r) &= \left(\tilde{h}_{0\,lm}(\omega,r), \tilde{h}_{1\,lm}(\omega,r)\right)\,. \end{aligned} \tag{15.68}$$

[3] Other gauge choices are possible, such as for instance the *Chandrasekhar-Ferrari gauge* [35], in which $H_{1\,lm} = 0$ while $G_{lm} \neq 0$.

In the following, we shall omit the tilde in the Fourier-transformed quantities for ease of notation. With this decomposition, in the Regge-Wheeler gauge the perturbations are described by the following scalar functions

$$H_{0\,lm}(\omega,r),\ H_{1\,lm}(r,\omega),\ H_{2\,lm}(\omega,r),\ K_{lm}(r,\omega) \qquad \text{(polar perturbations)}$$
$$h_{0\,lm}(r,\omega),\ h_{1\,lm}(r,\omega) \qquad \text{(axial perturbations)}\,.$$

15.2.4 Perturbations with $l = 0$ and $l = 1$

While the scalar spherical harmonics Y^{lm} are defined for all integer values of $l \geq 0$, the vector spherical harmonics Y^{lm}_a, S^{lm}_a only exist for $l \geq 1$, and the rank-two tensor spherical harmonics Z^{lm}_{ab}, S^{lm}_{ab} only exist for $l \geq 2$. Therefore, when $l = 0, 1$ some of the perturbations in the expansion 15.42 vanish. The equations for the $l = 0, 1$ perturbations should thus be derived separately.

On the other hand, as discussed in Sec. 13.2.3, the monopole and dipole moments of the source (corresponding to $l = 0, 1$ perturbations, respectively) are not associated to the emission of gravitational waves. Indeed, an $l = 0$ perturbation can be interpreted as a small change of the black hole mass, while an $l = 1$ perturbation can be interpreted as the black hole acquiring a small angular momentum. Therefore, in the remain of this chapter we shall focus on perturbations with $l \geq 2$.

15.2.5 Linearized Einstein's equations in vacuum

We are now ready to insert the harmonic expansion of the metric perturbations (Eq. 15.67) into the linearized Einstein equations in vacuum, Eq. 15.37:

$$2\delta R_{\mu\nu} = h^{\rho}{}_{\nu;\mu\rho} + h^{\rho}{}_{\mu;\nu\rho} - h^{\rho}{}_{\rho;\mu\nu} - g^{0\,\sigma\rho}h_{\mu\nu;\sigma\rho} = 0\,. \tag{15.69}$$

Since $\delta R_{\mu\nu}$ is a symmetric, rank-two tensor field on the background spacetime $\mathbf{M_2} \times \mathbf{S^2}$, its components can be split as in Eq. 15.41,

$$\delta R_{\mu\nu} = \begin{pmatrix} \delta R_{AB} & \delta R_{Aa} \\ \delta R_{aA} & \delta R_{ab} \end{pmatrix}, \tag{15.70}$$

and then expanded in tensor spherical harmonics Y^{lm}, Y^{lm}_a, S^{lm}_a, Z^{lm}_{ab}, S^{lm}_{ab}. With a lengthy but conceptually simple computation, one finds[4]:

$$2\delta R_{AB}(t,r,\theta,\varphi) = \int_{-\infty}^{+\infty} d\omega \begin{pmatrix} A^{(0)}_{lm}(\omega,r) & A^{(1)}_{lm}(\omega,r) \\ * & A^{(2)}_{lm}(\omega,r) \end{pmatrix} Y^{lm}(\theta,\varphi)e^{-i\omega t} = 0\,, \tag{15.71}$$

$$2\delta R_{Aa}(t,r,\theta,\varphi) = \int_{-\infty}^{+\infty} d\omega \left(\alpha^{(A)}_{lm}(\omega,r)Y^{lm}_a(\theta,\varphi) + \beta^{(A)}_{lm}(\omega,r)S^{lm}_a(\theta,\varphi)\right) e^{-i\omega t} = 0\,, \tag{15.72}$$

$$2\delta R_{ab}(t,r,\theta,\varphi) = \int_{-\infty}^{+\infty} d\omega \Big(A^{(3)}_{lm}(\omega,r)r^2\gamma_{ab}Y^{lm}(\theta,\varphi) + s_{lm}(\omega,r)Z^{lm}_{ab}(\theta,\varphi) + t_{lm}(\omega,r)S^{lm}_{ab}(\theta,\varphi)\Big) e^{-i\omega t} = 0\,, \tag{15.73}$$

[4]We use the notation of [70], with a different overall sign.

where $\beta^{(A)}_{lm}$ (with $A = 0, 1$) and t_{lm} are combinations of the axial functions $h_{A\,lm}(\omega, r)$ and of their derivatives with respect to r (which we denote, for brevity, with a prime):

$$
\begin{aligned}
\beta^{(0)}_{lm}(\omega,r) &= -f\left(h''_{0\,lm} + i\omega h'_{1\,lm}\right) - 2i\omega\frac{f}{r}h_{1\,lm} + \frac{l(l+1)r - 4M}{r^3}h_{0\,lm}\,, \qquad (15.74)\\
\beta^{(1)}_{lm}(\omega,r) &= f\left(i\omega h'_{0\,lm} - \omega^2 h_{1\,lm}\right) - 2i\omega\frac{f^{-1}}{r}h_{0\,lm} + \frac{(l-1)(l+2)}{r^2}h_{1\,lm}\,,\\
t_{lm}(\omega,r) &= i\omega f^{-1}h_{0\,lm} + fh'_{1\,lm} + \frac{2M}{r^2}h_{1\,lm}\,,
\end{aligned}
$$

and, similarly, $A^{(I)}_{lm}$ and $\alpha^{(A)}_{lm}$ (with $I = 0, \ldots, 3$, $A = 0, 1$) are combinations of the polar functions and of their derivatives:

$$
\begin{aligned}
A^{(0)}_{lm}(\omega,r) &= -2f^2K''_{lm} - \frac{2f}{r^2}(3r-5M)K'_{lm} + \frac{2f^2}{r}H'_{2\,lm} \qquad (15.75)\\
&\quad + \frac{(l-1)(l+2)f}{r^2}K_{lm} + \frac{f}{r^2}(l^2+l+2)H_{2\,lm}\\
A^{(1)}_{lm}(\omega,r) &= 2i\omega K'_{lm} + \frac{2i\omega f^{-1}}{r^2}(r-3M)K_{lm} - \frac{2i\omega}{r}H_{2\,lm} + \frac{l(l+1)}{r^2}H_{1,\,lm}\\
A^{(2)}_{lm}(\omega,r) &= 2\omega^2f^2K_{lm} + \frac{2f^{-1}}{r^2}(r-M)K'_{lm} - \frac{4i\omega f^{-1}}{r}H_{1\,lm} - \frac{2}{r}H'_{0\,lm}\\
&\quad - \frac{(l-1)(l+2)f^{-1}}{r^2}K_{lm} + \frac{l(l+1)f^{-1}}{r^2}H_{0\,lm} - \frac{2f^{-1}}{r^2}H_{2\,lm}\\
A^{(3)}_{lm}(\omega,r) &= -f(H''_{0\,lm} - K''_{0\,lm}) + \omega^2f^{-1}(K_{lm} + H_{2\,lm}) - 2i\omega H'_{1\,lm}\\
&\quad - \frac{r+M}{r^2}H'_{0\,lm} - \frac{r-M}{r^2}(H'_{2\,lm} - 2K'_{lm})\\
&\quad - \frac{2i\omega f^{-1}}{r^2}(r-M)H_{1\,lm} + \frac{l(l+1)}{2r^2}(H_{0\,lm} - H_{2\,lm})\\
\alpha^{(0)}_{lm}(\omega,r) &= fH'_{1\,lm} + i\omega(H_{2\,lm} + K_{lm}) + \frac{2M}{r^2}H_{1\,lm}\\
\alpha^{(1)}_{lm}(\omega,r) &= H'_{0\,lm} - K'_{lm} + i\omega f^{-1}H_{1\,lm} - \frac{f^{-1}}{r^2}(r-3M)(H_{0\,lm} - H_{2\,lm})\\
s_{lm}(\omega,r) &= \frac{1}{2}(H_{0\,lm} - H_{2\,lm})\,.
\end{aligned}
$$

15.2.6 Separation of the angular dependence

Owing to the radial and angular dependence, Eqs. 15.71-15.73 still form a system of *partial* differential equations for functions that depend on (r, θ, φ). It is however possible to separate the angular part by using the orthogonality properties of the spherical harmonics.

Eqs. 15.71 and the trace of Eq. 15.73, i.e.

$$2\gamma^{ab}\delta R_{ab} = 0 \qquad \rightarrow \qquad 4A^{(3)}_{lm}(\omega,r)r^2Y^{lm}(\theta,\varphi) = 0\,, \qquad (15.76)$$

have the form $\sum_{lm} A^{(I)}_{lm}Y^{lm} = 0$, with $I = 0, 1, 2, 3$. They can be separated simply by multiplying each of them by $Y^{*\,l'm'}$ and integrating over the sphere. Then, Eq. 15.9 yields immediately

$$\int d\Omega\, Y^{*lm}(\theta,\varphi)\delta R_{(I)}(\omega,r,\theta,\varphi) = 0 \quad \rightarrow A^{(I)}_{lm}(\omega,r) = 0\,, \qquad (15.77)$$

where $\delta R_{(I)}$ corresponds to δR_{tt}, δR_{tr}, δR_{rr}, and $\delta R_{\theta\theta} + \delta R_{\varphi\varphi}/\sin\theta^2$ for $I = 0, 1, 2, 3$, respectively. To hereafter, we shall not write explicitly the dependence on $(\omega, r, \theta, \varphi)$.

Similarly, Eqs. 15.72 can be decoupled using Eqs. 15.56, 15.57 as follows (the proof is left as an exercise)

$$\int d\Omega \left[Y^{*l'm'}_{,\theta}\,\delta R_{(J\theta)} + \frac{Y^{*l'm'}_{,\varphi}}{\sin\theta}\delta R_{(J\varphi)} \right] = 0 \quad \to l(l+1)\alpha_l^{(J)} = 0\,, \tag{15.78}$$

$$\int d\Omega \left[Y^{*l'm'}_{,\theta}\,\delta R_{(J\varphi)} - \frac{Y^{*l'm'}_{,\varphi}}{\sin\theta}\delta R_{(J\theta)} \right] = 0 \quad \to l(l+1)\beta_l^{(J)} = 0\,, \tag{15.79}$$

where $\delta R_{J\theta} = (\delta R_{t\theta}, \delta R_{r\theta})$ and $\delta G_{J\varphi} = (\delta R_{t\varphi}, \delta R_{r\varphi})$, for $J = 0, 1$, respectively. Finally, the angular dependence of Eqs. 15.71 (subtracting the trace part, Eq. 15.76) can be decoupled by using Eq. 15.58 and constructing the following relations

$$\begin{aligned} &\int d\Omega \frac{1}{l(l+1)-2} \left[W^{*l'm'}\delta R_{(-)} + \frac{X^{*l'm'}}{\sin\theta}\delta R_{\theta\varphi} \right] = 0 \quad \to l(l+1)s_l = 0\,, \\ &\int d\Omega \frac{1}{l(l+1)-2} \left[W^{*l'm'}\delta R_{\theta\varphi} - \frac{X^{*l'm'}}{\sin\theta}\delta R_{(-)} \right] = 0 \quad \to l(l+1)t_l = 0\,, \end{aligned} \tag{15.80}$$

where we defined $\delta R_{(-)} = \delta R_{\theta\theta} - \delta R_{\varphi}/\sin\theta^2$.

To summarize, our decoupling procedure allows to obtain a system of ten coupled, *ordinary* differential equations governing the radial dependence of the metric perturbations. Noteworthy, the axial and polar sectors completely decouple from each other and can be studied separately. Indeed, for the polar perturbations we obtain the following seven equations,

$$\begin{aligned} A^{(I)}_{lm}(\omega, r) &= 0 \\ \alpha^{(J)}_{lm}(\omega, r) &= 0 \\ s_{lm}(\omega, r) &= 0 \end{aligned} \tag{15.81}$$

($I = 0, \ldots, 3$ and $J = 0, 1$), whereas for the axial perturbations we obtain the following three equations,

$$\begin{aligned} \beta^{(J)}_{lm}(\omega, r) &= 0 \\ t_{lm}(\omega, r) &= 0\,. \end{aligned} \tag{15.82}$$

Not all ten linearized Einstein equations are independent and the coupled system can be simplified further, as we will show in the next sections.

These equations depend on the harmonic index l but not on the azimuthal number m. It can be shown that this property is a consequence of the fact that the background is spherically symmetric.

15.3 MASTER EQUATIONS FOR AXIAL AND POLAR PERTURBATIONS

We shall now show that the equations governing the perturbations of the Schwarzschild spacetime can be reduced to two decoupled second-order differential equations (sometimes called *master equations*), similar to the wave equation describing a test scalar field (Eq. 15.32).

15.3.1 The Regge-Wheeler equation for the axial perturbations

The equations for the axial perturbations can be found by replacing the expressions of $\beta^{(0)}_{lm}, \beta^{(1)}_{lm}$, and t_{lm} given in Eqs. 15.74 and 15.82,

$$-f\left(h''_{0\,lm} + i\omega h'_{1\,lm}\right) - 2i\omega\frac{f}{r}h_{1\,lm} + \frac{l(l+1)r - 4M}{r^3}h_{0\,lm} = 0\,, \tag{15.83}$$

$$f\left(i\omega h'_{0\,lm} - \omega^2 h_{1\,lm}\right) - 2i\omega\frac{f^{-1}}{r}h_{0\,lm} + \frac{(l-1)(l+2)}{r^2}h_{1\,lm} = 0\,, \tag{15.84}$$

$$i\omega f^{-1}h_{0\,lm} + fh'_{1\,lm} + \frac{2M}{r^2}h_{1\,lm} = 0\,. \tag{15.85}$$

In order to simplify the above equations, it is convenient to define the **Regge-Wheeler function** Q_{lm},

$$Q_{lm} \equiv f\frac{h_{1\,lm}}{r}\,. \tag{15.86}$$

Since

$$(Q_{lm}r)' = fh'_{1\,lm} + f'h_{1\,lm} = fh'_{1\,lm} + \frac{2M}{r^2}h_{1\,lm} \tag{15.87}$$

Eq. 15.85 yields

$$-if^{-1}\omega h_{0\,lm} = (Q_{lm}r)' \tag{15.88}$$

from which it follows

$$h_{0\,lm} = \frac{if}{\omega}\left(Q_{lm}r\right)'\,. \tag{15.89}$$

Substituting the expressions of h_0 and h_1 in terms of Q_{lm} in Eq. 15.84 we find the **Regge-Wheeler equation** [99]

$$\frac{d^2Q_{lm}}{dr_*^2} + \left(\omega^2 - V_l^{\text{axial}}\right)Q_{lm} = 0\,, \tag{15.90}$$

where we have defined

$$V_l^{\text{axial}}(r) = \left(1 - \frac{2M}{r}\right)\left[\frac{l(l+1)}{r^2} - \frac{6M}{r^3}\right]\,. \tag{15.91}$$

Eq. 15.83 is trivially satisfied, since it can be obtained from Eqs. 15.84 and 15.85. Once the Regge-Wheeler function Q_{lm} is found by solving Eq. 15.90, the axial functions $h_{0\,lm}$ and $h_{1\,lm}$ can be obtained from it by using Eqs. 15.86 and 15.89.

We note that the Regge-Wheeler equation has the same form as the Klein-Gordon equation on Schwarzschild spacetime (Eq. 15.32), i.e. it is a wave equation in the tortoise coordinate r_*, with a potential barrier. The two effective potentials are different but very similar (compare Eq. 15.91 with Eq. 15.29); indeed, V_l^{axial} has the same qualitative features shown in Fig. 15.1. In particular, it vanishes near the horizon and at infinity, and it has a maximum close to the light ring $r = 3M$. A comparison between the scalar and the gravitational potentials is shown in Fig. 15.2 below.

15.3.2 The Zerilli equation for the polar perturbations

The equations for the polar perturbations can be found by replacing the expressions of $(A^{(I)}_{lm}, \alpha^{(J)}_{lm})$ $(I = 0, \ldots, 3$ and $J = 0, 1)$ and of s_{lm}, given in Eqs. 15.75 and 15.81. To begin with, we note that the equation $s_{lm} = 0$ implies

$$H_{2\,lm} = H_{0\,lm}\,, \tag{15.92}$$

so that we can eliminate $H_{2,lm}$ from the remaining equations. We leave to the reader as an exercise to show that the remaining equations for the metric perturbations $H_{0\,lm}$, $H_{1\,lm}$, K_{lm} can be combined into three first-order ordinary differential equations:

$$K'_{lm} + \frac{f^{-2}}{r^2}(r-3M)K_{lm} - \frac{1}{r}H_{0\,lm} + \frac{l(l+1)}{2i\omega r^2}H_{1\,lm} = 0\,, \tag{15.93}$$

$$\frac{1}{i\omega}H'_{1\,lm} + f^{-2}(H_{0\,lm} + K_{lm}) + \frac{2Mf^{-2}}{i\omega r^2}H_{1\,lm} = 0\,,$$

$$H'_{0\,lm} + \frac{f^{-2}}{r^2}(r-3M)K_{lm} - \frac{f^{-2}}{r^2}(r-4M)H_{0\,lm} - \left(\omega^2 f^{-2} - \frac{l(l+1)}{2r^2}\right)\frac{1}{i\omega}H_{1\,lm} = 0\,,$$

and an algebraic relation:

$$\begin{aligned}&((l-1)(l+2)r + 6M)rH_{0\,lm} + \frac{1}{i\omega}\left(2\omega^2 r^3 - Ml(l+1)\right)H_{1\,lm}\\&+(2\omega^2 r^4 - (l-1)(l+2)r^2 - 2Mr + 2M^2 f^2)K_{lm} = 0\,.\end{aligned} \tag{15.94}$$

Using the algebraic relation to eliminate $H_{0\,lm}$, the above system can be reduced to two differential equations for $H_{1\,lm}$ and K_{lm}. Then, defining the **Zerilli function** Z_{lm} obtained solving the system

$$\begin{aligned}K_{lm} &= F_{11}Z_{lm} + F_{12}\frac{dZ_{lm}}{dr_*}\,,\\ H_{1\,lm} &= F_{21}Z_{lm} + F_{22}\frac{dZ_{lm}}{dr_*}\,,\end{aligned} \tag{15.95}$$

where

$$\begin{aligned}F_{11} &= \frac{6\left(l(l+1)-2\right)Mr + (l-1)l(l+1)(l+2)r^2 + 24M^2}{2r^2\left[\left(l(l+1)-2\right)r + 6M\right]}\,,\\ F_{12} &= 1\,,\\ F_{21} &= i\left[M\left(\frac{1}{r-2M} + \frac{6}{\left(l(l+1)-2\right)r + 6M}\right) - 1\right],\\ F_{22} &= -\frac{ir^2}{r-2M}\,,\end{aligned} \tag{15.96}$$

the two remaining equations reduce to a single second-order, differential equation

$$\frac{d^2 Z_{lm}}{dr_*^2} + \left(\omega^2 - V_l^{\text{polar}}\right)Z_{lm} = 0\,, \tag{15.97}$$

where

$$V_l^{\text{polar}}(r) = \left(1 - \frac{2M}{r}\right)\left[\frac{(l-1)(l+2)}{3}\left(\frac{1}{r^2} + \frac{2(l-1)(l+2)(l^2+l+1)}{\left(6M + r(l-1)(l+2)\right)^2}\right) + \frac{2M}{r^3}\right], \tag{15.98}$$

or, defining $2n = (l-1)(l+2)$,

$$V_l^{\text{polar}}(r) = \frac{2(r-2M)}{r^4(nr+3M)^2}[n^2(n+1)r^3 + 3Mn^2r^2 + 9M^2nr + 9M^3]\,. \tag{15.99}$$

Equation 15.97 is called the **Zerilli equation** [124] ; it has the same structure as the Regge-Wheeler equation, i.e. it is a wave equation in the tortoise coordinate r_*, with a potential barrier. Although the expression of the Zerilli potential $V_l^{\rm polar}$ differs from that of the Regge-Wheeler potential $V^{\rm axial}$, they are qualitatively very similar, as shown in Fig. 15.2.

Once the Zerilli function is known, $H_{1\,lm}$ and K_{lm} can be obtained in terms of Z_{lm} and its derivative with respect to r_* (see Eq. 15.95), and the polar function $H_{0\,lm}$ can be obtained by Eq. 15.94.

In conclusion, the gravitational perturbations of a Schwarzschild black hole are described by two Schrödinger-like equations, one for the polar, one for the axial perturbations, with a potential barrier which depends on the spacetime curvature.

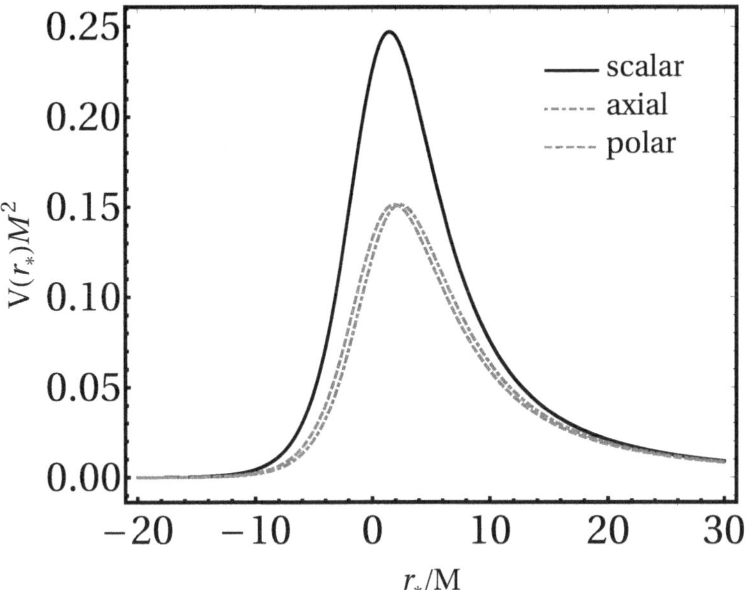

Figure 15.2: Comparison between the effective potentials governing scalar, axial gravitational, and polar gravitational perturbations of a Schwarzschild black hole with $l = 2$.

15.4 THE QUASI-NORMAL MODES OF A SCHWARZSCHILD BLACK HOLE

A powerful approach to study the behaviour of physical systems near equilibrium is to analyse their normal modes, i.e. the characteristic modes of oscillation in the absence of an external force. This approach is also effective to study the behaviour of black holes near equilibrium: if a black hole is perturbed by an external event, for instance a mass falling toward the horizon, after a transient it starts oscillating at some characteristic frequencies. A perturbed black hole can also be the leftover of the gravitational collapse of a massive star, or of the merger of a binary system (see Sec. 14.2). The newborn black hole is, in general, non-stationary but after some time it reaches a quasi-stationary configuration, and in this regime it oscillates with its characteristic frequencies. In all cases, the black hole emits gravitational waves with the same frequencies of its non-radial oscillations, until it becomes a static, or a stationary, black hole.

The non-radial oscillations are *damped*, because the black hole emits gravitational waves; consequently the mode eigenfrequencies are complex

$$\omega = \omega_R + i\,\omega_I\,, \tag{15.100}$$

since any quantity with a time dependence $\sim e^{-i\omega t}$ with $\omega_I < 0$ describes a damped oscillation

$$e^{-i\omega t} = e^{-i\omega_R t} e^{\omega_I t}, \tag{15.101}$$

where the *oscillation frequency* of the mode is $\nu = \omega_R/(2\pi)$, and the corresponding e-folding time, the *damping time*, is

$$\tau = -\frac{1}{\omega_I}. \tag{15.102}$$

The free oscillation modes of a perturbed black hole are called **quasi-normal modes**, since their eigenfrequencies are complex, at variance with the normal modes of non-dissipative systems, which have real eigenfrequencies. Note that we have assumed $\omega_I < 0$. If, instead, $\omega_I > 0$, the oscillation grows exponentially, with *growth time* $\tau = 1/\omega_I$.

In the previous sections we derived the Regge-Wheeler and the Zerilli equations 15.90, 15.97, which govern the perturbations of a Schwarzschild black hole, by expanding in tensor harmonics the perturbation $h_{\mu\nu}$, and taking its Fourier expansion, i.e.

$$h_{\mu\nu}(t,r,\theta,\varphi) = \int_{-\infty}^{+\infty} d\omega \begin{pmatrix} h_{AB\,lm}(\omega,r)Y^{lm}(\theta,\varphi) & h_{A\,lm}(\omega,r)S_a^{lm}(\theta,\varphi) \\ h_{A\,lm}(\omega,r)S_a^{lm}(\theta,\varphi) & r^2 K_{lm}(\omega,r)\gamma_{ab}Y^{lm}(\theta,\varphi) \end{pmatrix} e^{-i\omega t}. \tag{15.103}$$

To study the black hole oscillations, it is convenient to assume, as an ansatz, that the perturbation has the following time dependence

$$h_{\mu\nu}(t,r,\theta,\varphi) = \begin{pmatrix} h_{AB\,lm}(\omega,r)Y^{lm}(\theta,\varphi) & h_{A\,lm}(\omega,r)S_a^{lm}(\theta,\varphi) \\ h_{A\,lm}(\omega,r)S_a^{lm}(\theta,\varphi) & r^2 K_{lm}(\omega,r)\gamma_{ab}Y^{lm}(\theta,\varphi) \end{pmatrix} e^{-i\omega t}. \tag{15.104}$$

Since the oscillations are damped by the emission of gravitational waves, the frequency ω in Eq. 15.104 can be a complex number, whereas in Eq. 15.103 ω is a real number, since it is the variable of the Fourier transform. With the choice 15.104, the definition of the Regge-Wheeler and Zerilli functions remains the same as in Eqs. 15.86, 15.95, as well as the derivation which leads to the Regge-Wheeler and Zerilli equations. Consequently, Einstein's equations for the metric perturbations of a Schwarzschild black hole can be cast in the form

$$\frac{d^2\Psi_{lm}}{dr_*^2} + \left(\omega^2 - V_l\right)\Psi_{lm} = 0, \tag{15.105}$$

where ω is complex and Ψ is either the Regge-Wheeler or the Zerilli function. The explicit form of the potential is given in Eq. 15.91 and in Eq. 15.98, for the axial and polar perturbations, respectively.

As previously mentioned, Eq. 15.105 has the same form as the time-independent Schrödinger equation. Thus, as in quantum mechanics, it can be studied as a scattering problem, by imposing suitable boundary conditions. Since the potential vanishes both near the horizon and at infinity, corresponding to $r_* \to -\infty$ and to $r_* \to +\infty$, respectively, the asymptotic behaviour of the solution of Eq. 15.105 is simply that of a harmonic oscillator,

$$\frac{d^2\Psi_{lm}}{dr_*^2} + \omega^2\Psi_{lm} \simeq 0, \tag{15.106}$$

whose general solution when $r_* \to \pm\infty$ is

$$\Psi_{lm} \simeq A_{\text{in}} e^{-i\omega r_*} + A_{\text{out}} e^{i\omega r_*}. \tag{15.107}$$

Owing to the $e^{-i\omega t}$ time dependence, the asymptotic behaviour of Ψ_{lm} is

$$\Psi_{lm} e^{-i\omega t} \simeq A_{\text{in}} e^{-i\omega(r_*+t)} + A_{\text{out}} e^{i\omega(r_*-t)}. \tag{15.108}$$

The first term represents an *ingoing* wave with amplitude A_{in}, whereas the second represents an *outgoing* wave with amplitude A_{out}. If we are interested in the free oscillations of the black hole, we have to impose that no wave is incoming from infinity, i.e.

$$\Psi_{lm} \propto e^{i\omega r_*} \qquad r_* \to \infty \,. \tag{15.109}$$

As a consequence, $A_{\text{in}} = 0$ as $r_* \to \infty$. In addition, since nothing can escape from a black hole horizon, only ingoing waves are allowed when $r_* \to -\infty$, i.e.

$$\Psi_{lm} \propto e^{-i\omega r_*} \qquad r_* \to -\infty \,, \tag{15.110}$$

therefore, $A_{\text{out}} = 0$ as $r_* \to -\infty$. The boundary conditions (Eqs. 15.109 and 15.110) are shown in Fig. 15.3.

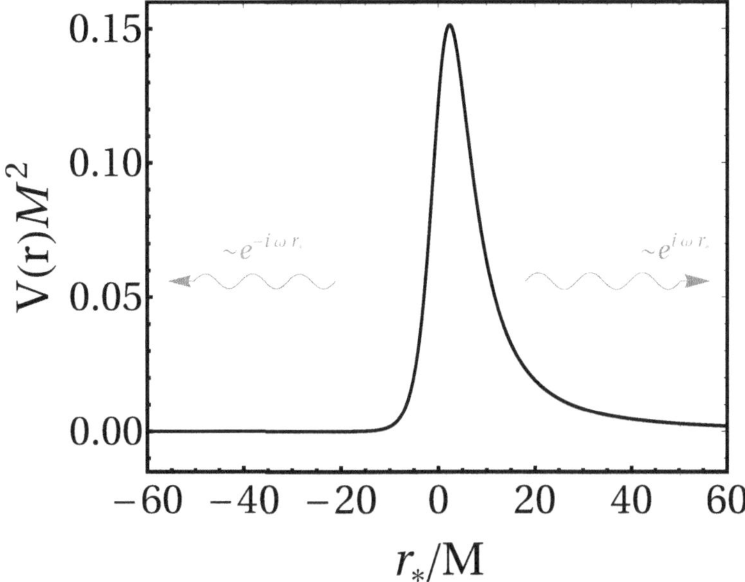

Figure 15.3: The boundary conditions defining the quasi-normal modes of a black hole: the wave is purely ingoing near the horizon ($r_* \to -\infty$) and purely outgoing at infinity ($r_* \to \infty$).

Equation 15.105 is a second-order differential equation and, together with the two boundary conditions, completely defines an eigenvalue problem for the frequency ω. This means that both boundary conditions are satisfied only for a *discrete* set of eigenfrequencies, which are the quasi-normal mode frequencies of the black hole. As expected, it turns out that these are complex frequencies as in Eq. 15.100, $\omega = \omega_R + i\omega_I$. Thus, the perturbation oscillates with frequency $\nu = \omega_R/(2\pi)$ and damping time $\tau = -1/\omega_I$.

For general black hole spacetimes these mode frequencies depend on the values of (l, m). However, for a spherically symmetric black hole the azimuthal number is degenerate, i.e. it does not affect the spectrum (this is not the case for spinning black holes, see Box 15-D below). Furthermore, since the quasi-normal mode frequencies form a discrete set, we can label them with an integer number n, such that $n = 0, 1, 2, 3, \ldots$ labels the fundamental mode ($n = 0$) and the overtones $n = 1, 2, 3, \ldots$.

We remark that, at variance with the ordinary normal modes of non-dissipative systems, the quasi-normal modes do not form a complete basis of the space of black hole perturbations. Indeed, it has been shown [78] that when a black hole is perturbed, after a transient the signal is described with good approximation by a combination of quasi-normal modes, but at late times it also includes a *power-law tail* $\sim t^{-2l+3}$.

In principle the quasi-normal modes of the axial and polar perturbations of a Schwarzschild black hole should be different. However, as shown by Chandrasekhar, the Regge-Wheeler and the Zerilli functions can be mapped one to another through a linear transformation which also leaves the boundary conditions unchanged [34]. Consequently, the polar and axial quasi-normal modes of a Schwarzschild black hole have the remarkable property to be *isospectral*. Since the Regge-Wheeler potential is simpler, it is sufficient to compute the quasi-normal mode frequencies using this potential to obtain the full spectrum. This can be done by solving numerically the Regge-Wheeler equation with the boundary conditions discussed above. Although these frequencies are known only numerically, they can be computed with arbitrary precision.

It has been proved rigorously that the Schwarzschild black hole is stable against linear perturbations; consequently $\omega_I < 0$ for all modes, and this guarantees that the perturbation decays exponentially in time.

The frequency of the fundamental mode (i.e. the one with the smallest imaginary part, which therefore has the longest damping time) for $l = 2$ reads

$$M\omega^{(0)} \simeq 0.3736 - i0.0890\,. \tag{15.111}$$

Note that, in geometrized units ($G = c = 1$), ω has dimensions of an inverse length, whereas M has dimensions of a length; hence $M\omega$ is dimensionless.

As discussed in Sec. 14.2.5, the fundamental mode dominates the part of the gravitational wave signal named "ringing tail" or ringdown, emitted in the merging of two black holes after the remnant black hole is formed. In the case of the event GW150914, the oscillation frequency of this mode has been indeed extracted from the signal.

We finally remark that, for simplicity, in this chapter we focused on the perturbations of static, spherically-symmetric black holes, described by the Schwarzschild metric. As will be discussed in Chapter 18, the most general stationary black-hole solution of Einstein's equations is axisymmetric, and is described by the *Kerr metric*. The equations governing the perturbations of the Kerr metric are much more involved than those described in this section, and will not be treated in this book. However, in Box 15-D we list their main properties. The interested reader can find a complete treatment in Chandrasekhar's monograph [34].

Box 15-D

Perturbations of a spinning black hole

We have so far focused on a static black hole, i.e. on the Schwarzschild solution. In Chapter 18 we shall introduce the Kerr solution which describes axisymmetric, spinning black holes. There are several qualitative and quantitative differences between the perturbations of non-rotating and rotating black holes. First of all, separating the radial and angular dependence of the perturbations of a spinning black hole is much more involved. The separability of the equations governing the scalar, electromagnetic, and gravitational perturbations of a Kerr black hole was achieved by Saul Teukolsky in the early 1970s [109] using a decomposition in the so-called spin-weighted spheroidal harmonics. It can be shown that the radial equations can be recast in a form like Eq. 15.105 but:

- the potential depends both on the mass and on the spin of the black hole;
- the potential depends both on the harmonic index l and on the azimuthal number m, thus breaking the azimuthal degeneracy; modes with $m > 0$ are called *co-rotating*, whereas those with $m < 0$ are called *counter-rotating*;
- the potential is complex, depends on the frequency, and does not vanish near the horizon;
- since the background spacetime is not spherically symmetric, the perturbations have not definite parity; consequently, they cannot be separated in axial and polar.

As we shall see, the Kerr metric depends only on two parameters: the mass and the spin of the black hole. Therefore, the frequencies of the whole quasi-normal mode spectrum depend only on these two quantities, pretty much like the vibration modes of a drum depend on the properties of the latter (its shape, composition, etc.). In all cases, $\omega_I < 0$ so that also spinning black holes are stable with respect to linear perturbations.

CHAPTER 16

Compact stars

In this chapter we shall discuss the structure of *compact stars*, namely *white dwarfs* and *neutron stars*, which are significantly more compact than an ordinary star like the Sun. Compact stars form at the end point of the evolution of sufficiently massive main-sequence stars. In order to understand how these objects can form, we shall briefly describe the main steps of stellar evolution. The interested reader can find a complete treatment of the subject in [69, 105].

16.1 STELLAR EVOLUTION IN A NUTSHELL

Stars are formed by the fragmentation of primordial hydrogen clouds. Under the action of gravity, the fragments contract and heat up; if the mass is larger than $\sim 0.08\ M_\odot$ the temperature in the core reaches values of the order of $T \sim 4 \times 10^6$ K, which are sufficient to ignite the fusion of hydrogen first to deuterium, and then to helium through the proton-proton chain reaction. This is the starting point of stellar evolution; the star enters the so-called *main sequence*, to which our Sun also belongs. During this phase the star is in equilibrium under the contrasting actions of the inward pressure exerted by gravity and of that of the outward thermal pressure provided by the nuclear reactions. The fact that the energy which sustains a star against collapse is provided by the fusion of hydrogen in helium was firstly understood by Arthur Eddington in 1920. Main sequence stars are very hot; the heat produced in the core propagates through the star and is radiated away through the surface. When the reservoir of hydrogen burning in the core is exhausted, the star exits from the main sequence and its subsequent evolution depends on its mass. If this is smaller than $\sim 0.5\ M_\odot$ a *white dwarf* forms directly. These objects will be described in detail in Sec. 16.2.

If the star has a larger mass, it contracts and heats up, reaching core temperatures of the order of $\sim 10^8$ K, at which a new chain of nuclear reactions starts, which transform three nuclei of Helium 4 (^{4}He) in one nucleus of carbon 12 (^{12}C). These reactions have a strong dependence on temperature and produce a large amount of energy in a relatively short time. The energy released is able to halt the collapse, and due to the strong radiation pressure, the outer layers expand; the radius reaches values of the order of hundred times the initial radius, and the star reaches a new equilibrium state as a *red giant*. The light emission from a red giant is more intense, but has a smaller frequency than that of a main-sequence star. In this phase, ^{12}C is produced in the core, whereas in the external layers hydrogen continues to burn into helium.

When the helium in the core is exhausted, the star starts contracting again and heating up, igniting nuclear processes which produce heavier elements. However, if the star has a mass in the range $\sim (0.5, 8-10)\ M_\odot$, the temperatures that can be reached in the core do not allow the formation of elements heavier than oxygen; in this case, when the oxygen core has formed and the nuclear fuel is exhausted, the thermal pressure is no longer sufficient to support the star and the hot core contracts until a white dwarf forms. It should be mentioned that during the entire course of its evolution, which typically takes several billion years, the star loses a major fraction of its original mass through stellar winds, whereas the outer layers of matter are expelled near the end of the nuclear burning process and create a planetary nebula.

White dwarfs have masses comparable to that of the Sun, but they are much more compact: their typical radius is comparable to that of Earth and their average density is of the order of $\bar{\rho} \sim 10^6$ g/cm^3 (for comparison, the average density of the Sun is $\bar{\rho}_\odot = 1.4$ g/cm^3). They are mainly composed of helium, carbon, and oxygen. If the progenitor has a mass in the range $\sim 8 - 10\ M_\odot$, oxygen-neon-magnesium stars can form, but they are rare. The number of white dwarfs in the Milky Way is of the order of 10^{10}. In Sec. 16.2 we shall show that white dwarfs of mass exceeding the *Chandrasekhar mass*, $M_{\rm Ch} \sim 1.4 M_\odot$, cannot exist.

If the mass of the progenitor star is in the range $\sim (8, 20-30)\ M_\odot$ the evolutionary path is different. Nuclear processes are able to burn elements heavier than carbon and oxygen, and exothermic nuclear reactions can proceed all the way to ^{56}Fe, which is the most stable element in nature; indeed, no element heavier than ^{56}Fe can be generated by the fusion of lighter elements through exothermic reactions. Thus, the evolving star has an onion-like layered structure, with the heavier elements in the core, and progressively lighter elements in the surrounding shells.

The process which produces ^{56}Fe starts with silicon burning, and goes this way:

$$
\begin{aligned}
{}^{28}\mathrm{Si} + {}^{28}\mathrm{Si} &\rightarrow {}^{56}\mathrm{Ni} + \gamma\,, && (16.1)\\
{}^{56}\mathrm{Ni} &\rightarrow {}^{56}\mathrm{Co} + e^+ + \nu_e\,, && (16.2)\\
{}^{56}\mathrm{Co} &\rightarrow {}^{56}\mathrm{Fe} + e^+ + \nu_e\,. && (16.3)
\end{aligned}
$$

It should be reminded that the atomic number Z (number of protons) and the mass number A (number of protons + neutrons) of the elements involved in the process are[1]:

- Si, $Z = 14$, $A = 28.086 \simeq 28$
- Ni, $Z = 28$, $A = 58.693 \simeq 59$
- Co, $Z = 27$, $A = 59.933 \simeq 60$
- Fe, $Z = 26$, $A = 55.845 \simeq 56$.

Atoms with a number of nucleons smaller than the mass number of are unstable. The ^{56}Ni which forms in the reaction 16.1 has tree neutrons less than required for stability; therefore, it decays as indicated in the reaction 16.2, forming ^{56}Co; this is again unstable, being in defect of four neutrons, and decays forming ^{56}Fe, which is stable.

Furthermore, as the core density increases, the inverse β-decay process,

$$e^- + p \rightarrow n + \nu_e\,, \tag{16.4}$$

through which electrons are captured by protons forming neutrons and neutrinos, becomes more and more efficient and nuclei richer of neutrons than ^{56}Fe can form, like ^{118}Kr. However, these heavy nuclei are formed through endothermic reactions and subtract energy to

[1]The mass numbers are not integers because in the stable state a gas of a given element is a mixture of isotopes of that element.

the star. Elements heavier than krypton can form during the subsequent phase of collapse, as explained below. As the iron core forms the pressure provided by nuclear reactions is able to maintain the star in equilibrium. However, several processes occur which act in the opposite direction:

- A large number of neutrinos are produced both in the reactions 16.2-16.3 and in the inverse β-decay process. Since neutrinos interact with matter very weakly, they diffuse through the star and escape through the surface, subtracting energy to the core.
- The production of elements heavier than Fe and Ni, which mainly occurs through neutron capture, subtracts energy to the star.
- In the reaction 16.1 photons are produced. At temperatures of the order of 10^{10} K, the number of high energy photons (> 8 MeV) is sufficient to ignite the iron photo-disintegration process,
$$\gamma +^{56} \mathrm{Fe} \rightarrow 13\ ^4\mathrm{He} + 4n\,. \tag{16.5}$$
This is an endothermic process which subtracts further energy to the core. In addition, this process produces a large number of neutrons.

All these processes tend to destabilize the star; when the core mass becomes larger than a critical mass the internal pressure gradient can no longer balance the gravitational attraction, and the core collapses reaching densities typical of atomic nuclei, $\sim 10^{14}$ g/cm^3, in a fraction of a second. The core is now composed mainly of neutrons, and it is so rigid that the infalling matter bounces back producing a violent shock wave that – in a spectacular explosion – ejects most of the material external to the core in the outer space. This phenomenon is called *supernova explosion*: the luminosity of the star suddenly increases to values of the order of $\sim 10^9\ L_\odot$, where $L_\odot = 10^{33}$ erg/s is the Sun luminosity; in this phase elements heavier than ^{56}Fe are created. The remnant of this explosion is a nebula, in the middle of which sits what remains of the core, i.e. a *neutron star*, whose structure and composition will be described in Sec. 16.3. These stars have mass in the range of $1 - 3$ solar masses, radius of the order of $10 - 15$ km, and their average density is $\bar{\rho} \sim 10^{14}$ g/cm^3, therefore they are extremely compact.

As for white dwarfs, a critical mass exists also for neutron stars. The absolute upper limit is in the range $\sim 2 - 3M_\odot$. The value of the critical mass depends on the equation of state which describes matter at the supranuclear densities prevailing in the core of these stars.

Neutron stars are often observed as *pulsars*, i.e. radio sources whose emission exhibits an extremely stable periodicity; pulsars are rapidly rotating neutron stars with strong magnetic fields ($B \sim 10^{11} - 10^{13}$ Gauss), which emit beams of radio waves; the observed periodicity is due to the fact that, since the star rotates and the magnetic field is in general not aligned with the rotation axis, the beam is visible only when it points in the direction of the detector. However, not all neutron stars are detectable as pulsars, since their beams may not point toward the Earth, or their magnetic field may not be sufficiently strong. Neutron stars can also be observed in other regions of the electromagnetic spectrum. For instance when a neutron star in a binary system accretes matter from a companion, the gravitational energy released by the infalling matter is converted in thermal energy, which is radiated away mainly in X-rays. Moreover, as mentioned in Chapter 14, the coalescence of a binary neutron star can be observed by detecting their gravitational wave emission.

The estimated number of neutron stars in the Milky Way is $\sim 10^9$. This number has been inferred from the total number of stars in our Galaxy at different evolutionary stages, using our understanding of stellar evolution.

Finally, studies of stellar evolution indicate that if the mass of the progenitor star is larger than $\sim 20 - 30\ M_\odot$, at the end of its thermonuclear evolution the star collapses to form a black hole. A detailed discussion of this important topic is beyond the scope of this book; we refer the reader to [58, 106] for a detailed treatment.

Black holes of astrophysical origin can have very different masses ranging from a few solar masses, for the "stellar black holes" born in the gravitational collapse of big stars or in the coalescence or accretion-driven processes in binary systems, to supermassive black holes, with masses in the range $\sim (10^5 - 10^{11})\ M_\odot$; the latter sit at the center of most (or all) galaxies. Astrophysical observations of black holes will be further discussed in Sec. 18.1.

16.2 WHITE DWARFS

In this section we shall focus on the study of white dwarfs. Since, as mentioned above, these objects have masses of the order of that of the Sun and radii of the order of 5000 km, their compactness is very small

$$\frac{GM}{c^2R} \sim \frac{M_\odot(\text{in km})}{R} \sim \frac{1.5\,\text{km}}{5000\,\text{km}} = 3 \times 10^{-4}\,. \tag{16.6}$$

Therefore, General Relativity effects on the structure of white dwarfs are negligible. As we shall show in the next sections, Special Relativity effects can be relevant in the microscopic description of the fluid composing these stars, since particles in the fluid can reach relativistic velocities. However, they are negligible in the macroscopic description of white dwarfs structure, since we consider static stars in which the fluid is at rest. For this reason, in Secs. 16.2.5 and 16.2.6 we will use the equations of Newtonian physics to determine the structure of these stars.

We shall start with some historical remarks on the discovery of white dwarfs and on the debate on their internal structure.

16.2.1 The discovery of white dwarfs

The first white dwarf was observed by William Herschel in 1783 in the triple system 40 Eridani, and in 1910 it was shown that 40 Eridani B was a white star (spectral type A). This type of star has an effective temperature in the range (7500, 10000) K. However it was only in 1915 – after the discovery of Sirius B, the companion of Sirius, by W.S. Adams – that the structure of white dwarfs started to be understood. Sirius is one of the closest stars[2] to the Earth, the distance being about ~ 8.6 ly (where 1 ly $= 9.46 \times 10^{17}$ cm, see Table A). Adams found that the spectrum of the stellar object orbiting around Sirius was that of a white star, not very different from the spectrum of Sirius. Since Sirius B is in a binary system, its mass could be estimated from the analysis of its orbital motion, and it was found to be in the range $(0.75, 0.95) M_\odot$. Moreover, knowing the distance of the system from Earth, from the apparent magnitude of this star it was possible to estimate its luminosity, while its effective temperature $T_{\rm eff} \sim 8000$ K was determined from the observed spectrum. Since for a black-body emission the luminosity depends on the effective temperature and on the stellar radius, $L \sim R^2 T_{\rm eff}^4$, it was then possible to estimate the radius of the star which was, surprisingly, $R_{\rm Sirius\,B} \sim 18800\,\text{km}$, much smaller than that of the Sun! The actual values of the mass and radius of Sirius B are $M_{\rm Sirius\,B} = (1.034 \pm 0.026)\, M_\odot$ and $R_{\rm Sirius\,B} = (0.0084 \pm 0.00025)\, R_\odot$, i.e. $R_{\rm Sirius\,B} \sim 5850\,\text{km}$.

At the time of the discovery this result came really as a surprise, because a star having a mass comparable to that of the Sun but a radius nearly forty times smaller had never been observed before. In addition, although the deflection of light predicted by Einstein's theory

[2]The star closest to the Earth is Proxima Centauri, at a distance of ~ 4.3 ly.

of Relativity had already been measured in the famous Eddington expedition in 1919, the redshift of the spectral lines of Sirius B measured by Adams in 1925 provided a much better verification of the theory.

The discovery of such an extremely dense star raised an outstanding question: how can this "white dwarf" (as it was named) support its matter against collapse? Indeed, if the matter composing the star were a perfect gas its temperature would be too low to prevent the collapse, i.e. the corresponding pressure gradient would not be sufficient to balance the gravitational attraction. About this problem, in his seminal monograph *The internal constitution of stars* [43], Eddington wrote:

"It seems likely that the ordinary failure of the gas laws due to finite sizes of molecules will occur at these high densities, and I do not suppose that the white dwarfs behave like perfect gas".

What is then that keeps white dwarfs in equilibrium? The answer to this question came a few years later when, in August 1926, Paul Dirac formulated the Fermi-Dirac statistics. Few months later (December 1926), Ralph H. Fowler identified the pressure holding up a white dwarf from collapsing with the *electron degeneracy pressure.* This was the crucial step toward the formulation of a consistent theory of these stars that led Subrahmanyan Chandrasekhar to predict the existence of a critical mass above which no stable white dwarf could exist.

16.2.2 Degenerate gas in quantum mechanics

A gas of particles is *degenerate* if its behaviour differs from the classical behaviour due to the quantum properties of the system. Let us discuss under which conditions degeneracy sets in.

Let us consider a gas composed of non-interacting, indistinguishable particles. In general, the properties of such a system will be completely described once the number of particles per unit phase-space volume, i.e. the number density in the phase space, is assigned

$$\frac{dN}{d^3x d^3p} = \frac{g}{h^3} f(\mathbf{x}, \mathbf{p})\,; \tag{16.7}$$

here h is Planck's constant, h^3 is the volume of a cell in the phase-space, $g = 2s + 1$ is the number of available states for a particle with spin s, position $\mathbf{x}$ and three-momentum $\mathbf{p}$, and $f(\mathbf{x}, \mathbf{p}) d^3x d^3p$ is the probability of finding a particle in the volume $d^3x d^3p$ around $(\mathbf{x}, \mathbf{p})$. The function $f(\mathbf{x}, \mathbf{p})$ is called *distribution function.*

Let m be the particle mass, $E = (p^2c^2 + m^2c^4)^{\frac{1}{2}}$ its total energy (see Box 16-A), and $E_c = E - mc^2$ its kinetic energy; the *total energy density* of the gas is

$$\mathcal{E} = \int E \frac{dN}{d^3x d^3p} d^3p = \frac{g}{h^3} \int E f(\mathbf{x}, \mathbf{p})\, d^3p\,. \tag{16.8}$$

For an ideal gas of fermions or bosons in equilibrium, the distribution function is

$$f = \frac{1}{e^{\frac{E-\mu}{k_B T}} \pm 1}\,, \tag{16.9}$$

where the plus sign holds for fermions (*Fermi-Dirac distribution*), whereas the minus sign holds for bosons (*Bose-Einstein distribution*). In Eq. 16.9 k_B is Boltzman's constant and μ is the chemical potential. This quantity is the energy needed to insert a new particle, in the same state, in the gas in equilibrium. Equation 16.9 shows that, since f must be positive, the chemical potential of fermions can take any real value, either positive or negative, whereas that of bosons is bounded to be $\mu < E$.

If the number density in the phase space is low (this occurs when the density of the gas is sufficiently low and its temperature is sufficiently high), quantum effects are negligible. Indeed, from Eq. 16.7 we see that $f \ll 1$, which implies $\frac{E-\mu}{k_B T} \gg 1$[3]. In this case both the Bose-Einstein and the Fermi-Dirac distributions reduce to the classical *Maxwell-Boltzmann distribution*

$$f \simeq e^{\frac{\mu - E}{k_B T}} . \tag{16.10}$$

Since the function f depends only on E, i.e. it depends only on the norm of the three-momentum p, the distribution of momenta is isotropic, and we can write

$$\int f(p) d^3p = 4\pi \int_0^\infty f(p) p^2 dp . \tag{16.11}$$

Therefore Eq. 16.8 becomes

$$\mathcal{E} = \frac{4\pi g}{h^3} \int_0^\infty \frac{E\, p^2\, dp}{e^{\frac{E-\mu}{k_B T}} \pm 1} . \tag{16.12}$$

Defining the *pressure*[4] as the momentum flux, we can write

$$P = \frac{1}{3} \int p\, v \frac{dN}{d^3x d^3p} d^3p = \frac{4\pi g}{3h^3} \int_0^\infty \frac{v\, p^3 dp}{e^{\frac{E-\mu}{k_B T}} \pm 1} , \tag{16.13}$$

where v is the particles velocity and the factor 1/3 takes into account the flux of momentum along one of the three spatial directions.

Furthermore, the *total number of particles* and the *internal (kinetic) energy* of the system are

$$N = \int \frac{dN}{d^3x d^3p} d^3x\, d^3p = \frac{4\pi g V}{h^3} \int_0^\infty \frac{p^2 dp}{e^{\frac{E-\mu}{k_B T}} \pm 1} , \tag{16.14}$$

$$U = \int E_c \frac{dN}{d^3x d^3p} d^3x\, d^3p = \frac{4\pi g V}{h^3} \int_0^\infty E_c \frac{p^2 dp}{e^{\frac{E-\mu}{k_B T}} \pm 1} , \tag{16.15}$$

where V is the total volume of the gas.

---

[3] In this limit both the chemical potential (in modulus) and the temperature are very large, but $\mu/(k_B T) \to -\infty$.

[4] In this section we denote the pressure with an uppercase P, in order to distinguish it from the three-momentum p.

Box 16-A

Some useful kinematical relations in Special Relativity

The four-momentum of a relativistic particle is

$$\vec{p} = (mc\gamma, \mathbf{p}) \,, \tag{16.16}$$

where $\mathbf{p} = m\gamma\mathbf{v}$ is the three-momentum. Moreover, the total energy of the particle is $E = p^0 c = \gamma mc^2$. Since $p_\alpha p^\alpha = -m^2c^2$, it follows that

$$-\frac{E^2}{c^2} + p^2 = -m^2c^2 \,, \tag{16.17}$$

where p is the norm of the three-momentum, and consequently the total energy of the particle can be written as

$$E = \sqrt{p^2c^2 + m^2c^4} \,. \tag{16.18}$$

Since $E = mc^2\gamma$, the above relation yields

$$\gamma = \frac{\sqrt{p^2c^2 + m^2c^4}}{mc^2} \,, \tag{16.19}$$

and since the norm of the particle velocity is $v = p/(m\gamma)$, we get

$$v = \frac{pc^2}{\sqrt{p^2c^2 + m^2c^4}} = \frac{pc^2}{E} \,. \tag{16.20}$$

16.2.3 A criterion for degeneracy

Let us consider the non-relativistic limit for $E_c \simeq \frac{1}{2}mv^2 = \frac{p^2}{2m}$. The latter expression can be found by expanding the definition of E when $pc \ll mc^2$. If we introduce the dimensionless variables

$$\zeta = e^{(\mu - mc^2)/(k_{\rm B}T)} \qquad \text{and} \qquad x^2 = \frac{p^2}{2mk_{\rm B}T} \,, \tag{16.21}$$

it is easy to see that Eqs. 16.14 reduce to

$$\begin{aligned} N &= \frac{4\pi gV}{h^3}\left(2mk_{\rm B}T\right)^{3/2} \int_0^\infty \frac{x^2 dx}{\zeta^{-1}e^{x^2} \pm 1} \,, \\ U &= \frac{4\pi gV}{h^3}(2m)^{3/2}\left(k_{\rm B}T\right)^{5/2} \int_0^\infty \frac{x^4 dx}{\zeta^{-1}e^{x^2} \pm 1} \,, \end{aligned} \tag{16.22}$$

where we have used the fact that $E - \mu \simeq \frac{p^2}{2m} - (\mu - mc^2)$.

In principle, these integrals can be solved in terms of special functions and the *degeneracy parameter* ζ can be found as a function of the thermodynamical variables. Here we shall consider explicitly the limit when $\zeta \ll 1$, in which case the integrals in Eqs. 16.22 simplify to

$$\begin{aligned} \int_0^\infty \frac{x^2 dx}{\zeta^{-1}e^{x^2} \pm 1} &\simeq \zeta \int_0^\infty x^2 e^{-x^2} dx \,, \\ \int_0^\infty \frac{x^4 dx}{\zeta^{-1}e^{x^2} \pm 1} &\simeq \zeta \int_0^\infty x^4 e^{-x^2} dx \,. \end{aligned} \tag{16.23}$$

Using these expressions the ratio between U and N becomes

$$\frac{U}{N} = k_{_B}T\frac{\int_0^\infty x^4 e^{-x^2}dx}{\int_0^\infty x^2 e^{-x^2}dx}\,, \tag{16.24}$$

and since $\int_0^\infty x^4 e^{-x^2}dx = \frac{3}{8}\sqrt{\pi}$ and $\int_0^\infty x^2 e^{-x^2}dx = \frac{1}{4}\sqrt{\pi}$, Eq. 16.24 gives $U = \frac{3}{2}Nk_{_B}T$, which is the expression of the internal energy of a classical perfect gas. Thus $\zeta \ll 1$ corresponds to the classical limit. In this limit, from the first of Eqs. 16.22 and Eqs. 16.23 we find

$$\zeta = \frac{nh^3}{g}\left(2\pi m k_{_B}T\right)^{-3/2}\,, \tag{16.25}$$

where n is the *number density* $n = N/V$, i.e. the number of particles per unit volume. The above equation can be rewritten as

$$\zeta = \left(\frac{T_{\rm deg}}{T}\right)^{3/2}\,, \tag{16.26}$$

where we have defined the *degeneracy temperature*

$$T_{\rm deg} = \frac{h^2}{2\pi m k_{_B}}\left(\frac{n}{g}\right)^{2/3}\,. \tag{16.27}$$

If $T \gg T_{\rm deg}$, then $\zeta \ll 1$ and the gas behaves as a classical gas. Conversely, if T is comparable or smaller than $T_{\rm deg}$ (i.e. ζ comparable or larger than one), the gas is said to be *degenerate.* When $h \to 0$, $T_{\rm deg}$ tends to zero, showing that the degeneracy of a gas is of a quantum nature. Furthermore, Eq. 16.27 shows that at a given density n, $T_{\rm deg}$ is higher for particles with smaller mass m. Strictly speaking, the degeneracy parameter ζ depends on both the density and the temperature of the gas: $\zeta \propto nT^{-3/2}$ (see Eq. 16.25); therefore, *degeneracy sets in at high densities or low temperatures.*

We note, however, that Eqs. 16.26, 16.27 have been derived under the assumption $\zeta \ll 1$. When ζ is comparable to one, Eq. 16.26 provides only a qualitative estimate of ζ.

It is useful to consider some examples:

- A hydrogen gas in normal condition, i.e. $T = 300$ K and $P = 1$ bar, has number density $n = 2.5 \times 10^{19}\text{cm}^{-3}$ and the degeneracy temperature (since $m_{H_2} = 3.3 \times 10^{-24}$ g) is $T_{\rm deg} \sim 0.13K$; therefore $\zeta \sim 8.8 \times 10^{-6}$ and the gas behaves as a classical perfect gas.
- For gases heavier than hydrogen ζ and $T_{\rm deg}$ are even smaller, and consequently at ordinary pressures and temperatures they are non-degenerate.
- A gas of photons is always degenerate because $m_\gamma = 0$, then $T_{\rm deg} \to \infty$.
- Electrons in metals are degenerate, due to their small mass ($m_e = 9.1 \times 10^{-28}$ g) and high density ($n \sim 10^{23}$ cm^{-3}). Indeed in this case $T_{\rm deg} \sim 7.5 \times 10^4$ K, and $\zeta \sim 4 \times 10^3$ (assuming for example room temperature, $T = 300$ K).

Let us now go back to white dwarfs. As previously mentioned, they are mainly composed of helium, carbon, and oxygen, with a small fraction of heavier elements in the inner core. When the nuclear material in the core has been burnt, the core contracts up to a point when the distance between two atoms becomes comparable to their size. For example, let us consider a white dwarf composed of carbon (the results are similar if one considers helium or oxygen stars). A typical white dwarf has mass $\sim M_\odot$ and radius ~ 5000 km, corresponding to a mean density $\bar{\rho} \sim 4 \times 10^6$ g/cm^3. Since the mass of an atom of carbon is $\sim 10^{-23}$ g, at these densities the average distance among atoms is $\sim 10^{-10}$ cm, which is smaller than its atomic radius $\sim 10^{-8}$ cm. In this situation, no more space is left for the external orbits of the electrons which are squeezed off the atoms, starting a pressure-driven ionization process; this proceeds as the density increases, progressively involving the innermost orbits. As a consequence, a dense core of nucleons forms, immersed in a gas of free electrons. Since, as shown in Box 16-B, $T \ll T_{\rm deg}$, the electron gas is *fully degenerate.*

At the same time the shells of lighter elements that surround the stellar core continue their nuclear evolution until all nuclear fuel is exhausted, and contraction and ionization processes take place also in the more exterior layers; the star then radiates its residual thermal energy and cools down. A more accurate description of white dwarfs should take into account other effects, like for example electrostatic corrections due to the fact that the positive charges are concentrated in individual nuclei rather than being uniformly distributed [5]. However, in what follows we shall neglect these effects and consider a simplified model of white dwarf at the endpoint of the evolution: we shall assume that the ionization process has been completed and that the thermal energy has been radiated away, so that the star is composed exclusively of a dense core of nucleons immersed in a gas of electrons, which behaves as a degenerate gas at zero temperature.

Box 16-B

Relation between the density of the star and the degeneracy temperature of the electron gas

Since electrons are much lighter than nucleons ($m_e = 9.109 \times 10^{-28}\,$g $\ll m_n = 1.675 \times 10^{-24}\,$g, where m_e, m_n are the masses of electrons and neutrons, see Table A) the density of a star is contributed mainly by the latter; therefore, if there are κ nucleons for each electron (for instance $\kappa \sim 2$ for stars that have used their hydrogen fuel and have formed a helium core) the mass density is

$$\rho = \kappa n m_N \,, \tag{16.28}$$

where n is the electron density and $m_N \simeq 1.674 \times 10^{-24}\,$g is the average nucleon mass in a white dwarf. Combining Eqs. 16.27 and 16.28, it is possible to establish a relation between the degeneracy temperature of the electron gas and the density of the star

$$T_{\rm deg} = \frac{h^2}{2\pi m k_{\rm B}} \left(\frac{1}{g m_N}\right)^{2/3} \left(\frac{\rho}{\kappa}\right)^{2/3} . \tag{16.29}$$

Recalling that the values of Planck's and Boltzman's constants are, respectively, $h = 6.626 \times 10^{-27}$ erg s and $k_{\rm B} = 1.381 \times 10^{-16}$ erg/K (see Table A), we find

$$T_{\rm deg} \sim \frac{2.4 \times 10^5}{\kappa^{2/3}} \left(\frac{\rho}{\rm g/cm^3}\right)^{2/3} \,{\rm K}\,. \tag{16.30}$$

Considering that the temperature of a white dwarf is of the order of $T \sim 10^4$ K, we see that at densities $\rho \sim 10^6$ g/cm^3, typical for these stars, the degeneracy temperature is $T_{\rm deg} \sim 2.4 \times 10^9 \left(\frac{1}{\kappa}\right)^{2/3}$ K and $T \ll T_{\rm deg}$, i.e. the electron gas behaves as a degenerate gas.

[5]Electrostatic corrections have been considered for the first time by Hamada and Salpeter in 1961 [54].

16.2.4 The equation of state of a fully degenerate gas of fermions

The structure of a white dwarf can be studied by solving the Newtonian equations of stellar structure for a fluid in hydrostatic equilibrium (see Box 16-C). In order to solve these equations it is necessary to know the **equation of state** of the matter describing the star, i.e. a relation among the thermodynamical quantities: the pressure, the density, and the temperature. We shall see that the equation of state (to hereafter EoS) of white dwarfs is *barotropic*, i.e. the pressure is a function only of the density, $P = P(\rho)$ (see Sec. 16.3.3).

Since the electron gas in a white dwarf is fully degenerate, $\zeta \gg 1$ and we cannot use the approximated relations derived in Sec. 16.2.3, which assumed $\zeta \ll 1$. Conversely, since $T \ll T_{\rm deg}$, we can consider the gas as having $T \simeq 0$ temperature.

In the limit $T \to 0$ the Fermi-Dirac distribution function, given by (see Eq. 16.9)

$$f(E) = \frac{1}{e^{\frac{E-\mu}{k_B T}} + 1}, \tag{16.31}$$

becomes

$$f(E) = \begin{cases} 1 & \text{for} \quad E \le E_F \quad (\text{or} \quad p \le p_F) \\ 0 & \text{for} \quad E > E_F\,, \quad (\text{or} \quad p > p_F) \end{cases} \tag{16.32}$$

where

$$E_F = \lim_{T\to 0} \mu \tag{16.33}$$

is the *Fermi energy* and p_F is the corresponding *Fermi momentum*, related to the Fermi energy by $E_F = \sqrt{m_e^2 c^4 + p_F^2 c^2}$ (see Box 16-A) [6]. We also define the *Fermi temperature* $T_F = (E_F - m_e c^2)/k$. Strictly speaking, the Fermi-Dirac distribution function has the form 16.32 not only when $T \to 0$ but whenever $T \ll T_F$. Note that the condition $T \ll T_F$ implies that the gas is degenerate; indeed, in this limit (see Eq. 16.21)

$$\zeta = e^{(\mu - m_e c^2)/(k_B T)} = e^{T_F/T} \gg 1\,. \tag{16.34}$$

This relation is different from Eq. 16.26, $\zeta = (T_{\rm deg}/T)^{3/2}$, which was obtained under the assumption $\zeta \ll 1$. However, it is possible to show that the degeneracy temperature $T_{\rm deg}$ defined in Eq. 16.27 is comparable with the Fermi temperature, $T_{\rm deg} \sim T_F$, therefore also the condition $T \ll T_{\rm deg}$ ensures that the gas is degenerate.

Since the temperature of the gas is negligible, the particles have negligible kinetic energy. In these conditions, if they were bosons they would occupy the lowest energy level $E = 0$, as it occurs in a Bose condensate. However fermions cannot do this, since Pauli's exclusion principle implies that in each energy level there can be at most two electrons, one with spin up and one with spin down. Thus, electrons will fill all states with energy up to E_F (note that $f = 0$ when $E > E_F$).

The Fermi momentum p_F can be expressed as a function of the density of the star as follows. The number of levels with momenta between p and $p + dp$ per unit volume in the phase space is [7]

$$d\chi = \frac{\text{number of levels}}{\text{unit of phase-space volume}} = \frac{4\pi p^2 dp}{h^3}\,. \tag{16.35}$$

[6] As $T \to 0$, the argument of the exponential in Eq. 16.31 tends to plus infinity if $E - E_F > 0$, to minus infinity if $E - E_F < 0$. Thus, in the former case $f(E) \to 0$, in the latter $f(E) \to 1$.

[7] Remember that in quantum mechanics the phase space is quantized; each energy level corresponds to a phase-space volume h^3.

Since each level can be occupied by only two electrons, the number of electrons per unit volume is

$$n = \frac{N}{V} = 2\int_0^{p_F} d\chi = \int_0^{p_F} \frac{8\pi p^2 dp}{h^3} = \frac{8\pi}{3h^3}p_F^3 \,. \tag{16.36}$$

Using the expression of the density given in Eq. 16.28, from Eq. 16.36 we find

$$p_F = h\left(\frac{3}{8\pi\kappa m_N}\rho\right)^{\frac{1}{3}} \,. \tag{16.37}$$

We shall now show that the thermodynamical quantities can be expressed in terms of the Fermi momentum. Then, using Eq. 16.37, the pressure will be expressed as functions of the density, obtaining the EoS $P(\rho)$.

The *kinetic energy-density* ϵ_c of the gas can be determined in terms of p_F as follows. Eq. 16.15 and Eq. 16.32 yield

$$\epsilon_c = \frac{U}{V} = \frac{8\pi}{h^3}\int_0^{p_F}\left[\left(p^2c^2 + m_e^2c^4\right)^{\frac{1}{2}} - m_ec^2\right]p^2dp; \tag{16.38}$$

moreover, the pressure P can be found using Eqs. 16.13, 16.20,

$$P = \frac{8\pi}{3h^3}\int_0^{p_F}\frac{p^4c^2}{\left[p^2c^2 + m_e^2c^4\right]^{\frac{1}{2}}}dp \,. \tag{16.39}$$

The evaluation of the integrals in Eqs. 16.38 and 16.39 is particularly simple in two opposite regimes: 1) the *non-relativistic limit* and 2) the *ultra-relativistic limit.* To this purpose, it is useful to define a critical density, ρ_{crit}, as the density at which the Fermi momentum becomes equal to m_ec; using Eq. 16.37 we find

$$\rho_{\text{crit}} = \kappa\frac{8\pi}{3}m_N\left(\frac{m_e\,c}{h}\right)^3 = 0.98\times 10^6\kappa\,\text{g/cm}^3 \,. \tag{16.40}$$

We can now study the two aforementioned regimes.

Non-relativistic regime: If $\rho \ll \rho_{\text{crit}}$, $p_F \ll m_ec$ and the electrons are non-relativistic. In this case Eq. 16.39 gives

$$P = \frac{8\pi}{3h^3}\int_0^{p_F}\frac{p^4}{m_e}dp = \frac{8\pi}{15h^3}\frac{p_F^5}{m_e} \,, \tag{16.41}$$

and using Eq. 16.37

$$P = \frac{h^2}{5m_e}\left(\frac{3}{8\pi}\right)^{2/3}\left(\frac{1}{\kappa m_N}\right)^{\frac{5}{3}}\rho^{\frac{5}{3}} \,. \tag{16.42}$$

Thus, the EoS of a non-relativistic, degenerate electron gas is *polytropic*, i.e. it has the form

$$P = K\rho^\gamma \,, \tag{16.43}$$

where

$$K = \frac{h^2}{5m_e}\left(\frac{3}{8\pi}\right)^{2/3}\left(\frac{1}{\kappa m_N}\right)^{\frac{5}{3}}, \qquad \text{and} \qquad \gamma = \frac{5}{3} \,. \tag{16.44}$$

Note that K depends only on fundamental quantities, namely Planck's constant, the electron mass, and the mass of the nucleons, and on the number of nucleons per electron κ. Moreover, from Eq. 16.38 the kinetic energy-density in the non-relativistic regime is

$$\epsilon_c = \frac{8\pi}{h^3}\int_0^{p_F}\left[m_ec^2\left(1+\frac{1}{2}\frac{p^2c^2}{m_e^2c^4}\right) - m_ec^2\right]p^2dp = \frac{4\pi}{h^3}\int_0^{p_F}\frac{p^4}{m_e}\,dp \,. \tag{16.45}$$

Using Eq. 16.41 we obtain

$$\epsilon_c = \frac{4\pi}{5h^3}\frac{p_F^5}{m_e} \qquad \rightarrow \qquad \epsilon_c = \frac{3}{2}P\,. \tag{16.46}$$

Ultra-relativistic limit: If $\rho \gg \rho_{\rm crit}$, $p_F \gg m_e c$ and the electrons are ultra-relativistic. In this case from Eq. 16.39 we find

$$P = \frac{8\pi}{3h^3}\int_0^{p_F} p^3 c dp = \frac{2\pi c}{3h^3}p_F^4\,, \tag{16.47}$$

and using Eq. 16.37

$$P = \frac{c\,h}{8}\left(\frac{3}{\pi}\right)^{1/3}\left(\frac{1}{m_N\kappa}\right)^{\frac{4}{3}}\rho^{\frac{4}{3}}\,. \tag{16.48}$$

Again, the EoS of a degenerate gas of electrons is a polytropic EoS

$$P = K\rho^\gamma\,, \tag{16.49}$$

where now

$$K = \frac{c\,h}{8}\left(\frac{3}{\pi}\right)^{1/3}\left(\frac{1}{m_N\kappa}\right)^{\frac{4}{3}}\,, \qquad \text{and} \qquad \gamma = \frac{4}{3}\,. \tag{16.50}$$

Note that again K depends only on fundamental quantities and on the number of nucleons per electron κ. Moreover

$$\epsilon_c = \frac{8\pi}{h^3}\int_0^{p_F} p^3 c dp\,, \tag{16.51}$$

thus, comparing with Eq. 16.47 we find

$$\epsilon_c = 3P\,. \tag{16.52}$$

General case: In general, Eqs. 16.38 and 16.39 can be integrated analytically. If we define the dimensionless Fermi momentum $x = p_F/(m_e c)$, Eq. 16.37 can be written as [32]

$$\rho(x) = \frac{8\pi m_e^3 c^3 \kappa m_N}{3h^3}x^3\,, \tag{16.53}$$

and the integral 16.39 can be solved explicitly, and gives

$$P(x) = \frac{\pi m_e^4 c^5 \kappa m_N}{3h^3}\left[x(2x^2-3)\sqrt{1+x^2} + 3\sinh^{-1}x\right]\,. \tag{16.54}$$

Eqs. 16.53 and 16.54 provide the white dwarf equation of state $P(\rho)$ in a parametric form, i.e. in terms of the implicit parameter x. Alternatively, one can solve Eq. 16.53 for x as a function of ρ and substitute it in Eq. 16.54 to find the barotropic EoS $P = P(\rho)$.

This function (which is monotonically increasing, since both Eqs. 16.53 and 16.54 are monotonically increasing functions of x) is valid at all densities between the non-relativistic and the ultra-relativistic limits. It can be checked that by expanding Eq. 16.54 for $x \ll 1$ and $x \gg 1$, the expression of $P(\rho)$ in the non-relativistic and relativistic regimes considered above are recovered.

From these expressions we see that, in a degenerate gas, the pressure only depends on the density, and increases as the density increases. Thus, when the electron gas in the contracting core of the star becomes degenerate, the pressure gradients become sufficient to support the equilibrium against gravitational collapse. This is true, as we shall later see, provided the mass does not exceed a critical value. We remark that, both in the non-relativistic and in the ultra-relativistic regime, a degenerate gas is described be a *polytropic* EoS.

The main results for the EoS of a fully degenerate electron gas are summarized in Box 16-D.

Box 16-C

The Newtonian equations of stellar structure

Let us consider a spherical shell of fluid of radius r and thickness dr; let $dV = dAdr$ be the volume of an element of the shell, where dA is its section (orthogonal to r), and $dm = \rho\, dV$ its mass. The forces acting on the fluid element are the gravitational attraction exerted by the sphere of mass $m(r)$ inside the shell, and the gradient of pressure across the shell, multiplied by the volume element; if the fluid element is in equilibrium, these two forces balance each other, i.e.

$$-\frac{dP}{dr}drdA = \frac{Gm(r)}{r^2}dm \qquad \rightarrow \qquad \frac{dP}{dr} = -\frac{Gm(r)\rho(r)}{r^2}\,. \tag{16.55}$$

The mass contained in a sphere of radius r is

$$m(r) = \int_0^r \rho(r)\, 4\pi r^2 dr\,, \qquad \rightarrow \qquad \frac{dm}{dr} = 4\pi r^2 \rho(r)\,. \tag{16.56}$$

Equations 16.55 and 16.56 can be solved only if we assign a further equation which relates pressure and density, i.e. an *equation of state*. We refer the reader to Sec. 16.3.3 for a more extended discussion on the EoS of compact stars. Here we only remark that in general the pressure is also related to other thermodynamical quantities, such as the temperature, the chemical potential, etc. However, as we have shown in Sec. 16.2.4, in the case of white dwarfs the EoS is barotropic, i.e. the pressure depends only on the density, $P = P(\rho)$. Thus, the closed system of equations which describe the equilibrium structure of a white dwarf is

$$\begin{cases} \dfrac{dm}{dr} = 4\pi r^2 \rho(r)\,, \\[2ex] \dfrac{dP}{dr} = -\dfrac{Gm(r)}{r^2}\rho(r)\,, \\[2ex] P = P(\rho)\,. \end{cases} \tag{16.57}$$

Box 16-D

Summary: the equation of state of a fully degenerate electron gas in the non-relativistic and in the ultra-relativistic regimes

We here summarize the results of this section for the EoS of a fully degenerate electron gas, in the non-relativistic and in the ultra-relativistic regimes. We have shown that in both cases the gas is governed by a polytropic EoS

$$P = K\rho^{\gamma}\,, \tag{16.58}$$

where K and γ depend on the regime. It is useful to define

$$\rho_{\rm crit} = \kappa\frac{8\pi}{3}m_N\left(\frac{m_e c}{h}\right)^3 = 0.98\times 10^6\kappa\,{\rm g/cm}^3\,, \tag{16.59}$$

in terms of which we have:

- non-relativistic regime, $\rho \ll \rho_{\rm crit}$; in this case,

$$\begin{aligned} \gamma &= \frac{5}{3} \\ K &= \frac{h^2}{5m_e}\left(\frac{3}{8\pi}\right)^{2/3}\left(\frac{1}{\kappa m_N}\right)^{\frac{5}{3}} = 9.92\times 10^{12}\kappa^{-5/3}\,{\rm erg}^2\ {\rm s}^2/{\rm g}^{8/3}\,, \end{aligned} \tag{16.60}$$

- ultra-relativistic regime, $\rho \gg \rho_{\rm crit}$; in this case

$$\begin{aligned} \gamma &= \frac{4}{3} \\ K &= \frac{ch}{8}\left(\frac{3}{\pi}\right)^{1/3}\left(\frac{1}{m_N\kappa}\right)^{\frac{4}{3}} = 1.23\times 10^{15}\kappa^{-4/3}\,{\rm erg\ cm/g}^{4/3}\,. \end{aligned} \tag{16.61}$$

When the conditions for the non-relativistic or for the ultra-relativistic regimes are not fulfilled, the EoS of a degenerate gas cannot be written as a polytropic as in Eq. 16.58. However, it has a barotropic form, $P = P(\rho)$, suitable for numerical integration of the equations of stellar structure. The analytic expression of this EoS, in terms of the implicit parameter $x = p_F/(m_e c)$, is

$$\begin{aligned} \rho(x) &= \frac{8\pi m_e^3 c^3 \kappa m_N}{3h^3}x^3\,, \\ P(x) &= \frac{\pi m_e^4 c^5 \kappa m_N}{3h^3}\left[x(2x^2-3)\sqrt{1+x^2} + 3\sinh^{-1}x\right]. \end{aligned} \tag{16.62}$$

Eqs. 16.62 reduce to the non-relativistic and ultra-relativistic cases in the corresponding regimes.

16.2.5 The structure of a white dwarf

In order to find the equilibrium configurations of a white dwarf, we shall solve the Newtonian equations of hydrostatic equilibrium given in Box 16-C using the results obtained in the previous section.

We shall assume that the EoS of matter inside the star is a polytropic one, and we shall use the values of the polytropic parameters given in Box 16-D, which correspond to the non-relativistic ($\rho \ll \rho_{\rm crit}$), and to the ultra-relativistic regime ($\rho \gg \rho_{\rm crit}$). Then, the complete set of equations which describe the structure of the star is

$$\begin{cases} \dfrac{dm}{dr} = 4\pi r^2 \rho \,, \\ \\ \dfrac{dP}{dr} = -\dfrac{Gm}{r^2}\rho \,, \\ \\ P = K\rho^\gamma \,. \end{cases} \tag{16.63}$$

These equations have to be solved by imposing that at $r = 0$ the density takes some assigned value ρ_c, and that at the surface of the star, $r = R$, the pressure vanishes, i.e.

$$\rho(0) = \rho_c, \qquad\qquad P(R) = 0\,. \tag{16.64}$$

By differentiating the second equation 16.63 and using the first equation, it is easy to show that the two equations can be combined into the following second-order, ordinary, differential equation:

$$\frac{1}{r^2}\frac{d}{dr}\left(\frac{r^2}{\rho}\frac{dP}{dr}\right) = -4\pi G\rho\,. \tag{16.65}$$

It is convenient to write the above equation in a dimensionless form as follows. We introduce the function $\Theta(r)$ such that

$$\begin{cases} \rho = \rho_c\,\Theta^n(r)\,, \\ \\ P = K\;\rho_c^{1+\frac{1}{n}}\;\Theta^{(n+1)}(r)\,, \end{cases} \tag{16.66}$$

where n is called *polytropic index*, and

$$\gamma = 1 + \frac{1}{n}\,. \tag{16.67}$$

In terms of the function $\Theta(r)$ Eq. 16.65 becomes

$$(n+1)K\;\rho_c^{(\frac{1}{n}-1)}\;\frac{1}{r^2}\frac{d}{dr}\left(r^2\frac{d\Theta}{dr}\right) = -4\pi G\;\Theta^n. \tag{16.68}$$

If we now introduce the following dimensionless, radial coordinate

$$\xi = \frac{r}{\alpha}, \qquad \text{where} \qquad \alpha = \left[\frac{(n+1)K\;\rho_c^{(\frac{1}{n}-1)}}{4\pi G}\right]^{\frac{1}{2}}, \tag{16.69}$$

Eq. 16.68 yields

$$\frac{1}{\xi^2}\frac{d}{d\xi}\left(\xi^2\frac{d\Theta}{d\xi}\right) = -\Theta^n\,. \tag{16.70}$$

This is known as the **Lane-Emden equation**; it is a dimensionless equation, which depends only on the polytropic index n. Being a second-order, ordinary, differential equation, to find a unique solution we need to specify the value of Θ and of its derivative at some point, typically the origin of the star, $r = 0$. Since $\rho = \rho_c\,\Theta^n$ and $\rho_c = \rho(0)$, the first boundary condition is $\Theta(0) = 1$; moreover, since the mass goes to zero as $m(r) \sim \frac{4\pi}{3}\rho_c r^3$, from Eq. 16.63 it follows that

$$\frac{dP}{dr} \sim -\frac{4\pi G}{3} r \rho_c^2 , \tag{16.71}$$

i.e. $\frac{dP}{dr}$ goes to zero as $\sim r$. From the EoS, $P = K\rho^\gamma$, we find

$$\frac{dP}{dr} = K\gamma\, \rho^{\gamma-1}\, \frac{d\rho}{dr} , \tag{16.72}$$

from which it follows that since $\frac{dP}{dr}$ tends to zero, $\frac{d\rho}{dr}$ must also tend to zero. Thus, a further condition to impose is $\Theta'(0) = 0$, where the prime indicates derivative with respect to ξ. In conclusion the Lane-Emden equation 16.70 must be integrated by imposing the following boundary conditions at the center of the star:

$$\begin{cases} \Theta(0) = 1 , \\ \Theta'(0) = 0 . \end{cases} \tag{16.73}$$

It can be shown that for $\gamma > \frac{6}{5}$ (or, equivalently, for $n < 5$) the solution $\Theta(\xi)$ is a monotonically decreasing function which vanishes for some $\xi = \xi_1$. When $\Theta = 0$ both the density and the pressure vanish, therefore ξ_1 corresponds to the boundary of the star; its value can be found by integrating the Lane-Emden equation numerically for a fixed value of n.

The procedure to find the stellar structure in Newtonian gravity can be summarized as follows:

1. Choose a value of γ, use the corresponding polytropic index $n = \frac{1}{\gamma-1}$, and integrate numerically Eq. 16.70 with the boundary conditions (Eqs. 16.73) up to the value $\xi = \xi_1$ where $\Theta = 0$. For instance, for $\gamma = \frac{5}{3}$ and $\gamma = \frac{4}{3}$ the numerical integration gives

$$\gamma = \tfrac{5}{3} \qquad n = \tfrac{3}{2} \quad \xi_1 = 3.65375 \qquad \xi_1^2\, \Theta'(\xi_1) = -2.71406 ,$$

$$\gamma = \tfrac{4}{3} \qquad n = 3 \quad \xi_1 = 6.89685 \qquad \xi_1^2\, \Theta'(\xi_1) = -2.01824 ; \tag{16.74}$$

   the combination $\xi_1^2\, \Theta'(\xi_1)$ is useful to determine the mass of the star, as will be explained below. It should be noted that since Θ is a monotonically decreasing function of ξ, at the boundary of the star its first derivative is negative.

2. Assign a value to κ, the number of nucleons per electron, then find K from Eq. 16.60 or Eq. 16.61. Choose a central density ρ_c. Knowing K and ρ_c the radius of the star can be found using the definition of ξ given in Eq. 16.69

$$R = \alpha\xi_1 \quad \rightarrow \quad R = \xi_1 \sqrt{\frac{(n+1)K}{4\pi G}\rho_c^{\frac{1-n}{2n}}} . \tag{16.75}$$

3. The mass of the star, $M = m(R)$, can be determined by integrating Eq. 16.56 as follows

$$\begin{aligned} M &= \int_0^R 4\pi r^2 \rho(r) dr = 4\pi\alpha^3\, \rho_c \int_0^{\xi_1} \xi^2\, \Theta^n\, d\xi \\ &= -4\pi\alpha^3\, \rho_c \int_0^{\xi_1} \frac{d}{d\xi}\left(\xi^2 \frac{d\Theta}{d\xi}\right) d\xi \\ &= -4\pi\alpha^3\, \rho_c\, \xi_1^2\, \Theta'(\xi_1) , \end{aligned}$$

where we have used Eq. 16.70. Finally, the value of M as a function of K and ρ_c can be found by using the expression of α given in Eq. 16.69

$$M = 4\pi\ \xi_1^2\ |\Theta'(\xi_1)|\ \left[\frac{(n+1)K}{4\pi G}\right]^{\frac{3}{2}} \rho_c^{\frac{3-n}{2n}}\,. \tag{16.76}$$

Let us define

$$A = \frac{(n+1)K}{4\pi G}\,, \qquad B = 4\pi\ \xi_1^2\ |\Theta'(\xi_1)|\,, \tag{16.77}$$

so that

$$R = \xi_1 A^{1/2} \rho_c^{\frac{1-n}{2n}}\,, \tag{16.78}$$

and

$$M = B A^{3/2} \rho_c^{\frac{3-n}{2n}}\,. \tag{16.79}$$

Combining Eqs. 16.78 and 16.79, we can finally find a relation between M and R:

$$M = \left(B A^{\frac{n}{n-1}} \xi_1^{\frac{n-3}{1-n}}\right) R^{\frac{3-n}{1-n}}\,. \tag{16.80}$$

The procedure outlined above shows that, once we fix the number of nucleons per electron κ and the polytropic index n, and integrate (numerically) the Lane-Emden equation to find ξ_1 and $\Theta'(\xi_1)$, we obtain a family of solutions parametrized by different values of the central density ρ_c, with radii and masses given by Eqs. 16.75 and 16.76.

Conversely, if we change the value of κ, the new configuration can easily be obtained by rescaling the various quantities in the following way

$$\begin{aligned} \hat{\rho} &= \tfrac{\hat{\kappa}}{\kappa}\rho\,, & \hat{P} &= P\,, \\ \hat{m}(r) &= \left(\tfrac{\kappa}{\hat{\kappa}}\right)^2 m(r)\,, & \hat{r} &= \frac{\kappa}{\hat{\kappa}} r\,. \end{aligned} \tag{16.81}$$

Box 16-E

Note on the numerical integration of the Lane-Emden equation

Since the Lane-Emden equation diverges in $\xi = 0$, we cannot integrate Eq. 16.70 starting from that point. This problem can be overcome starting the numerical integration at some small, but finite value of $\xi = \xi_{\rm start}$, and by Taylor-expanding the solution near the origin, writing it in terms of $\Theta(\xi_{\rm start})$ and $\Theta'(\xi_{\rm start})$.
Since we know from Eq. 16.73 that $\Theta(0) = 1$ and $\Theta'(0) = 0$, we can write the approximate solution near $\xi = 0$ as a power series

$$\Theta(\xi) \simeq 1 + \Theta_2\, \xi^2 + \Theta_3\, \xi^3 + \Theta_4\, \xi^4 + O(\xi^5) + \dots\,; \tag{16.82}$$

the truncation order of the series is arbitrary, but let us keep up to the fourth power of ξ as an example. To determine the values of the coefficients $\Theta_2, \Theta_3, \Theta_4$ we Taylor-expand the function Θ^n on the right-hand side of Eq. 16.70, i.e.

$$\Theta^n \simeq 1 + n\Theta_2\xi^2 + O(\xi^3); \tag{16.83}$$

by substituting Eqs. 16.82 and 16.83 in Eq. 16.70 we find

$$6\Theta_2 + 12\Theta_3\xi + 20\Theta_4\xi^2 + \dots = -[1 + n\Theta_2\, \xi^2] + \dots\,. \tag{16.84}$$

This equation is satisfied only if the coefficients of the same power of ξ vanish, i.e.

$$\begin{aligned} 1 &= -6\Theta_2 & &\to & \Theta_2 &= -\frac{1}{6}\,, \\ \Theta_3 &= 0 & & & & \\ 20\,\Theta_4 &= -n\Theta_2 & &\to & \Theta_4 &= \frac{n}{120}\,. \end{aligned}$$

Thus, the expansion of $\Theta(\xi)$ has only even powers of ξ (this property holds at any order since $\rho(r)$ must be an even function due to symmetry reasons), and the approximate solution for Θ and Θ' near the origin is

$$\begin{aligned} \Theta(\xi) &\simeq 1 - \frac{1}{6}\xi^2 + \frac{n}{120}\xi^4 + O(\xi^6) \\ \Theta'(\xi) &\simeq -\frac{1}{3}\xi + \frac{n}{30}\xi^3 + O(\xi^5)\,. \end{aligned} \tag{16.85}$$

We now have all we need to integrate the Lane-Emden equation numerically: we can start at, say, $\xi_{\rm start} = 10^{-4}$ using as initial values the functions in Eq. 16.85 computed at $\xi_{\rm start}$. As customary, it should always be checked that the solution does not depend on the choice of $\xi_{\rm start}$ within some numerical accuracy.

16.2.6 The Chandrasekhar limit

In Sec. 16.2.4 we have shown that if the density in the star is much smaller than the critical density $\rho_{\rm crit} = 0.98 \times 10^6\ \kappa$ g/cm^3, electrons behave as a polytropic gas with $\gamma = \frac{5}{3}$ and $n = \frac{3}{2}$, while if it is much larger than $\rho_{\rm crit}$, they behave as a polytropic gas with $\gamma = \frac{4}{3}$ and

$n = 3$. Moreover, Eq. 16.79 gives

$$M = BA^{3/2}\rho_c^{\frac{3-n}{2n}} = B\left(\frac{n+1}{4\pi G}\right)^{3/2} K^{3/2}\rho_c^{\frac{3-n}{2n}} . \tag{16.86}$$

There is an important difference between the non-relativistic and the ultra-relativistic regimes. In the non-relativistic case ($n = \frac{3}{2}$) the exponent of ρ_c in Eq. 16.86 is $\frac{3-n}{2n} = \frac{1}{2}$. Moreover, using the expressions of K and $\rho_{\rm crit}$ given in Eqs. 16.60 and 16.59, it follows that

$$K^{3/2} = const\ \frac{\kappa^{-2}}{\rho_{\rm crit}^{1/2}} , \tag{16.87}$$

where the constant depends only on fundamental constants (m_e, m_N, etc.). Therefore, Eq. 16.86 gives

$$M \propto \kappa^{-2}\left(\frac{\rho_c}{\rho_{\rm crit}}\right)^{1/2} . \tag{16.88}$$

Thus, in the non-relativistic regime the mass is an increasing function of the central density.

Conversely, in the ultra-relativistic regime ($n = 3$) the exponent of the central density in Eq. 16.86 vanishes, so that the mass *does not depend on the central density at all.*

An explicit computation (which we leave to the reader as an exercise; all the relevant fundamental constants are explicitly given in Table A) shows that, in the non-relativistic regime,

$$M = 2.72\ \kappa^{-2}\left(\frac{\rho_c}{\rho_{\rm crit}}\right)^{1/2} M_\odot , \tag{16.89}$$

whereas in the ultra-relativistic regime

$$M = M_{\rm Ch} = 5.73\ \kappa^{-2} M_\odot . \tag{16.90}$$

The quantity $M_{\rm Ch}$ is the critical mass above which no equilibrium configuration for a white dwarf can exist; it is called the *Chandrasekhar mass*, after Subrahmanyan Chandrasekhar who derived[8] it in 1931. It should be noted that the information on the internal composition of the star is contained entirely in the parameter κ. For a white dwarf made of helium, $\kappa = 2$ and

$$M_{\rm Ch} = 1.43\ M_\odot . \tag{16.91}$$

The fact that a critical mass should exist can also be understood from the following qualitative considerations, which were proposed by Lev Landau in 1932 [74, 73]. A given configuration of matter will be in equilibrium if the gradient of pressure is balanced by the gravitational attraction per unit of volume. In the non-relativistic case:

$$P \sim \rho^{\frac{5}{3}} \quad \rightarrow \quad P \sim \frac{M^{\frac{5}{3}}}{R^5} \quad \rightarrow \quad \frac{dP}{dr} \sim \frac{M^{\frac{5}{3}}}{R^6} . \tag{16.92}$$

In the ultra-relativistic case:

$$P \sim \rho^{\frac{4}{3}} \quad \rightarrow \quad P \sim \frac{M^{\frac{4}{3}}}{R^4} \quad \rightarrow \quad \frac{dP}{dr} \sim \frac{M^{\frac{4}{3}}}{R^5} . \tag{16.93}$$

The gravitational force per unit volume scales as

$$\frac{Gm(r)\rho}{r^2} \sim \frac{M^2}{R^5} , \tag{16.94}$$

[8]The concept of a limiting mass for white dwarfs was first introduced by Chandrasekhar in a paper published in 1931 [31]. The problem was subsequently investigated in a series of papers, and a complete account can be found in the book that Chandrasekhar wrote on the subject in 1939 [33].

and the star is in equilibrium if

$$\frac{dP}{dr} = -\frac{Gm(r)\rho}{r^2}\,. \tag{16.95}$$

Since in the non-relativistic case the gradient of pressure (Eq. 16.92) and the gravitational force per unit volume (Eq. 16.94) depend on the radius with a different power, for a given value of the mass the star can "adjust" the radius until the two forces balance each other. Conversely, in the ultra-relativistic case (Eq. 16.93) the gradient of pressure and the gravitational force per unit volume have the same dependence on the radius, and therefore the equilibrium is possible only for one value of the mass, i.e. for the critical mass. If $M > M_{\rm Ch}$ the gravitational attraction exceeds the gradient of pressure and equilibrium configurations are no longer possible.

An alternative version of this argument is given in Box 16-F.

Box 16-F

Chandrasekhar mass from energy considerations

Let us consider a star of radius R composed of N nucleons, each one of mass $m_N \simeq 1.674 \times 10^{-24}$ g (see Box 16-B). From Eq. 16.37, the Fermi energy of the system in the relativistic limit reads

$$E_F = cp_F = hc\left(\frac{3\rho}{8\pi\kappa m_N}\right)^{1/3} \simeq \frac{\hbar c}{R}\left(\frac{9\pi N}{4\kappa}\right)^{1/3}, \tag{16.96}$$

where $\hbar = h/(2\pi) = 1.055 \times 10^{-27}$ erg s (see Table A), and in the last step we approximated ρ/m_N with the mean nucleon number density, i.e. $\rho/m_N \simeq N/(4\pi R^3/3)$. Assuming that the star mass is $M \sim Nm_N$, the gravitational energy per nucleon is

$$E_G \simeq -\frac{GNm_N{}^2}{R}. \tag{16.97}$$

Thus, the star total energy is

$$E \equiv E_F + E_G = \frac{1}{R}\left[\hbar c\left(\frac{9\pi N}{4\kappa}\right)^{1/3} - GNm_N^2\right]. \tag{16.98}$$

A bound configuration requires $E < 0$ and therefore $E = 0$ corresponds to the critical configuration beyond which no equilibrium is possible. The condition $E = 0$ yields

$$N_{\text{max}} \simeq \sqrt{\frac{9\pi}{4\kappa}}\left(\frac{\hbar c}{Gm_N^2}\right)^{3/2}, \tag{16.99}$$

and therefore the maximum mass is of the order of

$$M_{\text{max}} = m_N N_{\text{max}} \simeq \frac{1}{m_N^2}\sqrt{\frac{9\pi}{4\kappa}}\left(\frac{\hbar c}{G}\right)^{3/2} \simeq 3.5\, M_\odot, \tag{16.100}$$

where in the last step we assumed $\kappa \simeq 2$. The above maximum mass is only a factor 2 larger than the precise value in Eq. 16.91. Note that, in addition to the proton mass, the above formula depends on several fundamental constants: the (reduced) Planck constant, the speed of light, and Newton's constant, showing that the Chandrasekhar limit arises from a combination of quantum, relativistic, and gravitational arguments.

Although nowadays the existence of a critical mass for white dwarfs seems an obvious consequence of the theory, it was not accepted when Chandrasekhar found it. The prejudice at that time was that white dwarfs were the unique final state of stellar evolution, and that they could have any mass (neutron stars were discovered much later in 1967). The famous astronomer Arthur Eddington was the strongest opponent to the new theory, and referred to it as "a stellar buffonery". Nobody, at that time, gave to Chandrasekhar any public support, although a few, as for example Rosenfeld, told him in private that they thought his result was correct[9]. However, Chandrasekhar understood that his discovery would have had important implications, and in 1934 he wrote: "The life history of a star of small mass

[9] An interesting account of the controversy between Eddington and Chandrasekhar on white dwarfs' maximum mass can be found in the book "*Chandra: a biography of S. Chandrasekhar*" [116].

must be essentially different from the life history of a star of large mass. For a star of small mass the natural white-dwarf stage is an initial step towards complete extinction. A star of large mass cannot pass into the white-dwarf stage and one is left speculating on other possibilities".

It should be stressed that the Chandrasekhar mass is a *static* limit, i.e. it refers to the equilibrium configuration. However, even if a star is in equilibrium it may be unstable against small perturbations. In this case the star would be *dynamically unstable* (see Sec. 16.3.9).

A second point which should be noted is that in the derivation of the critical mass, General Relativity plays no role. The basic ingredients are special relativity and the Fermi-Dirac statistics.

Box 16-G

Mass-radius diagram for a white dwarf

The mass-radius diagram of white dwarfs can be found by integrating the Newtonian equations of stellar structure given in Box 16-C, for different values of the central density and assuming the white dwarf EoS derived in Sec. 16.2.4 (Eq. 16.62). The result of the numerical integration, assuming $\kappa = 2$, is shown in Fig. 16.1. As expected, there exists a maximum mass, $M = M_{\rm Ch} \simeq 1.4\, M_\odot$, corresponding to a radius $R \sim 10^{-3} R_\odot \sim 700\,$km. The small discrepancies between the theoretical and observed values of the mass and radius of Sirius B are due to the assumption of zero temperature in the computation leading to Eq. 16.62, see Ref. [97] for a discussion. The configurations on the left of maximum are unstable, as will be explained in Sec. 16.3.9.

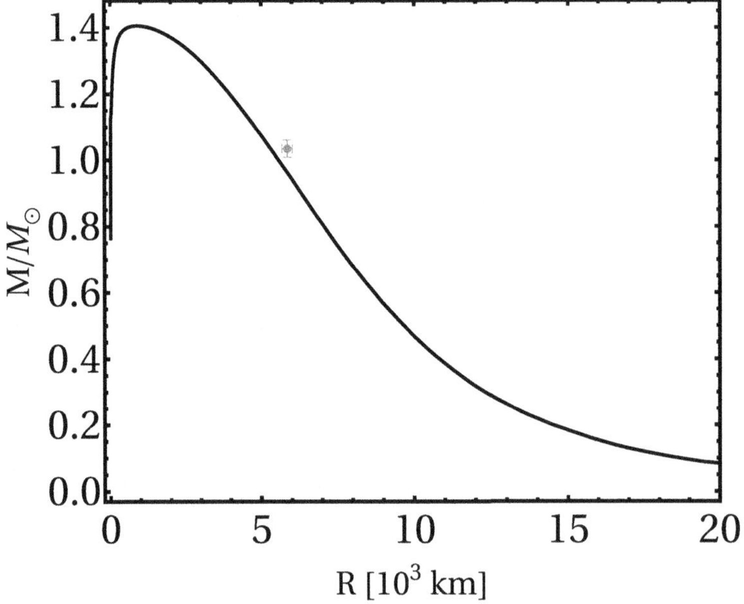

Figure 16.1: Mass-radius diagram of a zero-temperature white dwarf. Each point on the curve corresponds to a different value of the central density of the star. The marker with error bars is the mass and radius of Sirius B.

16.3 NEUTRON STARS

As discussed in Sec. 16.1, when the mass of the progenitor star is in the range $\sim (8, 20-30)\ M_\odot$, its thermonuclear evolution leads to a supernova explosion whose remnant is a neutron star. In this section, after a brief review on the discovery of these extremely compact objects, we shall discuss their internal composition and derive the general relativistic equations which allow to determine the internal structure and the gravitational field of these stars.

16.3.1 The discovery of neutron stars

The first explicit prediction of the existence of neutron stars was made by Walter Baade and Fritz Zwicky in 1934, two years after the discovery of the neutron [10, 12, 11]. They suggested that supernovae mark the transition from ordinary stars into neutron stars, and in [11] they wrote: "Such a star may possess a very small radius and an extremely high density. As neutrons can be packed much more closely than ordinary nuclei and electrons, the "gravitational packing" energy in a cold neutron star may become very large and, under certain circumstances, far exceed the ordinary nuclear packing fractions. A neutron star would therefore represent the most stable configuration of matter as such."

As mentioned in Sec. 16.1, neutron stars are often observed as pulsars, and it is interesting to follow the historical path that led to the discovery of pulsars and that allowed to establish the connection between supernova explosions and neutron stars. In 1942 Baade identified the Crab Nebula and the star in its center as the remnant of the supernova explosion that occurred in 1054, which was observed by the Chinese astronomer Yang Wei-te [13]. Twenty-five years later, in 1967, Jocelyn Bell, Antony Hewish, and collaborators discovered the first pulsar [62]. The observed radio pulses were separated by the very short time interval of 1.33 seconds and were extremely regular; it was soon realized that they could not be of human origin since they came from outside our solar system. In the same year the Italian astronomer Franco Pacini suggested that a rotating neutron star with a magnetic field would emit pulsed radiation [88]. However, the correlation between pulsars and supernovae was firmly established only when a pulsar was discovered in the remnant of the Vela supernova [76] and a very fast pulsar (with period $P = 0.033\,\mathrm{s}$) was discovered in the Crab Nebula [107]. Thus, more than thirty years later, the prediction of Baade and Zwicky was finally confirmed.

16.3.2 The internal structure of a neutron star

Current studies on the internal structure of neutron stars show that these stars can be modeled, as shown in Fig. 16.2, as a sequence of layers of different composition and thickness surrounding an innermost core. Proceeding from the exterior, we first encounter an outer crust, about 0.3 km thick, an inner crust, $\sim 0.5\,\mathrm{km}$ thick, and a core extending over about 10 km. We shall assume that the temperature of matter in the neutron star interior is $T = 0\,\mathrm{K}$. Indeed, shortly after their birth ($\sim$ a year after) neutron stars reach temperatures $T \lesssim 10^9\,\mathrm{K}$; since at the densities typical of neutron stars ($\bar{\rho} \sim 10^{14}\,\mathrm{g/cm^3}$) neutrons have a Fermi (kinetic) energy $E_F - m_n c^2 \sim 10^{-5}\,\mathrm{erg}$ (see Eq. 16.37), which corresponds to a Fermi temperature $T_F = (E_F - m_n c^2)/k \sim 10^{11}\,\mathrm{K}$, then $T \ll T_F$, which justifies our assumption (see the discussion in Sec. 16.2.4). In addition neutron stars are transparent to neutrinos, because the mean free path of neutrinos in nuclear matter at $T \lesssim 10^9\,\mathrm{K}$ is much larger than the typical radius of a neutron star, $R \sim (10-15)\,\mathrm{km}$.

- *The outer crust.*

  The matter density ranges from $\sim 10^7\ \mathrm{g/cm^3}$ to $4 \times 10^{11}\ \mathrm{g/cm^3}$. It is composed of a

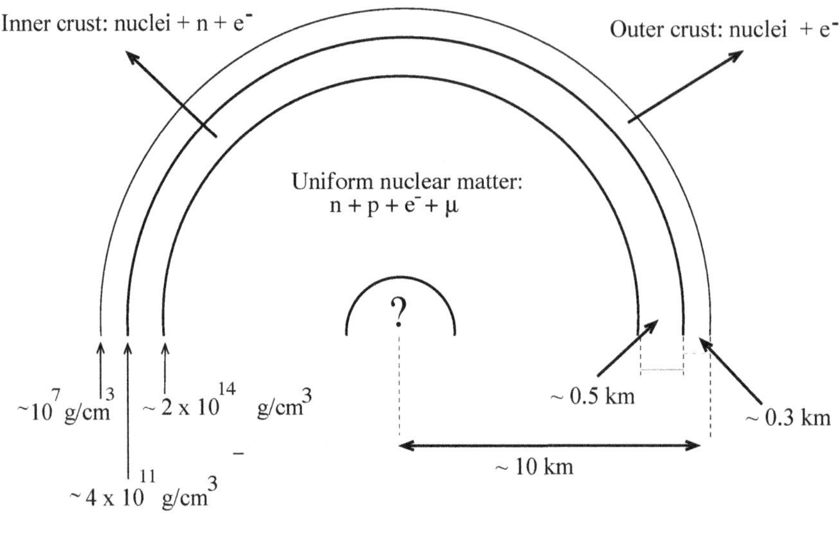

Figure 16.2: Schematic view of a neutron star internal structure (not in scale).

lattice of heavy nuclei immersed in an electron gas. Proceeding from the external to the internal boundary, as the density increases the inverse β-decay process becomes more and more efficient and neutrons are produced copiously according to Eq. 16.4

$$e^- + p \rightarrow n + \nu_e\,. \tag{16.101}$$

The produced neutrinos only weakly interact with the matter and leave the star. In the outer crust the pressure is mainly provided by the degenerate electron gas. At $\rho = \rho_\mathrm{d} = 4 \times 10^{11}\mathrm{g/cm}^3$ all bound states available in the nuclei for neutrons are filled. Neutrons can no longer live bound to the nuclei and start leaking out; this phenomenon is called *neutron drip.*

- *The inner crust.*
  In this region the density ranges between ρ_d and the nuclear density $\rho_0 = 2.67 \times 10^{14}\,\mathrm{g/cm}^3$, and the dominant contribution to pressure is due to the neutron gas. Matter is composed of a mixture of two phases: one, with density comparable to ρ_0, is rich of neutrons and is therefore called *neutron-rich matter* (NRM); the other is a *neutron gas* (NG). An electron gas is also present to ensure charge neutrality. In order to determine the fundamental state of matter in this region, one has to specify the density of the two phases, ρ_NRM and ρ_NG, which determines the fraction of volume each phase occupies, the proton fraction in the NRM, and the geometrical properties of the structures that are formed by the two phases; these strongly depend on surface effects at the interface between different phases. For $\rho_d \lesssim \rho \lesssim 0.35\ \rho_0$, the minimum energy configuration is formed by spherical drops of NRM, surrounded by the NG. For higher densities the separation between spheres decreases up to the touching limit. Consequently, for $0.35\ \rho_0 \lesssim \rho \lesssim 0.5\ \rho_0$ the spheres merge, forming bar-type structures, called "spaghetti phase", and for $0.5\ \rho_0 \lesssim \rho \lesssim 0.56\ \rho_0$ the bars merge to form slab-type structures, called "lasagna phase". For larger densities, the merged nuclei of NRM become a uniform fluid, with increasingly smaller contributions of "pasta phases" of

NG. When the density reaches the nuclear density ρ_0 the two phases are no longer separated and form a homogeneous fluid of protons, neutrons, and electrons.

- *The core.*
  For $\rho > \rho_0$, in the so-called *outer core* of the neutron star, matter is composed of a homogeneous fluid of p, n, e^-, in β-equilibrium, i.e. in equilibrium with respect to the neutron β-decay

$$n \to p + e^- + \bar{\nu}_e\,, \tag{16.102}$$

and to the inverse β-decay given in Eq. 16.101. By minimizing the free energy one finds that matter in the core is stable only if protons are up to about 10% of the total. It should be stressed that the main contribution to the pressure in the core comes from neutrons; since they are more massive than electrons, the total energy is also mostly provided by the neutrons themselves. Neutrons, as electrons, are fermions. However, the pressure they generate cannot be associated only to Pauli's exclusion principle, because at the core densities we can no longer treat neutrons as non-interacting particles, as we did for the degenerate electron gas in white dwarfs. Indeed, if we assume that the neutron star is composed of non-interacting neutrons, we find a maximum mass of 0.7 $M_\odot$, which is much lower than the mass of observed neutron stars [87].

In addition, several processes may develop at higher density ($\rho \gtrsim 2\rho_0$), in the *inner core*. For instance, as the density increases electrons become more energetic and their kinetic energy increases. So does the chemical potential which, we remind, is the energy needed to insert a new particle in the same state in a gas in equilibrium. Thus, at some density the chemical potential of the electrons becomes larger than the rest mass of the muon $m_{\mu^-} = 105\,\text{MeV}$. At this point the decay of a neutron through the reaction

$$n \to p + \mu^- + \bar{\nu}_\mu\,, \tag{16.103}$$

which creates the muon, becomes energetically more convenient than the neutron β-decay given in Eq. 16.102. Neutrinos escape through the surface, whereas n, p, e^-, μ^- remain trapped in the core. Furthermore, other particles may form in the core at high density. For instance in some models heavy baryons may form through the reaction

$$n + e^- \to \Sigma^- + \nu_e\,; \tag{16.104}$$

or, when $\rho \sim 2 - 3\ \rho_0$, π or K mesons may also appear; these particles are bosons and therefore not subjected to Pauli's exclusion principle; in this case a Bose-Einstein condensate may form in the innermost region of the star. Or, even further, since nucleons are known to be composite objects of size $\sim 0.5 - 1.0$ fm, corresponding to a density $\sim 10^{15}$ g/cm^3, if the density in the core reaches this value matter undergoes a transition to a new phase, in which quarks are no longer confined into nucleons or hadrons [10]. Thus, at high densities different particle content would lead to different equations of state.

There is a quite general consensus on the EoS of matter in the outer and inner crust, and in the outer part of the core, because at these densities the properties of matter are constrained by experimental data on neutron-rich nuclei. Conversely, densities and energies as those prevailing in the inner part of the core are presently unreachable in a laboratory; consequently, the variety of models of EoS at supranuclear densities proposed in the literature

[10]This phase transition may lead to the formation of *quark stars*, which are composed of deconfined quarks. It has been suggested that a quark star would be a sort of large "nucleus" formed of a mixture of up, down, and strange quarks, which would be the "ground state of matter", i.e. the lowest energy phase of matter in the Universe [23, 120].

are based on theoretical studies, and are only partially constrained by empirical data. Moreover, since hadronic matter at these densities is described by the non-perturbative regime of quantum chromo-dynamics, our theoretical understanding of its behaviour is also limited. Astrophysical and gravitational wave observations of neutron stars will provide information to constrain these models, and eventually understand the behaviour of matter in the inner core of these stars (see Box 16-I). A detailed description of the equations of state of neutron stars is beyond the scope of this textbook; the interested reader can find further information e.g. in [77, 53, 50].

The typical mass and radius predicted by the EoS proposed in the literature range roughly within $M \sim [1-3]\,M_\odot$ and $R \sim [10-15]$ km, and the stellar compactness is of the order of

$$\frac{GM}{c^2R} \sim 10^{-1} \tag{16.105}$$

(we remind that $\frac{GM_\odot}{c^2} \sim 1.47\,\text{km}$); therefore, to determine the structure of a neutron star General Relativity must be used. In what follows we shall first introduce the tools that are needed to describe perfect fluids in General Relativity; we shall then derive the equilibrium equations of a compact star, on the assumption that matter in the interior can be approximated as a perfect fluid.

In the rest of this chapter we shall use geometrized units $G = c = 1$.

16.3.3 Thermodynamics of perfect fluids in General Relativity

Let us consider a perfect fluid – i.e., a non-viscous fluid in which heat flux is absent – with fixed chemical composition and in thermodynamical equilibrium.

We can describe its (macroscopic) motion in terms of a stream of curves, which are worldlines of small fluid elements, i.e. in terms of a *four-velocity field* $u^\alpha(x)$ which is tangent in each point to these worldlines.

As in non-relativistic fluid dynamics and thermodynamics, the fluid elements are such that their three-volume is much larger than the scale of the microphysical interactions (for instance, nuclear interactions) [11], but it is small with respect to the macroscopic length-scales of the system. In particular, we require that the three-volume of the fluid element is small with respect to the length-scale of the gravitational interaction (i.e., the curvature radius) so that it can be covered by a single LIF. It is then possible to approximate each fluid element with a point particle located, for instance, at its center of mass.

Let us consider a fluid element located at the space point Q_0 at a time t_0. We can define a LIF centered in the event $\mathbf{q_0} = (t_0, Q_0)$, and such that the fluid element in $\mathbf{q_0}$ is at rest, i.e. the frame is *comoving* with the fluid element. Thus in this frame the four-velocity of the fluid in $\mathbf{q_0}$ is simply $u^\mu = (1, 0, 0, 0)$. In the following we shall indicate this frame as LICF (locally inertial comoving frame). We have assumed that the three-volume of the fluid element is small enough to be covered by the LICF and thus, for a short enough time interval, its four-volume is also covered by the LICF (see Fig. 16.3).

It is always possible to define a LICF, because the Equivalence Principle guarantees the existence of a LIF centered in $\mathbf{q_0}$; a Lorentz transformation (which maps a LIF into another LIF, see Sec. 1.5) can make it comoving with the fluid element in $\mathbf{q_0}$. We remark that the LICF should not be confused with the Fermi frame of the fluid element (see Sec. 3.9): the Fermi frame is comoving with the fluid element along its entire motion, but it is not locally inertial (unless the motion of the fluid is geodesic); conversely, the LICF is locally inertial,

[11] It is also required that the three-volume is large enough to contain a sufficiently large number of particles to properly define thermodynamical variables.

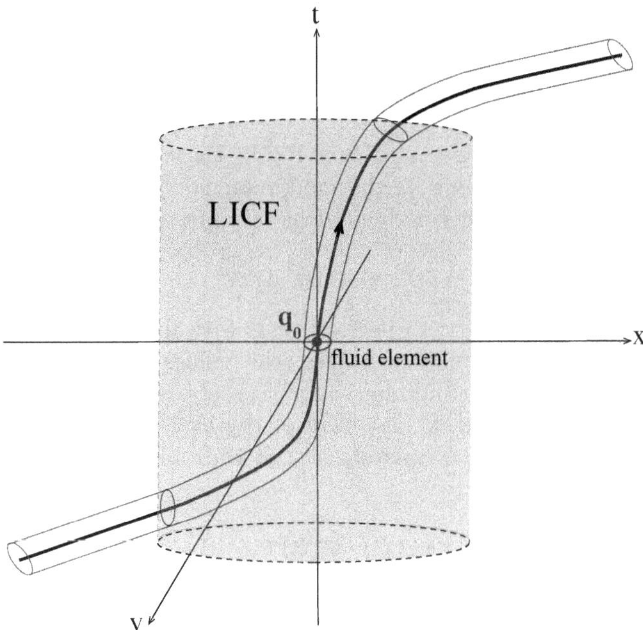

Figure 16.3: The LICF around an event $\mathbf{q_0} = (t_0, Q_0)$ in the fluid. For simplicity, we only show $2+1$ dimensions (two for space, one for time).

but it is only *momentarily* comoving with the fluid element, i.e. in a short time interval around t_0.

According to the Equivalence Principle, the fluid in the LICF is governed by the laws of Special Relativity for a fluid at rest. Since the fluid is in thermodynamical equilibrium, we can define the following thermodynamical quantities *measured in the LICF*:

- the number of baryons per unit of volume, or *baryon number density* n;
- the *energy density* ϵ, which includes the rest-mass energy and the potential energy of the interactions;
- the *pressure*[12] p;
- the *temperature* T;
- the *entropy per baryon* s.

Note that we define the baryon number density, and not the particle number density, because the number of particles can change when they interact, while the baryon number is conserved by all interactions. If we assume that the star does not contain antimatter and that the mesons content is negligible, the baryon number[13] coincides with the actual number of baryons. Since baryons are much heavier than electrons and neutrinos, the star *rest mass*

[12]To hereafter, at variance with Sec. 16.2, we denote the fluid pressure and energy density with a lowercase p and ϵ, respectively.

[13]The baryon number is $B = \dfrac{n_q - n_{\bar{q}}}{3}$ where n_q is the number of quarks, and $n_{\bar{q}}$ is the number of antiquarks, and is a conserved quantum number.

is due, with a good approximation, to baryons only. We also define the baryon rest-mass density as $\rho = m_n n$, where m_n is the neutron mass.

If we consider a different point Q of the fluid, and a different time t, we can always construct a LICF around the event $\mathbf{q} = (t, Q)$, and define the thermodynamical quantities n, ϵ, p, T, s in $\mathbf{q}$.

As discussed in Sec. 16.2.4 (see also Box 16-C), the *equation of state* of a thermodynamical system is a relation among its thermodynamical variables. In the case of a perfect fluid, the EoS gives one of the thermodynamical variables in terms of two of the others, for instance

$$\epsilon = \epsilon(p, s)\,. \tag{16.106}$$

This reflects a remarkable property of perfect fluids with fixed chemical composition: given the values of two thermodynamical variables, the values of all the others are uniquely determined. In other words, the thermodynamical state is completely determined by two thermodynamical variables. The explicit form of the EoS encodes the information on the microphysics of the system, and ultimately on the hadronic interactions occurring at high density.

Baryon number conservation law

The first of the equations describing fluid thermodynamics in General Relativity is the *baryon number conservation law*, also called *continuity equation*:

$$(nu^\alpha)_{;\alpha} = 0\,. \tag{16.107}$$

In order to derive Eq. 16.107 let us consider a fluid volume V passing through $\mathbf{q_0} = (t_0, Q_0)$, small enough to be contained in the domain of the LICF in $\mathbf{q_0}$. This volume contains a number $A = nV$ of baryons, which does not change with time due to the conservation of baryon number, i.e.

$$\frac{d}{d\tau}(nV) = 0 \tag{16.108}$$

where τ is the proper time of the fluid element at the LICF origin. This equation is not covariant, but in the following we shall show that it can be rewritten as Eq. 16.107, which is a tensor equation valid in any reference frame.

Let $\{x^\alpha\} = \{t, x^1, x^2, x^3\}$ be the coordinates of the LICF, and let us assume for simplicity that V is a cube of edges $\Delta x^1 = \Delta x^2 = \Delta x^3 = L$ and that the LICF origin, $Q_0 = (0, 0, 0)$, is chosen as in Fig. 16.4. In Q_0 the fluid is at rest, but inside the volume it has a small velocity (relative to Q_0)

$$v^i = \frac{dx^i}{dt}\,. \tag{16.109}$$

We expand Eq. 16.108 as follows

$$\frac{d}{d\tau}(nV) = u^\alpha\,(nV)_{,\alpha} = n_{,\alpha}Vu^\alpha + nV_{,\alpha}u^\alpha = 0\,. \tag{16.110}$$

Let us first evaluate the order of the various terms in Eq. 16.110. In the LICF (and thus in the small volume V) $g_{\mu\nu} = \eta_{\mu\nu} + O(x^2)$ (see Box 3-A); moreover, since the LICF is a comoving frame, $v^i(Q_0) = 0$ and thus

$$v^i = \frac{\partial v^i}{\partial x^j}x^j + O(x^2) = O(x)\,. \tag{16.111}$$

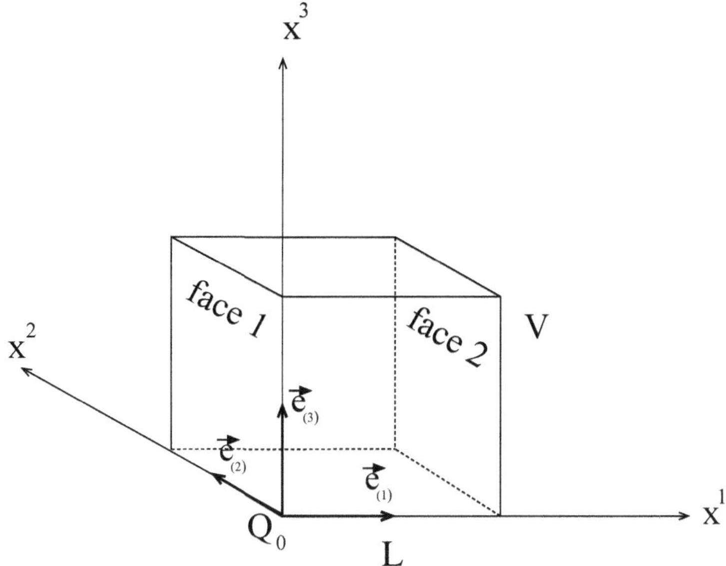

Figure 16.4: The fluid volume V in a LICF at some fixed instant of time.

Therefore,

$$d\tau = \sqrt{-g_{\alpha\beta}dx^\alpha dx^\beta} = dt\sqrt{-g_{00} + 2g_{0i}v^i + g_{ij}v^iv^j} = dt\sqrt{1 + O(x^2)} = dt + O(x^2)\,. \tag{16.112}$$

Consequently, the components of u^α read

$$\begin{aligned} u^0 &= \frac{dt}{d\tau} = 1 + O(x^2) \\ u^i &= \frac{dx^i}{d\tau} = v^i\left[1 + O(x^2)\right] = O(x)\,. \end{aligned} \tag{16.113}$$

The term $V_{,\alpha}u^\alpha$ in Eq. 16.110 can then be written as

$$V_{,\alpha}u^\alpha = V_{,0}u^0 + V_{,i}u^i = V_{,0} + O(x)\,. \tag{16.114}$$

In order to determine $V_{,0}$, we shall evaluate how the volume V changes in a time interval δt. The separation between the two faces orthogonal to the x^1 axis (see Fig. 16.4), Δx^1, change as

$$\delta(\Delta x^1) = \Delta(\delta x^1) = \left[\frac{dx^1}{dt}\delta t\right]_{\text{face 2}} - \left[\frac{dx^1}{dt}\delta t\right]_{\text{face 1}} = \Delta v^1\delta t = \frac{\partial v^1}{\partial x^1}\Delta x^1\,\delta t\,. \tag{16.115}$$

Similarly,

$$\begin{aligned} \delta(\Delta x^2) &= \frac{\partial v^2}{\partial x^2}\Delta x^2\,\delta t \\ \delta(\Delta x^3) &= \frac{\partial v^3}{\partial x^3}\Delta x^3\,\delta t\,, \end{aligned} \tag{16.116}$$

and the volume $V = \Delta x^1 \Delta x^2 \Delta x^3$ changes as

$$\begin{aligned} \delta V &= \delta(\Delta x^1)\Delta x^2 \Delta x^3 + \delta(\Delta x^2)\Delta x^3 \Delta x^1 + \delta(\Delta x^3)\Delta x^1 \Delta x^2 \\ &= \sum_{i=1}^{3} \frac{\partial v^i}{\partial x^i} L^3 \, \delta t \end{aligned} \tag{16.117}$$

so that (leaving implicit the sum on $i = 1, \ldots, 3$)

$$\frac{\partial V}{\partial t} = V \frac{\partial v^i}{\partial x^i} \,. \tag{16.118}$$

Then, Eq. 16.114 becomes

$$V_{\alpha} u^{\alpha} = V_{,0} u^0 + V_{,i} u^i = V \frac{\partial v^i}{\partial x^i} + O(x) \,, \tag{16.119}$$

and since $u^i = v^i \left[1 + O(x^2)\right]$ we find

$$V_{\alpha} u^{\alpha} = V \frac{\partial u^{\alpha}}{\partial x^{\alpha}} + O(x) \,. \tag{16.120}$$

By replacing this term in Eqs. 16.110 and evaluating it in Q_0 we finally find

$$\frac{d}{d\tau}(nV) = n_{,\alpha} V u^{\alpha} + nV u^{\alpha}_{,\alpha} = V(nu^{\alpha})_{,\alpha} = 0 \quad \rightarrow \quad (nu^{\alpha})_{;\alpha} = 0 \,, \tag{16.121}$$

where we have used the property that ordinary and covariant derivative coincide in a LIF. We have thus derived Eq. 16.107, which is a tensor equation, satisfied in any reference frame.

The first law of thermodynamics

Let us consider a fluid element (small enough to be covered by a LICF) with energy density ϵ and entropy per baryon s; a fluid element of volume V (measured in the LICF), consisting of a number $A = nV$ of baryons, has energy $E = V\epsilon$ and entropy $S = As$. The first law of thermodynamics,

$$dE = -pdV + TdS \,, \tag{16.122}$$

can then be written as

$$d\left(\frac{A}{n}\epsilon\right) = -pd\left(\frac{A}{n}\right) + Td(As) \,. \tag{16.123}$$

Multiplying by n/A, we get

$$d\epsilon = \frac{\epsilon + p}{n} dn + nTds \,. \tag{16.124}$$

Using an EoS for instance in the form

$$\epsilon = \epsilon(n, s) \,, \tag{16.125}$$

by differentiating with respect to the thermodynamical variables, we find

$$\left(\frac{\partial \epsilon}{\partial n}\right)_s = \frac{\epsilon + p}{n} \qquad \text{and} \qquad \left(\frac{\partial \epsilon}{\partial s}\right)_n = nT \,, \tag{16.126}$$

where the subscript in the brackets indicates that the partial derivative is taken keeping the quantity in the subscript constant. The pressure and the temperature of the fluid can then be expressed as

$$p(n,s) = n\left(\frac{\partial \epsilon}{\partial n}\right)_s - \epsilon \tag{16.127}$$

$$T(n,s) = \frac{1}{n}\left(\frac{\partial \epsilon}{\partial s}\right)_n . \tag{16.128}$$

Thus, as anticipated above, given an EoS – namely a relation between one thermodynamical variable (ϵ in the previous example) and two others (n and s) – the remaining thermodynamical variables (p and T in the example) can be determined using the first law of thermodynamics.

Another important function which describes the fluid is the *chemical potential* μ, which is the energy per baryon required to insert a small extra quantity of fluid composed by δA baryons, in the fluid volume in the same thermodynamical state. The volume variation due to the introduction of the extra baryons is $\delta V = \delta A/n$ (n does not change since the variation occurs in the same thermodynamical state). The energy needed to insert the δA baryons is the sum of the energy of the baryons, $\epsilon\delta V$, and of the energy needed to change the volume of the fluid by δV, i.e. the work $p\delta V$. Consequently

$$\mu = \frac{p\delta V + \epsilon\delta V}{\delta A} = \frac{p+\epsilon}{n}, \tag{16.129}$$

and, from Eq. 16.126, we get

$$\mu = \left(\frac{\partial \epsilon}{\partial n}\right)_s . \tag{16.130}$$

Barotropic equation of state

If the EoS does not depend on temperature or entropy, i.e. $\epsilon = \epsilon(n)$, it is named *barotropic* (see also Sec. 16.2.4). In this case, using Eq. 16.127 the EoS can also be written as

$$p = p(\epsilon) \qquad \text{or} \qquad \epsilon = \epsilon(p) \tag{16.131}$$

i.e., an equation relating the pressure and the energy density. Since $\left(\frac{\partial \epsilon}{\partial s}\right)_n = 0$, from Eq. 16.128 it follows that the temperature of a barotropic fluid is zero.

For a barotropic EoS, the first law of thermodynamics becomes

$$d\epsilon = \frac{\epsilon + p}{n} dn . \tag{16.132}$$

By differentiating the definition of the chemical potential, Eq. 16.129, and using Eq. 16.132 we find

$$d\mu = -\frac{\epsilon + p}{n^2}\, dn + \frac{d\epsilon + dp}{n} = \frac{dp}{n}; \tag{16.133}$$

therefore for a barotropic EoS we also have

$$dp = n d\mu . \tag{16.134}$$

The polytropic EoS which we have used to model white dwarfs is barotropic. As discussed in Sec. 16.3.2, a neutron star after about one year from birth reaches temperatures much smaller than the Fermi temperature of neutrons in the core; therefore they can be treated as having vanishing temperature and modeled using a barotropic EoS. Main-sequence stars (like our Sun), or newly born, hot neutron stars, conversely, have a non-barotropic EoS with non-trivial dependence on temperature and entropy.

16.3.4 The stress-energy tensor of a perfect fluid

In a perfect fluid, both viscosity and heat flow are absent. As explained in Chapter 5, in a LIF T^{00} is the energy density, T^{0i} $(i = 1, 2, 3)$ is the energy which flows per unit time across the unit surface orthogonal to the axis x^i, and T^{ij} is the amount of the i-th-component of momentum which flows per unit time across the unit surface orthogonal to the axis x^j. In the non-relativistic limit, this quantity is the i-th-component of the *force* per unit surface orthogonal to the axis x^j.

Let us consider a fluid element and the associated LICF. In this frame the fluid is at rest and the components of the stress-energy tensor are the following

- $T^{00} = \epsilon$, the energy density.
- $T^{0i} = 0$, indeed the fluid element does not exchange energy with its surroundings, because there is no heat flow.
- $T^{ij} = p\delta^{ij}$, indeed in a perfect fluid no tangential stresses are allowed, which means that the force exerted on the surface orthogonal to the axis x^j must be orthogonal to the surface, and this force per unit surface is, by definition, the pressure.

Thus, in the LICF the expression of the fluid stress-energy tensor is

$$T^{\mu\nu} = \begin{pmatrix} \epsilon & 0 & 0 & 0 \\ 0 & p & 0 & 0 \\ 0 & 0 & p & 0 \\ 0 & 0 & 0 & p \end{pmatrix}, \tag{16.135}$$

and, since in this frame the four-velocity of the fluid element is $u^\mu = (1, 0, 0, 0)$, it can also be written as

$$T^{\mu\nu} = (\epsilon + p)u^\mu u^\nu + p\eta^{\mu\nu} . \tag{16.136}$$

In addition, in a LICF $g^{\mu\nu} \equiv \eta^{\mu\nu}$, therefore we can also write

$$T^{\mu\nu} = (\epsilon + p)u^\mu u^\nu + pg^{\mu\nu} . \tag{16.137}$$

This is a tensorial expression and, by the General Covariance Principle, it must be valid in any other reference frame. Thus, Eq. 16.137 is the *covariant form for the stress-energy tensor of a perfect fluid in General Relativity.* Note that although Eq. 16.137 holds in any frame, ϵ and p are defined as the energy density and the pressure of the fluid element measured by a locally inertial, comoving observer. These quantities are, by definition, *scalar fields* on the manifold, and their value in a given point does not depend on the coordinate frame. Also note that according to Eq. 16.135, which follows from the assumption that viscosity and heat flow are absent, a comoving observer sees the fluid as isotropic.

Fluid equations from the stress-energy tensor

The stress-energy tensor 16.137 satisfies the divergence-free equation (see Chapter 5)

$$T^{\mu\nu}{}_{;\nu} = 0 . \tag{16.138}$$

Its contraction with u_μ yields

$$\begin{aligned} u_\mu T^{\mu\nu}{}_{;\nu} &= u_\mu u^\mu u^\nu(\epsilon + p)_{,\nu} + (\epsilon + p)(u_\mu u^\mu u^\nu{}_{;\nu} + u_\mu u^\mu{}_{;\nu} u^\nu) + u^\nu p_{,\nu} \\ &= -u^\nu \epsilon_{,\nu} - (\epsilon + p)u^\nu{}_{;\nu} = 0 , \end{aligned} \tag{16.139}$$

where we have used the relation

$$u_\mu (u^\mu)_{;\nu} = \frac{1}{2}(u_\mu u^\mu)_{;\nu} = 0\,. \tag{16.140}$$

Using the baryon number conservation 16.107, Eq. 16.139 gives

$$u^\nu \epsilon_{,\nu} = -(\epsilon + p) u^\nu{}_{;\nu} = \frac{\epsilon + p}{n} u^\nu n_{,\nu}\,, \tag{16.141}$$

i.e.

$$\frac{d\epsilon}{d\tau} = \frac{\epsilon + p}{n}\frac{dn}{d\tau}\,. \tag{16.142}$$

On the other hand, from the first law of thermodynamics (Eq. 16.124),

$$\frac{d\epsilon}{d\tau} = \frac{\epsilon + p}{n}\frac{dn}{d\tau} + nT\frac{ds}{d\tau}\,, \tag{16.143}$$

and the two equations are compatible only if

$$\frac{ds}{d\tau} = 0\,, \tag{16.144}$$

which reflects the fact that in a perfect fluid a fluid element does not exchange heat with its surroundings. Thus, the divergenceless equation of the stress-energy tensor and the baryon number conservation imply that a perfect fluid is *isentropic*. Note that we have not assumed that the EoS is barotropic; this property applies both to cold and warm perfect fluids.

To study the space components of Eq. 16.138 we define the projector onto the subspace orthogonal to u^μ,

$$P_{\mu\nu} = g_{\mu\nu} + u_\mu u_\nu\,. \tag{16.145}$$

This is a projector because $P^2 = P$, and it projects onto the subspace orthogonal to u^μ since $P_{\mu\nu}u^\nu = 0$. By applying $P_{\mu\nu}$ to Eq. 16.138 we find

$$\begin{aligned}
P_{\gamma\alpha}T^{\alpha\beta}{}_{;\beta} &= P_{\gamma\alpha}\left[(\epsilon + p)_{,\beta}u^\alpha u^\beta + (\epsilon + p)(u^\beta u^\alpha{}_{;\beta} + u^\alpha u^\beta{}_{;\beta}) + g^{\alpha\beta}p_{,\beta}\right] \\
&= (g_{\gamma\alpha} + u_\gamma u_\alpha)(\epsilon + p)u^\beta u^\alpha{}_{;\beta} + P_\gamma{}^\beta p_{,\beta} \\
&= (\epsilon + p)u^\beta u_{\gamma;\beta} + P_\gamma{}^\beta p_{,\beta} = 0\,,
\end{aligned} \tag{16.146}$$

where we have used Eq. 16.140. This equation gives

$$P_\gamma{}^\beta p_{,\beta} = -(\epsilon + p)u^\beta u_{\gamma;\beta}\,, \tag{16.147}$$

and shows that the pressure gradient projected on the subspace orthogonal to u^μ (that is, the space gradient of the pressure) is equal to the fluid acceleration, $u^\beta u_{\gamma;\beta}$, multiplied by the sum of the energy density and the pressure; this is the relativistic generalization of one of Euler's equations.

16.3.5 The equations of stellar structure in General Relativity

In this section we shall derive the equations describing the structure of a non-rotating star in static equilibrium in General Relativity. Since the spacetime generated by the star is static and spherically symmetric, as discussed in Chapter 9 the appropriate form of the metric is given by Eq. 9.15, i.e.

$$ds^2 = -e^{2\nu(r)}dt^2 + e^{2\lambda(r)}dr^2 + r^2(d\theta^2 + \sin^2\theta d\varphi^2)\,. \tag{16.148}$$

We shall assume that the interior of the star is described by a perfect fluid, with stress-energy tensor given by Eq. 16.137:

$$T^{\alpha\beta} = (\epsilon + p)u^\alpha u^\beta + p g^{\alpha\beta}, \tag{16.149}$$

where $u^\alpha = \frac{dx^\alpha}{d\tau}$ is the four-velocity of a fluid element, and p and ϵ are the pressure and the energy density, as discussed in the previous section. The pressure and energy density are related by an assigned EoS.

It should be stressed that ϵ is the *relativistic energy density* which, in the non-relativistic limit, reduces to the rest energy density ρc^2 (in physical units), where ρ is the mass density.

At this point some considerations about the dimensions of the physical quantities are in order. Since we are working in $G = c = 1$ units, $T_{\mu\nu}$ has the same dimensions as $G_{\mu\nu}$, i.e.

$$[T_{\mu\nu}] = (\text{length})^{-2}. \tag{16.150}$$

Consistently, both ϵ and p have $(\text{length})^{-2}$ dimensions. In physical units,

$$\epsilon = \frac{G}{c^4}\epsilon_{\text{phys}}, \qquad \text{and} \qquad p = \frac{G}{c^4}p_{\text{phys}}, \tag{16.151}$$

$[\epsilon_{\text{phys}}] = [p_{\text{phys}}] = (\text{mass})(\text{length})^{-1}(\text{time})^{-2}$.

Since by assumption the fluid is at rest, the only non-vanishing component of the velocity of the generic fluid element is u^0 and the normalization of u^μ gives

$$g_{\mu\nu}u^\mu u^\nu = -1 \qquad \to \qquad u^t = e^{-\nu}, \qquad u_t = -e^\nu. \tag{16.152}$$

Therefore, the non-vanishing covariant and contravariant components of the stress-energy tensor read

$$\begin{aligned} T_{tt} &= \epsilon\, e^{2\nu} & T_{\theta\theta} &= r^2\, p \\ T_{rr} &= p\, e^{2\lambda} & T_{\varphi\varphi} &= \sin^2\theta\, T_{\theta\theta}, \end{aligned} \tag{16.153}$$

and

$$\begin{aligned} T^{tt} &= \epsilon\, e^{-2\nu} & T^{\theta\theta} &= \tfrac{p}{r^2} \\ T^{rr} &= p\, e^{-2\lambda} & T^{\varphi\varphi} &= \tfrac{1}{\sin^2\theta}\, T^{\theta\theta}. \end{aligned} \tag{16.154}$$

Thus, the field equations are

$$G_{\mu\nu} = 8\pi T_{\mu\nu} \tag{16.155}$$

$$T^{\mu\nu}{}_{;\nu} = \frac{1}{\sqrt{-g}}\frac{\partial}{\partial x^\nu}\left(\sqrt{-g}T^{\mu\nu}\right) + \Gamma^\mu_{\lambda\nu}T^{\nu\lambda} = 0. \tag{16.156}$$

It should be noted that Eqs. 16.155, 16.156 are not independent. Indeed, as discussed in Chapter 6, the divergence-free equation satisfied by the stress-energy tensor is a consequence of the Bianchi identities satisfied by the Riemann tensor.

In order to write explicitly Eq. 16.156 we need the expression of Christoffel's symbols for the metric 16.148 which has been computed in Box 9-A, and we rewrite here for convenience:

$$\begin{aligned} &\Gamma^t_{tr} = \nu_{,r}, && \Gamma^r_{tt} = e^{2(\nu-\lambda)}\nu_{,r}, && \Gamma^r_{rr} = \lambda_{,r}, \\ &\Gamma^r_{\theta\theta} = -e^{-2\lambda}r, && \Gamma^r_{\varphi\varphi} = -e^{-2\lambda}r\sin^2\theta \\ &\Gamma^\theta_{r\theta} = \tfrac{1}{r}, && \Gamma^\theta_{\varphi\varphi} = -\cos\theta\sin\theta \\ &\Gamma^\varphi_{\theta\varphi} = \tfrac{\cos\theta}{\sin\theta}, && \Gamma^\varphi_{r\varphi} = \frac{1}{r}. \end{aligned} \tag{16.157}$$

In addition

$$\sqrt{-g} = r^2 e^{\nu+\lambda}\sin\theta. \tag{16.158}$$

The only non-trivial component of Eq. 16.156 is for $\mu = r$ and gives

$$\frac{1}{\sqrt{-g}}\frac{\partial}{\partial x^\nu}\left(\sqrt{-g}T^{r\nu}\right) + \Gamma^r_{\lambda\nu}T^{\nu\lambda} = 0\,, \tag{16.159}$$

i.e.

$$\begin{aligned} &\frac{1}{\sqrt{-g}}\frac{\partial}{\partial r}\left(\sqrt{-g}T^{rr}\right) + \Gamma^r_{tt}T^{tt} + +\Gamma^r_{rr}T^{rr} + \Gamma^r_{\theta\theta}T^{\theta\theta} + \Gamma^r_{\varphi\varphi}T^{\varphi\varphi} = \\ &\frac{e^{-(\nu+\lambda)}}{r^2}\left(r^2 e^{(\nu+\lambda)} p e^{-2\lambda}\right)_{,r} + e^{2(\nu-\lambda)}\nu_{,r}\epsilon e^{-2\nu} + e^{-2\lambda}\lambda_{,r}p - 2\frac{e^{-2\lambda}p}{r} = 0\,, \end{aligned} \tag{16.160}$$

which reduce to

$$\nu_{,r} = -\frac{p_{,r}}{\epsilon + p}\,. \tag{16.161}$$

Einstein's equations (see Eqs. 9.18,...,9.21) give

$$G_{tt} = 8\pi T_{tt} \rightarrow \quad \frac{1}{r^2}e^{2\nu}\frac{d}{dr}\left[r\left(1 - e^{-2\lambda}\right)\right] = 8\pi\epsilon\, e^{2\nu}\,, \tag{16.162}$$

$$G_{rr} = 8\pi T_{rr} \rightarrow \quad -\frac{1}{r^2}e^{2\lambda}\left(1 - e^{-2\lambda}\right) + \frac{2}{r}\nu_{,r} = 8\pi p\, e^{2\lambda}\,, \tag{16.163}$$

$$G_{\theta\theta} = 8\pi T_{\theta\theta} \rightarrow \quad r^2 e^{-2\lambda}\left[\nu_{,rr} + \nu_{,r}^2 + \frac{\nu_{,r}}{r} - \nu_{,r}\lambda_{,r} - \frac{\lambda_{,r}}{r}\right] = 8\pi r^2\, p\,. \tag{16.164}$$

If we define

$$m(r) \equiv \frac{1}{2}r\left(1 - e^{-2\lambda(r)}\right) \qquad \rightarrow \qquad e^{-2\lambda(r)} = 1 - \frac{2m(r)}{r}\,, \tag{16.165}$$

Eq. 16.162 becomes

$$\frac{dm}{dr} = 4\pi r^2\,\epsilon\,, \tag{16.166}$$

which is the generalization of the Newtonian equation 16.56. Furthermore, Eq. 16.163 can be rewritten as

$$\frac{\left(1 - e^{-2\lambda}\right)}{r^2} - \frac{2}{r}e^{-2\lambda}\nu_{,r} = -8\pi\, p\,, \tag{16.167}$$

which, using Eq. 16.165, becomes

$$\frac{2}{r}\left(1 - \frac{2m}{r}\right)\nu_{,r} = \frac{2m}{r^3} - 8\pi\, p\,, \tag{16.168}$$

which gives

$$\nu_{,r} = \frac{m + 4\pi r^3 p}{r\left(r - 2m\right)}\,. \tag{16.169}$$

In the Newtonian limit the pressure in geometrized units is small compared to the energy density. For example in the case of the Sun, the pressure-to-density ratio at the center is $\sim 10^{-6}$. In addition $m(r) \ll r$ and Eq. 16.169 reduces to

$$\nu_{,r} \simeq \frac{m(r)}{r^2}\,. \tag{16.170}$$

Recalling that in this limit $e^{2\nu} \rightarrow 1 + 2\Phi$, where Φ is the Newtonian potential and $\Phi_{,r} = \frac{m(r)}{r^2}$, Eq. 16.170 simply shows the Newtonian result that the gravitational force at r is exerted by the mass enclosed within a sphere of radius r. Eq. 16.169 shows that in General Relativity there is an additional contribution, $4\pi r^3 p$, which is due to the pressure: this term acts as an *effective mass* (note that p and ϵ have the same dimensions). This means that the gravitational attraction exerted on the mass shell between r and $r+dr$ is due to both contributions, and the pressure – which should contrast gravity – to some extent enhances its effects. This phenomenon is called *regeneration of the pressure.* Finally, Eqs. 16.161 and 16.169 can be combined, and the final set of equations, known as the **Tolman-Oppenheimer-Volkoff (TOV) equations**, is

$$\begin{cases} \dfrac{dm}{dr} = 4\pi r^2\,\epsilon \\ \\ \dfrac{dp}{dr} = -\dfrac{(\epsilon+p)[m(r)+4\pi r^3\,p]}{r[r-2m(r)]}\,. \end{cases} \tag{16.171}$$

The above system is closed if we assign a barotropic EoS of the form $p = p(\epsilon)$ (or $\epsilon = \epsilon(p)$). If this assumption is not satisfied, the pressure may depend on other quantities such as the entropy or the chemical composition; in this case the TOV equations have to be supplemented by further equations which describe the behaviour of the different quantities.

Boundary conditions of the TOV equations

The system of equations 16.171, supplemented with a barotropic EoS, consists of two first-order, ordinary differential equations. As such, it requires two boundary conditions at some point (respectively on $m(r)$ and $p(r)$), for instance at the center of the star. The first boundary condition is $m(r=0)=0$. Indeed, if we take a tiny sphere of radius $r = x$, the proper circumference is $2\pi x$ and the proper radius is

$$\int_0^x e^\lambda dr \simeq e^\lambda x\,, \tag{16.172}$$

hence their ratio is $2\pi e^{-\lambda}$. Since, by the Equivalence Principle, the spacetime is locally flat the ratio between the circumference of this infinitesimal sphere and the radius must be 2π. This implies that $e^\lambda \to 1$ as $r \to 0$. Since

$$e^{2\lambda} = \frac{1}{1-\frac{2m}{r}}\,, \tag{16.173}$$

it follows that $m(0) = 0$. This result can also be obtained by solving the first of the TOV equations using a Taylor expansion for $m(r)$ and $\epsilon(r)$ near $r \sim 0$, which shows that $m(r) \sim r^3$.

If the EoS is barotropic, for any assigned EoS in the form $p = p(\epsilon)$, we have a one-parameter family of solutions identified by the value of the energy density at $r = 0$, i.e. $\epsilon(r=0) = \epsilon_c$. Therefore, the second boundary condition is $p(r=0) = p(\epsilon_c)$ for a given EoS.

The functions $m(r)$, $p(r)$, and $\epsilon(r)$ can be determined by numerical integration; the radius of the star $r = R$ is, by definition, the surface where the pressure vanishes: $p(R) = 0$. Once $m(r)$, $p(r)$, $\epsilon(r)$, and R have been found, $e^{2\lambda(r)}$ follows from Eq. 16.173, whereas $\nu(r)$ can be found by integrating Eq. 16.161, which gives

$$\nu(r) = -\int_0^r \frac{p_{,r'}}{(\epsilon+p)}dr' + \nu_0\,. \tag{16.174}$$

The constant $\nu_0 = \nu(0)$ can be determined as follows. The solution of Eqs. 16.171, 16.173 and

16.174 describes the gravitational field and the distribution of pressure and energy density *inside the star*. Outside the star (i.e., at $r > R$) $p = \epsilon = 0$ and Einstein's equations reduce to those for a vacuum, static, spherically symmetric spacetime whose unique asymptotically flat solution is, by Birkhoff's theorem, the Schwarzschild metric (see Sec. 9.6). Thus, when $r = R$ the metric computed in the interior of the star must reduce to the Schwarzschild metric, and by imposing this condition the constant ν_0 in Eq. 16.174 can be found:

$$e^{2\nu(R)} \equiv e^{2\nu_0} e^{-2\int_0^R \frac{p_{,r}}{(\epsilon+p)} dr} = 1 - \frac{2m(R)}{R} \qquad \rightarrow \qquad e^{2\nu_0} = \frac{1 - \frac{2m(R)}{R}}{e^{2\int_0^R - \frac{p_{,r}}{(\epsilon+p)} dr}} . \tag{16.175}$$

The quantity

$$M \equiv m(R) = 4\pi \int_0^R r^2 \epsilon(r)\, dr \tag{16.176}$$

has the same expression as the mass in terms of the density in Newtonian theory, and is the *total mass-energy* inside the radius R. This interpretation will be further clarified in Chapter 17, where we will show that, using the far-field limit of an isolated object, M can be obtained as an integral, over a large space proper volume, of the conserved quantity $(-g)\left(T^{00} + t^{00}\right)$.

We can split M in three contributions

$$M = E_{\text{rest}} + E_{\text{int}} + E_B\,. \tag{16.177}$$

The first contribution, E_{rest}, is the *rest mass-energy* of the star, i.e. the integral over the proper volume element

$$dV_{\text{prop}} = \sqrt{g^{(3)}} dr d\theta d\varphi = e^{\lambda(r)} r^2 \sin\theta dr d\theta d\varphi \tag{16.178}$$

of the baryon rest-mass density $\rho = m_n n$ ($g^{(3)}$ is the determinant of the metric induced on the three-dimensional hypersurface $t = const$, see Box 7-A). We then get

$$\begin{aligned} E_{\text{rest}} &= \int_V m_n n(r)\, dV_{\text{prop}} \qquad (16.179)\\ &= \int_0^R e^{\lambda(r)} m_n n(r)\, r^2\, dr \int_0^\pi \sin(\theta)\, d\theta \int_0^{2\pi} d\varphi = 4\pi \int_0^R \frac{m_n n(r)\, r^2 dr}{\sqrt{1 - \frac{2m(r)}{r}}} . \end{aligned}$$

The function $n(r)$ in the above equation can be found using the pressure and energy density profiles, $p(r)$ and $\epsilon(r)$, and the first law of thermodynamics which, in the case of a barotropic EoS, gives (cf. Eqs. 16.124 and 16.144)

$$d\epsilon = \frac{\epsilon + p}{n} dn \qquad \rightarrow \qquad \frac{dn}{dr} = \frac{n}{\epsilon + p} \frac{d\epsilon}{dr} . \tag{16.180}$$

For a given barotropic EoS, $p(\epsilon)$, one can write

$$\frac{d\epsilon}{dr} = \frac{d\epsilon}{dp}\frac{dp}{dr} = \frac{1}{v_s^2}\frac{dp}{dr} , \tag{16.181}$$

where it can be shown that $v_s = \sqrt{dp/d\epsilon}$ is the speed of sound in the fluid. Substituting the second of the TOV equations 16.171 in the above equation, the explicit equation for $n(r)$ is

$$\frac{dn}{dr} = -\frac{n}{v_s^2} \frac{m + 4\pi r^3 p}{r[r - 2m]} . \tag{16.182}$$

The second term in Eq. 16.177, $E_{\rm int}$, is the *internal energy* (which includes various contributions, e.g. thermal, compressional, etc.), given by

$$E_{\rm int} = \int_V [\epsilon - m_n n] \, dV_{\rm prop} = 4\pi \int_0^R \frac{[\epsilon - m_n n] r^2 \, dr}{\sqrt{1 - \frac{2m}{r}}} \,. \tag{16.183}$$

The last contribution to the total mass-energy is the gravitational potential energy, i.e. the *binding energy* E_B, given by

$$E_B = m(R) - E_{\rm rest} - E_{\rm int} = 4\pi \int_0^R r^2 \epsilon [1 - e^{\lambda}] \, dr = 4\pi \int_0^R dr \, r^2 \epsilon \left[1 - \frac{1}{\sqrt{1 - \frac{2m}{r}}} \right] . \tag{16.184}$$

It is easy to see that $E_B < 0$, as required for a bound system.

Finally, we note that $r - 2m(r) > 0$ throughout the star. This follows from the second of the TOV equations 16.171,

$$\frac{dp}{dr} = -(\epsilon + p) \frac{m + 4\pi p r^3}{r(r - 2m)} \,. \tag{16.185}$$

Indeed, near the center of the star (i.e., at $r \simeq 0$) $r - 2m \simeq r - \frac{8}{3}\pi r^3 > 0$, and thus $\frac{dp}{dr} < 0$. As r increases, the pressure decreases, until it reaches $p = 0$ at $r = R$. Therefore, $p(r)$ is a positive, decreasing function of r. This implies that $r > 2m(r)$ at any point inside the star, because if at some point $r - 2m(r) = 0$, the pressure derivative diverges.

Consequently, $r > 2m$ and thus $e^{2\lambda} > 0$ in the entire star: at variance with black holes, a star does not have a horizon (see Chapter 9).

A note on the chemical potential

Let us consider a spherical star with a barotropic EoS $p = p(\epsilon)$; combining Eqs. 16.129 and 16.134 we find

$$dp = n d\mu = \frac{\epsilon + p}{\mu} d\mu \,. \tag{16.186}$$

By integrating this equation between two arbitrary radii (say r and r'), and using Eq. 16.174, we find

$$\int_{\mu(r)}^{\mu(r')} \frac{d\mu}{\mu} = \int_{p(r)}^{p(r')} \frac{dp}{\epsilon + p} = -\int_{\nu(r)}^{\nu(r')} d\nu = \nu(r) - \nu(r') \,. \tag{16.187}$$

Since the first integral gives $\log(\mu(r')/\mu(r))$, we obtain the relation

$$\mu(r) e^{\nu(r)} = \mu(r') e^{\nu(r')} \,, \tag{16.188}$$

which holds for any value of r and r'. Thus the chemical potential, corrected by the redshift factor e^{ν}, is a constant at any depth in the star. In particular, for any $r < R$ where R is stellar radius

$$\mu(r) e^{\nu(r)} = \left(1 - \frac{2M}{R}\right)^{1/2} \mu(R) \,. \tag{16.189}$$

16.3.6 The Schwarzschild solution for a homogeneous star

An analytic solution of the equations of stellar structure (Eqs. 16.171) can be obtained by considering the very simple EoS:

$$\epsilon = const \,. \tag{16.190}$$

This solution was found by Karl Schwarzschild in 1916, soon after he found the vacuum solution for a static, spherically symmetric spacetime.

Although homogeneous (i.e. uniform energy density) stars are unrealistic since the speed of sound $v_s = \left(\frac{dp}{d\epsilon}\right)^{1/2}$ diverges in their interior, they can be used as an approximation for the core of very dense stars, and this solution provides a simplified model to study the effects of gravity in a regime as strong as it can ever become under the condition of hydrostatic equilibrium. If $\epsilon = const$

$$m(r) = \frac{4}{3}\pi r^3 \epsilon\,, \tag{16.191}$$

and, from Eq. 16.165, one of the metric functions is immediately found to be

$$e^{2\lambda(r)} = \left(1 - \frac{2m(r)}{r}\right)^{-1} = \left(1 - \frac{8}{3}\pi\epsilon r^2\right)^{-1}. \tag{16.192}$$

Therefore, $m + 4\pi p r^3 = \frac{4\pi}{3}(\epsilon + 3p)r^3$, and the second of the TOV equations 16.171 reduces to

$$\frac{dp}{dr} = -\frac{4}{3}\pi r \frac{(\epsilon + p)(\epsilon + 3p)}{1 - \frac{8\pi}{3}r^2\epsilon}\,, \tag{16.193}$$

which gives

$$\frac{2\epsilon dp}{(\epsilon + p)(\epsilon + 3p)} = -\frac{8\pi}{3}\frac{\epsilon r dr}{1 - \frac{8\pi}{3}r^2\epsilon} = \frac{1}{2}\frac{d\left(1 - \frac{8\pi}{3}r^2\epsilon\right)}{1 - \frac{8\pi}{3}r^2\epsilon}\,. \tag{16.194}$$

Integrating between r and the radius of the star $r = R$, where the pressure has to vanish, we find the pressure as a function of r:

$$\log\left(\frac{\epsilon + 3p}{\epsilon + p}\right)\Bigg|_{p(r)}^{0} = \frac{1}{2}\log\left(1 - \frac{8\pi}{3}r^2\epsilon\right)\Bigg|_{r}^{R}. \tag{16.195}$$

This yields

$$\frac{\epsilon + p(r)}{\epsilon + 3p(r)} = \frac{y_1}{y(r)} \qquad \rightarrow \qquad p = \epsilon\frac{(y - y_1)}{(3y_1 - y)}\,, \tag{16.196}$$

where we have defined (remember that $M \equiv m(R)$)

$$y^2(r) = 1 - \frac{8\pi}{3}r^2\epsilon = 1 - \frac{2m(r)}{r}\,, \qquad \text{and} \qquad y_1^2 = y^2(R) = 1 - \frac{2M}{R}\,. \tag{16.197}$$

Eq. 16.196 evaluated in $r = 0$ gives the central pressure

$$p_c \equiv p(r = 0) = \epsilon\frac{1 - \sqrt{1 - \frac{2M}{R}}}{3\sqrt{1 - \frac{2M}{R}} - 1}\,. \tag{16.198}$$

Note that while the numerator of this expression is always finite and positive, the denominator can in principle be negative. Since the pressure can not be negative, homogeneous stars can exist only if the denominator of Eq. 16.198 is positive, i.e. for

$$3\sqrt{1 - \frac{2M}{R}} - 1 > 0 \qquad \rightarrow \qquad \frac{M}{R} < \frac{4}{9}\,, \tag{16.199}$$

or, equivalently,

$$R > \frac{9}{4}M\,. \tag{16.200}$$

This constraint sets a lower limit on the radius that a star of a given mass can have, and an upper limit on its *compactness* M/R (see Box 11-A), provided $\epsilon = const$. In the next section we will show that this result holds for a generic EoS under certain assumptions.

The radius of the star can be found by replacing $p(0) = p_c$ and $y(0) = 1$ in Eq. 16.196:

$$\frac{\epsilon + p_c}{\epsilon + 3p_c} = y_1 = \sqrt{1 - \frac{8\pi}{3}R^2\epsilon} \qquad \rightarrow \qquad \frac{8\pi}{3}R^2\epsilon = 1 - \frac{(\epsilon + p_c)^2}{(\epsilon + 3p_c)^2} \tag{16.201}$$

from which we find

$$R = \left[\frac{3}{8\pi\epsilon}\left(1 - \frac{(\epsilon + p_c)^2}{(\epsilon + 3p_c)^2}\right)\right]^{1/2}. \tag{16.202}$$

Thus, for any assigned value of ϵ and of $p_c > 0$, there exists a configuration of radius R given by 16.202 and mass $M = \frac{4\pi}{3}r^3\epsilon$.

To complete the solution we need to find the metric function $\nu(r)$, which can be determined from Eq. 16.174

$$\nu(r) = \nu_0 - \int_0^r \frac{p_{,r'}}{(\epsilon + p(r'))}dr' = \nu_0 - \int_{p_c}^{p(r)} \frac{dp}{\epsilon + p} = -\log\frac{\epsilon + p(r)}{\epsilon + p_c}; \tag{16.203}$$

since (see Eq. 16.196)

$$\epsilon + p = \epsilon\frac{2y_1}{(3y_1 - y)}, \tag{16.204}$$

Eq. 16.203 gives

$$\nu = \nu_0 - \log\frac{3y_1 - 1}{3y_1 - y} \qquad \rightarrow \qquad e^{2\nu} = e^{2\nu_0}\left(\frac{3y_1 - y}{3y_1 - 1}\right)^2. \tag{16.205}$$

At the boundary of the star $y(R) = y_1$ and

$$e^{2\nu(R)} = e^{2\nu_0}\frac{4y_1^2}{(3y_1 - 1)^2}; \tag{16.206}$$

by imposing that for $r \geq R$ the metric reduces to the Schwarzschild metric in vacuum, we find

$$e^{2\nu(R)} = 1 - \frac{2M}{R} \equiv y_1^2. \tag{16.207}$$

The value of the integration constant ν_0 is then found by equating Eq. 16.206 and Eq. 16.207

$$e^{2\nu_0} = \frac{(3y_1 - 1)^2}{4}, \tag{16.208}$$

and the solution for $\nu(r)$ is

$$e^{2\nu(r)} = \frac{(3y_1 - y)^2}{4}. \tag{16.209}$$

As required by Birkhoff's theorem, the metric functions 16.192 and 16.209 outside the star reduce to those of the Schwarzschild metric in vacuum.

16.3.7 Relativistic polytropes

In this section we shall generalize the Lane-Emden equation found in Sec 16.2.5 to General Relativity. To this purpose we extend the polytropic EoS introduced in Sec. 16.2.4, $p = K\rho^\gamma$, to the general relativistic expression

$$p = K\epsilon^\gamma \tag{16.210}$$

where ϵ is the relativistic energy density and $\gamma = const$ is the polytropic index[14]. Following what we did in Sec. 16.2.5, we shall show how to solve the relativistic equations of stellar structure (Eqs. 16.171) assuming that the EoS of matter inside the star is given by Eq. 16.210

$$\begin{cases} \dfrac{dm}{dr} = 4\pi r^2 \epsilon \\[2ex] \dfrac{dp}{dr} = -\dfrac{(\epsilon + p)[m + 4\pi r^3 p]}{r[r-2m]} \\[2ex] p = K\epsilon^{\gamma} . \end{cases} \tag{16.211}$$

The above system can be integrated numerically with standard methods. However, it is useful to introduce dimensionless quantities as done in Sec. 16.2.5. We shall make the ansatz:

$$\begin{cases} \gamma = 1 + \dfrac{1}{n} , \\[2ex] \epsilon = \epsilon_c \ \Theta^n(r) , \end{cases} \tag{16.212}$$

from which it follows that

$$p = K \ \epsilon_c^{1+\frac{1}{n}} \ \Theta^{(n+1)}(r) = p_c \ \Theta^{(n+1)}(r) , \qquad \text{where} \qquad p_c = K \ \epsilon_c^{1+\frac{1}{n}} . \tag{16.213}$$

With these substitutions Eqs. 16.211 become

$$\begin{cases} \dfrac{dm}{dr} = 4\pi\epsilon_c r^2 \Theta^n \\[2ex] \dfrac{d\Theta}{dr} = -\dfrac{\epsilon_c + p_c\Theta}{p_c(n+1)} \dfrac{m + 4\pi r^3 p_c \ \Theta^{(n+1)}}{r(r-2m)} , \end{cases} \tag{16.214}$$

which, setting

$$\alpha_0 = \frac{\epsilon_c}{p_c} , \tag{16.215}$$

yield

$$\begin{cases} \dfrac{dm}{dr} = 4\pi\epsilon_c r^2 \Theta^n \\[2ex] \dfrac{d\Theta}{dr} = -\dfrac{\alpha_0 + \Theta}{n+1} \dfrac{m + 4\pi r^3 \ \frac{\epsilon_c}{\alpha_0} \ \Theta^{(n+1)}}{r(r-2m)} . \end{cases} \tag{16.216}$$

As explained in Sec. 16.3.5, both ϵ and p have dimensions $(\text{length})^{-2}$, therefore the quantity $\sqrt{\epsilon_c}$ has dimension $(\text{length})^{-1}$ and we can use it to rescale the radial coordinate as follows. We define

$$\xi = r \ \sqrt{\epsilon_c}, \qquad \text{and} \qquad \mathcal{M} = \sqrt{\epsilon_c} \ m , \tag{16.217}$$

and rewrite Eqs. 16.216 in terms of the new variables (note that ξ and $\mathcal{M}(\xi)$ are dimensionless quantities)

$$\begin{cases} \dfrac{d\mathcal{M}}{d\xi} = 4\pi\xi^2 \Theta^n , \\[2ex] \dfrac{d\Theta}{d\xi} = -\dfrac{\alpha_0 + \Theta}{(n+1)} \dfrac{\mathcal{M} + 4\pi\xi^3 \ \frac{1}{\alpha_0} \ \Theta^{(n+1)}}{\xi \, (\xi - 2\mathcal{M})} . \end{cases} \tag{16.218}$$

[14]Another possible extension of polytropic EoS is $p = K\rho^{\gamma}$ where $\rho = \epsilon - p/(\gamma - 1)$ is the baryon rest-mass density. Both extensions reduce to $p = K\rho^{\gamma}$ in the Newtonian limit.

We may, at this point, multiply the second equation by ξ^2, differentiate it with respect to ξ and – upon substitution of $\frac{d\mathcal{M}}{d\xi}$ into the resulting equation – find a second-order differential equation for Θ in a form similar to the Lane-Emden equation (see Eq. 16.70); however, the equation we would get is much more complicated than Eq. 16.70, and it is much better to work with the system of equations 16.218.

Another important difference with the Newtonian equations is that, in that case, once we assign the value of the polytropic index n and integrate the Lane-Emden equation finding $\Theta(\xi)$ up to the stellar radius ξ_1, the function $\Theta(\xi)$ allows us to construct a family of solutions by assigning the value of K and of the central density ρ_c; no further integrations are needed and, for instance, the radius and the mass of the star can be found from Eqs. 16.75 and 16.76. This is not possible in the relativistic case, because to solve Eqs. 16.218 we need to assign *both* n *and* α_0, i.e. the ratio between the energy density and the pressure at $\xi = 0$. Therefore solutions with different values of the central density are not related by a simple rescaling.

Eqs. 16.218 can be integrated numerically with the initial condition $\mathcal{M}(0) = 0$ and $\Theta(0) = 1$; to avoid the singularity at $\xi = 0$ we can choose a starting point ξ very close to the origin, and use the expansion of the functions Θ and $\mathcal{M}$ discussed in Box 16-H, up to the point ξ_1 where the function Θ, and consequently the pressure, vanishes. As in the Newtonian case, this indicates that the boundary of the star has been reached. After computing $\Theta(\xi)$ numerically, from Eqs. 16.212 and 16.213 we get the energy density and the pressure profiles inside the star; in addition, we can compute the mass profile

$$\mathcal{M}(\xi) = \int_0^{\xi} 4\pi\epsilon_c \xi'^2 \Theta^n(\xi')\, d\xi' \qquad \to \qquad m(r) = \mathcal{M}(\xi)/\sqrt{\epsilon_c}\,, \tag{16.219}$$

and the rr-component of the metric,

$$e^{2\lambda} = \frac{1}{1 - \frac{2\mathcal{M}(\xi)}{\xi}} = \frac{1}{1 - \frac{2m(r)}{r}}\,. \tag{16.220}$$

The value of the stellar radius can be found using Eq. 16.217, and is

$$R = \xi_1/\sqrt{\epsilon_c}\,. \tag{16.221}$$

At the boundary of the star, the second of Eqs. 16.218 yields

$$\Theta'(\xi_1) = -\frac{\alpha_0}{(n+1)} \frac{\mathcal{M}(\xi_1)}{\xi_1(\xi_1 - 2\mathcal{M}(\xi_1))}\,, \tag{16.222}$$

from which we find the mass of the star [15]

$$\mathcal{M}(\xi_1) = -\frac{(n+1)\xi_1^2\Theta'(\xi_1)}{\alpha_0 - 2\xi_1(n+1)\Theta'(\xi_1)} \qquad \to \qquad M = \mathcal{M}(\xi_1)/\sqrt{\epsilon_c}\,, \tag{16.223}$$

where we remind that $\Theta'(\xi_1)$ is negative. The remaining metric function $e^{2\nu}$ can be found from Eq. 16.174 which now becomes

$$\begin{aligned} \nu(\xi) &= \nu_0 - \int_0^{\xi} \frac{p_{,\xi'}}{\epsilon + p} d\xi' = \nu_0 - \int_0^{\xi} \frac{p_c(\Theta^{(n+1)})_{,\xi'}}{\epsilon_c\Theta^n + p_c\Theta^{(n+1)}} d\xi' \\ &= \nu_0 - (n+1)\int_0^{\xi} \frac{\Theta'}{\alpha_0 + \Theta} d\xi' = \nu_0 + \log\left[\frac{\alpha_0 + 1}{\alpha_0 + \Theta(\xi)}\right]^{(n+1)}. \end{aligned} \tag{16.224}$$

[15] Clearly, the mass of the star can also be found from Eq. 16.219 integrated from 0 to ξ_1.

On the surface the metric must reduce to the Schwarzschild metric in vacuum, therefore (since $\Theta(\xi_1) = 0$)

$$e^{2\nu(\xi_1)} = e^{2\nu_0}\left(\frac{\alpha_0+1}{\alpha_0}\right)^{2(n+1)} = 1 - \frac{2\mathcal{M}(\xi_1)}{\xi_1}, \tag{16.225}$$

which, using Eq. 16.223, gives

$$e^{2\nu_0} = \left(\frac{\alpha_0}{\alpha_0+1}\right)^{2(n+1)} \frac{\alpha_0}{\alpha_0 - 2\xi_1(n+1)\Theta'(\xi_1)}. \tag{16.226}$$

Thus,

$$e^{2\nu(\xi)} = \left(\frac{\alpha_0}{\alpha_0+1}\right)^{2(n+1)} \frac{\alpha_0}{\alpha_0 - 2\xi_1(n+1)\Theta'(\xi_1)} \left[\frac{\alpha_0+1}{\alpha_0+\Theta(\xi)}\right]^{2(n+1)}, \tag{16.227}$$

and the solution is finally complete.

The EoS proposed by nuclear physicists are much more complex than the simple polytropic EoS discussed in this section (see Box 16-I). However, these are useful approximations, which capture several features of the more "realistic" EoS. In addition, any equation of state can be approximated as a sequence of polytropic EoS, and this method is largely used in the literature.

Box 16-H

A note on the numerical integration of Eqs. 16.218

As explained in Sec. 16.3.5, near the origin the mass goes to zero as $m(r) \sim r^3$; by Taylor-expanding both $\mathcal{M}(\xi)$ and $\Theta(\xi)$ and by substituting these expansions in Eqs. 16.218, it is easy to check that the coefficients of odd powers of ξ for Θ, and even powers for $\mathcal{M}$, vanish. Therefore, as in the Newtonian case, the appropriate expansions are

$$\begin{aligned} \Theta(\xi) &\sim 1 + \Theta_2\,\xi^2 + \Theta_4\,\xi^4 + O(\xi^6), \\ \mathcal{M}(\xi) &\sim m_3\xi^3 + m_5\xi^5 + O(\xi^7). \end{aligned} \tag{16.228}$$

By inserting these expansions in Eqs. 16.218 we find

$$\begin{aligned} 3m_3\xi^2 + 5m_5\xi^4 &= 4\pi\xi^2 + 4\pi n\,\Theta_2\,\xi^4 \\ 2\,\Theta_2\,\xi + 4\,\Theta_4\,\xi^3 &= -\frac{1}{n+1}\left\{\left(m_3 + \frac{4\pi}{\alpha_0}\right)\left[(1+\alpha_0)\xi + \Theta_2\,\xi^3\right]\right\} \end{aligned}$$

and, by equating the coefficients of the same power of ξ, we get

$$m_3 = \tfrac{4\pi}{3}, \qquad \Theta_2 = -2\pi\frac{(1+\alpha_0)(3+\alpha_0)}{3\alpha_0(n+1)}, \tag{16.229}$$

$$m_5 = \tfrac{4\pi n\Theta_2}{5}, \quad \Theta_4 = -\frac{\Theta_2}{2(n+1)}\left(m_3 + \frac{4\pi}{\alpha_0}\right). \tag{16.230}$$

With these values of the coefficients, the expansions 16.228 can be evaluated at any point ξ near the origin, to start the numerical integration of Eqs. 16.218.

Box 16-I

The mass-radius diagram for realistic neutron stars

As discussed in Sec. 16.3.2, the behaviour of matter in the core of neutron stars is largely unknown since the core density is higher than what can be probed in a laboratory. Thus, different nuclear-physics models provide EoS that agree at relatively small densities (in the crust region) but can significantly differ in the core. These EoS do not have a simple analytical polytropic expression, but are provided in *tabulated form*, i.e. as an ordered list of $\{\epsilon, p(\epsilon)\}$ computed using complex nuclear-physics models.

Nonetheless, a tabulated EoS is sufficient to integrate the equations of the stellar structure using the methods discussed in the main text. Fig. 16.5 shows the mass-radius diagram for a variety of tabulated EoS. As can be seen, this diagram, as well as the maximum mass and the minimum radius of a neutron star, depends significantly on the EoS. Future electromagnetic and gravitational wave observations hold the promise to provide data that will help measuring the mass and radius of a neutron star with sufficient accuracy to rule out certain EoS. For example, the observation of neutron stars with $M \simeq 2M_\odot$ with pulsar timing [41, 8] has ruled out those EoS that give a smaller maximum mass. Likewise, as discussed in Sec. 14.2.6, the recent gravitational wave detection of a binary neutron star coalescence, GW170817, has allowed a first measurement of the tidal deformability of the stars. The upper limit on the tidal Love number provided by LIGO and Virgo has ruled out those EoS which are not compatible with that limit [2]. Future observations will place even stronger constraints.

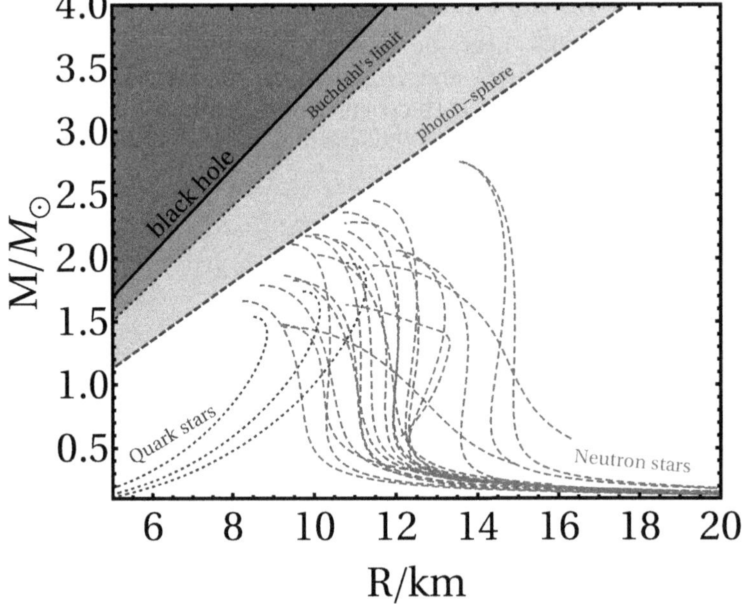

Figure 16.5: Mass-radius diagram of non-spinning neutron stars for several representative EoS [77, 87]. Each curve is computed in the stable branch (for smaller radii the equilibrium solution is unstable, see Sec. 16.3.9). The three straight lines correspond to $R = 2M$, $R = 9/4M$, and $R = 3M$, i.e. to the Schwarzschild radius, to the Buchdahl limit (Sec. 16.3.8), and to the light ring (Sec. 10.4). The dotted curves starting from the bottom-left part of the diagram correspond to quark stars (see Sec. 16.3.2), whereas the dashed curves starting on the bottom-right part of the diagram are ordinary neutron stars.

16.3.8 Buchdahl's theorem

In Sec. 16.3.6 we showed that the minimum value of the radius of a homogeneous star (i.e. one with constant energy density) is $R_{\rm min} = \frac{9}{4}M$. A theorem proved by Hans Buchdahl in 1959 [26] establishes that this result is much more general. The theorem is based on the assumptions that the star is static and spherically symmetric, its interior is described by a perfect fluid [16], with positive energy density and pressure, and that the energy density is a monotonically decreasing function of the radial coordinate, i.e.

$$\epsilon \geq 0\,, \qquad p \geq 0\,, \qquad \frac{d\epsilon}{dr} \leq 0\,. \tag{16.231}$$

No assumption is made on the EoS that relates ϵ and the pressure p. In order to prove the theorem, we shall use the rr- and $\theta\theta$-components of Einstein's equations, Eqs. 16.163, 16.164:

$$\begin{aligned} -\frac{e^{2\lambda}}{r^2}\left(1-e^{-2\lambda}\right)+\frac{2}{r}\nu_{,r} &= 8\pi p\, e^{2\lambda} \\ r^2 e^{-2\lambda}\left(\nu_{,rr}+\nu_{,r}^2+\frac{\nu_{,r}-\lambda_{,r}}{r}-\nu_{,r}\lambda_{,r}\right) &= 8\pi r^2\, p\,. \end{aligned} \tag{16.232}$$

By multiplying the second equation above by $\frac{e^{2\lambda}}{r^2}$ and subtracting the first equation we find

$$\nu_{,rr}+\nu_{,r}(\nu_{,r}-\lambda_{,r})-\frac{\nu_{,r}+\lambda_{,r}}{r}+\frac{e^{2\lambda}}{r^2}(1-e^{-2\lambda})=0\,. \tag{16.233}$$

Therefore,

$$\begin{aligned} \frac{d}{dr}\left[\frac{e^{-\lambda}}{r}\frac{de^{\nu}}{dr}\right] &= \frac{e^{\nu-\lambda}}{r}\left[\nu_{,rr}+\nu_{,r}(\nu_{,r}-\lambda_{,r})-\frac{\nu_{,r}}{r}\right] \\ &= \frac{e^{\nu-\lambda}}{r^2}\lambda_{,r}-\frac{e^{\nu+\lambda}}{r^3}(1-e^{-2\lambda}) = e^{\nu+\lambda}\frac{d}{dr}\left[\frac{1-e^{-2\lambda}}{2r^2}\right] \end{aligned} \tag{16.234}$$

and thus

$$\frac{d}{dr}\left[\frac{e^{-\lambda}}{r}\frac{de^{\nu}}{dr}\right] = e^{\nu+\lambda}\frac{d}{dr}\left[\frac{m(r)}{r^3}\right]\,, \tag{16.235}$$

where, as usual, $m(r) = 4\pi\int_0^r \epsilon r'^2 dr'$. For any r we can always define a density $\bar{\epsilon}_r$ such that

$$m(r) = \frac{4}{3}\pi\bar{\epsilon}_r r^3\,, \tag{16.236}$$

and since ϵ is a monotonically decreasing function of r, $\bar{\epsilon}_r$, and consequently $\frac{m(r)}{r^3}$, is also monotonically decreasing. Hence

$$\frac{d}{dr}\left[\frac{e^{-\lambda}}{r}\frac{de^{\nu}}{dr}\right] \leq 0\,. \tag{16.237}$$

It should be noted that the minimum value of $\bar{\epsilon}_r$ is attained at the boundary, i.e. $\bar{\epsilon}_{\rm min} = \bar{\epsilon}_R$. Thus

$$M = \frac{4}{3}\pi\bar{\epsilon}_{\rm min}R^3\,. \tag{16.238}$$

[16] Actually, Buchdahl's theorem can be generalized to mildly anisotropic (and thus non-perfect) fluids, for which the radial pressure is always larger than the tangential pressure.

Inside the star $\bar{\epsilon}_r \geq \bar{\epsilon}_{\min}$, and consequently

$$\frac{4}{3}\pi\bar{\epsilon}_r R^3 \geq \frac{4}{3}\pi\bar{\epsilon}_{\min} R^3 \,, \qquad \rightarrow \qquad m(r) \geq \frac{M}{R^3} r^3 \,, \tag{16.239}$$

i.e. $m(r)$ is always bigger than that of a homogeneous star with energy density equal to $\bar{\epsilon}_{\min}$.

From Eq. 16.237 it follows that

$$\frac{e^{-\lambda}}{r}\frac{de^{\nu}}{dr} \geq \left[\frac{e^{-\lambda}}{r}\frac{d}{dr}\left(e^{\nu}\right)\right]\bigg|_{r=R} . \tag{16.240}$$

When $r = R$ the metric reduces to the Schwarzschild metric in vacuum, therefore $e^{2\nu}\Big|_{r=R} = e^{-2\lambda}\Big|_{r=R} = 1 - \frac{2M}{R}$ and

$$\left[\frac{e^{-\lambda}}{r}\frac{de^{\nu}}{dr}\right]\bigg|_{r=R} = \frac{1}{2R}\frac{d}{dr}e^{2\nu}\bigg|_{r=R} = \frac{M}{R^3} \,, \tag{16.241}$$

therefore Eq. 16.240 gives

$$\frac{e^{-\lambda}}{r}\frac{de^{\nu}}{dr} \geq \frac{M}{R^3} \qquad \rightarrow \qquad \frac{de^{\nu}}{dr} \geq re^{\lambda}\frac{M}{R^3} . \tag{16.242}$$

By integrating Eq. 16.242 between 0 and R we find

$$e^{\nu(R)} - e^{\nu_0} \geq \frac{M}{R^3}\int_0^R re^{\lambda}dr \,, \tag{16.243}$$

and since $e^{-2\lambda} = 1 - \frac{2m(r)}{r}$

$$e^{\nu_0} \leq \sqrt{1 - \frac{2M}{R}} - \frac{M}{R^3}\int_0^R \frac{rdr}{\sqrt{1 - \frac{2m(r)}{r}}} . \tag{16.244}$$

We want to establish an upper bound for e^{ν_0}, therefore we need to determine when the right-hand side of Eq. 16.244 attains its maximum value. From Eq. 16.239 we know that $m(r) \geq \frac{M}{R^3}r^3$, and consequently

$$\sqrt{1 - \frac{2m(r)}{r}} \leq \sqrt{1 - \frac{2M}{R^3}r^2}, \qquad \rightarrow \qquad \int_0^R \frac{rdr}{\sqrt{1 - \frac{2m(r)}{r}}} \geq \int_0^R \frac{rdr}{\sqrt{1 - \frac{2M}{R^3}r^2}} . \tag{16.245}$$

Thus, Eq. 16.244 gives

$$e^{2\nu_0} \leq \sqrt{1 - \frac{2M}{R}} - \frac{M}{R^3}\int_0^R \frac{rdr}{\sqrt{1 - \frac{2M}{R^3}r^2}} = \frac{3}{2}\sqrt{1 - \frac{2M}{R}} - \frac{1}{2} . \tag{16.246}$$

In writing the 00-component of the metric as in Eq. 16.148, i.e.

$$g_{00} = -e^{2\nu} \,, \tag{16.247}$$

we have used one of the hypotheses of the theorem, namely the condition that the metric is static. Indeed, a static spacetime admits a timelike Killing vector, which must remain timelike in the interior of the star, i.e.

$$\vec{\xi} \cdot \vec{\xi} = g_{00}(\xi^0)^2 < 0 \,. \tag{16.248}$$

This condition is satisfied by Eq. 16.247 since $e^{2\nu}$ is positive. It follows from Eq. 16.246 that

$$\frac{3}{2}\sqrt{1 - \frac{2M}{R}} - \frac{1}{2} > 0 \,, \tag{16.249}$$

which yields

$$\frac{M}{R} < \frac{4}{9} \qquad \rightarrow \qquad R > \frac{9}{4}M \,, \tag{16.250}$$

and the theorem is proved.

Thus, the radius of a static, spherically symmetric, perfect-fluid star in hydrostatic equilibrium must be larger than $\frac{9}{4}M$, and it should be noted that this value is larger than the Schwarzschild radius $R_S = 2M$. In other words, the compactness $C = \frac{M}{R}$ of a static, spherically symmetric star (see Box 11-A) can not exceed $C \simeq 0.44$. In practice, current models of the neutron star EoS predict that $R \gtrsim 3M$, i.e. $C \lesssim 0.33$ (see Box 16-I).

16.3.9 Stability of a compact star

A solution of the TOV equations 16.171, satisfying the appropriate boundary conditions discussed in Sec. 16.3.5, describes a stellar configuration in hydrostatic equilibrium. This equilibrium can, in principle, be either stable or unstable. In this section we will discuss the conditions for stability under radial perturbations, starting with a qualitative argument.

A qualitative argument for the stability of a compact star

Let us consider a sequence of equilibrium configurations obtained by integrating the TOV equations with an assigned EoS, for different values of the central energy density ϵ_c. The gravitational mass is thus a function of ϵ_c, i.e. $M = M(\epsilon_c)$.

The typical form of the profile $M(\epsilon_c)$ obtained by numerical integration is sketched in Fig. 16.6. Each point of this curve represents an equilibrium configuration, i.e. a solution of the TOV equations. Given a star in the equilibrium configuration A, if a small radial perturbation *reduces* its central energy density to a value, say, ϵ_{A1} leaving the mass unchanged, the new (non-equilibrium) configuration will be represented by point A_1. This is *above* the equilibrium curve, therefore the perturbed star has a mass which is *larger* than that corresponding to ϵ_{A1} in equilibrium. Since the mass is larger than the equilibrium mass, gravity exceeds the pressure gradient needed for equilibrium, and the star contracts; its central energy density consequently increases until the star goes back to the equilibrium configuration A.

In a similar way, if a perturbation *increases* the central energy density to ϵ_{A2} leaving the mass unchanged, the new configuration A_2 is a point *below* the curve shown in Fig. 16.6. The star in A_2 has mass *smaller* than that of the equilibrium configuration corresponding to ϵ_{A2}; in this case gravity is weaker than the pressure gradient needed to keep the star in equilibrium, and the star expands reducing its central energy density to return to the equilibrium configuration A. Thus, A is a *stable* equilibrium configuration, and the condition for stability is

$$\frac{dM}{d\epsilon_c} > 0 \,. \tag{16.251}$$

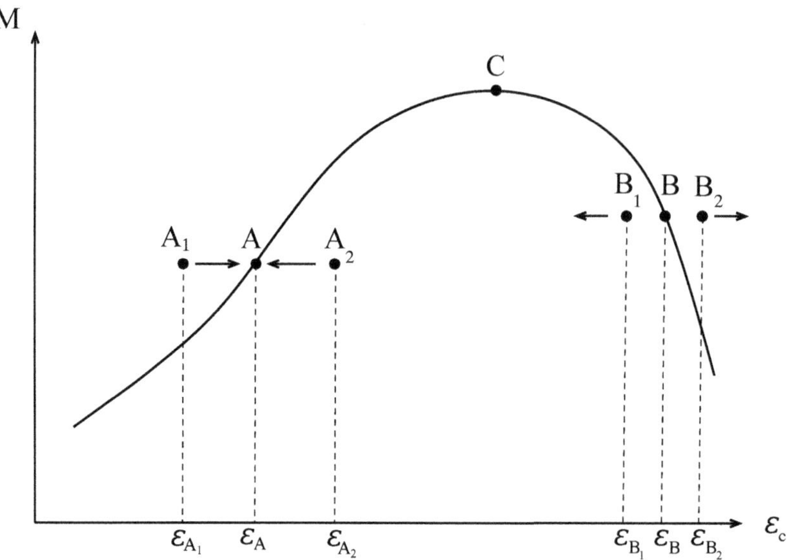

Figure 16.6: Gravitational mass of equilibrium stellar configurations as a function of the central energy density, near a relative maximum.

Conversely, a similar discussion about the point B in Fig. 16.6, where

$$\frac{dM}{d\epsilon_c} < 0\,, \tag{16.252}$$

shows that a displacement to B_1 brings the star to a configuration where gravity is weaker than the pressure gradient so that the star expands, further reducing the central energy density. Similarly, a displacement to B_2 brings the star to a configuration where gravity exceeds the pressure gradient, leading the star to contract and to further increase the central energy density: the equilibrium in B is *unstable*.

In Fig. 16.6 the branch of the curve on the left of the maximum C corresponds, in principle, to stable configurations, whereas that on the right to unstable configurations. The point C is the configuration of *maximum mass*.

An example is provided by Newtonian polytropic stars: the function $M(\epsilon_c)$, given in Eq. 16.76, which we rewrite here for convenience:

$$M = 4\pi\ \xi_1^2\ |\Theta'(\xi_1)|\ \left[\frac{(n+1)K}{4\pi G}\right]^{\frac{3}{2}}\ \rho_c^{\frac{3-n}{2n}}\,, \tag{16.253}$$

shows that M is an increasing function of the central density ρ_c for $n < 3$, it is maximum for $n = 3$, and decreasing for $n > 3$; therefore our qualitative argument suggests that the polytropic star is stable only if $n < 3$.

If we consider the stellar mass as a function of the radius, we find that since

$$\frac{dM}{d\epsilon_c} = \frac{dM}{dR}\frac{dR}{d\epsilon_c}\,, \tag{16.254}$$

the stability criterion 16.251 is satisfied in two cases

$$dR/d\epsilon_c > 0 \quad \text{and} \qquad dM/dR > 0$$
$$dR/d\epsilon_c < 0 \quad \text{and} \qquad dM/dR < 0\,.$$

In most cases the radius of the star *decreases* as the central density increases, and the stable branches of the function $M(R)$ are those for which

$$\frac{dM}{dR} < 0\,. \tag{16.255}$$

This is the case, for instance, of most of the neutron star curves shown in the $M - R$ diagram in Fig. 16.5. Conversely, for quark stars, the radius *increases* as the central density increases. In this case the stable branches are those for which $\frac{dM}{dR} > 0$. Indeed, all curves shown in Fig. 16.5 correspond to *stable* solutions.

Is the condition $\frac{dM}{d\epsilon_c} > 0$ sufficient for stability?

We now wish to discuss whether the condition $\frac{dM}{d\epsilon_c} > 0$ is sufficient for the stability of a static, spherically symmetric star.

The answer is negative, and the reason can be understood by considering the full theory of radial perturbations of a star. A detailed treatment is outside the scope of this book, therefore we shall just sketch the main results and give the basic notions needed to understand them (for a more detailed analysis see e.g. [113, 38]). Note that, by Birkhoff's theorem, if the perturbed star is spherically symmetric the gravitational field in the exterior is static, i.e., *a radially oscillating star does not emit gravitational waves.*

A star in equilibrium has an infinite set of radial, proper oscillation modes, labelled by an integer, positive index $n = 0, 1, 2, \ldots$; when the star oscillates in the n-th mode, each fluid element is displaced from the equilibrium position by a radial displacement

$$\xi_n(r,t) = u_n(r)e^{-i\omega_n t}\,, \tag{16.256}$$

where ω_n is the mode frequency and $u_n(r)$ is the mode amplitude. The mode number n corresponds to the number of nodes that $u_n(r)$ has inside the star: $n = 0$ for zero nodes, $n = 1$ for one node, etc. The mode frequencies are ordered as

$$\omega_0^2 < \omega_1^2 < \omega_2^2 < \ldots\,, \tag{16.257}$$

and the mode corresponding to ω_0 is called the *fundamental mode.* The latter corresponds to a global oscillation of the star, which is either expanding or contracting; for this reason it is also called the "breathing mode".

If $\omega_n^2 > 0$, the fluid element oscillates about the equilibrium position and the corresponding mode is stable; conversely, if $\omega_n^2 < 0$ then ω_n is purely imaginary, the radial displacement in Eq. 16.256 grows exponentially for the root with positive imaginary part, and the mode is unstable. The existence of at least one unstable mode implies that the equilibrium configuration of the star is unstable.

In the presence of multiple maxima and minima of the curve $M(\epsilon_c)$, as in Fig. 16.7, the qualitative discussion about stability presented in the previous section is not applicable. In this case the theory of radial pulsations shows the following.

Given a sequence of stellar configurations, differing for the value of ϵ_c and having the same EoS, for each value of ϵ_c we can compute the mass of the star $M(\epsilon_c)$ and the frequency of the various radial modes. If for some value of ϵ_c, say $\epsilon_c^{\rm crit}$, the curve $M(\epsilon_c)$ has an extremal point, i.e.

$$\left.\frac{dM}{d\epsilon_c}\right|_{\epsilon_c^{\rm crit}} = 0\,, \tag{16.258}$$

then for $\epsilon_c = \epsilon_c^{\rm crit}$ the square of the frequency of one of the modes crosses the real axis, i.e.

$$\omega_{i,\epsilon_c^{\rm crit}}^2 = 0\,, \tag{16.259}$$

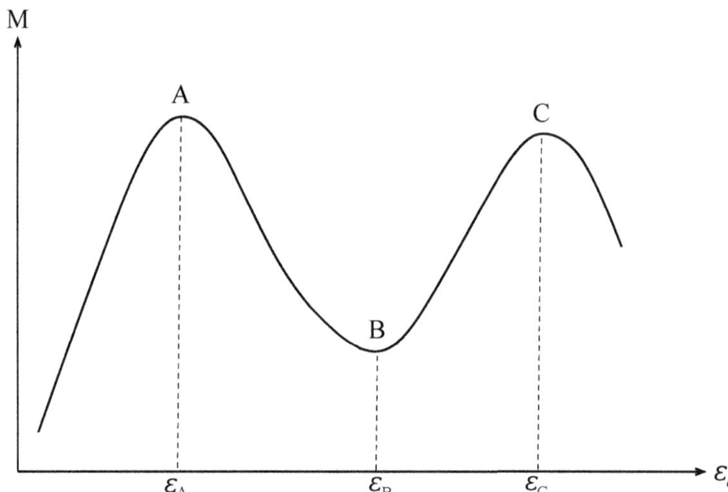

Figure 16.7: Mass of equilibrium stellar configurations as a function of the central density. Example of diagram with multiple maxima and minima.

and changes sign. This means the the i-th mode becomes unstable. Given Eq. 16.257, the $n = 0$ mode (which is the one with lowest frequency) is the first to become unstable.

Now consider the curve shown in Fig. 16.7 as an example, and suppose that for $\epsilon_c < \epsilon_A$ the fundamental mode has frequency $\omega_0^2 > 0$, i.e. it is stable. A is an extremal point, therefore in A $\omega_0^2 = 0$, and all configurations belonging to the branch AB will be unstable because their fundamental mode has $\omega_0^2 < 0$. Increasing the central energy density we reach the second extremal point B. Here two things may happen:

1. ω_0^2 changes sign again becoming positive. In this case the equilibrium configuration corresponding to B and all configurations of the branch BC would be stable for radial perturbations.

2. ω_0^2 remains negative (i.e. the fundamental mode remains unstable). Then, from Eq. 16.257, it must happen that the frequency of the $n = 1$ radial mode changes sign, and consequently the $n = 1$ mode also becomes unstable. In this case all configurations of the branch BC would be unstable.

This example clearly shows that the condition $\frac{dM}{d\epsilon_c} > 0$ provides a *necessary*, but *not sufficient* condition for stability.

Using linear perturbation theory, it can be shown that also the behavior of $dR/d\epsilon_c$ is needed to assess the (in)stability of a specific stellar configuration [113, 38, 105]. Indeed, if $dR/d\epsilon_c > 0$ at the critical point, the lowest radial mode with *odd* value of n must change sign, whereas if $dR/d\epsilon_c < 0$ the same is true for the lowest mode with *even* value of n. This general property – whose proof is beyond the scope of this book – can be used to establish which branch of solutions between two stationary points of the function $M(\epsilon_c)$ is stable.

Finally, we remark that we have only considered *radial* perturbations. If an equilibrium configuration of the star is stable under these perturbations, it can still be unstable under non-radial perturbations, the study of which is much more involved, since they are associated to the emission of gravitational radiation (see e.g. Chapter 15, where they are studied for black holes).

CHAPTER 17

The far-field limit of an isolated, stationary object

In this chapter we shall derive the metric which describes the gravitational field generated by an *isolated, stationary* object at large distance from the source. This will be useful to show how the angular momentum of a rotating body affects the spacetime.

Since the source is isolated, in the exterior the stress-energy tensor vanishes, and it is reasonable to assume that sufficiently "far away" from the source the gravitational field vanishes, i.e. that the spacetime is asymptotically flat. A spacetime is said *asymptotically flat* if it is possible to define, in an appropriate frame, a coordinate r such that [1]

$$\lim_{r\to\infty} g_{\mu\nu} = \eta_{\mu\nu} \,. \tag{17.1}$$

The coordinate r has to be a space coordinate, i.e. the vector $\frac{\partial}{\partial r}$ has to be spacelike (at least for sufficiently large r).

We call *far-field limit* the region of spacetime where $r \gg R$, given R a length-scale characteristic of the source. In this region we can expand all quantities in powers of $\frac{R}{r} \ll 1$; for simplicity, we shall denote the terms of n-th order in this expansion as $O(1/r^n)$. Since the metric is asymptotically flat, in the far-field limit (and in an appropriate reference frame) it can be written as

$$g_{\mu\nu} = \eta_{\mu\nu} + O\left(\frac{1}{r}\right) \,. \tag{17.2}$$

Since the source is stationary, we assume that the metric is also stationary; therefore it admits a timelike Killing vector field and, by a suitable choice of coordinates, it can be made independent of time (see Chapter 8).

We shall show that the metric of a stationary, axisymmetric source in the far-field limit, up to terms of order higher than $1/r$ in the expansion 17.2, in geometrized units can be written as

$$\begin{aligned} ds^2 &= -\left(1-\frac{2M}{r}\right)dt^2 + \left(1+\frac{2M}{r}\right)dr^2 + r^2(d\theta^2 + \sin^2\theta d\varphi^2) \\ &\quad -\frac{4J}{r}\sin^2\theta dt d\varphi + \text{ higher-order terms in } 1/r \,, \end{aligned} \tag{17.3}$$

[1] The definition 17.1 of asymptotic flatness is adequate for the discussion in this book, but it is not rigorous enough for other applications. Indeed, it does not allow to define the limit $r \to \infty$ in a coordinate-independent way, and technical problems may arise in computations when exchanging limits and derivatives. A more precise definition of asymptotic flatness is based on the concept of conformal rescalings, and is beyond the scope of this book. We refer the interested reader to Chapter 11 of [114] and references therein.

where M is the total mass-energy of the source mass and J is its angular momentum.

Let us write the expansion 17.2, which holds at large distance from the source, in a perturbative form

$$g_{\mu\nu} = \eta_{\mu\nu} + h_{\mu\nu} \tag{17.4}$$

with $|h_{\mu\nu}| \ll 1$, neglecting $O(h^2)$ terms. The perturbation $h_{\mu\nu}$ is a solution of the linearized Einstein equations in vacuum (see Chapter 12, Eq. 12.33)

$$\Box_F \bar{h}_{\mu\nu} = 0 \tag{17.5}$$

$$\bar{h}^{\mu}{}_{\nu,\mu} = 0\,, \tag{17.6}$$

where we remind that $\Box_F$ is the d'Alembert operator of the flat spacetime

$$\Box_F = \eta^{\alpha\beta}\frac{\partial}{\partial x^\alpha}\frac{\partial}{\partial x^\beta} = -\frac{\partial^2}{\partial t^2} + \nabla^2\,,$$

and

$$\bar{h}_{\mu\nu} \equiv h_{\mu\nu} - \frac{1}{2}\eta_{\mu\nu}h^{\alpha}{}_{\alpha}\,. \tag{17.7}$$

Since we are assuming that the spacetime is stationary, Eqs. 17.5 and 17.6 reduce to

$$\nabla^2 \bar{h}_{\mu\nu} = 0\,, \tag{17.8}$$

$$\bar{h}^{i}{}_{\nu,i} = 0\,. \tag{17.9}$$

We stress that Eqs. 17.8 and 17.9 hold only in the far-field limit $r \gg R$.

In the following, we shall first derive Eq. 17.3 in the simple case when the gravitational field generated by the source is weak everywhere, i.e. inside and outside the source. We shall then show that the Eq. 17.3 is far more general, since it holds also when the field near the source is strong. As in Chapters 12 and 13, in this chapter we shall use Latin indices $i, j, \ldots$ for the space components $1, 2, 3$, and Greek indices $\mu, \nu, \ldots$ for the spacetime components $0, 1, 2, 3$.

17.1 THE WEAK-FIELD CASE

If the gravitational field of the source is weak, the metric can be expressed as a perturbation of flat spacetime (Eq. 17.4) everywhere, not only in the far-field limit. As discussed in Chapter 13, in this case we can treat the metric perturbation $h_{\mu\nu}$ as a (tensor) field living in Minkowski's spacetime. Therefore, we raise and lower the indices of $h_{\mu\nu}$ by using Minkowski's metric (e.g., $h_{i\mu} = h^{i}{}_{\mu}$, $h_{0\mu} = -h^{0}{}_{\mu}$). Moreover, since the stress-energy tensor vanishes in the background (Minkowski's metric is solution of Einstein's equations in vacuum), the source term $T^{\mu\nu}$ is of order $O(h)$ and, neglecting $O(h^2)$ term, it satisfies the laws of Special Relativity such as e.g. the conservation law $T^{\mu\nu}{}_{,\nu} = 0$.

As shown in Chapter 12, in the weak-field approximation Einstein's equations for the metric perturbation $\bar{h}_{\mu\nu}$ become

$$\Box_F \bar{h}_{\mu\nu} = -16\pi T_{\mu\nu} \tag{17.10}$$

$$\bar{h}^{\mu}{}_{\nu,\mu} = 0\,,$$

whose general solution is Eq. 12.35, which we rewrite for convenience:

$$\bar{h}_{\mu\nu}(t, \mathbf{x}) = 4\int_V \frac{T_{\mu\nu}(t - |\mathbf{x}-\mathbf{x}'|, \mathbf{x}')}{|\mathbf{x}-\mathbf{x}'|}\, d^3x'\,; \tag{17.11}$$

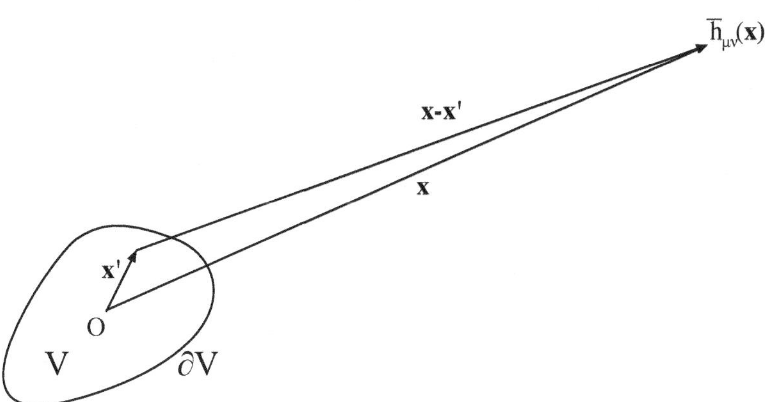

Figure 17.1: Metric perturbation $\bar{h}_{\mu\nu}$ produced at $\mathbf{x}$ by a stationary, isolated source enclosed in the volume V.

here V is the source three-volume (see Fig. 17.1). On the source boundary ∂V and outside the source the stress-energy tensor vanishes.

In Chapters 12 and 13 we were interested in the time-dependent part of the solution 17.11, which is associated to gravitational wave emission. Conversely, we here consider a stationary source, for which $T_{\mu\nu} = T_{\mu\nu}(\mathbf{x}')$ and Eq. 17.11 reduces to

$$\bar{h}_{\mu\nu}(\mathbf{x}) = 4\int_V \frac{T_{\mu\nu}(\mathbf{x}')}{|\mathbf{x}-\mathbf{x}'|}d^3x' \,. \tag{17.12}$$

17.1.1 The multipolar expansion

We choose a reference frame centered at the center of mass of the source. Let us consider a generic point of the source, with position three-vector $\mathbf{x}'$, and a point far away from the source, with position three-vector $\mathbf{x}$, such that $|\mathbf{x}| \gg |\mathbf{x}'|$ (see Fig. 17.1). The Taylor expansion of the quantity $1/|\mathbf{x}-\mathbf{x}'|$ around $|\mathbf{x}'| = 0$ is called *multipolar expansion*:

$$\frac{1}{|\mathbf{x}-\mathbf{x}'|} = \frac{1}{r} + \frac{x^i x'^i}{r^3} + O\left(\frac{1}{r^3}\right) , \tag{17.13}$$

where $r \equiv |\mathbf{x}|$. As mentioned above, since we raise/lower indices with Minkowski's metric, e.g. $h_{i\mu} = h^i{}_\mu$, in the following we shall not distinguish covariant space indices from contravariant ones. In addition, we remind that $x^i x'^i = \sum_{i=1}^3 x^i x'^i$. Note that the term $\frac{x^i x'^i}{r^3}$ is of order $O\left(\frac{1}{r^2}\right)$ because x^i is of the same order as r.

Substituting Eq. 17.13 in Eq. 17.12 we find

$$\bar{h}_{\mu\nu}(\mathbf{x}) = \frac{4}{r}\int_V T_{\mu\nu}d^3x' + \frac{4x^i}{r^3}\int_V T_{\mu\nu}x'^i d^3x' + O\left(\frac{1}{r^3}\right) . \tag{17.14}$$

We shall now compute the different components of $\bar{h}_{\mu\nu}$.

- *00-component.*
  In the weak-field limit $T_{00} = \rho$ (see Box 5-C and recall that we are using $G = c = 1$ units), therefore the first integral in Eq. 17.14 gives

$$\int_V T_{00} d^3x' = M\,. \tag{17.15}$$

The 00-component of the second integral in Eq. 17.14 gives the position of the source center of mass $x'^i_{\rm CM}$ which, by assumption, coincides with the origin of the coordinate frame, i.e.

$$\int_V T_{00} x'^i d^3x' = M x'^i_{\rm CM} = 0\,. \tag{17.16}$$

Therefore

$$\bar{h}_{00} = \frac{4M}{r} + O\left(\frac{1}{r^3}\right)\,. \tag{17.17}$$

- *μi-components (general properties).*
  The stress-energy tensor satisfies the conservation equation in flat space $T^{\mu\nu}{}_{,\nu} = 0$ which, in the stationary case, becomes

$$T^{\mu\nu}{}_{,\nu} = T^{\mu 0}{}_{,0} + T^{\mu i}{}_{,i} = T^{\mu i}{}_{,i} = 0\,. \tag{17.18}$$

Using Eq. 17.18 and the property $\frac{\partial x^i}{\partial x^j} = \delta^i_j$, we find

$$\begin{aligned} \int_V T^{\mu i} d^3x' &= \int_V T^{\mu k} \delta^i_k d^3x' = \int_V T^{\mu k} \frac{\partial x'^i}{\partial x'^k} d^3x' \\ &= -\int_V \left(\frac{\partial T^{\mu k}}{\partial x'^k}\right) x'^i d^3x' = 0\,, \end{aligned} \tag{17.19}$$

where we have integrated by parts; note that the surface terms do not contribute, because on the boundary of V, $T_{\mu\nu} = 0$. Thus the first integral in the multipolar expansion 17.14 vanishes:

$$\int_V T^{\mu i} d^3x' = 0\,. \tag{17.20}$$

Using the conservation equation satisfied by $T^{\mu\nu}$ it is easy to show that:

$$\begin{aligned} &\int_V \left(T^{\mu i} x'^j + T^{\mu j} x'^i\right) d^3x' = \int_V T^{\mu k} \left(\frac{\partial x'^i}{\partial x'^k} x'^j + \frac{\partial x'^j}{\partial x'^k} x'^i\right) d^3x' \\ &= \int_V T^{\mu k} \frac{\partial}{\partial x'^k} \left(x'^i x'^j\right) d^3x' = -\int_V x'^i x'^j T^{\mu k}{}_{,k} d^3x' = 0\,, \end{aligned} \tag{17.21}$$

from which it follows this remarkable property:

$$\int_V T^{\mu i} x'^j d^3x' = -\int_V T^{\mu j} x'^i d^3x'\,, \tag{17.22}$$

i.e. the second integral in Eq. 17.14 is antisymmetric in the last two indices.

- *ij-components.*
  Eq. 17.20 (with $\mu = j$) implies that $\int_V T^{ij} d^3x' = 0$. Moreover, the integral

$$\int_V T^{jk} x'^i d^3x' \tag{17.23}$$

is symmetric in the first two indices, and antisymmetric in the last two, because of Eq. 17.22; consequently

$$\int_V T^{ki}x'^j d^3x' = -\int_V T^{kj}x'^i d^3x' = -\int_V T^{jk}x'^i d^3x' = \int_V T^{ji}x'^k d^3x'$$
$$= \int_V T^{ij}x'^k d^3x' = -\int_V T^{ik}x'^j d^3x' = -\int_V T^{ki}x'^j d^3x' \quad (17.24)$$

i.e.

$$\int_V T^{ki}x'^j d^3x' = -\int_V T^{ki}x'^j d^3x' \quad \rightarrow \quad \int_V T^{ki}x'^j d^3x' = 0\,. \quad (17.25)$$

Summarizing, from Eqs. 17.14, 17.20, and 17.25 we find

$$\bar{h}_{ik} = O\left(\frac{1}{r^3}\right)\,. \quad (17.26)$$

- $0i$-*components.*
Eq. 17.20 (with $\mu = 0$) implies that $\int_V T^{0i}d^3x' = 0$. Moreover, from Eq. 17.22 it follows that the second integral in Eq. 17.14 is antisymmetric:

$$\int_V T^{0i}x'^j d^3x' = -\int_V T^{0j}x'^i d^3x'\,. \quad (17.27)$$

The components T^{0i} are the density of the i-th component of the momentum of the source (see Box 5-C),

$$T^{0i} = \mathcal{P}^i\,. \quad (17.28)$$

A matter element with volume d^3x' and position $\mathbf{x}'$ has momentum $\boldsymbol{\mathcal{P}}(\mathbf{x}')d^3x'$. Its angular momentum is $d\mathbf{J} = \mathbf{x}' \times \boldsymbol{\mathcal{P}}d^3x'$, where $\times$ indicates the vector product [2]. Thus, the source angular momentum is

$$\mathbf{J} = \int_V \mathbf{x}' \times \boldsymbol{\mathcal{P}}\, d^3x'\,. \quad (17.29)$$

The components of $\mathbf{J}$ can be written as follows

$$J^i = -\epsilon_{ijk}\int_V T^{0j}x'^k d^3x'\,, \quad (17.30)$$

where ϵ_{ijk} is the three-dimensional Levi-Civita symbol (see Box 8-C) which, in flat spacetime, coincides with the Levi-Civita tensor. We remind that the Levi-Civita is completely antisymmetric, and thus its non-vanishing components are only those for which the three indices are different. Moreover

$$\epsilon_{123} = 1\,. \quad (17.31)$$

As shown in Box 17-A, given an antisymmetric three-dimensional tensor $B^{ij} = -B^{ji}$, if a vector A^i is defined as

$$A^i = \epsilon_{ijk}B^{jk}\,, \quad (17.32)$$

[2] Strictly speaking, in Special Relativity this is the angular momentum seen by a static observer (see Sec. 10.3).

then B^{ij} can be written in terms of A^i as follows:

$$B^{ij} = \frac{1}{2}\epsilon_{ijk}A^k . \tag{17.33}$$

Thus, if we define $B^{jk} = -\int_V T^{0j}x'^k d^3x'$, from Eqs. 17.30 and 17.33 it follows

$$-\int_V T^{0i}x'^j d^3x' = \frac{1}{2}\epsilon_{ijk}J^k , \tag{17.34}$$

and the second integral in Eq. 17.14 can be written as

$$\frac{4x^j}{r^3}\int_V T_{0i}x'^j d^3x' = -\frac{4x^j}{r^3}\int_V T^{0i}x'^j d^3x' = \frac{2x^j}{r^3}\epsilon_{ijk}J^k . \tag{17.35}$$

From Eqs. 17.14, 17.20, and 17.35 it follows that the $0i$-components of $h_{\mu\nu}$ are

$$\bar{h}_{0i} = \frac{2}{r^3}\epsilon_{ijk}x^j J^k + O\left(\frac{1}{r^3}\right) . \tag{17.36}$$

In summary, the multipolar expansion 17.14 gives

$$\begin{aligned} \bar{h}_{00} &= \frac{4M}{r} + O\left(\frac{1}{r^3}\right) \\ \bar{h}_{0i} &= \frac{2}{r^3}\epsilon_{ijk}x^j J^k + O\left(\frac{1}{r^3}\right) \\ \bar{h}_{ij} &= O\left(\frac{1}{r^3}\right) . \end{aligned} \tag{17.37}$$

In terms of $h_{\mu\nu}$, since[3]

$$h_{\mu\nu} = \bar{h}_{\mu\nu} - \frac{1}{2}\eta_{\mu\nu}\bar{h}^\alpha{}_\alpha \tag{17.38}$$

and $h^\alpha{}_\alpha = -4M/r^2$, we find

$$\begin{aligned} h_{00} &= \frac{2M}{r} + O\left(\frac{1}{r^3}\right) \\ h_{0i} &= \frac{2}{r^3}\epsilon_{ijk}x^j J^k + O\left(\frac{1}{r^3}\right) \\ h_{ij} &= \frac{2M}{r}\delta_{ij} + O\left(\frac{1}{r^3}\right) . \end{aligned} \tag{17.39}$$

This is the solution of the linearized Einstein equations in the weak-field limit. If we consider the full, non-linear Einstein equations and include terms of order $O(h^2)$ in the weak-field expansion, the solution has corrections $\sim M^2/r^2$ in h_{00} and h_{ij}. We neglect these terms, which are of order $O(1/r^2)$ in the far-field expansion. Therefore, the solution of the fully non-linear Einstein equations expanded in the far-field limit reads

$$\begin{aligned} h_{00} &= \frac{2M}{r} + O\left(\frac{1}{r^2}\right) \\ h_{0i} &= \frac{2}{r^3}\epsilon_{ijk}x^j J^k + O\left(\frac{1}{r^3}\right) \\ h_{ij} &= \frac{2M}{r}\delta_{ij} + O\left(\frac{1}{r^2}\right) . \end{aligned} \tag{17.40}$$

[3] Eq. 17.38 can be found by replacing $\bar{h}^\alpha{}_\alpha = -h^\alpha{}_\alpha$ (Eq. 12.34) in Eq. 17.7.

Box 17-A

Vector representation of a three-dimensional antisymmetric tensor

Let us consider a generic antisymmetric tensor in the three-dimensional Euclidean space $B^{ij} = -B^{ji}$, and let

$$A^k = \epsilon_{klm} B^{lm} . \tag{17.41}$$

We shall show that

$$B^{ij} = \frac{1}{2}\epsilon_{ijk} A^k . \tag{17.42}$$

To prove Eq. 17.42 we contract Eq. 17.41 with $\frac{1}{2}\epsilon_{ijk}$:

$$\frac{1}{2}\epsilon_{ijk} A^k = \frac{1}{2}\epsilon_{ijk}\epsilon_{klm} B^{lm} . \tag{17.43}$$

The following equality is easy to prove:

$$\epsilon_{ijk}\epsilon_{klm} = \delta_{il}\delta_{jm} - \delta_{im}\delta_{jl} . \tag{17.44}$$

Indeed, the non-vanishing components of ϵ_{ijk} are those for which the indices are 123 with some ordering. Thus in the contraction $\epsilon_{ijk}\epsilon_{klm}$, the couples of indices ij and lm have to be the same, either $ij = lm$ or $ij = ml$. If $ij = lm$, then $\epsilon_{ijk}\epsilon_{klm} = 1$; if $ij = ml$, then $\epsilon_{ijk}\epsilon_{klm} = -1$. Consequently, $\epsilon_{ijk}\epsilon_{klm} = \delta_{il}\delta_{jm} - \delta_{im}\delta_{jl}$.
Finally, replacing Eq. 17.44 in Eq. 17.43 we find

$$\frac{1}{2}\epsilon_{ijk} A^k = \frac{1}{2}\left(\delta_{il}\delta_{jm} - \delta_{im}\delta_{jl}\right) B^{lm} = B^{ij} . \tag{17.45}$$

17.1.2 The metric of the far-field limit in polar coordinates

Let us transform the solution 17.40 in polar coordinates

$$\begin{aligned} x^1 &= r\sin\theta\cos\varphi \\ x^2 &= r\sin\theta\sin\varphi \\ x^3 &= r\cos\theta . \end{aligned} \tag{17.46}$$

Using Eqs. 17.40 we find

$$h_{\mu\nu}dx^\mu dx^\nu = \frac{2M}{r}dt^2 + \frac{4}{r^3}\epsilon_{ijk}x^j J^k\, dtdx^i + \frac{2M}{r}\delta_{ij}dx^i dx^j + \text{ higher-order terms in } 1/r . \tag{17.47}$$

Since $\delta_{ij}dx^i dx^j = dr^2 + r^2 d\theta^2 + r^2\sin^2\theta d\varphi^2$, the last term in Eq. 17.47 is

$$h_{ij}dx^i dx^j = \frac{2M}{r}\left(dr^2 + r^2 d\theta^2 + r^2\sin^2\theta d\varphi^2\right) . \tag{17.48}$$

In order to transform $h_{0i}dtdx^i$, we choose the frame orientation such that the angular momentum is directed along the z-axis, i.e. $\mathbf{J} = (0, 0, J)$. Therefore,

$$\begin{aligned} h_{0i}dtdx^i &= \left(\frac{2}{r^3}dt\right)\epsilon_{ijk}x^j J^k dx^i = -\left(\frac{2}{r^3}dt\right)J(x^1dx^2 - x^2dx^1) \\ &= -\frac{2J}{r}\sin^2\theta dtd\varphi, \end{aligned} \tag{17.49}$$

where the equality $x^1dx^2 - x^2dx^1 = r^2\sin^2\theta$ can easily be proved by differentiating Eqs. 17.46.

Finally, since $\eta_{\mu\nu}dx^\mu dx^\nu = -dt^2 + dr^2 + r^2d\theta^2 + r^2\sin^2\theta d\varphi^2$, the line element is

$$\begin{aligned} ds^2 = (\eta_{\mu\nu} + h_{\mu\nu})dx^\mu dx^\nu &= -\left(1 - \frac{2M}{r}\right)dt^2 + \left(1 + \frac{2M}{r}\right)\left[dr^2 + r^2(d\theta^2 + \sin^2\theta d\varphi^2)\right] \\ &- \left(\frac{4J}{r}\sin^2\theta\right)dtd\varphi + \text{ higher-order terms in } 1/r\,. \end{aligned} \tag{17.50}$$

Moreover, if we redefine the radial coordinate as:

$$r \to r - M \tag{17.51}$$

and neglect contributions of order $O(1/r^2)$, the only term which changes in the metric is

$$\begin{aligned} &\left(1 + \frac{2M}{r}\right)r^2(d\theta^2 + \sin^2\theta d\varphi^2) \to \left(1 + \frac{2M}{r}\right)(r - M)^2(d\theta^2 + \sin^2\theta d\varphi^2) \\ &= r^2\left(1 + O\left(\frac{1}{r}\right)\right)(d\theta^2 + \sin^2\theta d\varphi^2)\,. \end{aligned} \tag{17.52}$$

Therefore, with the redefinition 17.51 the metric 17.50 becomes

$$\begin{aligned} ds^2 &= -\left(1 - \frac{2M}{r}\right)dt^2 + \left(1 + \frac{2M}{r}\right)dr^2 + r^2(d\theta^2 + \sin^2\theta d\varphi^2) \\ &\quad -\frac{4J}{r}\sin^2\theta dtd\varphi + \text{ higher-order terms in } 1/r\,, \end{aligned}$$

which coincides with Eq. 17.3.

17.2 THE STRONG-FIELD CASE

In this section we shall consider the general case in which the field near and on the source can be strong; the weak-field approximation still holds in the far-field limit, but it is no longer applicable to the source and its surroundings.

As discussed in the introduction of this chapter, in the far-field limit the metric can be written as a perturbation of Minkowski's spacetime

$$g_{\mu\nu} = \eta_{\mu\nu} + h_{\mu\nu}\,, \tag{17.53}$$

and we neglect terms of order $O(h^2)$. In this case Einstein's equations, linearized in the perturbation, yield

$$\nabla^2\bar{h}_{\mu\nu} = 0\,, \tag{17.54}$$

$$\bar{h}^i{}_{\nu,i} = 0\,, \tag{17.55}$$

where ∇^2 is the Laplace operator in the three-dimensional, Euclidean space. Since the metric perturbation vanishes as $r \to \infty$, we look for a solution to Eqs. 17.54, 17.55 of the form

$$\bar{h}_{\mu\nu} = \frac{a_{\mu\nu}(\theta,\varphi)}{r} + \frac{b_{\mu\nu}(\theta,\varphi)}{r^2} + O\left(\frac{1}{r^3}\right) . \tag{17.56}$$

Note that the coefficients $a_{\mu\nu}$, $b_{\mu\nu}$ depend only on the angular variables θ,φ. Laplace's operator in spherical coordinates has the form (see Box 3-E)

$$\nabla^2 = \frac{1}{r^2}\frac{\partial}{\partial r}r^2\frac{\partial}{\partial r} + \frac{I\!L}{r^2} \tag{17.57}$$

where $I\!L$ is the operator acting on the angular variables defined in Eq. 3.88

$$I\!L \equiv \frac{\partial^2}{\partial\theta^2} + \cot\theta\frac{\partial}{\partial\theta} + \sin^{-2}\theta\frac{\partial^2}{\partial\varphi^2} . \tag{17.58}$$

By substituting Eq. 17.56 in Eq. 17.54, and imposing that each term of the $1/r$ expansion vanishes independently, we find

$$I\!L a_{\mu\nu}(\theta,\varphi) = 0 \tag{17.59}$$
$$I\!L b_{\mu\nu}(\theta,\varphi) = -2b_{\mu\nu}(\theta,\varphi) . \tag{17.60}$$

The eigenfunctions of the operator $I\!L$ are the *spherical harmonics* $Y^{lm}(\theta,\varphi)$, with $l = 0, 1, \ldots$ and $m = -l, -l+1, \ldots, l-1, l$ (see Box 15-A). They are defined by the property

$$I\!L Y^{lm} = -l(l+1)Y^{lm} . \tag{17.61}$$

Eq. 17.59 implies that $a_{\mu\nu}$ is proportional to Y^{00}, which does not depend on θ,φ, while Eq. 17.60 shows that $b_{\mu\nu}$ is a linear combination of the spherical harmonics with $l = 1$, which are (see Eq. 15.15)

$$Y^{11} = -\sqrt{\frac{3}{8\pi}}\sin\theta e^{\mathrm{i}\varphi} , \qquad Y^{10} = \sqrt{\frac{3}{4\pi}}\cos\theta , \qquad Y^{1-1} = \sqrt{\frac{3}{8\pi}}\sin\theta e^{-\mathrm{i}\varphi} . \tag{17.62}$$

The $l = 1$ spherical harmonics are linear combinations of the direction cosines $n^i = x^i/r$:

$$n^1 = \frac{x^1}{r} = \sin\theta\cos\varphi = \sqrt{\frac{8\pi}{3}}\frac{-Y^{11}+Y^{1-1}}{2} \tag{17.63}$$
$$n^2 = \frac{x^2}{r} = \sin\theta\sin\varphi = \sqrt{\frac{8\pi}{3}}\frac{-Y^{11}-Y^{1-1}}{2\mathrm{i}}$$
$$n^3 = \frac{x^3}{r} = \cos\theta = \sqrt{\frac{4\pi}{3}}Y^{10} ;$$

therefore the functions $b_{\mu\nu}(\theta,\varphi)$ are linear combinations of n^1, n^2, and n^3:

$$b_{\mu\nu}(\theta,\varphi) = b_{\mu\nu i}n^i(\theta,\varphi) , \tag{17.64}$$

and the expansion 17.56 can be written as

$$\bar{h}_{\mu\nu} = \frac{a_{\mu\nu}}{r} + \frac{b_{\mu\nu i}x^i}{r^3} + O\left(\frac{1}{r^3}\right) , \tag{17.65}$$

with $a_{\mu\nu}$, $b_{\mu\nu i}$ constant coefficients, symmetric in the indices μ, ν.

We now impose on Eq. 17.65 the harmonic gauge condition 17.55, $\bar{h}^{\mu i}{}_{,i} = 0$ which, using $\partial_j 1/r^n = -nx^j/r^{n+2}$, gives

$$\bar{h}_{\mu j,j} = -\frac{a_{\mu j}x^j}{r^3} + \frac{b_{\mu ji}\left(\delta^{ij}r^2 - 3x^i x^j\right)}{r^5} = 0\,. \tag{17.66}$$

Eq. 17.66 has to be satisfied for all (large) values of r and for all values of $n^i = x^i/r$. As before, this occurs if each coefficient in the $1/r$-expansion vanishes, i.e.

$$a_{\mu j} = 0 \tag{17.67}$$

$$\left(\delta^{ij} - 3n^i n^j\right) b_{\mu ij} = 0\,. \tag{17.68}$$

Note that these algebraic equations do not involve a_{00} and b_{00i}, which are free constants of the general solutions. In order to simplify the notation, we redefine the latter as:

$$\begin{aligned} a &\equiv a_{00} \\ b_i &\equiv b_{00i}\,. \end{aligned} \tag{17.69}$$

Eq. 17.67 implies that all the constants $a_{\mu\nu}$ except a_{00} vanish. Eq. 17.68 can be rewritten as:

$$H^{ij}b_{0ij} = 0 \tag{17.70}$$

$$H^{ij}b_{kij} = 0\,, \tag{17.71}$$

where we have defined $H^{ij} \equiv \delta^{ij} - 3n^i n^j$. The general solution of Eqs. 17.70, 17.71 is

$$b_{0ij} = b\delta_{ij} + c_k \epsilon_{ijk} \tag{17.72}$$

$$b_{kij} = d_k\delta_{ij} + d_i\delta_{kj} - d_j\delta_{ki}\,, \tag{17.73}$$

where b, c_k, d_k are constants.

A rigorous proof of 17.72, 17.73 would require the use of the structures of Group Theory, which goes beyond the scope of this book. Here we will only provide an intuitive, non-rigorous proof of the first solution, 17.72. First of all note that H^{ij} depends on the angles, while b_{0ij} – which are the components of rank-two tensors on the three-dimensional Euclidean space – are constants. Eq. 17.70 must be satisfied for any value of n^i, i.e. for any value of the angular variables θ, φ. This is possible only if the left-hand side of Eq. 17.70 *vanishes identically for symmetry reasons*: since H^{ij} is *symmetric and traceless* ($H^{ij} = H^{ji}$, $\delta_{ij}H^{ij} = 0$), the combination $H^{ij}(\theta,\varphi)b_{0ij}$ vanishes for all values of θ, φ when $b_{0ij} = -b_{0ji}$ and when $b_{0ij} \propto \delta_{ij}$. There exist only two Euclidean, three-dimensional constant tensors with these properties: the Levi-Civita tensor, ϵ_{ijk}, and the Kronecker delta, δ_{ij}; thus b_{0ij} must be a linear combination of these two tensors, as in Eq. 17.72. The proof that Eq. 17.73 is solution of Eq. 17.71 follows from similar, albeit more involved, arguments.

In summary, by imposing the harmonic gauge condition on the expansion 17.65 we get

$$\begin{aligned} \bar{h}_{00} &= \frac{a}{r} + \frac{b_i x^i}{r^3} + O\left(\frac{1}{r^3}\right) \\ \bar{h}_{0i} &= \frac{bx^i}{r^3} + \epsilon_{ijk}\frac{x^j c_k}{r^3} + O\left(\frac{1}{r^3}\right) \\ \bar{h}_{ij} &= \frac{1}{r^3}\left(d_i x^j + d_j x^i - \delta_{ij}d_k x^k\right) + O\left(\frac{1}{r^3}\right)\,; \end{aligned} \tag{17.74}$$

this solution depends on the constants a, b_i, b, c_k, d_k.

The constants b, d_k can be eliminated with an appropriate infinitesimal coordinate transformation $x^\mu \to x^\mu + \epsilon^\mu$ (see Sec. 12.2):

$$\epsilon^\mu = \frac{1}{r}\left(-b, d^i\right) . \tag{17.75}$$

Note that this transformation can only be defined in the far-field limit. As a consequence, the change in the metric neglecting terms of order $O(h^2)$ is (see Eq. 12.24)

$$\delta h_{\mu\nu} = \eta_{\mu\alpha}\epsilon^\alpha{}_{,\nu} + \eta_{\nu\alpha}\epsilon^\alpha{}_{,\mu} \tag{17.76}$$

and thus, in terms of the perturbation $\bar{h}_{\mu\nu}$,

$$\delta\bar{h}_{\mu\nu} = \delta h_{\mu\nu} - \frac{1}{2}\eta_{\mu\nu}\eta^{\alpha\beta}\delta h_{\alpha\beta} = \eta_{\mu\alpha}\epsilon^\alpha{}_{,\nu} + \eta_{\nu\alpha}\epsilon^\alpha{}_{,\mu} - \eta_{\mu\nu}\epsilon^\alpha{}_{,\alpha} . \tag{17.77}$$

Since $\epsilon^\mu{}_{,0} = 0$, $\epsilon^0{}_{,i} = \frac{bx^i}{r^3}$, and $\epsilon^k{}_{,i} = -\frac{d^k x^i}{r^3}$, we find

$$\begin{aligned}
\delta\bar{h}_{00} &= -\eta_{00}\epsilon^k{}_{,k} + O\left(\frac{1}{r^3}\right) = -\frac{d^k x^k}{r^3} + O\left(\frac{1}{r^3}\right) \\
\delta\bar{h}_{0i} &= \eta_{00}\epsilon^0_{,i} + O\left(\frac{1}{r^3}\right) = -\frac{bx^i}{r^3} + O\left(\frac{1}{r^3}\right) \\
\delta\bar{h}_{ij} &= \epsilon^i{}_{,j} + \epsilon^j{}_{,i} - \delta_{ij}\epsilon^k{}_{,k} = -\frac{1}{r^3}\left[d^i x^j + d^j x^j - \eta_{ij} d^k x^k\right] .
\end{aligned} \tag{17.78}$$

Thus, the metric perturbation becomes

$$\begin{aligned}
\bar{h}_{00} &= \frac{a}{r} + \frac{\tilde{b}_i x^i}{r^3} + O\left(\frac{1}{r^3}\right) \\
\bar{h}_{0i} &= \epsilon_{ijk}\frac{x^j c_k}{r^3} + O\left(\frac{1}{r^3}\right) \\
\bar{h}_{ij} &= O\left(\frac{1}{r^3}\right) ,
\end{aligned} \tag{17.79}$$

where we have defined $\tilde{b}_i \equiv b_i - d_i$. Finally we can get rid of $\tilde{b}_i$ by performing a translation

$$x^i \to x^i + \frac{\tilde{b}_i}{a} , \tag{17.80}$$

which produces the following change in the a/r term:

$$\begin{aligned}
\frac{a}{r} = a\left[(x^i)^2\right]^{-1/2} &\to a\left[\left(x^i + \frac{\tilde{b}_i}{a}\right)^2\right]^{-1/2} = a\left[r^2\left(1 + 2\frac{\tilde{b}_i x^i}{r^2 a}\right)\right]^{-1/2} + O\left(\frac{1}{r^3}\right) \\
&= \frac{a}{r}\left(1 - \frac{\tilde{b}_i x^i}{r^2 a}\right) + O\left(\frac{1}{r^3}\right) = \frac{a}{r} - \frac{\tilde{b}_i x^i}{r^3} + O\left(\frac{1}{r^3}\right) .
\end{aligned} \tag{17.81}$$

Therefore,

$$\begin{aligned}
\bar{h}_{00} &= \frac{a}{r} + O\left(\frac{1}{r^3}\right) \\
\bar{h}_{0i} &= \epsilon_{ijk}\frac{x^j c_k}{r^3} + O\left(\frac{1}{r^3}\right) \\
\bar{h}_{ij} &= O\left(\frac{1}{r^3}\right) .
\end{aligned} \tag{17.82}$$

Eq. 17.82 coincides with Eq. 17.37, derived in Sec. 17.1.1 under the assumption of weak field on the source, with the identifications

$$a = 4M \qquad c^k = 2J^k\,. \tag{17.83}$$

Thus, repeating the derivation in the last part of Sec. 17.1, i.e. moving from $\bar{h}_{\mu\nu}$ to $h_{\mu\nu}$ and from Cartesian to polar coordinates, and shifting the radial coordinate by M, we find again the metric of the far-field limit 17.3, i.e.

$$\begin{aligned} ds^2 &= -\left(1-\frac{2M}{r}\right)dt^2+\left(1+\frac{2M}{r}\right)dr^2+r^2(d\theta^2+\sin^2\theta d\varphi^2) \\ &\quad -\frac{4J}{r}\sin^2\theta dt d\varphi + \text{ higher-order terms in } 1/r\,. \end{aligned} \tag{17.84}$$

Thus, the metric 17.84 describes the far-field limit of a stationary, isolated source even when the gravitational field is strong near the source.

17.3 MASS AND ANGULAR MOMENTUM OF AN ISOLATED OBJECT

As shown in Sec. 17.2, if the gravitational field is weak everywhere, close to and far from the source, the constants M and J emerge as the mass and angular momentum of the source as defined in Special Relativity. If the field is *not* weak on the source and in its neighborhood, then M and J arise as integration constants of the general solution of the far-field equations. In this case, we need to assess the physical meaning of these constants.

One possibility is to provide an *operational definition* of mass and angular momentum for an isolated, stationary object, based on the following argument. Suppose that a test body is moving in the spacetime described by the metric 17.84, far away from the source; the study of its motion does not allow to distinguish whether the gravitational field is weak or strong near the source, because the metric in the far-field limit is the same in both cases. Thus, if the source is a strong-gravity one, we can *define* its mass and angular momentum as the quantities M and J, related to the integration constants appearing in the far-field metric by Eq. 17.83. The mass M can be measured from the orbital frequency of the test mass through Kepler's third law, and the angular momentum J can be inferred by measuring the precession of gyroscopes orbiting around the source (see Sec. 17.4 below).

A more rigorous way to assess the physical meaning of M and J for a strong-gravity source uses the stress-energy pseudo-tensor $t^{\mu\nu}$, which we defined in Chapter 13. We remind that $t^{\mu\nu}$ describes the energy and momentum carried by the gravitational field, and satisfies, together with the stress-energy tensor of matter and fields $T^{\mu\nu}$, the conservation law

$$\left[(-g)(T^{\mu\nu}+t^{\mu\nu})\right]_{,\nu} = 0\,. \tag{17.85}$$

As shown in Sec. 13.6.1, the quantity $(-g)(T^{\mu\nu}+t^{\mu\nu})$ can be defined in terms of the spacetime metric as follows:

$$(-g)(T^{\mu\nu}+t^{\mu\nu}) = \frac{\partial\zeta^{\mu\nu\alpha}}{\partial x^\alpha} \tag{17.86}$$

where

$$\zeta^{\mu\nu\alpha} = \frac{\partial\lambda^{\mu\nu\alpha\beta}}{\partial x^\beta} \tag{17.87}$$

and

$$\lambda^{\mu\nu\alpha\beta} \equiv \frac{1}{16\pi}\left[(-g)(g^{\mu\nu}g^{\alpha\beta}-g^{\mu\alpha}g^{\nu\beta})\right]\,. \tag{17.88}$$

Since we are considering a stationary spacetime, Eq. 17.86 becomes

$$(-g)(T^{\mu\nu}+t^{\mu\nu})=\frac{\partial\zeta^{\mu\nu k}}{\partial x^k}\,,\qquad k=1,2,3\,,\tag{17.89}$$

and $\zeta^{\mu\nu\alpha}=\frac{\partial\lambda^{\mu\nu\alpha i}}{\partial x^i}$.

Let us now consider a spherical, three-dimensional volume V centered on the source, with radius r much larger than the source size (we choose an asymptotically flat coordinate frame (t,x^i)). As discussed in Sec. 13.6.1, the total four-momentum P^μ enclosed in the volume V is given by

$$P^\mu=\int_V d^3x(-g)(T^{0\mu}+t^{0\mu})\tag{17.90}$$

(see Eq. 13.120) and it is contributed *both by the source and by the gravitational field.* By substituting Eq. 17.89 in Eq. 17.90, P^μ can be expressed as

$$P^\mu=\int_V d^3x\frac{\partial\zeta^{0\mu k}}{\partial x^k}\,.\tag{17.91}$$

Using Gauss' theorem (see Box 5-D) the integral of a three-divergence of a vector over the volume V can be written as the flux of the vector across the spherical surface ∂V surrounding the volume

$$P^\mu=\int_{\partial V}\zeta^{0\mu k}dS_k\,,\tag{17.92}$$

where $dS_k=n^k dS=n^k r^2 d\Omega$ and $n^k=\frac{x^i}{r}$ is the unit vector orthogonal to the surface element. In Sec. 17.1 we computed the mass and the angular momentum of a stationary, isolated source – on the assumption that the gravitational field it generates is weak – by integrating suitable components of the stress-energy tensor on the volume of the source (see Eqs. 17.15 and 17.30). We are now considering a stationary, isolated source whose gravitational field is not weak and, in order to compute the *total mass-energy* and the *total angular momentum* which include the contribution of the gravitational field, we need to evaluate the integral in Eq. 17.91 over a volume much larger than the source volume. We have shown that this volume integral reduces to the surface integral 17.92 on the boundary ∂V which is in the far-field region, where the metric is known; thus, $\zeta^{0\mu k}$ can be computed using Eqs. 17.87 and 17.88. The total mass-energy is

$$M_{\rm tot}=P^0=\int_{\partial V}\zeta^{00k}n^k r^2 d\Omega\,.\tag{17.93}$$

As discussed in Sec. 17.1, the three-momentum of the volume element d^3x located at a point of coordinates x^i is $\mathcal{P}^i d^3x$, and the angular momentum of the same element is $dJ^i=(\mathbf{x}\times\mathcal{P})^i\,d^3x=-\epsilon_{ijk}\mathcal{P}^j x^k d^3x$. In the present case $\mathcal{P}^j$ is [4]

$$\mathcal{P}^j=(-g)(T^{0j}+t^{0j})=(-g)(T^{j0}+t^{j0})=\frac{\partial}{\partial x^l}\zeta^{j0l}\,,\tag{17.94}$$

therefore the total angular momentum which generalizes Eq. 17.30 and includes the contribution of the gravitational field is

$$\begin{aligned}J^i_{\rm tot}&=-\epsilon_{ijk}\int_V d^3x\frac{\partial\zeta^{j0l}}{\partial x^l}x^k=-\epsilon_{ijk}\int_V d^3x\left[\frac{\partial(\zeta^{j0l}x^k)}{\partial x^l}-\zeta^{j0l}\frac{\partial x^k}{\partial x^l}\right]\\&=-\epsilon_{ijk}\int_V d^3x\left[\frac{\partial(\zeta^{j0l}x^k)}{\partial x^l}-\zeta^{j0k}\right]\,,\end{aligned}\tag{17.95}$$

[4]Note that ${\zeta^{0jl}}_{,l}={\zeta^{j0l}}_{,l}$ but $\zeta^{0jl}\neq\zeta^{j0l}$; we write Eq. 17.94 in terms of ζ^{j0l} because in this way the subsequent computations are simpler.

where the latter equality holds in stationary spacetimes, and V is assumed to be larger than the source volume. By replacing Eq. 17.87 in Eq. 17.95 we find

$$J^i_{\text{tot}} = -\epsilon_{ijk}\int_V d^3x \frac{\partial}{\partial x^l}\left(\zeta^{j0l}x^k - \lambda^{j0kl}\right), \tag{17.96}$$

and, by Gauss' theorem, the components of the total angular momentum of the source are

$$J^i_{\text{tot}} = -\epsilon_{ijk}\int_{\partial V}\left(\zeta^{j0l}x^k - \lambda^{j0kl}\right) n^l r^2 d\Omega\,. \tag{17.97}$$

Box 17-B

Summary: total mass-energy and angular momentum of a stationary, isolated body using the stress-energy pseudo-tensor

The total mass and angular momentum of a stationary, isolated source, which includes the contribution of the gravitational field it generates, can be written in terms of the stress-energy pseudo-tensor as follows:

$$M_{\text{tot}} = \int_V d^3x(-g)(T^{0\mu} + t^{0\mu}) = \int_{\partial V}\zeta^{00i}n^i r^2 d\Omega \tag{17.98}$$

$$J^i_{\text{tot}} = \epsilon_{ijk}\int_V(-g)(T^{0k} + t^{0k})x^j d^3x = -\epsilon_{ijk}\int_{\partial V}\left(\zeta^{j0l}x^k - \lambda^{j0kl}\right) n^l r^2 d\Omega\,, \tag{17.99}$$

where $\zeta^{\alpha\mu\nu}$, $\lambda^{\mu\nu\alpha\beta}$ are given in Eqs. 17.87, 17.88, respectively, and V is much larger than the source size, so that the surface ∂V is located in the far-field region.

Let us now explicitly compute Eqs. 17.98 and 17.99 assuming that the surface enclosing the source is located in the far-field region; in this case the metric to be used to evaluate the surface integrals is $g_{\mu\nu} = \eta_{\mu\nu} + h_{\mu\nu}$, where $h_{\mu\nu}$ is given by Eqs. 17.40:

$$\begin{aligned} h_{00} &= \frac{2M}{r} + O\left(\frac{1}{r^2}\right) \\ h_{0i} &= \frac{2}{r^3}\epsilon_{ijk}x^j J^k + O\left(\frac{1}{r^3}\right) \\ h_{ij} &= \frac{2M}{r}\delta_{ij} + O\left(\frac{1}{r^2}\right). \end{aligned} \tag{17.100}$$

We first recall that (see Sec. 6.1)

$$g^{\mu\nu} = \eta^{\mu\nu} - h^{\mu\nu} + O(h^2)\,, \tag{17.101}$$

where the indices of $h_{\mu\nu}$ have been raised with Minkowski's metric. The determinant of $g_{\mu\nu}$ is

$$g = (-1 + h_{00})(1 + h_{ii}) = -(1 + \frac{4M}{r}) + O\left(\frac{1}{r^2}\right). \tag{17.102}$$

Note that in this expression we have neglected the term J/r^3 with respect to M/r, since we are in the far-field limit.

The quantity ζ^{00i}, which appears in Eq. 17.98, can be evaluated using Eqs. 17.87, 17.88 and the metric 17.100; neglecting terms $O\left(1/r^2\right)$, we find

$$\begin{aligned}
\zeta^{00i}n^i &= \frac{1}{16\pi}n^i\frac{\partial}{\partial x^j}\left[(-g)\left(g^{00}g^{ij}-g^{0i}g^{0j}\right)\right] = \frac{1}{16\pi}n^i\frac{\partial}{\partial x^j}\left[(-g)g^{00}g^{ij}\right]+O(h^2)\\
&= \frac{1}{16\pi}n^i\frac{\partial}{\partial x^j}\left[\left(1+\frac{4M}{r}\right)\left(-1-\frac{2M}{r}\right)\left(1-\frac{2M}{r}\right)\delta_{ij}\right]+O\left(\frac{1}{r^2}\right)\\
&= -\frac{1}{16\pi}n^i\frac{\partial}{\partial x^j}\frac{4M}{r}\delta_{ij}+O\left(\frac{1}{r^2}\right). \qquad (17.103)
\end{aligned}$$

Since $\frac{\partial}{\partial x^j}\frac{1}{r}=-\frac{n^j}{r^2}$

$$\zeta^{00i}n^i=\frac{1}{4\pi}\frac{M}{r^2}n^in^i=\frac{1}{4\pi}\frac{M}{r^2}, \qquad (17.104)$$

and finally

$$M_{\text{tot}}=P^0=\int_{\partial V}\zeta^{00i}n^ir^2d\Omega=M. \qquad (17.105)$$

Thus, *the constant M appearing in the metric of the far-field limit is the total mass-energy of the source.*

Let us now prove that the components of the angular momentum defined in Eq. 17.99 coincide with the constants J^i which appear in the metric of far-field limit 17.100. Using Eq. 17.87, and the relation $x^k=rn^k$, Eq. 17.99 becomes

$$\begin{aligned}
J^i_{\text{tot}} &= -\epsilon_{ijk}\int_{\partial V}\left(\zeta^{j0l}x^k-\lambda^{j0kl}\right)dS_l \qquad (17.106)\\
&= -\epsilon_{ijk}\int_{\partial V}\left(n^kn^lr^3\frac{\partial}{\partial x^m}\lambda^{j0lm}-n^lr^2\lambda^{j0kl}\right)d\Omega.
\end{aligned}$$

Using Eqs. 17.88, 17.102, and

$$g^{0j}=-h^{0j}=h_{0j}=\frac{2}{r^3}\epsilon_{jrs}x^rJ^s+O\left(\frac{1}{r^3}\right), \qquad (17.107)$$

neglecting terms $O(h^2)$ we find

$$\begin{aligned}
\lambda^{j0lm} &= \frac{1}{16\pi}(h_{0j}\delta_{lm}-h_{0m}\delta_{jl})+O\left(\frac{1}{r^3}\right) \qquad (17.108)\\
&= \frac{1}{8\pi r^3}(\epsilon_{jrs}\delta_{lm}-\epsilon_{mrs}\delta_{jl})x^rJ^s+O\left(\frac{1}{r^3}\right).
\end{aligned}$$

Let us compute the first term in Eq. 17.106. Neglecting higher-order terms in $1/r$ we obtain

$$-\epsilon_{ijk}n^kn^lr^3\frac{\partial}{\partial x^m}\lambda^{j0lm}=-\frac{1}{8\pi}\epsilon_{ijk}n^kn^lr^3\frac{\partial}{\partial x^m}\left(\frac{\epsilon_{jrs}\delta_{lm}-\epsilon_{mrs}\delta_{jl}}{r^3}x^r\right)J^s; \qquad (17.109)$$

since $\delta_{jl}\epsilon_{ijk}n^kn^l=\epsilon_{ijk}n^kn^j=0$ and

$$n^m\frac{\partial}{\partial x^m}\left(\frac{x^r}{r^3}\right)=\frac{\delta_{rm}-3n^mn^r}{r^3}n^m=-2\frac{n^r}{r^3}, \qquad (17.110)$$

using the property $\epsilon_{ijk}\epsilon_{jrs}=\delta_{kr}\delta_{is}-\delta_{ks}\delta_{ir}$ (see Eq. 17.44), we find

$$\begin{aligned}
-\epsilon_{ijk}n^kn^lr^3\frac{\partial}{\partial x^m}\lambda^{0jlm} &= -\frac{1}{8\pi}\epsilon_{ijk}\epsilon_{jrs}J^sn^kn^mr^3\frac{\partial}{\partial x^m}\left(\frac{x^r}{r^3}\right) \qquad (17.111)\\
&= \frac{1}{4\pi}\epsilon_{ijk}\epsilon_{jrs}J^sn^kn^r=\frac{1}{4\pi}(\delta_{kr}\delta_{is}-\delta_{ks}\delta_{ir})J^sn^kn^r\\
&= \frac{1}{4\pi}(J^i-J^kn^kn^i).
\end{aligned}$$

For the second term in Eq. 17.106, using the property $\epsilon_{ijk}\delta_{jk} = 0$, we find

$$\begin{aligned}\epsilon_{ijk}n^l r^2 \lambda^{0jkl} &= \frac{n^l}{8\pi}\epsilon_{ijk}(\epsilon_{jrs}\delta_{kl} - \epsilon_{lrs}\delta_{jk})n^r J^s \\ &= \frac{1}{8\pi}\epsilon_{ijk}\epsilon_{jrs}n^k n^r J^s = \frac{1}{8\pi}(\delta_{kr}\delta_{is} - \delta_{ks}\delta_{ir})n^k n^r J^s = \frac{1}{8\pi}(J^i - J^k n^k n^i)\,.\end{aligned} \tag{17.112}$$

Replacing in Eq. 17.106,

$$\begin{aligned}J^i_{\text{tot}} &= -\epsilon_{ijk}\int_{\partial V}\left(n^k n^l r^3 \frac{\partial}{\partial x^m}\lambda^{0jlm} - n^l r^2 \lambda^{j0kl}\right) d\Omega \\ &= \frac{3}{2}\frac{1}{4\pi}\int_{\partial V}(J^i - J^k n^k n^i)d\Omega = \frac{3}{2}J^k \frac{1}{4\pi}\int_{\partial V}(\delta_{ik} - n^i n^k)d\Omega = J^i\,,\end{aligned} \tag{17.113}$$

where we have used $\int_{\partial V} d\Omega n^i n^k = (4\pi/3)\delta_{ik}$ (see Eq. 13.145).

We can conclude that the integration constants M and J appearing in the metric of the far-field limit of a stationary, isolated source given in Eq. 17.3 can be correctly interpreted as the mass-energy and the angular momentum of the system. For a weakly gravitating source, the contribution of the gravitational field is negligible; if the gravitational field of the source is strong, it contributes to these quantities through the stress-energy pseudo-tensor $t^{\mu\nu}$.

17.4 PRECESSION OF A GYROSCOPE IN A GRAVITATIONAL FIELD

A *gyroscope* is a body with an intrinsic angular momentum which is not subjected to external torques; this means that non-gravitational forces, if present, act on its center of mass. We can model a gyroscope in General Relativity as a point particle with four-velocity $\vec{u}$, and with an *intrinsic spin four-vector* $\vec{S}$.

In a LIF the laws of Special Relativity hold and, if the body moves on a geodesic, absence of external torques implies[5]

$$\frac{dS^\mu}{d\tau} = 0\,, \tag{17.114}$$

which can be written as $u^\alpha S^\mu{}_{,\alpha} = u^\alpha S^\mu{}_{;\alpha} = 0$. By definition, the intrinsic spin vector in Special Relativity – like the orbital angular momentum of a moving particle – is orthogonal to the four-velocity of the body: $S_\mu u^\mu = 0$ (see Eq. 10.38); this condition is equivalent to the requirement that the spin vector has vanishing time component in the comoving frame. Since these are tensor equations, they hold in any coordinate frame:

$$u^\alpha S^\mu{}_{;\alpha} = 0 \tag{17.115}$$

$$S_\mu u^\mu = 0\,. \tag{17.116}$$

If the gyroscope does not move on a geodesic, i.e. non-gravitational forces act on its center of mass, it has a non-vanishing four-acceleration $a^\mu = u^\nu u^\mu{}_{;\nu}$. It can be shown that the four-acceleration induces a rotation of the spin vector in the plane containing the four-velocity and the four-acceleration:

$$u^\alpha S^\mu{}_{;\alpha} = -u^\mu a^\alpha S_\alpha\,. \tag{17.117}$$

This phenomenon, also present in Special Relativity, is called *Thomas precession* and will not be discussed in this book.

[5] In Newtonian physics the angular momentum $\mathbf{L}$ satisfies the equation $d\mathbf{L}/dt = \mathbf{M}$, where $\mathbf{M}$ is the external torque; if $\mathbf{M} = 0$ the angular momentum is conserved. Eq. 17.114 is the special relativistic generalization of the conservation of angular momentum.

17.4.1 Gyroscopes in the gravitational field of a rotating body: the Lense-Thirring precession

Let us consider a gyroscope in the gravitational field generated by an isolated, stationary object with non-vanishing angular momentum J^i. Let $\{x^\mu\}$ be the coordinate frame in which $g_{\mu\nu,0}=0$ and, in the far-field limit, $g_{\mu\nu}=\eta_{\mu\nu}+h_{\mu\nu}$ with $h_{\mu\nu}$ given by Eqs. 17.100. We shall show that, due to the coupling with the angular momentum of the central body, the spin of the gyroscope precesses. This effect, called **Lense-Thirring precession**, is a remarkable prediction of General Relativity.

Let us first consider, for simplicity, a gyroscope which does not move in space, i.e. such that $u^i=0$. Therefore, Eq. 17.116 gives $S_0=0$, and thus $S^0=g^{0\mu}S_\mu=-h^{0i}S_i=O(h)$. The space components of Eq. 17.115 reduce to[6]

$$u^\mu S^i{}_{;\mu}=u^0 S^i{}_{;0}=u^0 S^i{}_{,0}+\Gamma^i_{0j}S^j+\Gamma^i_{00}S^0=0 \tag{17.118}$$

and therefore, neglecting $O(h^2)$ terms,

$$\frac{dS^i}{d\tau}=u^0S^i{}_{,0}=-\Gamma^i_{0j}S^j+\Gamma^i_{00}S^0=-\frac{1}{2}(h_{0i,j}-h_{0j,i})S^j=-2B_{ij}S^j \tag{17.119}$$

where we have used the fact that $\Gamma^i_{00}S^0=O(h^2)$, and we have defined

$$B_{ij}=-B_{ji}=\frac{1}{4}(h_{0i,j}-h_{0j,i})\,. \tag{17.120}$$

As shown in Box 17-A, if we define $\omega^k=\epsilon^{kij}B_{ij}$, it follows that $B_{ij}=\frac{1}{2}\epsilon_{ijk}\omega^k$. Hence, Eq. 17.119 can be written as

$$\frac{dS^i}{d\tau}=-\epsilon_{ijk}\omega^kS^j=\epsilon_{ijk}\omega^jS^k\,,\qquad \text{i.e.}\qquad \frac{d\mathbf{S}}{d\tau}=\boldsymbol{\omega}\times\mathbf{S}: \tag{17.121}$$

the three-vector $\mathbf{S}=\{S^i\}$ rotates in space with angular velocity $\boldsymbol{\omega}=\{\omega^i\}$, which can be found using Eq. 17.120,

$$\omega^k=\frac{1}{2}\epsilon^{kij}h_{0i,j}\,. \tag{17.122}$$

Since the gyroscope is moving in the far-field region of a stationary, isolated object, h_{0i} is given by Eqs. 17.100:

$$h_{0i}=\frac{2}{r^3}\epsilon_{ijk}x^jJ^k\,. \tag{17.123}$$

The angular velocity of the spin precession is then

$$\omega^k=\frac{1}{2}\epsilon^{ijk}h_{0i,j}=\epsilon^{ijk}\epsilon_{ilm}\left(\frac{x^lJ^m}{r^3}\right)_{,j}=\frac{J^m}{r^3}\left(\delta^j_l\delta^k_m-\delta^j_m\delta^k_l\right)\left(\delta^l_j-\frac{3x^lx^j}{r^2}\right), \tag{17.124}$$

where we have used the relation $r^n{}_{,i}=nx^ir^{n-2}$ and the property 17.44, $\epsilon^{ijk}\epsilon_{klm}=\delta^i_l\delta^j_m-\delta^i_m\delta^j_l$. Then, since $\delta^j_l(\delta^l_j-3x^lx^j/r^2)=0$,

$$\omega^k=-\frac{J^j}{r^3}\left(\delta^k_j-\frac{3x^kx^j}{r^2}\right)=\frac{1}{r^3}\left(-J^k+3\frac{J^jx^jx^k}{r^2}\right), \tag{17.125}$$

[6]Strictly speaking, a static body in a gravitational field has a non-vanishing acceleration, and thus the general formula 17.117 should be used: $u^\alpha S^\mu{}_{;\alpha}=-u^\mu a^\alpha S_\alpha$. However, since we are only interested in the equations for the space components S^i and $u^i=0$, the acceleration term disappears and Eq. 17.118 is correct.

i.e.

$$\omega = \omega_{\rm LT} \equiv \frac{1}{r^3}\left(-\mathbf{J} + 3\frac{\mathbf{J}\cdot\mathbf{x}}{r^2}\mathbf{x}\right) . \tag{17.126}$$

The precession frequency $\boldsymbol{\omega}_{\rm LT}$ defined in Eq. 17.126 is called *Lense-Thirring frequency.*

The Lense-Thirring precession is an example of *dragging of inertial frames* (which will be also discussed in Sec. 18.5): as the central body rotates, it drags the inertial frames in the surrounding spacetime, because the angular momentum of the central body couples with the spin of the gyroscope. Thus, although the gyroscope keeps its direction fixed with respect to the LIF, it rotates with respect to the asymptotically flat frame. In practice, the gyroscope rotates with respect to the direction of far-away stars (which – neglecting for simplicity the effects of cosmology – have "fixed" directions in the asymptotically flat frame).

17.4.2 Moving gyroscopes: geodesic precession

In the previous section we considered a gyroscope at rest, i.e. with $u^i = 0$. If the gyroscope is in motion, its spin undergoes a further precession (distinct from the Lense-Thirring one) due to the coupling between the orbital angular momentum of the gyroscope and its spin. This phenomenon is called **geodesic precession** or *de Sitter precession*, and is typically much larger than the Lense-Thirring precession.

In order to study the gyroscope motion it is convenient to consider the comoving frame, i.e. the local Fermi frame $\{x^{*\mu}\}$ adapted to the source (see Sec. 3.9). For simplicity we shall assume that the gyroscope moves along a geodesic.

We remind that the Fermi frame $\{\vec{e}^{\,*}_{(\mu)}\}$ is an orthonormal basis ($\vec{e}^{\,*}_{(\mu)} \cdot \vec{e}^{\,*}_{(\nu)} = \eta_{\mu\nu}$) such that $\vec{e}^{\,*}_{(0)} = \vec{u}$ is the four-velocity of the body (in this case, the gyroscope), and $\{\vec{e}^{\,*}_{(i)}\}$ are three spacelike vectors orthogonal to $\vec{u}$. With this definition the spatial orientation of the vectors $\{\vec{e}^{\,*}_{(i)}\}$ can be arbitrarily chosen; we choose them to have the same space orientation as the coordinate basis vectors associated to the coordinates (r, θ, φ) of the asymptotically flat metric.

The motion of a gyroscope moving on a general geodesic orbit is quite involved. We shall here consider the simple case of a *circular, equatorial* orbit around a spherically symmetric object (thus neglecting its angular momentum). Obviously, in this case the Lense-Thirring precession will not be present. As discussed in Chapter 9, the spacetime outside the central body is described by the Schwarzschild metric,

$$ds^2 = -\left(1 - \frac{2M}{r}\right)dt^2 + \left(1 - \frac{2M}{r}\right)^{-1} dr^2 + r^2(d\theta^2 + \sin^2\theta d\varphi^2) , \tag{17.127}$$

where we remind that M is the mass of the central body in geometrized units. We shall first determine how the spin of the gyroscope evolves in the frame $\{x^\mu\} = (t, r, \theta, \varphi)$, and then derive the corresponding equations in the local Fermi frame.

Since the gyroscope is a (massive) body in circular equatorial motion around the central object [7], $u^\mu = u^t(1, 0, 0, \omega)$ where (see Sec. 10.5)

$$\omega = \frac{d\varphi}{dt} = \frac{u^\varphi}{u^t} = \sqrt{\frac{M}{r^3}} \tag{17.128}$$

is the Keplerian orbital frequency of the body. Moreover, since $u^\mu u_\mu = -1$ and $\theta = \pi/2$,

$$g_{\mu\nu}u^\mu u^\nu = (u^t)^2\left[-\left(1 - \frac{2M}{r}\right) + \omega^2 r^2\right] = -(u^t)^2\left(1 - \frac{3M}{r}\right) = -1 ,$$

[7] We assume that the mass of the gyroscope is much smaller than that of the central body.

therefore

$$u^t = \left(1 - \frac{3M}{r}\right)^{-1/2} . \tag{17.129}$$

Eq. 17.116, $S_\mu u^\mu = 0$, yields

$$g_{\mu\nu} S^\mu u^\nu = -\left(1 - \frac{2M}{r}\right) S^t u^t + r^2 S^\varphi \omega u^t = 0 , \tag{17.130}$$

hence

$$S^t = r^2 \left(1 - \frac{2M}{r}\right)^{-1} \omega S^\varphi . \tag{17.131}$$

The evolution equation of the spin vector is given by Eq. 17.115, $u^\alpha S^\mu{}_{;\alpha} = 0$, i.e.

$$\frac{dS^i}{d\tau} = u^\alpha S^i{}_{,\alpha} = -\Gamma^i_{\alpha\beta} u^\alpha S^\beta . \tag{17.132}$$

The non-vanishing Christoffel symbols of the Schwarzschild metric are given in Box 9-A; on the equatorial plane they become

$$\begin{array}{lll} \Gamma^t_{tr} = \Gamma^t_{rt} = \frac{M}{r^2}\left(1 - \frac{2M}{r}\right)^{-1} & \Gamma^r_{tt} = \frac{M}{r^2}\left(1 - \frac{2M}{r}\right) & \Gamma^r_{rr} = -\frac{M}{r^2}\left(1 - \frac{2M}{r}\right)^{-1} \\ \Gamma^r_{\theta\theta} = -(r - 2M) & \Gamma^r_{\varphi\varphi} = -(r - 2M) & \Gamma^\varphi_{r\varphi} = \Gamma^\varphi_{\varphi r} = \Gamma^\theta_{r\theta} = \Gamma^\theta_{\theta r} = \frac{1}{r} . \end{array} \tag{17.133}$$

Using these expressions, the θ-component of Eq. 17.132 gives

$$\frac{dS^\theta}{d\tau} = -\Gamma^\theta_{\alpha\beta} u^\alpha S^\beta = -\Gamma^\theta_{r\theta}(u^r S^\theta + u^\theta S^r) = 0 , \tag{17.134}$$

since $u^r = u^\theta = 0$. Thus, S^θ is constant; let us assume for simplicity that $S^\theta = 0$, i.e. that the spin of the gyroscope lies in the orbital plane.

The r-component of Eq. 17.132 is (using $u^\varphi = \omega u^t$ and Eq. 17.131)

$$\begin{aligned} \frac{dS^r}{d\tau} &= -\Gamma^r_{\alpha\beta} u^\alpha S^\beta = -\Gamma^r_{tt} u^t S^t - \Gamma^r_{rr} u^r S^r - \Gamma^r_{\theta\theta} u^\theta S^\theta - \Gamma^r_{\varphi\varphi} u^\varphi S^\varphi & (17.135) \\ &= -u^t \left(\Gamma^r_{tt} S^t + \Gamma^r_{\varphi\varphi} \omega S^\varphi\right) = -\omega u^t S^\varphi \left[\Gamma^r_{tt} r^2 \left(1 - \frac{2M}{r}\right)^{-1} + \Gamma^r_{\varphi\varphi}\right] & (17.136) \\ &= -\omega u^t S^\varphi \left[(M - (r - 2M)\right] = \omega u^t (r - 3M) S^\varphi . \end{aligned}$$

The φ-component is

$$\frac{dS^\varphi}{d\tau} = -\Gamma^\varphi_{\alpha\beta} u^\alpha S^\beta = -\Gamma^\varphi_{\varphi r} u^\varphi S^r = -\omega \frac{u^t}{r} S^r . \tag{17.137}$$

Finally, in terms of the global time coordinate t, we find the system of equations

$$\begin{aligned} \frac{dS^r}{dt} &= \frac{1}{u^t}\frac{dS^r}{d\tau} = \omega r \left(1 - \frac{3M}{r}\right) S^\varphi & (17.138) \\ \frac{dS^\varphi}{dt} &= \frac{1}{u^t}\frac{dS^\varphi}{d\tau} = -\omega \frac{1}{r} S^r . \end{aligned}$$

The Fermi frame with spacelike vectors oriented as those of the coordinate basis associated to (r, θ, φ), i.e. $\{\vec{e}_{(\mu)}\} = \left\{\frac{\partial}{\partial x^\mu}\right\}$, in the equatorial plane is

$$\begin{aligned}
\vec{e}^{\,*}_{(0)} &= \vec{u} = u^t\left(\vec{e}_{(0)} + \omega \vec{e}_{(3)}\right) \\
\vec{e}^{\,*}_{(1)} &= \left(1 - \frac{2M}{r}\right)^{1/2} \vec{e}_{(1)} \\
\vec{e}^{\,*}_{(2)} &= \frac{1}{r}\vec{e}_{(2)} \\
\vec{e}^{\,*}_{(3)} &= u^t\left[\omega r\left(1 - \frac{2M}{r}\right)^{-1/2} \vec{e}_{(0)} + \left(1 - \frac{2M}{r}\right)^{1/2} \frac{1}{r}\vec{e}_{(3)}\right] .
\end{aligned} \tag{17.139}$$

Note that the spacelike vector $\vec{e}^{\,*}_{(3)}$ is a linear combination of $\vec{e}_{(3)}$ and $\vec{e}_{(0)}$ (the latter is needed to have $\vec{e}^{\,*}_{(3)} \cdot \vec{u} = 0$), and it is orthogonal to the other space vectors $\vec{e}_{(1)}$ and $\vec{e}_{(2)}$. It is simple to check, using the property $\vec{e}_{(\mu)} \cdot \vec{e}_{(\nu)} = g_{\mu\nu}$ and Eq. 17.129, that $\vec{e}^{\,*}_{(\mu)} \cdot \vec{e}^{\,*}_{(\nu)} = \eta_{\mu\nu}$ (we leave the proof as an exercise).

The spin vector is

$$\vec{S} = S^\mu \vec{e}_{(\mu)} = S^{*\mu} \vec{e}^{\,*}_{(\mu)} ; \tag{17.140}$$

therefore, recalling that $\vec{e}_{(1)} \cdot \vec{e}^{\,*}_{(3)} = 0$ and using Eq. 17.131, we find that the components of the spin of the gyroscope in the Fermi frame are

$$\begin{aligned}
S^{*t} &= -\vec{S} \cdot \vec{e}^{\,*}_{(0)} = -\vec{S} \cdot \vec{u} = 0 \\
S^{*r} &= \vec{S} \cdot \vec{e}^{\,*}_{(1)} = S^r \vec{e}_{(1)} \cdot \vec{e}^{\,*}_{(1)} = \left(1 - \frac{2M}{r}\right)^{-1/2} S^r \\
S^{*\theta} &= \vec{S} \cdot \vec{e}^{\,*}_{(2)} = 0 \\
S^{*\varphi} &= \vec{S} \cdot \vec{e}^{\,*}_{(3)} = \left(S^t \vec{e}_{(0)} + S^r \vec{e}_{(1)} + S^\varphi \vec{e}_{(3)}\right) \cdot \vec{e}^{\,*}_{(3)} \\
&= \left[r^2\left(1 - \frac{2M}{r}\right)^{-1} \omega S^\varphi \vec{e}_{(0)} + S^\varphi \vec{e}_{(3)}\right] \\
&\quad \cdot u^t\left[\omega r\left(1 - \frac{2M}{r}\right)^{-1/2} \vec{e}_{(0)} + \left(1 - \frac{2M}{r}\right)^{1/2} \frac{1}{r}\vec{e}_{(3)}\right] \\
&= u^t\left[-\omega^2 r^3\left(1 - \frac{2M}{r}\right)^{-1/2} + r\left(1 - \frac{2M}{r}\right)^{1/2}\right] S^\varphi \\
&= u^t r\left(1 - \frac{2M}{r}\right)^{-1/2}\left(1 - \frac{2M}{r} - \omega^2 r^2\right) S^\varphi \\
&= r\left(1 - \frac{2M}{r}\right)^{-1/2}\left(1 - \frac{3M}{r}\right)^{1/2} S^\varphi .
\end{aligned} \tag{17.141}$$

Then, multiplying Eqs. 17.138 by $(1 - 2M/r)^{-1/2}$ and $r(1 - 3M/r)^{1/2}(1 - 2M/r)^{-1/2}$, respectively, we obtain

$$\begin{aligned}
\frac{dS^{*r}}{dt} &= \omega r\left(1 - \frac{3M}{r}\right)\left(1 - \frac{2M}{r}\right)^{-1/2} S^\varphi = \omega\left(1 - \frac{3M}{r}\right)^{1/2} S^{*\varphi} \\
\frac{dS^{*\varphi}}{dt} &= -\omega\left(1 - \frac{3M}{r}\right)^{1/2} S^{*r} .
\end{aligned} \tag{17.142}$$

Defining

$$\omega' = \left(1 - \frac{3M}{r}\right)^{1/2} \omega \,, \tag{17.143}$$

Eqs. 17.142 become

$$\begin{aligned} \frac{dS^{*r}}{dt} &= \omega' S^{*\varphi} \\ \frac{dS^{*\varphi}}{dt} &= -\omega' S^{*r} \,. \end{aligned} \tag{17.144}$$

These equations show that the spin vector rotates with frequency ω' in the comoving frame $\{\vec{e}^{\,*}_{(r)}, \vec{e}^{\,*}_{(\varphi)}\}$. If, for instance, at $t = 0$ the spin points towards the radial direction, $S^{*r}(0) = S_0$, $S^{*\varphi}(0) = 0$, then $S^{*r}(t) = S_0 \cos\omega' t$ and $S^{*\varphi}(t) = -S_0 \sin\omega' t$.

We now determine the spin of the gyroscope with respect to the asymptotically flat directions, i.e. the directions which, in the far-field limit, correspond to the Cartesian coordinate lines of the Minkowski metric $\eta_{\mu\nu}$. These are often called "fixed stars directions" (neglecting the effect of the cosmological expansion).

To this aim, reminding that we have the freedom to rotate the spatial vectors of the Fermi frame, we change the basis vectors $\{\vec{e}^{\,*}_{(r)}, \vec{e}^{\,*}_{(\varphi)}\}$ to $\{\vec{e}^{\,*}_{(x)}, \vec{e}^{\,*}_{(y)}\}$, parallel to the Cartesian coordinate axes in the far-field limit, defined as follows:

$$\begin{aligned} \vec{e}^{\,*}_{(x)} &= \cos\varphi \vec{e}^{\,*}_{(r)} - \sin\varphi \vec{e}^{\,*}_{(\varphi)} \\ \vec{e}^{\,*}_{(y)} &= \sin\varphi \vec{e}^{\,*}_{(r)} + \cos\varphi \vec{e}^{\,*}_{(\varphi)} \,. \end{aligned} \tag{17.145}$$

These expressions can be obtained from the rotation $x = r\cos\varphi$, $y = r\sin\varphi$, by inverting Eq. 3.136. We stress that the Cartesian vectors $\vec{e}^{\,*}_{(x)}$, $\vec{e}^{\,*}_{(y)}$ are still vectors of the local Fermi frame of the gyroscope, with a different space orientation.

The spin components along the Cartesian directions are

$$\begin{aligned} S^{*x} &= \vec{S} \cdot \vec{e}^{\,*}_{(x)} = (S^{*r} \vec{e}^{\,*}_{(r)} + S^{*\varphi} \vec{e}^{\,*}_{(\varphi)}) \cdot (\cos\varphi \vec{e}^{\,*}_{(r)} - \sin\varphi \vec{e}^{\,*}_{(\varphi)}) \\ &= \cos\varphi S^{*r} - \sin\varphi S^{*\varphi} \\ S^{*y} &= \vec{S} \cdot \vec{e}^{\,*}_{(y)} = (S^{*r} \vec{e}^{\,*}_{(r)} + S^{*\varphi} \vec{e}^{\,*}_{(\varphi)}) \cdot (\sin\varphi \vec{e}^{\,*}_{(r)} + \cos\varphi \vec{e}^{\,*}_{(\varphi)}) \\ &= \sin\varphi S^{*r} + \cos\varphi S^{*\varphi} \,, \end{aligned} \tag{17.146}$$

and since $S^{*r}(t) = S_0 \cos\omega' t$, $S^{*\varphi} = -S_0 \sin\omega' t$, and and $\varphi = \omega t$ along the circular orbit,

$$\begin{aligned} S^{*x} &= S_0(\cos\omega t \cos\omega' t + \sin\omega t \cos\omega' t) = S_0 \cos(\omega - \omega')t \\ S^{*y} &= S_0(\sin\omega t \cos\omega' t - \cos\omega t \sin\omega' t) = -S_0 \sin(\omega - \omega')r \,. \end{aligned} \tag{17.147}$$

In the Newtonian limit $\omega = \omega'$ and thus (S^{*x}, S^{*y}) are constant: as it is well known, a moving gyroscope keeps its direction fixed with respect to far-away stars. In General Relativity, instead, the gyroscope precesses with respect to the "fixed stars directions", with the *geodesic precession frequancy*

$$\omega_{\rm GP} = \omega - \omega' = \left[1 - \left(1 - \frac{3M}{r}\right)^{1/2}\right] \omega \simeq \frac{3M}{2r}\omega = \frac{3M}{2r}\sqrt{\frac{M}{r^3}} \,, \tag{17.148}$$

where we have neglected higher-order terms in $1/r$ while using Eq. 17.143.

When the central body rotates, both the geodesic precession and the Lense-Thirring precession are present. It can be shown that (in the weak-field limit) the spin, measured in the local frame with the orientation of the asymptotically Cartesian directions, satisfies the equation

$$\frac{d\mathbf{S}^*}{dt} = (\boldsymbol{\omega}_{\rm GP} + \boldsymbol{\omega}_{\rm LT}) \times \mathbf{S}^* \tag{17.149}$$

where $\boldsymbol{\omega}_{\rm GP}$ is the geodesic precession frequency (which reduces, in the case of circular equatorial motion, to Eq. 17.148) and $\boldsymbol{\omega}_{\rm LT}$ is the Lense-Thirring frequency given in Eq. 17.126.

As mentioned above, the geodesic precession frequency is typically much larger than the Lense-Thirring frequency. Let us consider, for instance, a gyroscope in a spacecraft moving on a circular, equatorial orbit 600 km above the surface of Earth (i.e. about 7000 km from the center of Earth). In this case the Earth angular momentum is orthogonal to the position vector $\mathbf{x}$ of the gyroscope, hence Eqs. 17.126, 17.148 give (in physical units, see Box 9-A)

$$|\omega_{\rm LT}| = \frac{GJ}{c^2 r^3} \tag{17.150}$$

$$|\omega_{\rm GP}| = \frac{3GM}{2c^2 r}\sqrt{\frac{GM}{r^3}} \tag{17.151}$$

where $M = M_\oplus = 5.9 \times 10^{27}$ g, $r = 7.0 \times 10^8$ cm, and, with the rough approximation of a uniform-density Earth, $J \sim \frac{2}{5} M_\oplus R_\oplus^2 \Omega_\oplus$. Since the rotation frequency of the Earth is $\Omega_\oplus = 2\pi/86400 = 7.3 \times 10^{-5}$ s^{-1}, its radius is $R_\oplus = 6.378 \times 10^8$ cm and $G = 6.674 \times 10^{-8}$ cm^3 g^{-1} s^{-2}, $c = 2.998 \times 10^{10}$ cm/s (see Table A), we find $J \sim 7 \times 10^{40}$ g cm^2 s^{-1}; hence $|\omega_{\rm LT}| \sim 1.5 \times 10^{-14}$ rad/s ~ 0.1 arcsec/year, and $|\omega_{\rm GP}| \sim 10^{-12}$ rad/s ~ 6 arcsec/year.

The configuration discussed above maximizes the Lense-Thirring effect. If the angular momentum of the central object is not orthogonal to the orbital plane (as in the case of an orbit along a meridian circle of the Earth, see Sec. 17.4.3) the Lense-Thirring frequency is even smaller, as shown by Eq. 17.126, while the geodesic precession frequency is the same.

17.4.3 Measurement of geodesic and Lense-Thirring frequencies: Gravity Probe B

The idea of an experiment to measure the geodesic and Lense-Thirring precessions dates back to 1960, when Shiff [103] noted that a gyroscope on a spacecraft orbiting around Earth would undergo geodesic and Lense-Thirring precessions with respect to the directions of "fixed stars" at infinity (see also [98] for a similar, independent proposal). An estimate of the order of magnitude of these effects showed that they would be "difficult, but not impossible, to observe".

The realization of this experiment – called Gravity Probe B (GPB) – has been extremely difficult. The satellite was launched in 2004 and the mission ended in 2005; after that, six more years were needed to analyse the data and understand all sources of error. In 2011 the final results were published [45]: the geodesic precession was measured with an accuracy of $\simeq 0.3\%$, and the more elusive Lense-Thirring precession was measured with an accuracy of $\simeq 20\%$. The observed values are compatible with the predictions of General Relativity, and can be considered a further *kinematical test of General Relativity*, which adds to those discussed in Chapter 11.

As discussed above the geodesic precession frequency is much larger than the Lense-Thirring frequency. Thus if $\boldsymbol{\omega}_{\rm GP}$ is parallel to $\boldsymbol{\omega}_{\rm LT}$, the Lense-Thirring frequency is "buried" in the geodesic precession frequency, increasing the difficulty of the measurement. To overcome this problem the orbit of GPB was chosen to be orthogonal to the equatorial plane (see Fig. 17.2), and the spin of the gyroscope was oriented orthogonally to both the angular momenta of the Earth and of the orbit. Thus, in the Cartesian frame $(Oxyz)$ in Fig. 17.2, the angular momentum of the Earth is $J^i = (0, 0, J)$, the orbit lies in the $x - z$ plane, and the spin of the gyroscope is $S^i = (S, 0, 0)$. With this configuration, the geodesic precession

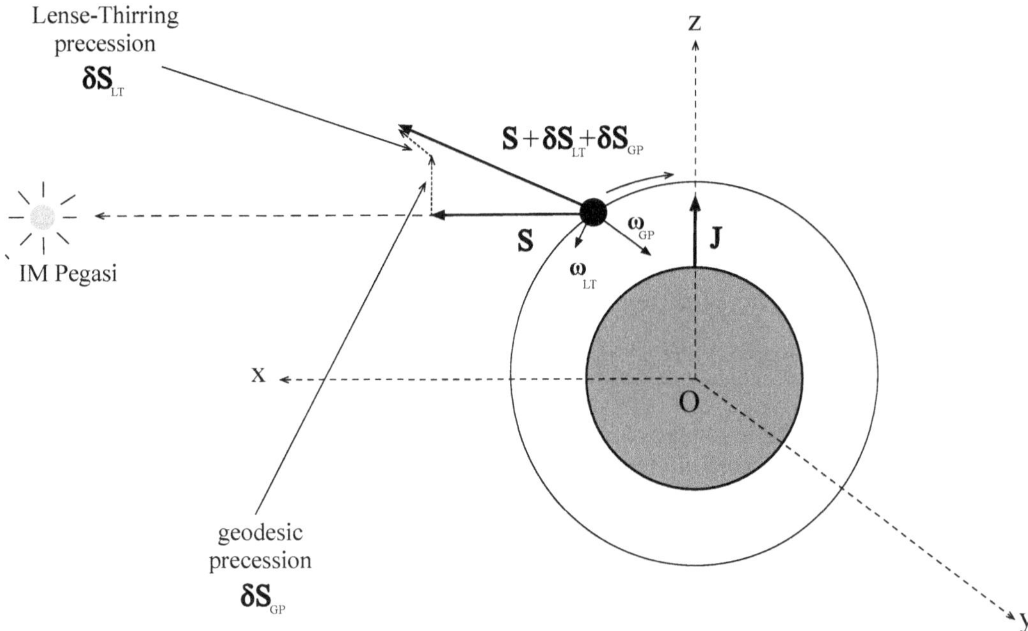

Figure 17.2: Schematic description of the GPB experiment. The spin of the satellite has a geodesic precession from South to North, and a Lense-Thirring precession from East to West.

frequency – which is always orthogonal to the orbital plane – is parallel to the y-axis, and determines a precession of the spin in the $x - z$ plane, i.e. in the North-South direction. The Lense-Thirring frequency 17.126, instead, lies in the orbital plane $x - z$ (because both the Earth angular momentum $\mathbf{J}$ and the radial direction $\mathbf{n}$ belong to that plane); since the spin $\mathbf{S}$ is parallel to the x-axis, the Lense-Thirring precession 17.126 – $\delta\mathbf{S} \sim \boldsymbol{\omega}_{\rm LT} \times \mathbf{S} = (0, \omega^z_{\rm LT} S, 0)$ – is parallel to the y-axis, i.e. in the West-East direction. It was thus possible to measure the two effects independently.

The actual value of the Earth moment of inertia (known from geophysical modelling of the Earth) is $I_\oplus = 8.0 \times 10^{44}$ g cm^2. Eq. 17.151 yields $|\omega_{\rm GP}| = 6.6$ arcsec/year. To compute the Lense-Thirring frequency 17.126 we note that $\mathbf{n} = (\cos\Omega_\oplus t, 0, \sin\Omega_\oplus t)$ and $\mathbf{J} = (0, 0, J)$, therefore

$$-\mathbf{J} + 3(\mathbf{J}\cdot\mathbf{n})\mathbf{n} = (3J\sin\Omega_\oplus t\cos\Omega_\oplus t, 0, -J(1 - 3\sin^2\Omega_\oplus t)) \tag{17.152}$$

hence, averaging the z-component of the Lense-Thirring frequency on an orbit with period $T = 2\pi/\Omega_\oplus$,

$$\langle\omega^z_{\rm LT}\rangle = \frac{1}{T}\int_0^T J\left(3\sin^2\frac{2\pi}{T}t - 1\right)dt = \frac{J}{2\pi}\int_0^{2\pi}\left(3\sin^2\alpha - 1\right)d\alpha = \frac{1}{2}J\,, \tag{17.153}$$

which, in physical units, gives

$$|\omega^z_{\rm LT}| = \frac{GI\Omega_\oplus}{2c^2r^3} = 0.04\,\text{arcsec/year}\,. \tag{17.154}$$

GPB contained four gyroscopes, which were electrostatically suspended, rotating spheres. Each sphere contained a superconducting loop producing a magnetic dipole moment parallel to the spin direction. A magnetometer measured the direction of the magnetic moment, and then of the spin, of each gyroscope. These directions were compared with the direction of the far-away star IM Pegasi, determined by a telescope on the spacecraft. The results obtained by combining the observations of the four gyroscopes are shown in Table 17.1, and compared with the theoretical predictions of General Relativity.

Table 17.1: Precession frequency measured by Gravity Probe B along the North-South direction (Lense-Thirring frequency) and along the West-East direction (geodesic precession frequency), compared with the theoretical values predicted by General Relativity.

	$\omega_{\rm LT}$ (milli-arcsec/year)	$\omega_{\rm GP}$ (milli-arcsec/year)
General Relativity prediction	-39.2	-6606.1
GPB	-37.2 ± 7.2	-6601.8 ± 18.3

Remarkably, when the results of this experiment were released the geodesic and Lense-Thirring effect of the Earth had already been measured by analysing the motion of a pair of satellites, whose position was constantly monitored by sending laser impulses from the Earth to the satellites. This experiment, called LAGEOS, required the subtraction of the non-spherical component of the Earth gravitational field, which had previously been determined by a similar experiment, GRACE. In 2004 the LAGEOS experiment measured the geodesic precession frequency with an accuracy of $\simeq 0.7\%$, and the Lense-Thirring frequency with an accuracy of $\simeq 10\%$. The accuracy of these measurements has significantly been improved with a recent analysis of the LAGEOS data [37, 80].

We also note that while the Lense-Thirring effect was firstly observed by the GPB and LAGEOS experiments, the geodesic precession had been measured decades earlier studying the motion of binary pulsars such as PSR 1913+16 (see e.g. [118] and references therein).

CHAPTER 18

The Kerr solution

As shown in Chapter 9, the solution of Einstein's equations describing the exterior of an isolated, spherically symmetric, static object is quite simple. Indeed, the Schwarzschild solution was found in 1916, immediately after the publication of Einstein's equation. Finding the solution for the gravitational field of a rotating body (all astrophysical objects do rotate to some extent) is a much more difficult problem; indeed we do not know any exact solution describing the exterior of a rotating star, and only some approximate solutions are known. However, there exists an exact solution of Einstein's equations in vacuum ($T_{\mu\nu} = 0$), which describes a rotating, stationary, axially symmetric *black hole*. It was derived in 1963 by Roy Kerr [68], and it is known as the **Kerr solution**. As for the Schwarzschild solution, it describes the spacetime generated by a curvature singularity concealed by a *horizon*.

We stress that while, owing to Birkhoff's theorem (Sec. 9.6), the Schwarzschild metric for $r > 2M$ also describes the exterior of *any* spherically symmetric, static, isolated object (a star, a planet, etc.), the Kerr metric outside the horizon can only describe the exterior of a black hole [1].

18.1 OBSERVATIONAL EVIDENCE FOR ROTATING BLACK HOLES

Black holes are solutions of Einstein's equations *in vacuum*. Since these equations do not contain dimensionful parameters (see Sec. 6.2) the mass M, which arises as a dimensionful integration constant, must be a *free parameter*. This is at variance with the mass of a stellar object (e.g., a neutron star), since the microphysical properties of stars determine a maximum mass (see Chapter 16). Therefore, as mathematical solutions of Einstein's equations, black holes can have *any* mass. However, this does not necessarily imply that black holes exist with any mass in nature. Indeed, the mass of *astrophysical* black holes is determined, and limited, by their formation mechanism and their evolutionary path.

A growing amount of observational evidence suggests that astrophysical black holes exist (at least) in two categories:

- *Stellar-mass black holes.* These are either the outcome of the gravitational collapse of sufficiently massive stars or, as discussed in Chapter 14, of the merger of compact objects (neutron stars or black holes). The former have masses in the range from a few to several tens of solar masses, and have been indirectly observed through the X-ray emission from their accretion disk, whose inner edge can be associated to the ISCO of the black hole (see Sec. 10.5); the latter can reach masses up to $\sim 10^2\,M_\odot$, and the gravitational wave signal emitted when they form has been detected by interferometric

[1] Actually, a rigorous proof that a stellar model matching with Kerr's metric at its surface does not exist is still lacking.

detectors. The heaviest black hole detected during the first two observational runs of the LIGO and Virgo interferometers is the remnant of GW170729, with mass $M \simeq 80.3^{+14.6}_{-10.2} M_\odot$ (see Table 14.1). At the time of writing, the events detected during the third observational run in 2019 are still being analyzed. Binary black hole mergers have been detected on a weekly basis and the heaviest one is the remnant of GW190521 with mass $M \simeq 142^{+28}_{-16} M_\odot$ [5].

- *Supermassive black holes.* These are much heavier black holes, whose masses range approximately from $10^5 M_\odot$ to $10^{11} M_\odot$. They sit at the center of most galaxies and play a crucial role in galaxy evolution. The center of the Milky Way hosts Sgr A*, a supermassive black hole with $M \approx 4 \times 10^6 M_\odot$. During the cosmic evolution, supermassive black holes are expected to merge as a result of galaxy mergers, thus producing heavier black holes. Nonetheless, the origin of the heaviest supermassive black holes which have been observed at high redshift is still unclear. To date supermassive black holes have been observed indirectly by monitoring the motions of stars in close orbit around them (as in the case of Sgr A* [48]), or by measuring the X-ray and infrared spectrum from accreting supermassive compact objects. In 2019, the Event Horizon Telescope Collaboration provided the first radio image of the "shadow" (see Sec. 11.5) of the supermassive black hole in the galaxy M87, also measuring its mass, $M \approx 6.5 \times 10^9 M_\odot$ [111]. So far the only observational evidence for supermassive black holes comes from electromagnetic measurements. Detecting the gravitational waves emitted during the coalescence of two supermassive black holes is one of the main science goals of the Laser Interferometer Space Antenna (LISA), a European/American space mission expected to be launched in 2034 [7] (see Sec. 14.2.6).

There are observational and theoretical hints that *intermediate-mass black holes*, with masses filling the gap between stellar-mass and supermassive black holes, should exist. The strongest evidence to date comes from the detection of GW190521 [5].

Finally, it has been speculated that *primordial black holes* – with masses below the solar mass – might be formed in the early universe (see Sec. 20.4). Although the search for these objects is very active, to date their existence remains hypothetical.

18.2 THE KERR METRIC IN THE BOYER-LINDQUIST COORDINATES

At variance with the Schwarzschild metric, the derivation of the Kerr solution is a formidable task which goes beyond the scope of this book. We refer the interested reader to the original paper by Kerr [68] or to Chandrasekhar's monograph [34]. Here we simply give the explicit form of the Kerr metric (we shall use geometrized units $G = c = 1$, unless otherwise specified):

$$ds^2 = -dt^2 + \Sigma\left(\frac{dr^2}{\Delta} + d\theta^2\right) + (r^2 + a^2)\sin^2\theta d\varphi^2 + \frac{2Mr}{\Sigma}(a\sin^2\theta d\varphi - dt)^2\,, \qquad (18.1)$$

where

$$\begin{aligned} \Delta(r) &\equiv r^2 - 2Mr + a^2\,, \\ \Sigma(r,\theta) &\equiv r^2 + a^2\cos^2\theta\,, \end{aligned} \qquad (18.2)$$

and M and a are constants with the dimensions of a length. The coordinates (t, r, θ, φ), in terms of which the metric has the form given by Eq. 18.1, are called the *Boyer-Lindquist coordinates* [24].

The Kerr metric depends on two parameters, M and a; comparing Eq. 18.1 with the metric of the far-field limit of an isolated object (Eq. 17.3), we see that M is the black hole mass, and $J \equiv Ma$ its angular momentum.

Some properties of the Kerr metric can be directly deduced from the line element in Eq. 18.1:

- It is *axisymmetric*, since it does not depend on the azimuthal angle φ.
- It is not static, since it is not invariant for time reversal $t \to -t$.
- It is *stationary*, since it does not depend on time.
- It is invariant for the simultaneous inversion of t and φ,

$$t \to -t \qquad \varphi \to -\varphi \,. \tag{18.3}$$

  This property can be understood by noting that the time reversal of a rotating object corresponds to the same object rotating in the opposite direction.

- It is also invariant for the simultaneous inversion of φ and a,

$$a \to -a \qquad \varphi \to -\varphi \,. \tag{18.4}$$

  Note that $a \to -a$ corresponds to an object that rotates in the opposite direction. Using this property, in the following we shall focus on $a \geq 0$ without loss of generality.

- In the limit $r \to \infty$, the metric in Eq. 18.1 reduces to Minkowski's metric in polar coordinates; therefore, the Kerr spacetime is *asymptotically flat.*
- In the limit $a \to 0$ (with $M \neq 0$), $\Delta \to r^2 - 2Mr$, $\Sigma \to r^2$, and Eq. 18.1 reduces to the Schwarzschild metric

$$ds^2 \to -\left(1 - \frac{2M}{r}\right) dt^2 + \left(1 - \frac{2M}{r}\right)^{-1} dr^2 + r^2(d\theta^2 + \sin^2\theta d\varphi^2) \,. \tag{18.5}$$

- In the limit $M \to 0$ (with $a \neq 0$), Eq. 18.1 reduces to

$$ds^2 = -dt^2 + \frac{r^2 + a^2\cos^2\theta}{r^2 + a^2} dr^2 + (r^2 + a^2\cos^2\theta)d\theta^2 + (r^2 + a^2)\sin^2\theta d\varphi^2 \,, \tag{18.6}$$

  which is the metric of flat spacetime in spheroidal coordinates

$$\begin{aligned} x &= \sqrt{r^2 + a^2}\sin\theta\cos\varphi \\ y &= \sqrt{r^2 + a^2}\sin\theta\sin\varphi \\ z &= r\cos\theta \,. \end{aligned} \tag{18.7}$$

  Indeed,

$$\begin{aligned} dx &= \frac{r}{\sqrt{r^2 + a^2}}\sin\theta\cos\varphi dr + \sqrt{r^2 + a^2}\cos\theta\cos\varphi d\theta - \sqrt{r^2 + a^2}\sin\theta\sin\varphi d\varphi \,, \\ dy &= \frac{r}{\sqrt{r^2 + a^2}}\sin\theta\sin\varphi dr + \sqrt{r^2 + a^2}\cos\theta\sin\varphi d\theta + \sqrt{r^2 + a^2}\sin\theta\cos\varphi d\varphi \,, \\ dz &= \cos\theta dr - r\sin\theta d\theta \,; \end{aligned} \tag{18.8}$$

  thus

$$\begin{aligned} ds^2 = \; & -dt^2 + dx^2 + dy^2 + dz^2 = -dt^2 + \left(\frac{r^2}{r^2 + a^2}\sin^2\theta + \cos^2\theta\right) dr^2 \\ & + \left[(r^2 + a^2)\cos^2\theta + r^2\sin^2\theta\right] d\theta^2 + (r^2 + a^2)\sin^2\theta d\varphi^2 \\ & = -dt^2 + \frac{r^2 + a^2\cos^2\theta}{r^2 + a^2} dr^2 + (r^2 + a^2\cos^2\theta)d\theta^2 + (r^2 + a^2)\sin^2\theta d\varphi^2 \,. \end{aligned} \tag{18.9}$$

- The metric in Eq. 18.1 is singular when $\Delta = 0$ and when $\Sigma = 0$. Since the curvature invariant $R_{\mu\nu\alpha\beta}R^{\mu\nu\alpha\beta}$ (see Sec. 9.3) is singular on $\Sigma = 0$, this is a true, curvature singularity of the manifold, the structure of which will be discussed in Sec. 18.7. All curvature invariants are, instead, regular on $\Delta = 0$. As we shall show in Sec. 18.4.2, the roots of $\Delta = 0$ are *two* horizons of the Kerr metric, r_+ and r_-. These are coordinate singularities, which can be removed through an appropriate coordinate transformation.

  Note that in the Schwarzschild limit ($a = 0$), $\Sigma = r^2 = 0$ is the curvature singularity, while (for $r \neq 0$) $\Delta = r(r - 2M) = 0$ is the coordinate singularity corresponding to the black hole horizon of the Schwarzschild metric.

The metric has the form

$$g_{\mu\nu} = \begin{pmatrix} g_{tt} & 0 & 0 & g_{t\varphi} \\ 0 & \frac{\Sigma}{\Delta} & 0 & 0 \\ 0 & 0 & \Sigma & 0 \\ g_{t\varphi} & 0 & 0 & g_{\varphi\varphi} \end{pmatrix}, \tag{18.10}$$

with

$$\begin{aligned} g_{tt} &= -\left(1 - \frac{2Mr}{\Sigma}\right), \qquad g_{rr} = \frac{\Sigma}{\Delta}, \\ g_{t\varphi} &= -\frac{2Mr}{\Sigma} a \sin^2\theta, \qquad g_{\theta\theta} = \Sigma, \\ g_{\varphi\varphi} &= \left(r^2 + a^2 + \frac{2Mra^2}{\Sigma}\sin^2\theta\right)\sin^2\theta. \end{aligned} \tag{18.11}$$

Note also that, using $\Sigma = r^2 + a^2\cos^2\theta$,

$$\begin{aligned} g_{\varphi\varphi} &= (r^2 + a^2)\sin^2\theta + \frac{2Mr}{\Sigma}a^2\sin^4\theta \\ &= \frac{r^2 + a^2}{\Sigma}(r^2 + a^2 - a^2\sin^2\theta)\sin^2\theta + \frac{2Mr}{\Sigma}a^2\sin^4\theta \\ &= \frac{1}{\Sigma}\left[(r^2 + a^2)^2 - (r^2 + a^2 - 2Mr)a^2\sin^2\theta\right]\sin^2\theta \\ &= \frac{(r^2 + a^2)^2 - \Delta a^2\sin^2\theta}{\Sigma}\sin^2\theta. \end{aligned} \tag{18.12}$$

To compute the inverse metric $g^{\mu\nu}$, we only need to invert the $(t - \varphi)$ block, while the inversion of the $(r - \theta)$ part, which is diagonal, is trivial. The $(t - \varphi)$ block is

$$\tilde{g}_{ab} = \begin{pmatrix} g_{tt} & g_{t\varphi} \\ g_{t\varphi} & g_{\varphi\varphi} \end{pmatrix}, \tag{18.13}$$

and its determinant is

$$\begin{aligned} \tilde{g} &= g_{tt}g_{\varphi\varphi} - g_{t\varphi}^2 \\ &= -\left(1 - \frac{2Mr}{\Sigma}\right)\left(r^2 + a^2 + \frac{2Mra^2}{\Sigma}\sin^2\theta\right)\sin^2\theta - \frac{4M^2r^2a^2}{\Sigma^2}\sin^4\theta \\ &= -(r^2 + a^2)\sin^2\theta + 2Mr\sin^2\theta = -\Delta\sin^2\theta. \end{aligned} \tag{18.14}$$

Therefore

$$\tilde{g}^{ab} = -\frac{1}{\Delta\sin^2\theta}\begin{pmatrix} g_{\varphi\varphi} & -g_{t\varphi} \\ -g_{t\varphi} & g_{tt} \end{pmatrix} \tag{18.15}$$

and

$$g^{\mu\nu} = \begin{pmatrix} g^{tt} & 0 & 0 & g^{t\varphi} \\ 0 & \frac{\Delta}{\Sigma} & 0 & 0 \\ 0 & 0 & \frac{1}{\Sigma} & 0 \\ g^{t\varphi} & 0 & 0 & g^{\varphi\varphi} \end{pmatrix}, \tag{18.16}$$

with

$$\begin{aligned} g^{tt} &= -\frac{1}{\Delta}\left(r^2 + a^2 + \frac{2Mra^2}{\Sigma}\sin^2\theta\right), \\ g^{t\varphi} &= -\frac{2Mra}{\Sigma\Delta}, \\ g^{\varphi\varphi} &= \frac{\Delta - a^2\sin^2\theta}{\Sigma\Delta\sin^2\theta}, \end{aligned} \tag{18.17}$$

where we have used the following equality

$$\frac{\Sigma - 2Mr}{\Sigma\Delta\sin^2\theta} = \frac{r^2 + a^2\cos^2\theta - 2Mr}{\Sigma\Delta\sin^2\theta} = \frac{\Delta - a^2\sin^2\theta}{\Sigma\Delta\sin^2\theta}. \tag{18.18}$$

18.3 SYMMETRIES

Being stationary and axisymmetric, the Kerr metric admits two Killing vector fields:

$$\vec{k} \equiv \frac{\partial}{\partial t}, \qquad \vec{m} \equiv \frac{\partial}{\partial\varphi}, \tag{18.19}$$

or equivalently, in coordinates (t, r, θ, φ),

$$k^\mu \equiv (1, 0, 0, 0) \qquad m^\mu \equiv (0, 0, 0, 1). \tag{18.20}$$

As a consequence, there are two conserved quantities associated to particles in geodesic motion:

$$E \equiv -u^\mu k_\mu = -u_t \qquad L \equiv u^\mu m_\mu = u_\varphi, \tag{18.21}$$

where u^μ is the particle four-velocity. As discussed in Sec. 10.3 in the case of the Schwarzschild spacetime, for massive particles the constants E and L are the energy and azimuthal angular momentum of the particle per unit mass, measured by a static observer. For massless particles, we can choose the affine parameter such that the four-momentum coincides with the four-velocity, i.e. $p^\mu = u^\mu$; with this choice, for massless particles E and L are the energy and angular momentum of the particle as measured by a static observer.

It can be shown that $\vec{k}$, $\vec{m}$ are the only independent Killing vector fields admitted by the Kerr metric; thus, any Killing vector field is a linear combination of them.

18.4 BLACK HOLE HORIZONS

In this section we will show that the equation $\Delta = 0$ identifies *two* black hole horizons of the Kerr metric, r_+ and r_-, and we shall discuss their structure. Furthermore, we will show that the singularities on the surfaces $r = r_+$ and $r = r_-$ are coordinate singularities, which can be removed by an appropriate coordinate transformation.

18.4.1 Horizon structure

Let us consider the submanifold

$$\Delta = r^2 + a^2 - 2Mr = 0\,, \tag{18.22}$$

where the Kerr metric, written in the Boyer-Lindquist coordinates, has a coordinate singularity.

When $a^2 \leq M^2$, Eq. 18.22 has two roots:

$$\begin{aligned} r_+ &\equiv M + \sqrt{M^2 - a^2}\,, \\ r_- &\equiv M - \sqrt{M^2 - a^2}\,, \end{aligned} \tag{18.23}$$

and Δ can be written as

$$\Delta(r) = (r - r_+)(r - r_-)\,. \tag{18.24}$$

The surfaces where there is a coordinate singularity are then $r = r_+$ and $r = r_-$. Note that $\Delta < 0$ for $r_- < r < r_+$, and $\Delta > 0$ for $r < r_-$ and $r > r_+$.

When $a^2 > M^2$, Eq. 18.22 has no real solution, and the Kerr metric does not describe a black hole, since the horizon is absent; in this case the singularity at $\Sigma = 0$ is "naked", i.e. it is not concealed by a horizon (see the discussion in Sec. 9.5.4). According to Penrose's cosmic censorship conjecture, a naked singularity cannot form in realistic situations. Numerical simulations of astrophysical processes leading to black hole formation provide strong support to the hypothesis that the final object cannot have $a > M$. In addition, theoretical studies on the mathematical structure of spacetime indicate that when $a^2 > M^2$ there are several pathologies, whose discussion is beyond the scope of this book. Thus, in general the solution with $a > M$ is considered unphysical, although we remark that this is still an open issue. To hereafter, we will restrict our analysis to the case

$$a^2 \leq M^2\,. \tag{18.25}$$

In the limiting case $a^2 = M^2$, the two roots coincide ($r_+ = r_-$) and the Kerr solution is said to describe an *extremal black hole.*

In Sec. 9.4 we showed how to establish whether a hypersurface is spacelike, timelike, or null, by studying the norm of its normal vector. We remind that in the Schwarzschild case the null hypersurface $r = 2M$ separates regions of spacetime where $r = const$ are timelike hypersurfaces (which can be crossed in both directions), from regions where $r = const$ are spacelike hypersurfaces (which can be crossed in one direction only); thus an object crossing the null hypersurface can never go back, and for this reason $r = 2M$ is called *event horizon.*

Given the vector n^μ normal to the hypersurfaces $r = const$, i.e. $n_\mu = (0, 1, 0, 0)$, using the Kerr metric given in Eq. 18.16 we find

$$n_\mu n^\mu = n_\mu n_\nu g^{\mu\nu} = g^{rr} = \frac{\Delta}{\Sigma}\,. \tag{18.26}$$

Thus, the vector normal to the surfaces $r = r_+$ and $r = r_-$, where $\Delta = 0$, is a null vector, and these are *null* surfaces which, as shown in Sec. 9.4, can be crossed in one direction only. As in the case of the Schwarzschild spacetime (see Sec. 9.5.4), the surfaces $r = r_+$ and $r = r_-$ in the future of the events outside the black hole are crossed inwards by timelike and null geodesics, as we will also show in Sec. 18.8.1. Therefore, these surfaces are *horizons*: any body or signal crossing one of them cannot come back. Since $r_- < r_+$, the surface $r = r_+$ is called *outer horizon*, and $r = r_-$ is called *inner horizon.*

The two horizons separate the spacetime in three regions:

I. $r > r_+$. Here the $r = const$ hypersurfaces are timelike, and thus can be crossed both inwards and outwards. The asymptotic limit $r \to \infty$, where the metric becomes flat, is in this region, which can be considered as the black hole exterior.

II. $r_- < r < r_+$. Here the $r = const$ hypersurfaces are spacelike, and thus can be crossed in one direction only. An object which falls inside the outer horizon can only continue falling inwards, until it reaches the inner horizon and passes to region III.

III. $r < r_-$. Here the $r = const$ hypersurfaces are timelike and then, as in region I, they can be crossed in both directions. This region contains the $\Sigma = 0$ singularity, which will be studied in Sec. 18.7.

In the case of extremal black holes, when $a^2 = M^2$, the two horizons coincide, and region II disappears.

Roughly speaking, the outer horizon r_+ can be considered as a sort of "black hole surface", although obviously such surface is immaterial. In the non-spinning case $a = 0$, the outer horizon coincides with the Schwarzschild one, $r_+ = 2M$, whereas the inner horizon tends to the curvature singularity of the Schwarzschild metric, $r_- = 0$.

18.4.2 How to remove the singularity at the horizons

In order to show that $\Delta = (r - r_+)(r - r_-) = 0$ is a coordinate singularity, we make a coordinate transformation that brings the metric into a form which is not singular on that surface. Following the same procedure used in Chapter 9 for the Schwarzschild spacetime, we look for a family of null geodesics crossing the $\Delta = 0$ surface from the exterior, and choose a coordinate system in which these geodesics are coordinate lines. In the case of Kerr's geometry, the spacetime cannot be decomposed in the product of two-dimensional manifolds, thus the study of null geodesics is more complex than in the Schwarzschild case. The Kerr metric admits two special families of null geodesics, named *principal null geodesics*, with tangent vector

$$u^\mu = \frac{dx^\mu}{d\lambda} = \left(\frac{dt}{d\lambda}, \frac{dr}{d\lambda}, \frac{d\theta}{d\lambda}, \frac{d\varphi}{d\lambda}\right) = \left(\frac{r^2 + a^2}{\Delta}, \pm 1, 0, \frac{a}{\Delta}\right), \tag{18.27}$$

where the plus (minus) sign corresponds to outgoing (ingoing) geodesics. In the Schwarzschild limit these are the usual outgoing and ingoing geodesics, with $\frac{dt}{dr} = \pm\frac{r^2}{\Delta} = \pm\frac{r}{r-2M}$ (see Sec. 9.9); in the Kerr case Eq. 18.27 shows that u^μ acquires an angular component $d\varphi/d\lambda$ proportional to a and diverging on $\Delta = 0$.

The proof that these worldlines are geodesics, i.e. that they satisfy the geodesic equation $u^\nu u^\mu{}_{;\nu} = 0$, is shown in Box 18-A. Here we show that u^μ is a null vector, i.e.

$$g_{\mu\nu} u^\mu u^\nu = 0\,. \tag{18.28}$$

We have

$$\begin{aligned} g_{\mu\nu}\frac{dx^\mu}{d\lambda}\frac{dx^\nu}{d\lambda} &= -\left(\frac{dt}{d\lambda}\right)^2 + \Sigma\left[\frac{1}{\Delta}\left(\frac{dr}{d\lambda}\right)^2 + \left(\frac{d\theta}{d\lambda}\right)^2\right] \\ &\quad +(r^2 + a^2)\sin^2\theta\left(\frac{d\varphi}{d\lambda}\right)^2 + \frac{2Mr}{\Sigma}\left(a\sin^2\theta\frac{d\varphi}{d\lambda} - \frac{dt}{d\lambda}\right)^2 . \end{aligned} \tag{18.29}$$

Note that

$$\frac{dt}{d\lambda} - a\sin^2\theta\frac{d\varphi}{d\lambda} = \frac{r^2 + a^2 - a^2\sin^2\theta}{\Delta} = \frac{\Sigma}{\Delta}, \tag{18.30}$$

hence

$$
\begin{aligned}
g_{\mu\nu}u^\mu u^\nu &= -\frac{(r^2+a^2)^2}{\Delta^2} + \frac{\Sigma}{\Delta} + (r^2+a^2)\sin^2\theta\frac{a^2}{\Delta^2} + \frac{2Mr\Sigma}{\Delta^2} \\
&= \frac{1}{\Delta^2}\left[-(r^2+a^2)(r^2+a^2) + (r^2+a^2\cos^2\theta)(r^2+a^2-2Mr)\right. \\
&\quad \left. + \sin^2\theta a^2(r^2+a^2) + (r^2+a^2\cos^2\theta)2Mr\right] = 0\,. \qquad (18.31)
\end{aligned}
$$

Consequently, the vectors given in Eq. 18.27 (with both signs) are null.

We remark that although these geodesics can in principle describe the motion of a massless particle, here we introduce them with a different aim, i.e. to find a coordinate transformation which allows for the extension of the spacetime across the coordinate singularity $\Delta = 0$. Therefore, the choice of the affine parameter λ in Eq. 18.27 is different from that of Sec. 18.3 and of Chapter 19: $u^0 = (r^2+a^2)/\Delta$ is *not* the energy of a massless particle moving on a principal null geodesic (see Box 19-A).

Let us consider the ingoing geodesics, and indicate their tangent vector as

$$l^\mu = \left(\frac{r^2+a^2}{\Delta}, -1, 0, \frac{a}{\Delta}\right)\,; \qquad (18.32)$$

as clearly shown by Eq. 18.32, these geodesics are parametrized by $\lambda = -r$. Therefore,

$$\frac{dt}{dr} = -\frac{r^2+a^2}{\Delta} \qquad \frac{d\varphi}{dr} = -\frac{a}{\Delta}\,. \qquad (18.33)$$

We want these geodesics to be coordinate lines of our new system; choosing one of our coordinates to be r, the other coordinates have to be constant along each geodesic belonging to the family. One of these, θ, is constant because $d\theta/dr = 0$. The remaining two are given by

$$
\begin{aligned}
v &\equiv t + T(r)\,, \qquad (18.34)\\
\bar{\varphi} &\equiv \varphi + \Phi(r)\,,
\end{aligned}
$$

where $T(r)$ and $\Phi(r)$ are solutions of[2]

$$
\begin{aligned}
\frac{dT}{dr} &= \frac{r^2+a^2}{\Delta}\,, \qquad (18.35)\\
\frac{d\Phi}{dr} &= \frac{a}{\Delta}\,,
\end{aligned}
$$

so that, along a geodesic of the family,

$$\frac{dv}{dr} = 0\,, \qquad \frac{d\bar{\varphi}}{dr} = 0\,, \qquad (18.36)$$

and the tangent vector of the ingoing principal null geodesics in Eq. 18.32, in the new coordinates $(v, r, \theta, \bar{\varphi})$, is

$$l^\mu = (0, -1, 0, 0)\,. \qquad (18.37)$$

We can now compute the metric tensor in the new frame. We recall that, in the Boyer-Lindquist coordinates,

$$ds^2 = -dt^2 + \Sigma\left(\frac{dr^2}{\Delta} + d\theta^2\right) + (r^2+a^2)\sin^2\theta d\varphi^2 + \frac{2Mr}{\Sigma}(a\sin^2\theta d\varphi - dt)^2\,. \qquad (18.38)$$

[2]Note that Eqs. 18.36 are first-order differential equations for $T(r)$ and $\Phi(r)$. As such, they depend on a free constant each, and this is related to the choice of the origin of v and $\bar{\varphi}$.

From Eqs. 18.34 we have

$$\begin{aligned} dv &= dt + \frac{r^2+a^2}{\Delta}dr \quad , \quad dt = dv - \frac{r^2+a^2}{\Delta}dr\,, \\ d\bar{\varphi} &= d\varphi + \frac{a}{\Delta}dr \quad , \quad d\varphi = d\bar{\varphi} - \frac{a}{\Delta}dr\,, \end{aligned} \tag{18.39}$$

hence

$$\begin{aligned} -dt^2 &= -dv^2 - \frac{(r^2+a^2)^2}{\Delta^2}dr^2 + 2\frac{r^2+a^2}{\Delta}dvdr\,, \\ (r^2+a^2)\sin^2\theta d\varphi^2 &= (r^2+a^2)\sin^2\theta d\bar{\varphi}^2 + (r^2+a^2)\frac{a^2}{\Delta^2}\sin^2\theta dr^2\,, \\ &\quad -2(r^2+a^2)\frac{a}{\Delta}\sin^2\theta dr d\bar{\varphi}\,. \end{aligned} \tag{18.40}$$

The quantity $\frac{\Sigma}{\Delta}dr^2 + \Sigma d\theta^2$ does not change (r, θ are also coordinates in the new frame), and the parenthesis in the last term of Eq. 18.38 reduces to

$$\begin{aligned} dt - a\sin^2\theta d\varphi &= dv - a\sin^2\theta d\bar{\varphi} - \frac{r^2+a^2-a^2\sin^2\theta}{\Delta}dr \\ &= dv - a\sin^2\theta d\bar{\varphi} - \frac{\Sigma}{\Delta}dr\,, \end{aligned} \tag{18.41}$$

thus

$$\begin{aligned} \frac{2Mr}{\Sigma}(dt - a\sin^2\theta d\varphi)^2 &= \frac{2Mr}{\Sigma}dv^2 + \frac{2Mr}{\Sigma}a^2\sin^4\theta d\bar{\varphi}^2 \\ &\quad + \frac{2Mr\Sigma}{\Delta^2}dr^2 - \frac{4Mr}{\Sigma}a\sin^2\theta dv d\bar{\varphi} \\ &\quad - \frac{4Mr}{\Delta}dvdr + \frac{4Mra\sin^2\theta}{\Delta}d\bar{\varphi}dr\,. \end{aligned} \tag{18.42}$$

Finally, collecting all terms, we find

$$\begin{aligned} ds^2 &= -\left(1-\frac{2Mr}{\Sigma}\right)dv^2 + 2dvdr + \Sigma d\theta^2 + \frac{(r^2+a^2)^2 - \Delta a^2\sin^2\theta}{\Sigma}\sin^2\theta d\bar{\varphi}^2 \\ &\quad -2a\sin^2\theta dr d\bar{\varphi} - \frac{4Mra}{\Sigma}\sin^2\theta dv d\bar{\varphi}\,. \end{aligned} \tag{18.43}$$

The coordinates $(v, r, \theta, \bar{\varphi})$ are called the *Kerr coordinates.* In this frame, the metric is not singular on $\Delta = 0$. This means that, while the Boyer-Lindquist coordinates are defined in $\Delta \neq 0$ and $\Sigma \neq 0$[3], the Kerr coordinates can be also defined in the submanifold $\Delta = 0$ (whereas $\Sigma = 0$ is a true curvature singularity, see Sec. 18.7). Then, after changing coordinates to the Kerr frame, $(v, r, \theta, \bar{\varphi})$, we can *extend* the manifold, to include the submanifold $\Delta = 0$.

[3]To be precise, one should subtract also the extrema of the domain of angular coordinates, $\theta = 0, \pi$, $\varphi = 0, 2\pi$, as usual when polar-like coordinates are considered. Anyway, as discussed in footnote 6 of Chapter 9, the related "pathologies" are much easier to cure by simple coordinate redefinitions; therefore we will limit our analysis to the "pathologies" on $\Delta = 0$ and $\Sigma = 0$.

Box 18-A

Principal null geodesics of the Kerr metric

In Boyer-Lindquist coordinates (t, r, θ, φ), the principal null geodesics have tangent vectors given by Eq. 18.27:

$$l^\mu = \left(\frac{r^2+a^2}{\Delta}, -1, 0, \frac{a}{\Delta}\right) \quad \text{(ingoing)} \tag{18.44}$$

$$n^\mu = \left(\frac{r^2+a^2}{\Delta}, +1, 0, \frac{a}{\Delta}\right) \quad \text{(outgoing)}\,. \tag{18.45}$$

In order to prove that these worldlines are indeed geodesics, we need to show that l^μ and n^μ satisfy the geodesic equation: $l^\nu l^\mu{}_{;\nu} = n^\nu n^\mu{}_{;\nu} = 0$.
Let us first consider the ingoing geodesics. The proof is simpler if we use the Kerr frame $(v, r, \theta, \bar{\varphi})$, in which the metric is (see Eq. 18.43)

$$g_{\mu\nu} = \begin{pmatrix} -\left(1-\frac{2Mr}{\Sigma}\right) & 1 & 0 & -\frac{2Mra}{\Sigma}\sin^2\theta \\ 1 & 0 & 0 & -a\sin^2\theta \\ 0 & 0 & \Sigma & 0 \\ -\frac{2Mra}{\Sigma}\sin^2\theta & -a\sin^2\theta & 0 & \frac{(r^2+a^2)^2-\Delta a^2\sin^2\theta}{\Sigma}\sin^2\theta \end{pmatrix} \tag{18.46}$$

and, since the ingoing principal null geodesics are coordinate lines in this frame, the tangent vector l^μ has components (see Eq. 18.37)

$$l^\mu = (0, -1, 0, 0)\,. \tag{18.47}$$

The geodesic equation in this frame reduces to

$$l^\nu l^\mu{}_{;\nu} = l^\nu l^\alpha \Gamma^\mu_{\nu\alpha} = \Gamma^\mu_{rr} = \frac{1}{2} g^{\mu\beta}(2g_{\beta r,r} - g_{rr,\beta})\,. \tag{18.48}$$

Since in the Kerr frame (see Eq. 18.46) $g_{rr} = 0$ and $g_{r\mu,r} = 0$, all terms in Eq. 18.48 vanish, therefore $l^\nu l^\mu{}_{;\nu} = 0$.
In order to prove that the vector n^μ satisfies the geodesic equation, we consider a coordinate frame in which the outgoing principal null geodesics are coordinate lines. Along these geodesics

$$\frac{dt}{dr} = \frac{r^2+a^2}{\Delta} \qquad \frac{d\varphi}{dr} = \frac{a}{\Delta} \tag{18.49}$$

and, by defining $u = t - T(r)$ and $\hat{\varphi} = \varphi - \Phi(r)$, with $T(r)$, $\Phi(r)$ solutions of Eqs. 18.36, we find a coordinate frame $(u, r, \theta, \hat{\varphi})$ in which

$$n^\mu = (0, 1, 0, 0)\,. \tag{18.50}$$

It is simple to show that in the coordinates $(u, r, \theta, \hat{\varphi})$, $g_{rr} = g_{r\theta} = 0$, $g_{ru} = -1$, and $g_{r\hat{\varphi}} = a\sin^2\theta$, therefore $n^\nu n^\mu{}_{;\nu} = n^\nu n^\alpha \Gamma^\mu_{\nu\alpha} = \Gamma^\mu_{rr} = \frac{1}{2} g^{\mu\beta}(2g_{\beta r,r} - g_{rr,\beta}) = 0$.

For later use we note that, since

$$g_{vr} = 1 \qquad g_{rr} = g_{\theta r} = 0 \qquad g_{\bar{\varphi} r} = -a\sin^2\theta\,, \tag{18.51}$$

the components of the one-form $\tilde{l}$ obtained lowering the index of the vector $\vec{l}$ are

$$l_\mu = (-1, 0, 0, a\sin^2\theta) . \tag{18.52}$$

We also note that (see Eq. 18.12)

$$g_{\bar{\varphi}\bar{\varphi}} = \frac{(r^2+a^2)^2 - \Delta a^2 \sin^2\theta}{\Sigma}\sin^2\theta = (r^2+a^2)\sin^2\theta + \frac{2Mr}{\Sigma}a^2\sin^4\theta = g_{\varphi\varphi} , \tag{18.53}$$

i.e. $g_{\bar{\varphi}\bar{\varphi}}$ in Kerr coordinates has the same expression as $g_{\varphi\varphi}$ in the Boyer-Lindquist coordinates. Moreover

$$\frac{2Mr}{\Sigma}(dv - a\sin^2\theta d\bar{\varphi})^2 = \frac{2Mr}{\Sigma}\left[dv^2 + a^2\sin^4\theta d\bar{\varphi}^2 - 2a\sin^2\theta dv d\bar{\varphi}\right] , \tag{18.54}$$

therefore the metric in Kerr coordinates can also be written in the simpler form

$$\begin{aligned} ds^2 &= -dv^2 + 2dvdr + \Sigma d\theta^2 + (r^2+a^2)\sin^2\theta d\bar{\varphi}^2 - 2a\sin^2\theta dr d\bar{\varphi} \\ &+\frac{2Mr}{\Sigma}(dv - a\sin^2\theta d\bar{\varphi})^2 . \end{aligned} \tag{18.55}$$

We can define a coordinate which (outside the horizon, see Sec. 18.4.1) behaves as a sort of "time" coordinate:

$$\bar{t} \equiv v - r , \tag{18.56}$$

so that Eq. 18.55 becomes

$$\begin{aligned} ds^2 &= -d\bar{t}^2 + dr^2 + \Sigma d\theta^2 + (r^2+a^2)\sin^2\theta d\bar{\varphi}^2 - 2a\sin^2\theta dr d\bar{\varphi} \\ &+\frac{2Mr}{\Sigma}(d\bar{t} + dr - a\sin^2\theta d\bar{\varphi})^2 . \end{aligned} \tag{18.57}$$

18.5 FRAME DRAGGING

Let us consider an observer, with timelike four-velocity u^μ, which falls toward the black hole from infinity with zero angular momentum: in the Boyer-Lindquist coordinates,

$$L = u_\varphi = 0 . \tag{18.58}$$

An observer with $L = 0$ is conventionally named ZAMO, which stands for *zero angular momentum observer.*

In general, for an observer with four-velocity u^μ, the angular velocity is defined as

$$\Omega \equiv \frac{d\varphi}{dt} = \frac{\frac{d\varphi}{d\tau}}{\frac{dt}{d\tau}} = \frac{u^\varphi}{u^t} . \tag{18.59}$$

For a ZAMO, Eq. 18.58 implies that, since the metric becomes flat as $r \to \infty$, $u^\varphi = \eta^{\varphi\mu}u_\mu \propto u_\varphi = 0$ in this limit. Thus, the angular velocity of a ZAMO vanishes as $r \to \infty$. At finite values of r, instead,

$$u^\varphi = g^{\varphi t}u_t \neq 0 \qquad \longrightarrow \qquad \Omega \neq 0 . \tag{18.60}$$

In other words, while falling toward the black hole, the observer acquires an angular velocity. To compute Ω in terms of the metric 18.1, we use the condition 18.58:

$$u_\varphi = g_{\varphi\varphi}u^\varphi + g_{\varphi t}u^t = 0 \quad \to \quad u^\varphi = -\frac{g_{\varphi t}}{g_{\varphi\varphi}}u^t , \tag{18.61}$$

from which it follows that

$$\Omega = \frac{u^\varphi}{u^t} = -\frac{g_{\varphi t}}{g_{\varphi\varphi}}\,. \tag{18.62}$$

Thus, since

$$g_{\varphi t} = -\frac{2Mra}{\Sigma}\sin^2\theta\,, \tag{18.63}$$

and (see Eq. 18.12)

$$g_{\varphi\varphi} = \frac{\sin^2\theta}{\Sigma}\left[(r^2+a^2)^2 - a^2\sin^2\theta\Delta\right]\,, \tag{18.64}$$

the ZAMO angular velocity is

$$\Omega = \frac{2Mar}{(r^2+a^2)^2 - a^2\Delta\sin^2\theta}\,. \tag{18.65}$$

As expected, $\Omega \to 0$ as $r \to \infty$. Furthermore, since

$$(r^2+a^2)^2 > a^2\sin^2\theta(r^2+a^2-2Mr)\,, \tag{18.66}$$

the denominator in Eq. 18.65 is positive, and therefore $\Omega/(Ma) > 0$ for all values of M and a; this means that the angular velocity has the same sign of the black hole angular momentum Ma.

Therefore, an observer which moves toward a Kerr black hole starting at radial infinity with zero angular momentum (which, at infinity, implies zero angular velocity) is *dragged* by the black hole gravitational field, and acquires an angular velocity which forces the ZAMO to *corotate* with the black hole. We discussed a similar phenomenon in Sec. 17.4.1.

On the outer horizon, $r = r_+$, $\Delta = 0$, and the ZAMO angular velocity becomes

$$\Omega = \frac{2Mar_+}{(r_+^2+a^2)^2} \equiv \Omega_H\,, \tag{18.67}$$

or, since $\Delta(r_+) = 0$,

$$r_+^2 + a^2 = 2Mr_+\,, \tag{18.68}$$

and thus

$$\Omega_H = \frac{a}{2Mr_+} = \frac{a}{r_+^2+a^2}\,. \tag{18.69}$$

As mentioned above, $r = r_+$ can be considered as a sort of "black hole surface", therefore Ω_H can be considered as the *black hole angular velocity.* Since Ω_H is constant, i.e. it does not depend on t, θ, and φ, we can say that a black hole *rotates rigidly.*

As we shall discuss in Chapter 19, the frame dragging plays an important role for the geodesics of the Kerr metric.

18.6 THE ERGOSPHERE

While in Schwarzschild's spacetime the horizon is also the surface where g_{tt} changes sign, in Kerr's spacetime these surfaces do not coincide. Indeed

$$\begin{aligned} g_{tt} &= -1 + \frac{2Mr}{\Sigma} = -\frac{1}{\Sigma}\left(r^2 - 2Mr + a^2\cos^2\theta\right) \\ &= -\frac{1}{\Sigma}(r - r_{S+})(r - r_{S-}) = 0\,, \end{aligned} \tag{18.70}$$

with

$$r_{S\pm} \equiv M \pm \sqrt{M^2 - a^2\cos^2\theta}\,. \tag{18.71}$$

These surfaces are called *infinite redshift surfaces*, although this name may be misleading, as will be discussed in Sec. 19.3.3.

Since the coefficient of r^2 in Eq. 18.70 is negative, $g_{tt} < 0$ outside the interval $[r_{S_-}, r_{S_+}]$, and $g_{tt} > 0$ inside. In addition, given $\sqrt{M^2 - a^2\cos^2\theta} \geq \sqrt{M^2 - a^2}$, the horizons $r_\pm = M \pm \sqrt{M^2 - a^2}$ (see Eq. 18.23) lie inside the interval $[r_{S_-}, r_{S_+}]$, i.e.

$$r_{S-} \leq r_- \leq r_+ \leq r_{S+}\,. \tag{18.72}$$

At $\theta = 0$ and $\theta = \pi$, i.e. on the symmetry axis,

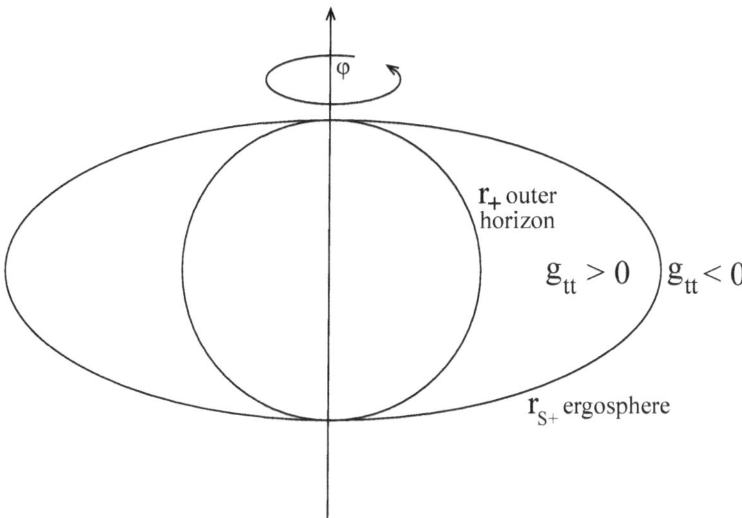

Figure 18.1: Schematic illustration of the ergosphere and of the outer horizon of the Kerr metric.

$$r_{S-} = r_- \qquad \text{and} \qquad r_{S+} = r_+\,; \tag{18.73}$$

on the equatorial plane

$$r_{S+} = 2M \qquad \text{and} \qquad r_{S-} = 0\,. \tag{18.74}$$

Therefore, there is a region *outside the outer horizon* where $g_{tt} > 0$ [4]. This region, i.e.

$$r_+ < r < r_{S+} \tag{18.75}$$

is called **ergoregion**, and its outer boundary, the surface $r = r_{S+}$, is called **ergosphere**. Since the ergoregion is located outside the outer horizon, an object arriving from outside may cross the ergosphere, enter in the ergoregion, and then cross again the ergosphere without crossing the horizon, and escape to infinity.

In the ergoregion the Killing vector field $k^\mu = (1, 0, 0, 0)$ is spacelike:

$$k^\mu k^\nu g_{\mu\nu} = g_{tt} > 0\,. \tag{18.76}$$

[4] This does not happen in Schwarzschild spacetime, where $g_{tt} > 0$ only inside the horizon.

18.6.1 Static and stationary observers

As discussed in Sec. 3.9, an observer in General Relativity is characterized by a timelike curve. We define as *static observer* one with tangent vector (i.e. with four-velocity u^μ) proportional to $k^\mu = (1,0,0,0)$. Thus, on the worldlines of a static observer the coordinates r,θ,φ are constant. Since $u^\mu u_\mu = -1$, we have $u^\mu = k^\mu/|\vec{k}|$ (where $|\vec{k}| = \sqrt{|k^\mu k_\mu|}$).

Since inside the ergosphere k^μ is spacelike, *in that region static observers cannot exist.* In other words, an observer inside the ergosphere cannot stay still, and is forced to move.

A *stationary observer* is one who does not see the metric changing while moving. Then, its tangent vector must be a Killing vector, i.e. it must be a linear combination of the two independent Killing vectors of the Kerr metric, $\vec{k} = \partial/\partial t$ and $\vec{m} = \partial/\partial\varphi$,

$$u^\mu = \frac{k^\mu + \omega m^\mu}{|\vec{k} + \omega\vec{m}|} = (u^t, 0, 0, u^\varphi) = u^t(1,0,0,\omega)\,, \tag{18.77}$$

where ω is the angular velocity of the observer

$$\omega \equiv \frac{d\varphi}{dt} = \frac{u^\varphi}{u^t}\,. \tag{18.78}$$

Thus, the worldline of a stationary observer has constant r and θ. (S)he can only move along circles with angular velocity ω, because on such orbits (s)he does not see the metric changing, since the spacetime is axially symmetric.

A stationary observer can exist provided

$$u^\mu u^\nu g_{\mu\nu} = (u^t)^2\left[g_{tt} + 2\omega g_{t\varphi} + \omega^2 g_{\varphi\varphi}\right] = -1\,, \tag{18.79}$$

i.e.

$$\omega^2 g_{\varphi\varphi} + 2\omega g_{t\varphi} + g_{tt} < 0\,. \tag{18.80}$$

To find the regions where Eq. 18.80 is satisfied, let us consider the equation

$$\omega^2 g_{\varphi\varphi} + 2\omega g_{t\varphi} + g_{tt} = 0\,, \tag{18.81}$$

with solutions

$$\omega_\pm = \frac{-g_{t\varphi} \pm \sqrt{g_{t\varphi}^2 - g_{tt}g_{\varphi\varphi}}}{g_{\varphi\varphi}} = \frac{-g_{t\varphi} \pm \sin\theta\sqrt{\Delta}}{g_{\varphi\varphi}}\,, \tag{18.82}$$

where we have used Eq. 18.14 to compute $g_{t\varphi}^2 - g_{tt}g_{\varphi\varphi}$. From this equation we see that stationary observers cannot exist when $\Delta \le 0$, i.e. in the region between the two horizons. Since (see Eqs. 18.64, 18.66)

$$g_{\varphi\varphi} = \frac{\sin^2\theta}{\Sigma}\left[(r^2+a^2)^2 - a^2\sin^2\theta\Delta\right] > 0\,, \tag{18.83}$$

the coefficient of ω^2 in Eq. 18.80 is positive; consequently the inequality 18.80 is satisfied outside the outer horizon (where $r > r_+$, $\Delta > 0$ and thus $\omega_- < \omega_+$), for

$$\omega_- < \omega < \omega_+\,. \tag{18.84}$$

Therefore, a stationary observer cannot have arbitrary angular velocity, since the constraint $u^\mu u_\mu = -1$ forces ω to be in the range 18.84.

Note that on the outer horizon $r = r_+$, $\Delta = 0$, and $\omega_- = \omega_+$; therefore Eq. 18.81 has coincident solutions,

$$\omega_\pm = -\frac{g_{t\varphi}(r_+)}{g_{\varphi\varphi}(r_+)}\,. \tag{18.85}$$

This is the angular velocity of the ZAMO at the horizon, i.e. the angular velocity of the black hole (see Eqs. 18.62, 18.67):

$$\omega = \Omega(r_+) = \Omega_H \,. \tag{18.86}$$

Thus, while the ZAMO satisfies Eq. 18.81 on the horizon, it does not satisfy the inequality 18.80: strictly speaking, the ZAMO on the horizon is not an observer, because its four-velocity is a null vector.

On the ergosphere, $g_{tt} = 0$ and (being $g_{t\varphi} < 0$)

$$\omega_- = \frac{-g_{t\varphi} - \sqrt{g_{t\varphi}^2}}{g_{\varphi\varphi}} = 0\,. \tag{18.87}$$

Outside the ergosphere, i.e. for $r \geq r_{S_+}$,

$$g_{tt} < 0, \qquad g_{\varphi\varphi} > 0, \qquad \text{therefore} \qquad \omega_- < 0 \quad \text{and} \quad \omega_+ > 0\,. \tag{18.88}$$

Thus, outside the ergosphere a stationary observer can be either co-rotating or counter-rotating with the black hole. Conversely, in the ergoregion, where $r_+ < r < r_{S_+}$,

$$g_{tt} > 0, \qquad g_{\varphi\varphi} > 0, \qquad \text{therefore} \qquad \omega_- > 0 \quad \text{and} \quad \omega_+ > 0\,: \tag{18.89}$$

a stationary observer in the ergosphere does necessarily corotate with the black hole.

18.7 THE SINGULARITY OF THE KERR SPACETIME

The Kerr spacetime has a curvature singularity on the surface

$$\Sigma = r^2 + a^2\cos^2\theta = 0\,, \tag{18.90}$$

i.e. for $r = 0$ *and* $\theta = \pi/2$. Remarkably, the spacetime is *not* singular for $r = 0$ and $\theta \neq \pi/2$, since in this case $\Sigma \neq 0$ and – as we are going to show – the metric has just a coordinate singularity. This may come as a surprise: in flat spacetime (or even in the Euclidean space) when polar coordinates are used, and in the Schwarzschild spacetime, $r = 0$ is a *single* point, regardless of the values of the angular variables. The behaviour of the Boyer-Lindquist coordinates (r, θ, φ) near $r = 0$ (and of the Kerr coordinates $(r, \theta, \bar{\varphi})$ as well) is quite different: the properties of the $r \sim 0$ region depend on the values of the angular coordinates.

18.7.1 The Kerr-Schild coordinates

In order to understand the structure of the singularity at $(r, \theta) = (0, \pi/2)$, we need to introduce the so-called *Kerr-Schild coordinates*. Let us start with the metric in Kerr coordinates $(\bar{t}, r, \theta, \bar{\varphi})$, given in Eq. 18.57:

$$\begin{aligned} ds^2 &= -d\bar{t}^2 + dr^2 + \Sigma d\theta^2 + (r^2 + a^2)\sin^2\theta d\bar{\varphi}^2 - 2a\sin^2\theta dr d\bar{\varphi} \\ &\quad + \frac{2Mr}{\Sigma}(d\bar{t} + dr - a\sin^2\theta d\bar{\varphi})^2\,. \end{aligned} \tag{18.91}$$

The Kerr-Schild coordinates $(\bar{t}, x, y, z)$ are defined by

$$\begin{aligned} x &= \sqrt{r^2 + a^2}\sin\theta\cos\left(\bar{\varphi} + \arctan\frac{a}{r}\right) \\ y &= \sqrt{r^2 + a^2}\sin\theta\sin\left(\bar{\varphi} + \arctan\frac{a}{r}\right) \\ z &= r\cos\theta\,. \end{aligned} \tag{18.92}$$

The metric expressed in these coordinates will be derived in the next section; here we use this frame to describe the Kerr spacetime near the singularity.

In Figs. 18.2 and 18.3 we respectively show the $r = const$ and the $\theta = const$ surfaces in the $(\bar{t}, x, y, z)$ frame. Here, for the purpose of visualization, (x, y, z) are represented as Cartesian coordinates in the three-dimensional Euclidean space, and (r, θ) are considered as functions of (x, y, z). The surfaces shown in these figures are ellipsoids and hyperboloids, respectively. Indeed, from Eqs. 18.92 we get

$$\begin{aligned} x^2 + y^2 &= (r^2 + a^2) \sin^2 \theta \\ z^2 &= r^2 \cos^2 \theta \, , \end{aligned} \tag{18.93}$$

hence

$$\frac{x^2 + y^2}{r^2 + a^2} + \frac{z^2}{r^2} = 1 \, ; \tag{18.94}$$

therefore, the surfaces with constant r are ellipsoids, shown in Fig. 18.2. Likewise, since

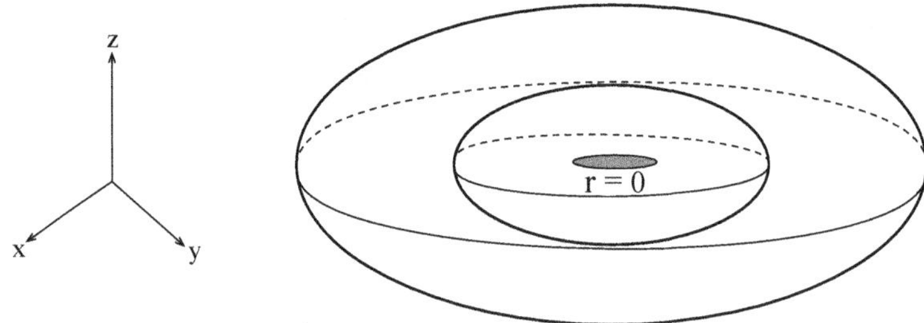

Figure 18.2: $r = const$ ellipsoidal surfaces in the Kerr-Schild frame. The shadowed disk is the $r = 0$ surface.

$$\frac{x^2 + y^2}{a^2 \sin^2 \theta} - \frac{z^2}{a^2 \cos^2 \theta} = 1 \, , \tag{18.95}$$

the surfaces with constant θ are hyperboloids, shown in Fig. 18.3. Strictly speaking, the θ-constant surfaces are *half*-hyperboloids. Indeed, since $z = r \cos \theta$ (see Eq. 18.92), $z > 0$ for $0 < \theta < \pi/2$ and the surfaces $\theta = const$ correspond to the upper half-hyperboloids; conversely, $z < 0$ for $\pi/2 < \theta < \pi$, and the surfaces $\theta = const$ correspond to the lower half-hyperboloids.

It should be noted that for r sufficiently large the r, θ coordinates behave like ordinary polar coordinates. However, near the black hole they behave differently: for $r = 0$ and $0 < \theta < \pi$ Eq. 18.93 describes a disk of radius a in the equatorial plane

$$x^2 + y^2 = a^2 \sin^2 \theta \, , \qquad z = 0 \, , \tag{18.96}$$

thus $r = 0$ is not a single point, it is a disk. Each value of θ corresponds to the circle $x^2 + y^2 = a^2 \sin^2 \theta < a^2$, which is the intersection of the $r = 0$ disk with the half-hyperboloid corresponding to that value of θ. In particular, the singularity

$$r = 0 \qquad \theta = \frac{\pi}{2} \tag{18.97}$$

corresponds to the *ring*

$$x^2 + y^2 = a^2 \, , \qquad z = 0 \, . \tag{18.98}$$

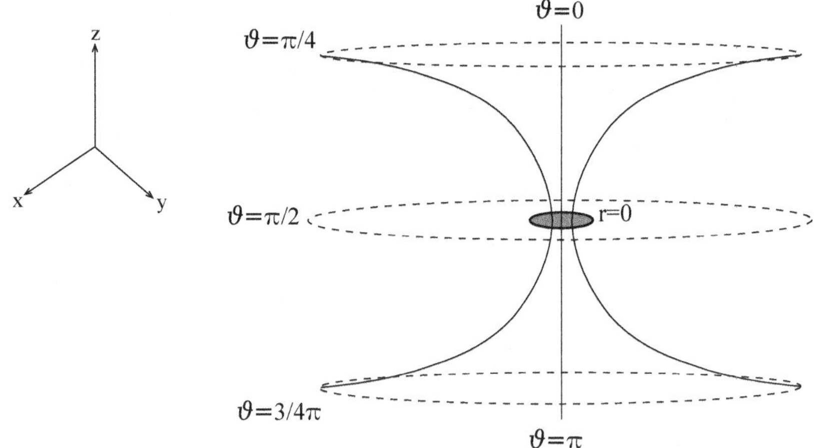

Figure 18.3: $\theta = const$ half-hyperboloidal surfaces in the Kerr-Schild frame; the upper (lower) branch of each hyperboloid corresponds to $0 < \theta < \pi/2$ ($\pi/2 < \theta < \pi$). The thick ring represents the $r = 0$, $\theta = \pi/2$ singularity, and the disk inside the ring corresponds to $r = 0$, $\theta \neq \pi/2$. The equatorial plane outside the thick ring corresponds to $r > 0$, $\theta = \pi/2$.

On this ring $\Sigma = 0$, and the Kretschmann scalar diverges (see Sec. 9.3). Therefore $\Sigma = 0$ is a true singularity of the Kerr spacetime and it is called *ring singularity*.

The interior disk $r = 0$, $\theta \neq \pi/2$ is instead a coordinate singularity in the Boyer-Lindquist and in the Kerr coordinates; indeed, when a timelike or null geodesic crosses the disk, the coordinate θ has a discontinuity ($\theta < \pi/2$ on the top, $\theta > \pi/2$ on the bottom). This coordinate singularity can be removed in the Kerr-Schild coordinates by extending the Kerr metric, as will be discussed in Sec. 18.8.1.

18.7.2 The metric in Kerr-Schild's coordinates

Let us define the quantity $\alpha = \arctan a/r$. Since $\tan^2\alpha = a^2/r^2$, it follows that $r^2\sin^2\alpha = a^2\cos^2\alpha$. Adding $r^2\cos^2\alpha$ and $a^2\sin^2\alpha$ to both sides of this expression we get

$$\begin{aligned} r &= \sqrt{r^2+a^2}\cos\alpha\,, \\ a &= \sqrt{r^2+a^2}\sin\alpha\,, \end{aligned} \tag{18.99}$$

respectively. Therefore, rewriting Eq. 18.92 as

$$\begin{aligned} x &= \sin\theta\sqrt{r^2+a^2}(\cos\bar{\varphi}\cos\alpha - \sin\bar{\varphi}\sin\alpha) \\ y &= \sin\theta\sqrt{r^2+a^2}(\sin\bar{\varphi}\cos\alpha + \cos\bar{\varphi}\sin\alpha) \\ z &= r\cos\theta \end{aligned} \tag{18.100}$$

we find

$$\begin{aligned} x &= \sin\theta(r\cos\bar{\varphi} - a\sin\bar{\varphi}) \\ y &= \sin\theta(r\sin\bar{\varphi} + a\cos\bar{\varphi}) \\ z &= r\cos\theta\,. \end{aligned} \tag{18.101}$$

By differentiating these expressions

$$\begin{aligned} dx &= \cos\theta(r\cos\bar\varphi - a\sin\bar\varphi)d\theta + \sin\theta\cos\bar\varphi dr - \sin\theta(r\sin\bar\varphi + a\cos\bar\varphi)d\bar\varphi \\ dy &= \cos\theta(r\sin\bar\varphi + a\cos\bar\varphi)d\theta + \sin\theta\sin\bar\varphi dr + \sin\theta(r\cos\bar\varphi - a\sin\bar\varphi)d\bar\varphi \\ dz &= -r\sin\theta d\theta + \cos\theta dr\,, \end{aligned} \tag{18.102}$$

then

$$\begin{aligned} dx^2 + dy^2 + dz^2 &= dr^2 + \left(r^2\sin^2\theta + (r^2+a^2)\cos^2\theta\right)d\theta^2 \\ &\quad +(r^2+a^2)\sin^2\theta d\bar\varphi^2 - 2\sin^2\theta a dr d\bar\varphi \\ &= dr^2 + \Sigma d\theta^2 + (r^2+a^2)\sin^2\theta d\bar\varphi^2 - 2a\sin^2\theta dr d\bar\varphi\,. \end{aligned} \tag{18.103}$$

Thus, the metric in Eq. 18.91 can be written as the Minkowski metric plus the term

$$\frac{2Mr}{\Sigma}(d\bar t + dr - a\sin^2\theta d\bar\varphi)^2\,. \tag{18.104}$$

Since

$$\Sigma = r^2 + a^2\cos^2\theta = r^2 + \frac{a^2z^2}{r^2}\,, \tag{18.105}$$

the factor $2Mr/\Sigma$ is easily expressed in the Kerr-Schild coordinates:

$$\frac{2Mr}{\Sigma} = \frac{2Mr^3}{r^4 + a^2z^2}\,. \tag{18.106}$$

The one-form $d\bar t + dr - a\sin^2\theta d\bar\varphi$ is more complicated to transform. We will prove that

$$d\bar t + dr - a\sin^2\theta d\bar\varphi = d\bar t + \frac{r(xdx+ydy) - a(xdy-ydx)}{r^2+a^2} + \frac{zdz}{r}\,. \tag{18.107}$$

First of all, let us express the differentials in Eq. 18.102 as

$$\begin{aligned} dx &= \frac{\cos\theta}{\sin\theta}xd\theta + \sin\theta\cos\bar\varphi dr - yd\bar\varphi \\ dy &= \frac{\cos\theta}{\sin\theta}yd\theta + \sin\theta\sin\bar\varphi dr + xd\bar\varphi \\ dz &= -r\sin\theta d\theta + \cos\theta dr\,. \end{aligned} \tag{18.108}$$

We have

$$\begin{aligned} xdx + ydy &= \frac{\cos\theta}{\sin\theta}(x^2+y^2)d\theta + \sin\theta(x\cos\bar\varphi + y\sin\bar\varphi)dr \\ &= \sin\theta\cos\theta(r^2+a^2)d\theta + \sin^2\theta r dr\,, \end{aligned} \tag{18.109}$$

$$\begin{aligned} ydx - xdy &= -(x^2+y^2)d\bar\varphi + \sin\theta(y\cos\bar\varphi - x\sin\bar\varphi)dr \\ &= -(r^2+a^2)\sin^2\theta d\bar\varphi + \sin^2\theta a dr\,, \end{aligned}$$

$$zdz = -r^2\sin\theta\cos\theta d\theta + r\cos^2\theta dr\,,$$

then

$$
\begin{aligned}
&(xdx + ydy)\frac{r}{r^2+a^2} + (ydx - xdy)\frac{a}{r^2+a^2} + \frac{zdz}{r} \qquad (18.110)\\
= &\left(r\sin\theta\cos\theta d\theta + \frac{r^2}{r^2+a^2}\sin^2\theta dr\right)\\
&+\left(-a\sin^2\theta d\bar{\varphi} + \frac{a^2}{r^2+a^2}\sin^2\theta dr\right) + \left(-r\sin\theta\cos\theta d\theta + \cos^2\theta dr\right)\\
&= dr - a\sin^2\theta d\bar{\varphi}\,,
\end{aligned}
$$

which proves Eq. 18.107. The Kerr metric in Kerr-Schild coordinates is then

$$
\begin{aligned}
ds^2 &= -d\bar{t}^2 + dx^2 + dy^2 + dz^2 \qquad (18.111)\\
&+\frac{2Mr^3}{r^4+a^2z^2}\left[d\bar{t} + \frac{r(xdx+ydy) - a(xdy - ydx)}{r^2+a^2} + \frac{zdz}{r}\right]^2.
\end{aligned}
$$

The above metric depends on a function $r(x, y, z)$, which is defined implicitly by

$$r^4 - (x^2 + y^2 + z^2 - a^2)r^2 - a^2z^2 = 0\,. \qquad (18.112)$$

Indeed, by combining Eqs. 18.93 we find $r^2 - (x^2 + y^2 + z^2 - a^2) = a^2\cos^2\theta = z^2a^2/r^2$, which is equivalent to Eq. 18.112.

Note that the metric 18.111 has the form

$$g_{\mu\nu} = \eta_{\mu\nu} + Hl_\mu l_\nu \qquad (18.113)$$

with

$$H \equiv \frac{2Mr^3}{r^4+a^2z^2} \qquad (18.114)$$

and, in Kerr-Schild coordinates,

$$l_\mu dx^\mu = -\left(d\bar{t} + \frac{r(xdx+ydy) - a(xdy-ydx)}{r^2+a^2} + \frac{zdz}{r}\right), \qquad (18.115)$$

while in Kerr coordinates

$$l_\alpha dx^\alpha = -d\bar{t} - dr + a\sin^2\theta d\bar{\varphi} = -dv + a\sin^2\theta d\bar{\varphi}\,; \qquad (18.116)$$

thus l_μ is exactly the null vector given in Eq. 18.52, i.e. the generator of the principal null geodesics which have been used to define the Kerr coordinates.

18.8 THE INTERIOR OF AN ETERNAL KERR BLACK HOLE

While the Kerr metric describes the exterior – i.e., the region outside the outer horizon $r = r_+$ – of a stationary, astrophysical black hole formed in the gravitational collapse of a star or in the coalescence of compact bodies, it cannot describe its interior, i.e. the region $r < r_+$ (see also Sec. 9.5.4). Strictly speaking, the Kerr metric (which includes the external and the internal regions) describes an *eternal* Kerr black hole, i.e. one which exists for $t \in (-\infty, +\infty)$. In this section we shall discuss some peculiar properties of the interior of an eternal Kerr black hole, in particular of the region close to $r = 0$. In reality, the interior of an astrophysical black hole is not empty but contains the matter fields that underwent the gravitational collapse. This might change the inner structure significantly and resolve some of the potential pathologies that we are going to discuss[5].

[5] The curvature singularities that exist in the Schwarzschild and Kerr metric are not a prerogative of these solutions. Indeed, a remarkable series of results due to Penrose and Hawking – the so-called "singularity theorems" – show that in General Relativity curvature singularities emerge generically as the outcome of a gravitational collapse (see [57] for a rigorous discussion).

18.8.1 Extensions of the Kerr metric

Let us consider the metric in the Kerr-Schild coordinates $\{\bar{t}, x, y, z\}$ (Eq. 18.111)

$$\begin{aligned} ds^2 &= -d\bar{t}^2 + dx^2 + dy^2 + dz^2 \\ &+ \frac{2Mr^3}{r^4 + a^2z^2}\left[d\bar{t} + \frac{r(xdx + ydy) - a(xdy - ydx)}{r^2 + a^2} + \frac{zdz}{r}\right]^2 . \end{aligned} \tag{18.117}$$

The function $r(x, y, z)$ is given by Eq. 18.112 which, for each value of x, y, z, has two real solutions (besides two unphysical complex conjugate solutions), one positive and one negative. We could be tempted to discard the $r < 0$ solution as unphysical, but in this way – although the coordinates $\{\bar{t}, x, y, z\}$ are continuous across the disk $r = 0$, $\theta \neq 0$ – the coordinate singularity on the disk would not be removed, as we are going to show.

Let us consider, for simplicity, the $x = y = 0$ submanifold, where the metric 18.111 reduces to

$$ds^2 = -d\bar{t}^2 + dz^2 + \frac{2Mr^3}{r^4 + a^2z^2}\left[d\bar{t} + \frac{zdz}{r}\right]^2 , \tag{18.118}$$

and Eq. 18.112 reduces to $r^4 - (z^2 - a^2)r^2 - z^2a^2 = 0$, whose solutions are $r(z) = \pm z$. If we require that the radial coordinate is positive, $r = |z|$ and Eq. 18.118 becomes

$$ds^2 = -d\bar{t}^2 + dz^2 + \frac{2M|z|}{z^2 + a^2}\left[d\bar{t} + \frac{zdz}{|z|}\right]^2 , \tag{18.119}$$

which is continuous but not differentiable at $z = 0$. A computation of the Christoffel symbols shows that $\Gamma^z_{\bar{t}\bar{t}}$ is discontinuous across the disk and thus, for a timelike geodesic with tangent vector $\frac{dx^\mu}{d\lambda}$ crossing the disk, $\frac{d^2z}{d\lambda^2}$ would also be discontinuous.

These problems arise because we have forced r to be positive, but there is no fundamental reason for this assumption. If we allow r to have negative values, when an observer crosses the disk the coordinate r changes sign. For instance, in the $x = y = 0$ submanifold, we can choose the solution $r = z$ of Eq. 18.112 along the entire axis. The metric in this submanifold is then

$$ds^2 = -d\bar{t}^2 + dz^2 + \frac{2Mz}{z^2 + a^2}\left[d\bar{t} + dz\right]^2 , \tag{18.120}$$

which is regular at $z = 0$. Note that this choice also "cures" the discontinuity of $dr/d\lambda$ across the disk, and the discontinuity of $\theta(\lambda)$ as well, since (given $z = r\cos\theta$) $\theta = 0$ along the entire axis.

In order to extend the spacetime across the $r = 0$ disk we have to consider a manifold formed by at least two copies of the spacetime described by Eq. 18.111: one with $r > 0$, the other with $r < 0$. The $r < 0$ spacetime is also asymptotically flat, but it has no horizons. If the top of the disk of the $r > 0$ spacetime is identified with the bottom of the disk with $r < 0$ spacetime and vice versa (see Fig. 18.4), the worldlines crossing the disk move from one copy to the other. In this way, the metric is regular across the disk, and the coordinate singularity is removed[6]. Note, however, that there is no reason to assume that two observers in the $r > 0$ spacetime, one crossing the disk from the top, the other from the bottom, reach the same $r < 0$ spacetime, as in Fig. 18.4. A larger spacetime would consist of different copies of the same manifold, such that the two observers crossing the disk from different sides end up in different $r < 0$ manifolds.

This is not the maximal extension of the Kerr metric. A detailed analysis of geodesic

[6]This extension is analogous to the extension of the complex plane to Riemann surfaces for the representation of multi-valued functions of a complex variable.

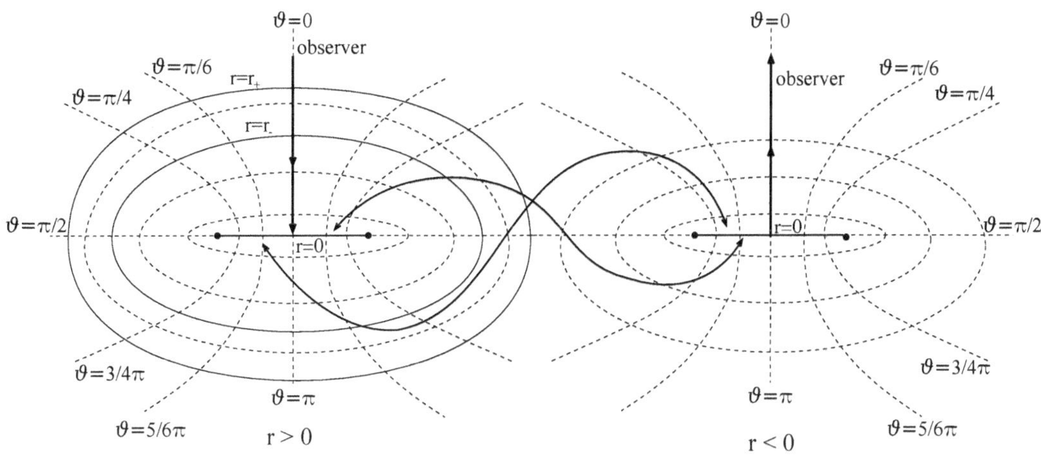

Figure 18.4: Two copies of the spacetime, one with $r > 0$, one with $r < 0$, are patched together, identifying the top of the $r > 0$ disc with the bottom of the $r < 0$ disk, and vice versa. The $r > 0$ spacetime contains the $r = r_\pm$ horizons. An observer enters in the disk from the top of the $r > 0$ space and emerges from the top of the $r < 0$ space.

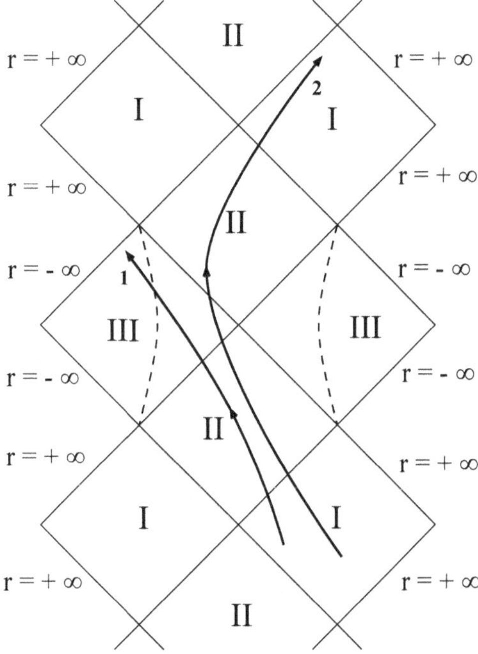

Figure 18.5: Schematic representation of the maximal extension of the Kerr metric, along the $\theta = 0$ axis. The dashed hyperbolic curves correspond to $r = 0$. We indicate with I the exterior of the black hole, with II the regions between the inner and outer horizons, with III the inner regions, which include the center of the $r = 0$ disk and the asymptotically flat region with $r < 0$. The solid lines represent two timelike observers escaping to different asymptotically flat spaces.

completeness – which is extremely involved and would go far beyond the scope of this book, see e.g. [57] – shows that if we require that the spacetime is inextendible, i.e. that all timelike or null geodesics either hit the ring singularity, or can be extended to arbitrarily large values of the affine parameter (see Box 9-C and Sec. 9.5.4), we have to patch together an infinite number of spaces like those in Fig. 18.4. Here we only describe the maximal extension of the $x = y = 0$ submanifold, i.e. of the $\theta = 0$ axis. The structure of this spacetime is shown in Fig. 18.5, where the regions I,II,III correspond to:

$$\begin{array}{llll} I & : & r_+ < r < +\infty & \text{(exterior of the black hole, asymptotically flat)} \\ II & : & r_- < r < r_+ & \text{(where the } r = const. \text{ surfaces are spacelike)} \\ III & : & -\infty < r < r_- & \text{(contains the asymptotically flat space with } r < 0). \end{array}$$

The dashed hyperbolic curves correspond to $r = 0$. Note that, as in the Schwarzschild spacetime (see Sec. 9.5.4), the outer horizon $r = r_+$ in the future of region I is crossed inwards by timelike and null worldlines, which also cross inwards the inner horizon $r = r_-$ in the future of regions I and II.

Remarkably, in this spacetime an observer can travel through different asymptotically flat regions. This is different from the maximal extension of the Schwarzschild spacetime discussed in Sec. 9.5.4, where the observers falling in the black hole necessarily hit the singularity, and the two asymptotically flat regions (indicated as I and IV in Fig. 9.10) are causally disconnected. In the maximally extended Kerr spacetime, instead, once an observer crosses the inner horizon and enters in region III, (s)he can either cross the disk $r = 0$, and escape to the asymptotically flat region $r < 0$ (trajectory 1 in Fig. 18.5), or move outwards [7] and cross the inner horizon again (trajectory 2 in Fig. 18.5). In this case the observer, after leaving region III, would enter into a different copy of region II, where (s)he can keep moving to increasing values of r, and finally enter into a different copy of region I.

This fascinating structure, however, is unlikely to be realized in actual astrophysical objects. As we noted at the beginning of this section, the Kerr metric and its extensions only describe *eternal black holes* (similarly to case of the Schwarzschild metric in Kruskal's coordinates discussed in Sec. 9.5.4). In an astrophysical black hole formed in a gravitational collapse or in the coalescence of compact bodies, the region I cannot receive signals from the region II, because these signals would come from $t \to -\infty$, when the black hole was not born yet. This prevents the formation of the multiple copies shown in Figs. 18.4 and 18.5.

Moreover, it has been shown that the inner horizon $r = r_-$ is unstable: a small perturbation produced by mass accretion would grow up [94], potentially leading to drastic changes in the structure of the $r \leq r_-$ region. However, the nature of this instability is still unclear (see e.g. [15] or, more recently, [39]).

18.8.2 Causality violations

Let us consider a curve γ on the equatorial plane, consisting in a ring just outside the curvature singularity ring, in the spacetime with $r < 0$:

$$\gamma: \quad \left\{\bar{t} = const,\, r = const,\, \theta = \frac{\pi}{2},\, 0 \leq \bar{\varphi} \leq 2\pi\right\} \quad \text{with } |r| \ll M,\, r < 0\,. \qquad (18.121)$$

The curve γ belongs to region III of the black hole, and can be reached by an observer from positive, large values of r who crosses the two horizons, passes through the $r = 0$ ring, and turns around it up to the $z = 0$ plane, just outside the ring (see Fig. 18.6).

[7] We remind that in region III the $r = const$ surfaces are timelike and can be crossed in both directions.

The tangent vector to the curve γ is the Killing vector $\vec{m}$, and its norm is

$$m^\mu m^\nu g_{\mu\nu} = g_{\bar{\varphi}\bar{\varphi}} = (r^2 + a^2)\sin^2\theta + \frac{2Mr}{\Sigma} a^2 \sin^4\theta = r^2 + a^2 + \frac{2Ma^2}{r}\,, \tag{18.122}$$

where we used Eq. 18.53 evaluated in $\theta = \pi/2$. Since $r < 0$ and $|r| \ll M$, the term $2Ma^2/r$ is negative and dominates the others, therefore $m^\mu m^\nu g_{\mu\nu} < 0$. The curve γ is then a timelike curve, and can be interpreted as the worldline of an observer; however, it is a closed curve and its existence may violate causality: the observer moving on a *closed timelike curve* (CTC) [8] would meet him-/herself in its own past. However, it has recently been argued that this paradox might be resolved including thermodynamical considerations; a causality violation would require not only a particle, but a thermodynamical system – for instance, a clock keeping track of time – meeting itself in its own future (for details see [101]). There are reasons to believe that in a rotating black hole born in a gravitational

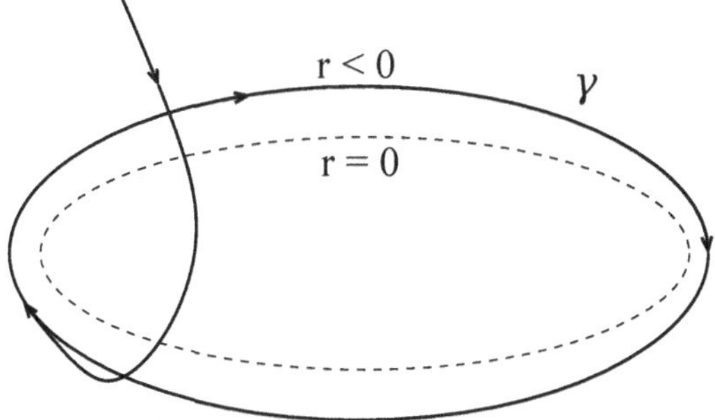

Figure 18.6: Closed timelike curve in Kerr spacetime.

collapse or in a compact binary coalescence, the structure of the ring singularity (and then the occurrence of CTCs) would be destroyed by the presence of the matter and/or by the instability of the inner horizon. However, there is no clear proof that this is the case.

A possible point of view to interpret these troublesome features of Kerr's spacetime is that causality violations, together with the existence of singularities, are *inconsistencies* of the theory of General Relativity, which would disappear once a more fundamental theory unifying General Relativity with Quantum Mechanics will be discovered (see footnote 8 in Chapter 9). Indeed, quantum gravity effects are expected to be significant near the singularities.

A different point of view is that these features are not problematic since they can not be observed, because they are hidden behind horizons. This is a further motivation for the cosmic censorship conjecture discussed in Sec. 9.5.4.

18.9 GENERAL BLACK HOLE SOLUTIONS

When a black hole forms in the gravitational collapse of a sufficiently massive star or in the coalescence of compact binaries, the violent oscillations that follow are damped by gravitational wave emission and by other dissipative processes. We expect that, after some time related to the damping time of the black hole quasi-normal modes (see Chapter 15), the newborn black hole settles down to a *stationary* configuration.

[8]The occurrence of closed timelike curves was first found by Kurt Gödel in an exact solution of Einstein's equations, which is considered to be unphysical.

In addition to the Schwarzschild and Kerr solutions, there exists a stationary, axisymmetric solution of the Einstein-Maxwell equations [9] known as the **Kerr-Newman solution** [85]; the metric, in the Boyer-Lindquist coordinates, is

$$ds^2 = -\frac{\Delta}{\Sigma}(dt - a\sin^2\theta d\varphi)^2 + \frac{\sin^2\theta}{\Sigma}[adt - (r^2+a^2)d\varphi]^2 + \frac{\Sigma}{\Delta}dr^2 + \Sigma d\theta^2 \,, \tag{18.123}$$

where

$$\begin{aligned} \Delta(r) &\equiv r^2 - 2Mr + a^2 + Q^2 \,, \\ \Sigma(r,\theta) &\equiv r^2 + a^2\cos^2\theta \,, \end{aligned} \tag{18.124}$$

and Q is the *black hole electric charge*; here we have used units [10] such that $4\pi\epsilon_0 = 1$ (see also Box 18-B). It is easy to check that if $Q = 0$, the solution 18.123 reduces to the Kerr metric.

In the case of zero spin, $a = 0$ and the Kerr-Newman solution reduces to the **Reissner-Nordström solution**,

$$ds^2 = -\left(1 - \frac{2M}{r} + \frac{Q^2}{r^2}\right)dt^2 + \left(1 - \frac{2M}{r} + \frac{Q^2}{r^2}\right)^{-1} dr^2 + r^2 d\Omega^2 \,, \tag{18.125}$$

which describes a spherically symmetric, electrically charged black hole in Einstein-Maxwell theory. However, as discussed in Box 18-B, the charge of astrophysical black holes is expected to be negligibly small and therefore astrophysical black holes are perfectly described by the Kerr solution.

There are some remarkable theorems [100] on stationary black holes, derived by S. Hawking, W. Israel, B. Carter, and others, which prove the following:

- A stationary, asymptotically flat black hole must be axially symmetric (while, as we know from Birkhoff's theorem, a static black hole is necessarily spherically symmetric).
- Any stationary, asymptotically flat black hole, with vanishing electric charge, is described by the Kerr solution.
- Any stationary, asymptotically flat black hole is described by the Kerr-Newman solution, and it is therefore characterized by *only three parameters*: the mass M, the angular momentum aM, and the charge Q.

Besides the mass, angular momentum, and electric charge, all other features that a star possessed before collapsing, such as a particular structure of the magnetic field, departure from axisymmetry, matter currents, differential rotation, etc., disappear when the final black hole forms. This result, which goes under the name of *no-hair theorem*, has been nicely summarized with the sentence: "*A black hole has no hair*" [11].

---

[9] The equations which couple the electromagnetic field to gravity can be derived from a variational principle, as shown in Chapter 7, by adding Maxwell's action to the Einstein-Hilbert action.

[10] In geometrized ($G = 1$) and unrationalized Gaussian ($4\pi\epsilon_0 = 1$) units, the ratio Q/M is dimensionless.

[11] The quote is attributed to John Archibald Wheeler who, in turn, attributed it to Jacob Bekenstein.

Box 18-B

The charge of astrophysical black holes

Astrophysical black holes are considered to be electrically neutral as a consequence of various physical processes: quantum discharge effects, electron-positron pair production, and charge neutralization by astrophysical plasmas. Without entering into the details, these arguments rely — one way or another — on the huge charge-to-mass ratio of the electron. The simplest argument is purely kinematical. Let us consider a black hole with mass M and electric charge Q and a low-energy electron in radial motion with a small, initial radial velocity. For the electron to be absorbed by the black hole, the magnitude of the electric (Coulomb) force

$$F_C = \frac{Qe}{4\pi\epsilon_0 r^2} \tag{18.126}$$

(here ϵ_0 is the vacuum dielectric constant) must be smaller than the gravitational force (we use a Newtonian approximation for the sake of simplicity)

$$F_N = \frac{GMm_e}{r^2} \,. \tag{18.127}$$

The condition $F_C < F_N$ implies

$$eQ \leq 4\pi\epsilon_0 GMm_e \,. \tag{18.128}$$

Note that the quantity $\sqrt{4\pi\epsilon_0 G}M$ has the dimensions of a charge. It is convenient to use geometrized ($G = 1$) and unrationalized Gaussian ($4\pi\epsilon_0 = 1$) units, in which the charge-to-mass ratio is dimensionless. In these units 1 C $= 1.16 \times 10^{13}$ g and the charge of the electron is (see Table A) $e = 1.602 \times 10^{-19}$ C $\sim 2 \times 10^{-6}$ g $\sim 2 \times 10^{21} m_e$. Thus, Eq. (18.128) can be written as

$$\frac{Q}{M} \leq \frac{m_e}{e} \sim 5 \times 10^{-22} \,. \tag{18.129}$$

Therefore, due to the tiny mass-to-charge ratio of the electron, the dimensionless parameter Q/M is extremely small.

Similarly to the spin parameter of a Kerr black hole (which is limited by $|a|/M \leq 1$) also the charge of a Reissner-Nordström black hole is limited, $Q^2/M^2 \leq 1$. For $Q = M$ the black hole is said to be *extremal*, whereas for $Q > M$ there is no horizon and a naked singularity appears. In the $Q/M \ll 1$ limit one recovers the Schwarzschild solution and Eq. 18.129 implies that Q must be negligibly small in units of the black hole mass.

The above argument does not apply if the initial radial velocity of the electron is very large. However, more sophisticated arguments (e.g. charge neutralization by surrounding plasma) show that – even when the electrons have large velocities – the charge of astrophysical black holes is always incredibly small and can be neglected.

CHAPTER 19

Geodesic motion in Kerr's spacetime

In this chapter we shall study the geodesic motion of massive and massless particles in Kerr's spacetime. We shall only consider the motion outside the outer horizon ($r \geq r_+$ in Boyer-Lindquist coordinates), because this is the region relevant for astrophysical observations.

We shall look for a set of four algebraic equations for the components of the particle four-velocity, i.e. of the tangent vector to the geodesic

$$u^\mu = \frac{dx^\mu}{d\lambda} \equiv \dot{x}^\mu \,, \tag{19.1}$$

where λ is the affine parameter. These equations can be integrated to find the worldlines $x^\mu(\lambda)$ of massive and massless particles in a *closed form*, in terms of integrals of functions of the coordinates.

As discussed in Sec 18.3, the Kerr metric admits only two Killing vector fields $\vec{k} = \frac{\partial}{\partial t}\,, \vec{m} = \frac{\partial}{\partial \varphi}$; we shall show that, as in the Schwarzschild case, two algebraic equations for $\dot{t}$ and $\dot{\varphi}$ can be found by exploiting these spacetime symmetries and the associated constants of motion E and L. In addition, an algebraic equation for $\dot{r}$ can be obtained from

$$g_{\mu\nu} u^\mu u^\nu = \kappa \,, \tag{19.2}$$

with $\kappa = -1$ or $\kappa = 0$ for timelike and null geodesics, respectively.

In the Schwarzschild case, a fourth equation was provided by the planarity of the orbit: in Sec. 10.2, assuming that $\theta(\lambda = 0) = \frac{\pi}{2}$ and that $\dot{\theta}(\lambda = 0) = 0$, we showed that $\dot{\theta} = 0$ along the entire particle trajectory, i.e. the orbit is planar; due to the spherical symmetry, this property is generic, since we can always rotate the spatial frame and make an arbitrarily oriented initial velocity coincident with the initial conditions mentioned above, leaving the metric unchanged.

Since the Kerr spacetime is axisymmetric the orbits are generically non-planar; the only orbits which are planar are those along the rotational axis with initial velocity parallel to it, and those which start in the equatorial plane with $\dot{\theta}(\lambda = 0) = 0$. In this case a rotation of the spatial frame would not leave the metric unchanged; therefore, at variance with the Schwarzschild case, these orbits are not generic and are restricted to the equatorial plane.

However, there exists an additional constant of motion, the *Carter constant*, which, as we shall show, is crucial to find algebraic equations for $\dot{r}$ and $\dot{\theta}$. The existence of this constant, which is not associated to Killing vectors, is a highly nontrivial property featured by the Kerr metric.

We shall derive the geodesic equations in the general case using the Hamilton-Jacobi approach, and show how the Carter constant emerges in this framework. The resulting equations are quite complicated and need to be solved numerically. Since the equations on the equatorial plane are much simpler, we shall discuss them in detail, in particular for null geodesics. A more detailed discussion of the non-equatorial geodesics of the Kerr metric can be found in Chandrasekhar's monograph [34].

In the following we shall use Boyer-Lindquist's coordinates and write the Kerr metric as (see Sec. 18.2)

$$ds^2 = -dt^2 + \Sigma\left(\frac{dr^2}{\Delta} + d\theta^2\right) + (r^2 + a^2)\sin^2\theta d\varphi^2 + \frac{2Mr}{\Sigma}(a\sin^2\theta d\varphi - dt)^2\,, \tag{19.3}$$

where we remind that $\Delta = r^2 - 2Mr + a^2$ and $\Sigma = r^2 + a^2\cos^2\theta$.

19.1 THE EQUATIONS FOR $\dot{t}$ AND $\dot{\varphi}$

In Sec. 18.3 we showed that, since the spacetime is stationary and axisymmetric, geodesic motion is characterized by two constants

$$E \equiv -u^\mu k_\mu = -u_t \qquad L \equiv u^\mu m_\mu = u_\varphi\,, \tag{19.4}$$

where $\vec{k}$ and $\vec{m}$ are the two Killing vector fields. From these, using Eqs. 18.11, we find

$$\begin{aligned} E &= -g_{t\mu}u^\mu = \left(1 - \frac{2Mr}{\Sigma}\right)\dot{t} + \frac{2Mr}{\Sigma}a\sin^2\theta\,\dot{\varphi}\,, \\ L &= g_{\varphi\mu}u^\mu = -\frac{2Mr}{\Sigma}a\sin^2\theta\,\dot{t} + \left(r^2 + a^2 + \frac{2Mra^2}{\Sigma}\sin^2\theta\right)\sin^2\theta\,\dot{\varphi}\,. \end{aligned} \tag{19.5}$$

These equations can easily be solved for $\dot{t}$ and $\dot{\varphi}$; let us define

$$\begin{aligned} A &\equiv -g_{tt} = 1 - \frac{2Mr}{\Sigma}\,, \\ B &\equiv -g_{t\varphi} = \frac{2Mra}{\Sigma}\sin^2\theta\,, \\ C &\equiv g_{\varphi\varphi} = \left(r^2 + a^2 + \frac{2Mra^2}{\Sigma}\sin^2\theta\right)\sin^2\theta\,, \end{aligned} \tag{19.6}$$

and write Eqs. 19.5 as

$$E = A\dot{t} + B\dot{\varphi}\,, \tag{19.7}$$

$$L = -B\dot{t} + C\dot{\varphi}\,. \tag{19.8}$$

Furthermore,

$$AC + B^2 = -g_{tt}g_{\varphi\varphi} + g_{t\varphi}^2 = \Delta\sin^2\theta \tag{19.9}$$

as shown in Eq. 18.14. Therefore,

$$\begin{aligned} \dot{t} &= \frac{1}{\Delta\sin^2\theta}(CE - BL) \\ &= \frac{1}{\Delta}\left[\left(r^2 + a^2 + \frac{2Mra^2}{\Sigma}\sin^2\theta\right)E - \frac{2Mra}{\Sigma}L\right]\,, \\ \dot{\varphi} &= \frac{1}{\Delta\sin^2\theta}(AL + BE) = \frac{1}{\Delta}\left[\left(1 - \frac{2Mr}{\Sigma}\right)\frac{L}{\sin^2\theta} + \frac{2Mra}{\Sigma}E\right]\,. \end{aligned} \tag{19.10}$$

These are algebraic expressions for $\dot{t}$ and $\dot{\varphi}$ in terms of the two constants of motion, E and L, and of the coordinates.

19.2 THE EQUATIONS FOR $\dot{r}$ AND $\dot{\theta}$

We now look for algebraic expressions for the components $\dot{r}$ and $\dot{\theta}$ of the particle four-velocity. To this aim, it is convenient to use the *Hamilton-Jacobi approach* which we shall briefly review.

Given the Lagrangian of the particle (see Sec. 10.1)

$$\mathcal{L}(x^\mu, \dot{x}^\mu) = \frac{1}{2} g_{\mu\nu} \dot{x}^\mu \dot{x}^\nu \tag{19.11}$$

and given the conjugate momenta[1]

$$p_\mu = \frac{\partial \mathcal{L}}{\partial \dot{x}^\mu} = g_{\mu\nu} \dot{x}^\nu , \tag{19.12}$$

which coincide with the covariant components of the four-velocity $u^\mu = \dot{x}^\mu$, i.e.

$$u^\mu = g^{\mu\nu} p_\nu , \tag{19.13}$$

we can define the *Hamiltonian* of the particle

$$H(x^\mu, p_\nu) = p_\mu \, \dot{x}^\mu(p_\nu) - \mathcal{L}\left(x^\mu, \dot{x}^\mu(p_\nu)\right) = \frac{1}{2} g^{\mu\nu} p_\mu p_\nu \, . \tag{19.14}$$

Since, as shown in Sec. 10.1, the geodesic equations are equivalent to the Euler-Lagrange equations for the Lagrangian 19.11, they are also equivalent to the Hamilton equations

$$\begin{aligned} \dot{x}^\mu &= \frac{\partial H}{\partial p_\mu} , \\ \dot{p}_\mu &= -\frac{\partial H}{\partial x^\mu} . \end{aligned} \tag{19.15}$$

Solving Eqs. 19.15 presents the same difficulties as solving Euler-Lagrange's equations. However, in the Hamilton-Jacobi approach the existence of a further constant of motion emerges quite naturally, as we shall now show.

We look for a function of the coordinates and of the parameter λ on the curve,

$$S = S(x^\mu, \lambda) \tag{19.16}$$

which is the solution to the *Hamilton-Jacobi equation*

$$H\left(x^\mu, \frac{\partial S}{\partial x^\mu}\right) + \frac{\partial S}{\partial \lambda} = 0. \tag{19.17}$$

If the solution S depends on *four independent constants of motion*, it is called a *complete integral.* It can be shown that, if S is a complete integral, then

$$\frac{\partial S}{\partial x^\mu} = p_\mu \, . \tag{19.18}$$

Therefore, once Eq. 19.17 is solved, the expressions of the conjugate momenta (and of $\dot{x}^\mu$) follow in terms of the four constants, and allow to write the solutions of the geodesic equations in closed form, in terms of simple integrals.

[1] Not to be confused with the four-momentum of the particle.

In order to find the complete integral of Eq. 19.17, we can exploit the three constants of motion we already know, i.e.

$$\begin{aligned} H &= \frac{1}{2}g^{\mu\nu}p_\mu p_\nu = \frac{1}{2}\kappa = const \\ p_t &= -E = const \\ p_\varphi &= L = const\,. \end{aligned} \tag{19.19}$$

It is convenient to write S as

$$S = -\frac{1}{2}\kappa\lambda - Et + L\varphi + S^{(r\theta)}(r,\theta)\,, \tag{19.20}$$

where $S^{(r\theta)}$ is a function of r and θ to be determined. Indeed, in this way Eq. 19.18 is automatically satisfied for $\mu = t, \varphi$, and

$$\frac{\partial S}{\partial \lambda} = -\frac{1}{2}\kappa \tag{19.21}$$

which is consistent with Eq. 19.17, with $H = \kappa/2$.

We look for a *separable* solution, $S^{(r\theta)}(r,\theta) = S^{(r)}(r) + S^{(\theta)}(\theta)$, therefore [2]

$$S = -\frac{1}{2}\kappa\lambda - Et + L\varphi + S^{(r)}(r) + S^{(\theta)}(\theta)\,. \tag{19.22}$$

By substituting Eq. 19.22 into Eq. 19.17, and using the expression 18.16 for the inverse metric, we find

$$\begin{aligned} -\kappa + g^{\mu\nu}\frac{\partial S}{\partial x^\mu}\frac{\partial S}{\partial x^\nu} &= -\kappa + \frac{\Delta}{\Sigma}\left(\frac{dS^{(r)}}{dr}\right)^2 + \frac{1}{\Sigma}\left(\frac{dS^{(\theta)}}{d\theta}\right)^2 \\ &-\frac{1}{\Delta}\left[r^2 + a^2 + \frac{2Mra^2}{\Sigma}\sin^2\theta\right]E^2 + \frac{4Mra}{\Sigma\Delta}EL + \frac{\Delta - a^2\sin^2\theta}{\Sigma\Delta\sin^2\theta}L^2 = 0\,. \end{aligned} \tag{19.23}$$

Using the relation 18.64 which we rewrite for convenience

$$(r^2 + a^2) + \frac{2Mra^2}{\Sigma}\sin^2\theta = \frac{1}{\Sigma}\left[(r^2+a^2)^2 - a^2\sin^2\theta\Delta\right]\,, \tag{19.24}$$

and multiplying by $\Sigma = r^2 + a^2\cos^2\theta$, we get

$$\begin{aligned} &-\kappa(r^2 + a^2\cos^2\theta) + \Delta\left(\frac{dS^{(r)}}{dr}\right)^2 + \left(\frac{dS^{(\theta)}}{d\theta}\right)^2 \\ &-\left[\frac{(r^2+a^2)^2}{\Delta} - a^2\sin^2\theta\right]E^2 + \frac{4Mra}{\Delta}EL + \left(\frac{1}{\sin^2\theta} - \frac{a^2}{\Delta}\right)L^2 = 0\,, \end{aligned} \tag{19.25}$$

i.e.

$$\begin{aligned} &\Delta\left(\frac{dS^{(r)}}{dr}\right)^2 - \kappa r^2 - \frac{(r^2+a^2)^2}{\Delta}E^2 + \frac{4Mra}{\Delta}EL - \frac{a^2}{\Delta}L^2 \\ &= -\left(\frac{dS^{(\theta)}}{d\theta}\right)^2 + \kappa a^2\cos^2\theta - a^2\sin^2\theta E^2 - \frac{1}{\sin^2\theta}L^2\,. \end{aligned} \tag{19.26}$$

[2]This assumption will be verified *a posteriori*, once we show that a solution in this form satisfies the Hamilton-Jacobi equation.

Eq. 19.26 can be rearranged by adding to both sides the constant quantity $a^2E^2 + L^2$:

$$\Delta\left(\frac{dS^{(r)}}{dr}\right)^2 - \kappa r^2 - \frac{(r^2+a^2)^2}{\Delta}E^2 + \frac{4Mra}{\Delta}EL - \frac{a^2}{\Delta}L^2 + a^2E^2 + L^2 \qquad (19.27)$$
$$= -\left(\frac{dS^{(\theta)}}{d\theta}\right)^2 + \kappa a^2\cos^2\theta + a^2\cos^2\theta E^2 - \frac{\cos^2\theta}{\sin^2\theta}L^2 .$$

In Eq. 19.27, the left-hand side depends only on r, whereas the right-hand side depends only on θ; this equation is satisfied only if the right- and the left-hand sides are equal to the same constant $\mathcal{C}$, named the Carter constant after its discoverer Brandon Carter [30]. Therefore we get

$$\left(\frac{dS^{(\theta)}}{d\theta}\right)^2 - \cos^2\theta\left[(\kappa+E^2)a^2 - \frac{1}{\sin^2\theta}L^2\right] = \mathcal{C} \qquad (19.28)$$
$$\Delta\left(\frac{dS^{(r)}}{dr}\right)^2 - \kappa r^2 - \frac{(r^2+a^2)^2}{\Delta}E^2 + \frac{4Mra}{\Delta}EL - \frac{a^2}{\Delta}L^2 + E^2a^2 + L^2$$
$$= \Delta\left(\frac{dS^{(r)}}{dr}\right)^2 - \kappa r^2 + (L-aE)^2 - \frac{1}{\Delta}\left[E(r^2+a^2) - La\right]^2 = -\mathcal{C} .$$

Note that in rearranging the terms in the last two lines, we have used the relation

$$-2aLE + 2aLE\frac{r^2+a^2}{\Delta} = -\frac{4aMr}{\Delta}LE . \qquad (19.29)$$

If we define the functions $R(r)$ and $\Theta(\theta)$ as

$$\Theta(\theta) \equiv \mathcal{C} + \cos^2\theta\left[(\kappa+E^2)a^2 - \frac{1}{\sin^2\theta}L^2\right], \qquad (19.30)$$
$$R(r) \equiv \Delta\left[\kappa r^2 - (L-aE)^2 - \mathcal{C}\right] + \left[E(r^2+a^2) - La\right]^2 , \qquad (19.31)$$

then

$$\left(\frac{dS^{(\theta)}}{d\theta}\right)^2 = \Theta , \qquad (19.32)$$
$$\left(\frac{dS^{(r)}}{dr}\right)^2 = \frac{R}{\Delta^2} ,$$

and the solution of the Hamilton-Jacobi equation takes the form

$$S = -\frac{1}{2}\kappa\lambda - Et + L\varphi + \int\frac{\sqrt{R}}{\Delta}dr + \int\sqrt{\Theta}d\theta . \qquad (19.33)$$

Thus, the Carter constant $\mathcal{C}$ emerges as a separation constant and, together with E, L, and κ, allows to solve completely the geodesic problem in Kerr's spacetime. *We stress that, unlike E and L, this constant is not associated to a spacetime isometry*[3].

Given the solution of the Hamilton-Jacobi equations depending on the four constants

[3]The existence of the Carter constant for the Kerr metric is associated with the so-called Killing-Yano tensor, from which other geometrical properties of the spacetime can be derived. The existence of this tensor is sometimes called a *"hidden symmetry"* of the Kerr metric.

$(\kappa, E, L, \mathcal{C})$, it is possible to find algebraic expressions for the components $\dot{r}, \dot{\theta}$ of the four-velocity. Indeed, from Eqs. 19.12, 19.18 the conjugate momenta squared are

$$\begin{aligned} p_\theta^2 &= (\Sigma\dot{\theta})^2 = \Theta(\theta)\,, \qquad (19.34) \\ p_r^2 &= \left(\frac{\Sigma}{\Delta}\dot{r}\right)^2 = \frac{R(r)}{\Delta^2}\,, \end{aligned}$$

therefore

$$\dot{\theta} = \pm\frac{1}{\Sigma}\sqrt{\Theta}\,, \qquad (19.35)$$

$$\dot{r} = \pm\frac{1}{\Sigma}\sqrt{R}\,, \qquad (19.36)$$

where the two choices $\pm$ depend on the sign of the velocity along the angular and radial direction, respectively. These equations, together with Eqs. 19.10, are four algebraic equations for the components of the four-velocity $\vec{u}$, and give the particle trajectory in terms of simple integrals.

To integrate Eqs. 19.35, 19.36 it is convenient to define a new variable $\tilde{\lambda}$ to parametrize the geodesic, the so-called "Mino time" [84]:

$$d\lambda = \Sigma d\tilde{\lambda} = (r^2 + a^2\cos^2\theta)d\tilde{\lambda}\,. \qquad (19.37)$$

In terms of this parameter the equations for $r(\tilde{\lambda})$ and $\theta(\tilde{\lambda})$ are

$$\begin{aligned} \frac{dr}{d\tilde{\lambda}} &= \pm\sqrt{R(r)}\,. \qquad (19.38) \\ \frac{d\theta}{d\tilde{\lambda}} &= \pm\sqrt{\Theta(\theta)}\,. \end{aligned}$$

These equations, unlike Eqs. 19.35, 19.36, are *decoupled*: the equation for $r(\tilde{\lambda})$ depends only on r, while the equation for $\theta(\tilde{\lambda})$ depends only on θ. An important consequence of this property is that for bound orbits, which satisfy $R(r) > 0$ in the interval $r \in [r_1, r_2]$ where r_1 and r_2 are the turning points, and $\Theta(\theta) > 0$ in an interval $\theta \in [\theta_1, \theta_2]$, the functions $r(\tilde{\lambda})$ and $\theta(\tilde{\lambda})$ are periodic:

$$\begin{aligned} r(\tilde{\lambda}) &= r(\tilde{\lambda} + n\Lambda_r) \qquad (19.39) \\ \theta(\tilde{\lambda}) &= \theta(\tilde{\lambda} + k\Lambda_\theta) \end{aligned}$$

with n, k integer numbers and periods

$$\Lambda_r = 2\int_{r_1}^{r_2}\frac{dr}{\sqrt{R}}\,, \qquad \Lambda_\theta = 2\int_{\theta_1}^{\theta_2}\frac{d\theta}{\sqrt{\Theta}}\,. \qquad (19.40)$$

The orbits, still, are very complicated, since the periods in r and θ are different, and φ is not periodic at all: a generic bound orbit is very different from an ellipse. An example of these trajectories is shown in Fig. 19.1 for the values of the constants $E = 0.969$, $L = 2.539$, $\mathcal{C} = 6.470$.

19.3 EQUATORIAL GEODESICS

In this section we shall study the geodesic motion of a particle on the equatorial plane of the Kerr metric in Boyer-Lindquist coordinates. Replacing $\theta = \frac{\pi}{2}$ in Eq. 19.30 we find

$$\Theta\left(\frac{\pi}{2}\right) = \mathcal{C}\,; \qquad (19.41)$$

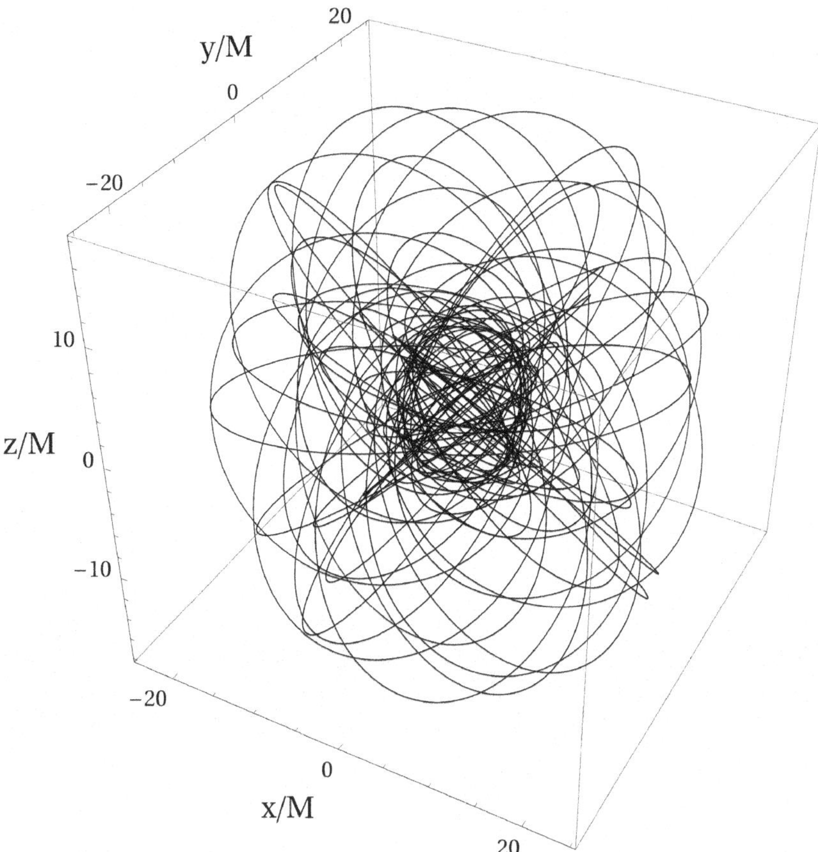

Figure 19.1: Example of a trajectory in Kerr's geometry with $a = 0.9M$, in Cartesian coordinates obtained from the Boyer-Lindquist coordinates (r, θ, φ). The constants of motion are $E = 0.969$, $L = 2.539$, $\mathcal{C} = 6.470$.

consequently, recalling that on the equatorial plane $\Sigma = r^2$, Eq. 19.35 gives

$$\dot{\theta} = \pm\frac{1}{r^2}\sqrt{\mathcal{C}}\,. \tag{19.42}$$

Thus, among all geodesics starting on the equatorial plane, only those with $\mathcal{C} = 0$ have zero initial derivative along the θ direction and remain on that plane at later times. As discussed above, in the Schwarzschild spacetime, due to the spherical symmetry, it is possible to generalize the result to any orbit, and to prove that all geodesics are planar. This generalization is not possible for the Kerr metric which is axially symmetric: in this case only the equatorial geodesics with $\mathcal{C} = 0$ are planar.

Let us now derive the equations for $\dot{t}, \dot{\varphi}, \dot{r}$ for a geodesic on the equatorial plane. Eqs. 19.10 evaluated at $\theta = \frac{\pi}{2}$ give

$$\begin{aligned} \dot{t} &= \frac{1}{\Delta}\left[\left(r^2 + a^2 + \frac{2Ma^2}{r}\right)E - \frac{2Ma}{r}L\right], \\ \dot{\varphi} &= \frac{1}{\Delta}\left[\left(1 - \frac{2M}{r}\right)L + \frac{2Ma}{r}E\right]. \end{aligned} \tag{19.43}$$

It may be noted that the quantities A, B and C, defined in Eqs. 19.5, for $\theta = \frac{\pi}{2}$ reduce to

$$\begin{aligned} A &\equiv -g_{tt} = 1 - \frac{2M}{r}, \\ B &\equiv -g_{t\varphi} = \frac{2Ma}{r}, \\ C &\equiv g_{\varphi\varphi} = r^2 + a^2 + \frac{2Ma^2}{r}, \end{aligned} \tag{19.44}$$

and Eq. 19.9 reduces to $AC + B^2 = \Delta$. Moreover, replacing $\theta = \pi/2$ in Eq. 18.12 we find

$$C = \frac{(r^2 + a^2)^2 - a^2\Delta}{r^2}. \tag{19.45}$$

The equation for $\dot{r}$ can be found by evaluating Eqs. 19.31 and 19.36 for $\theta = \frac{\pi}{2}$ and $\mathcal{C} = 0$

$$\begin{aligned} \dot{r}^2 &= \frac{R(r)}{r^4} = \frac{1}{r^4}\left(\Delta\kappa r^2 - \Delta(L - aE)^2 + \left[E(r^2 + a^2) - La\right]^2\right) \\ &= \frac{1}{r^2}\left(CE^2 - 2BLE - AL^2\right) + \frac{\kappa\Delta}{r^2}. \end{aligned} \tag{19.46}$$

This equation can also be derived from $g_{\mu\nu}u^\mu u^\nu = \kappa$ (Eq. 19.2), using Eqs. 19.43, 19.44, $g_{rr} = r^2/\Delta$ and $\dot{\theta} = 0$.

The polynomial $P(E) \equiv CE^2 - 2BLE - AL^2$ has two roots

$$V_\pm = \frac{BL \pm \sqrt{B^2L^2 + ACL^2}}{C} = \frac{L}{C}\left(B \pm \sqrt{\Delta}\right); \tag{19.47}$$

consequently, Eq. 19.46 can be written as

$$\dot{r}^2 = \frac{C}{r^2}(E - V_+)(E - V_-) + \frac{\kappa\Delta}{r^2}. \tag{19.48}$$

By defining

$$\begin{aligned} V_{\rm r}(r) &= \frac{C}{r^2}(E - V_+)(E - V_-) + \frac{\kappa\Delta}{r^2} \\ &= \frac{(r^2 + a^2)^2 - a^2\Delta}{r^4}(E - V_+)(E - V_-) + \frac{\kappa\Delta}{r^2} \end{aligned} \tag{19.49}$$

(where we have used Eq. 19.45), we get

$$\dot{r}^2 = V_{\rm r}(r)\,. \tag{19.50}$$

Moreover, we can write Eq. 19.47 as

$$V_\pm = \frac{2Mar \pm r^2\sqrt{\Delta}}{(r^2+a^2)^2 - a^2\Delta}L\,. \tag{19.51}$$

In the Schwarzschild limit $a \to 0$ and

$$V_+ + V_- \propto a \to 0\,, \qquad V_+V_- \to \frac{L^2\Delta}{r^4}\,; \tag{19.52}$$

therefore, if we define $V \equiv -V_+V_-$, Eqs. 19.50 and 19.51 reduce to the equations discussed in Sec. 10.2:

$$\dot{r}^2 = E^2 - V(r), \qquad \text{where} \qquad V(r) = -\frac{\kappa\Delta}{r^2} + \frac{L^2\Delta}{r^4} = \left(1 - \frac{2M}{r}\right)\left(-\kappa + \frac{L^2}{r^2}\right)\,, \tag{19.53}$$

where we recall that $\kappa = -1$ for timelike geodesics and $\kappa = 0$ for null geodesics.

19.3.1 Potentials for equatorial geodesics

In principle, when studying the potential of the radial equation given in Eq. 19.49 we have four possibilities, corresponding to L positive and negative and a positive and negative. In practice, the cases obtained by a simultaneous exchange of the signs of a and L,

$$a \to -a\,, \qquad L \to -L\,, \tag{19.54}$$

are equivalent; indeed the physical processes of interest are those with $La > 0$ or with $La < 0$, in which the test particle is, respectively, corotating and counter-rotating with the black hole. Note, however, that by changing the signs of a and L the potentials $V_\pm$ interchange (V_+ becomes V_- and vice versa); to avoid this ambiguity, it is better to redefine the potentials as follows

$$V_\pm = \frac{2MLar \pm |L|r^2\sqrt{\Delta}}{(r^2+a^2)^2 - a^2\Delta}\,. \tag{19.55}$$

With this definition the potentials V_+ and V_- are also invariant under the symmetry 19.54. Moreover, the following inequality is always true:

$$V_+ \geq V_-\,. \tag{19.56}$$

In the following we shall assume $a > 0$. The case $a < 0$ can be obtained by changing the sign of L, and using the symmetry 19.54.

In general, the following properties apply:

- $V_\pm$ vanish in the limit $r \to \infty$.
- V_+ and V_- coincide for $\Delta = 0$, i.e. at the horizon

  $$r = r_+ = M + \sqrt{M^2 - a^2}\,, \tag{19.57}$$

  and

  $$V_+(r_+) = V_-(r_+) = \frac{2Mr_+La}{(r_+^2 + a^2)^2}\,, \tag{19.58}$$

  which is positive if $La > 0$ and negative if $La < 0$. Note that since the angular velocity of the black hole is $\Omega_H = 2Mar_+/(r_+^2 + a^2)^2$ (Eq. 18.67), then

  $$V_+(r_+) = V_-(r_+) = \Omega_H L\,. \tag{19.59}$$

- If $La > 0$ (corotating orbits), the potential V_+ is positive definite; V_- is positive at r_+, and vanishes when

$$r\sqrt{\Delta} = 2Ma \quad \rightarrow \quad r^2(r^2 - 2Mr + a^2) = 4M^2a^2\,, \tag{19.60}$$

which gives

$$r^4 - 2Mr^3 + a^2r^2 - 4M^2a^2 = (r - 2M)(r^3 + a^2r + 2Ma^2) = 0\,; \tag{19.61}$$

thus V_- vanishes at $r = 2M$, which is the location of the ergosphere on the equatorial plane.

- If $La < 0$ (counter-rotating orbits), the potential V_- is negative definite and V_+, which is negative at r_+, vanishes at $r = 2M$.

- The study of the derivatives of $V_\pm$ shows that both potentials, V_+ and V_-, have only one stationary point, which is a local maximum for V_+ and a local minimum for V_-.

In summary, $V_+(r)$ and $V_-(r)$ have the shapes shown in Fig. 19.2, where the upper and lower panels refer to the $La > 0$ and $La < 0$ cases, respectively.

19.3.2 Null geodesics

Let us consider a massless particle, say a photon, moving on a null geodesic on the equatorial plane. In this case $\kappa = 0$ and the radial equation 19.48 becomes

$$\dot{r}^2 = V_{\rm r} = \frac{(r^2 + a^2)^2 - a^2\Delta}{r^4}(E - V_+)(E - V_-)\,. \tag{19.62}$$

Since $\dot{r}^2$ must be positive, and given $(r^2 + a^2)^2 - a^2\Delta > 0$ (Eq. 18.66), massless particles can only move on those geodesics with constant of motion E which satisfies the following inequalities

$$E < V_- \qquad \text{or} \qquad E > V_+\,. \tag{19.63}$$

Thus, the region $V_- < E < V_+$, corresponding to the shadowed region in the two panels of Fig. 19.2, is forbidden.

In order to study the equatorial orbits allowed to massless particles, it is useful to compute the radial acceleration. By differentiating Eq. 19.62 with respect to the affine parameter λ, we find

$$2\dot{r}\ddot{r} = V_{\rm r}'\dot{r} \tag{19.64}$$

where the prime indicates differentiation with respect to r; using Eq. 19.49,

$$\ddot{r} = \frac{1}{2}V_{\rm r}' = \frac{1}{2}\left(\frac{C}{r^2}\right)'(E - V_+)(E - V_-) - \frac{C}{2r^2}\left[V_+'(E - V_-) + V_-'(E - V_+)\right]\,. \tag{19.65}$$

Let us evaluate $\ddot{r}$ in a point where the radial velocity $\dot{r}$ is zero, i.e. when $E = V_+$ or $E = V_-$:

$$\ddot{r} \;=\; = -\frac{C}{2r^2}V_+'(V_+ - V_-) \qquad \text{if } E = V_+\,, \tag{19.66}$$

$$\ddot{r} \;=\; = -\frac{C}{2r^2}V_-'(V_- - V_+) \qquad \text{if } E = V_-\,. \tag{19.67}$$

Since

$$V_+ - V_- = \frac{2|L|r^2\sqrt{\Delta}}{(r^2 + a^2)^2 - a^2\Delta} = \frac{2|L|\sqrt{\Delta}}{C}\,, \tag{19.68}$$

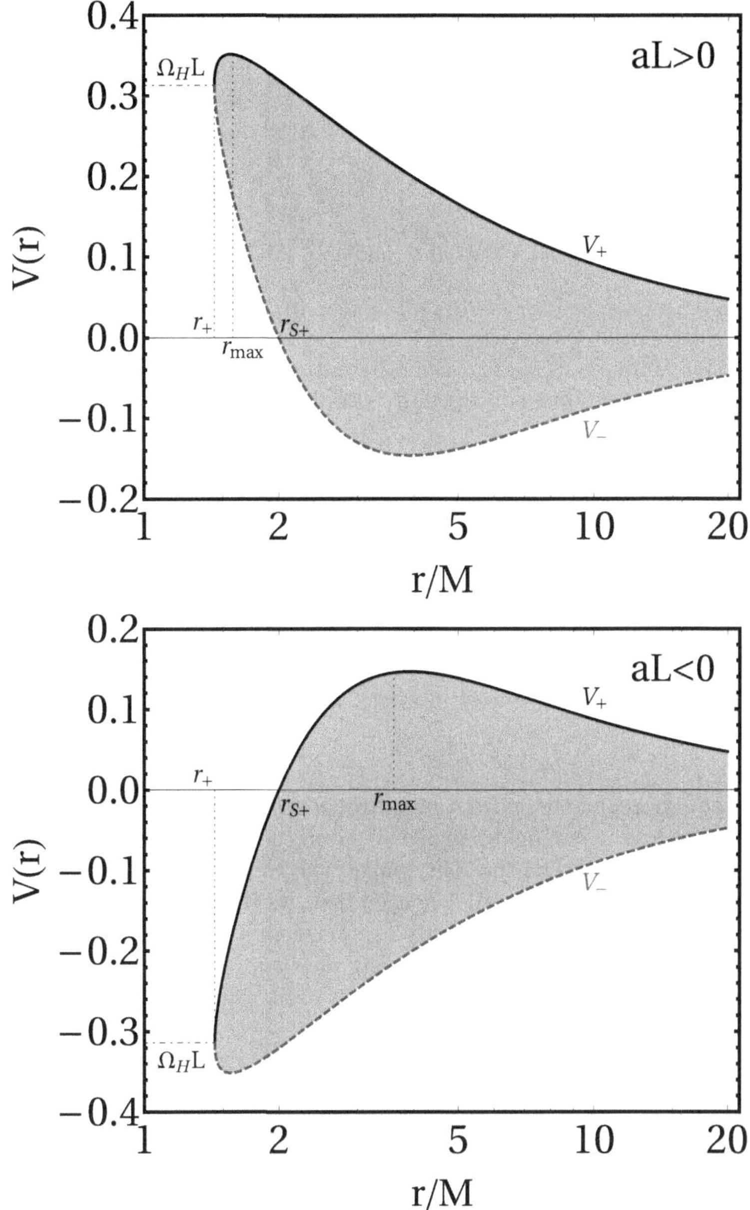

Figure 19.2: The potentials $V_+(r)$ and $V_-(r)$, for corotating ($aL > 0$, top panel) and counter-rotating ($aL < 0$, bottom panel) equatorial orbits around a Kerr black hole with spin parameter $a = 0.9M$. The shadowed region is not accessible to the motion of massless particles. r_{S+} is the radius of the ergosphere on the equator, $\theta = \pi/2$.

the radial acceleration at these points is

$$\ddot{r} = -\frac{|L|\sqrt{\Delta}}{r^2} V'_{\pm} \qquad \text{if } E = V_{\pm}\,. \tag{19.69}$$

We can now classify null, equatorial geodesics depending on the value of the constant of motion E. Here we only consider the cases with $E > V_+$. The cases with $E < V_-$ (see Eq. 19.63) will be discussed in Sec. 19.3.3. Moreover, we do not consider here the geodesics with $L = aE$, which have special properties and will be discussed in Box 19-A.

Unstable circular orbits

For circular orbits $r = r_{\max}$, $\dot{r} = \ddot{r} = 0$ and, since $\dot{r}^2 = V_{\rm r}$, $\ddot{r} = \frac{1}{2}V'_{\rm r}$,

$$V_{\rm r} = 0 \qquad V'_{\rm r} = 0\,. \tag{19.70}$$

The first condition imposes $E = V_+$. By differentiating the right-hand side of Eq. 19.62 with respect to r, a lengthy but conceptually simple computation yields

$$V'_{\rm r} = \frac{2(L - aE)(L(r - 3M) + aE(r + 3M)}{r^4} = 0\,. \tag{19.71}$$

Excluding the special root $L = aE$ discussed in Box 19-A, the solution of the above equation is

$$\frac{L}{E} = -a\frac{r + 3M}{r - 3M}\,, \tag{19.72}$$

so that orbits with $r < 3M$ are corotating ($aL > 0$) whereas orbits with $r > 3M$ are counter-rotating ($aL < 0$). After some algebra, Eq. 19.72, together with $E = V_+$, can be cast in the following form

$$r(r - 3M)^2 - 4Ma^2 = 0\,. \tag{19.73}$$

This is a cubic equation and therefore it has three roots. One root is always smaller than the outer horizon and therefore can be discarded. We dub the two physical roots as $r = r^i_{\rm LR}$, with $i = 1, 2$, since they correspond to the *light ring* of the Kerr metric. Indeed, since at $r = r^i_{\rm LR}$ both the radial velocity and the radial acceleration vanish, these are circular null orbits. However, as can easily be verified by studying the second derivative of $V_{\rm r}$, all null circular orbits are unstable, exactly as for the light ring at $r = 3M$ in the Schwarzschild case (see Sec. 10.4). The two roots of Eq. 19.73 can be most easily found by defining $a = \cos(\phi/2)M$, where $\phi \in [0, \pi]$.

$$\begin{aligned} r^{(1)}_{\rm LR}/M &= 2 - \cos(\phi/3) + \sqrt{3}\sin(\phi/3)\,, \\ r^{(2)}_{\rm LR}/M &= 4\cos(\phi/6)^2\,. \end{aligned} \tag{19.74}$$

Note that $r^{(1)}_{\rm LR}$ is a decreasing function of a such that $1 \leq r^{(1)}_{\rm LR}/M \leq 3$. Hence, from Eq. 19.72, it corresponds to (unstable) null circular orbits corotating with the black hole. Conversely, $r^{(2)}_{\rm LR}$ is an increasing function of a such that $3 \leq r^{(2)}_{\rm LR}/M \leq 4$ and it therefore corresponds to counter-rotating orbits. The dependence of $r^{(i)}_{\rm LR}$ on the spin of the black hole is shown in Fig. 19.3. In particular,

$$\begin{aligned} r^{(1,2)}_{\rm LR} &= 3M && \text{for } a = 0\,, \\ r^{(1)}_{\rm LR} &= M && \text{for } a = M\,, \\ r^{(2)}_{\rm LR} &= 4M && \text{for } a = M\,. \end{aligned} \tag{19.75}$$

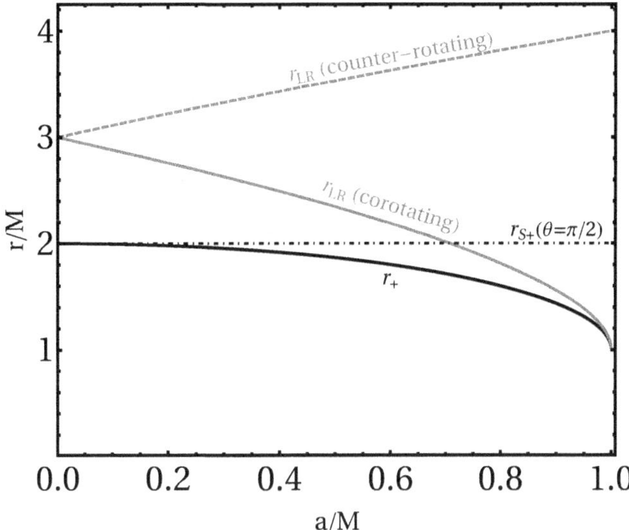

Figure 19.3: Location of the light ring of a Kerr black hole in Boyer-Lindquist's coordinates as a function of the spin. Continuous (dashed) curves refer to corotating (counter-rotating) orbits. We also show the location of the outer horizon r_+ and the boundary of the ergoregion on the equator ($r = 2M$). Note that at high spin some corotating circular orbits are within the ergoregion and that $r_{\rm LR} = r_+ = M$ when $a = M$.

Therefore, while for a Schwarzschild black hole the unstable circular orbit of a photon (i.e., the light ring) is at $r = 3M$, for a Kerr black hole it can be much closer to the horizon; in particular, at high spin it can be inside the ergosphere and, in the extremal case $a = M$, for corotating orbits $r_{\rm max} = M$ coincides with the outer horizon, $r_{\rm LR} = r_+ = M$.

Particle capture

A photon falling from infinity with constant of motion $E > V_+(r_{\rm max})$ and $\dot{r}(\infty) < 0$ will move towards the black hole with decreasing radial velocity up to $r_{\rm max}$, and then with increasing radial velocity, until it crosses the horizon (see Fig. 19.2); the number of turns it makes around the black hole will depend on the value of L [4].

Particle deflection

If the massless particle starts it motion with $\dot{r}(\infty) < 0$ and $0 < E < V_+(r_{\rm max})$, it reaches the turning point where $E = V_+(r)$ and $\dot{r} = 0$. Since at the turning point $V_+' < 0$, according to Eq. 19.69 $\ddot{r} > 0$, i.e. the radial acceleration is positive. Therefore the particle reverses its motion and escapes at infinity.

[4]Photons with $E < V_+(r_{\rm max})$ can also be captured, as long as they are created near the black hole. Assuming $E > 0$, a corotating photon ($aL > 0$) can be created in the region $r_+ < r < r_{\rm max}$ if $V_+(r_+) < E < V_+(r_{\rm max})$, while a counter-rotating photon ($aL < 0$) can be created in the region $r_{S+} < r < r_{\rm max}$ if $0 < E < V_+(r_{\rm max})$.

Box 19-A

Principal null geodesics in the equatorial plane

We shall discuss here the null equatorial geodesics with $L = aE$. By replacing $L = aE$ in Eqs. 19.43 we find

$$\dot{t} = \frac{dt}{d\lambda} = \frac{r^2 + a^2}{\Delta} E , \qquad \dot{\varphi} = \frac{d\varphi}{d\lambda} = \frac{a}{\Delta} E , \tag{19.76}$$

whereas Eq. 19.46 gives

$$\dot{r}^2 = E^2 \qquad \to \qquad \dot{r} = \frac{dr}{d\lambda} = \pm E . \tag{19.77}$$

These geodesics coincide with the principal null geodesics discussed in Sec. 18.4.2 and in Box 18-A, restricted to the equatorial plane, using, however, a different parametrization. Indeed, rescaling the affine parameter as $\lambda = \hat{\lambda}/E$, we find

$$\frac{dx^\alpha}{d\hat{\lambda}} = \left(\frac{r^2 + a^2}{\Delta}, \pm 1, 0, \frac{a}{\Delta} \right) , \tag{19.78}$$

which coincides with the tangent vector given in Eq. 18.27.
The angular momentum of the massless particle is $L = aE$ while the angular momentum of the black hole is $J = aM$, therefore

$$\frac{L}{E} = \frac{J}{M} . \tag{19.79}$$

Thus, these geodesics have a special feature: the angular momentum per unit energy of the particle and that per unit mass of the black hole coincide. Using Eqs. 19.76, 19.77 we find

$$\frac{dt}{dr} = \pm \frac{r^2 + a^2}{\Delta} , \tag{19.80}$$

the solution of which is

$$t = \pm r_* + const , \tag{19.81}$$

where r_* is the *tortoise coordinate of the Kerr metric*, defined by

$$\frac{dr_*}{dr} = \frac{r^2 + a^2}{\Delta} . \tag{19.82}$$

The explicit expression of $r_*(r)$ is

$$r_* = r + \left(M + \frac{M^2}{\sqrt{M^2 - a^2}} \right) \log \left(\frac{r - r_+}{2M} \right) + \left(M - \frac{M^2}{\sqrt{M^2 - a^2}} \right) \log \left(\frac{r - r_-}{2M} \right) . \tag{19.83}$$

We refer the interested reader to [34] for further details on principal null geodesics.

19.3.3 Energy of a particle in Kerr's spacetime

So far we have assumed that the constant of motion E associated to the timelike Killing vector is positive. In the following we shall discuss whether this quantity can take negative values. This will turn out to be useful to study the cases $E < V_-$, which are formally allowed by the condition 19.63.

In order to establish whether the constant of motion E can assume negative values, we need to remind whether, and when, it can be related to the energy of the particle, following the discussion in Sec. 10.3.

The energy of a particle is an observer-dependent quantity, and in General Relativity it is defined by Eq. 10.32, which we rewrite here for convenience:

$$\mathcal{E}^{(U)} = -g_{\mu\nu}U^\mu p^\nu = -U^\mu p_\mu \,, \tag{19.84}$$

where $\vec{p}$ is the particle four-momentum, and $\vec{U}$ is the four-velocity of the observer. As discussed in Sec. 10.3, the energy $\mathcal{E}^{(U)}$ measured by any observer $\vec{U}$, which is also the energy measured in a LIF comoving with the observer $\vec{U}$, *must be positive.* On the other hand, the constant of motion E is given by Eq. 19.4 and reads $E = -u_t$, where $\vec{u}$ is the particle four-velocity.

Let us now consider a static observer in Kerr's spacetime, with $U^\mu_{st} = (1,0,0,0)$, located outside the ergosphere, where such observer can exist. According to the definition 19.84, the energy measured by this observer is $\mathcal{E}^{(U_{st})} = -p_t$. If the particle is massless we can always parametrize the geodesic in such a way that $p_t \equiv u_t$. Thus:

$$\mathcal{E}^{(U_{st})} = -p_t = E \,. \tag{19.85}$$

We conclude that *for a particle which happens to be outside the ergoregion during its motion, the constant E can be interpreted as the particle energy measured by a static observer located outside the ergosphere*[5]. For such particles, orbits with negative values of E are not allowed (see Sec. 10.3). Thus, referring to Fig. 19.2, corotating or counter-rotating orbits with $E < V_-$ impinging from radial infinity are forbidden, even though for such orbits $\dot{r}^2 > 0$ (see Eq. 19.62).

Let us now consider a massless particle which starts its motion *within* the ergoregion, i.e. between r_+ and r_{S+}, and never crosses the ergosphere. In this region static observers cannot exist; therefore we need to consider a different observer, for instance the ZAMO introduced in Sec. 18.5, i.e. a stationary observer with $L = u_\varphi = 0$ and angular velocity on the equatorial plane given by (see Eq. 18.65)

$$\Omega = \frac{2Mar}{(r^2+a^2)^2 - a^2\Delta} \,. \tag{19.86}$$

Since the ZAMO is a stationary observer, as explained in Sec. 18.6.1, its four-velocity is

$$U^\mu_Z = const\,(1,0,0,\Omega) \,. \tag{19.87}$$

The constant can be found by imposing $g_{\mu\nu}U^\mu_Z U^\nu_Z = -1$. Note that this constant must be positive, otherwise the observer would move backwards in time (see discussion in Sec. 10.3). The particle energy measured by the ZAMO is

$$\mathcal{E}^Z = -p_\mu U^\mu_Z = const\,(E - \Omega L) \,, \tag{19.88}$$

where we remind that $p_\mu = (-E, p_r, 0, L)$. Thus, the requirement $\mathcal{E}^Z > 0$ is equivalent to

$$E > \Omega L \,. \tag{19.89}$$

[5]Similarly, for massive particles E is the energy per unit mass as measured by a static observer outside the ergosphere, i.e. $\mathcal{E}^{(U_{st})} = -p_t = m_{\rm P}E$, where $m_{\rm P}$ is the particle mass.

By comparing Eq. 19.86 with the expression of the potentials $V_\pm$ given by Eq. 19.55, we find that

$$V_- \leq \Omega L \leq V_+ \,, \tag{19.90}$$

where the equal sign holds for $\Delta = 0$, i.e. at the horizon. This implies that the null geodesics with

$$E > V_+ \tag{19.91}$$

satisfy the positive energy condition 19.89 and are allowed, whereas those with $E < V_-$ are forbidden.

Thus, referring to Fig. 19.2:

- a corotating particle ($La > 0$) can move within the ergoregion only if the constant of motion E is positive and is in the range

  $$V_+(r_+) < E < V_+(r_{\rm max}) \,. \tag{19.92}$$

  If $E > V_+(r_{\rm max})$ the particle can cross the ergosphere and escape at infinity.

- For counter-rotating particles ($La < 0$), since in the ergoregion V_+ is negative the requirement $E > V_+$ (necessary and sufficient to ensure that $\mathcal{E} > 0$) *allows for negative values of the constant of motion* E, provided

  $$V_+(r_+) = V_-(r_+) < E < 0 \,. \tag{19.93}$$

  The consequences of this possibility will be discussed in Sec. 19.4.

  It should be stressed that the fact that E is negative is not in contradiction with the requirement that the particle energy must be positive. Indeed these two quantities coincide only when the particle energy is measured by a static observer, but these observers cannot exist inside the ergosphere; the geodesics with $E < 0$ we are considering never escape from the ergoregion.

19.3.4 Timelike geodesics

The study of equatorial, timelike geodesics is much more involved than the study of null geodesics. Indeed, when $\kappa = -1$, Eq. 19.46 can be written as

$$\dot{r}^2 = V_{\rm r} \equiv \frac{1}{r^4}\left(\left[E(r^2+a^2) - aL\right]^2 - \Delta\left[r^2 + (L-aE)^2\right]\right) \,, \tag{19.94}$$

and it does not allow for a simple qualitative study as in the null case. Therefore, here we only report some results of a detailed study of the geodesic equations in this general case. We refer the interested reader to [17, 34] for further information.

On a circular orbit

$$V_{\rm r} = 0 \qquad V_{\rm r}' = 0 \,. \tag{19.95}$$

These two equations can be solved for E and L and give [17]

$$E = \frac{r^{3/2} - 2Mr^{1/2} \pm aM^{1/2}}{r^{3/4}\left(r^{3/2} - 3Mr^{1/2} \pm aM^{1/2}\right)^{1/2}} \,, \tag{19.96}$$

$$L = \pm M^{1/2} \frac{r^2 \mp 2aM^{1/2}r^{1/2} + a^2}{\left(r^{3/2} - 3Mr^{1/2} \pm aM^{1/2}\right)^{1/2}} \,. \tag{19.97}$$

As discussed in Sec. 18.2 we can choose $a \geq 0$ without loss of generality, so that $L > 0$

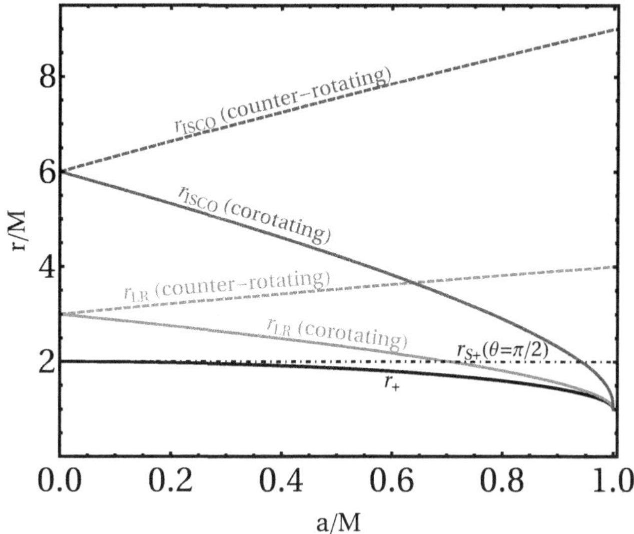

Figure 19.4: Location of the ISCO and of the light ring in the equatorial plane of a Kerr black hole in Boyer-Lindquist coordinates as a function of the spin. Continuous (dashed) curves refer to corotating (counter-rotating) orbits. We also show the horizon location and the boundary of the ergoregion on the equatorial plane ($r = 2M$). Note that at high spin some corotating stable circular orbits are within the ergoregion.

($L < 0$) corresponds to corotating (counter-rotating) orbits. It is easy to see that the term $\left(r^2 \mp 2aM^{1/2}r^{1/2} + a^2\right)$ in the numerator of Eq. 19.97 is always positive when $r \geq r_+ = M + \sqrt{M^2 - a^2}$.[6] Therefore, the upper (lower) sign in Eq. 19.97 (and in the following equations) corresponds to $L > 0$ ($L < 0$).

Furthermore, from Eqs. 19.96, 19.97 we see that circular orbits exist only if the denominators in the expressions of E and L are real, i.e. if

$$r^{3/2} - 3Mr^{1/2} \pm aM^{1/2} \geq 0\,, \tag{19.98}$$

where the limiting case corresponds to infinite energy per unit mass, i.e. to a photon circular orbit.

If we further require that the circular orbit is *stable*, we need to impose that $\frac{d^2V_r}{dr^2} \leq 0$, and this gives the following equivalent conditions

$$\begin{aligned} 1 - E^2 &\geq \frac{2}{3r}M\,, \\ r^2 - 6Mr \pm 8aM^{1/2}r^{1/2} - 3a^2 &\geq 0\,, \end{aligned} \tag{19.99}$$

or

$$r \geq r_{\rm ISCO}\,, \tag{19.100}$$

where $r_{\rm ISCO}$ is the radius of the innermost (marginally) stable circular orbit, whose analytical expression reads [17]

$$r_{\rm ISCO} = M\left(3 + Z_2 \mp [(3 - Z_1)(3 + Z_1 + 2Z_2)]^{1/2}\right)\,, \tag{19.101}$$

[6]Indeed, the square of the inequality $r - M \geq \sqrt{M^2 - a^2}$ implies $r^2 + a^2 \geq 2Mr$, and thus $r^2 - 2aM^{1/2}r^{1/2} + a^2 \geq 2Mr - 2aM^{1/2}r^{1/2} = 2\sqrt{Mr}(\sqrt{Mr} - a)$, which is always positive because $r \geq M \geq a$.

where

$$Z_1 \equiv 1 + \left(1 - \frac{a^2}{M^2}\right)^{1/3} \left[\left(1 + \frac{a}{M}\right)^{1/3} + \left(1 - \frac{a}{M}\right)^{1/3}\right], \qquad (19.102)$$

$$Z_2 \equiv \left(3\frac{a^2}{M^2} + Z_1\right)^{1/2}.$$

For $a = 0$, the two solutions corresponding to corotating and counter-rotating orbits coincide and give $r_{\rm ISCO} = 6M$, as expected (see Sec. 10.5); by increasing a, the ISCO moves closer to the black hole for corotating orbits, and farther for counter-rotating orbits. In the extremal case ($a = M$), the corotating ISCO coincides with the outer horizon, $r_{\rm ISCO} = r_+ = M$, whereas the counter-rotating ISCO is at $r = 9M$. This behaviour is similar to that we have already seen in the case of unstable circular orbits for null geodesics.

In Fig. 19.4 we show the locations of the innermost stable circular orbits for timelike geodesics, together with the unstable circular orbits for null geodesics discussed in the previous section.

Box 19-B

The shadow of a Kerr black hole

Earlier in Fig. 11.9 we showed the first direct radio image of a black hole, taken by the Event Horizon Telescope Collaboration at the center of galaxy M87 [111]. Although spin measurements for this supermassive ($M \simeq 6.5 \times 10^9 M_\odot$) black hole are not yet accurate enough, it is likely that this black hole is rotating, i.e. it is a Kerr black hole. In this case, its shadow and radio image are more involved than what was discussed in Sec. 11.5.

First of all, we remind that the shadow is the "silhouette" of the black hole, which can be obtained by ray-tracing the photons emitted by a light source at infinity and deflected by the black hole. On the other hand, Fig. 11.9 mostly shows the radiation emitted by the accretion disk around the black hole. As discussed in Sec. 11.5.1, a thin, equatorial accretion disk typically extends up to the ISCO, whose location depends on the spin. As shown in Fig. 19.4, for highly-spinning black holes the ISCO is closer to the horizon and therefore the radiation comes from regions where the gravitational field is stronger and is more redshifted than in the Schwarzschild case.

Furthermore, the spin breaks the spherical symmetry of the system. This implies that the contour of the shadow is not spherical as in the Schwarzschild case. In that case the profile of the shadow is a sphere of radius $3\sqrt{3}M$, which coincides with the critical impact parameter of photons coming from a source at infinity. In the Kerr case the shadow contour depends on the spin of the black hole and on the inclination angle of the observer relative to the black hole rotation axis [122]. In particular, due to the frame-dragging effect, the shadow is mostly distorted in the plane orthogonal to the rotation axis.

Besides deforming the shadow, the frame dragging is responsible for a distortion of the image of the accretion disk which also depends on the viewing angle. The image shown in Fig. 11.9 is brighter in the south part. This effect is mostly due to *Doppler beaming* of material that is rotating in the same direction of the black hole, i.e. clockwise with respect to the celestial plane of the observer. Indeed, the velocity of corotating material is higher due to the frame dragging, and therefore the relativistic beaming of the emitted radiation is stronger in one direction relative to the opposite one.

It is worthwhile to note that – owing to the opposite contributions of the frame dragging and of the non-zero quadrupole moment of the Kerr metric – the shape of the shadow depends much more mildly on the spin as compared to, e.g., the ISCO. It is therefore quite challenging to measure the black hole spin through the observation of its silhouette, whereas such measurement might be possible through an accurate modelling of the accretion flow. Quantitative studies are based on sophisticated general-relativistic, magneto-hydrodynamical simulations and accurate astrophysical modeling; the interested reader can find more details in [111] and in the references therein.

19.3.5 Emission of light from the ergosphere

In Sec. 18.6 we mentioned that the ergosphere is sometimes called *infinite redshift surface* because by definition $g_{tt} = 0$ when $r = r_{S+}$. This might suggest that a signal is infinitely redshifted when emitted from the ergosphere despite the fact that the ergosphere is outside the horizon of the Kerr metric. However, the name "infinite redshift surface" is misleading because the signals emitted on the ergosphere are *not* infinitely redshifted (otherwise no signal from the ergosphere could reach an observer at infinity and this surface would be an event horizon!). Indeed, the argument presented in Sec. 11.1 for the gravitational redshift assumes that the source and the observer are *at fixed positions*, but we now know that this assumption cannot be fulfilled by a source in the ergoregion, since it is forced to corotate with the black hole.

To understand the properties of a light signal emitted at the ergosphere, let us extend the discussion of Sec. 10.6 – about a spaceship emitting an SOS signal while falling radially towards a Schwarzschild black hole – to the case of a spaceship falling into a Kerr black hole. For simplicity, we focus on equatorial orbits with $L = aE$ for both the spaceship and the SOS signals. Note that this implies that the signals travel along equatorial principal null geodesics, as those considered in Box 19-A.

For the spaceship orbit, replacing $L = aE$ in Eq. 19.43 gives

$$\frac{dt}{d\tau} = \frac{r^2 + a^2}{\Delta} E \,, \tag{19.103}$$

and Eqs. 19.31, 19.36 give (since $\mathcal{C} = 0$ and $\kappa = -1$)

$$R(r) = -\Delta r^2 + E^2 r^4 \tag{19.104}$$

and

$$\frac{dr}{d\tau} = -\frac{1}{r^2}\sqrt{R} = -\sqrt{E^2 - \frac{\Delta}{r^2}} \,. \tag{19.105}$$

Assuming that the spaceship starts its motion at rest at infinity, $E = 1$ and the geodesic equations are

$$\frac{dt}{d\tau} = \frac{r^2 + a^2}{\Delta} \,, \tag{19.106}$$

$$\frac{dr}{d\tau} = -\sqrt{\frac{2M}{r} - \frac{a^2}{r^2}} \,. \tag{19.107}$$

As expected, when $a = 0$ these equations reduce to Eqs. 10.69 and 10.70 for the Schwarzschild metric.

As in Sec. 11.1 the spaceship, while falling towards the Kerr black hole horizon, sends an SOS in the form of a sequence of equally spaced electromagnetic pulses, and these signals are received by a static observer at radial infinity. Since we assumed that the signal travels along an equatorial null geodesic with $L = aE$ then, as shown in Box 19-A,

$$t = \pm r_* + const \,, \tag{19.108}$$

where r_* is the tortoise coordinate of the Kerr metric given in Eq. 19.83, satisfying

$$\frac{dr_*}{dr} = \frac{r^2 + a^2}{\Delta} \,. \tag{19.109}$$

Following the same steps that lead to Eq. 10.85, we consider two pulses emitted from the spaceship at $t = t_1$ and $t = t_2$, and received from an observer at infinity at $t^{obs} = t_1^{obs}$ and $t^{obs} = t_2^{obs}$. Since for each pulse $t - r_*$ is constant along the null geodesic,

$$t_2^{obs} - t_1^{obs} = (t_2^{obs} - r_*^{obs}) - (t_1^{obs} - r_*^{obs}) = (t_2 - r_{*2}) - (t_1 - r_{*1}) \,. \tag{19.110}$$

If the two pulses are emitted at a very short time interval, $dt^{obs} = dt - dr_*$, and

$$\frac{dt^{obs}}{d\tau} = \frac{dt}{d\tau} - \frac{dr_*}{d\tau} = \frac{dt}{d\tau} - \frac{dr_*}{dr}\frac{dr}{d\tau} = \frac{r^2+a^2}{\Delta}\left(1+\sqrt{\frac{2M}{r}-\frac{a^2}{r^2}}\right), \tag{19.111}$$

where we have used Eqs. 19.106, 19.107, and 19.109.

The above equation shows that the time interval between pulses as detected by the static observer at infinity diverges as the spaceship approaches the horizon ($\Delta = 0$). However, the same quantity is *finite* on the ergosphere. Indeed, on the equatorial plane $r_{S+} = 2M$ and Eq. 19.111 reduces to

$$\left.\frac{dt^{obs}}{d\tau}\right|_{r=r_{S+}} = \frac{a^2+4M^2}{a^2}\left(1+\sqrt{1-\frac{a^2}{4M^2}}\right). \tag{19.112}$$

19.3.6 Generalization of Kepler's third law in Kerr's spacetime

As shown in Sec. 10.5, Kepler's third law is satisfied not only in Newtonian gravity, but also in General Relativity for circular, timelike geodesics around a Schwarzschild black holes. We want to generalize this law to the case of a particle moving on a circular, timelike geodesic in the equatorial plane of a Kerr black hole. Given $g_{r\mu} = 0$ if $\mu \neq r$, the r-component of the Euler-Lagrange equation,

$$\frac{d}{d\lambda}\frac{\partial\mathcal{L}}{\partial\dot{r}} = \frac{\partial\mathcal{L}}{\partial r} \tag{19.113}$$

with $\mathcal{L} = \frac{1}{2}g_{\mu\nu}\dot{x}^\mu\dot{x}^\nu$, gives

$$\frac{d}{d\lambda}(g_{rr}\dot{r}) = \frac{1}{2}g_{\mu\nu,r}\dot{x}^\mu\dot{x}^\nu\,. \tag{19.114}$$

On circular orbits $\dot{r} = \ddot{r} = 0$, and this equation reduces to

$$g_{tt,r}\dot{t}^2 + 2g_{t\varphi,r}\dot{t}\dot{\phi} + g_{\varphi\varphi,r}\dot{\varphi}^2 = 0\,. \tag{19.115}$$

The angular velocity of the particle is $\omega = \dot{\varphi}/\dot{t}$, thus

$$g_{\varphi\varphi,r}\omega^2 + 2g_{t\varphi,r}\omega + g_{tt,r} = 0\,. \tag{19.116}$$

We remind that on the equatorial plane

$$\begin{aligned} g_{tt} &= -\left(1-\frac{2M}{r}\right), \\ g_{t\varphi} &= -\frac{2Ma}{r}, \\ g_{\varphi\varphi} &= r^2+a^2+\frac{2Ma^2}{r}, \end{aligned} \tag{19.117}$$

therefore Eq. 19.116 becomes

$$(r^3 - Ma^2)\omega^2 + 2Ma\omega - M = 0\,. \tag{19.118}$$

The solution for ω is

$$\begin{aligned} \omega_\pm &= \frac{-Ma \pm \sqrt{Mr^3}}{r^3 - Ma^2} = \pm\sqrt{M}\,\frac{r^{3/2}\mp a\sqrt{M}}{r^3-Ma^2} \\ &= \pm\sqrt{M}\frac{r^{3/2}\mp a\sqrt{M}}{(r^{3/2}+a\sqrt{M})(r^{3/2}-a\sqrt{M})} \\ &= \pm\frac{\sqrt{M}}{r^{3/2}\pm a\sqrt{M}}, \end{aligned} \tag{19.119}$$

where again the $\pm$ signs refer to corotating and counter-rotating orbits. Eq. 19.119 is the *generalization of Kepler's third law in Kerr's spacetime.* For $a = 0$, the above expression reduces to the standard Keplerian frequency

$$\omega_{\pm} = \pm\sqrt{\frac{M}{r^3}}\,. \tag{19.120}$$

19.4 ENERGY EXTRACTION FROM BLACK HOLES

In this section we shall show that, in principle, it is possible to extract energy from a rotating black hole. There are two phenomena in which this can occur: *Penrose's process*, involving particles, and *superradiant scattering*, involving waves. As we are going to show, these are different realizations of the same physical process.

19.4.1 The Penrose process

In the following we will use a slightly different notation for the constants of motion E, L, which have been defined in Sec. 18.3 to be the energy and angular momentum *per unit mass*, for massive particles, and the energy and angular momentum for massless particles, as measured by a static observer outside the ergoregion, i.e. $E = -k^\mu u_\mu$, $L = m^\mu u_\mu$ (Eqs. 18.21). Here we define E and L to be the energy and angular momentum at infinity, both for massive and massless particles, i.e.

$$E = -k^\mu p_\mu\,, \qquad L = m^\mu p_\mu \tag{19.121}$$

where $\vec{p}$ is the particle four-momentum. In other words, in the case of massive particles, E and L are redefined by multiplying them by the particle mass. Again we shall assume $a > 0$ without loss of generality.

Since particles with negative E can exist in the ergoregion, as discussed in Sec. 19.3.3, we can imagine a process through which it may be possible to extract rotational energy from a Kerr black hole; this is named **Penrose's process** [91], and it works as follows.

Suppose that we shoot a massive particle with energy E and angular momentum L from infinity, so that it falls towards the black hole on the equatorial plane. The covariant components of the particle four-momentum initially read

$$p_\mu(\lambda = 0) = (-E, p_r^0, 0, L)\,; \tag{19.122}$$

$p_r^0 = p_r(\lambda = 0)$, λ is the affine parameter of the geodesic. During the particle motion its four-momentum changes, but the covariant components $p_t = k^\mu p_\mu = -E$ and $p_\varphi = m^\mu p_\mu = L$ remain constant, i.e.,

$$p_\mu(\lambda) = (-E, p_r(\lambda), 0, L)\,. \tag{19.123}$$

Let us suppose that when the particle enters the ergoregion it decays in two photons (see Fig. 19.5), with momenta

$$p_{1\,\mu} = (-E_1, p_{1r}, 0, L_1) \qquad p_{2\,\mu} = (-E_2, p_{2r}, 0, L_2)\,. \tag{19.124}$$

Since the four-momentum is conserved in this decay, we have

$$p_\mu = p_{1\mu} + p_{2\mu}\,, \tag{19.125}$$

from which it follows that

$$E = E_1 + E_2\,, \qquad L = L_1 + L_2\,. \tag{19.126}$$

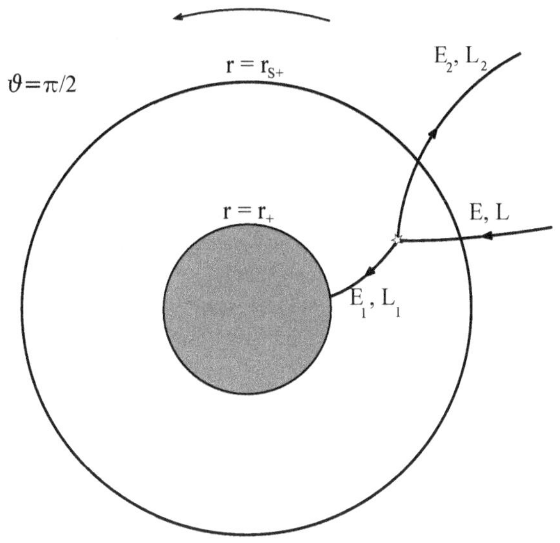

Figure 19.5: Schematic representation of Penrose's process. A massive particle moving on the equatorial plane enters in the ergoregion of a Kerr black hole and decays in two massless particles, one falling into the horizon, the other escaping back to infinity.

Let us assume that $\dot{r}_1 < 0$, so that photon 1 falls into the black hole, and that it has negative constants of motion, i.e. $E_1 < 0$ and $L_1 < 0$, with (see Eq. 19.93)

$$V_+(r_+) = V_-(r_+) < E_1 < 0 \,.$$

We further assume that $\dot{r}_2 > 0$, i.e. photon 2 goes back to infinity. Note that, as explained in Sec. 19.3.3, this is possible only if

$$E_2 > V_+(r_{\rm max}) \,.$$

Its energy and angular momentum are

$$\begin{aligned} E_2 &= E - E_1 > E \\ L_2 &= L - L_1 > L \,, \end{aligned} \tag{19.127}$$

thus, *at the end of the process the particle at infinity is more energetic than the one initially sent to the black hole!*

Since $E_1 < 0, L_1 < 0$, the capture of photon 1 by the black hole reduces the black hole mass-energy M and the angular momentum $J = Ma$; indeed their values $M_{\rm fin}$, $J_{\rm fin}$ after absorbing photon 1 are, respectively,

$$M_{\rm fin} = M + E_1 < M \tag{19.128}$$

$$J_{\rm fin} = J + L_1 < J \,. \tag{19.129}$$

To prove the relation 19.128 we note that, as shown in Chapter 17, the total mass-energy of the system is

$$P^0_{\rm tot} = \int_V d^3x(-g)(T^{00} + t^{00}) \,, \tag{19.130}$$

where V is the volume of a $t = const$ three-surface. If we neglect the gravitational field generated by the particle, t^{00} is due to the black hole gravitational field only, thus (see Eq. 17.93)

$$\int_V d^3x(-g)t^{00} = M\,, \tag{19.131}$$

and

$$P^0_{\text{tot}} = \int_V d^3x(-g)T^{00}_{\text{particle}} + M\,. \tag{19.132}$$

Let us compute the integral when the process starts, i.e. at a time when the massive particle is shot with energy E into the black hole; the spacetime near the particle is flat and the 00-component of the stress-energy tensor, in Minkowskian coordinates, is

$$T^{00}_{\text{particle}} = E\delta^3(\mathbf{x} - \mathbf{x}(t))\,. \tag{19.133}$$

Thus Eq. 19.132 gives

$$P^0_{\text{tot in}} = E + M_{\text{in}}\,. \tag{19.134}$$

Repeating the computation at the end of the process, namely when the photon 2 reaches infinity, we find

$$P^0_{\text{tot fin}} = E_2 + M_{\text{fin}}\,. \tag{19.135}$$

If we neglect the outgoing gravitational flux generated by the particle, P^0_{tot} is a conserved quantity; therefore by equating the initial and final momenta we find

$$P^0_{\text{tot in}} = P^0_{\text{tot fin}} \quad \to \quad M_{\text{fin}} = M_{\text{in}} + (E - E_2) \quad \to \quad M_{\text{fin}} = M_{\text{in}} + E_1 < M_{\text{in}}\,. \tag{19.136}$$

This proves the relation 19.128. The corresponding relation for the angular momentum, Eq. 19.129, can be proved in a similar way.

In conclusion, by this process one can *extract* rotational energy from the black hole. Note that this phenomenon is possible due to the presence of an ergoregion: it is not possible to extract energy from a static black hole at the *classical* level. This is not anymore true when quantum effects are taken into account, as discussed in the next chapter.

19.4.2 Superradiant scattering

Penrose's process involves particle decays; however, there exists an analog process, which involves *waves* scattered off a rotating black hole. The corresponding energy extraction process is called *superradiance* (for a detailed analysis we refer the interested reader to the monograph [25]). Superradiance can occur in several physical systems; in the context of black hole physics it was discovered by Zeldovich in 1971 [123] and studied in detail by Press and Teukolsky later on [95, 110]. It can occur for any *bosonic* wave, i.e., for scalar, electromagnetic, or even gravitational waves.

Studying the superradiant scattering off a rotating black hole requires the study of the perturbations of the Kerr metric, which goes beyond the scope of this book (for a comprehensive treatment, we refer the reader to Chandrasekhar's book [34]). Nonetheless, superradiance can be proved on very generic grounds. It can be shown that the dynamics of perturbations of the Kerr metric with a time dependence $\sim e^{-i\omega t}$ can be described in terms of a Schrödinger-like equation as in the Schwarzschild case (see Box 15-D)

$$\frac{d^2\psi}{dr_*^2} + V_{\text{eff}}\psi = 0\,, \tag{19.137}$$

where ψ is some function that describes the radial behavior of the perturbations, the tortoise coordinate r_*, defined in Eq. 19.83, maps the region $r \in [r_+, \infty)$ to the entire real axis, and the potential $V_{\rm eff}$ depends also on the frequency ω. Its precise form might be complicated but, for our purposes, it is sufficient to know that the effective potential is *real*[7] and that it has the following asymptotic behavior:

$$V_{\rm eff} \to \begin{cases} k_H \equiv \omega - m\Omega_H & r* \to -\infty \\ \omega & r* \to +\infty \end{cases} , \tag{19.138}$$

where m is the azimuthal number of the perturbation (see Box 15-D).

Let us consider the scattering of a monochromatic wave with frequency ω and azimuthal number m, incident from infinity with amplitude $\mathcal{I}$. Since $V_{\rm eff}$ is constant at the boundaries (see Eq. 19.138), the solution to Eq. 19.137 near the boundaries is a superposition of plane waves, namely

$$\psi \sim \begin{cases} \mathcal{T} e^{-ik_H r_*} & \text{as } r* \to -\infty\,, \\ \mathcal{R} e^{i\omega r_*} + \mathcal{I} e^{-i\omega r_*} & \text{as } r* \to +\infty\,, \end{cases} \tag{19.139}$$

where $\mathcal{T}$ and $\mathcal{R}$ are the transmission and reflection coefficients, respectively. Indeed, owing to the time dependence $\sim e^{-i\omega t}$ of the wave, terms like $e^{i\omega r_*}$ (with $\omega > 0$) describe outgoing waves, and vice versa for $e^{-i\omega r_*}$ (see Chapter 15). Note that the presence of a horizon imposes that waves at $r* \to -\infty$ (i.e., $r \sim r_+$) are purely ingoing.

Since the effective potential is real, ψ^* (the complex conjugate of ψ) is still a solution of Eq. 19.137 which satisfies the complex conjugate boundary conditions, namely

$$\psi^* \sim \begin{cases} \mathcal{T}^* e^{ik_H r_*} & \text{as } r* \to -\infty\,, \\ \mathcal{R}^* e^{-i\omega r_*} + \mathcal{I}^* e^{i\omega r_*} & \text{as } r* \to +\infty\,. \end{cases} \tag{19.140}$$

Furthermore, since the background is stationary, the field equations are invariant under the transformations $t \to -t$, $a \to -a$, and $\omega \to -\omega$, which map the conjugate boundary conditions 19.140 into a form analog to Eq. 19.139. Therefore, the solutions ψ and ψ^* satisfy the same differential equation with the same form of boundary conditions and, in addition, are linearly independent. The standard theory of ordinary differential equations tells us that their Wronskian,

$$W = \frac{d\psi}{dr_*}\psi^* - \frac{d\psi^*}{dr_*}\psi\,, \tag{19.141}$$

is independent of r_*. By evaluating the Wronskian near the horizon and near infinity using Eq. 19.139 we obtain

$$W = \begin{cases} -2ik_H |\mathcal{T}|^2 & \text{as } r* \to -\infty \\ 2i\omega(|\mathcal{R}|^2 - |\mathcal{I}|^2) & \text{as } r* \to +\infty \end{cases} , \tag{19.142}$$

respectively. Since W is constant, the two expressions above must be equal to each other; therefore

$$|\mathcal{R}|^2 = |\mathcal{I}|^2 - \frac{k_H}{\omega}|\mathcal{T}|^2 = |\mathcal{I}|^2 - \frac{\omega - m\Omega_H}{\omega}|\mathcal{T}|^2\,, \tag{19.143}$$

independently of the details of the potential in the wave equation. Thus, for waves satisfying the *superradiant condition*

$$\omega(\omega - m\Omega_H) < 0\,, \tag{19.144}$$

we obtain $|\mathcal{R}|^2 > |\mathcal{I}|^2$, i.e. the scattered wave is *amplified* since the reflected energy is *larger* than the incident one. The energy excess comes from the rotational energy of the black hole, which slows down, just like in Penrose's process.

[7] In general, the effective potential for perturbations of the Kerr metric is complex. However, it can be shown that a redefinition of the perturbation fields satisfies an equation like Eq. 19.137 with $V_{\rm eff}$ real [42].

CHAPTER 20

Black hole thermodynamics

The study of the mathematical properties of black holes has revealed the existence of an unexpected link between black hole mechanics and thermodynamics. This result, due to Bekenstein, Hawking, Bardeen, Carter, Christodoulou, and others, was totally unexpected, since black holes – as discussed in previous chapters – are extremely simple objects, only characterized by their mass and angular momentum, while thermodynamics involves, by its own nature, systems with a very large number of degrees of freedom.

In this chapter we shall present a brief introduction to black hole thermodynamics. We shall see that it is possible to associate an entropy to a black hole, and that – when quantum mechanics is taken into account – a black hole is not totally black, but rather emits radiation as a black body. The interested reader can find an exhaustive discussion of these issues in [114, 115, 40, 89] and references therein.

20.1 IRREDUCIBLE MASS AND BLACK HOLE AREA THEOREM

Let us consider a stationary black hole which, owing to the no-hair theorems discussed in Sec. 18.9, is uniquely described by the Kerr metric with mass M and angular momentum $J = Ma$. To hereafter we shall assume $a \geq 0$ without loss of generality. If the black hole interacts with matter and energy outside the horizon it becomes non-stationary, unless the process is slow enough that it can be considered as a sequence of infinitesimal interaction processes. In this case – within a good approximation – the black hole goes through a sequence of stationary states (similarly to a fluid which undergoes a quasi-static transformation through a sequence of equilibrium configurations). Each of these processes can be thought as the absorption of a (massive or massless) particle with energy $E \ll M$ and angular momentum $L \ll J$[1].

As shown in Sec. 19.4.1, when the particle crosses the black hole horizon the change in the black hole mass is equal to the energy of the particle, $\delta M = E$ (see Eq. 19.136). Similarly, it can be shown that the change in the black hole angular momentum is equal to the angular momentum of the particle, $\delta J = L$. Note that in Sec. 19.4.1 we considered a Penrose process in which the particle crossing the horizon has negative energy and angular momentum, but in fact the result $\delta M = E$ (and $\delta J = L$) applies in general, also to particles with positive energy and/or positive angular momentum.

Let us reconsider the Penrose process: a massive particle with energy E and angular momentum L falls from infinity into the black hole on the equatorial plane; in the ergoregion

[1] For the constants of geodesic motion we follow the same notation used in Sec. 19.4.1, i.e. $E = -k^\mu p_\mu$, $L = m^\mu p_\mu$, where p^μ is the particle four-momentum.

it decays in two (massless) particles, with

$$\begin{aligned} E &= E_1 + E_2 \\ L &= L_1 + L_2 \end{aligned} \tag{20.1}$$

and

$$V_+(r) < E_1 < 0 \qquad L_1 < 0\,, \tag{20.2}$$

where V_+ is defined in Eq. 19.55. The particle 2 goes back to infinity, while the particle 1 falls into the black hole. Thus, the mass and the angular momentum of the black hole change by

$$\begin{aligned} \delta M &= E_1 = E - E_2 < 0 \\ \delta J &= L_1 = L - L_2 < 0\,, \end{aligned} \tag{20.3}$$

that is, the black hole mass and angular momentum decrease.

We shall now show that the energy which can be extracted in a Penrose process, $|\delta M| = |E_1|$, is limited by the extracted angular momentum, $|\delta J| = |L_1|$. Indeed, since the particle 1 is massless and moves in the ergoregion towards the black hole (see the lower panel of Fig. 19.2), we have

$$E_1 \geq V_+(r) \geq V_+(r_+)\,, \tag{20.4}$$

where r is the radial coordinate at which the particle is created. Moreover (see Eq. 19.59)

$$V_+(r_+) = \Omega_H L_1 = \Omega_H \delta J\,, \tag{20.5}$$

where

$$\Omega_H = \frac{2Mar_+}{(r_+^2 + a^2)^2} = \frac{a}{r_+^2 + a^2} \tag{20.6}$$

is the black hole angular velocity defined in Eq. 18.67 (see also Eq. 18.68). Summarizing, $E_1 \geq V_+(r_+)$ implies that

$$\delta M \geq \Omega_H \delta J\,. \tag{20.7}$$

Since in the Penrose process $\delta M < 0$ and $\delta J < 0$, we can write this inequality as

$$|\delta M| \leq \Omega_H |\delta J|\,, \tag{20.8}$$

which shows that the extracted energy is always smaller than, or equal to, the extracted angular momentum multiplied by the black hole angular velocity.

The inequality 20.7 becomes an equality if and only if the two inequalities in Eq. 20.4 are saturated, i.e. if and only if $E_1 = V_+(r)$ and $r = r_+$. Note that the condition $E_1 = V_+(r)$ is equivalent to $\dot{r} = 0$ because $\dot{r}^2 \propto E_1 - V_+(r)$ (see Eq. 19.62). We conclude that $\delta M = \Omega_H \delta J$ if and only if

$$\dot{r} = 0 \quad \text{and} \quad r = r_+\,, \tag{20.9}$$

i.e. if the particle is emitted at $r = r_+$ with vanishing radial velocity. Conversely, if the conditions 20.9 are not satisfied, then $|\delta M| < \Omega_H |\delta J|$ and the energy extracted is lower than the extracted angular momentum multiplied by Ω_H.

The condition 20.7 is also satisfied when the angular momentum and/or the energy of the particle swallowed by the black hole are positive. Indeed, as discussed in Sec. 19.3.3, the energy E of a massless particle on an equatorial orbit around a Kerr black hole always satisfies the inequality $E > V_+(r)$. Moreover, looking at both panels of Fig. 19.2 it is clear that if the particle has been created at the radial coordinate r inside the ergosphere, $V_+(r) > V_+(r_+)$; if, instead, the particle 1 originates from outside the ergosphere and reaches the horizon, $E_1 > V_{\rm max} > V_+(r_+)$. The limiting case $E_1 = V_+(r_+) = \Omega_H L_1$ is possible only when the particle is created in the ergosphere, and the two conditions 20.9 are satisfied. This can occur either with $L_1 < 0$ and $E_1 < 0$, as in Penrose's process (lower panel of Fig. 19.2), or with $L_1 > 0$ and $E_1 > 0$ (upper panel of Fig. 19.2).

It can be shown that the inequality 20.7 also applies to massive particles, and to particles in non-equatorial orbits. In other words, *when a Kerr black hole captures a particle, the change in its mass and angular momentum always satisfies the inequality* $\delta M \geq \Omega_H \delta J$. It can also be shown that this inequality is always saturated under the conditions 20.9, i.e. when the particle is released from the horizon with $\dot{r} = 0$. These statements (whose proofs in the general case are quite involved and beyond the scope of this book) have far-reaching consequences, which we now discuss.

Given $J = Ma$, Eq. 20.7 yields

$$M\delta M \geq \frac{J\delta J}{r_+^2 + a^2}\,, \tag{20.10}$$

which implies (see Box 20-B)

$$\delta\left(M^2 - \frac{J^2}{r_+^2 + a^2}\right) \geq 0\,. \tag{20.11}$$

Therefore in any process in which a particle is swallowed by a Kerr black hole, the quantity $M^2 - J^2/(r_+^2 + a^2)$ *never decreases.* This quantity can be expressed in a simpler way, by using Eq. 18.68, $r_+^2 + a^2 = 2Mr_+$:

$$M^2 - \frac{J^2}{r_+^2 + a^2} = M^2\left(1 - \frac{a^2}{r_+^2 + a^2}\right) = \frac{M^2 r_+^2}{r_+^2 + a^2} = \frac{r_+^2 + a^2}{4}\,; \tag{20.12}$$

therefore, as noted by Christodoulou [36], if we define the **irreducible mass**

$$M_{\rm irr} \equiv \frac{1}{2}\sqrt{r_+^2 + a^2}\,, \tag{20.13}$$

from Eq. 20.11 it follows that

$$\delta M_{\rm irr} \geq 0\,, \tag{20.14}$$

i.e., *when a Kerr black hole captures a particle, its irreducible mass never decreases.* As discussed above, any infinitesimal interaction process can be modeled as the capture of a particle with $|\delta M| \ll M$ and $|\delta J| \ll J$; thus in these processes the irreducible mass cannot decrease.

For later use, we provide here some other useful relations:

$$M_{\rm irr}^2 = \frac{Mr_+}{2} = \frac{M^2 + \sqrt{M^4 - J^2}}{2}\,, \tag{20.15}$$

and

$$M_{\rm irr}^2 = M^2 - \frac{J^2}{4M_{\rm irr}^2}\,. \tag{20.16}$$

The latter follows from Eq. 20.12. Note also that if $J = 0$, $M_{\rm irr} = M$.

The mere existence of a quantity – the irreducible mass – which cannot decrease leads to the notion of reversible and irreversible processes. Indeed, let us consider a process in which a particle with energy E_1 and angular momentum L_1 is sent into a black hole with mass M and angular momentum J, leading to a black hole with mass $M + E_1$ and angular momentum $M + L_1$. If the irreducible mass is constant, this process is called *reversible*, because it is possible to perform a process in which a particle with energy $-E_1$ and angular momentum $-L_1$ is sent into the black hole under the same conditions, bringing back the black hole to mass M and angular momentum J. If instead the irreducible mass increases, $\delta M_{\rm irr} > 0$, the process is *irreversible*: the "reverse" cannot occur, because it would have $\delta M_{\rm irr} < 0$, violating the condition 20.14. Note that a similar argument holds in thermodynamics, where the existence of a quantity which cannot decrease (in isolated systems), the *entropy*, leads to the notion of irreversible processes and thus to the second law of thermodynamics.

We remark that the condition for reversible processes, $\delta M_{\rm irr} = 0$, is equivalent to say that $\delta M = \Omega_H \delta J$. We have shown that in a Penrose process with a given δJ, the extracted energy $|\delta M| \leq \Omega_H |\delta J|$ is maximum when this inequality is saturated, i.e. when the process is reversible. Therefore, if we want to extract the maximum amount of energy from a black hole with mass M and angular momentum J, we must perform a sequence of reversible Penrose processes, until all angular momentum has been extracted and we end up with a Schwarzschild black hole. The mass of the final, non-rotating black hole is equal to its irreducible mass, which has not changed:

$$M^2_{\rm fin} = M^2_{\rm irr} = \frac{Mr_+}{2}\,, \tag{20.17}$$

where r_+ is the horizon radius of the *initial* black hole. The relative change in the mass therefore is

$$\frac{\Delta M}{M} = \frac{M - M_{\rm fin}}{M} = 1 - \sqrt{\frac{r_+}{2M}}\,. \tag{20.18}$$

If the processes are not reversible, $M_{\rm irr}$ increases, $M_{\rm fin} = M_{\rm irr}$ is larger than $Mr_+/2$, and the extracted energy is smaller than in the reversible case. Since r_+ is a decreasing function of a, the extracted energy increases with the initial black hole angular momentum. If $a = 0$ (i.e. the initial black hole is Schwarzschild), $r_+ = 2M$ and $\Delta M/M = 0$; conversely, if $a = M$ (i.e. the initial black hole is extremal), $r_+ = M$ and

$$\frac{\Delta M}{M} = 1 - \frac{1}{\sqrt{2}} = 0.29\,. \tag{20.19}$$

This equation shows that the Penrose process can be extremely efficient, extracting up to 29% of the mass-energy of a rotating black hole. This efficiency is by far larger than that of any known astrophysical process in the Universe.

We can gain further insight into the meaning of the irreducible mass by determining the *black hole area*, i.e. the area of the two-surface corresponding to the horizon evaluated at a constant "time". Note that the Boyer-Lindquist coordinates are not well defined at $r = r_+$; to properly describe the horizon we use the Kerr coordinates $(v, r, \theta, \bar{\phi})$ defined in Sec. 18.4.2, and consider the two-surface $v = const$ and $r = r_+$. The Kerr metric 18.43, restricted to this surface, is

$$d\sigma^2 = (r_+^2 + a^2\cos^2\theta)d\theta^2 + \frac{(r_+^2 + a^2)^2\sin^2\theta}{r_+^2 + a^2\cos^2\theta}d\bar{\phi}^2\,, \tag{20.20}$$

and has determinant $(r_+^2 + a^2)^2\sin^2\theta$. Therefore, the black hole area is

$$A = \int (r_+^2 + a^2)\sin\theta d\theta d\bar{\phi} = 4\pi(r_+^2 + a^2) = 16\pi M^2_{\rm irr}\,. \tag{20.21}$$

Thus, the square of the irreducible mass is a measure of the black hole area.

We can conclude that, when a particle falls into a Kerr black hole, the area of the latter can only increase or, if the process is reversible, remain constant; namely,

$$\delta A \geq 0\,. \tag{20.22}$$

Box 20-A

Summary: reversible and irreversible processes

Let us consider a particle, either massive or massless, falling into a Kerr black hole. The black hole is stationary both before and after the capture. The process is **reversible**, if

$$\delta M = \Omega_H \delta J\,; \tag{20.23}$$

in this case, the particle is released from the outer horizon with

$$\dot{r} = 0 \quad \text{and} \quad r = r_+\,, \tag{20.24}$$

and

$$\delta M_{\rm irr} = 0 \qquad \text{i.e.} \qquad \delta A = 0\,. \tag{20.25}$$

The process is **irreversible** if

$$\delta M > \Omega_H \delta J\,; \tag{20.26}$$

in this case, the particle does not satisfy either one of the conditions 20.24 or both, and

$$\delta M_{\rm irr} > 0 \qquad \text{i.e.} \qquad \delta A > 0\,. \tag{20.27}$$

The result $\delta A \geq 0$ holds not only for infinitesimal interactions of stationary black holes, as those discussed above and summarized in Box 20-A, but also for any process – not necessarily stationary – involving one or more black holes. Indeed, the **black hole area theorem**, proved by Hawking (see e.g. [57] and references therein), states that *in any spacetime with black holes, the total area of the black holes' surface can only increase (or remain constant in reversible processes).*

This theorem is extremely powerful; for instance, it implies that in the coalescence of two black holes such as those discussed in Chapter 14, the area of the final black hole is always larger than the sum of the areas of the initial black holes.

Box 20-B

Proof of the inequality in Eq. 20.11

Here we show that Eq. 20.10,

$$M\delta M \geq \frac{J\delta J}{r_+^2 + a^2}\,, \tag{20.28}$$

implies Eq. 20.11,

$$\delta\left(M^2 - \frac{J^2}{r_+^2 + a^2}\right) \geq 0\,. \tag{20.29}$$

Indeed, since $r_+^2 + a^2 = 2Mr_+$,

$$\delta\left(M^2 - \frac{J^2}{r_+^2 + a^2}\right) = 2\left(M\delta M - \frac{J\delta J}{r_+^2 + a^2}\right) + \frac{2J^2}{(2Mr_+)^2}\delta(Mr_+)\,, \tag{20.30}$$

and

$$\begin{aligned} \delta(Mr_+) &= \delta(M^2 + \sqrt{M^4 - J^2}) = 2M\delta M + \frac{2M^3\delta M - J\delta J}{\sqrt{M^4 - J^2}} \qquad (20.31) \\ &= 2M\delta M + \frac{2M^2\delta M - J\delta J/M}{r_+ - M} = \frac{2Mr_+\delta M - J\delta J/M}{r_+ - M} \\ &= \frac{2r_+}{r_+ - M}\left(M\delta M - \frac{J\delta J}{2Mr_+}\right); \end{aligned}$$

it follows

$$\delta\left(M^2 - \frac{J^2}{r_+^2 + a^2}\right) = \left(2 + \frac{4J^2 r_+}{(r_+ - M)(2Mr_+)^2}\right)\left(M\delta M - \frac{J\delta J}{r_+^2 + a^2}\right) \geq 0\,. \tag{20.32}$$

20.2 THE LAWS OF BLACK HOLE THERMODYNAMICS

As remarked in Sec. 20.1 the notion of reversible and irreversible processes, which naturally arises in the study of black hole dynamics, has been introduced in analogy with thermodynamical processes. Going further along these lines, it is possible to draw an analogy – which, at this stage of the discussion, is just a formal correspondence – between Eq. 20.22 and the second law of thermodynamics. This analogy can be further extended to other laws of thermodynamics, as we are going to discuss.

In thermodynamics, a fluid in equilibrium is described by a set of state variables, such as the internal energy U, the volume V, the temperature T, the entropy S, the pressure p, etc. As discussed in Chapter 16, a state of a perfect fluid is fully identified by the values of two of these variables; the equation of state provides the remaining variables in terms of these two. Similarly, a stationary black hole is characterized by two "state variables": the mass M and the angular momentum J. A further state function is the area A which – like the entropy S of an isolated fluid – cannot decrease.

Each of these quantities can be expressed in terms of the other two. The black hole analog of the equation of state of a fluid is then the expression of the black hole area in terms of its mass and angular momentum, $A = A(M, J) = 16\pi M_{\rm irr}^2 = (M^2 + \sqrt{M^4 - J^2})/2$, or, equivalently, the inverse function $M = M(A, J)$ which follows from Eq. 20.16:

$$M^2 = M_{\rm irr}^2 + \frac{J^2}{4M_{\rm irr}^2} = \frac{A}{16\pi} + \frac{4\pi J^2}{A}. \tag{20.33}$$

The first law of thermodynamics establishes how the internal energy of a fluid changes in an infinitesimal transformation between equilibrium states [2]:

$$\delta U = -p\delta V + T\delta S. \tag{20.34}$$

The term $-p\delta V$ describes the work done *on* the fluid (the work done *by* the fluid has the opposite sign), while $T\delta S$ is the heat received by the fluid. We remark that other contributions to the work term have to be included if, for instance, the fluid is in global motion. In particular, if the fluid is rigidly rotating with angular velocity Ω and angular momentum J, the work term associated to a change in the angular momentum, δJ, is $\Omega\delta J$.

Let us now consider an infinitesimal transformation of a black hole between stationary configurations. The mass change can be found by differentiating Eq. 20.33 (see Box 20-C):

$$\delta M = \Omega_H \delta J + \frac{\kappa}{8\pi}\delta A \tag{20.35}$$

where

$$\kappa = \frac{\sqrt{M^2 - a^2}}{r_+^2 + a^2} \tag{20.36}$$

is called **surface gravity** of the black hole (see Box 20-D).

Eq. 20.35 is similar to the first law of thermodynamics: δU corresponds to δM, $\Omega_H \delta J$ can be interpreted as the work done on the black hole to change its angular momentum by δJ, and corresponds to $-p\delta V$; the term $\kappa\delta A/(8\pi)$ corresponds to the heat received by the fluid, $T\delta S$. Moreover, the area law $\delta A \geq 0$ (Eq. 20.22), which shows that the black hole area always increases, is similar to the second law of thermodynamics, $\delta S \geq 0$.

This analogy suggests that the black hole area can be considered as a sort of *black hole entropy*: we define

$$S_{\rm BH} \equiv \alpha A, \tag{20.37}$$

where, choosing units in which the Boltzmann constant $k_{\rm B}$ is unity, the entropy is dimensionless and α is a constant with the dimension of an inverse square length. If we also define the *temperature* of the black hole as

$$T_{\rm BH} \equiv \frac{\kappa}{8\pi\alpha}, \tag{20.38}$$

the change of mass between two neighbouring stationary configurations of the black hole can be written as

$$\delta M = \Omega_H \delta J + T_{\rm BH}\delta S_{\rm BH}. \tag{20.39}$$

This is the **first law of black hole thermodynamics**, while

$$\delta S_{\rm BH} \geq 0 \tag{20.40}$$

is the **second law of black hole thermodynamics**.

[2] To adopt a uniform notation, we describe the small changes of a thermodynamical variable "X" as the variation δX between two neighbouring states, rather than as the differential dX.

Thus, for instance, a process in which the black hole angular momentum J decreases corresponds to the expansion of a fluid; if J is reduced through a reversible process (in which $S_{\rm BH}$ is constant), this corresponds to an adiabatic expansion of a fluid, which determines a reduction of the internal energy – and indeed the black hole mass decreases, too. If, instead, J is reduced through a non-reversible process, $S_{\rm BH}$ increases, and the thermodynamical analogous process is the expansion of a fluid receiving heat; in this case the reduction of internal energy (like the reduction of the black hole mass) is smaller than in the reversible case. Another relevant example is the process in which two (initially stationary) black holes coalesce to a single stationary black hole with a larger area: it corresponds to the process in which two thermodynamical systems merge, increasing their total entropy.

We mention that this analogy has been shown to extend to the third law of thermodynamics, too: it is impossible to reach $T_{\rm BH} = 0$ by any sequence of physical processes, i.e. by accreting matter or energy onto the black hole. Thus, since $T_{\rm BH} \propto \kappa$ and the surface gravity κ vanishes for extremal Kerr black holes (see Eq. 20.36), we can conclude that astrophysical black holes cannot be spun up to extremality by accretion or, more generally, by any interaction with matter and energy surrounding the black hole.

Box 20-C

Proof of Eq. 20.35

To prove Eq. 20.35 we differentiate Eq. 20.33,

$$M^2 = \frac{A}{16\pi} + \frac{4\pi J^2}{A}\,. \tag{20.41}$$

We have

$$\frac{\partial M}{\partial J} = \frac{1}{2M}\frac{\partial M^2}{\partial J} = \frac{J}{M}\frac{4\pi}{A} = \frac{a}{r_+^2 + a^2} = \Omega_H \tag{20.42}$$

and

$$\begin{aligned}\frac{\partial M}{\partial A} &= \frac{1}{2M}\frac{\partial M^2}{\partial A} = \frac{1}{2M}\left(\frac{1}{16\pi} - \frac{4\pi J^2}{A^2}\right) = \frac{1}{32\pi M}\left(1 - \frac{4M^2a^2}{(r_+^2+a^2)^2}\right)\\ &= \frac{1}{32\pi M}\left(1 - \frac{a^2}{r_+^2}\right) = \frac{1}{32\pi M}\left(2 - \frac{r_+^2+a^2}{r_+^2}\right)\\ &= \frac{1}{32\pi M}\left(2 - \frac{2M}{r_+}\right) = \frac{1}{8\pi}\frac{r_+ - M}{2Mr_+} = \frac{1}{8\pi}\frac{\sqrt{M^2-a^2}}{r_+^2+a^2} = \frac{\kappa}{8\pi}\,.\end{aligned} \tag{20.43}$$

Therefore,

$$\delta M = \frac{\partial M}{\partial J}\delta J + \frac{\partial M}{\partial A}\delta A = \Omega_H \delta J + \frac{\kappa}{8\pi}\delta A\,. \tag{20.44}$$

Box 20-D

The surface gravity

As discussed in Sec. 18.6.1, in Kerr spacetime stationary observers with four-velocity

$$u^\mu = u^0(1, 0, 0, \omega)\,, \tag{20.45}$$

and ω constant, only exist for $r > r_+$. On the horizon, the vector u^μ cannot be timelike: if $\omega = -g_{t\phi}/g_{\phi\phi}$, it is a null vector, otherwise it is spacelike. Indeed, using Eqs. 18.14 and 18.79, it is easy to show that the time component of the four-velocity, u^0, diverges when $r \to r_+$ as $u^0 \to (\Delta \sin^2\theta/g_{\phi\phi})^{-1/2}$. In the same limit the norm of the four-acceleration of the observer $a^\mu = u^\nu u^\mu{}_{;\nu}$ diverges as well. This has a clear physical interpretation: the acceleration needed to keep a particle in stationary motion diverges as the particle orbit approaches the horizon.

It turns out that the ratio between the norm of the four-acceleration of a stationary observer and the time component of the four-velocity of the same observer is *finite* on the horizon. This quantity,

$$\kappa \equiv \lim_{r\to r_+} \frac{\sqrt{\vec{a}\cdot\vec{a}}}{u^0}\,, \tag{20.46}$$

is called **surface gravity** of the black hole.

Let us firstly compute the surface gravity of a Schwarzschild black hole. Consider a static observer, whose four-velocity is $u^\mu = (u^0, 0, 0, 0)$, where $u^0 = (-g_{00})^{-1/2}$; its four-acceleration is

$$a^\alpha = u^\mu u^\alpha{}_{;\mu} = u^0 u^\nu{}_{;0} = (u^0)^2\, \Gamma^\alpha_{00}\,. \tag{20.47}$$

Since $\Gamma^\alpha_{00} = -\frac{1}{2}g^{\alpha\beta}g_{00,\beta}$, the metric is diagonal and g_{00} only depends on r, it follows that $a^0 = a^2 = a^3 = 0$, and

$$a^1 = -\frac{1}{2}(u^0)^2 g^{11} g_{00,r} = -\frac{1}{2}\left(1 - \frac{2M}{r}\right)^{-1}\left(1 - \frac{2M}{r}\right)\left(-\frac{2M}{r^2}\right) = \frac{M}{r^2}\,. \tag{20.48}$$

The norm of the acceleration is $\sqrt{g_{\mu\nu}a^\mu a^\nu} = \sqrt{\left(1 - \frac{2M}{r}\right)^{-1}(a^1)^2}$. When divided by $u^0 = \left(1 - \frac{2M}{r}\right)^{-1/2}$ and evaluated at the horizon, $R_S = 2M$, it gives the surface gravity of a Schwarzschild black hole

$$\kappa = \lim_{r\to R_S} \frac{\sqrt{g_{\mu\nu}a^\mu a^\nu}}{u^0} = \lim_{r\to R_S} \frac{M}{r^2} = \frac{M}{R_S^2} = \frac{1}{4M}\,. \tag{20.49}$$

It is worth noting that a spherical star with radius R is also described, in the exterior, by the Schwarzschild metric, and its surface gravity is $\kappa = \lim_{r\to R} M/r^2 = M/R^2$, which coincides with the well-known surface gravity in Newtonian physics (in physical units, GM/R^2).

In the case of a Kerr black hole, a long but conceptually simple computation gives

$$\kappa = \lim_{r\to r_+} \frac{\sqrt{g_{\mu\nu}a^\mu a^\nu}}{u^0} = \frac{\sqrt{M^2 - a^2}}{r_+^2 + a^2}\,, \tag{20.50}$$

which coincides with Eq. 20.36. The surface gravity is a *decreasing* function of a. Indeed $r_+ = M + \sqrt{M^2 - a^2}$ decreases with a, and since κ can be written as $\kappa = (r_+ - M)/(2Mr_+) = 1/(2M) - 1/(2r_+)$, it also decreases as a increases. For extremal black holes ($a^2 = M^2$), Eq. 20.50 yields $\kappa = 0$.

20.3 THE GENERALIZED SECOND LAW OF THERMODYNAMICS

When it was first proposed by Bardeen, Carter, and Hawking [16], the correspondence between the quantities $S_{\rm BH}$, $T_{\rm BH}$ and the thermodynamical entropy and temperature seemed to be just a formal analogy. Then, Bekenstein conjectured [18, 19] that this correspondence is based on real physical motivations.

The starting point of Bekenstein's argument is the following. The entropy of a thermodynamical system in equilibrium measures the fact that – in our macroscopic description in terms of state variables – we cannot distinguish between different microstates corresponding to the same macroscopic state. This can be rephrased by saying that, in our thermodynamical description, we *do not know* part of the information on the physical state of the system, and the entropy (which is proportional to the logarithm of the number of microstates corresponding to a given macroscopic state) is a measure of this "ignorance". Similarly, due to the no-hair theorems discussed in Sec. 18.9, it is impossible to distinguish between stationary black holes formed in different astrophysical processes (e.g., from the collapse of stars with different composition), as long as they have the same masses and angular momenta. Bekenstein argued that the entropy $S_{\rm BH}$ of a stationary black hole measures the inaccessibility of the information on the particular configuration corresponding to the black hole mass and angular momentum.

In order to make this statement quantitative, let us consider a stationary black hole with mass M and angular momentum J, and an "object" (a fluid, particles, radiation, i.e. anything accreting onto the black hole) with energy $E \ll M$, angular momentum $L \ll J$, and thermodynamical entropy S, falling into it.

The entropy of the object disappears when it crosses the horizon, because the information on its structure becomes inaccessible; since the final black hole is characterized by mass and angular momentum only, the total entropy of the universe would then decrease, violating the second law of thermodynamics. However, when the object is swallowed by the black hole, as discussed in Sec. 20.2 the black hole entropy, $S_{\rm BH} = \alpha A$, increases. Thus, the second law of thermodynamics can be recovered in an extended form if we include this contribution:

$$\delta S + \delta S_{\rm BH} \geq 0\,. \tag{20.51}$$

This is the **generalized second law of thermodynamics**. Hence, the entropy $S_{\rm BH}$ can be considered as a measure of the information stored inside the black hole which cannot be seen from outside. This conjecture does not tell us which is the value of the constant α, but some further reasoning suggests at least its order of magnitude.

Let us consider, for instance, a *massive* particle which falls into a Kerr black hole. As shown in Sec. 20.1, in this process $\delta A \geq 0$, which implies $\delta S_{\rm BH} \geq 0$. Furthermore, $\delta S_{\rm BH} = 0$ (i.e. $\delta A = 0$) if and only if the particle is released from the horizon with zero radial velocity (see Box 20-A). However, due to Heisenberg's principle we cannot locate a particle with $\dot{r} = 0$ exactly at $r = r_+$: there is an uncertainty in the position, with associated proper length[3] of the order of the Compton wavelength

$$\lambda = \frac{\hbar}{m_{\rm p}}\,, \tag{20.52}$$

where $m_{\rm p}$ is the mass of the particle[4]. Therefore, the conditions $\dot{r} = 0$ and $r = r_+$ cannot be

[3]Strictly speaking, the Compton wavelength λ is the *coordinate* separation between two events at the same time, as measured in the LIF comoving with the particle when the latter is emitted. As discussed in Box 3-A, the spacetime coordinate separation in a LIF coincides with the *proper length*, at leading order in the coordinate separation (see Eq. 3.19). Therefore, for two simultaneous events their *space* coordinate separation measured in a LIF and the proper length, to leading order, coincide.

[4]Note that although the particle has $\dot{r} = 0$, it moves along the azimuthal direction with angular velocity

simultaneously satisfied, since the uncertainty on the particle position implies that $r > r_+$; consequently the process *can not be reversible*, the area increases by an amount δA, and so does the entropy $\delta S_{\rm BH} = \alpha\delta A$. If we identify the minimum amount of entropy increase $\delta S^{\rm min}_{\rm BH} = \alpha\delta A^{\rm min}$ with the minimum entropy of the particle itself, which is of the order of $\ln 2$ (i.e. a bit of information [5]), and impose the condition $\delta S_{\rm BH} \geq \delta S^{\rm min}_{\rm BH}$, we find that $\alpha \sim \ln 2/\delta A^{\rm min}$. Therefore, in order to estimate the value of α we need to estimate $\delta A^{\rm min}$. To this purpose, we assume that a particle with mass $m_{\rm p}$ is released with $\dot{r} = 0$ from a point at $r_+ + \delta r$, such that the proper distance between this point and the horizon is of the order of the Compton wavelength λ given in Eq. 20.52. To begin with, we shall estimate the proper distance

$$d = \int_{r_+}^{r_+ + \delta r} \sqrt{g_{rr}} dr \,. \tag{20.53}$$

For simplicity, we assume that the particle moves with $L = 0$ in the equatorial plane, where

$$g_{rr} = \frac{\Sigma}{\Delta} = \frac{r^2}{(r - r_+)(r - r_-)} \,. \tag{20.54}$$

Defining $r = r_+ + x$ (with $x \ll r_+$),

$$g_{rr} = \left(\frac{r_+^2}{r_+ - r_-}\right)\frac{1}{x} + \dots \tag{20.55}$$

thus

$$d = \frac{r_+}{\sqrt{r_+ - r_-}} \int_0^{\delta r} \frac{dx}{\sqrt{x}} + \dots = 2\sqrt{\delta r}\frac{r_+}{\sqrt{r_+ - r_-}} + \dots \tag{20.56}$$

and

$$\delta r = \frac{1}{4}\frac{r_+ - r_-}{r_+^2} d^2 + O(d^3) \,. \tag{20.57}$$

Since $L = 0$, $V_\pm = 0$ (see Eq. 19.51) and the condition $\dot{r} = 0$ gives (see Eq. 19.48) [6]

$$\dot{r}^2 = \frac{C}{r^2}\left(\frac{E}{m_{\rm p}}\right)^2 - \frac{\Delta}{r^2} = 0 \,, \tag{20.58}$$

where $C = r^2 + a^2 + 2Ma^2/r$. Therefore, the mass of the black hole increases by

$$\delta M = E = m_{\rm p}\sqrt{\frac{\Delta}{C}} \,. \tag{20.59}$$

By replacing $r = r_+ + \delta r$ and Eq. 20.57 in Δ and C we find

$$\Delta = (r - r_+)(r - r_-) = (r_+ - r_-)\delta r + O(\delta r^2) = \left(\frac{r_+ - r_-}{2r_+}\right)^2 d^2 + O(d^3) \tag{20.60}$$

Ω_H. This determines a Lorentz contraction of the Compton wavelength in the φ-direction, but it does not affect the Compton wavelength in the radial direction, which is the quantity relevant for this argument.

[5] We can add a bit of information to a system by merging it with another system having two possible configurations, e.g. spin up and spin down. The number N of possible states of the final system is doubled, and the logarithm of N is increased by $\ln 2$.

[6] Note that in Eq. 20.58, differently from Eq. 19.48, E is divided by the particle mass $m_{\rm p}$. This is due to the different normalization conventions: here E is the particle energy, while in Sec. 19.3 E was the energy per unit of mass.

and

$$C = r^2 + a^2 + \frac{2Ma^2}{r} = r_+^2 + a^2 + \frac{2Ma^2}{r_+} + O(d^2)$$
$$= 2Mr_+ + \frac{2Ma^2}{r_+} + O(d^2) = \frac{2M}{r_+}(r_+^2 + a^2) + O(d^2) = 4M^2 + O(d^2) \tag{20.61}$$

where we used the relation 18.68, $r_+^2 + a^2 = 2Mr_+$. Thus,

$$\delta M \simeq m_{\rm p}\frac{r_+ - r_-}{4Mr_+}d = \frac{1}{2}\frac{r_+ - r_-}{r_+^2 + a^2}m_{\rm p}d = \frac{\sqrt{M^2 - a^2}}{r_+^2 + a^2}m_{\rm p}d = \kappa m_{\rm p}d\,. \tag{20.62}$$

Finally, by requiring that the proper distance between the horizon and the point in which the particle is released equates the Compton wavelength 20.52, we find

$$d = \lambda = \frac{\hbar}{m_{\rm p}} \qquad \rightarrow \qquad \delta M \simeq \kappa\hbar\,. \tag{20.63}$$

From Eq. 20.35, since $L = 0$ and then $\delta J = 0$, we find

$$\delta A^{\rm min} = \frac{8\pi}{\kappa}\delta M \tag{20.64}$$

which, using Eq. 20.63, yields

$$\delta A^{\rm min} \simeq 8\pi\hbar\,, \tag{20.65}$$

and consequently

$$\alpha \sim \frac{\ln 2}{\delta A_{\rm min}} \simeq \frac{\ln 2}{8\pi\hbar} \sim \frac{1}{\hbar}\,. \tag{20.66}$$

Note that in the units $c = G = 1$, the Planck constant has dimension [7] $[\hbar] = ({\rm length})^2$. Since, as discussed above, the constant α has dimensions $({\rm length})^{-2}$, and the only fundamental dimensionful constant in the theory is $\hbar$, it is not surprising that $\alpha \sim \hbar^{-1}$.

In terms of the *Planck length* (see Box 9-A and Table A)

$$l_{\rm P} = \sqrt{\hbar} = 1.6 \times 10^{-33}\,{\rm cm} \tag{20.67}$$

we have $\alpha \sim 1/l_{\rm P}^2$. If we introduce the dimensionless constant

$$\hat{\alpha} = \hbar\alpha\,, \tag{20.68}$$

the black hole entropy can be written as

$$S_{\rm BH} = \hat{\alpha}\frac{A}{\hbar} = \hat{\alpha}\frac{A}{l_{\rm P}^2}\,. \tag{20.69}$$

The estimate in Eq. 20.66 thus implies that the (still undetermined) constant $\hat{\alpha}$ is of the order of unity.

Finally, the black hole temperature can be written in terms of $\hat{\alpha}$ as

$$T_{\rm BH} \equiv \frac{\hbar\kappa}{8\pi\hat{\alpha}}\,. \tag{20.70}$$

[7] It should be stressed that different choices of units are possible. An alternative choice consists in setting $\hbar = c = 1$; in this case G has dimensions $({\rm length})^2$. The units with $G = c = 1$, used in this book, are common in the context of gravitational physics, while the units with $\hbar = c = 1$ are often used in the context of particle physics.

Note that Bekenstein's conjecture provides a thermodynamical interpretation for $S_{\rm BH}$, but not for the temperature $T_{\rm BH}$: how is it possible to assign a temperature to an object which, by definition, has no emission? Indeed, the generalized second law of thermodynamics presents a problem. If a black hole with mass M and temperature $T_{\rm BH}$ is surrounded by radiation with energy U and temperature T, and some radiation enters into the black hole with zero angular momentum, the fluid internal energy and the black hole mass change, respectively, by

$$\delta U = T\delta S\,, \qquad \delta M = T_{\rm BH}\delta S_{\rm BH}\,, \tag{20.71}$$

with $\delta U = -\delta M$. The generalized second law of thermodynamics (Eq. 20.51) states that

$$\delta S_{\rm BH} + \delta S = \frac{\delta M}{T_{\rm BH}} + \frac{\delta U}{T} = \left(\frac{1}{T_{\rm BH}} - \frac{1}{T}\right)\delta M \geq 0\,. \tag{20.72}$$

Thus, if the black hole is "colder" than the surrounding radiation, $T_{\rm BH} < T$ and $\delta M \geq 0$, as expected. If, instead, the black hole is "warmer" than the surrounding radiation, $T_{\rm BH} > T$ and $\delta M \leq 0$, i.e. the black hole should emit energy. In other words, the generalized second law of thermodynamics implies that black holes must satisfy a fundamental property of thermodynamical systems: any black hole with finite temperature which is warmer than the surrounding has to radiate. But how can a black hole radiate energy?

20.4 THE HAWKING RADIATION

A decisive proof of Bekenstein's conjecture, together with a precise determination of the constant $\hat{\alpha}$, came from the discovery, due to Hawking [55], that if quantum mechanics is taken into account [8] a black hole emits the so-called **Hawking radiation**, whose distribution function (see Sec. 16.2.2) is

$$f = \frac{1}{e^{2\pi\omega/\kappa} - 1}\,, \tag{20.73}$$

where ω is the frequency of the emitted radiation and κ is the surface gravity given in Eq. 20.36. The above equation has the same form as the black-body distribution function

$$f = \frac{1}{e^{\hbar\omega/T} - 1} \tag{20.74}$$

with temperature

$$T = T_{\rm BH} \equiv \frac{\hbar\kappa}{2\pi} \tag{20.75}$$

(we remind that we have set Boltzmann's constant to 1). This result provides a convincing interpretation of the black hole temperature, and determines the precise value of the constant $\hat{\alpha}$, which, as supposed by Bekenstein, is of the order of unity:

$$\hat{\alpha} = \frac{1}{4}\,. \tag{20.76}$$

Consequently, the black hole entropy $S_{\rm BH} = \alpha A = \hat{\alpha}A/\hbar$ (see Eqs. 20.37 and 20.68) is

$$S_{\rm BH} = \frac{1}{4}\frac{A}{\hbar}\,. \tag{20.77}$$

[8]It should be stressed that in the derivation of Hawking's radiation, the radiation is treated as a quantum field while the gravitational field is treated classically. This approximation is accurate as long as the black hole radius is much larger than the Planck length $l_{\rm P}$. Presently there is no way to overcome this approximation: the quantum description of gravity is still an open problem in physics (see footnote 8 in Chapter 9).

Summarizing, if quantum mechanics is taken into account, black holes radiate with a black-body spectrum at temperature $T_{\rm BH}$. When matter and energy enter into a black hole, their entropy becomes inaccessible from the exterior, but it shows up as an increase of the black hole area, according to Eq. 20.77, and the generalized second law of thermodynamics holds:

$$\delta S_{\rm BH} + \delta S \geq 0\,. \tag{20.78}$$

We shall now estimate the temperature of astrophysical black holes. To this aim, we convert Eq. 20.75 in physical units:

$$T_{\rm BH} = \frac{\hbar\kappa}{2\pi c k_{\rm B}} \tag{20.79}$$

where (see Table A) $\hbar = 1.055 \times 10^{-27}$ erg s, $c = 2.998 \times 10^{8}$ m/s, $k_{\rm B} = 1.381 \times 10^{-16}$ erg/K is Boltzmann's constant, and the surface gravity has the dimensions of an acceleration. The surface gravity of a Kerr black hole is of the same order as that of a Schwarzschild black hole (Eq. 20.49), unless the black hole is close to extremality. In physical units, it is

$$\kappa \simeq \frac{GM}{R_S^2} = \frac{GM}{(2GM/c^2)^2} = \frac{c^4}{4GM}\,, \tag{20.80}$$

where $G = 6.674 \times 10^{-8}$ cm^3/(g s^2). Replacing in Eq. 20.79 we have

$$T_{\rm BH} \simeq \frac{\hbar c^3}{8\pi G k_{\rm B}}\frac{1}{M}\,. \tag{20.81}$$

Remarkably, the formula for black hole temperature contains the fundamental constants $\hbar, G, k_{\rm B}$. This is why it is often seen as an important bridge between quantum theory (governed by $\hbar$), gravity (governed by G), relativity (governed by c), and thermodynamics (governed by $k_{\rm B}$).

By multiplying and dividing the above equation for the mass of the Sun, $M_\odot = 1.989 \times 10^{33}$ g, we get

$$T_{\rm BH} \simeq \frac{\hbar c^3}{8\pi G k_{\rm B} M_\odot}\frac{M_\odot}{M} \simeq 10^{-7}\frac{M_\odot}{M}\,\text{K}. \tag{20.82}$$

Since the present universe is filled with the cosmic microwave background at temperature $T_{\rm CMB} \sim 3$ K, we can conclude that astrophysical black holes are *colder* than their environment and do not emit Hawking radiation. Only extremely small black holes (with masses $M \lesssim 3 \times 10^{-8} M_\odot$) could have $T_{\rm BH} > T_{\rm CMB}$.

Although astrophysical black holes are too heavy to emit, it has been conjectured that primordial black holes small enough to be warmer than the surrounding radiation could have been formed in the aftermath of the big bang. In this case, the Hawking emission would determine a decrease of their mass, and thus of their area. This would not violate the generalized second law of thermodynamics 20.78 because the *total* entropy – including that of the emitted radiation – would increase. The black hole would then become hotter and hotter, until complete evaporation. The final outocome of the Hawking evaporation is unknown. In the stages immediately before evaporating completely, the black hole would be extremely small and hot and its curvature at the horizon would be enormous. In these conditions quantum gravity effects become important and only a quantum theory of gravity can explain the final state.

It should be mentioned that the possibility – even as a matter of principle – that black holes evaporate leads to a serious conceptual problem also known as **Hawking's information loss paradox**: the quantum information of anything crossing the black hole horizon would disappear once the black hole evaporates completely, but this "information loss" would be in conflict with the unitarity of the evolution in quantum mechanics, which is a fundamental principle, necessary for the very consistency of any quantum theory. Whether black hole evaporation leads to information loss is still a matter of debate.

Bibliography

[1] B. P. Abbott et al. Observation of Gravitational Waves from a Binary Black Hole Merger. *Phys. Rev. Lett.*, 116(6):061102, 2016.

[2] B.P. Abbott et al. GW170817: Measurements of neutron star radii and equation of state. *Phys. Rev. Lett.*, 121(16):161101, 2018.

[3] B.P. Abbott et al. GWTC-1: A Gravitational-Wave Transient Catalog of Compact Binary Mergers Observed by LIGO and Virgo during the First and Second Observing Runs. *Phys. Rev. X*, 9(3):031040, 2019.

[4] B.P. Abbott et al. GW190425: Observation of a Compact Binary Coalescence with Total Mass $\sim 3.4 M_{\odot}$. *Astrophys. J. Lett.*, 892(1):L3, 2020.

[5] R. Abbott et al. GW190521: A Binary Black Hole Merger with a Total Mass of 150 $M_{\odot}$. *Phys. Rev. Lett.*, 125(10):101102, 2020.

[6] Miguel Alcubierre. *Introduction to 3+1 numerical relativity.* International series of monographs on physics. Oxford Univ. Press, Oxford, 2008.

[7] Pau Amaro-Seoane et al. Laser Interferometer Space Antenna. *arXiv e-prints*, page arXiv:1702.00786, February 2017.

[8] John Antoniadis et al. A Massive Pulsar in a Compact Relativistic Binary. *Science*, 340:6131, 2013.

[9] Neil Ashby. Relativity in the global positioning system. *Living Reviews in Relativity*, 6(1):1, Jan 2003.

[10] W. Baade and F. Zwicky. Cosmic Rays from Super-novae. *Physical Reviews*, 45:138, 1934.

[11] W. Baade and F. Zwicky. Cosmic Rays from Super-novae. *Proceedings of the National Academy of Science*, 20:259–263, May 1934.

[12] W. Baade and F. Zwicky. On Super-novae. *Proceedings of the National Academy of Science*, 20:254–259, May 1934.

[13] Walter Baade. The crab nebula. *The Astrophysical Journal*, 96:188, 1942.

[14] John G. Baker, Joan Centrella, Dae-Il Choi, Michael Koppitz, and James van Meter. Gravitational wave extraction from an inspiraling configuration of merging black holes. *Phys. Rev. Lett.*, 96:111102, 2006.

[15] R. Balbinot, P. R. Brady, W. Israel, and E. Poisson. How singular are black hole interiors? *Phys. Lett.*, A161(3):223–226, 1991.

[16] James M. Bardeen, B. Carter, and S. W. Hawking. The Four laws of black hole mechanics. *Commun. Math. Phys.*, 31:161–170, 1973.

[17] James M. Bardeen, William H. Press, and Saul A. Teukolsky. Rotating Black Holes: Locally Nonrotating Frames, Energy Extraction, and Scalar Synchrotron Radiation. *The Astrophysical Journal*, 178:347–370, Dec 1972.

[18] Jacob D. Bekenstein. Black holes and entropy. *Phys. Rev.*, D7:2333–2346, 1973.

[19] Jacob D. Bekenstein. Generalized second law of thermodynamics in black hole physics. *Phys. Rev.*, D9:3292–3300, 1974.

[20] Joel Bergé, Philippe Brax, Gilles Métris, Martin Pernot-Borràs, Pierre Touboul, and Jean-Philippe Uzan. MICROSCOPE Mission: First Constraints on the Violation of the Weak Equivalence Principle by a Light Scalar Dilaton. *Phys. Rev. Lett.*, 120(14):141101, 2018.

[21] Emanuele Berti et al. Testing General Relativity with Present and Future Astrophysical Observations. *Class. Quant. Grav.*, 32:243001, 2015.

[22] B. Bertotti, L. Iess, and P. Tortora. A test of general relativity using radio links with the Cassini spacecraft. *Nature*, 425:374–376, 2003.

[23] A.R. Bodmer. Collapsed nuclei. *Phys. Rev. D*, 4:1601–1606, 1971.

[24] Robert H. Boyer and Richard W. Lindquist. Maximal analytic extension of the Kerr metric. *J. Math. Phys.*, 8:265, 1967.

[25] Richard Brito, Vitor Cardoso, and Paolo Pani. Superradiance. *Lect. Notes Phys.*, 906:pp.1–237, 2015.

[26] Hans A. Buchdahl. General Relativistic Fluid Spheres. *Phys. Rev.*, 116:1027, 1959.

[27] M. Burgay, N. D'Amico, A. Possenti, R. N. Manchester, A. G. Lyne, B. C. Joshi, M. A. McLaughlin, M. Kramer, J. M. Sarkissian, F. Camilo, V. Kalogera, C. Kim, and D. R. Lorimer. An increased estimate of the merger rate of double neutron stars from observations of a highly relativistic system. *Nature*, 426:531–533, December 2003.

[28] Manuela Campanelli, C. O. Lousto, P. Marronetti, and Y. Zlochower. Accurate evolutions of orbiting black-hole binaries without excision. *Phys. Rev. Lett.*, 96:111101, 2006.

[29] Sean Carrol. *Spacetime and Geometry.* Addison-Wesley, 2003.

[30] Brandon Carter. Hamilton-Jacobi and Schrödinger Separable Solutions of Einstein's Equations. *Communications in Mathematical Physics*, 10(4):280–310, Dec 1968.

[31] S. Chandrasekhar. The Maximum Mass of Ideal White Dwarfs. *The Astrophysical Journal*, 74:81, July 1931.

[32] S. Chandrasekhar. The highly collapsed configurations of a stellar mass (Second paper). *Mon.Not.Roy.Astron.Soc.*, 95:207–225, 1935.

[33] S. Chandrasekhar. *An introduction to the study of stellar structure.* Chicago, Ill., The University of Chicago Press, 1939.

[34] Subrahmanyan Chandrasekhar. *The mathematical theory of black holes.* Oxford University Press, 1983.

[35] Subrahmanyan Chandrasekhar and V. Ferrari. On the non-radial oscillations of a star. *Proc. Roy. Soc. Lond.*, A432:247–279, 1991.

[36] D. Christodoulou. Reversible and irreversible transformations in black hole physics. *Phys. Rev. Lett.*, 25:1596–1597, 1970.

[37] Ignazio Ciufolini, Antonio Paolozzi, Erricos C. Pavlis, Giampiero Sindoni, John Ries, Richard Matzner, Rolf Koenig, Claudio Paris, Vahe Gurzadyan, and Roger Penrose. An Improved Test of the General Relativistic Effect of Frame-Dragging Using the LARES and LAGEOS Satellites. *Eur. Phys. J.*, C79(10):872, 2019.

[38] John P. Cox. *Theory of Stellar Pulsation*, volume 2. Princeton University Press, 2017.

[39] Mihalis Dafermos and Jonathan Luk. The interior of dynamical vacuum black holes I: The C^0-stability of the Kerr Cauchy horizon. *arXiv e-prints*, page 1710.01722, 2017.

[40] Thibault Damour. The entropy of black holes: A primer. *Prog. in Math. Phys.*, 38:227, 2004.

[41] Paul Demorest, Tim Pennucci, Scott Ransom, Mallory Roberts, and Jason Hessels. Shapiro Delay Measurement of A Two Solar Mass Neutron Star. *Nature*, 467:1081–1083, 2010.

[42] S. Detweiler. On resonant oscillations of a rapidly rotating black hole. *Proceedings of the Royal Society of London Series A*, 352:381–395, July 1977.

[43] A.S. Eddington. The internal constitution of the stars. *The Observatory*, 43:341–358, 1920.

[44] A. Einstein. Über Gravitationswellen. *Sitzungsberichte der Königlich Preußischen Akademie der Wissenschaften (Berlin), Seite 154-167.*, 1918. English translation: "The Collected Papers of Albert Einstein - Volume 6: The Berlin Years: Writings, 1914-1917", translated by Alfred Engel.

[45] C. W. F. Everitt et al. Gravity Probe B: Final Results of a Space Experiment to Test General Relativity. *Phys. Rev. Lett.*, 106:221101, 2011.

[46] U.H. Gerlach and U.K. Sengupta. Gauge invariant perturbations on most general spherically symmetric space-times. *Phys. Rev. D*, 19:2268–2272, 1979.

[47] R. Geroch. Singularities. In M. Carmeli et al., editor, *Relativity*, pages 259–291. Springer, 1970.

[48] A. M. Ghez et al. Measuring Distance and Properties of the Milky Way's Central Supermassive Black Hole with Stellar Orbits. *Astrophys. J.*, 689:1044–1062, 2008.

[49] G. W. Gibbons and S. W. Hawking. Action integrals and partition functions in quantum gravity. *Phys. Rev. D*, 15:2752–2756, May 1977.

[50] Norman K. Glendenning. *Compact stars: Nuclear physics, particle physics and general relativity.* Springer Science & Business Media, 2012.

[51] E. Gourgoulhon and S. Bonazzola. Gravitational waves from isolated neutron stars. In *International Conference on Gravitational Waves: Sources and Detectors*, pages 51–60, 3 1996.

[52] J. B. Griffiths. *Colliding Plane Waves in General Relativity.* Oxford Mathematical Monographs, 1991.

[53] P. Haensel, A. Y. Potekhin, and D. G. Yakovlev. Neutron stars 1: Equation of state and structure. *Astrophys. Space Sci. Libr.*, 326:pp.1–619, 2007.

[54] T. Hamada and E. E. Salpeter. Models for Zero-Temperature Stars. *The Astrophysical Journal*, 134:683, November 1961.

[55] S. W. Hawking. Particle Creation by Black Holes. *Commun. Math. Phys.*, 43:199–220, 1975.

[56] Stephen Hawking. Singularities and the geometry of spacetime. *European Physical Journal H*, 39(4):413–503, Nov 2014.

[57] S.W. Hawking and G.F.R. Ellis. *The Large Scale Structure of Space-Time.* Cambridge Monographs on Mathematical Physics, 1975.

[58] Alexander Heger, C. L. Fryer, S. E. Woosley, N. Langer, and D. H. Hartmann. How massive single stars end their life. *Astrophys. J.*, 591:288–300, 2003.

[59] Friedrich W. Hehl, J. Dermott McCrea, Eckehard W. Mielke, and Yuval Ne'eman. Metric affine gauge theory of gravity: Field equations, Noether identities, world spinors, and breaking of dilation invariance. *Phys. Rept.*, 258:1–171, 1995.

[60] Carlos A. R. Herdeiro and José P. S. Lemos. The black hole fifty years after: Genesis of the name. *Gazeta de Fisica*, pages 41(2), 2, 2018.

[61] S. Herrmann, A. Senger, K. Mohle, M. Nagel, E. V. Kovalchuk, and A. Peters. Rotating optical cavity experiment testing Lorentz invariance at the 10^{-17} level. *Phys. Rev.*, D80:105011, 2009.

[62] Antony Hewish, S. Jocelyn Bell, John D.H. Pilkington, Paul Frederick Scott, and Robin Ashley Collins. Observation of a rapidly pulsating radio source. *Nature*, 217(5130):709, 1968.

[63] David Hilbert. Die grundlagen der physik.(erste mitteilung.). *Nachrichten von der Gesellschaft der Wissenschaften zu Göttingen, Mathematisch-Physikalische Klasse*, 1915:395–408, 1915. English translation in: "The genesis of general relativity". Springer, Dodrecht, 2007. p. 1925-1938.

[64] R. A. Hulse and J. H. Taylor. Discovery of a pulsar in a binary system. *The Astrophysics Journal Letters*, 195:L51–L53, January 1975.

[65] Richard A. Isaacson. Gravitational Radiation in the Limit of High Frequency. I. The Linear Approximation and Geometrical Optics. *Phys. Rev.*, 166:1263–1271, 1968.

[66] Richard A. Isaacson. Gravitational radiation in the limit of high frequency. II. Nonlinear terms and the effective stress tensor. *Physical Review*, 166(5):1272, 1968.

[67] J. D. Jackson. *Classical Electordynamics.* John Wiley & Sons, 1999.

[68] Roy P. Kerr. Gravitational field of a spinning mass as an example of algebraically special metrics. *Phys. Rev. Lett.*, 11:237–238, Sep 1963.

[69] R. Kippenhahn, A. Weigert, and A. Weiss. *Stellar structure and evolution.* Springer-Verlag, 2012.

[70] Y. Kojima. Equations governing the nonradial oscillations of a slowly rotating relativistic star. *Phys. Rev.*, D46:4289–4303, 1992.

[71] E.W. Kolb and M.S. Turner. *The Early Universe.* Addison-Wesley - Redwood City, 1990.

[72] M. Kramer et al. Tests of general relativity from timing the double pulsar. *Science*, 314:97–102, 2006.

[73] L. D. Landau. Origin of Stellar Energy. *Nature*, 141:333–334, 1938.

[74] L. D. Landau. 8 - on the theory of stars. In D. Ter Haar, editor, *Collected Papers of L.D. Landau*, pages 60 – 62. Pergamon, 1965.

[75] Lev Davidovich Landau and Evgenii Mikhailovich Lifshitz. *The classical theory of fields.* Pergamon, 1971.

[76] M. I. Large, A. E. Vaughan, and B. Y. Mills. A Pulsar Supernova Association? *Nature*, 220(5165):340–341, Oct 1968.

[77] James M. Lattimer and Maddapa Prakash. Neutron Star Observations: Prognosis for Equation of State Constraints. *Phys. Rept.*, 442:109–165, 2007.

[78] Edward W. Leaver. Spectral decomposition of the perturbation response of the Schwarzschild geometry. *Phys. Rev.*, D34:384–408, 1986.

[79] D. R. Lorimer and M. Kramer. *Handbook of Pulsar Astronomy.* October 2012.

[80] David M. Lucchesi, Massimo Visco, Roberto Peron, Massimo Bassan, Giuseppe Pucacco, Carmen Pardini, Luciano Anselmo, and Carmelo Magnafico. An improved measurement of the Lense-Thirring precession on the orbits of laser-ranged satellites with an accuracy approaching the 1% level. *arXiv e-prints*, page arXiv:1910.01941, 2019.

[81] A. G. Lyne, M. Burgay, M. Kramer, A. Possenti, R. N. Manchester, F. Camilo, M. A. McLaughlin, D. R. Lorimer, N. D'Amico, B. C. Joshi, J. Reynolds, and P. C. C. Freire. A Double-Pulsar System: A Rare Laboratory for Relativistic Gravity and Plasma Physics. *Science*, 303:1153–1157, February 2004.

[82] Michele Maggiore. *Gravitational waves: Volume 1: Theory and experiments*, volume 1. Oxford University Press, 2008.

[83] Jing Ming et al. Results from an Einstein@Home search for continuous gravitational waves from Cassiopeia A, Vela Jr. and G347.3. *Phys. Rev.*, D100(2):024063, 2019.

[84] Yasushi Mino. Perturbative approach to an orbital evolution around a supermassive black hole. *Phys. Rev.*, D67:084027, 2003.

[85] E. T. Newman, R. Couch, K. Chinnapared, A. Exton, A. Prakash, and R. Torrence. Metric of a Rotating, Charged Mass. *J. Math. Phys.*, 6:918–919, 1965.

[86] M. Ostrogradsky. Memoires de l'Academie Imperiale des Science de Saint-Petersbourg. 4:385, 1850.

[87] Feryal Ozel and Paulo Freire. Masses, Radii, and the Equation of State of Neutron Stars. *Ann. Rev. Astron. Astrophys.*, 54:401–440, 2016.

[88] Franco Pacini. Energy emission from a neutron star. *Nature*, 216(5115):567, 1967.

[89] Don N. Page. Hawking radiation and black hole thermodynamics. *New Journal of Physics*, 7(1):203, 2005.

[90] R. Penrose. Republication of: Conformal treatment of infinity. *Gen. Rel. Grav.*, 43:901–922, 2011.

[91] Roger Penrose. *Nuovo Cimento.J.*, Serie 1:252, 1969.

[92] Planck Collaboration. Planck 2018 results. VI. Cosmological parameters. *arXiv e-prints*, page arXiv:1807.06209, July 2018.

[93] E. Poisson and C.M. Will. *Gravity: Newtonian, Post-Newtonian, Relativistic.* Cambridge University Press, Cambridge, UK, 1953.

[94] Eric Poisson and W. Israel. Inner-horizon instability and mass inflation in black holes. *Phys. Rev. Lett.*, 63:1663–1666, 1989.

[95] William H. Press and Saul A. Teukolsky. Floating Orbits, Superradiant Scattering and the Black-hole Bomb. *Nature*, 238:211–212, 1972.

[96] Frans Pretorius. Evolution of binary black hole spacetimes. *Phys. Rev. Lett.*, 95:121101, 2005.

[97] J. L. Provencal, H. L. Shipman, E. Høg, and P. Thejll. Testing the White Dwarf Mass-Radius Relation with HIPPARCOS. *The Astrophysical Journal*, 494:759–767, February 1998.

[98] George E. Pugh. Proposal for a Satellite Test of the Coriolis Predictions of General Relativity. In *Nonlinear Gravitodynamics: The Lense-Thirring Effect*, pages 414–426. World Scientific, 2003.

[99] Tullio Regge and John A. Wheeler. Stability of a Schwarzschild singularity. *Phys. Rev.*, 108:1063–1069, 1957.

[100] D. Robinson. *Four decades of black holes uniqueness theorems.* Cambridge University Press, 2009.

[101] Carlo Rovelli. Can we travel to the past? Irreversible physics along closed timelike curves. *arXiv e-prints*, page arXiv:1912.04702, 12 2019.

[102] Barbara Ryden. *Introduction to cosmology.* Cambridge University Press, 2017.

[103] L. I. Schiff. Possible New Experimental Test of General Relativity Theory. *Phys. Rev. Lett.*, 4:215–217, 1960.

[104] Bernard F. Schutz. Determining the Hubble Constant from Gravitational Wave Observations. *Nature*, 323:310–311, 1986.

[105] Stuart L. Shapiro and Saul A. Teukolsky. *Black holes, white dwarfs, and neutron stars: The physics of compact objects.* John Wiley & Sons, 2008.

[106] Stephen J. Smartt. Progenitors of core-collapse supernovae. *Ann. Rev. Astron. Astrophys.*, 47:63–106, 2009.

[107] David H. Staelin and Edward C. Reifenstein. Pulsating radio sources near the crab nebula. *Science*, 162(3861):1481–1483, 1968.

[108] László B. Szabados. Quasi-Local Energy-Momentum and Angular Momentum in General Relativity. *Living Rev. Rel.*, 12:4, 2009.

[109] S. A. Teukolsky. Rotating black holes - separable wave equations for gravitational and electromagnetic perturbations. *Phys. Rev. Lett.*, 29:1114–1118, 1972.

[110] S.A. Teukolsky and W.H. Press. Perturbations of a rotating black hole. III - Interaction of the hole with gravitational and electromagnetic radiation. *Astrophys.J.*, 193:443–461, 1974.

[111] The Event Horizon Telescope Collaboration. First M87 Event Horizon Telescope Results. I. The Shadow of the Supermassive Black Hole. *ApJL*, 877:1, 2019.

[112] Kip S. Thorne, S.W. Hawking, and W. Israel. Three hundred years of gravitation. *Gravitational Radiation.*

[113] W. Unno, Y. Osaki, H. Ando, H. Saio, and H. Shibahashi. *Nonradial oscillations of stars.* 1989.

[114] Robert M. Wald. *General Relativity.* Chicago University Press, Chicago, USA, 1984.

[115] Robert M. Wald. The thermodynamics of black holes. *Living Rev. Rel.*, 4:6, 2001.

[116] Kameshwar C. Wali. *Chandra: A biography of S. Chandrasekhar.* Chicago, Ill., The University of Chicago Press, 1991.

[117] Steven Weinberg. *Gravitation and Cosmology.* John Wiley & Sons, Inc., 1972.

[118] Joel M. Weisberg and Joseph H. Taylor. Relativistic binary pulsar B1913+16: Thirty years of observations and analysis. *ASP Conf. Ser.*, 328:25, 2005.

[119] Clifford M. Will. The Confrontation between General Relativity and Experiment. *Living Rev. Rel.*, 17:4, 2014.

[120] Edward Witten. Cosmic Separation of Phases. *Phys. Rev. D*, 30:272–285, 1984.

[121] James W. York. Role of conformal three-geometry in the dynamics of gravitation. *Phys. Rev. Lett.*, 28:1082–1085, Apr 1972.

[122] Peter J. Young. Capture of particles from plunge orbits by a black hole. *Phys. Rev.*, D14:3281–3289, 1976.

[123] Ya. B. Zel'dovich. *Pis'ma Zh. Eksp. Teor. Fiz.*, 14:270 [JETP Lett. **14**, 180 (1971)], 1971.

[124] Frank J. Zerilli. Effective potential for even parity Regge-Wheeler gravitational perturbation equations. *Phys.Rev.Lett.*, 24:737–738, 1970.

Index

Printed in the United States
By Bookmasters